Major Biological Events

	Millions of Years Ago	
	0.135	Modern *Homo sapiens* arises.
	2	Genus *Homo* arises.
	4	Australopithecines present.
Gymnosperms, angiosperms widespread. Temperate grasslands, forests expand.	65	Mammals diversify. Primates arise.
Angiosperms arise and diversify.		Major extinction event. Most large reptiles, ancient birds extinct.
		Teleost fish diversify. Dinosaurs dominant. Modern crustaceans common.
Gymnosperms, ferns dominant.		Dinosaur ancestors common. First mammals, birds.
Conifers appeared.		Mammallike reptiles common. Major extinction of invertebrates, amphibia.
Forests widespread. Coal deposits form.		First reptiles. Amphibia diversify. Major extinction event.
First forests. Vascular plants and seeds present.		First insects, sharks, amphibians. Fish diversify.
Green, red, brown algae common.		First land arthropods. Jawed fish arise.
First vascular land plants probably appeared.		Second major extinction event. Jawless fish diversify; large invertebrates present; mollusks diversify. First tracks left by land animals.
Algae dominant.		First major extinction event. Trilobites common; onychophorans, first jawless fish at end of period. Evolution of many phyla.
Algae abundant. Multicellular organisms: algae, fungi. Cyanobacteria diversified. Eukaryotes present: green algae, protists.		Wormlike animals, cnidarians present.
Photosynthetic cells liberate oxygen.		
Origin of life. Prokaryotic heterotrophs. Chemical evolution.		

errors

fig 39. a mislabeled
hydrophobic tail

fig 20.2 left brackets misdrawn

fig 20.3 not printed in the book

p351 prevented
no diagrams/paintings of blood cells

p336 no elastic fiber in
labeled sketch — too hard to
see

p460 ~ intestine & H_2O absorption.

ABOUT THE AUTHORS

John H. Postlethwait, professor of biology at the University of Oregon, holds a B.A. degree from Purdue University and a Ph.D. from Case Western Reserve University and did post-doctoral research at Harvard University. He was visiting Research Scientist for a year at the Institut für Molekular Biologie of the Austrian Academy of Sciences; the Laboratoire de Génétique Moléculaire, Faculté de Médecine, in Strasbourg, France; and the Imperial Cancer Research Fund in Oxford, England. He has received the Ersted Award for Distinguished Teaching at the University of Oregon. Author of *The Nature of Life,* with Janet Hopson, he has written more than one hundred research papers on genetics and developmental biology.

Janet L. Hopson is a freelance science writer and a lecturer for the Science Communication Program, University of California at Santa Cruz. She has also taught writing courses at the University of California at Berkeley and Mills College. She holds B.A. and M.S. degrees from Southern Illinois University and the University of Missouri. Coauthor of three other biology textbooks for McGraw-Hill, *Biology* and *Essentials of Biology,* with Norman Wessells, and *The Nature of Life,* with John Postlethwait, she has also written a trade book on the human sense of smell and dozens of articles in national magazines and newspapers.

Ruth C. Veres is a science writer and editor with eighteen years of experience in textbook publishing. She holds a B.A. degree from Swarthmore College and M.A. degrees from Columbia University and Tufts University. She has developed and edited more than twenty texts, including the introductory biology text *The Nature of Life,* by John Postlethwait and Janet Hopson. She coauthored a book on the immune system and has taught writing and languages at the University of California at Berkeley.

≋ BIOLOGY!
BRINGING SCIENCE TO LIFE

JOHN H. POSTLETHWAIT
University of Oregon

JANET L. HOPSON
University of California, Santa Cruz

RUTH C. VERES

McGRAW-HILL, INC.

New York   St. Louis   San Francisco   Auckland   Bogotá   Caracas
Hamburg   Lisbon   London   Madrid   Mexico   Milan   Montreal   New Delhi
Paris   San Juan   São Paulo   Singapore   Sydney   Tokyo   Toronto

For Nita, Holly, Lance, Heather, Joe, and Jordan.—J.H.P.
For Alan, Benjamin, Gillian, James, John, Julie, Madelynn, and
Steven Hopson.—J.L.H.
For Peter, Maia, Michael, Kati, Bernard, and in memory of
Anne.—R.C.V.

BIOLOGY!
BRINGING SCIENCE TO LIFE

1 2 3 4 5 6 7 8 9 0 VNH VNH 9 5 4 3 2 1

ISBN 0-07-050631-0

Library of Congress Cataloging-in-Publication Data

Postlethwait, John H.
    Biology! bringing science to life / John H. Postlethwait,
Janet L. Hopson, Ruth C. Veres.
        p.   cm.
    Includes bibliographical references and index.
    ISBN 0-07-050631-0
    1. Biology.   I. Hopson, Janet L.   II. Veres, Ruth C.
    III. Title.
QH308.2.P66   1991   90-24326
574—dc20   CIP

Sponsoring Editors: Denise Schanck, June Smith
Senior Associate Editor: Mary Eshelman
Editing Supervisors: Phyllis Niklas, Alice Mace Nakanishi
Production Supervisor: Pattie Myers
Designer: Janet Bollow
Developmental Art Consultants: Peter Veres, Cherie Wetzel
Illustrators: Martha Blake, Wayne Clark, Cindy Clark-Huegel,
Cecile Duray-Bito, Paula McKenzie, Linda McVay, Elizabeth
Morales-Denney, Victor Royer, Carla Simmons, John Waller,
Judith Waller
Box Logo Illustrator: Steve Madson
Photo Researcher: Stuart Kenter
Copyeditor: Janet Greenblatt
Indexer: Barbara Littlewood
Compositor: Graphic Typesetting Service
Color Separator: Black Dot Graphics
Printer and Binder: Von Hoffmann Press
Cover Photo: Queensland harlequin bugs (*Tectocoris dioph-
thalmus*) by David Maitland/Planet Earth Pictures
Cover Color Separator: Color Tech
Cover Printer: Phoenix Color Corp.
Production assistance by: Betty Duncan-Todd, Marian
Hartsough, Lorna Lo, Jane Moorman, Edie Williams

BRIEF CONTENTS

FULL CONTENTS

PART ONE   LIFE'S FUNDAMENTALS

  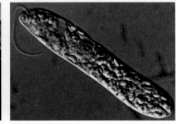 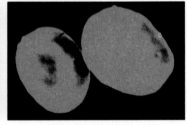

PART TWO PERPETUATION OF LIFE

  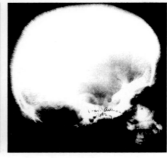 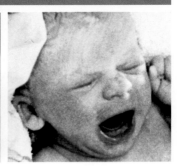

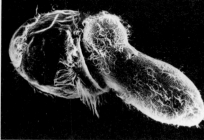

PART FOUR   HOW ANIMALS SURVIVE

CHAPTER 19   A STEADY STATE: THE KEY TO ANIMAL SURVIVAL   334

CHAPTER 20   CIRCULATION: THE INTERNAL TRANSPORT SYSTEM   346

CHAPTER 21   RESPIRATION: GAS EXCHANGE WITH THE ENVIRONMENT   360

CHAPTER 22   THE IMMUNE SYSTEM: DEFENSE FROM DISEASE   371

CHAPTER 29   PLANT FORM AND FUNCTION   472

CHAPTER 30   HOW PLANTS GROW   488

PART SIX   INTERACTIONS: ORGANISMS AND ENVIRONMENT

CHAPTER 32   THE GENETIC BASIS FOR EVOLUTION   518

PREFACE

In recent years, several groups have tested the scientific literacy of American adults, and the results have been startling.

■ Nearly a quarter of Americans think the sun revolves around the earth.
■ Forty percent can't place the Pacific Ocean on an unlabeled map.
■ One third couldn't say what a molecule is.
■ Five out of six don't know the basics of genetic engineering.
■ More than half failed a Kinsey Institute quiz on sex and reproduction.
■ Even our best high school biology students (those in advanced placement courses) scored lowest, on average, among their peers from thirteen other industrialized nations in a recent exam.

While there are several, often conflicting, explanations for our national illiteracy in biology and other sciences, surveys reveal unambiguously negative attitudes among the very people who have the most to learn. The public, to put it simply, doesn't like science very much. Most students think the subjects are difficult, only for top scholars, and all-around just plain boring.

Required to memorize biological details such as the life cycles of the slime molds, the molecular structure of a ribosome, or the anatomy of a worm, students will quite logically wonder: What's this got to do with me and my life? Texts and courses have frequently overlooked the need for a clear framework showing how details like these can fit into the broader themes of biology and everyday life, not to mention underemphasizing health, fitness, food, sex, the environment, and other topics of intrinsic interest to students.

Somehow, despite our nation's superiority in science and technology, students have missed out on the excitement of scientific discovery and the pleasure of exploring and understanding the physical world. Their natural curiosity about nature has all too often been stifled by the time they reach college, and thus they are unaware of the vital role that biology and other sciences play in our survival, now and in the future. People today are faced with dozens of important choices about what to eat or avoid; what to buy or not buy; what to support politically or vote down; how to protect themselves against AIDS, heart disease, obesity, and other conditions; and what they can do personally to prevent the global environment from being further denuded, planted, paved, and polluted. Quite clearly, we face a dilemma of serious proportions: an entire generation of unprepared students who nevertheless need to understand biological science as never before.

It was against this challenging backdrop that we planned and produced *Biology! Bringing Science to Life*. We wanted to create a learning tool that was not simply a shortened version of our well-received book *The Nature of Life*. Instead, we designed a book that could provide any beginning college student, regardless of previous biology courses or attitudes toward science, with the information he or she needs to make informed personal and societal choices as a citizen of the late twentieth century. To do this, the book would have to selectively present those supporting facts and concepts needed for clear understanding, place them in a broad context, and deliver them in a way that would interest, entertain, and provoke the reader. While the student was acquiring the basic knowledge to make informed life choices, he or she could also pick up a heightened appreciation along the way for the natural world, and for the ways biologists think, work, and discover. Beginning students and their professors have repeatedly called for such a book, and we hope *Biology!* does the job.

Each chapter begins with a story—a dilemma, really.

■ How can people, all of whom are 65 percent water, live and thrive in the desert?
■ How does solar energy (indirectly, of course) power a hiker up a mountain trail?
■ Why do some diseases resist treatment with antibiotics?
■ Why can't some people drink diet sodas?
■ How can a climber reach the top of Mt. Everest without an oxygen tank?
■ How does cocaine damage the nerves and brain of an unborn baby?
■ Is our environment heating up, and how will that affect us?

Each chapter then provides the necessary background facts and concepts to answer the question, as well as raising and addressing related ones, many of direct relevance to the students.

■ Should you catch some rays?
■ Should you undergo genetic screening?
■ What can you do to help prevent global warming and ozone depletion?

We pose many of these questions in the boxes called Health Choices, Societal Choices, and Environmental Choices. These essays always provide evidence about various sides of the issues, and encourage students to formulate their own opinions and plans of action. The essays end with lists of additional sources of information, and the lists represent our response to an interesting point raised by surveys of science literacy. The vast majority of respondents consider it far more important for a citizen to be able to locate, read, and apply scientific information than to memorize science facts and concepts and spout them at will. The much sought-after educational commodity called "critical thinking" depends on this information gathering and processing. The boxed essays called Personal Impact are also designed to stimulate critical thinking by applying basic material from the surrounding chapter to topics of student interest.

The book further encourages critical thinking through another series of boxes called Discovery. These boxes are models of the scientific process that show how biologists study the world—how they make observations, pose specific questions, and test possible answers. The boxes are designed to be more than a recitation of the scientific method. They are intended to show how scientists think and what processes they use—processes which we all employ every day—in solving problems.

The contemporary graphic and art styles we have chosen for *Biology!* are intended to invite and appeal to readers, as well as to illustrate the material in a simple, clear way. An innovative feature called At a Glance provides an illustrated summary at the end of each chapter. Other study tools in each chapter include underlined take-home messages, a Connections section to review and interrelate concepts, a New Terms list, Study Questions, and a For Further Reading list. Our approach is deliberately visual and oriented toward high student interest and easy review. We hope it accomplishes our goals and will help in some small way to close the worrisome gap between society's science illiteracy and its desperate need to understand and apply science facts and ideas to daily life and global directions.

≋ SUPPLEMENTARY MATERIALS

A comprehensive and completely integrated package of supplementary materials accompanies *Biology! Bringing Science to Life.*

- **Instructor's Manual and Resource Guide**
  Dennis Todd, University of Oregon
- **LecturePak** (transparency masters of lecture outlines)
  Dennis Todd, University of Oregon
- **Study Guide**
  Deborah M. Brosnan, Oregon State University
- **Test Bank**
  Dennis Todd, University of Oregon
- **Laboratory Manual**
  Eileen Jokinen, University of Connecticut, Storrs
  Theodore Taigen, University of Connecticut, Storrs
  Thomas Terry, University of Connecticut, Storrs
  David Wagner, University of Connecticut, Storrs
- **Laboratory Preparator's Guide**
  Eileen Jokinen, University of Connecticut, Storrs
  Theodore Taigen, University of Connecticut, Storrs

Thomas Terry, University of Connecticut, Storrs
David Wagner, University of Connecticut, Storrs
- Computerized Instructor's Manual (available in IBM, Macintosh, Apple)
- Computerized Test Bank (available in IBM, Macintosh, Apple)
- *BioPartner* (Computerized Study Guide; available in IBM, Macintosh)
- Videos
- Biology Slides and Acetate Package
- Videodisk
- HyperMedia Software (Macintosh):
  Lecture Planner (to accompany the Videodisk)
    Daniel Udovic, University of Oregon
  Biological Diversity (interactive software)
    Daniel Udovic, University of Oregon

For further information regarding the supplements available, please contact your local McGraw-Hill representative.

≋ ACKNOWLEDGMENTS

We have sought the advice of hundreds of instructors around the country to help us create a textbook that would meet the unique needs of the introductory biology market. Our sincere thanks are extended to the following individuals who responded to our market questionnaires:

Dr. Laura Adamkewicz, George Mason University; **Olukemi Adewusi**, Ferris State University; **Kraig Adler**, Cornell University; **Dr. John U. Aliff**, Glendale Community College; **Joanna T. Ambron**, Queensborough Community College; **Steven Austad**, Harvard University; **Robert J. Baalman**, California State University, Hayward; **Stuart S. Bamforth**, Tulane University; **Sarah F. Barlow**, Middle Tennessee State University; **R. J. Barnett**, California State University, Chico; **Joseph A. Beatty**, Southern Illinois University, Carbondale; **Nancy Benchimol**, Nassau Community College; **Dr. Rolf W. Benseler**, California State University, Hayward; **Gerald Bergtrom**, University of Wisconsin, Milwaukee; **Dr. Dorothy B. Berner**, Temple University; **Dr. A. K. Boateng**, Florida Community College, Jacksonville; **William S. Bradshaw**, Brigham Young University; **Jonathan Brosin**, Sacramento City College; **Howard E. Buhse, Jr.**, University of Illinois, Chicago; **John Burger**, University of New Hampshire; **F. M. Butterworth**, Oakland University; **Guy Cameron**, University of Houston; **Ian M. Campbell**, University of Pittsburgh; **John L. Caruso**, University of Cincinnati; **Brenda Casper**, University of Pennsylvania; **Doug Cheeseman**, De Anza College; **Dr. Gregory Cheplick**, University of Wisconsin; **Dr. Joseph P. Chinnici**, Virginia Commonwealth University; **Carl F. Chuey**, Youngstown State University; **Dr. Simon Chung**, Northeastern Illinois University; **Norman S. Cohn**, Ohio University; **Paul Colinvaux**, Ohio State University; **Scott L. Collins**, University of Oklahoma; **Dr. August J. Colo**, Middlesex County College; **G. Dennis Cooke**, Kent State; **Jack D. Cote**, College of Lake

County; **Gerald T. Cowley**, University of South Carolina; **Louis Crescitelli**, Bergen Community College; **Orlando Cuellar**, University of Utah; Thomas Daniel, University of Washington; **J. Michael DeBow**, San Joaquin Delta College; **Loren Denny**, Southwest Missouri State University; **Ron DePry**, Fresno City College; **Dr. Kathryn Dickson**, California State University, Fullerton; **Patrick J. Doyle**, Middle Tennessee State University; **Dr. David W. Eldridge**, Baylor University; **Lynne Elkin**, California State University, Hayward; **Paul R. Elliott**, Florida State University; **Eldon Enger**, Delta College; **Gauhari Farooka**, University of Nebraska, Omaha; **Marvin Fawley**, North Dakota State University; **Ronald R. Fenstermacher**, Community College of Philadelphia; **Edwin Franks**, Western Illinois University; **C. E. Freeman**, University of Texas, El Paso; **Lawrence D. Friedman**, University of Missouri, St. Louis; **Dr. Ric A. Garcia**, Clemson University; **Wendell Gauger**, University of Nebraska, Lincoln; **Dr. S. M. Gittleson**, Fairleigh Dickinson University; **E. Goudsmit**, Oakland University; **John S. Graham**, Bowling Green State University; **Shirley Graham**, Kent State University; **Thomas Gregg**, Miami University; **Alan Groeger**, Southwest Texas State University; **Thaddeus A. Grudzien**, Oakland University; **James A. Guikema**, Kansas State University; **Robert W. Hamilton**, Loyola University of Chicago; **Richard C. Harrel**, Lamar University; **T. P. Harrison**, Central State University; **Maurice E. Hartman**, Palm Beach Community College; **Dr. Karl H. Hasenstein**, University of Southwestern Louisiana; **Martin A. Hegyi**, Fordham University; **Dr. John J. Heise**, Georgia Institute of Technology; **H. T. Hendrickson**, University of North Carolina, Greensboro; **T. R. Hoage**, Sam Houston State University; **Kurt G. Hofer**, Florida State University; **Dr. Rhodes B. Holliman**, Virginia Polytechnic Institute and State University; **Harry L. Holloway**, University of North Dakota; **E. Bruce Holmes**, Western Illinois University; **Jerry H. Hubschman**, Wright State University; **Hadar Isseroff**, State University of New York College, Buffalo; **Dr. Ira James**, California State University, Long Beach; **Dr. Wilmar B. Jansma**, University of Northern Iowa; **Dr. Margaret Jefferson**, California State University, Los Angeles; **Dr. Ira Jones**, California State University, Long Beach; **Dr. Patricia P. Jones**, Stanford University; **Dr. Craig T. Jordan**, University of Texas, San Antonio; **Maurice C. Kalb**, University of Wisconsin, Whitewater; **Bonnie Kalison**, Mesa College; **Judy Kandel**, California State University, Fullerton; **Arnold Karpoff**, University of Louisville; **L. G. Kavaljian**, California State University, Sacramento; **Donald R. Kirk**, Shasta College; **R. Koide**, Pennsylvania State University; **Mark Konikoff**, University of Southwestern Louisiana; **Barbara S. Lake**, Central Piedmont Community College; **Jim des Lauvérs**, Chaffey College; **Tami Levitt-Gilmarr**, Pennsylvania State University; **Daniel Linzer**, Northwestern University; **J. R. Loewenberg**, University of Wisconsin, Milwaukee; **Dr. Robert Lonard**, University of Texas, Pan American; **Sharon R. Long**, Stanford University; **Carmita E. Love**, Community College of Philadelphia; **C. E. Ludwig**, California State University, Sacramento; **Dr. Ann S. Lumsden**, Florida State University; **Dr. Bonnie Lustigman**, Montclair State College; **Edward B. Lyke**, California State University, Hayward; **Douglas Lyng**, Indiana University–Purdue University, Ft. Wayne; **George**

L. Marchin, Kansas State University; **Philip M. Mathis**, Middle Tennessee State University; **Mrs. Margaret L. May**, Virginia Commonwealth University; **Edward McCrady**, University of North Carolina, Greensboro; **Bruce McCune**, Oregon State University; **Dr. John O. Mecom**, Richland College; **Tekié Mehary**, University of Washington; **Richard L. Miller**, Temple University; **Phyllis Moore**, University of Arkansas, Little Rock; **Carl Moos**, State University of New York, Stony Brook; **Doris Morgan**, Middlesex County College; **Donald B. Morzenti**, Milwaukee Area Technical College; **Steve Murray**, California State University, Fullerton; **Robert Neill**, University of Texas, Arlington; **Paul Nollen**, Western Illinois University; **Kenneth Nuss**, University of Northern Iowa; **William D. O'Dell**, University of Nebraska, Omaha; **Dr. Joyce K. Ono**, California State University, Fullerton; **James T. Oris**, Miami University; **Clark L. Ovrebo**, Central State University, Edmond; **Charles Page**, El Camino College; **Kay Pauling**, Foothill College; **Dr. Chris E. Petersen**, College of DuPage; **Richard Petersen**, Portland State University; **Jeffrey Pommerville**, Glendale Community College; **David I. Rasmussen**, Arizona State University; **Daniel Read**, Central Piedmont Community College; **Dr. Don Reinhardt**, Georgia State University; **Louis Renaud**, Prince George's Community College; **Jackie Reynolds**, Richland College; **Jennifer H. Richards**, Florida International University; **Thomas L. Richards**, California State Polytechnic University; **Tom Rike**, Glendale Community College, California; **C. L. Rockett**, Bowling Green State University; **Hugh Rooney**, J. S. Reynolds Community College; **Wayne C. Rosing**, Middle Tennessee State University; **Frederick C. Ross**, Delta College; **A. H. Rothman**, California State University, Fullerton; **Mary Lou Rottman**, University of Colorado, Denver; **Dr. Donald J. Roufa**, Kansas State University; **Dr. Michael Rourke**, Bakersfield College; **Chester E. Rufh**, Youngstown State University; **Mariette Ruppert**, Clemson University; **Charles L. Rutherford**, Virginia Polytechnic Institute and State University; **Dr. Milton Saier**, University of California, San Diego; **Lisa Sardinia**, San Francisco State University; **A. G. Scarbrough**, Towson State University; **Dan Scheirer**, Northeastern University; **Randall Schietzelf**, Harper College; **Robert W. Schuhmacher**, Kean College of New Jersey; **Joel S. Schwartz**, College of Staten Island; **Roger S. Sharpe**, University of Nebraska, Omaha; **Stanley Shostak**, University of Pittsburgh; **J. Kenneth Shull, Jr.**, Appalachian State University; **C. Steven Sikes**, University of South Alabama; **Christopher C. Smith**, Kansas State University; **John O. Stanton**, Monroe Community College; **D. R. Starr**, Mt. Hood Community College; **Dr. Ruth B. Thomas**, Sam Houston State University; **Nancy C. Tuckman**, Loyola University of Chicago; **Dr. Spencer Jay Turkel**, New York Institute of Technology; **William A. Turner**, Wayne State University; **C. L. Tydings**, Youngstown State University; **John Tyson**, Virginia Polytechnic Institute and State University; **Richard R. Vance**, University of California, Los Angeles; **Harry van Keulen**, Cleveland State University; **Roy M. Ventullo**, University of Dayton; **Judith A. Verbeke**, University of Illinois, Chicago; **Dr. Ronald B. Walter**, Southwest Texas State University; **Stephen Watts**, University of Alabama, Birmingham; **Dr. Joel D. Weintraub**, California State University, Fullerton; **Marion R. Wells**,

Middle Tennessee State University; **James White**, New York City Technical College; **Joe Whitesell**, University of Arkansas, Little Rock; **Fred Whittaker**, University of Louisville; **Roberta Williams**, University of Nevada, Las Vegas; **Chuck Wimpee**, University of Wisconsin, Milwaukee; **Mala Wingerd**, San Diego State University; **Richard Wise**, Bakersfield College; **Gary Wisehart**, San Diego City College; **Dan Wivagg**, Baylor University; **Richard P. Wurst**, Central Connecticut State University; **Edward K. Yeargers**, Georgia Institute of Technology; **Linda Yasui**, Northern Illinois University

There would be no *Biology! Bringing Science to Life* without the talented leadership and steady guidance of Eirik Borve, June Smith, and Denise Schanck, as well as the support of McGraw-Hill. We are also indebted to the dedicated professionals who produced this complex work, and who assisted in various other ways: Blake Edgar, Mary Eshelman, Karen Judd, Stuart Kenter, Alice Mace Nakanishi, Pattie Myers, Phyllis Niklas, Francis Owens, Phyllis Snyder, and Lesley Walsh.

We are grateful to Professor Nancy Parker, Southern Illinois University, Edwardsville, who helped steer us to the most up-to-date information, presented in the most accurate manner possible.

We would also like to thank the following individuals who contributed material and advice, which has been incorporated in this text: Charles L. Aker, The National Autonomous University of Nicaragua; Russ Fernald, University of Oregon; Craig Heller, Stanford University; Kent Holsinger, University of Connecticut; V. Pat Lombardi, University of Oregon; Christopher Stringer, British Museum, London; Daniel Udovic, University of Oregon; and Lenny Vincent, Fullerton College.

We would like to warmly acknowledge Peter Veres for his conceptualization of the At a Glance feature.

We greatly appreciate the many scientists and artists whose photographs and drawings illustrate these pages. Their names are listed on pages iv and A-13 to A-16.

Finally, we thank each other for the spirit of cooperation, mutual respect, and support that has made every step of this project enjoyable. If our student readers enjoy it half as much, we will be well satisfied!

John Postlethwait
Janet Hopson
Ruth Veres

WHAT IS LIFE?

LIFE HISTORY OF THE STRAWBERRY FROG

Several times a year, in the lush, misty cloud forests of Central America, small, flamboyantly scarlet strawberry frogs produce and care for a new generation. The young emerge from the fertilized eggs as dark, wriggling tadpoles with yolk from the egg still filling much of their bodies. After six weeks, the tadpoles sprout arms and legs, lose their tails, and emerge from the water as brilliant red adult frogs with deadly poisonous skin glands (Figure 1.1), ready to mate, brood, and care for new young.

The strawberry frog has a number of unusual characteristics. The most obvious is its dazzling coloration. But also unique is the manner in which

these frogs care for their offspring. The female transports newly hatched tadpoles piggyback from a rolled-up dry leaf to a basin of clear water trapped in a tropical plant. When she returns to feed her young, they signal their presence by stiffening their bodies and vibrating their tails. She then deposits a few infertile eggs for them to eat.

The strawberry frog's life history provides a dramatic illustration of the two most basic problems every living thing must solve. First, living things must take in energy to maintain their internal order and organization, because everything in the universe, living and nonliving, tends naturally toward a state of

disorder. This disorganization is apparent from the fact that strawberry frog tadpoles die and disintegrate unless the mother deposits sterile eggs into their basin as a source of food energy.

Second, individual living organisms face the certainty of death. Despite its bright scarlet hue, which serves to forewarn enemies, and its poison glands, which discourage or kill attackers, a strawberry frog will usually survive only about three years. If groups of strawberry frogs are to exist beyond an individual's normal life span, they must not only take in energy, but also pair up and reproduce. Through reproduction, each kind of living thing continues even though individuals die.

Our major goal in this book is to explore **biology,** the science of living things. We will repeatedly encounter the twin challenges of disorder and death in our discussions and see that living things share a set of fundamental characteristics that help them meet those challenges. We will also discuss fascinating experiments to see how biologists study the living world. Finally, we will consider many of the challenges now facing our planet and its organisms.

Chapter 1 outlines the general characteristics of the living state. And like each chapter that follows, it addresses several questions, which we list as an advance organizer. Here are the questions we will answer in Chapter 1:

- Why study biology?
- What nine characteristics define a living thing?
- How do people learn about the living world?
- How can biology help us solve some of our world's problems?

FIGURE 1.1 *Strawberry Frog in a Cloud Forest. Dendrobates pumilio*, a tiny resident of the Central American jungle, has flaming red skin and highly poisonous skin glands. The frog carries a newly hatched tadpole on its back to a pool of water trapped in a leaf. This behavior ultimately helps ensure the survival of new generations of strawberry frogs.

※ WHY STUDY BIOLOGY?

Students sometimes wonder how life science—the realm of fungi and fruit flies and DNA—relates to them and why they should work hard to study and understand it. In fact, biology is one of the most relevant subjects in a modern university. Here are just a few of the reasons why (Figure 1.2):

1. If you're reading this, you are alive! We humans are living mammals: We are born, we eat, we grow, we have sex, we reproduce, we age, and we die. We live in a range of the earth's ecological zones, from deserts to polar caps, from mountains and prairies to forests and coastlines, and our bodies serve as habitats for a profusion of bacteria, protozoa, fungi, and viruses. Studying biology allows us to understand our place

in the earth's environments, our unwitting hospitality to so many microorganisms, and our relationship to life forms here and perhaps elsewhere in the universe.

2. As conscious beings, we make daily choices about how to maintain our own life and health. Each of us decides what to eat; when and how much to sleep; how much and what types of exercise we need; whether or not to smoke, drink, or use birth control; how to avoid sexually transmitted diseases; and so on. A course in biology helps us understand the basic mechanisms and effects of eating, sleeping, exercising, smoking, and other common activities, and how to make the best choices for ourselves.

3. As finely tuned biological entities, we sometimes get out of balance and become ill. Biology focuses not only on normal life processes, but on the ways they

(a)

(b)

(c)

FIGURE 1.2 *Why Study Biology?* (a) This Eskimo family lives year round in an arctic environment of extreme cold, where there are short summers of continuous day and long winters of continuous night. (b) Vigorous exercise is one means of staying fit. (c) A genetic counselor (right) can help prospective parents, or young parents considering having more children, make informed reproductive decisions based on their families' genetic history and the available options. (d) The computerized tomography (CAT) scanner (on right in background) can reconstruct images of internal organs, such as the brain, and display them on a television screen. With this high-tech tool, physicians can locate tumors or blood clots without surgery or other invasive procedures and gather precise information for evaluating medical emergencies. (e) Every car on the crowded freeways of Los Angeles produces polluting exhaust fumes that contribute to the heavy smog of the area. During smog alerts, it is not safe to breathe the air outdoors.

can go awry in disease states such as cancer, AIDS, heart attack, Rh incompatibility during pregnancy, and cystic fibrosis and other genetic conditions. Understanding the mechanisms of health and disease improves our decision making as medical consumers and as citizens who vote on issues like abortion, human gene therapy, and animal experimentation.

4. We live in an era of science and technology, with a constant stream of complex, sometimes confusing new developments modifying our lives in both large and small ways. Without a knowledge of biology, significant developments, such as gene splicing and genetic screening, would seem threatening rather than the positive tools they have already become for medicine, agriculture, and scientific research.

5. We also live in an era of human overpopulation, and our species' burgeoning numbers have led to an unprecedented assault on the air, water, land, climate, and other living organisms of our planet. Studying life science can help us interpret the sobering news published daily, and it can help us make personal and political choices that will protect our global life-support systems as well as the other living things that share our world.

6. Finally, a solid background in life science is required for many careers, including work in the allied health professions, agriculture, food service, parks and recreation, and education. What's more, occupations with a scientific and technological component will be increasingly available in the future.

This list shows how important biology is to a college student in today's world. And beyond all these reasons, it is simply a fascinating subject. Let's dig in and see.

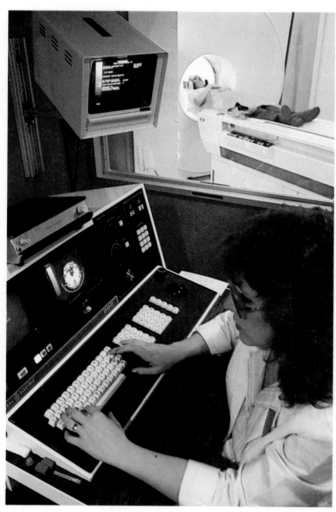

(d)

(e)

✳ TO BE ALIVE: THE SHARED FEATURES OF LIVING THINGS

What is a living thing? When is it alive or dead? What, in fact, is the nature of life? These questions have never been more important, because today's biologists and physicians have unprecedented abilities to sustain the human body on life-support machines; to freeze or dry living materials for later use; and to change and merge the traits of particular microbes, plants, and animals. In the broadest sense, biological science probes the question, *What is life?*, but biologists have not yet formulated a simple definition. Despite current sophisticated knowledge and thousands of years of observations and descriptions of the natural world, life scientists are not able to define the essence of life in a single statement that most can agree on and that is not a circular definition such as, "Life is the quality of living things that separates them from nonliving things."

Instead, biologists usually sidestep the thorny issue by defining life through a set of nine observable characteristics of living things, listed in Table 1.1. The first five traits in the table relate to an individual's day-to-day survival, and they address the universal tendency toward disorganization. They are (1) *order,* a precise arrangement of structural units and activities; (2) *adaptations,* structures and activities that allow an individual organism to use its environment better; (3) *metabolism,* the chemical breakdown, conversion, and use of energy-rich compounds; (4) *movement,* the self-generated motion of an individual or its parts; and (5) *responsiveness,* the ability to sense and react to surroundings. Without these trademarks, most living things would perish rapidly.

The next three characteristics relate to the persistence of groups of very similar living things over time, even though individuals die: (6) *Reproduction* is the process that gives rise to offspring; (7) *development* is the orderly sequence of physical and behavioral changes (such as growth or sexual maturation) that occur during an organism's life cycle; and (8) *genes* are the units of inheritance, passed from parent to offspring, that control many daily functions as well as the way an organism looks.

Finally, populations of living things adjust to environmental variations through (9) *evolution,* a change in the traits of a group over time. Ultimately, evolution provides new ways for organisms to overcome disorganization and death.

Many nonliving entities have one or more of these characteristics; for example, waves move, flames use energy, and crystals grow. Only living organisms, however, display *all* the characteristics simultaneously or at some point during their individual life cycle or group history.

The third column in Table 1.1 explains how the strawberry frogs of Central America's cloud forests manifest each life characteristic. Throughout the rest of the book, you will find hundreds more examples of how living things display the collective traits that distinguish a rock from a rockrose and water from a water lily.

In contrast to individual living things, life as a whole on the planet earth can be characterized by two additional qualities—*unity* and *diversity*—both of which are consequences and products of evolutionary change. Living things on earth share the same hereditary language, or genetic code. The singular nature of the genetic code reveals the historical relatedness of all living things. In addition, living organisms are fantastically varied in shape, size, color, and life habits, with at least 5 million and perhaps upward of 10 million separate species now alive. This stunning diversity is not a prerequisite of life; a planet could easily be inhabited by just a few kinds of simple organisms, as the earth itself most likely was 4 billion years ago. But constant changes in global climate and the massive shifting of continents and bodies of water during the earth's long history opened up new sources of energy and presented new opportunities for change within living species. Unity and diversity are thus an outgrowth of life's history on a changing planet and evidence of evolution at work.

Let us consider the defining characteristics of life one by one now and see how they allow living things to overcome disorganization and death.

✳ ORDER AND DISORDER: LIFE TRAITS THAT COMBAT DISORGANIZATION

Living organisms are highly ordered. Strawberry frogs, for example, have two large dark eyes, muscles in certain places with certain dimensions that allow the legs to paddle and hop, and a dazzling red skin that covers every bit of the animal's exterior. As time passes, some of this organization starts to deteriorate. While climbing a tree one day in search of ants, a frog may slip and crash to the forest floor, puncturing its skin, bruising its legs, and becoming more disorganized.

TABLE 1.1 ≋ NINE CHARACTERISTICS OF LIFE

	EXPLANATION	EXAMPLE
Day-to-Day Survival: Overcoming Disorganization		
1. Order	There is a precise arrangement of structures and activities in living things, and each has a specific relation to all the others.	Frogs have one eye on each side of the head.
2. Adaptations	Specific structures and behaviors suit organisms to their environments.	Frogs have a long, sticky tongue that entraps insects and other food items.
3. Metabolism	Organized chemical steps break down molecules and convert them into products that build body parts or make energy available.	A frog digests an ant into small chemicals, then reorganizes them into new body parts.
4. Movement	Using their own power, organisms are able to move themselves or their body parts through space.	A frog's tongue flips out to snag a fly.
5. Responsiveness	Organisms perceive the environment and react to it.	When a frog sees an ant, it flips out its tongue.
Survival Over Time: Continuing the Group Despite the Death of Individuals		
6. Reproduction	Organisms give rise to others of the same type.	Frogs produce more frogs.
7. Development	Ordered sequences of progressive changes result in increased complexity.	An egg becomes a tadpole, which becomes a frog.
8. Genes	Organisms have units of inheritance that control physical, chemical, and behavioral traits.	Genes control whether a frog will be red or black.
Group Changes Over Time: Adjusting to Environmental Change		
9. Evolution	Through evolution, species acquire new ways to survive, to obtain and use energy, and to reproduce.	Bright red frogs with poison skin survive and reproduce effectively in the cloud forest environment.

This kind of wear and tear is not unique to life: Rocks, stars, rivers, and mountain ranges all tend to become more disordered as time goes by. In an injured frog, however, broken skin is replaced and wounded muscle heals when the animal's body extracts energy and materials from the insects it catches and eats, then uses the energy and materials to generate new skin and healthy muscle.

In general, living things counteract the disorder that comes with time by taking energy and materials from their surroundings and employing them for maintenance, growth, and other survival activities.

(a)

(b)

(c)

FIGURE 1.3 *Order in Life and Nonlife.* (a) A fruit fly's eye and (b) the tightly packed seeds of a sunflower form geometric arrays that reveal the order inherent in living things. (c) A translucent, angular quartz crystal is also geometric and constructed from orderly arrays of molecules. There are, however, essential differences between living and nonliving things, and the shared characteristics of the fly and the flower form a useful working definition of life.

LIFE CHARACTERISTIC 1: ORDER WITHIN LIVING THINGS

The workings of a Swiss clock are marvelously intricate, and the parts of a race car are beautifully engineered for high performance. But living organisms possess a degree of **order**—a structural and behavioral complexity and regularity far greater than anything in the nonliving world. The eye of a fly, for example, and the spiral-packed seeds of a sunflower head consist of highly organized units repeated and arranged in precise geometric arrays (Figure 1.3a and b). In fact, the most highly organized struc-

ture yet discovered in the universe is the human brain, such as the one allowing you to read and understand this page. And while glittering quartz crystals are a highly ordered latticework of silicon dioxide molecules (Figure 1.3c), crystals and organisms grow in fundamentally different ways: Living things take energy and raw materials *inside* the body and reorganize them into body parts; crystals grow when chemicals are passively added to the *outside* surfaces, layer upon identical layer.

■ **Order Revealed in the Hierarchy of Life** Life's organization is obvious in whole organs such as eyes or flowers, but it occurs at the microscopic level as well and at the level of large living groups in their environments. Consider, for example, the African savanna (Figure 1.4). The levels of biological order and organization it represents reveal a fascinating **hierarchy of life** (Figure 1.5). The savanna is a collection of elephants, egrets, acacia trees, tussock grasses, clouds, arid plains, and other bio-

FIGURE 1.4 *Order Reigns at Every Level in the Living World.* An elephant herd in the shadow of Mt. Kilimanjaro on the African savanna symbolizes a hierarchy of organization. The elephants' bodies are made up of highly ordered cell parts, cells, tissues, organs, and organ systems that function together smoothly. The herd members are part of a population within a diverse community, including the birds picking insects off their thick hides and the grasses they graze and trample. The community, in turn, is part of the savanna ecosystem, with its expansive plain, snowcapped mountain range, and arid climate.

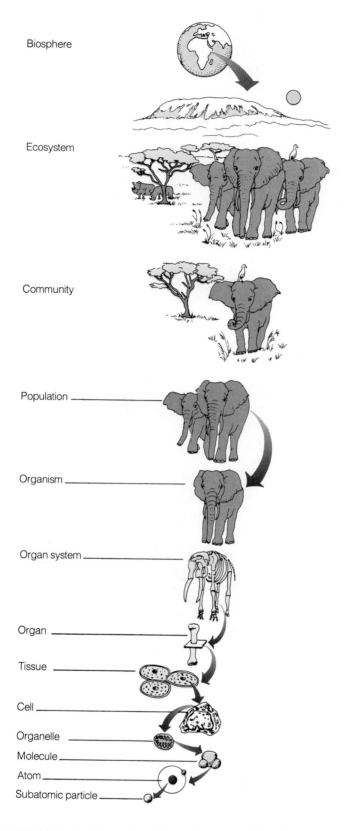

Biosphere

Ecosystem

Community

Population

Organism

Organ system

Organ

Tissue

Cell

Organelle

Molecule

Atom

Subatomic particle

FIGURE 1.5 *The Hierarchy of Life and Its Levels of Organization.* As the text explains, an elephant, as a whole organism, lies midway on a continuum of life's inherent hierarchy of order, stretching from subatomic particles to the biosphere.

logical and environmental elements. Although it is vast and complex, it is but a small portion of the *biosphere:* our entire planet earth and all its living inhabitants. The biosphere consists of many *ecosystems:* living things in particular regions and their nonliving physical surroundings. The savanna ecosystem includes elephants, the egrets that pick insects off their skin, and the coarse grass they chew and trample, as well as the water in the clouds, the sandy soil underfoot, and the hot African sun overhead. The living part of an ecosystem is the *community:* an assemblage of interacting organisms living in a particular place—in this case, the plants, animals, and other organisms of the savanna.

Communities are made up of *populations:* groups of individuals of a particular type that live in the same area and actively interbreed with one another, such as an elephant herd or a field of grass. Populations are made up of *organisms,* the next lowest level of organization. Organisms are independent individuals that express life's characteristics. A tuskless, old female elephant, the matriarch and leader in this particular population, is a single organism (see Figure 1.4). Each organism, in turn, is made up of *organ systems:* sets of body parts that carry out general functions within the organism. The skeletal system, for example, supports the elephant's body.

Moving further down the hierarchy, organ systems are composed of *organs:* units that perform specialized functions, such as a single bone that supports part of an elephant's leg or a single blade of grass that produces sugars for the grass plant.

Each organ, in turn, is made up of *tissues:* groups of similar cells. *Cells* are the simplest entities in the hierarchy that have all the properties of life listed in Table 1.1; they are the least complicated units that we can say are truly alive. Within cells are *organelles:* structures that perform specialized functions for the cell, such as the information-storing cell nucleus. Organelles are made up of *molecules,* which are clusters of atoms; and *atoms* are the smallest units of matter that still have distinct chemical properties. Atoms, finally, are composed of *subatomic particles,* fundamental units of energy and matter.

The hierarchy of life is important because it identifies the various levels at which we can examine and understand biology. For example, to understand the significance of a strawberry frog's fiery coloration (organism level), a biologist wants to know how the skin makes its brilliant colors (levels of organ systems and cells); how the bright colors affect the frog's reproductive activities (population level); and how these colors help the frog avoid being eaten (community level). Biologists delight in discovering new phenomena (like the strawberry frog's piggyback tadpole transport) and in explaining their origin and utility at various levels in the hierarchy of life.

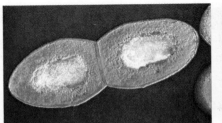

(a) Monera

(b) Protista

(c) Fungi

(d) Plantae

(e) Animalia

FIGURE 1.6 *Life's Five Kingdoms.* (a) This *Streptococcus* cell, a microscopic bacterium magnified 20,750 times, represents the kingdom Monera. The earliest life forms on earth were monerans. Today, there are more than 2500 bacterial species, living in air, soil, and water. (b) The kingdom Protista also contains microscopic single-celled organisms, but they are generally much larger and more complex than monerans, and they appeared on earth about 2 billion years after the monerans. This *Paramecium* cell, with its exterior covering of beating hairlike organelles, is a protist. (c) Fungi, such as this *Hygrophorus conicus* mushroom decomposing leaf litter on the forest floor, make up a third kingdom. Fungi live symbiotically with many plants and help expand the uptake of water and nutrients. Fungi are also used in making drugs, cheeses, and soy sauce. (d) The plant kingdom includes a large group of many-celled organisms that generate their own food molecules and are often conspicuous and graceful, such as this pine. Plants are the major source of food for animals—the fifth kingdom—including humans. (e) The largest of the five kingdoms contains over a million species of animals, including sponges, jellyfish, worms, insects, fishes, frogs, birds, and mammals. Most of these lack backbones, but some, like this Arctic loon, have an internal skeleton and complex behavior patterns.

Life's levels of organization provide a general outline for the inquiry in this book. Part One probes the nature of atoms and molecules, because all living things are built of them and are governed by the same rules of matter and energy. Part Two discusses the rules of heredity and development common to all organisms. Part Three describes the full range of living species, from bacteria to animals. Parts Four and Five explain how these organisms function and survive. And Part Six covers the highest levels of organization and the interactions of living things with one another and with their environment.

■ **Order Revealed in the Diversity of Life** In addition to the natural order apparent from the hierarchical levels of life, biologists have a system for categorizing organisms that separates living things into groups according to their similarities. Strawberry frogs, for example, are more similar to bullfrogs than to elephants, but frogs and elephants are more similar to each other than to tussock grass. To make sense of life's wonderful diversity on earth, biologists divide organisms into *species*, sets of structurally similar individuals that all descend from the same initial group and that have the potential to successfully breed with one another. Several related and similar species may make up a *genus* (plural, genera). Just as a person has a family name and a given name, biologists refer to each species by its genus name followed by its species name. For example, the strawberry frog is called *Dendrobates pumilio*, while the closely related "play-actor" frog is called *Dendrobates histrionicus* (see Figure 18.9); *D. histrionicus* has different color markings and does not breed with *D. pumilio* (note that the genus can be designated by its first letter). Just as organisms are grouped into species and species into genera, similar organisms are arranged by broader and broader criteria until ultimately, the millions of life forms on earth are assigned to just five **kingdoms** (Figure 1.6). The simplest kingdom of life is *Monera*, mostly single-celled organisms called bacteria, including the microscopic entities that turn milk into yogurt or that cause strep throat. The next kingdom, *Protista*, is also made up of single-celled organisms, but they are larger and more complex and include the interesting species that swim around in drops of pond water viewed through a microscope. Kingdom *Fungi* consists of decomposers: some single-celled organisms, including the yeasts, and many multicellular (many-celled) forms like mushrooms. Fungi absorb their energy and materials after decomposing other living or dead organisms or their parts. The fourth kingdom, *Plantae*, includes ferns, maple trees, and other members that are usually green and multicellular and that generate their own food from air, water, and sunlight. The fifth and largest kingdom, *Animalia*, includes the world's most complex organisms—earthworms, pigs, snails, people, and more than a million other species. Animals are multicellular and get their energy by ingesting other organisms, alive or dead.

Many types of evidence suggest that the five kingdoms of organisms are related to each other. Monerans form the most ancient kingdom, and their descendants, the protists, probably gave rise independently to the fungal, plant, and animal kingdoms.

LIFE CHARACTERISTIC 2: ADAPTATIONS

Because the physical world tends to become more disordered, the intricate organization within a frog, flower, or other living thing will begin to disappear without a constant input of useful energy. Different species have different ways of extracting energy and materials from their specific environments, and since these different methods *adapt* the organisms to their own special ways of life, they are called **adaptations.** Not all adaptations relate to the intake of energy and materials. Some improve the organism's ability to grow; others, to reproduce, or move, or live in a group, or attract a mate. All adaptations, however, increase the individual's chances of combating disorganization or improve the species' chance of continuing for further generations.

Different kingdoms of life have different adaptations for obtaining energy. The fungus in Figure 1.6c has adaptations of structure and activity that allow it to secrete substances into the forest soil, rich with rotting leaves and other plant matter. The fungal substances degrade some of that plant matter, and the fungal cells can then absorb nutrient molecules and use them for energy and materials. A frog has a different set of feeding adaptations, including a long, gluey tongue folded over at the tip (Figure 1.7, page 10). When an ant or fly ventures near, the frog quickly unfurls its tongue, mires the victim in a sticky coating, retracts its tongue, and devours the morsel. Plants do not move around in search of food; instead, they generate their own by trapping the energy of sunlight and converting molecules from air and water into sugars. Adaptations originate only over many generations as the environment allows better-adapted individuals to reproduce more effectively.

LIFE CHARACTERISTIC 3: METABOLISM

Once an organism has taken in energy-containing food molecules or has manufactured its own food using sunlight and gases, an organized series of chemical steps called **metabolism** breaks down the molecules. Other

FIGURE 1.7 *Sticky Tongues: A Feeding Adaptation in Frogs.* An adult frog has a long, sticky tongue that it flips forward with lightning speed to entrap insects for food.

metabolic processes then rearrange these simpler molecules into products that are useful for repairing old structures or building new ones and unlock energy necessary to transport and organize materials within the organism's body. Light shining on a palm leaf (Figure 1.8), for example, fuels a series of metabolic steps that causes the plant to construct high-energy compounds, move organelles around inside cells, construct new leaves, and break down worn-out cells and organs.

FIGURE 1.8 *Metabolism: Conversion of Energy and Materials.* The spoked leaf of a fan palm acts as a solar collector. Leaf cells take in the sun's energy and trap it in a chemical form that metabolic processes employ for growth, repair, and movement—activities that help maintain order within the plant.

FIGURE 1.9 *Poetry in Motion: Animal Movement Can Be Swift and Elegant.* These impalas, residents of savanna and woodland, cavort in the African sunshine.

LIFE CHARACTERISTIC 4: MOVEMENT

Energy unlocked by the steps of metabolism allows the movement of materials to areas where they are used to replace worn parts, help increase size, and generally combat disorganization. Self-propelled motion, or **movement,** is a diagnostic feature of life. Even simple organisms can move on their own; the tiny "hairs" on a *Paramecium*, for example, can propel it toward concentrations of food particles (see Figure 1.6b). Organisms such as plants, which may not seem to move about, nonetheless have a constant streaming rotation of fluid within each cell. What's more, whole plant organs, such as leaves, usually track the sun's path through the sky each day. Animals, of course, rely on movement in their pursuit of food, their escape from enemies, their social relations (Figure 1.9)—in general, in their avoidance of disorganization or death.

LIFE CHARACTERISTIC 5: RESPONSIVENESS

To maintain a well-ordered body, an organism must be responsive to its environment. It must exhibit **responsiveness**—detect the presence of food or enemies, sun or cold, water or dry land, and then react in ways that help maintain its body's organization or avoid its destruction. Responsiveness can be instantaneous, as when a small moth hears the high-pitched whine of a swooping, hungry bat (Figure 1.10), or it can be gradual and seasonal, as when trumpeter swans detect the shortening days of autumn and respond by flying to wetlands in warmer southern regions. Plants, too, must respond to their environments, conserving water during times of drought and capturing sunlight when it is most readily available.

THE FLOW OF ENERGY AND CYCLING OF MATERIALS: DEFEATING DISORGANIZATION

The five characteristics of life we have just described—order, adaptations, metabolism, movement, and responsiveness—depend on a flow of energy and a cycling of materials through many organisms and their environments. Although a few kinds of organisms live quite successfully in total isolation, most depend on other organisms for their energy and materials. Consider the greatest assemblage of grazing animals on our planet, the hooved animals of the African savanna (Figure 1.11a), and their sprawling, arid environment. The green grasses that carpet the plain take in energy from sunlight and absorb materials from water, air, and soil. Mixed herds of graceful zebras and bearded wildebeests graze on the grass, and their bodies use the plants' stored energy for growth, activity, and maintenance of their own internal organization. A warm-blooded wildebeest loses much of the energy it consumes to the cool African night air and loses more in the carbon dioxide it exhales and the wastes it eliminates. When the shaggy wildebeest dies, fungi and bacteria decompose the carcass, use energy and materials trapped in the body, and release additional heat and chemicals into the environment. Thus, the struggle to overcome disorganization is a collective, interdependent one; energy flows and materials cycle through living communities and their environments.

In the savanna, as in other ecosystems, energy and materials follow two separate but overlapping pathways (Figure 1.11b). Energy enters from the sun, flows through the plants, animals, fungi, and microbes, and exits as heat lost to the air and eventually to space. In contrast, materials continually cycle from plants to animals, fungi, and microbes, to the soil and air, and back to plants. While the flow of energy and the cycling of materials

(a)

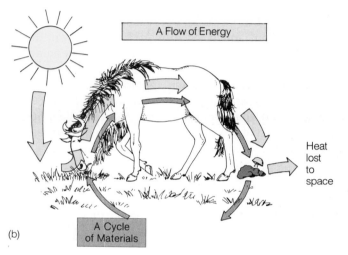
(b)

FIGURE 1.11 *Great Beasts of the African Plain: A Flow of Energy and a Cycling of Materials.* (a) Grazing wildebeest herds in the Serengeti plain of Kenya and Tanzania are an awesome reminder of nature's life-giving flux of energy and materials. (b) Energy flows and materials cycle. Energy from the sun is trapped by grass, then enters the animal's body cells after the animal eats the grass. Some energy is lost from the animal as heat, and some remains in the animal's solid wastes, but fungi, bacteria, and other soil organisms can use some of this energy. Additional energy escapes as heat from these decomposers in a one-way flow, forever lost back into space. In contrast, minerals and other materials cycle from the air and soil through the grass, the animal's body cells, its wastes, and back to the air and soil for eventual reuse by living things.

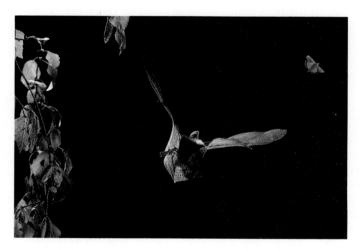

FIGURE 1.10 *Responsiveness: The Life Characteristics of Sensing and Reacting.* This moth sensed the approach of its enemy, the long-eared bat, and reacted by carrying out specific escape behaviors.

allow individuals to survive, disorganization inevitably gets the upper hand. Individuals appear to be programmed to age and die, and eventually all do. Since life has existed for billions of years, however, this tendency toward death has clearly been overcome by a second set of life characteristics.

≋ LIFE CHARACTERISTICS THAT PERPETUATE A POPULATION

Among the millions of species alive today, we human beings, with our typical life span of 70 years, are fairly long-lived. A few other species, however, live for much longer periods, the classic example being the 4000- to 5000-year-old bristlecone pine trees standing stalwart but gnarled on windswept limestone ridges in the White Mountains of California and Nevada. Despite their tenure throughout much of human history, even these astonishing survivors will eventually die. By generating similar copies of themselves, living entities overcome the inevitability of dying. When metabolism, movement, responsiveness, and the other characteristics of life cease in the original, they continue in the "copies." The mechanism by which organisms give rise to others of the same type is called *reproduction*.

LIFE CHARACTERISTIC 6: REPRODUCTION

Reproduction can be as simple as the splitting of a single-celled bacterium into two daughter cells (Figure 1.12). More complex organisms, however, usually undergo sexual reproduction, during which two specialized cells

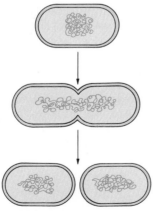

FIGURE 1.12 *Some Single-Celled Organisms Reproduce by Simple Division.* Bacteria and many other members of the kingdoms Monera and Protista divide in two, and the daughter cells grow into new bacteria or protists.

FIGURE 1.13 *Botanical Want Ad: The Penstemon Flower's Brilliant Red Color and Sweet Nectar Attract Hummingbirds and Ensure the Plant's Reproduction.* This penstemon flower is visited by a broad-tailed hummingbird. The animal laps up nectar with a long tongue, gains energy, and inadvertently carries pollen grains to the next flower it samples.

called the egg and sperm unite and form the first cell of a new individual. Many sexually reproducing organisms have curious reproductive adaptations. A male strawberry frog, for example, or a dazzling golden toad increases his chances of reproducing if he clearly advertises his location and amorous intent by a distinctive song attractive to females of his species. In another adaptation, flaming red tube-shaped penstemon flowers produce stores of energy-rich nectar, and both the flowers and the nectar attract hummingbirds (Figure 1.13). As each iridescent visitor drinks in some of the energy it needs to survive, it inadvertently assists in the flower's reproduction. Its feathers pick up sticky pollen grains, which contain the flower's sperm, and when the bird flies to the next plant, these grains can reach the eggs of another flower and lead to the formation of a new plant generation.

LIFE CHARACTERISTIC 7: DEVELOPMENT

When an organism reproduces, the young always start out smaller and usually much simpler in form than the parent(s). The offspring then grow in size and increase in complexity, eventually becoming parents themselves. This process is called **development.**

Although the details vary from one species to the next, consider the sequence of events in the development of a common bullfrog. With the union of the father's sperm and mother's egg, a new individual begins as a single, very large cell (Figure 1.14a). The newly formed cell divides and redivides into many smaller cells, which rapidly become organized into an embryo with eyes, internal organs, and a tail (Figure 1.14b). The tadpole hatches

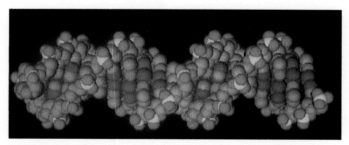

(a)

(b)

(c)

FIGURE 1.14 *Development: A Sequence of Growth, a Consequence of Reproduction.* (a) A European common frog (*Rana temporaria*) begins as a fertilized egg cell. (b) After a few hours, divisions in that original cell produce an embryo. Development continues during metamorphosis, a complete physical transformation that results in (c) the adult form.

and commences its main activity—eating—then grows larger, sprouts legs, resorbs its tail, and finally matures into an adult frog (Figure 1.14c). Although an adult frog must also eat to survive, its job, in the broadest sense, is to reproduce. Various adaptations, including its voice, sex organs, and innate behaviors for mating and parental care, aid the reproductive process.

Development is the orderly production and unfolding of structures and activities that allow each organism to cope successfully with its own special environment. One of the most intriguing questions in all biology is, How does a fertilized egg cell become the millions of cells of various types that will function as a viable organism? In other words, what makes an egg develop into a straw-

berry frog instead of a strawberry? The answer is the remarkable molecule of inheritance called DNA.

LIFE CHARACTERISTIC 8: HERITABLE UNITS OF INFORMATION, OR GENES

Identical twins make a convincing argument that some kind of information must direct the development of each individual in a fairly precise manner so that two very similar organisms result. When one compares a set of twins with their different-looking brothers and sisters, however, one can see that there are also variations in hereditary information. The units of inheritance that guide heredity and control the development of specific physical, chemical, or behavioral traits in a living organism are the **genes.** Specific genes, inherited from parents, determine whether a strawberry frog has big, bright red splashes of color or small, pale pink patches, whether a honeysuckle flower produces a lot of nectar or only a little, and whether a person has blue, green, brown, or hazel eyes. Likewise, genes control whether the brilliant harlequin bugs, shown on the cover of this book, are electric orange with black spots or iridescent deep purple with orange spots.

Genes are made of a remarkable and beautiful molecule called DNA (Figure 1.15). Each of your cells contains 46 of these twisting, threadlike molecules, and each molecule contains about 2000 genes. While a DNA molecule may be transmitted from parent to offspring without change, variations occasionally arise in its structure. Such changes are called *mutations*, and a familiar mutation that prevents blood from clotting leads to a disease called hemophilia (bleeder's disease) in people who inherit only a mutated copy of the gene.

As in hemophilia, many mutations in DNA are harmful. But occasionally, mutations occur that allow an organism to combat disorganization and death in some new and better way. The gradual accumulation of such mutations over many generations tends to change organisms, and the amassing of such changes in a population is called *evolution*.

FIGURE 1.15 *Multicolor Computer Graphic of the DNA Molecule.* DNA is the repository of genetic information for reproduction and cell function in all living cells.

EVOLUTION BY NATURAL SELECTION: THE KEY TO UNDERSTANDING UNITY AND DIVERSITY

LIFE CHARACTERISTIC 9: EVOLUTION

Living organisms change over time. We know this partly from the fossilized imprints of early organisms (Figure 1.16). The older a fossil, the less similar it is to present-day forms, and this provides good evidence of the constant change in living species over millennia. Using fossils, analysis of DNA, and other kinds of evidence, biologists can trace the history of an individual species. Living species are like the youngest branches of a family tree. For example, the strawberry frog (see Figure 1.1) and the play-actor frog (see Figure 18.9) are closely related to each other and somewhat more distantly related to bullfrogs. Reaching farther back in time, frogs and salamanders would share a common ancestor, and even farther back in the history of this lineage, frogs would be related to reptiles, fishes, and marine worms. Tracing the evolutionary tree of life to ever more remote periods, all organisms would eventually share a common ancestor in the ancient past. This family tree concept helps explain the origin of species: All living things are descended from a common ancestor and arose as the result of genetic modification in species that lived before them; this process is called **evolution.**

THE UNITY AND DIVERSITY OF LIFE

Evolution provides a framework for understanding life's unity and diversity. Life's **unity** is evident in the fact that all independently living creatures have hereditary material made up of the same type of molecule, DNA. Unity is also evident in the very similar ways that all organisms form new proteins, the basic building blocks of cell parts. Biologists have capitalized on that very similar machinery to trick bacterial cells into manufacturing useful human proteins, such as insulin. The insulin made in an engineered bacterial cell can prolong the lives of many diabetics.

No less amazing than life's unity is its **diversity.** Think of the dazzling variety evident in the colorful songbirds of eastern forests, the butterflies of midwestern weed fields, and the wildflowers of the Rocky Mountains. The earth, in fact, is covered with immensely varied communities of living species, all products of evolution. The evolutionary mechanism that brings about the diversity of new species from ancestral lines is **natural selection,** a major guiding principle for the study of life.

NATURAL SELECTION: A MECHANISM OF EVOLUTION

The graceful, towering giraffe is one of nature's magnificent products, and its extremely long neck and legs have been the objects of human curiosity for several centuries. Early naturalists noted that the giraffe's long neck allows it to browse on leaves from high branches that are inaccessible to wildebeests, zebras, elephants, and other inhabitants of the African savanna. We know today that these adaptations help the giraffe overcome disorder by collecting energy and materials and that the world's tallest animal evolved from shorter ancestors. At the turn of the nineteenth century, however, curious biologists could only speculate about how the giraffe got its wondrous neck and legs.

In 1809, French naturalist Jean Baptiste Lamarck suggested that early giraffes must have stretched their necks trying to graze on the leaves of high branches and that the long neck that an individual acquired through such stretching was passed along to its young. Experiments eventually showed that Lamarck's theory, the so-called inheritance of acquired characteristics, is incorrect and is not the mechanism underlying evolution. Consider just one example that clearly disproves Lamarck's idea: Many human families have for thousands of years and hundreds of generations removed the foreskins of infant males. Despite this deliberate alteration, however, every baby boy in each new generation is born with a foreskin—clear evidence that the acquired characteristic is not inherited. It remained for two English naturalists working in the mid-1800s, Charles Darwin and Alfred Russel Wallace, to devise an alternative explanation for how living things evolve (Figure 1.17).

In the 1830s, young Charles Darwin sailed around the world, investigating nature's diversity. Darwin agreed with Lamarck and others that evolution occurs, but

(a)   (b)

FIGURE 1.16 *Floral Ancestor in Stone: Fossil Gentian Compared with Modern Flower.* (a) This 50-million-year-old fossil flower bears a strong resemblance to (b) a modern gentian. Biologists use such comparisons in the study of evolution.

(a)   (b)

FIGURE 1.17 *The Authors of Evolution by Natural Selection.*
(a) Charles Darwin (1809–1882) and (b) Alfred Russel Wallace
(1823–1913) proposed remarkably similar theories to explain how
life's diversity evolved.

remained dissatisfied with existing explanations for how species change over time. Eventually, Darwin drew together two indisputable facts based on his own observations and synthesized a far-reaching conclusion:

FACT 1: Individuals in a population vary in many ways, and some of these variations are heritable.

FACT 2: Populations have the inherent ability to produce many more offspring than the environment's food, space, and other assets can possibly support. As a consequence, individuals of the same population compete with each other for limited resources.

DARWIN'S CONCLUSION:

Individuals equipped with traits that allow them to cope efficiently with the local environment leave more offspring than individuals with less adaptive traits. As a result, certain heritable variations become more common in succeeding generations.

Darwin used the term *natural selection* to describe the greater reproductive success of those individuals with adaptive characteristics as compared with members of the same species lacking the adaptations. He chose this term because nature "selects" the parents for the next generation.

The principle of natural selection, now widely accepted as a main mechanism behind evolution in nature, explains adaptations such as the long necks of giraffes and the brilliant color of strawberry frogs. If we begin with a population of giraffes browsing on trees in the savanna many thousands of years ago (Figure 1.18), we can imagine how some of the giraffes would have long necks and others short necks, just as some people have longer necks than others. The hereditary units called genes help determine neck length in both giraffes and people. Now, if on the savanna there were too few low-hanging leaves to feed all the giraffes in the population, then the long-necked giraffes could reach more food, harvest more

energy and materials, and survive to produce more offspring—many, like their parents, possessing the genes for long necks. With proportionately more long-necked genes around, the average neck length in the giraffe population would increase over time to present-day lengths. It is important to note, however, that natural selection is not the cause of variations within the population. Short necks, long necks, and other variations preexist in the population as a result of gene mutations. Natural selection simply chooses the best-adapted, best-competing individuals to be parents for the next generation.

Consider two basic facts from another example of natural selection. First, some strawberry frogs have brilliant red skin, while others have less color (Figure 1.19, page 16). Second, the Central American jungle lacks enough food and nesting sites to support all the strawberry tadpoles that hatch, since a single female can lay 600 eggs a year. Darwin might have concluded that only a few strawberry frogs from each clutch of eggs survive to

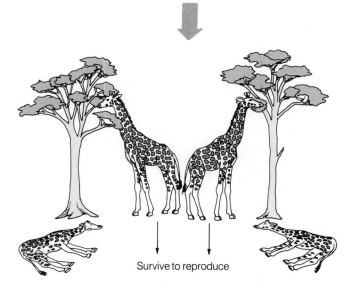

Survive to reproduce

FIGURE 1.18 *Natural Selection on the Savanna: Giraffes with Genetically Determined Long Necks Have Advantages for Survival and Reproduction.* Amid tall trees and long-necked neighbors, short-necked giraffes would have died and left few offspring, while long-necked ones would have survived to reproduce more often.

FIGURE 1.19 *Natural Variation in the Strawberry Frog.* As these photos show, not all strawberry frogs are a vivid scarlet. (a) Some have red bodies and large dark spots; and (b) some have yellowish bodies and small spots. Natural selection acts on such variation, "selecting" those individuals best equipped to survive in the local environment as parents for the next generation.

reproduce and that this natural selection of survivors is based on adaptations such as crimson skin color and the skin's content of powerful poison. (Such frog poisons serve as a hunting resource for South American Indians; see Figure 1.20.) A bird or snake that seizes a strawberry frog in its mouth immediately experiences burning, numbing, and foul taste and tends to drop the prey. After one disagreeable nibble on a poison strawberry frog, most predators learn to leave the frogs alone. The more brilliantly colored the frog, the more easily an enemy recognizes it as poisonous. Over time, more frogs belonging to brightly colored families survive and thereby produce more offspring, and genes for bright color increase in frequency in the population along with genes for poison production.

Darwin argued that evolution by natural selection, working over long stretches of time, could have produced the millions of living forms we now observe in all their splendid diversity (see Figure 1.21 for another fascinating example). Natural selection fine-tunes each species to its own environment by selecting genes that, in the broadest sense, help individual organisms overcome disorder and death. Evolution by natural selection is so grand an organizing principle for all of biology that it will resurface time and again throughout this book.

Like Darwin, Alfred Russel Wallace devised a theory of evolution by natural selection to explain what he saw in his travels (even choosing the term *natural selection* independently!). By 1858, Wallace had published four papers containing parts of a theory of evolution that included the mechanism of natural selection, the notion of descent with modification, and the idea of survival of the fittest. Wallace sent Darwin his fourth paper in 1858, and Darwin—realizing that his life's work was being scooped—rushed to prepare an abstract. The following year, Darwin released his monumental and well-docu-

mented book, *The Origin of Species.* The intellectual stimulation each scientist gave the other provided generations of biologists with the broadest conceptual framework for understanding the living world.

Before a biologist can apply the principle of evolution by natural selection to unraveling life's secrets, however, he or she must observe the natural world, suggest mechanisms for how it operates, and test those suggested mechanisms by experiment. This organized pattern of investigation is called the scientific method.

THE SCIENTIFIC METHOD: ORGANIZED COMMON SENSE

A biologist is rarely happy just to describe a curious event, such as the disappearance of a tadpole's tail as the young frog becomes an adult. Scientists instead want to learn what causes the event. To investigate the natural world in the most organized way, scientists use the **scientific method:**

1. They ask a *question* or identify a problem to be solved based on observations of the natural world.
2. They propose a **hypothesis,** a possible answer to the question or a potential solution to the problem.
3. They make a **prediction** of what they will observe in a specific situation if the hypothesis is correct.
4. They *test* the prediction by performing an experiment.

FIGURE 1.20 *Poison-Tipped Dart and Blowgun.* At some point in their cultural history, Indian tribes living in the jungles of Peru, South America, learned that strawberry frogs and similar tropical species make powerful poisons that will kill animals. In one type of frog, the poison from a single individual can kill 20 adult people! Since such poisons block the function of nerve and muscle cells, researchers also use them to investigate how nerves and muscles do their jobs.

FIGURE 1.21 *The Giant Panda's Thumb: An Odd Evolutionary Arrangement.* The handsome, thick-coated panda bear lives in bamboo forests along the Tibetan plateau in eastern China. The most curious of the panda's traits is its front paws: Instead of having five digits like other bears, dogs, and people, they alone have *six* on each paw. The panda's five "fingers" plus one opposable "thumb" are fully adapted for holding bamboo stalks, their favorite food. The "thumb" is actually an enlarged wrist bone with attached muscles and tendons. Sometime in the early evolutionary history of the giant panda, mutations occurred giving some individuals an enlarged wrist bone with muscles and tendons properly connected for opposable movement. These mutant bears must have been able to gather more bamboo, survive in greater numbers than their "thumbless" cohorts, and leave more offspring, because today, all surviving pandas have the opposable sixth digit.

If the hypothesis predicts the results correctly, the scientist makes other predictions based on the same hypothesis and tests them. Scientific research requires logic, analytical skills, and perseverance. Only after researchers create and test several likely hypotheses and find one that consistently predicts what they see in nature will they tentatively accept a hypothesis as correct.

While these steps may sound very regimented, they are really little more than an organized commonsense approach—one you use regularly in your own life. Let's say that one evening, you observe that your desk lamp stops working. You would pose a question (step 1): "What made my desk lamp go out?" And you would probably create a hypothesis (step 2): "Maybe the light bulb burned out." Next you would make a prediction (step 3): "If the bulb burned out, then when I replace it with a working bulb, the lamp should light." Finally, you would perform an experiment (step 4): You would remove a bulb from a floor lamp that works, screw it into your desk lamp, and watch the result. When you performed the test, you would include what scientists call a *control*, a check that

all factors of the experiment are the same except for the one in question. Here, the control is the borrowed bulb that works in the floor lamp. If the borrowed bulb fails to work in the desk lamp, you could conclude, based on your control, that a burned-out bulb is not the problem. You would next discard the faulty-bulb hypothesis and ask new questions: "Is the lamp itself broken? Is something wrong with the wiring to the wall socket?" You could then make new hypotheses, new predictions, and perform new tests until you discovered why your lamp went out.

While an orderly approach to exploring nature, the scientific method is by no means rigid, rote, or unimaginative in practice. Once a scientist has observed a curious phenomenon, it takes creativity to dream up a clear, testable hypothesis. It also takes logic, talent, experience, imagination, and intuition to follow through with cleverly designed experiments and alternative hypotheses. Finally, it takes an ability to communicate clearly through writing and speaking to share results with others.

No tool is more powerful for understanding the natural world than the scientific method. It does not apply, however, to matters of religion, politics, culture, ethics, or art. These valuable ways of approaching the world and its problems proceed along different lines of inquiry and experience. Nevertheless, many of the world's current problems have underlying biological bases, and thus they mainly demand biological solutions.

☰ HOW BIOLOGICAL SCIENCE CAN HELP SOLVE WORLD PROBLEMS

Anyone who reads a daily newspaper is well acquainted with world problems: overpopulation, famine, violent crime, territorial aggression, drug addiction, AIDS, cancer, heart disease, pollution, ozone depletion, acid rain, changes in climate, species extinction. While these problems tend to have both social and biological roots, attacking their biological bases may help ease their associated pressures on society. For this reason, citizens in all fields can benefit from an understanding of the biological bases of the world's problems.

Many of our most vexing problems stem from our species' enormous and burgeoning population. Five billion people are currently straining our planet's environmental resources, and by the year 2000, we will number 6 billion. Some observers believe that our future security and quality of life are threatened less by war among nations than by the burden we place on natural systems and resources with the sheer crush of humanity.

(a)

(b)

FIGURE 1.22 *Deforestation in Madagascar: Landscapes Effaced, Healing Plants Lost.* (a) Hills in Madagascar's central plateau lie denuded of their original forests, with precious topsoil washing away each time it rains. The jungles were cleared for firewood and for the creation of new farmland, but after a few years, the land's productivity plummeted, and farmers cleared new jungle. Thousands of species can be lost to this deforestation process. (b) For many years, people have collected and exported the Madagascar periwinkle, a native plant containing a compound that can help fight childhood leukemia. Will the potential for this or other life-saving drugs be lost along with Madagascar's forests?

Take, for example, the plight of Madagascar, an island about 1000 miles long off the southeast coast of Africa. In the past 35 years, half of Madagascar's forests have been leveled to provide fuel and uncover farmland for the impoverished and rapidly growing population. This has led to serious erosion of the rich topsoil (Figure 1.22a). In the place of mature trees, farmers have planted clove tree seedlings as a cash crop to help satisfy the world's appetite for vanilla ice cream, cola drinks, and other foods that contain extracts from clove buds. Yet their own children go hungry because cloves are not an adequate food source.

The increase in Madagascar's human population and the destruction of native forests also caused the loss of hundreds of species, including plants that produce potentially life-saving drugs. One such plant is a pink-petaled periwinkle flower (Figure 1.22b) discovered in Madagascar's forests and exported to pharmaceutical companies in Europe and North America to make drugs for children with leukemia. Many observers fear that plants like this will perish along with Madagascar's tropical forests.

Two decades ago, one might have predicted a similar fate for the Central American nation of Costa Rica, but by replanting deforested hillsides with fast-growing tropical hardwood trees, growing nutritionally improved food crops, setting aside up to 15 percent of their land as natural preserves for their tropical species, and instituting up-to-date health care and family-planning practices whenever possible, Costa Ricans are helping to save their land and their people. Their success is a hopeful sign for the future: biological solutions for social problems of biological origin.

In the chapters that follow, you will find many discussions of how biology is helping to solve social problems.

We are, in fact, in the midst of a revolution in the biological sciences, with exciting new information surfacing weekly in the fight against cancer, heart disease, AIDS, infertility, and obesity. Researchers are making rapid advances in gene manipulation to create new drugs, crops, and farm animals; in sports physiology to improve human performance; in the diagnosis of genetic diseases; and in the transplantation of organs, including brain tissue. Across all frontiers of biological science, at all levels of life's organization—from molecules to the biosphere—scientists are learning the most profound secrets of how living things use energy to overcome disorganization and reproduce to overcome death. In studying the living world, you are embarking on an adventure of discovery that will not only excite your imagination and enrich your appreciation of the natural world, but will provide a basis on which you can contribute intelligently to the difficult choices society must make in the future.

☰ CONNECTIONS

While a strawberry frog captures ants and recycles [uses] the energy stored in the insect's muscles, a periwinkle plant captures energy directly from the sun and absorbs materials from the air and soil. As different as these solutions may be, animals, plants, and all other organisms share a basic set of characteristics that allow them to meet the challenges of survival, and these attest to the common origin and unity of all living things, from bacteria to people.

To save ourselves and our world from ecological disaster, biologists and citizens in all fields must learn about life processes and the interactions of organisms with their environment. Our next few chapters examine the underlying principles of biological science, beginning with the laws of chemistry and physics and how they govern the atoms and molecules that make up living things.

(1) Biology is the study of living things and their environment.

(2) All living things face the universal tendencies toward disorder and death.

(3) The nine characteristics of life help organisms overcome disorder and death.

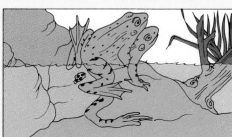

- ORDER
- ADAPTATION
- METABOLISM
- MOVEMENT
- RESPONSIVENESS

Help organisms to survive day by day

- REPRODUCTION
- DEVELOPMENT
- GENES

Enable species to persist over time

- EVOLUTION

Enables populations of living things to adjust to environmental change

(4) Natural selection, a driving force of evolution, helps explain the unity and diversity of life.

(5) The four steps of the scientific method provide a powerful tool for learning about the natural world:

| QUESTION | → | HYPOTHESIS | → | PREDICTION | → | TEST |

(6) Biologists have applied their knowledge of the natural world to help fight disease; create pest-resistant crops; and sustain productive ecosystems.

NEW TERMS

adaptation, page 9
biology, page 1
development, page 12
diversity, page 14
evolution, page 14
gene, page 13
hierarchy of life, page 6
hypothesis, page 16
kingdom, page 9

metabolism, page 9
movement, page 10
natural selection, page 14
order, page 6
prediction, page 16
reproduction, page 12
responsiveness, page 10
scientific method, page 16
unity, page 14

STUDY QUESTIONS

REVIEW WHAT YOU HAVE LEARNED

1. Which life characteristics are related to overcoming disorder? To overcoming death?
2. How do unity and diversity relate to evolution?
3. Define organism.
4. List the levels of hierarchy of biological organization on earth.
5. Once an organism takes in energy and materials, what basic process breaks them down and rearranges them into useful products?
6. What are the two basic types of reproduction?
7. What do genes do in organisms? What is the result of a mutation?
8. What is the most important mechanism leading to evolutionary change?

APPLY WHAT YOU HAVE LEARNED

1. A biology student wonders whether her philodendron plant really needs its weekly dose of nitrogen fertilizer. Using your knowledge of the scientific method, explain how this student can satisfy her curiosity.

FOR FURTHER READING

Darwin, C. *On the Origin of Species: A Facsimile of the First Edition.* Cambridge, MA: Harvard University Press, 1975.

Gould, S. J. *The Panda's Thumb.* New York: Norton, 1980.

Jolly, A. "Madagascar: A World Apart." *National Geographic* 171 (February 1987): 149–183.

Myers, C. W., and J. W. Daly, "Dart-Poison Frogs." *Scientific American* 248 (February 1983): 120–133.

Wassersug, R. "Why Tadpoles Love Fast Food." *Natural History* 93 (April 1984): 60–69.

Wells, K. D. "Courtship and Parental Behavior in a Panamanian Poison-Arrow Frog (*Dendrobates auratus*)." *Herpetologica* 34 (1978): 148–155.

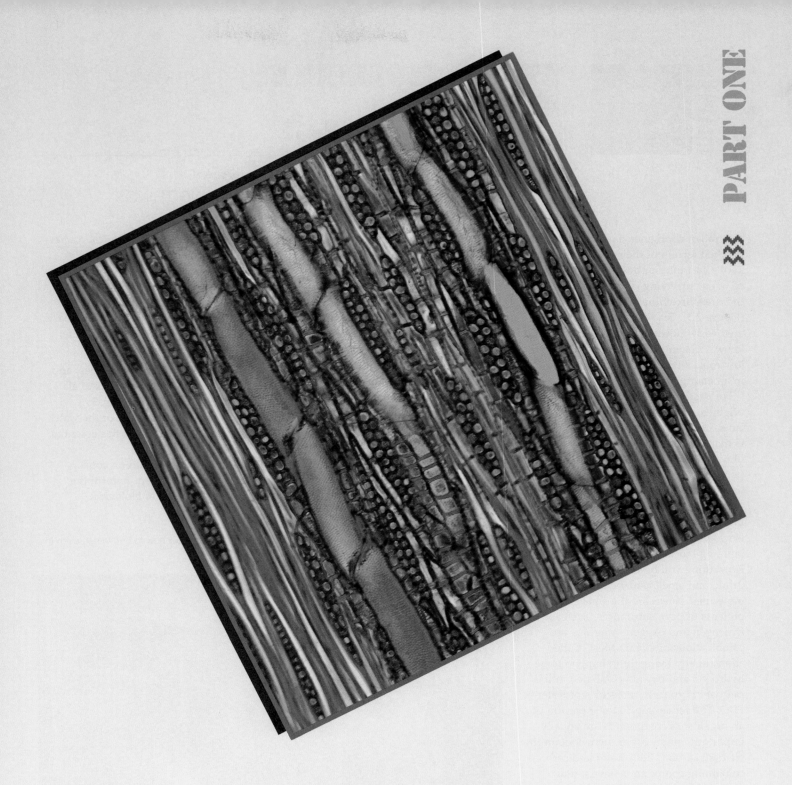

>>>

PART ONE

LIFE'S FUNDAMENTALS

ATOMS, MOLECULES, AND LIFE

WATER, SURVIVAL, AND THE CHEMISTRY OF LIVING THINGS

Australian aborigines have inhabited the parched terrain of their island continent's vast interior desert for thousands of years, surviving largely because of their resourcefulness at locating water (Figure 2.1). Traditional aborigines gathered early morning dew on grasses and knew a dozen plant species whose shallow roots store water that can be tapped in an emergency.

For desert inhabitants such as the aborigines, and in fact for all living organisms, life depends on water. One reason is that living things are made up mostly of water molecules. A tree is about 50 percent water, and most animals (including humans) are about 65 percent water.

This chapter lays the groundwork for understanding living cells and organisms by discussing their fundamental building blocks—atoms and molecules. Everything you breathe, touch, eat, drink, and smell is made of atoms and molecules. *Atoms* are the smallest components of pure substances, such as hydrogen and oxygen. *Molecules* are combinations of atoms. Most of the atoms in any living thing happen to be hydrogen and oxygen combined in the ratio of 2:1 to form molecules of water—H_2O. The remaining 10 to 50 percent of an organism's compounds are equally important, and most of them contain the element carbon. Both water and carbon-containing compounds have unique chemical and physical properties that make life possible.

Think of how thirsty you get on a hot summer day when you lose water through perspiration. Like the Australian aborigines and all other living organisms, you cannot survive without water. Nor can you get along without

the many carbon-containing compounds in your body. They store the energy you use to walk, sleep, or think; they store the genetic information you could pass on to your children. Every time you eat a sandwich or drink a glass of milk, you take in carbon-containing compounds.

By understanding chemistry, you will understand life, because the behavior of atoms and molecules underlies and explains the behavior of living cells. This fact forms a major unifying theme of this chapter. Whether cells are breaking down food molecules, building new cell parts, dividing, or moving, atoms and molecules are involved. A second unifying theme is related to the first: The physical structure of atoms and molecules determines their chemical properties and hence the roles they play in cells. The close relationship between structure and function will be evident throughout our study of biology.

In this chapter, we answer three basic questions about the chemistry of life:

- What are atoms and molecules, and how do they form the units of all living and nonliving things?
- How does the structure of the water molecule give it properties essential to life?
- What property of carbon compounds allows them to form the four main types of biological molecules?

FIGURE 2.1 *Life Depends on Water.* And Australian aborigines know how to find water, even in the continent's parched desert areas.

≋ ATOMS AND MOLECULES

THE NATURE OF MATTER: ATOMS AND ELEMENTS

Ancient Greek philosophers, keen observers of the natural world, recorded many basic truths about common materials. They noticed that some materials, such as rock, wood, and soil, are made up of mixtures of substances, while other materials, such as chunks of iron, gold, silver, and sulfur, are not mixtures and cannot be decomposed any further into constituents by chemical processes. Scientists later studied the pure substances and called them elementary substances, or **elements,** and named the mixtures of two or more elements **compounds.** By the early nineteenth century, scientists had concluded that each element is composed of identical particles, and they called the particles **atoms,** after the Greek word *atomos,* meaning "indivisible." Atoms are the smallest particles into which an element can be divided and yet still display the properties of that element. What's more, all the atoms in a sample of gold, for example, are identical to each other but different from the atoms in a piece of iron. Other scientists have since confirmed this *atomic theory.* In addition, chemists have discovered 92 naturally occurring elements and have created 13 more in the laboratory. These elements are marvelously different. For example, in its pure state, helium is a colorless, odorless gas, lithium a soft, whitish metal, calcium a white powder, and copper a reddish solid.

Why do these different elements have different properties? The explanation is based on a fascinating idea—the concept of *emergent properties.* In chemistry as well as biology, the whole (the whole atom, the whole molecule, the whole cell, the whole organism) is always more than the sum of its parts. The collective properties of the whole emerge not just from the properties of the individual parts, but from the precise way those parts are arranged and interact (Figure 2.2). For example, all of the 92 naturally occurring chemical elements are made up of the same three kinds of particles, but the properties of powdery white calcium, let's say, are very different from the properties of metallic copper because of the way the particles are arranged in calcium and copper atoms. Likewise, the body of a desert-dwelling aborigine contains the same six major chemical elements as the woody tissue in the eucalyptus tree he or she rests beneath for shade. But those six elements are arranged into different sorts of molecules in a human and a eucalyptus tree. Hence, the properties of person and tree emerge from how their constituent parts are arranged.

Although the earth contains 92 elements, scientists have found that living things contain just a few. In fact, earth and the life it supports have radically different compositions. Ninety-eight percent of all the atoms in

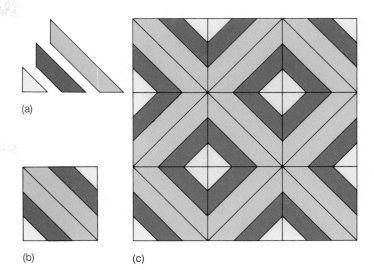

FIGURE 2.2 *Emergent Properties.* The whole is more than the sum of its parts. In a quilt, small geometric units (a) can be combined into a block to form a pattern (b). When sets of these blocks are arranged systematically, a higher-order pattern emerges (c). In a similar way, the properties of a living thing emerge from the precise arrangement of component parts: atoms, molecules, cell parts, cells, and so on.

our planet's surface layer belong to the eight elements listed in Figure 2.3a (page 24), headed by oxygen (O), silicon (Si), and aluminum (Al). In a typical organism, however, 99 percent of all the atoms belong to the six elements sulfur (S), carbon (C), hydrogen (H), nitrogen (N), oxygen (O), and phosphorus (P). (A convenient way to remember the six basic elements in living things is to think of the acronym SCHNOP, formed from the symbols for each of the elements.) In addition to these six basic elements, Figure 2.3b also names the nine other elements that occur in trace amounts in living organisms. The properties of elements are based on the internal organization of their atoms. Let's see why that is true.

ORGANIZATION WITHIN THE ATOM

All atoms are composed of the same three types of subatomic particles: **protons,** or positively charged particles; **neutrons,** or neutral particles with no electrical charge; and **electrons,** or negatively charged particles. Each atom is like a tiny solar system with a sun at its center. This "sun" is the atomic **nucleus,** and it contains a set number of protons and neutrons that account for most of the atom's mass. A specific number of electrons, equal to the number of protons, orbit about the nucleus at a relatively

(a) Composition of the earth

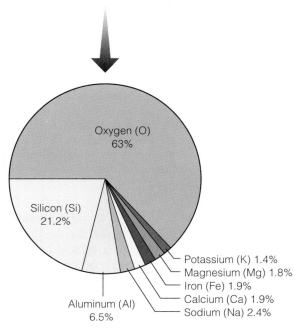

Oxygen (O)
63%

Silicon (Si)
21.2%

Potassium (K) 1.4%
Magnesium (Mg) 1.8%
Iron (Fe) 1.9%
Calcium (Ca) 1.9%
Sodium (Na) 2.4%

Aluminum (Al)
6.5%

(b) Composition of living things

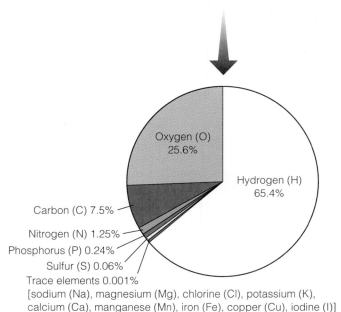

Oxygen (O)
25.6%

Hydrogen (H)
65.4%

Carbon (C) 7.5%

Nitrogen (N) 1.25%
Phosphorus (P) 0.24%
Sulfur (S) 0.06%
Trace elements 0.001%
[sodium (Na), magnesium (Mg), chlorine (Cl), potassium (K),
calcium (Ca), manganese (Mn), iron (Fe), copper (Cu), iodine (I)]

FIGURE 2.3 *Elements in Earth and Living Organisms.* If analyzed at the level of atoms, our planet has a very different composition from the living organisms that inhabit it. (a) The crust at our planet's surface, containing rock, soil, sand, and other materials, is mostly oxygen, silicon, and aluminum atoms in various compounds. (b) The people, mushrooms, trees, and other organisms that live on that crust are mostly hydrogen, oxygen, and carbon atoms. The unique properties of living things emerge from their combination of elements and the precise arrangement of those elements.

great distance. Figure 2.4 shows the structure of hydrogen, the simplest atom, with its single proton, single electron, and no neutron, as well as the structures of carbon (six protons, six neutrons, six electrons) and oxygen (eight protons, eight neutrons, eight electrons). If the nucleus of a hydrogen atom were magnified to the size of an orange, the single electron would orbit that enlarged nucleus at a distance of one-third of a mile. Of course, in reality, an atom is minute: More than 3 million hydrogen atoms could sit side by side on the period ending this sentence.

Every element has its own specific *atomic number,* equal to the number of protons in the nuclei of its atoms. Physicians have recently devised a way to use the physical properties of protons to view human tissues and detect disease (see the box on page 26).

The combined number of protons and neutrons in the nucleus is the *atomic mass,* a convenient measure of how heavy each atom is. Biologists use their knowledge of atomic mass to identify the various atoms and molecules in living organisms. There is often (but not always) an equal number of protons and neutrons. For example, the nucleus of the carbon atom contains six protons and six neutrons; therefore, the atomic number is 6 and the atomic mass (protons plus neutrons) is 12. Table 2.1 lists the atomic numbers for the six most common elements in living things.

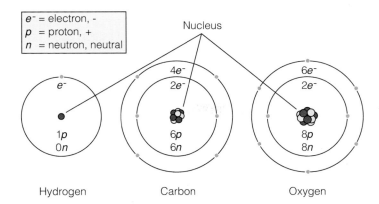

Hydrogen   Carbon   Oxygen

FIGURE 2.4 *Structure of the Atom.* In each atom, negatively charged electrons (here, small gold dots) orbit the positively charged nucleus, which contains one proton (brown spheres) for each electron and a set of neutrons (tan spheres), usually close to the number of protons. A balance between opposing physical forces keeps the electrons moving at set distances around the nucleus. The system of static rings (to represent energy levels) and dots (to represent electrons around the nucleus) is called the Bohr model. It helps us visualize how the parts interact, but it is obviously not drawn to scale. The electrons are actually much smaller than shown, orbit much farther from the nucleus, and do not travel in rigid circles.

Because the nucleus contains both positively charged protons and chargeless neutrons, it has an overall positive charge; by contrast, the orbiting electrons have a negative charge. In fact, each proton has a charge of $+1$, while each electron is -1. The attraction between these positive and negative charges pulls the electrons toward the nucleus, but the outward pushing force of the rapidly orbiting electrons tends to throw them outward, away from the nucleus, like the force of a rock tied to a twirling string. A balance reached between these forces holds the electrons in orbit in a set pattern around the nucleus.

TABLE 2.1 〰 THE MOST COMMON ELEMENTS IN LIVING THINGS

ELEMENT (SYMBOL)	ATOMIC NUMBER	WHERE FOUND
Hydrogen (H)	1	Water, many organic molecules
Carbon (C)	6	All organic molecules
Nitrogen (N)	7	Proteins, nucleic acids, urea
Oxygen (O)	8	Water, many organic molecules
Phosphorus (P)	15	Nucleic acids, bones
Sulfur (S)	16	Proteins

There are two types of exceptions to this standard structure. While the number of protons in the nucleus of an atom of a given element always remains the same, the number of neutrons sometimes varies from the normal number. Thus, the atomic number remains the same, but the atomic mass may change. Atoms with the same number of protons but with different numbers of neutrons are called **isotopes.** A good example of an isotope is carbon-14 (^{14}C), which has two extra neutrons in the nucleus and thus an atomic mass of 14 instead of 12 (Figure 2.5a). Isotopes like carbon-14 are *radioactive;* their nucleus is unstable and emits energy as it loses the extra

(a) Isotopes: more or fewer neutrons

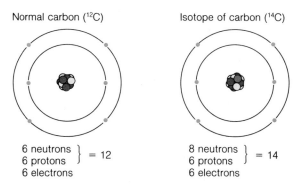

Normal carbon (^{12}C)   Isotope of carbon (^{14}C)

6 neutrons ⎫
6 protons ⎬ = 12
6 electrons ⎭

8 neutrons ⎫
6 protons ⎬ = 14
6 electrons ⎭

(b) Ions: more or fewer electrons

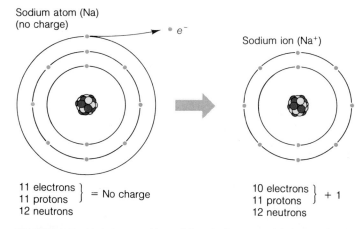

Sodium atom (Na) (no charge)   • e^- Sodium ion (Na$^+$)

11 electrons ⎫
11 protons ⎬ = No charge
12 neutrons ⎭

10 electrons ⎫
11 protons ⎬ + 1
12 neutrons ⎭

FIGURE 2.5 *Variations on Normal Atomic Structure.* (a) An atomic isotope has more or fewer neutrons than normal atoms of the same element. Here, a normal carbon atom, ^{12}C, is compared with one of its isotopes, ^{14}C. The radioactive energy given off by isotopes as the nucleus decays makes isotopes highly useful as markers in biological experiments. (b) Ions are another variation on atomic structure. Ions have more or fewer orbiting electrons than normal atoms of the same element. The change gives the atom an overall positive or negative charge. A normal sodium atom has 11 electrons, just one of which is in the outer shell. If this electron leaves the atom, the remaining sodium ion will have one more proton than electrons and thus a positive charge ($+1$).

MRI AND THE SECRETS IN HYDROGEN

Doctors are now using a powerful diagnostic tool called magnetic resonance imaging (MRI) based on the special properties of protons. Chemists discovered many decades ago that in certain atoms, such as hydrogen and phosphorus, the protons whirl like tops and create small magnetic fields in the nucleus that can be altered with external magnets and radio signals and then detected electronically. More recently, medical engineers created a diagnostic instrument out of a huge, ring-shaped magnet that can encompass the human body and generate a magnetic field 40,000 times greater than the earth's (Figure 1). The magnet is strong enough to jerk a 2 lb wrench from your hands, and yet the force fields it generates do not appear to harm the body in any way. The magnets cause protons in the body's hydrogen atoms (and in certain other types of atoms) to align and spin in the same direction. Then, when a second magnetic field is applied, the spinning protons flip back to their normal alignment; this flipping gives off a signal that can be recorded and used to reveal the presence and concentration of the atoms. Physicians can interpret the signal and use it to diagnose diseases.

Because MRI can detect protons in the nuclei of hydrogen atoms, doctors can now observe blood (which is 90 percent water) flowing through various tissues. This allows them to peer into the brain tissue of stroke victims to see which vessels were blocked during the stroke and which still allow the free flow of blood (Figure 2). Doctors are

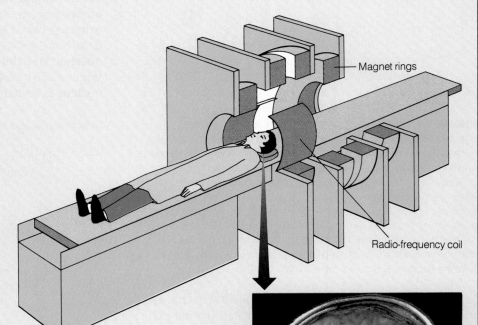

FIGURE 1 *Medical Diagnostic Device That Employs MRI.*

also using MRI to pinpoint the exact location of blockages caused by brain tumors, to locate sources of pain in the spine and limb joints, and to diagnose broken bones anywhere in the body, since blood seeps into the fracture. MRI may also replace some angiogram tests (which involve inserting tubes into the body) by producing a picture of the fatty tissue blocking blood vessels in the heart.

MRI is part of a new generation of medical technology that is replacing the doctor's black bag and the medical laboratory. Other techniques include computed axial tomography, or CAT scans, which combine X-rays and computers, and positron emission tomog-

FIGURE 2 *MRI Reveals Vivid Details of the Brain's Soft Tissue.*

raphy, or PET, which detects electrons given off by certain isotopes injected into the patient. While each of these life-saving techniques is unique, all three are based, ultimately, on our knowledge of the structure and properties of atoms.

neutrons and changes to a more stable form. Because carbon-14 decays to carbon-12 in a predictable way over time, biologists can calculate the age of early human bones and other remains by calculating the ratios of the carbon isotopes they contain. Isotopes are also used in research and medicine to mark and trace molecules that contain them.

In the second exception to an element's standard atomic structure, the number of orbiting electrons decreases or increases so that the whole atom, now called an **ion,** has a positive or negative charge. For example, if a sodium atom loses an electron, it will become an ion with 10 electrons and 11 protons (-10 and $+11 = +1$) and an overall positive charge of 1 (Figure 2.5b). It is designated Na^+ and has different properties than a sodium atom.

To recap the key points of an atom's internal organization: The number of protons in the nucleus, called the atomic number, defines each element; the number of electrons orbiting the nucleus equals the number of protons; and the number of protons plus the number of neutrons is the atomic mass. Exceptions to the standard structure can make an atom radioactive or give it an electrical charge.

ELECTRONS IN ORBIT: ATOMIC PROPERTIES EMERGE

With this picture of the atom in mind, we are ready to address the basic question, What is it about the structure of an atom of gold or sulfur or oxygen that gives each element its unique properties—its powdery or hard texture, its green or metallic color, its acrid or pleasant smell?

The answer lies, in large part, with the number and arrangement of the atom's electrons. The number of protons determines the number of electrons, and the number of electrons, in turn, dictates the volume, "shape," and tendency of the atom to react with other atoms. Electrons become arrayed in certain predictable ways based on the following principles: (1) Electrons are attracted by the nucleus but repelled by each other; (2) they move about the nucleus in *energy levels,* or zones, at given distances from the nucleus; (3) each level can hold up to a certain number of electrons—the innermost level up to two, the next several levels up to eight each; (4) an energy level that is completely filled with electrons (i.e., that contains the maximum number of two or eight) is more stable and less reactive than one that is partially filled; (5) atoms have a tendency to establish a filled outer level by gaining or losing electrons.

Now, returning to the original question of how atomic structure gives each element its unique properties, we can see why the electrons hold the key. The outermost electrons are "in contact" with the world, and their number and position largely determine the atom's chemical behavior. For example, atoms of the colorless, odorless gas neon (Ne) are very stable—so stable, in fact, that they are *inert;* they do not react easily with other atoms because their outer energy level is already filled with eight electrons (Figure 2.6a). In contrast, atoms of the black, solid element carbon are missing four electrons in their outer energy level (Figure 2.6b). To achieve a filled outer energy level, an atom of carbon must gain four electrons from other atoms. As a result of its "electron dearth," carbon tends to combine (share electrons) with other carbons and many other elements in a huge number of compounds, including those that make up living cells, and to take part in the chemical activities of the life process. Hydrogen, which needs one electron to fill its outer level, and oxygen, which has two unfilled slots in its outer level, also react and form compounds with each other (including H_2O—two hydrogen atoms and one oxygen atom) and with other elements.

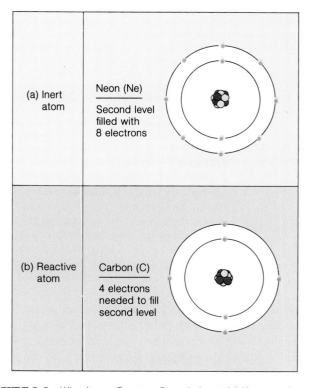

FIGURE 2.6 *Why Atoms React or Remain Inert.* (a) If an atom's outermost energy level is filled with the maximum number of electrons, the atom remains inert, or unreactive with other atoms. Neon is inert for this reason, having filled the first and second levels with two and eight electrons, respectively. (b) Carbon, however, is reactive. It is missing four electrons in its outer energy level.

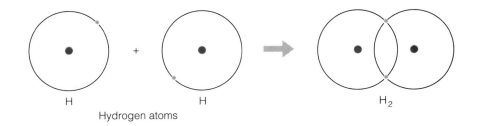

(a) Nonpolar covalent bonds

H   H H_2

Hydrogen atoms

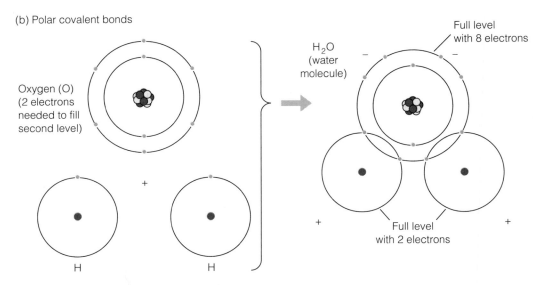

(b) Polar covalent bonds

Oxygen (O) (2 electrons needed to fill second level)

H   H

H_2O (water molecule)

Full level with 8 electrons

Full level with 2 electrons

FIGURE 2.7 *Covalent Bonds.* (a) As hydrogen atoms approach each other, their individual electron orbits overlap to form a covalent bond. Because the distribution of charge is symmetrical, the bond is called a nonpolar covalent bond, and the molecule is electrically neutral. (b) In the H_2O molecule, the oxygen nucleus with its six protons attracts the two hydrogen electrons more strongly than do the hydrogen nuclei. As a result, the shared electrons spend more time orbiting the oxygen atom than the two hydrogens. This causes the oxygen atom to act as if it were slightly negatively charged and the two hydrogen atoms to act as if they were slightly positively charged. This kind of bonding is called a polar covalent bond because regions (poles) of the molecule have slight positive or negative charges.

The structure and energy patterns of an atom determine how it reacts with other atoms and also whether the element will be an odorless gas, a black solid, a shiny metal, or some other physical form. In fact, all the properties of the elements emerge from both the structure of the parts (i.e., the number of electrons, protons, and neutrons) and the way the parts (particularly the electrons) are arranged. With this in mind, we can examine the tendency of atoms to react and combine with other atoms into molecules. This potential for combining into molecules is perhaps the most important aspect of atoms we will discuss, because the organelles and cells of living things are composed of molecules.

MOLECULES: ATOMS LINKED WITH "ENERGY GLUE"

Molecules are groups of identical or different atoms that are linked by a kind of energy glue called *molecular bonds*. Bonds are not actual physical objects like the couplings between railroad cars, nor are they solid like hardened glue; instead, they are *links of pure energy* usually based on shared or donated electrons. Bonds act like invisible springs; once a bond between two atoms is formed, it requires energy to pull the atoms apart or push them closer together.

■ **Covalent Bonds** The most common type of bond forms when a pair of electrons is shared by two or more atoms. A shared pair of electrons makes up a **covalent bond.** Hydrogen atoms provide a good example. Two hydrogen atoms can join to form one hydrogen molecule (H_2), with the pair of atoms sharing a pair of electrons, one from each atom (Figure 2.7a). (The subscript in H_2 shows that *two* hydrogens are bonded together in this relationship.) The sharing of two electrons, represented by H—H, creates a *single bond* (the single line represents a single bond). However, some atoms, such as carbon, can share four electrons and form *double bonds*, designated C=C. Whether bonds are single or double, the total number of bonds formed by an atom is generally equal to the number of electrons needed to fill its outer energy level. As Figure 2.7b shows, an oxygen atom has two fewer electrons than a full set of eight in its outer level; it can fill this outer level by bonding with two hydrogen atoms (each sharing one electron). The hydrogen atom, in contrast, can form a bond with just one other atom, since one shared pair of electrons fills its outer energy level. And carbon, the workhorse of atoms in your body, generally makes four bonds to fill its outer level.

In a molecule like H_2, the electrons spend as much time orbiting one nucleus as the other. In a case like this, the distribution of charge is symmetrical, and the bond is called a *nonpolar covalent bond* because it does not create oppositely charged *poles*, or ends, of the molecule. This equal sharing of electrons produces a molecule that is electrically balanced and neutral as a whole (see Figure 2.7a). In a molecule like H_2O, however, where two hydrogen atoms combine with one oxygen atom, the electrons spend more time orbiting the oxygen nucleus than the hydrogen nuclei. The negative electrical charge from the cloud of moving electrons is thus asymmetrical, and the bond is called a *polar covalent bond* (see Figure 2.7b).

■ **Hydrogen Bonds** Molecules with polar covalent bonds can interact with other such molecules because of the slight charges at their poles. In liquid water, for example, a hydrogen from one water molecule can form a bond with an oxygen from a different water molecule, a result of the attraction of opposite charges. This sort of bond between the negative pole of a polar molecule and the weak positive charge on a hydrogen atom is called a **hydrogen bond** (Figure 2.8a). "Weak" here means that the bond is easily broken and re-formed. Some of water's unusual properties (such as the tendency to form droplets) are based on hydrogen bonds (Figure 2.8b), and some important biological molecules (such as the molecule of inheritance called DNA) have millions of these bonds, contributing to the shape and function of the molecule.

■ **Ionic Bonds** Another type of molecular bond depends on the complete transfer of electrons from one atom to another, resulting in positively and negatively charged ions that attract each other. In sodium chloride (NaCl, common table salt), for example, sodium atoms lose the one electron from their outer energy level, and chlorine

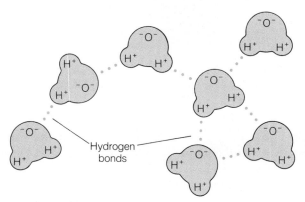

(a) Water molecules

(b) Water drop

FIGURE 2.8 *Hydrogen Bonds and Water Droplets.* (a) A hydrogen atom can be shared between the negative regions of two polar molecules to form a hydrogen bond. Such bonds form between the negative ($^-O^-$) regions of water molecules. This kind of bond is weak and easily broken, but it figures prominently in the molecular interactions within living organisms. These water molecules are drawn according to the "cloud model," which shows the molecules' three-dimensional "electrical" shape by filling in the cloudlike regions occupied by orbiting electrons. (b) Photo of a water drop, magnified many times.

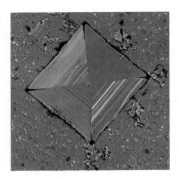

FIGURE 2.9 *Ionic Bonds.* A regular latticework of sodium chloride molecules (NaCl) underlies the cubic structure of a salt crystal, enlarged 14 times in this photograph.

atoms gain one. *Ionic bonds* form between the sodium ions (Na^+) and chlorine ions (Cl^-), resulting in a crystal of salt (Figure 2.9).

Our discussion of bonding—covalent, hydrogen, and ionic—helps show how the properties of molecules emerge from the internal organization and electrical activities of their constituent atoms. The water molecule is an excellent example of such emergent properties. It is a single remarkable chemical entity that makes up most of the matter in most living organisms, and its structure and bonding properties make it uniquely suited to biological systems.

≋ LIFE AND THE CHEMISTRY OF WATER

Our planet and its life forms are practically synonymous with water. Fully three-quarters of the earth's surface is covered by ocean (Figure 2.10). Life began in this aquatic realm, and hundreds of thousands of species still live in water today. Since water is transparent, light can penetrate tens of meters. This allows photosynthesis to take place, supporting most aquatic life. What's more, organisms themselves—whether in the sea or on land—are 50 to 90 percent water.

HYDROGEN BONDS AND THE PROPERTIES OF WATER

What makes water so special chemically and so necessary to life? The answer is that hydrogen bonding between water molecules and other substances makes life possible. Recall that water is a polar molecule with positive

and negative ends. When the charged pole of one molecule attracts the oppositely charged portion of a neighboring molecule, a hydrogen bond forms. Hydrogen bonding helps explain several of water's important properties relating to temperature. Water remains liquid throughout most of the earth's temperature range because it resists heating; that is, its temperature is slow to rise. Water has a high boiling point—100° Celsius (C), or 212° Fahrenheit (F)—and a low freezing point—0° C, or 32° F. This is because when heat is supplied to water, the heat energy must first break the hydrogen bonds between water molecules before it can increase the motion of the individual molecules and thus raise the temperature.

Water's resistance to heating is important to living things because it helps maintain the relatively constant external and internal environments they need. Oceans, large lakes, and rivers are all slow to change temperature, and thus aquatic life forms do not have to cope with rapid temperature changes in their surroundings. The fact that living things are 50 to 90 percent water means that body temperature, especially of large organisms, also changes slowly.

Because water temperature is slow to rise, the evaporation of water (the change from the liquid to the gaseous phase) requires a great deal of heat energy; as water evaporates, it absorbs heat from its surroundings, carries it away, and thus cools the surroundings. For a living thing, this property provides a natural cooling system and explains why many mammals—including our own species—have sweat glands. Sweat pouring from the

FIGURE 2.10 *Earth, Our Blue Planet.* We live on a water-covered planet, as this satellite photo from space reveals.

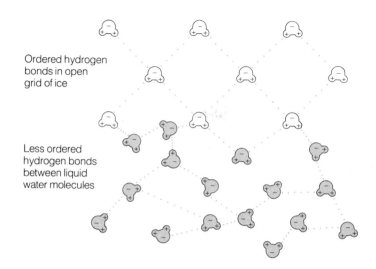

Ordered hydrogen bonds in open grid of ice

Less ordered hydrogen bonds between liquid water molecules

FIGURE 2.11 *The Crystalline Lattice of Ice.* The hydrogen bonds in ice are rigid and create an open grid.

glands in the skin forms a thin layer over the body surface. Only after large amounts of body heat transfer from the body to the skin and break hydrogen bonds in the water molecules is the water free to evaporate.

Finally, water tends to remain as a liquid because to form the rigid latticework of the solid state (ice), water's hydrogen bonds must become rigidly organized, and this only happens when the temperature drops to 0° C (32° F) or below (Figure 2.11).

Hydrogen bonding also explains water's mechanical properties, and those, too, are important to living things.

Water molecules exhibit *cohesion*, or the tendency of *like* molecules to cling to each other. The clinging of water molecules to each other and not to air creates *surface tension* and results in the elastic "skin" around water droplets, the surface on still water that is exploited by water striders and other insects (Figure 2.12a), and the moist surfaces inside the lungs (Figure 2.12b).

Water molecules also exhibit *adhesion*, the tendency of unlike molecules to cling to each other (e.g., water to paper, soil, or glass; see Figure 2.13, page 32). Adhesion and cohesion together account for **capillarity**, the tendency of a liquid substance to move upward (as when water moves upward inside the narrow "pipelines" in a plant's stem; see Figure 2.13).

Hydrogen bonding also creates a fairly rigid, open latticework in ice (review Figure 2.11), with the result that ice is less dense than liquid water and floats in it (Figure 2.14, page 32). Because ice is less dense, it remains at the top of lakes in winter, insulating the lower depths, and allowing plants and animals to survive winter in the chilly—but liquid—water beneath the ice.

The hydrogen bonds in water also help explain its most significant chemical property—the ability to dissolve other molecules, or act as a **solvent.** With some dissolved sub-

stances, or **solutes,** hydrogen bonds form between water and the substance. This happens when sugar dissolves in a cup of tea. With other solutes, such as salt dissolving in soup, water's polarity is of primary importance. The component ions of the salt dissociate, and each becomes

(a)

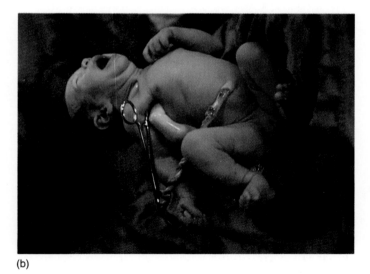

(b)

FIGURE 2.12 *Surface Tension.* (a) Water striders literally walk on water, exploiting the "elastic skin" that forms at the water's surface as a result of the cohesion of water molecules to each other. (b) A baby's first breath is facilitated by a special coating called a surfactant, which the lungs secrete. This material acts much like a detergent to decrease the surface tension of the fluid layer lining the lungs. Without the surfactant, hydrogen bonds in the water lining the small sacs of the lungs would pull water molecules together so tightly that the sacs would collapse. Babies born unable to produce enough surfactant can suffer collapsed lungs and die of respiratory failure.

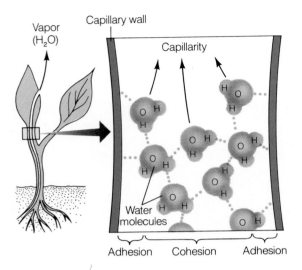

FIGURE 2.13 *Water's Mechanical Properties: Cohesion, Adhesion, and Capillarity.* As a result of hydrogen bonding, water molecules cling to other water molecules—a property of like bonding to like called cohesion. Water molecules also form hydrogen bonds with unlike molecules, the property known as adhesion. Finally, water tends to move upward through a narrow space because water molecules adhere to the walls above and other molecules below. This is called capillarity. Water movement in plants reflects all three mechanical properties.

surrounded by an oriented cloud of water molecules (Figure 2.15). Compounds such as sugar and salt that dissolve readily in water are called *hydrophilic,* or "water-loving," compounds. In contrast, such compounds as oils, waxes, and most plastics do not dissolve readily and are called *hydrophobic,* or "water-fearing," compounds (discussed later).

ACIDS AND BASES

Water has yet another chemical property—a slight tendency to fall apart—again with significant implications for biology. Water molecules themselves separate into positively charged hydrogen ions (H^+) and negatively charged hydroxide ions (OH^-):

$$H_2O \rightarrow H^+ + OH^-$$

This falling apart is a relatively rare event in pure water; only about two out of every billion water molecules tend to dissociate at a given time. Even so, the ions produced when water dissociates play a key role in the chemistry of life. Plants use the ions during the harvest of solar energy (see Chapter 6), and our bodies depend on them

for breaking apart food molecules during digestion (see Chapter 23). What's more, fluctuations of hydrogen ion concentration, caused by dissolved substances that give off or pick up H^+ in water, have broad biological consequences: They alter the acidity of a solution and with it the rate of cellular activities.

By definition, an **acid** is a substance that *gives off* hydrogen ions when dissolved in water and thereby increases the H^+ concentration of the solution. Your stomach produces a very strong acid (hydrochloric acid or HCl) that aids in food digestion. Too much acid in the stomach, however, can result in acid indigestion. The opposite of an acid is a **base,** any substance that *accepts* hydrogen ions when dissociated in water. Bicarbonate is a base. It is produced by cells in your pancreas and counteracts the acid made by your stomach before it reaches your small intestine. Pills to ease acid indigestion, such as Tums and Alka Seltzer, contain bicarbonate.

The most common measure of a solution's H^+ concentration is called the **pH scale,** which ranges from 1 to 14. There are just two important things to remember about the pH scale. First, like the Richter scale that measures an earthquake's magnitude, the pH scale is logarithmic; that is, *every step represents a tenfold change in the number of hydrogen ions in solution.* Second, because of the way the scale is derived, *the higher the H^+ concentration (and thus the more acidic the solution), the lower the number on the scale.* Water has the neutral pH value of 7, right

FIGURE 2.14 *Why Ice Floats.* Many organisms would perish if it weren't for the simple fact that ice floats, insulating the liquid lower depths. Frozen water is less dense than liquid water because the hydrogen bonds in ice create a rigid, open latticework. Hence, this less dense material floats in the more dense water.

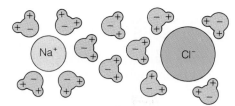

FIGURE 2.15 *Water, the Universal Solvent.* When salt (sodium chloride) is dissolved in water, the ionic bonds between the sodium and chloride ions break, and the water molecules surround the ions, with positive poles orienting toward the chloride ions and negative poles orienting toward the sodium ions. Water's polarity makes it the universal solvent, capable of orienting in either direction to surround ions with positive or negative charges.

in the middle of the scale. Figure 2.16 shows the pH of some common materials, while the box on page 42 examines the issue of acid rain.

Acids and bases are important to living things because living cells and tissues are extremely sensitive to small changes in hydrogen ion concentration. The pH inside most cells stays fairly neutral, between about 6.5 and 7.5, and it is only within this narrow range that many vital cellular reactions take place at optimal speed (see Chapters 4 and 5). Fluids such as blood or plant sap that surround most cells inside living things are also usually fairly neutral in pH. However, certain cells must withstand an acidic or basic environment. Cells lining the stomach, for example, survive in their highly acidic environment of stomach fluid (pH 2) only by secreting a protective mucous layer; and if the pancreas does not make enough bicarbonate to neutralize your stomach acid, you may get a sore, or ulcer, in the upper part of the small intestine. Human sperm, on the other hand, swim in a basic environment (pH 7.8).

The nearly neutral pH of both blood and the cell's interior is maintained largely by substances called *buffers*. Buffers regulate pH by "soaking up" or "doling out" hydrogen ions as needed. Bicarbonate in the blood functions as a buffer; it helps maintain blood pH at about 7.2 by soaking up hydrogen ions when the blood is too acidic and doling them out when the blood is too basic.

We have seen in this section that water—the sparkling substance of ocean waves, waterfalls, snow drifts, clouds, and icebergs—is also the major ingredient in living things. And we've also seen that the molecular properties of water, based on the atomic structures of oxygen and hydrogen and on the versatile hydrogen bond, make it possible for people to stay cool even in the desert, for plants to grow tall, and for life to occur in its myriad forms. Before we can truly understand the chemistry of life, however, we must consider another set of materials that makes up living things: compounds containing carbon.

≋ THE STUFF OF LIFE: COMPOUNDS CONTAINING CARBON

Despite the large amount of oxygen and hydrogen in living organisms, fully 18 percent of a person's weight comes from carbon atoms, and for a large tree, that figure

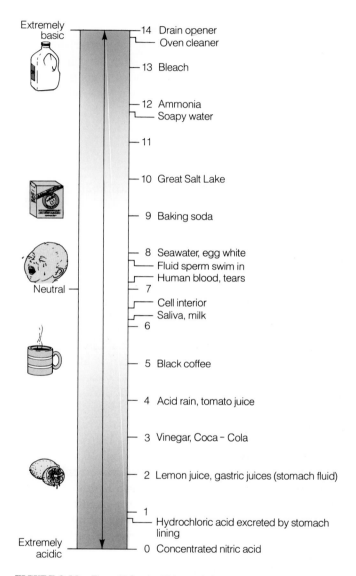

FIGURE 2.16 *The pH Scale.* This scale is a representation of the hydrogen ion concentration in common biological and nonbiological substances. The more hydrogen ions, the more acidic a substance but the lower the number on the pH scale; the fewer hydrogen ions, the more basic and the higher the number on the scale.

can approach 50 percent. So prevalent and so important are carbon atoms to living things that the study of compounds containing carbon is called *organic* ("related to organisms") *chemistry*, leaving to *inorganic* ("lifeless") *chemistry* all other substances. Interestingly, of the two divisions, organic chemistry deals with a far larger number of compounds. Why? Because the structure of the carbon atom is such that carbon can form bonds with up to four other atoms at a time; as a result, it can form millions of different combinations with other atoms—ten times more than all the compounds formed by the dozens of other elements put together. This bonding versatility is carbon's key characteristic, and it explains why crustal elements such as silicon and aluminum, which form relatively few compounds, could never have formed the basis of life with all its many manifestations.

CARBON: COMPOUNDS AND CHARACTERISTICS

Many organic compounds are large (compared to inorganic molecules such as water or salt) and have a "backbone" of carbon atoms bonded to other carbon atoms in long straight chains, branched chains, or rings (Figure 2.17). In living things, the complex compounds known as *biological molecules* belong to this type of carbon compound.

■ **Biological Molecules: Long Chains Composed of Subunits** There are four types of biological molecules: carbohydrates, lipids, proteins, and nucleic acids. They are vital to the structure and function of living things and together with a few other materials account for the diverse shapes, colors, textures, and characteristics of organisms. All are **polymers** ("many parts"), or long chains, which like strings of beads are made up of smaller subunits, the **monomers** ("single parts"), linked together by covalent bonds. Each class of biological molecule is built of a distinctive type of subunit: **Carbohydrates** are built of sugars, or **saccharides; lipids** are composed of **fatty acids; proteins** are long chains of **amino acids;** and **nucleic acids** are long strings of **nucleotides.** Figure 2.18 shows the sugar glucose, a carbohydrate monomer, and the polymer formed by joining many subunits. Life processes, such as the harvesting of energy and the building of new cell parts, depend on the continual construction and destruction of biological polymers.

BIOLOGICAL MOLECULES: THE FOUR CLASSES

Although the four main types of biological molecules all have a string-of-beads polymer organization as well as a backbone of carbon atoms, the particular arrangement of subunits in each type of molecule causes the four types to have very different properties and thus to carry out very different tasks in the living cell.

■ **Carbohydrates** Plants and fungi are mostly water and carbohydrates. Considering the earth's great forested expanses, its grassy plains, and all the aquatic plants of lake and sea, by far the most abundant carbon compounds in living organisms are carbohydrates, and they serve both as structural components of cells and as an energy reserve to fuel life processes. The term *carbohydrate* comes from "hydrate (water) of carbon," and indeed, carbohydrates are variations on the chemical theme of carbon plus water. The monomers in carbohydrates are sugar molecules called *monosaccharides* ("single sugars"); some have a sweet taste. They are linked into two-unit molecules called *disaccharides* or into polymers called **polysaccharides** ("many sugars").

The monosaccharides glucose and fructose are important carbohydrate monomers, since they make up many of the complex carbohydrates in starch, wood, and other biological materials. Glucose is the universal cellular fuel, broken down by virtually all living things to release the energy stored in its bonds. Glucose and its conversion products will figure prominently in our discussions of energy in Chapter 5. Fructose is called fruit sugar because it is the compound that gives many kinds of fruit their sweet flavor (Figure 2.19a, page 36). Glucose and fructose share the molecular formula $C_6H_{12}O_6$, which means they contain six carbon atoms, twelve hydrogen atoms, and six oxygen atoms. The different properties of the two sugars emerge from their different arrangements of the same atoms.

Disaccharides are the common form in which sugars are transported inside plants. Sucrose, or table sugar, is a disaccharide composed of glucose plus fructose. Sucrose is abundant in the saps of sugarcane, maple trees, and sugar beets—our major sources of sugar for refining (Figure 2.19b). Honey is also a disaccharide made up of glucose and fructose, and it has virtually no more nutrient value than refined sugar.

Polysaccharides either store energy or provide structural support. The polysaccharides **starch** and **glycogen** are molecular storage bins that serve as the primary energy reserves of plants and animals, respectively. Most humans consume a diet that is mainly starch gathered from the seeds of rice, wheat, corn, and other cereal plants grown agriculturally. People who lived by hunting and gathering food, such as the traditional Australian aborigines,

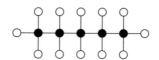

Unbranched chain
with single bonds

Unbranched chain with
single and double bonds

Branched chain

Ring

FIGURE 2.17 *Carbon Chains: Backbones of Life.* Carbon can form millions of organic compounds because each atom can form covalent bonds with up to four other atoms. Many organic compounds have unbranched, branched, or circular (ring) carbon "backbones," with hydrogens or other atoms projecting. Note that we depict these backbones by a ball-and-stick model in two dimensions. Throughout the remainder of this chapter, we will mainly use the two-dimensional ball-and-stick molecular models for clarity, with each colored ball (see code) representing an atom and each stick a covalent bond. Double sticks represent double bonds.

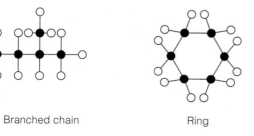

Color Code

Sulfur (S)	○	Yellow
Phosphorus (P)	◐	Green
Oxygen (O)	●	Red
Nitrogen (N)	●	Blue
Carbon (C)	●	Black
Hydrogen (H)	○	White

(a) Monomer

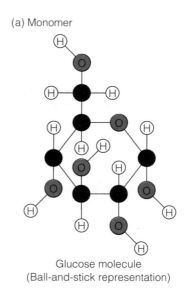

Glucose molecule
(Ball-and-stick representation)

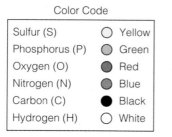

A corner occupied by an atom
other than carbon is indicated
by its symbol (here, O for oxygen)

A simple corner
is occupied by
carbon

At the end of a line is a hydrogen,
unless another atom or group
of atoms is shown. This hydrogen
is above the level of the plane.

CH₂OH

HO   OH   OH

OH

A thickened line suggests
that this part of a planar
molecule lies closest to the reader.

Glucose molecule
(Simplified representation)

FIGURE 2.18 *Biological Molecules: Chains of Subunits.* (a) Biological molecules are made up of monomers, or simple units, such as the glucose molecule depicted here. Even for a simple monomer, the ball-and-stick model depicting every atom can become unwieldy. Thus, we use an even simpler version, with the carbons of the ring omitted but implied by corners. (b) A polymer, such as the starch chain shown here, contains many glucose subunits. The unmarked corners contain a carbon, while the corners marked by an O hold an oxygen atom.

(b) Polymer

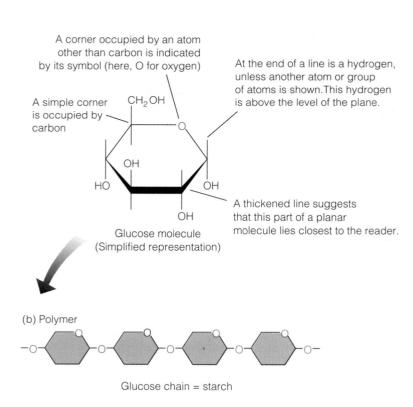

Glucose chain = starch

(a)

(b)

FIGURE 2.19 *Simple Carbohydrates: Monosaccharides and Disaccharides.* (a) Monosaccharides like glucose and fructose are small, simple molecules. Fructose makes peaches and other fruits taste sweet. (b) Sucrose, common table sugar, is a combination of glucose and fructose. Sugar beets, the enlarged roots of the sugar beet plant, store sucrose in special cells. Sugar beets are a major source of the millions of tons of sucrose Americans consume each year.

also depended heavily on starchy roots and seeds. In people and other animals, however, energy in the muscles and liver is stored in the form of glycogen (Figure 2.20).

Cellulose is a tough, fibrous material (the stuff of wood and paper) that gives form and rigidity to the cell walls in crisp leaves and woody stems (see Figure 2.20). Since people cannot digest cellulose, it provides us with no calories. Scientists have recently learned to modify cellulose so it can be mixed with starch and baked into cakes and bread with a fraction of the normal calories.

In later chapters, we will talk more about the great importance of carbohydrates—as structural materials in plant cell walls, as nutrient storage compounds in most kinds of organisms, and as a medium for energy exchange in all forms of life.

■ **Lipids** The term *lipid* may not have a familiar ring, but some of the compounds classed as lipids probably do: Lipids include the **fats,** such as bacon fat, lard, and butter; the **oils,** such as corn, coconut, and olive oils; the *waxes,* like beeswax and earwax; the **phospholipids,** which are important components of cell membranes; and the **steroids,** including certain vitamins, hormones, and cholesterol (the heart and blood vessel clogger). Like the carbohydrates, lipids can serve as energy storage molecules; they also provide waterproof coverings around cells.

Fats and oils have similar basic structures and serve as rich energy storage molecules. The monomers of fats and oils are fatty acids. Because there are three fatty acids in each molecule, fats and oils are usually called **triglycerides.** Triglycerides can be solid, as in animal fats, or liquid, as in vegetable oils (Figure 2.21, page 38). Because of their molecular structure, fats and oils provide more calories than an equivalent amount of sugar or polysaccharide, even though the energy release from lipids is usually slower. One gram (g) of carbohydrate provides 4 kilocalories (kcal) of food energy, while 1 g of fat provides 9 kcal. (One calorie [cal] is the amount of energy required to raise 1 g of water 1° C; 1 kcal = 1000 cal.) When we consume more calories than we burn, our bodies store the extra energy in concentrated form as fat. (We'll discuss food and calories at greater length in Chapter 23.)

Wax—the stuff of candles, honeycombs, and earwax—is another type of lipid. Like other lipids, wax molecules are insoluble in water and serve as important waterproof coatings on leaves, bark, and some fruits.

Another class of lipids, the phospholipids, has properties essential to the living cell. The phospholipid subunit can be visualized as a soluble ball (that includes the phosphate group) on an insoluble stick (two fatty acid chains; Figure 2.22a, page 38). When added to water, phospholipid subunits align to form single layers, double layers, or spheres, with the "balls" oriented toward the water and the tails oriented away from it (Figure 2.22b). This characteristic behavior is crucial to living things, for every cell is surrounded by a double layer of phospholipids (Figure 2.22c), and this vital barrier allows the cell to maintain its watery contents and integrity as a living unit while still exchanging materials with the fluid environment surrounding it.

A fourth class of lipids is the steroids. Steroid molecules are insoluble in water, but they can dissolve in oils or in lipid membranes. Examples of steroids include the sex hormones, estrogen and testosterone (see Chapter 12), and vitamins A, D, and E (see Chapter 23). Figure

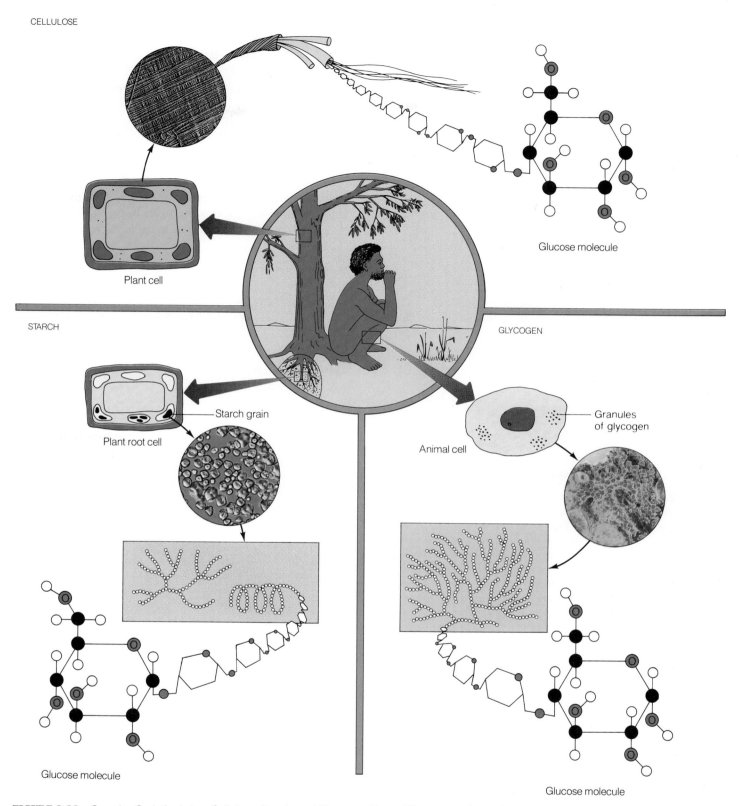

CELLULOSE

Glucose molecule

STARCH

GLYCOGEN

Plant cell

Plant root cell

Starch grain

Animal cell

Granules
of glycogen

Glucose molecule

Glucose molecule

FIGURE 2.20 *Complex Carbohydrates: Cellulose, Starch, and Glycogen.* Three different complex carbohydrates are represented in this scene of an Australian aborigine sitting by a tree. The roots of trees and other plants harvested for food contain cells that store starch grains. The grains are made up of branched or coiled starch molecules, which are themselves long polymers of glucose subunits. The tree contains plant cells with walls made up of cellulose, a tough, fibrous structural material. Cellulose occurs in bundles made up of cords that contain slender strands. Each small strand is composed of cellulose molecules, long polymers of glucose subunits that do not coil or twist. The aborigine's muscles and liver cells contain granules of glycogen, which is very highly branched and can be broken down rapidly to supply quick energy.

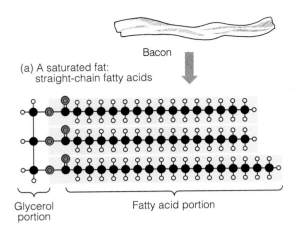

(a) A saturated fat:
straight-chain fatty acids

Bacon

Glycerol portion

Fatty acid portion

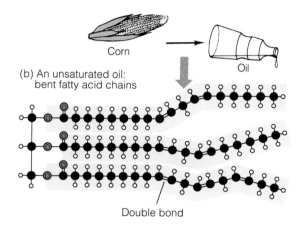

Corn

Oil

(b) An unsaturated oil:
bent fatty acid chains

Double bond

FIGURE 2.21 *Triglycerides: Fats and Oils.* Each triglyceride molecule contains a glycerol subunit and three fatty acids. (a) Saturated lipids are usually whitish solids (fats) because their fatty acids, mostly straight chains, lie close and thus can pack together into solid configurations. People with heart or circulatory problems should cut back on their intake of saturated fats. (b) Unsaturated triglycerides are usually golden-colored oils because their double bonds cause a bend in the fatty acid subunits that keeps the kinked chains from lying close together. Thus, they are slippery liquids.

2.23a shows another common steroid, cholesterol; it is a component of cell membranes. In some people, however, a diet high in saturated fats from animal products can lead to a buildup of cholesterol deposits in blood vessels, which in turn can lead to higher blood pressure and the risk of strokes and heart attacks (Figure 2.23b and c).

■ **Proteins: Key to Life's Diversity** The complexities of our bodies and the rich diversity of life—the millions of organisms of different shapes, colors, textures, sizes, and life-styles—could never be accounted for by carbohydrates and lipids alone. These classes of compounds are important for cell structure and for energy storage and use, but they simply do not contain enough different types of molecules to account for life's vast diversity. Proteins, on the other hand, come in such a wide variety of forms (at least 10 to 100 million different kinds in the spectrum of the earth's organisms) that they can easily explain the myriad forms and functions of living things. Moreover, the specialized shapes and functions of different cell types depend on protein, and the proteins called **enzymes** facilitate virtually all the life processes that go on in cells.

The building blocks of proteins are *amino acids.* Twenty different types of amino acids make up most of the proteins in plants, animals, and other life forms. Each amino acid has four key groups of atoms bound to one central carbon atom (Figure 2.24a). Each group imparts specific

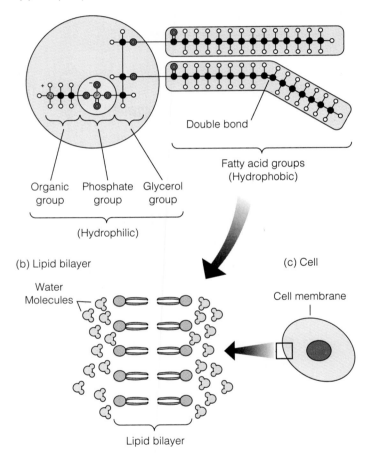

(a) Phospholipid

Organic group   Phosphate group   Glycerol group

(Hydrophilic)

Double bond

Fatty acid groups
(Hydrophobic)

(b) Lipid bilayer

Water Molecules

Lipid bilayer

(c) Cell

Cell membrane

FIGURE 2.22 *Phospholipids as Waterproof Barriers in Living Organisms.* (a) Each phospholipid has a head (containing a phosphate group) and a tail (made up of two fatty acids). The head is water-soluble (hydrophilic), but the tail is not. (b) When phospholipid molecules are surrounded by water, they can form double layers (bilayers) with their heads oriented toward the water molecules and their tails oriented away from the water, toward the inside of the "lipid sandwich." (c) The membranes surrounding living cells contain a double layer of phospholipids.

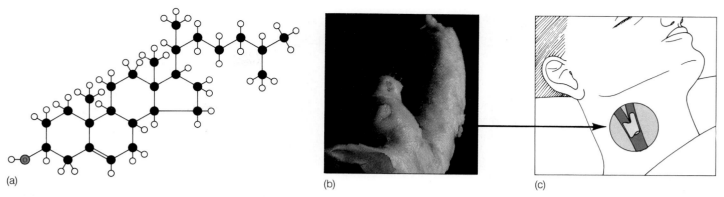

(a)   (b)   (c)

FIGURE 2.23 *Cholesterol and Plugged Arteries.* Steroids are a class of lipid made up of four interconnected rings of carbons with groups of atoms attached. Steroids can pass through phospholipid layers; therefore, hormones—important chemical messengers and regulators—are often steroids. (a) Cholesterol is a common steroid that occurs within cell membranes. But cholesterol deposits can build up into solid clots and plugs that block arteries and lead to serious diseases of the heart and blood vessels. (b) A plug, or plaque, of solid cholesterol and other materials was removed from (c) a patient's major neck artery to allow greater blood flow to the brain.

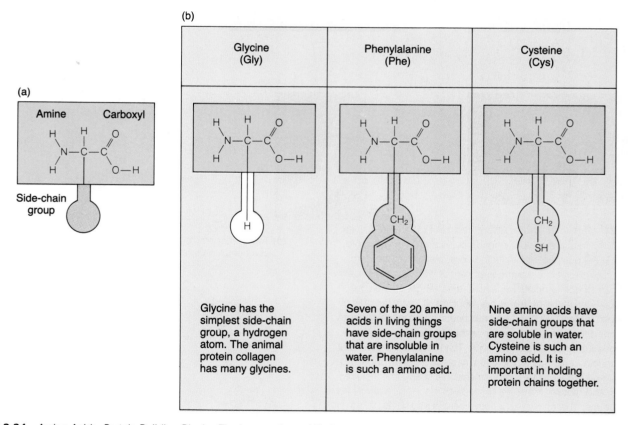

FIGURE 2.24 *Amino Acids: Protein Building Blocks.* The term *amino acid* is based on the structure of these protein subunits. (a) Each amino acid has an amine group (NH₂) on one end and an acidic carboxyl group (COOH) on the other end. The third group is always a hydrogen atom, shown here at the top of the molecule. Only the fourth group, designated R, varies from one amino acid to another. The four groups are all attached to a central carbon atom. The R group can be as simple as the single hydrogen atom found in the amino acid glycine or as complex as a double-ring structure or chain. (b) Representative amino acids.

(a) Amino acids join together to form a polypeptide.

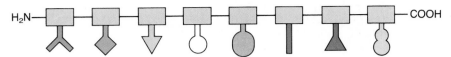

(b) Portions of polypeptides form coils, folded sheets, or random forms.

Coils

Folded sheets

(c) Coils, sheets, and random forms fold together into a three-dimensional shape.

Random form

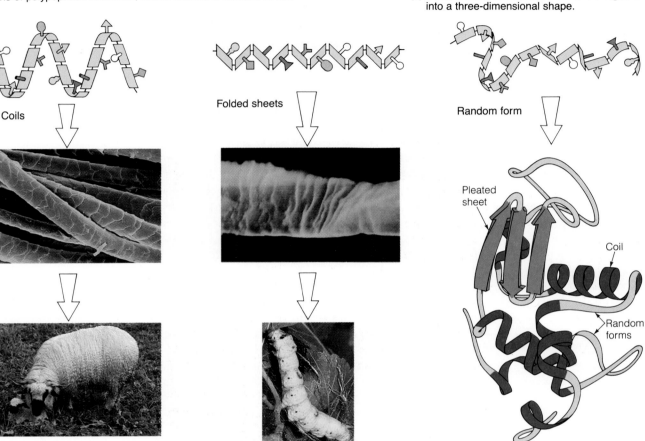

Pleated sheet

Coil

Random forms

FIGURE 2.25 *Protein Structure.* All proteins have (a) a unique linear sequence of amino acids in a polypeptide chain and (b) regular folding patterns in the polypeptide, including coils, pleats, and random forms. The protein keratin, a major component of wool, contains many coils. Wool can stretch to nearly twice its original length as the coils straighten out. Fibroin, the protein in silk, consists of long stretches of pleated sheets. In each silk fiber, several sheets stack on top of each other and impart flexibility; however, since the sheets are nearly completely extended, silk cannot stretch much. Thaumatin is a protein that has many random forms and is 100,000 times sweeter than sugar. (c) All proteins also have a unique three-dimensional shape that includes coils, sheets, and random forms like those seen in this molecule of the enzyme lysozyme. This enzyme is a constituent of egg white; it can break down bacterial cell walls and hence protect the embryo from infection. (d) Some proteins have two or more polypeptide chains that fit together like the pieces of a three-dimensional puzzle. Hemoglobin, the globular protein that transports oxygen in the blood, has four polypeptide chains, two called α chains, and two called β chains. Coils are a prominent part of each chain's structure. The various coils fold around a disk-shaped, iron-containing heme group, and the four chains come together to make the functional protein.

(d) Two or more folded chains fit together like pieces of a complex three-dimensional puzzle.

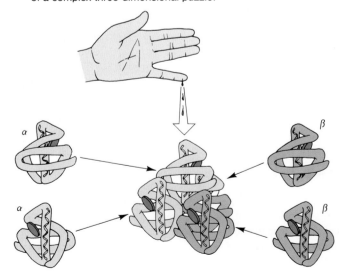

α

β

α

β

chemical behavior to the molecules to which it is attached. In fact, the term *amino acid* comes from two of the attached groups, the amine group and the acidic carboxyl group. Three of the 20 amino acids common in proteins are shown in Figure 2.24b. In a protein, amino acids are joined by covalent bonds, and long chains of many amino acid subunits are called **polypeptides.** Each protein molecule is composed of one or more polypeptides.

The 20 types of amino acids function as subunits in a biological alphabet, "spelling out" complex proteins much as the 26 letters of our alphabet can be combined into a nearly infinite array of words.

The human body contains about 50,000 different proteins, and biologists are now attempting to study the structure and function of many of them. Many observers believe that most genetic and some infectious diseases will eventually be traced to deficiencies in proteins.

Each protein has its own complex shape, and this shape determines how it functions. The shape itself is based on four factors: the order of amino acids, which then determines regions of localized bending or pleating, a three-dimensional folding of the entire polypeptide chain, and, in some proteins, the fitting together of two or more chains (Figure 2.25).

Proteins are chains usually composed of 20 basic types of amino acids; the chains can be 100 to 10,000 amino acids long. Thus, from only 20 subunits, it is possible to build tens of millions of unique amino acid sequences, each "spelling out" a specific type of protein and together making possible life's remarkable diversity (see Figure 2.25a). The chains of amino acids in a polypeptide can fold into local patterns, like the coils of a telephone cord, the pleats in the bellows of an accordion, or more random forms (see Figure 2.25b). For example, *keratin*, the protein in hair, fingernails, horns, claws, and feathers, has many coiled regions. The fact that weak hydrogen bonds stabilize the coils explains why hair can be stretched or curled when warm; heat disrupts the bonds and allows the coils to unfold and the polypeptide chain to stretch out. When the protein cools, coils re-form and the hair takes on the new shape of the brush, roller, or curling iron it was curled around. In contrast, folded sheets form much of *fibroin*, the protein in silk. The many hydrogen bonds cross-linking adjacent folded sheets in fibroin result in a very strong fiber. A recent exciting find was a protein from an African berry. This protein, called thaumatin, is now considered the world's sweetest substance, and chemists believe that regions of random forms may be responsible for triggering the sweet taste on our tongues.

The coils and sheets and random forms of polypeptides fold together in space to give each kind of protein molecule its unique three-dimensional shape (see Figure 2.25c) and, in turn, its function in the cell. Proteins assume their three-dimensional shapes automatically, given the right environment. Some proteins are long fibers. Collagen, the most abundant protein in the animal king-

dom, forms fibers. Collagen gives strength, flexibility, and shape to skin, tendons, ligaments, cartilage, and bones. Some physicians now inject collagen to "fill in" the scars, dents, and wrinkles left by acne, accidents, and aging. In contrast to long fibrous proteins, hemoglobin—the protein that makes blood red—is globular, that is, shaped like a ball. Figure 2.25d shows how the polypeptides of hemoglobin fold in three dimensions.

As already mentioned, some proteins have two or more folded polypeptide chains that fit together like the pieces of a three-dimensional puzzle. In fact, hemoglobin exhibits such a structure, and Figure 2.25d shows how hemoglobin's four chains come together. Hemoglobin's ability to pick up oxygen in the lungs and release it elsewhere in the body arises from its structure. Slight changes in that structure alter the molecule's chemical attraction for oxygen.

To summarize, in every protein, a unique sequence of amino acids determines the shape of the protein and hence the job that it does. Whether a protein serves a structural role, aids movement, transports oxygen, speeds chemical reactions, or attacks foreign invaders depends on its shape and solubility, and those biochemical "meanings" emerge from the language written in amino acids.

■ **Nucleic Acids and Related Molecules: Information Storage and Energy Transfer** Among the outstanding members of the final class of biological molecules are the nucleic acids, which carry the chemical "code of life" and bear genetic information from one generation to the next. This last group also includes transport compounds involved in critical energy transformations in every cell. All the nucleic acid molecules are built of monomers called nucleotides, which are composed of a nitrogen-containing base, a five-carbon sugar (either ribose or deoxyribose), and a phosphate group. The transport compounds are single nucleotides. An example of a nucleotide is shown in Figure 2.26a (page 43).

The nucleic acids DNA (for deoxyribonucleic acid) and RNA (for ribonucleic acid) are long polymers of nucleotide subunits (Figure 2.26b). Each DNA molecule has an elegant double helix shape, while RNA molecules are smaller and simpler. The order of the bases in DNA and RNA carries crucial information for constructing and maintaining cells. What is important here is that the sequence of nucleotides in DNA and RNA codes the sequence of amino acids in proteins (which, as we saw, determines protein shape and function).

Several other important biological molecules are based on modified nucleotide monomers but are not themselves polymers. One class, the *adenosine phosphates,*

ACID RAIN

Rain, cooling and earth scented, can fall in a fine mist, a light shower, or a pelting downpour, but it has always represented a return of life-sustaining moisture to the planet's surface and its organisms. Over the last few decades, however, our human activities have slowly changed a phenomenon as timeless and generative as the rain: In many places now, the falling precipitation corrodes, acidifies, and even kills (Figure 1).

In the northeastern United States, for example, rain is four times more acidic today than in 1900. And acid precipitation (rain, snow, fog, and dew) falls all across the upper Midwest, several southeastern states, the mountainous regions of the western United States, much of Canada, Europe, and the United Kingdom, and in the industrialized areas of India, Africa, Asia, and South America.

Over the decades, this corrosive cascade has caused enormous damage to freshwater habitats, forests, and marine environments, as well as to human-built structures:

- Acid rain has lowered the pH in thousands of northern and mountain lakes and killed or stunted young fish and other organisms.
- It has stripped the leaves and needles from millions of acres of trees and even killed whole forests by altering soil chemistry and lowering nutrient availability to the trees' roots.
- Acid rain has carried an immeasurable tonnage of nitrogen compounds into shallow coastal waters such as the Chesapeake Bay and caused overgrowths of algae that starve and choke out other marine life.
- It relentlessly etches away the surfaces of buildings, machinery, and

statuary, resulting in billions of dollars of repair costs annually.
- Finally, acid precipitation has aggravated many cases of human lung disease, and by lowering the pH in urban water supplies, it causes the leaching of toxic metals from pipes into our drinking water.

Scientists are quite certain about the basic causes of acid rain. The burning of oil and coal to generate electricity and propel vehicles throws massive clouds of sulfur dioxide (SO_2), nitrogen oxides (primarily NO and NO_2), and other gases into the atmosphere. Here they combine with water vapor to form nitric acid (HNO_3) and sulfuric acid (H_2SO_4) and are ferried down in normal precipitation. The smelting of ores contributes additional sulfur dioxide, while the burning of grasslands and forests to expose land for agriculture vaults more nitrogen oxides into the sky, as does (indirectly) the application of fertilizers.

Scientists are also quite sure that restrictions on these activities would have proportionately positive results: A 50 percent reduction in sulfur or nitrogen air pollution, for example, would bring about a 50 percent decrease in acid rain. But is society ready to take the necessary steps to solve this global environmental problem?

- Will people put up with substantially higher power bills so that utilities can build cleaner, more modern plants and install effective smoke stack scrubbers at older facilities?
- Will people vote to raise taxes on gasoline to discourage unnecessary driving and to fund new public transportation systems?
- Will people voluntarily drive less? Turn their thermostats down in winter and up in summer? Use fewer

FIGURE 1 *Effects of Acid Rain.*

disposable metals (tin cans, foil, aluminum containers)?
- Will they support conservation organizations that fight forest and grassland burning? Will they inform themselves about the perils of acid rain and function as an active constituency for clean air?

The decisions and actions rest with you, your family, friends, and neighbors, and other citizens in the community.

Here are places where you can learn more about the acid rain issue:

Ecology departments and courses at colleges and universities
U.S. Environmental Protection Agency
Local air quality management districts
The Sierra Club, the Environmental Defense Fund, and other environmental organizations
Schindler, D. W. "Effects of Acid Rain on Freshwater Ecosystems." *Science* 239 (January 1980): 149–154.
Schulze, E. D. "Air Pollution and Forest Decline in a Spruce *Picea abies* Forest." *Science* 244 (May 1989): 776–783.
Shabecoff, P. "Acid Rain Called Peril to Sea Life on Atlantic Coast," *New York Times*, 25 April 1988, p. 1.

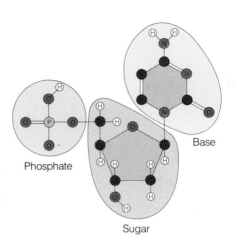

(a) Nucleotide: subunit of DNA

N H
H
N
N
N
H
H
H
O
H
H
H

Phosphate

O
P
O
O
O⁻
H

Sugar

Base

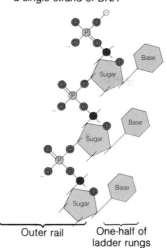

(b) Arrangement of subunits in a single strand of DNA

P
O
Sugar
Base
P
Sugar
Base
P
Sugar
Base
Sugar
Base

Outer rail — One-half of ladder rungs

(c)

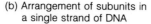

FIGURE 2.26 *DNA: A Double Helix Polymer Made Up of Nucleotides.* (a) A nucleotide has a phosphate group, a ribose or deoxyribose sugar (here we show deoxyribose), and a nitrogen-containing base. There are five types of bases, each with slightly different chemical compositions. DNA uses four of the five; RNA uses a different set of four. (b) A single-stranded DNA molecule consists of many nucleotides joined together in a long chain. (c) Two such chains twist about each other to form a double-stranded DNA molecule, or double helix. The double helix depicted here is a computer-generated color graphic that looks like a work of art. The precise order of the bases carries information for the amino acid sequence of enzymes and other proteins. In turn, enzymes act on raw materials to build all four kinds of biological molecules—carbohydrates, lipids, proteins, and nucleic acids—as well as the cell parts constructed of them. DNA is also responsible for the passing of hereditary information from one cell generation to the next, as we will see later. RNA molecules, most of which are single chains of nucleotides, act as intermediaries in the building of proteins. They help translate the hereditary information contained in DNA to an amino acid sequence in a protein (see Chapter 10).

includes ATP, an energy-carrying molecule that acts as a form of "currency" whose "expenditure" (chemical breakdown) releases energy and enables the cell to accomplish most of its tasks. We will discuss ATP and related compounds in detail in Chapters 4 and 5. Nucleotides called *coenzymes* are transport compounds necessary for energy harvest and the building of new cellular structures. We will become well acquainted with them in Chapters 5 and 6.

In summary, nucleotides provide the immediate energy source for most activities of living cells and join together in long chains to form nucleic acids, the reservoir of hereditary information in the cell.

※ CONNECTIONS

We have seen in this chapter that living things—from individual cells to indigenous peoples dwelling in the great Australian outback—are largely made up of water and depend utterly on this small molecule with its spe-

cial chemical properties. We have also seen that most of the other compounds in a living organism are carbon-based. Moreover, all of the biologically important molecules, whether as simple as H_2O or as complex as nucleic acid, owe their characteristic shapes and activities to the atoms within them. Once you begin to think of living things as incredibly complex and precisely arranged collections of atoms and molecules, then the fundamental importance of water and carbon compounds, the relationships of structure to function, and the concept of emergent properties all unite to help explain a great deal that we see in our daily lives: how physicians can diagnose life-threatening diseases with giant magnetic instruments; how sweating helps cool you off; why organisms store energy as fats and oils; and the list goes on. We will encounter atoms and molecules again and again in our exploration of biology. Chapter 3 examines how the life process emerges from the arrangement of molecules in living cells.

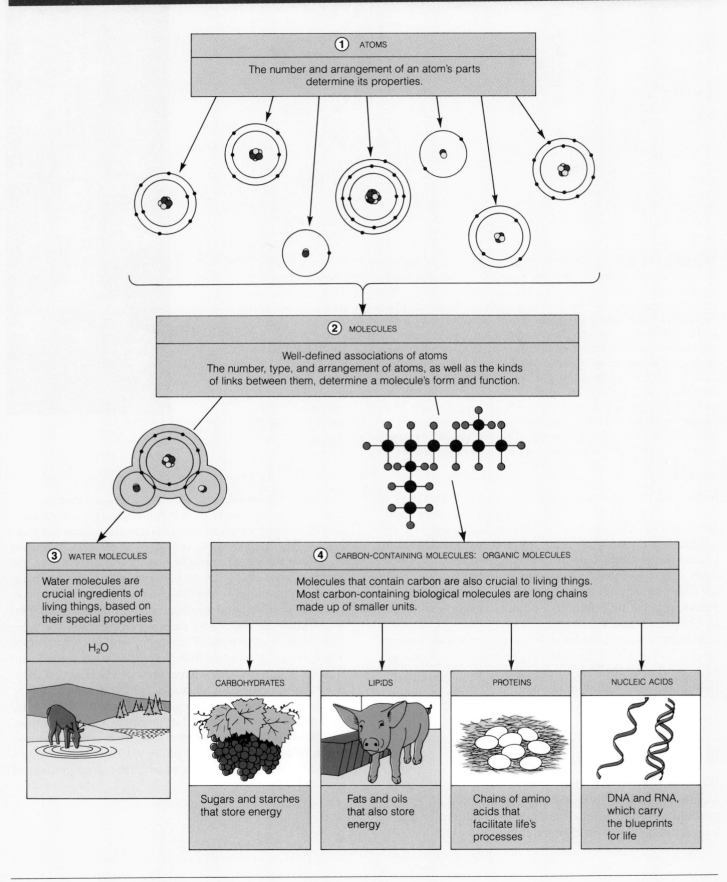

① ATOMS

The number and arrangement of an atom's parts determine its properties.

② MOLECULES

Well-defined associations of atoms
The number, type, and arrangement of atoms, as well as the kinds of links between them, determine a molecule's form and function.

③ WATER MOLECULES

Water molecules are crucial ingredients of living things, based on their special properties

H_2O

④ CARBON-CONTAINING MOLECULES: ORGANIC MOLECULES

Molecules that contain carbon are also crucial to living things. Most carbon-containing biological molecules are long chains made up of smaller units.

CARBOHYDRATES

Sugars and starches that store energy

LIPIDS

Fats and oils that also store energy

PROTEINS

Chains of amino acids that facilitate life's processes

NUCLEIC ACIDS

DNA and RNA, which carry the blueprints for life

NEW TERMS

acid, page 32
amino acid, page 34
atom, page 23
base, page 32
capillarity, page 31
carbohydrate, page 34
compound, page 23
covalent bond, page 29
electron, page 23
element, page 23
enzyme, page 38
fat, page 36
fatty acid, page 34
glycogen, page 34
hydrogen bond, page 29
ion, page 27
isotope, page 25
lipid, page 34
molecule, page 28

monomer, page 34
neutron, page 23
nucleic acid, page 34
nucleotide, page 34
nucleus, page 23
oil, page 36
pH scale, page 32
phospholipid, page 36
polymer, page 34
polypeptide, page 41
polysaccharide, page 34
protein, page 34
proton, page 23
saccharide, page 34
solute, page 31
solvent, page 31
starch, page 34
steroid, page 36
triglyceride, page 36

STUDY QUESTIONS

REVIEW WHAT YOU HAVE LEARNED

1. Although atoms of different elements contain the same basic particles (protons, neutrons, and electrons), they nevertheless have different properties. Explain.
2. What is the difference between a covalent bond and an ionic bond? Give an example of each.
3. While most atoms of an element normally have the same number of protons, neutrons, and electrons, there are exceptions. Name two types.
4. What does the pH scale measure? What makes water neutral on that scale?
5. List the different types of carbohydrates, and describe how they differ from one another. Give an example of each.
6. Name the different classes of lipids and their characteristics.
7. Describe the various aspects of protein structure, and give an example of each.
8. How are nucleotides important to living things?

APPLY WHAT YOU HAVE LEARNED

1. A water strider can walk on the surface of water. Which property of water allows this activity?
2. Cholesterol, which our cells manufacture, serves a structural role in cell membranes. Yet, a doctor may place a patient on a low-cholesterol diet. Why?

FOR FURTHER READING

Atkins, P. W. *Molecules.* New York: Scientific American Books, 1987.

CELLS: THE BASIC UNITS OF LIFE

EUGLENA: AN EXEMPLARY LIVING CELL

If you were to collect a water sample from almost any small pond, stream, or swamp and then view a few drops under a microscope, you might see the bright grassy green, spindle-shaped *Euglena* (Figure 3.1). This organism belongs to neither the plant nor the animal kingdom, but is instead a single-celled life form in the kingdom Protista (see Chapter 1). *Euglena* is a versatile single cell that makes a good case study of what cells are and how they work.

Members of the genus *Euglena* are rather complicated single cells. Like a plant, each one contains small green organelles, called chloroplasts, that convert solar energy into food energy. But like a tiny animal, the cell can move about by means of a whiplike organelle called a flagellum, which enables it to pursue sunlight, just as a lizard seeks the sun's warming rays. In short, a *Euglena* cell exhibits all the characteristics of life (growth, metabolism, reproduction, and so on) that a larger being with many cells displays. And although each *Euglena* species is unique, every individual illustrates important unifying themes that relate to all organisms. First, all cells carry out certain basic functions responsible for keeping them alive. They take in and convert energy to usable form; they build proteins, grow, and reproduce; and the extent and timing of all these activities are controlled. When that control goes awry in one of our cells, cancer can result.

Second, all cells share certain physical structures that might collectively be called the machinery of life. These structures include an outer envelope, or *plasma membrane;* a control center called the *nucleus* (or nucleoid); a jellylike inner substance, the *cytoplasm;* and often several other small internal structures. The fact that *Euglena* and most other cells share these structures is a strong argument for the *unity of life,* the evolutionary relatedness of all living things.

Third, each cell's specialization—how it lives and what it can do—depends on its shape, its size, and the particular organelles it contains. *Euglena,* for example, is a single cell specialized to move toward sunlight and harvest solar energy. Its green chloroplasts and whiplike flagellum are the specialized organelles that allow it to do these things. Similarly, your red blood cells contain specialized organelles that allow them to carry oxygen to all parts of your body; and your egg cells or sperm are specialized to give rise to a new individual.

How cells may have originated is the subject of Chapter 14. Here, in Chapter 3, we concentrate on what cells are, how they live, and why they are so important to an understanding of biology in general and to your health and well-being in particular. This chapter answers several questions:

- What is a cell?
- What are the main cell types and how are they organized?
- What common functions are carried out by cells?
- What specializations distinguish cells from one another?

FIGURE 3.1 Euglena: *A Versatile Single Cell.* Members of the genus *Euglena* are usually bright green and move about by means of a long, lashing "tail," a flagellum.

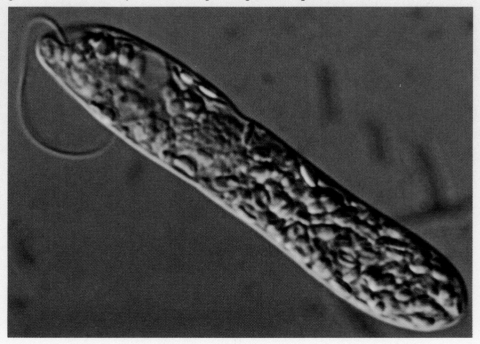

THE DISCOVERY AND BASIC THEORY OF CELLS

Robert Hooke, one of England's greatest scientists, was probably the first person to discover cells. In the 1660s, Hooke wondered why corks, which are solid, could float so easily. He searched for the answer by peering at a thin slice of cork through his ornate microscope, and he saw what he called "pores" or "cells" that reminded him of the little rooms monks inhabit (Figure 3.2). We know today that Hooke was seeing not living cells but their remains—specifically, plant cell walls, composed of cellulose and other molecules deposited outside each plant cell and remaining even after the cell dies.

Anton van Leeuwenhoek, a Dutch cloth merchant, was a contemporary of Hooke. Working in Delft, Holland, also in the 1600s, he had far greater success at seeing living cells such as *Euglena* in action, despite his smaller, simpler, hand-held microscopes (Figure 3.3). You might be able to see some of the creatures he observed by looking at a drop of water through a modern microscope (see Figure 1 on page 61).

It took more than two centuries of study before biologists could extend the work of the early microscopists

(a)　　　　　　　　　　　　　　　　(b)

FIGURE 3.3 *Anton van Leeuwenhoek: Microscopist Extraordinaire.* (a) Anton van Leeuwenhoek (1632–1723). (b) One of Leeuwenhoek's small, simple, hand-held microscopes is still capable of magnifying objects up to 500 times.

past simple observation and comprehend the true significance of cells to the living state. This significance is stated in the **cell theory,** a major doctrine of biology. According to modern cell theory:

1. All living things are made up of one or more cells.
2. Cells are the basic living units within organisms, and the chemical reactions of life take place within cells.
3. All cells arise from preexisting cells.

These tenets may seem obvious today, but in an era with few scientists, few microscopes, and millions of species to explore, they were profound revelations that led to a far greater understanding of the structures and processes that make up the living state.

 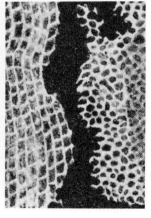

(a)　　　　　　　　　　　　　　　　(b)

FIGURE 3.2 *Discovery of the Cell.* (a) Robert Hooke, a seventeenth-century British scientist, used a simple microscope to explore common objects in a new way. (b) Hooke's 300-year-old drawing of a slice of cork, with its small chambers, is an accurate rendering. These subunits of a formerly living thing he called "cells," and they were the first cells to be discovered.

THE UNITS OF LIFE: AN OVERVIEW

Saying that cells are the basic units of life does not tell much about the units themselves. What are the major kinds of cells? How are they arranged in living things? What shapes and sizes are most cells? What tasks do cells carry out, and how do they do it? Let's look at the answers to these questions.

THE TWO MAJOR KINDS OF CELLS: PROKARYOTES AND EUKARYOTES

Recall from Chapter 1 that biologists divide the millions of species of living things into five kingdoms: monerans, protists, fungi, plants, and animals. Within these kingdoms, there are two major types of cells: **prokaryotic** and **eukaryotic**. The kingdom Monera contains more than 3000 species of single-celled organisms (mostly bacteria), and each individual is a prokaryotic cell. In contrast, the millions of species in the other four kingdoms are each made up of one or more eukaryotic cells.

The term *prokaryotic* means "before nucleus," and perhaps the most obvious thing about prokaryotes is that each prokaryotic cell lacks a distinct nucleus, or region of genetic material bounded by a membrane (Figure 3.4a and c). Prokaryotes do possess DNA, but it occurs in a naked circular strand that, although unprotected by a membrane, is concentrated into an unbounded region called the **nucleoid**. Prokaryotes also lack the other membrane-bounded organelles that eukaryotes possess. They do have some membranes, however, including one that forms an outer envelope around the cell. This **cell membrane** (also called the **plasma membrane**) encloses a clear

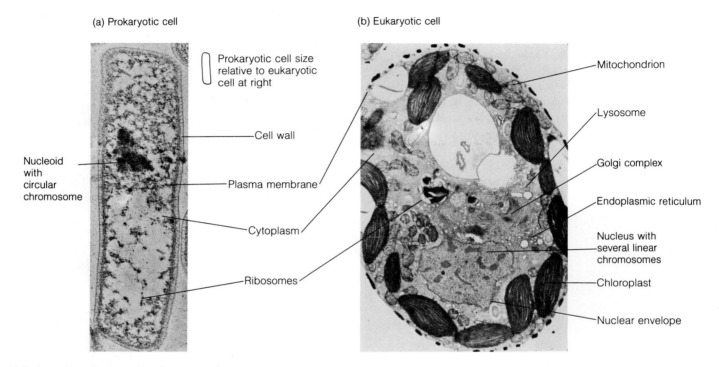

(a) Prokaryotic cell

Prokaryotic cell size relative to eukaryotic cell at right

Cell wall

Nucleoid with circular chromosome

Plasma membrane

Cytoplasm

Ribosomes

(b) Eukaryotic cell

Mitochondrion

Lysosome

Golgi complex

Endoplasmic reticulum

Nucleus with several linear chromosomes

Chloroplast

Nuclear envelope

(c) Prokaryotic and eukaryotic cells compared

	PROKARYOTIC	EUKARYOTIC
KINGDOM	(1) Monerans (mostly bacteria)	(2)　　(3)　　(4)　　(5) Protists, Fungi, Plants, Animals
DISTINGUISHING FEATURES OF CELL TYPE	■ No membrane-bounded nucleus ■ Circular strand of DNA ■ Few cell organelles ■ Single-celled organisms	■ Nucleus bounded by a membrane ■ DNA in several linear chromosomes ■ Many specialized organelles ■ Single-celled organisms or can coexist as subunits of multicellular organism

FIGURE 3.4 *Prokaryotic and Eukaryotic Cells Compared.* (a) A prokaryotic cell such as a bacterium, with major cellular components labeled. (b) A eukaryotic cell with its components labeled. Eukaryotic cells are usually over ten times bigger than the simpler prokaryotic cells. (c) The presence or absence of a true nucleus, the type of DNA packaging, and the number of different cell organelles are the main features that distinguish prokaryotes from eukaryotes.

TABLE 3.1 | COMPONENTS OF PROKARYOTIC AND EUKARYOTIC CELLS

STRUCTURE	WHERE FOUND			FUNCTION
	Bacteria	Plant Cells	Animal Cells	
Plasma membrane	✓	✓	✓	Protection; communication; regulates passage of materials
DNA	✓	✓	✓	Contains genetic information
Nucleus with nuclear envelope		✓	✓	Encloses genetic material
Nucleoid with circular chromosome	✓			Contains DNA
Several linear chromosomes	✓	✓		DNA wrapped around proteins
Cytoplasm	✓	✓	✓	Gel-like interior of cell
Cytoskeleton		✓	✓	Cell support and movement
Endoplasmic reticulum		✓	✓	Intracellular transport
Golgi complex		✓	✓	Packages materials
Ribosomes	✓	✓	✓	Manufacture proteins
Lysosomes		✓	✓	Contain enzymes for cellular digestion
Microbodies		✓	✓	Break down damaging compounds
Mitochondria		✓	✓	Provide cellular energy
Chloroplasts		✓		Produce cellular energy from sunlight
Central vacuole		✓		Maintains cell shape; stores materials and water
Flagella	✓	✓	✓	Cell movement
Cilia			✓	Cell movement
Cell wall	✓	✓		Protects cell; maintains cell shape
Extracellular meshwork			✓	Surrounds and protects cell
Intercellular links — Pili	✓			Exchange DNA
Junctions		✓	✓	Cell-to-cell communication; hold cells together

watery mass called the **cytoplasm.** Finally, most prokaryotic cells are surrounded by a **cell wall** just outside the cell membrane that can be either rigid or flexible, but that always gives shape and support to the cell. Figure 3.4c lists the basic distinguishing traits of prokaryotes, while Table 3.1 shows the components of this cell type. Prokaryotic organisms (largely bacteria) occur in soil, air, and water, as well as on other organisms; grow and reproduce rapidly; and, as a group, are able to use a huge spectrum of organic and inorganic materials as sources of energy.

Eukaryotic cells make up the individual cells of multicellular plants, animals, and fungi as well as the single-celled protists (such as *Euglena*). Like prokaryotes, eukaryotic cells are surrounded by a *cell membrane* (Figure 3.4b). In contrast to prokaryotes, however, eukaryotic cells have a membrane-bounded nucleus (the word

eukaryotic means "true nucleus"; see Figure 3.4b and c). Eukaryotic cells also have other membrane-bounded organelles and are generally 10 to 25 times larger than most prokaryotes. Table 3.1 lists the eukaryotic organelles, many of which are labeled in Figure 3.4b. We will discuss each of these in detail in the second half of this chapter. Eukaryotes also have a highly organized cytoplasm, the clear watery mass contained by the cell membrane and surrounding the nucleus. Crisscrossing this mass is an internal latticework called the *cytoskeleton* that is important to the structural support and movement of the cell and its parts.

Eukaryotic cells have evolved an important trait not shared with the simpler prokaryotic cells: a capacity to coexist and cooperate as subunits of multicellular organ-

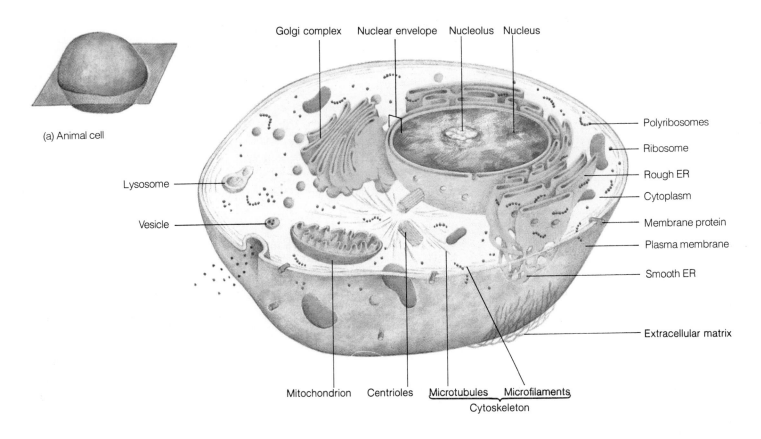

(a) Animal cell

Golgi complex   Nuclear envelope   Nucleolus   Nucleus

Polyribosomes
Ribosome
Rough ER
Cytoplasm
Membrane protein
Plasma membrane
Smooth ER

Extracellular matrix

Lysosome

Vesicle

Mitochondrion   Centrioles   Microtubules   Microfilaments
                              Cytoskeleton

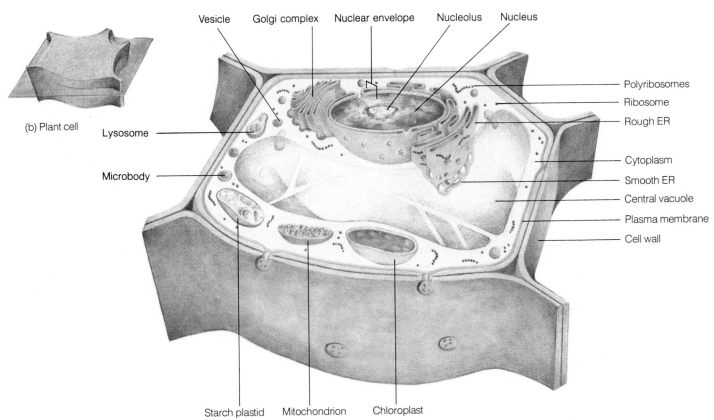

(b) Plant cell

Vesicle   Golgi complex   Nuclear envelope   Nucleolus   Nucleus

Polyribosomes
Ribosome
Rough ER

Cytoplasm
Smooth ER
Central vacuole
Plasma membrane
Cell wall

Lysosome

Microbody

Starch plastid   Mitochondrion   Chloroplast

FIGURE 3.5 *Animal and Plant Cells Compared.* These generalized drawings show the cellular components of (a) an animal cell and (b) a plant cell.

isms. Each multicellular organism depends on a *division of labor* in which groups of cells accomplish specific tasks. Thus, one subset of cells is involved in reproduction, another in digestion or photosynthesis, yet another in mechanical support, and so on.

The evolution of cellular cooperation and coexistence reached a zenith in complex plants and animals. Figure 3.5 shows generalized animal and plant cells with their internal organelles. A quick tour of these cells will help orient you for the more detailed discussions in the rest of the chapter. Notice that the two cell types share many structural features, including a *nucleus* for controlling cell functions and housing the genetic material, *mitochondria* for producing usable energy, *ribosomes* for making proteins, the *endoplasmic reticulum* and *Golgi complex* for processing and transporting proteins, and *microtubules*, which together with *microfilaments* maintain cell shape and allow cell movements. However, plant and animal cells also have obvious structural differences associated with their very different life-styles. For example, an animal cell is surrounded only by an outer or *plasma membrane* and can have a flagellum, which helps produce cell motility. In a plant cell, however, the plasma membrane lies just inside a tough supportive *cell wall* made of cellulose. Plant cells also have a large, usually clear (under the microscope) storage organelle, the *vacuole*, filling most of the cell, as well as one or more bright green *chloroplasts*, which convert solar energy into food energy in the process known as photosynthesis.

THE NUMBERS AND SIZES OF CELLS

When Robert Hooke first discovered the basic units of life in 1663, he also noticed how very numerous they are in a given piece of tissue—over 1 billion, he calculated, in a "cubick inch" of cork. One encounters such astonishing numbers frequently in the study of cell biology. A newborn human baby contains 2 trillion cells; an adult, 60 trillion. When you donate blood, you give away 5.4 billion cells—and scarcely miss them. Each day, in fact, your body sloughs off and replaces 1 percent of its cells, or about 600 billion. (No wonder you don't miss a mere 5.4 billion!)

For a body to contain so many cells, the cells must be extremely small. But just how small? Prokaryotes are the smallest cells, and some bacteria are no more than 0.2 micrometer in length. (A micrometer, designated μm, is one-millionth of a meter; see inside back cover.) An average-sized animal cell is about ten times larger than an average bacterial cell, and many plant cells are a bit larger still (Figure 3.6). Even very large cells, like some *Euglena* species, are still only as wide as the thickness of one page of this book.

Why are cells so small? Why don't we have just a few extremely large cells instead of trillions of microscopic

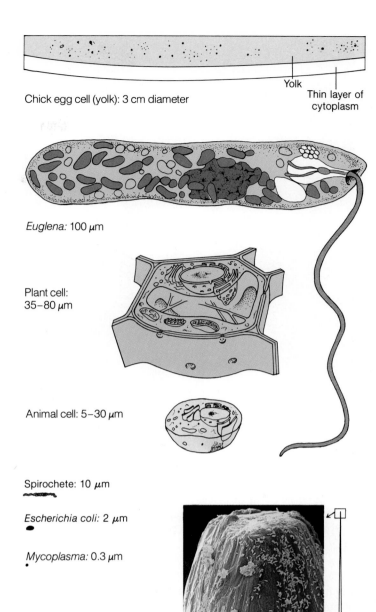

Chick egg cell (yolk): 3 cm diameter

Yolk

Thin layer of cytoplasm

Euglena: 100 μm

Plant cell: 35–80 μm

Animal cell: 5–30 μm

Spirochete: 10 μm

Escherichia coli: 2 μm

Mycoplasma: 0.3 μm

FIGURE 3.6 *Cell Sizes Can Differ Widely.* If a typical bacterium such as *Escherichia coli* were the size of the drawing at the bottom of this figure, then the animal and plant cells would be considerably larger: A *Euglena* cell would look positively whalelike, and the yolk of a chicken egg, a single cell 3 cm in diameter, would look like a small planet 50 m across! In reality, of course, thousands of *E. coli* cells can fit on the sharp tip of a pin and cannot be seen without a powerful microscope.

Cells with similar functions have similar sizes

FIGURE 3.7 *Of Mice and Elephants: Surface-to-Volume Ratio Determines Cell Size.* The life of a cell depends on the exchange of materials through its surface. The greater the cell volume, the more surface area required. Most cells are microscopic, and their surface-to-volume ratios are favorable. That's why a large organism has more cells than a small organism, but the cells are roughly the same size.

ones? The critical dimensions of any cell are its *surface area* and its *volume*. The life of a cell depends on exchanges of materials (ions, gases, nutrients, and wastes) with the environment, and the inward and outward movement of all these materials takes place through the cell's surface. The larger a cell's volume, the greater the amount of material to be exchanged. Since this exchange depends on the cell's surface area, the *surface-to-volume ratio* imposes limits on cell size. An analogy is a pile of wet laundry. If left in a heap with little exposed surface area, it will take a long time to dry; but with each item separated and hung on a line, the exposed surface area increases, water evaporates to the environment, and the laundry dries rapidly. Biological needs depending on surface area and volume thus determine how large a cell can be. A large cell with no special modifications would exchange mate-

rials with the environment too slowly to survive, but many small cells have enough surface area for the rapid exchanges that sustain life. That's why a liver cell in an elephant is the same size as a liver cell in a mouse; the elephant has trillions more cells than the mouse, not the same number of extremely large cells (Figure 3.7).

The one way around this surface-to-volume constraint on size is through altered cell shape, and there are a few types of truly colossal cells. A long, thin cell, such as the nerve cell that extends more than 3 m down a giraffe's leg, or an extremely flat one, such as the thin layer of active cytoplasm that surrounds the 7 cm (70,000 μm) yolk of an ostrich egg, can have the same volume as a round or cube-shaped cell but a greatly expanded surface area. If you take a water balloon and pull it into a long shape or squash it into a flat one, the volume of water inside will not change, but the rubber will have stretched considerably and the surface area will be much larger. Figure 3.8 shows a surface-expanding shape found in the cells that line the human small intestine. Numerous fingerlike extensions allow these cells to carry out their major task—absorption of nutrients—with speed and efficiency.

THE COMMON FUNCTIONS AND STRUCTURES OF ALL CELLS

The survival activities carried out by living cells are essentially synonymous with the life process itself. To carry out survival tasks, cells maintain a set of basic structures, and this section considers them one by one, starting at the cell's outer boundary, or plasma membrane, then moving inward to the most visible cell part, the nucleus, and finally considering the cytoplasm and all the various organelles within it.

Surveying cell components one by one necessarily eliminates the next higher level of organization, the *life* of the cell. In this chapter, try to imagine the "music" of the life process humming quietly but continuously in the

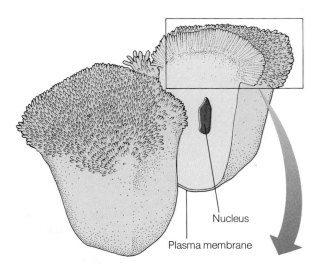

Nucleus

Plasma membrane

FIGURE 3.8 *One Way to Expand the Cell Surface.* High-powered magnification of the surface of human intestinal cells reveals microscopic, fingerlike projections called microvilli. These expand the surface area, so it can absorb nutrients quickly.

background as we focus on each cell part, each "molecular instrument," separately.

THE CELL'S DYNAMIC BOUNDARY: THE PLASMA MEMBRANE

A euglena gliding about in a pond can be easily distinguished from the pond water itself, even though the cell's composition is nearly 80 percent water. This is because a cell's content is delineated from the world around it by a flexible sheet of fatty material called the *plasma membrane* (see Figure 3.5). The outer boundary does not, however, form a tight seal around the cell. On the contrary, the membrane regulates a constant traffic of materials into and out of the cell, allowing water, ions, and certain organic molecules to pass through the boundary and allowing toxic or useless by-products of cellular metabolism to exit. At the same time, it keeps unneeded materials from entering and useful cell contents from oozing out.

The sheet of the plasma membrane is formed by the group of fatty compounds called phospholipids (see Figures 2.22, and 3.9a, page 54). Each molecule of this type has a head and a tail. When such compounds are surrounded by water, they align in a characteristic two-layered sheet (a **lipid bilayer**) with the "water-fearing" (hydrophobic) tails pointing inward, the "water-loving" (hydrophilic) heads pointing outward, and water excluded from the middle (Figure 3.9b). Since the cytoplasm is watery and since cells must be bathed in fluid on the outside (for reasons explored later), the phospholipids naturally assume this two-tiered configuration. And it is this lipid bilayer that is largely responsible for the membrane's barrier functions; many electrically charged substances, as well as most water-loving substances such as sugars, cannot pass through it without the aid of special passageways. The membranes surrounding internal cell organelles have essentially the same structure.

What is there in the structure of the plasma membrane that enables it to regulate the cell's exchanges with the outside world? The answer is that membrane-bound proteins form passageways for materials. Cell biologists have shown that the cell membrane is a *semipermeable* lipid bilayer (one that allows passage of certain materials but not others) studded with proteins that regulate the flow of materials into and out of the cell. This image is the **fluid mosaic model** of membrane structure (Figure 3.9c). The membrane has the fluid consistency of butter on a warm day rather than the solidity of lard. The term *mosaic* refers to the fact that the proteins are scattered about as in a tile mosaic and can move about the fluid plane like floating icebergs.

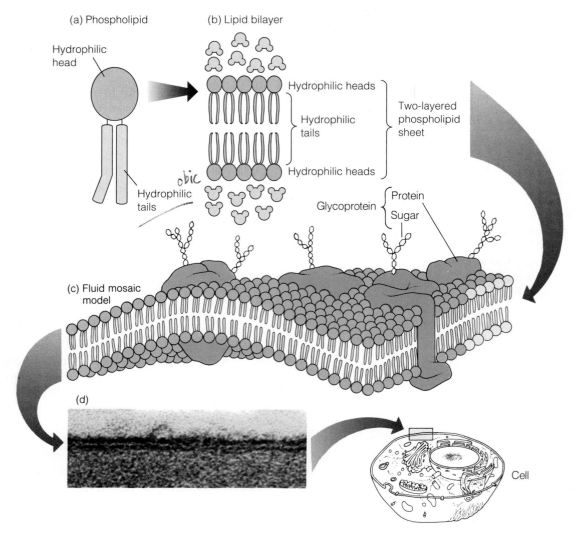

(a) Phospholipid

Hydrophilic head

Hydrophilic tails

(b) Lipid bilayer

Hydrophilic heads

Hydrophilic tails

Hydrophilic heads

Two-layered phospholipid sheet

obic

Glycoprotein { Protein, Sugar }

(c) Fluid mosaic model

(d)

Cell

FIGURE 3.9 *The Plasma Membrane: The Cell's Outer Boundary.* The plasma membrane is a flexible fatty sheet that protects the cell and controls the traffic of materials into and out of the cell. (a, b) Phospholipid molecules align with their heads oriented outward and tails oriented inward to form a lipid bilayer. (c) In the fluid mosaic model, the plasma membrane is pictured as a fluid plane with floating "icebergs"—proteins that can extend through the membrane and project from either or both sides. Special cell surface molecules, made up of sugars and proteins, look like antennae and act as cellular labels and as receptors for incoming chemical signals. (d) Electron micrograph of a sperm cell's plasma membrane.

oligosaccharides

Some membrane proteins in the cell recognize specific materials and either allow them to pass through the membrane or actively transport them across. One such protein is responsible for the most common inherited disease among white Americans, *cystic fibrosis*. In this disease, the tubes leading to the lungs and some other organs become clogged; most victims die before reaching college age. Further knowledge about how membrane proteins work may provide a cure for this devastating condition (Figure 3.10).

Different types of cells contain different populations of membrane proteins. For example, your red blood cells have special membrane proteins that label the cell, giving it your A, B, AB, or O blood type. Other molecular labels lead to either acceptance or rejection of a transplanted kidney, heart, or other organ.

Although many substances enter or leave cells through membrane pores, two methods of transport involve the infolding or outpocketing of the membrane itself and the taking in or exporting of materials (Figure 3.11). Interestingly, in some people, heart attacks are associated with a deficit in this mechanism. These people have an inher-

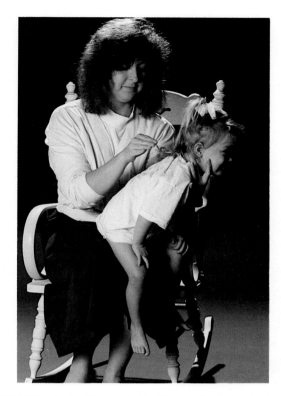

FIGURE 3.10 *Cystic Fibrosis: A Life-Threatening Defect in the Cell Membrane.* Before 1940, 80 percent of children with cystic fibrosis died before they turned five. Today, with better understanding and treatment, at least half of the youngsters with this lung-clogging condition survive into their 20s and beyond. Treatment includes antibiotics to get rid of bacteria that invade weak lungs, physical therapies like back-clapping to dislodge mucus, breathing exercises to increase the flow of oxygen, and a special diet rich in fats and carbohydrates, which provide extra energy.

ited condition that slows down the rate at which cells take in cholesterol circulating in the blood. The fatty material thus builds up outside the cells, clogs the blood vessels, and can lead to heart attacks.

THE NUCLEUS: CONTROL CENTER FOR THE CELL

In a eukaryotic cell, the largest organelle and often the most conspicuous when viewed through the microscope is the **nucleus,** the "brain" of the cellular operation. This roughly spherical structure contains DNA, the genetic information that controls a cell's activities. It also makes and exports RNA, the nucleic acids that relay genetic "orders" to the cytoplasm. These orders are patterns for building proteins—sometimes functional proteins (enzymes) and sometimes structural proteins (like those in the cell's skeleton). The information molecules issuing from the nucleus direct most of a cell's day-to-day activities.

The nucleus is surrounded by a **nuclear envelope.** This envelope is made up of *two* lipid bilayer membranes separated by a space, and the entire envelope is perforated at dozens of points by **nuclear pores,** giving the organelle the appearance of a golf ball (Figure 3.12, page 56). Each pore is a cluster of proteins that form a channel that somehow regulates the passage of RNA to the cell's cytoplasm.

Within the nucleus, proteins bind to DNA to form **chromosomes** (literally, "colored bodies"), structures that carry hereditary information and are visible under the microscope (see Chapters 7 and 9). There are one or more

(a) Infolding: an amoeba ingests a *Euglena* cell during phagocytosis

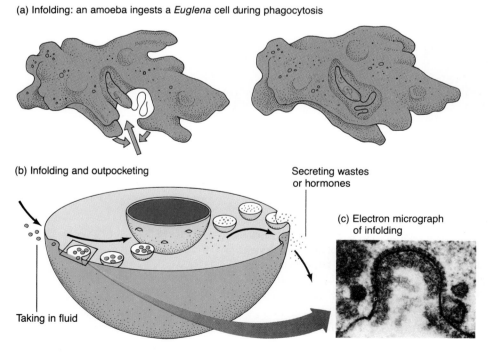

(b) Infolding and outpocketing

Secreting wastes or hormones

(c) Electron micrograph of infolding

Taking in fluid

FIGURE 3.11 *Infolding and Outpocketing of the Cell Membrane.* Material can be drawn into or expelled from the cell in a membranous pocket. (a) In the process of infolding, the cell membrane forms a pocket around a cluster of molecules or a food particle, sometimes around an entire cell. When the cell takes in food, the type of infolding is called *phagocytosis.* Here, an amoeba ingests a *Euglena* cell through phagocytosis. (b) When the cell takes in fluid, the membrane indents around the fluid and dissolved proteins (designated here by blue dots) and pinches off inside the cell. By means of outpocketing, a process that is the reverse of infolding, the cell can secrete pockets of material—indigestible wastes or products such as hormones for export to other parts of a multicellular organism. In outpocketing, tiny membranous spheres packed with the material to be jettisoned fuse with the plasma membrane and then open to release their contents to the outside. (c) Electron micrograph of infolding in a hen's egg cell.

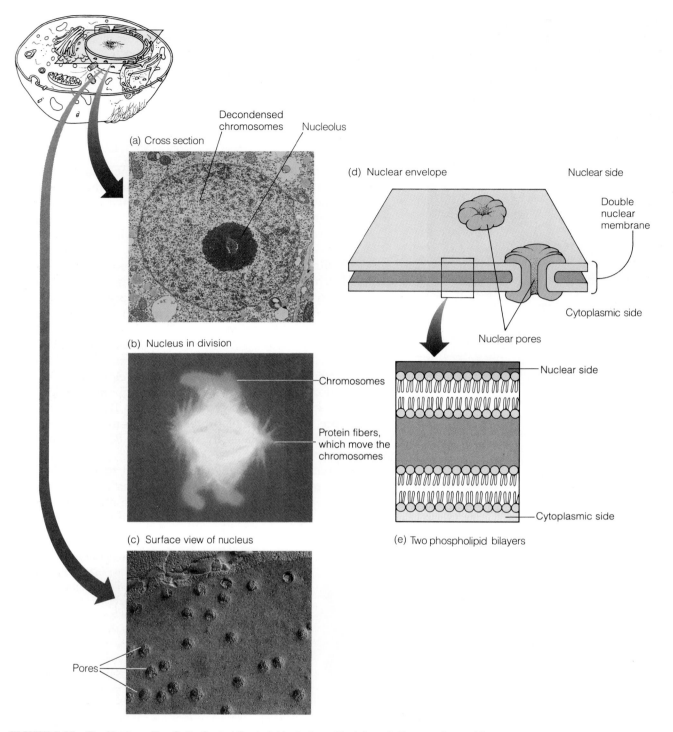

(a) Cross section

Decondensed chromosomes

Nucleolus

(b) Nucleus in division

Chromosomes

Protein fibers, which move the chromosomes

(c) Surface view of nucleus

Pores

(d) Nuclear envelope

Nuclear side

Double nuclear membrane

Cytoplasmic side

Nuclear pores

Nuclear side

Cytoplasmic side

(e) Two phospholipid bilayers

FIGURE 3.12 *The Nucleus: The Cell's Central Control.* Most of a cell's daily activities are directed by information molecules that issue from the nucleus. (a) A cross section of a nucleus reveals decondensed chromosomes. A dark-staining area, the nucleolus, is the site of ribosomal RNA production. (b) When the cell divides, the cell's hereditary material condenses into chromosomes. They appear here as brightly stained reddish orange rods surrounded by protein fibers (stained yellowish green), which move the chromosomes during cell division. (c) The often spherical nucleus looks a bit like a golf ball with its double membrane perforated by dozens of pores. (d) The nuclear envelope is perforated by pores, channels made up of proteins. The pore regulates the passage of RNA. (e) The nuclear envelope is composed of two phospholipid bilayers with a space in between.

dark-staining regions called **nucleoli** (singular, nucleolus, or "little nucleus"; see Figure 3.12a). The nucleolus synthesizes a special type of RNA called **ribosomal RNA,** which joins a few specific proteins and is then exported through the nuclear pores to the cytoplasm. There, the RNA joins more proteins and forms tiny, beadlike units called **ribosomes** that are involved in protein manufacture (Figure 3.13); this process is described fully in Chapter 10. Cells, such as egg cells, that are actively making proteins, have large nucleoli or several of them.

CYTOPLASM AND THE CYTOSKELETON: THE DYNAMIC BACKGROUND

The plasma membrane and nucleus account for only a small portion of the cell's mass. Most of the cell is made up of cytoplasm, a semifluid, highly organized ground substance that acts as a pool of raw materials and contains an internal lattice, the **cytoskeleton.** This weblike structure surrounds and suspends several kinds of cytoplasmic organelles. About 70 percent of the fluid portion of the cytoplasm consists of water molecules, and about 15 to 20 percent is made up of 10,000 different kinds of protein molecules, with 10 billion or so molecules occurring in an average cell!

Through the light microscope, the cytoplasm looks granular and gel-like. However, the electron microscope reveals that the cytoskeleton's three-dimensional latticework of protein filaments permeates the cytoplasm like a spider's web, suspending the organelles in proper spatial relationships to each other and allowing precisely regulated movement of cell parts.

Figure 3.14 depicts the three elements that make up the cytoskeleton: the *microfilaments*, *microtubules*, and

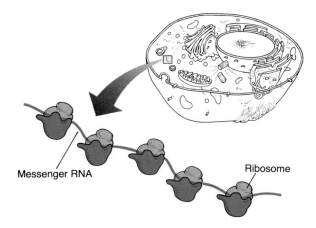

FIGURE 3.13 *Ribosomes: Sites of Protein Manufacture.* Ribosomes are dense, beadlike clusters of ribosomal RNA and certain proteins. They can be seen throughout the cell cytoplasm, either associated with various membrane surfaces, or in chains called polyribosomes. In a polyribosome, a string of ribosomes becomes associated with a ribbonlike molecule of a type of RNA called messenger RNA.

(a) Microfilaments

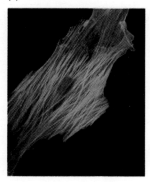

(b) Microtubules

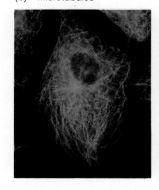

(c) Intermediate filaments

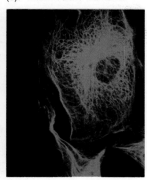

FIGURE 3.14 *The Cytoskeleton: Cell Support and Movement.* The cytoskeleton acts as an internal skeleton, supporting the cell's shape and activities. The photos here are analogous to X-rays of bones; each particular cytoskeletal element appears a different color. (a) Microfilaments, or stress fibers, are made mostly of *actin* proteins. They assist in many kinds of intracellular movements. Actin proteins can rapidly assemble or disassemble from the filament, and this dynamic activity causes the filament to lengthen or shorten, moving organelles along with it. For example, in plant cells, microfilaments attach to chloroplasts, organelles that collect sunlight and carry out photosynthesis. The microfilaments serve as running tracks that transport the chloroplasts toward the brightest side of the cell. (b) Microtubules, a second cytoskeletal element, are hollow cylindrical tubes 20–25 nm in diameter. They are composed of subunits of the globular protein *tubulin,* which, like actin, assemble and disassemble rapidly to lengthen or shorten the microtubules. A microtubule acts as an intracellular engine, moving particles inside cells in two directions at once. Microtubules are major structural elements in two kinds of organelles that generate cell movement, flagella and cilia; they also help establish the shape of many cells. A third component of the cytoskeleton is (c) the intermediate filaments, which are made up of *keratin* and other proteins. These are about 10 nm across, or midway in size between microfilaments and microtubules. Unlike the other cytoskeletal elements, these filaments appear to be incapable of dynamic assembly and disassembly. Instead, they seem to be involved in maintaining cell shape, acting as girders or tension-bearing cables that stabilize the cell's perimeter.

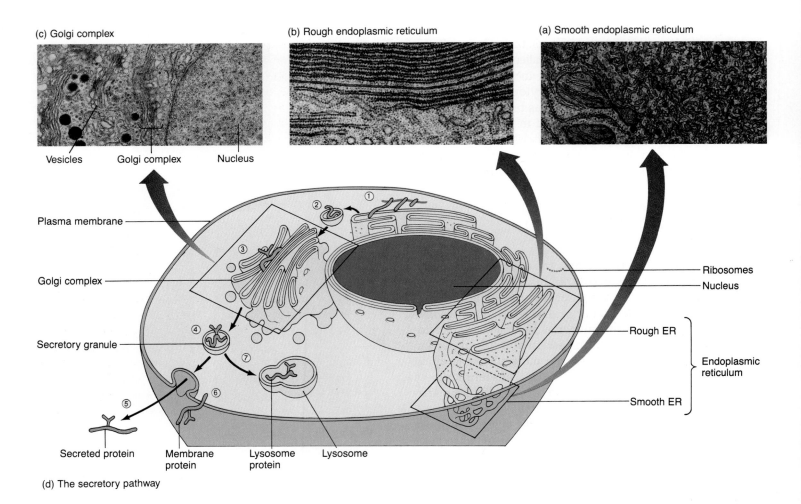

(c) Golgi complex

Vesicles   Golgi complex   Nucleus

(b) Rough endoplasmic reticulum

(a) Smooth endoplasmic reticulum

Plasma membrane

Golgi complex

Secretory granule

Secreted protein   Membrane protein   Lysosome protein   Lysosome

(d) The secretory pathway

Ribosomes
Nucleus

Rough ER

Endoplasmic reticulum

Smooth ER

FIGURE 3.15   *Function of the Endoplasmic Reticulum and Golgi Complex: The Production and Packaging of Proteins.* The endoplasmic reticulum (ER) is a network of membranes forming an interconnected set of channels throughout the cell. (a) Certain areas of the ER have a smooth appearance and are called *smooth ER.* Smooth ER is most common in cells that produce, store, and secrete nonprotein substances such as steroid hormones and other lipids. (b) The exterior surfaces of some ER are studded with ribosomes, giving them a grainy appearance when viewed with the electron microscope. These areas make up the *rough ER,* which is especially prominent in cells exporting or secreting proteins that do their jobs outside the cell of origin. (c) The Golgi complex consists of stacks of saucer-shaped membranes surrounded by vesicles. Biological molecules enter the Golgi complex from the ER through interconnecting channels or in small transport vesicles that pinch off from smooth or rough ER and move toward the Golgi complex. (d) Proteins that will end up in specific compartments in the cell often enter the rough endoplasmic reticulum as they are formed (1). Vesicles containing the newly formed proteins pinch off from the endoplasmic reticulum (2) and move to the Golgi complex, where the proteins may be modified (3). From the Golgi, vesicles pinch off and transport the modified proteins to some final location (4). Certain proteins, such as hormones, may exit the cell (5); others, like the proteins involved in rejecting transplanted organs, may be inserted in the cell membrane (6); still others, such as a few kinds of digestive enzymes, may enter lysosomes (7).

intermediate filaments. Most of the movements of living things, such as the beating of hearts, the running, flying, or swimming of whole organisms, and the tiny movements of structures within the cell, are based on the activities of these "engines" and "skeletal" parts. Microfilaments and microtubules also act as internal girders and cables that help maintain a cell's shape; they also transport cell organelles, much as a gymnast climbs up a stationary rope. The box on page 60 shows how a researcher applied the scientific method to uncovering the role of microfilaments in cell division.

A SYSTEM OF INTERNAL MEMBRANES FOR SYNTHESIS, STORAGE, AND EXPORT

Suspended within the cytoplasm is an interconnected system of organelles involved in the synthesis, transport, degradation, and export of proteins and other cell products. Interconnected membranous channels that are found throughout the cell form a major part of this system. The system of channels is called the **endoplasmic reticulum** (literally, "network within the cell," abbreviated ER), and its convoluted passageway extends from the nuclear envelope to the plasma membrane. Biologists call some areas of this organelle *smooth ER* because the membrane is folded into smooth sheets and tubes (Figure 3.15a). The smooth ER detoxifies poisonous chemicals and also manufactures lipids. For example, the smooth ER in the skin converts cholesterol (review Figure 2.23) into the lipid compound called vitamin D whenever sunlight strikes the skin; this vitamin helps maintain strong, healthy bones. North African women of the Bedouin tribe, who wear dark, full-length gar-

FIGURE 3.16 *Some Bedouin Women Have a Smooth ER Problem.* Because this woman's clothing leaves little or no skin exposed to sunlight, her smooth ER may not be able to make enough of the vitamin D necessary to maintain strong, healthy bones.

ments (Figure 3.16), get very little exposure to sunlight, and thus the smooth ER in their skin cells cannot make vitamin D. As a result, Bedouin women sometimes develop soft, weak bones.

While the smooth ER is involved in lipid synthesis, another part of the endoplasmic reticulum, called the *rough ER*, is involved in protein synthesis. Beginning with the nucleus, we can follow a protein through this set of membranous organelles. The nucleus generates information for making proteins, which moves to the cytoplasm as RNA. There, the information joins the small, beadlike ribosomes—biochemical anvils on which protein molecules will be forged (see Figure 3.13). Ribosomes can attach to the rough ER, and in fact, it is their presence there that makes this part of the ER look rough (Figure 3.15b). Proteins assembled on rough ER (Figure 3.15d, step 1) enter the ER cavity, are modified as they move along through the channels, and are eventually pinched off in little sacs, or vesicles (step 2).

Most of the sacs pinched off from the endoplasmic reticulum enter another membrane system, the **Golgi complex** (Figure 3.15c and Figure 3.15d, step 3), where proteins in the sacs are further modified. The modified proteins leave the Golgi complex in vesicles (Figure 3.15d, step 4) that can either fuse with the cell's plasma membrane and export the protein from the cell (step 5), or insert the protein in the cell's surface membrane (step 6), or fuse with the cell's digestive sacs (lysosomes) (step 7). You can think of the Golgi complex as the cell's traffic controller, directing proteins to their proper destinations. Golgi-packaged proteins and lipids repair the plasma membrane itself when it is damaged, and in plants, the precursors to the cellulose that forms the outer cell wall are packaged and exported from *dictyosomes*, the name given to the Golgi complex in plants.

While many vesicles that pinch off from the Golgi complex leave the cell, two main types—lysosomes and microbodies—take up permanent residence in the cytoplasm. **Lysosomes** are spherical vesicles within the cell that contain powerful digestive enzymes. These enzymes can help recycle worn-out cell parts. Biologists know that this recycling by lysosomes is crucial to human survival because babies born with defective lysosomes have Tay-Sachs disease. With this condition, lysosomes in brain cells fail to recycle certain carbohydrates, allowing them to accumulate, block brain development, and result in mental retardation and death.

Besides carrying out recycling, lysosomes can act like minute cellular stomachs. When a protozoan, such as *Euglena,* or one of your white blood cells engulfs a bacterium, lysosomes will merge with the membrane-encapsulated "meal" and release their enclosed digestive

(a)

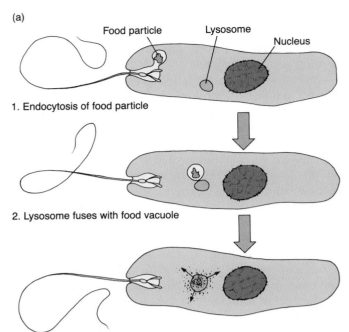

Food particle   Lysosome   Nucleus

1. Endocytosis of food particle

2. Lysosome fuses with food vacuole

3. Absorption of small molecules

(b)

FIGURE 3.17 *Lysosomes at Work.* (a) *Euglena* engulfing a food particle. 1. When a euglena happens upon a food particle of the right size and composition, the cell membrane forms a pocket around it and engulfs it (the process of endocytosis). 2. A lysosome then fuses with the food vacuole, and digestive enzymes break down the food particle. 3. Small nutrient molecules then pass through the lysosome membrane into the cell cytoplasm to nourish the cell. (b) Lysosomes at work help transform caterpillars into butterflies. Before this silver-washed fritillary butterfly (*A. paphia*) emerged from its cocoon, enzymes in lysosomes destroyed the old larval cells. Materials released by this breakdown were used to construct the new adult.

DISCOVERY

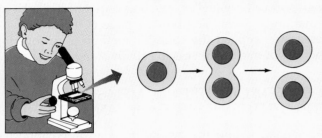

1.   OBSERVATION: A dividing animal cell pinches in two.

2.   QUESTION: I wonder what causes the cell to pinch in two during division?

3.   HYPOTHESIS: Maybe microfilaments close in like a tightening belt and split the cell.

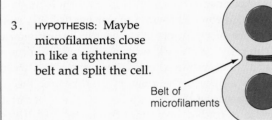

Belt of microfilaments

4.   EXPERIMENT AND PREDICTION: If I block the action of microfilaments with a chemical,* then the cell should not split.

5.   RESULTS: Blocking microfilaments really did prevent cell splitting.

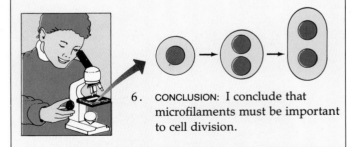

6.   CONCLUSION: I conclude that microfilaments must be important to cell division.

*The chemical cytochalasin b is known to disrupt microfilaments. Adapted from T. E. Schroeder, *J. Cell Biol.* **53** (1972): 419–434.

The researcher pictured in this box is using a light microscope. Figure 1 on page 61 shows three types of microscopes and an example of what you can see through each type.

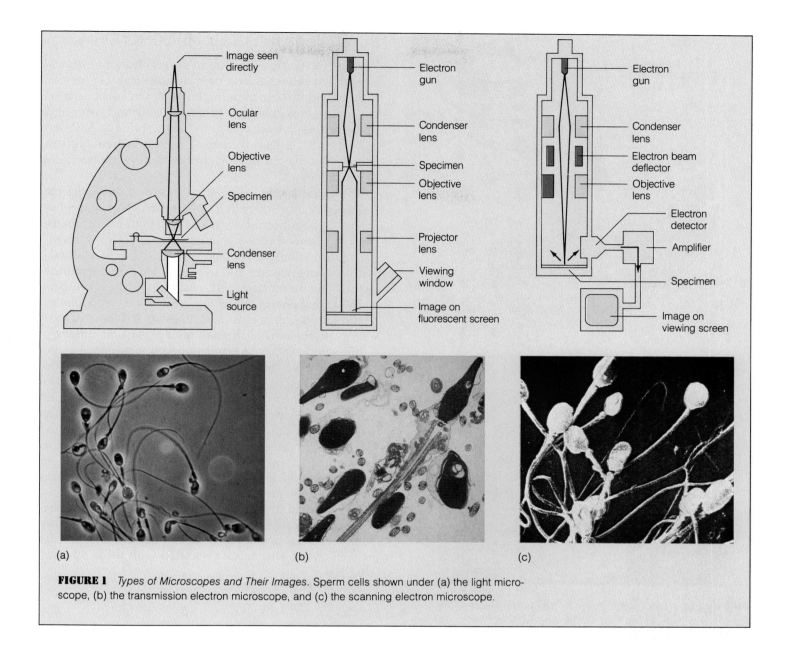

FIGURE 1 *Types of Microscopes and Their Images.* Sperm cells shown under (a) the light microscope, (b) the transmission electron microscope, and (c) the scanning electron microscope.

enzymes, breaking down the particle to its constituent molecules (Figure 3.17a). These molecules then pass through the lysosomal membrane into the cytoplasm, where they can be used as raw materials for producing energy or building new cell parts. This is one of the mechanisms that helps white blood cells clean out the bacteria infecting a cut in your finger.

Lysosomes also sometimes act as "suicide bags," breaking open and spilling their contents and literally digesting entire damaged or aged cells from the inside out. This process is especially dramatic in insects that are transformed within cocoons from caterpillars to moths or butterflies. One by one, the caterpillar cells die and are self-digested to make way for the new moth cells (Figure 3.17b). It is still not clear why the enzymes in the suicide bags don't digest the bags themselves, but it is

clear they can't break down every type of material to constituent molecules. In fact, aging eukaryotic organisms often accumulate brown-pigmented granules in their lysosomes. Older people often develop brownish "age spots" on their skin from these same pigments.

Microbodies are the other class of membranous vesicles that reside permanently in the cell. Microbodies called *peroxisomes* contain enzymes that break down damaging compounds. Peroxisomes within liver and kidney cells, for example, break down and detoxify fully half of the alcohol a person drinks.

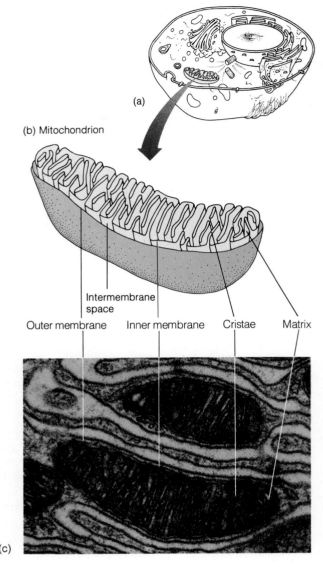

(a)

(b) Mitochondrion

Intermembrane
space

Outer membrane   Inner membrane   Cristae   Matrix

(c)

FIGURE 3.18 *Mitochondria: Power for the Cell.* (a) Mitochondria look like tiny grains of rice filled with threads. (b) Mitochondria have a smooth *outer membrane* and an *inner membrane* folded into a complex of overlapping sheets, or *cristae*, like the teeth of enmeshed gears. The membranes are lipid bilayers like the plasma membrane, but contain a different set of embedded proteins. Between the two membranes is the *intermembrane space*, a compartment that contains several kinds of enzymes. Inside the inner membrane, there is another compartment, called the *matrix*, that contains the genetic material DNA, ribosomes, and many different enzymes involved in producing ATP for energy. (c) Electron micrograph of kidney cell mitochondria.

POWER FOR THE CELL: MITOCHONDRIA

Within eukaryotic cells, one to several hundred organelles called **mitochondria** provide chemical fuel for cellular activities. Mitochondria break down small carbon-containing molecules into carbon dioxide and water and in the process release energy, which is stored in ATP (see Chapter 2). These "energy packets" then diffuse throughout the cell and fuel the biochemical transactions of life processes.

The activities of each mitochondrion are possible because of the organelle's unique structure. Like the nucleus, the mitochondrion has two lipid bilayer membranes: a smooth outer membrane and an inner membrane thrown into folds (Figure 3.18). At least 60 different types of enzymes and other proteins are embedded in the inner membrane folds, and many of these proteins help harvest energy in aerobic respiration (see Chapter 5). Prokaryotic cells lack mitochondria, but they convert energy on the inner surface of their plasma membranes.

Mitochondria have a semiautonomous existence in the cell: They have their own DNA that directs production of some of their component proteins, and they can divide in half and thus reproduce independently of the cell's normal cell division cycle. Surprisingly, mitochondria are passed to an animal only by the mother, since mitochondria are present in eggs but not in the part of the sperm that enters the egg. Thus, people can trace their mitochondria back to their mothers, grandmothers, great-grandmothers, and so on.

The cellular structures considered in this section are present in nearly all eukaryotic cells. The next section considers tasks carried out by some cells but not by others, as well as the structures associated with those specialized tasks.

≋ SPECIALIZED FUNCTIONS AND STRUCTURES IN CELLS

This chapter began with *Euglena,* a cell specialized to pursue sunlight by means of particular organelles that "outfit" the organism to move and to trap solar energy. Like *Euglena,* most cells have specialized organelles in addition to their nucleus, mitochondria, and other organelles of life support.

PLASTIDS: ORGANELLES OF PHOTOSYNTHESIS AND STORAGE

Plants and some protozoa have several types of oval organelles, collectively called **plastids,** that harvest solar energy and produce and store food. Like mitochondria, plastids have two lipid bilayer membranes, and the most important and widespread of the plastids, the **chloroplasts,** also carry out critical energy conversions (Figure 3.19). As the organelles of photosynthesis, chloroplasts trap the sun's energy in a chemical form (as carbohydrate molecules) that is ultimately used by nearly every organism on earth to power the myriad activities of life. Chloroplasts contain the green, light-absorbing pigment *chlorophyll* and sometimes brownish and yellowish pigments as well. Through photosynthesis, plant cells manufacture their own organic nutrients instead of absorbing ready-made nutrients from ingested food, as animal and fungal cells and most protozoa must do. Like mitochondria, chloroplasts have their own DNA, which encodes *some* of the organelle's proteins.

Two other types of plastids are common in plant cells. *Chromoplasts* ("colored plastids") store yellow, orange, and red energy-trapping pigments and give color to many fruits and flower petals. *Leucoplasts* ("white plastids"), on the other hand, are colorless organelles that often store starch granules. Potatoes, for example, contain billions of starch-storing leucoplasts.

VACUOLES: NOT-SO-EMPTY VESICLES

Through the microscope, most plant cells look strangely hollow, with what appears to be empty space filling most of the cell. Actually, the "space" is an important organelle called the **central vacuole** (from the Latin word for "empty"). This organelle contains water and various storage products, has a single membrane, and can occupy from 5 to 95 percent of the total cell volume.

The central vacuole fills up a plant cell with what amounts to a bag of water. This keeps the cells plump, giving firm shape to the leaves, stems, and other plant parts. It also presses a small amount of cytoplasm and all the cell's organelles into a thin layer just below the cell membrane (see Figure 3.5b), which makes the surface-to-volume ratio quite favorable. The next time you neglect a houseplant and it wilts, you'll know that its vacuoles need refilling. Although the central vacuole contains mostly water, it can also store sugars, proteins, poisonous substances to protect the plant from hungry animals, or specialized products such as rubber and opium.

Single-celled organisms that live in fresh water—*Euglena* is again a good example—have another type of vacuole that "bails out" excess water: the *contractile vacuole*. In *Euglena*, water collects quickly in the vacuole, is squeezed out, and then leaves the cell through a canal (Figure 3.20, page 64).

CELLULAR MOVEMENTS: CILIA, FLAGELLA, AND MORE

Many cells that pursue sunlight or food particles possess either of two specialized organelles of movement: fla-

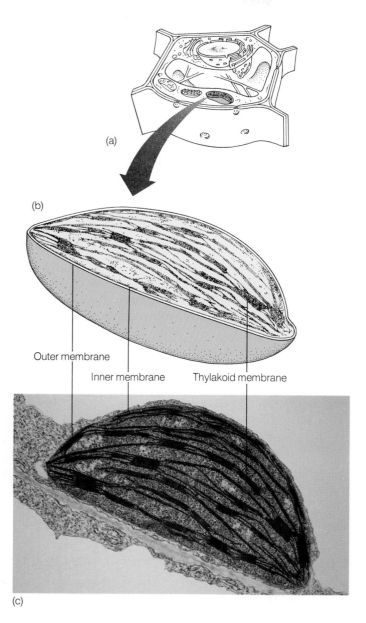

(a)

(b)

Outer membrane

Inner membrane   Thylakoid membrane

(c)

FIGURE 3.19 *Chloroplasts: Organelles of Photosynthesis.* (a) Chloroplasts are elongated organelles that contain green-colored pigments that trap the energy from sunlight. Biochemical processes within the chloroplast or on its membranes then convert and store that energy in the chemical bonds of carbohydrate molecules. (b) Chloroplast pigments are embedded in superthin membranes that arise from the inner membrane and form stacks of flattened disklike sacs called the *thylakoids.* The thylakoids trap sunlight and transfer the energy to ATP molecules. These pass to the surrounding space, where they fuel the production of sugar and starch (see Chapter 6). (c) Electron micrograph of a chloroplast (magnification 40,000×).

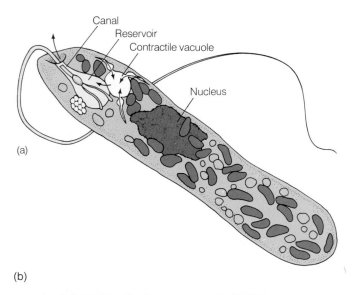

FIGURE 3.20 *The Contractile Vacuole: A Cellular Pump.* (a) *Euglena* and many other large, single-celled organisms living in fresh water have a contractile vacuole, an organelle that collects excess water. (b) When *Euglena*'s vacuole contracts, it pumps out water. The water then exits via a canal. The electron micrographs show filled and contracted vacuoles.

gella or cilia. Free-living cells like *Euglena* can have a fine whiplike organelle called a **flagellum** that extends from the cell surface and spins about like a corkscrew, pulling the cell forward through the water. Specialized cells within multicellular organisms can also have flagella: Each sperm has a flagellum that undulates like a lashing whip, pushing the cell toward its target, the egg (Figure 3.21).

Certain protozoa bear not one flagellum but thousands of similar shorter projections called **cilia** (singular, cilium) all over the cell surface (Figure 3.22). The cilia beat in concert like the oars of a medieval galley ship and allow the cell to swim quickly. In some protozoa and in the cells of certain multicellular organisms, cilia can have a different function: to sweep fluid and particles across the stationary cell. In cells that line the airways leading to our lungs, for example, cilia sweep dust particles out toward the air passages to eventually be expelled in mucus or swallowed. While cilia and flagella differ in length and motion, they have the same internal structure—a beautiful symmetrical pattern derived from structures called centrioles. This symmetrical pattern is described in Figure 3.23 (see also Figure 3.5a).

Some cells can creep along a flat surface by the activity of the cytoskeleton. The classic example is the bloblike single-celled hunter *Amoeba*, but some animal cells can also locomote slowly, either in the body or when removed and kept alive in a glass Petri dish (Figure 3.24, page 66).

CELL COVERINGS

The outer boundary of the cell cannot be said to fall strictly at the plasma membrane, because virtually all cells secrete coverings—either strong cell walls or fluffy coatings—that protect the delicate membrane and confer

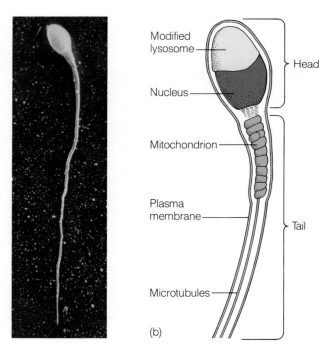

FIGURE 3.21 *Human Sperm with Its Long Flagellum.* (a) As you can see in this electron micrograph, the human sperm is a streamlined "missile" specialized to carry a payload—its nucleus—to the egg cell. (b) A cluster of mitochondria to provide power rings the sperm's long tail, and a flagellum contains microtubules to provide locomotion.

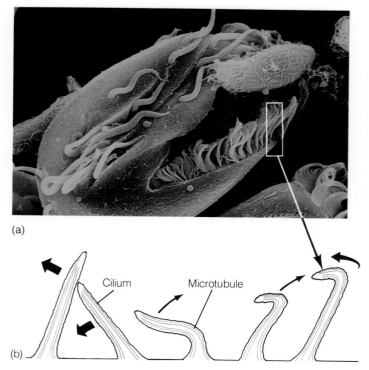

(a)

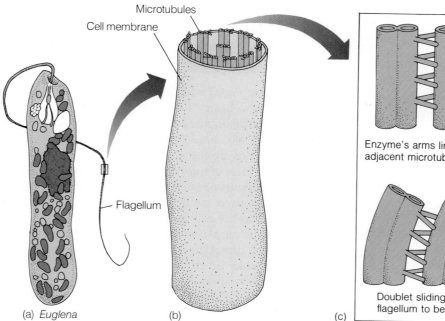

Cilium Microtubule

(b)

FIGURE 3.22 *Cilia: Hairlike Projections That Beat in Concert.* Cilia tend to occur in vast numbers, like short, movable hairs over the cell surface. (a) The slipper-shaped *Paramecium* cell in the upper right corner can swim quickly forward or backward by the wavelike motions that sweep continuously and sequentially across its thousands of surface cilia. The much larger *Euplotes* cell, with its own fringe of waving cilia, is attempting to consume the *Paramecium.* (b) As in flagella, microtubules span the interior length of each cilium and enable it to beat.

other advantages as well. *Cell walls* made largely of cellulose surround plant cells on all sides (Figure 3.25, page 66). These walls are porous, allowing water, gases, and solid materials to pass through to the plasma membrane. The cell wall is the major component of wood. Some fungi and protists have cellulose in their cell walls as well as a wide variety of other molecules. Bacteria also secrete cell walls, but these are made up of a complex material that includes sugars, lipids, and amino acids rather than cellulose. Be sure not to confuse the cell wall with the cell membrane. All cells have a lipid bilayer, or membrane; but only some cells, such as plant, fungal, and bacterial cells, have cell walls outside their cell membrane.

Virtually all animal cells secrete and become surrounded by a meshwork of molecules that serves as a scaffold and intercellular glue (Figure 3.26, page 66). The most common component of this meshwork is the fibrous protein called *collagen*, which has stiff, ropelike polypeptide chains wound around each other into fibrils. Because of the meshwork surrounding all animal cells, collagen molecules are the most plentiful protein in a typical mammal. Among other things, they are the main component of tendons, which act like pulleys and cables that enable muscles to move bones.

LINKS BETWEEN CELLS

Cells in multicellular organisms are not only embedded in a meshwork outside the cell, they are also attached to neighboring cells by physical linkages. Like the mesh-

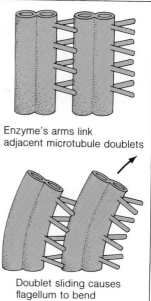

Microtubules

Cell membrane

Enzyme's arms link adjacent microtubule doublets

Doublet sliding causes flagellum to bend

Flagellum

(a) *Euglena* (b) (c)

FIGURE 3.23 *How Flagella and Cilia Move.* Cilia and flagella, organelles of cell movement, have a distinctive internal structure that helps explain how they bend and lash. For example, a *Euglena* cell's flagellum (a) has the cross-sectional arrangement shown in (b): nine pairs of microtubules arranged in a circle, with two microtubules in the center. The microtubules contain an enzyme that brings about the release of energy from ATP, the molecular fuel for flagellar and ciliary motion. As the enzyme splits ATP molecules and releases energy, the microtubules slide past each other. This sliding causes the cilia or flagella to bend (c).

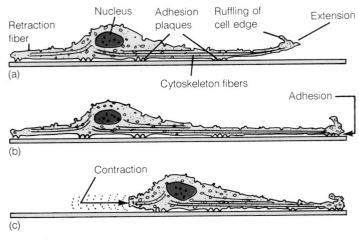

FIGURE 3.24 *Cell Movements: Creeping, Ruffling, and Inching.* The changeable shape of an animal cell, as well as the ability to creep and glide forward, is maintained by the dynamic cytoskeleton. As an animal cell in a Petri dish creeps forward, its leading edge extends (a), and sticky areas (adhesion plaques) on the lower surface cling to the substrate below (b). Eventually, the cell contracts, the trailing edge snaps forward in inchworm fashion, and the cell slides toward the ruffling edge (c).

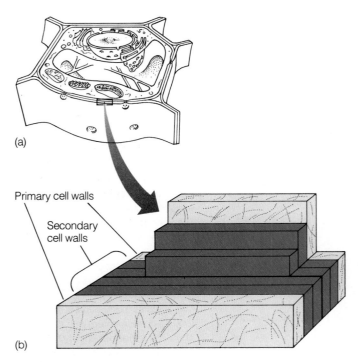

FIGURE 3.25 *Plant Cell Walls.* (a) Plant cells are surrounded by walls made up mostly of cellulose, the most abundant organic molecule on earth. (b) As a newly divided plant cell matures, it lays down a primary cell wall just outside the plasma membrane that remains flexible until a secondary cell wall, usually more rigid than the first, is deposited inside the primary wall, close to the plasma membrane.

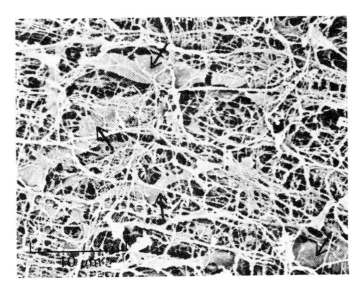

FIGURE 3.26 *A Meshwork Outside the Cell.* Most animal cells are surrounded by a meshwork of protein and polysaccharide molecules that cushions the cell, strengthens the tissue the cell is part of, and helps establish and maintain the cell's shape. The most common extracellular material in animals is collagen. This scanning electron micrograph reveals the collagen fibers surrounding the eye cells from a chick embryo. The black arrows point out the eye cells; the thin, white network is the collagen fibers.

work materials, these junctions help weld cells together into functional tissues and organs. They also allow the cells to communicate freely and to coordinate the activities of various cell types. While some linkages allow materials to flow between cells, others prevent fluid leaks between cells. Figures 3.27 and 3.28 describe the kinds of links found in plant and animal cells.

≋ CONNECTIONS

This chapter explored what cells are and how they work, focusing on the basic "housekeeping" tasks that keep cells alive and the organelles associated with those tasks. The structures of organelles are based on the same biological molecules discussed in Chapter 2: Lipids make up the membranes of many organelles; specific proteins, often embedded in or surrounded by membranes, give the organelles their particular function; carbohydrates give strength to the cell wall or provide energy in mitochondria and chloroplasts; and nucleic acids in the nucleus and ribosomes direct the activities of the cell. In eukaryotes, which make up the vast majority of cells, specialized organelles carry out specialized tasks; these include chloroplasts that allow for photosynthesis and cilia and flagella that make movement possible. Deficiencies in

FIGURE 3.27 *Junctions and Links Between Animal Cells.* Within a multicellular organism, most cells are connected to each other by a range of junctions that provide impermeability, adherence, and communication. An impermeable junction (1) results from a band around the cell where the plasma membrane touches the membrane of an adjacent cell, forming a tight seal that prevents molecules from passing between cells. In organs that store fluid, such as the bladder, these junctions act like rubber gaskets to keep urine from seeping out of the bladder and into the body. Cells adhere to each other by junctions that either run in bands around the outer edges of cells (2) or form a spot-weld-like link between the cell skeletons of adjacent cells (3) in tissues subject to heavy stretching and mechanical wear and tear, such as human skin. Another type of junction attaches a cell to the fibrous meshwork of proteins and carbohydrates the cell sits on (4). Cells can communicate through junctions made of protein-lined pores through which small molecules can travel from cell to cell (5). In the heart, such junctions allow for the passage of electrical currents between cells and thus the precise timing of the heartbeat.

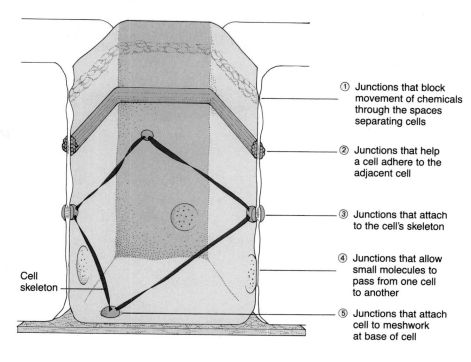

Cell skeleton

① Junctions that block movement of chemicals through the spaces separating cells

② Junctions that help a cell adhere to the adjacent cell

③ Junctions that attach to the cell's skeleton

④ Junctions that allow small molecules to pass from one cell to another

⑤ Junctions that attach cell to meshwork at base of cell

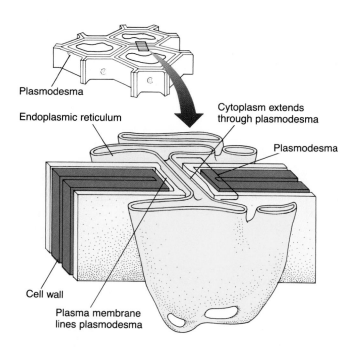

Plasmodesma

Endoplasmic reticulum

Cytoplasm extends through plasmodesma

Plasmodesma

Cell wall

Plasma membrane lines plasmodesma

FIGURE 3.28 *Connections in Plant Cells.* In plant cells, encasement by the cell wall would isolate and block intercellular coordination if it were not for junctions called *plasmodesmata.* These plant cell connectors are delicate bridges of cytoplasm that pass through the walls of adjacent cells and link them, allowing the rapid exchange of materials and sometimes electrical signals.

organelle structure and function can lead to brain disease, heart attack, soft bones, sterility, and other serious conditions.

This chapter also revealed the biologist's dilemma: One must look at each cell part individually to learn how it works, but must also realize that once dismantled, the cell will never again function. The life in that living unit is an emergent property based on the physical relationships and integrated activities of the cell parts—a property that must be passed down intact from earlier generations of cells to present and future ones. And the similarity of cells and their organelles shows an evolutionary relatedness that biologists refer to as the unity of life.

The next four chapters examine in some detail how cells obtain energy and reproduce. Our early focus on cells provides a basis for understanding the more complex living systems—the many-celled organisms and groups—we study in the remaining chapters of the book; for as the pioneers of cell biology discovered, all living things contain cells, and all the biochemical processes of life go on inside them. Thus, cells are truly "life modules," and to understand their activities is to understand much about life itself.

(1) All living things are made up of cells.

(2) All cells arise from preexisting cells.

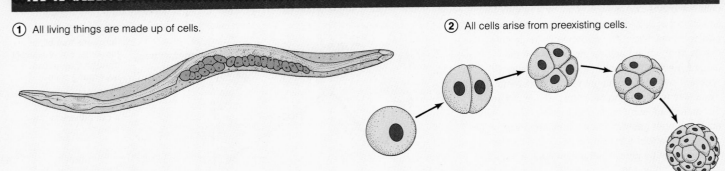

(3) There are two major cell types: Prokaryotic and Eukaryotic

PROKARYOTIC CELL

Most prokaryotes are single-celled bacteria (monerans) that have no membrane-bounded nucleus.

EUKARYOTIC CELL

Eukaryotes have a true membrane-enclosed nucleus and can exist either as single cells (the protists) or as large communities of cells (fungi, plants, and animals).

(4) All cells must:

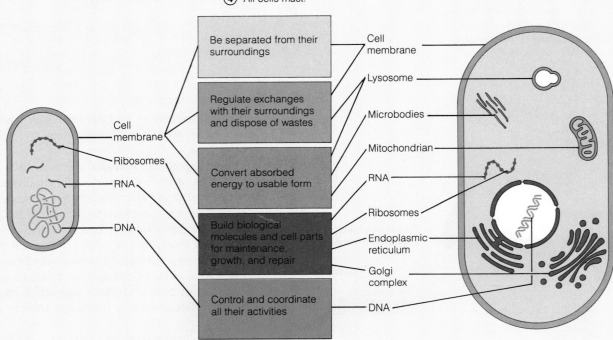

Be separated from their surroundings

Regulate exchanges with their surroundings and dispose of wastes

Convert absorbed energy to usable form

Build biological molecules and cell parts for maintenance, growth, and repair

Control and coordinate all their activities

Cell membrane

Ribosomes

RNA

DNA

Cell membrane

Lysosome

Microbodies

Mitochondrian

RNA

Ribosomes

Endoplasmic reticulum

Golgi complex

DNA

(5) Among eukaryotic cells particular specializations distinguish plant and animal cells.

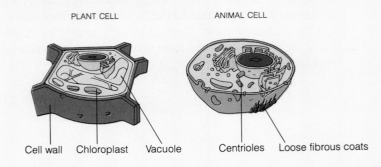

PLANT CELL

ANIMAL CELL

Cell wall   Chloroplast   Vacuole

Centrioles   Loose fibrous coats

(6) The life of a cell is an emergent property that depends on the smooth coordination of all activities.

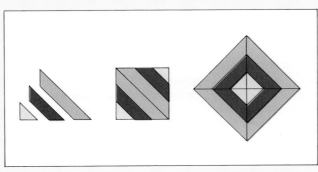

NEW TERMS

cell membrane, page 48
cell theory, page 47
cell wall, page 49
central vacuole, page 63
chloroplast, page 63
chromosome, page 55
cilium, page 64
cytoplasm, page 49
cytoskeleton, page 57
endoplasmic reticulum,
  page 59
eukaryotic cell, page 48
flagellum, page 64
fluid mosaic model, page 53

Golgi complex, page 59
lipid bilayer, page 53
lysosome, page 59
mitochondrion, page 62
nuclear envelope, page 55
nuclear pore, page 55
nucleoid, page 48
nucleolus, page 57
nucleus, page 55
plasma membrane, page 48
plastid, page 62
prokaryotic cell, page 48
ribosomal RNA, page 57
ribosome, page 57

STUDY QUESTIONS

REVIEW WHAT YOU HAVE LEARNED

1. How did Robert Hooke and Anton van Leeuwenhoek add to our knowledge of cells?
2. State the cell theory.
3. How are prokaryotic cells structurally different from eukaryotic cells?
4. What are the functions of a cell's plasma membrane, nucleus, and cytoplasm?
5. Although a cell is tiny, it contains a number of organelles, including ribosomes, endoplasmic reticulum, Golgi complex, lysosomes, and mitochondria. What role does each of these organelles play in the life of the cell?
6. True or false: Since bacteria do not contain a nucleus, they do not possess DNA. Explain your answer.
7. How are plant and animal cells similar? How are they different?

APPLY WHAT YOU HAVE LEARNED

1. A science fiction writer describes an amoeba growing to the size of an elephant, then devouring horses and cows. From your knowledge of surface-to-volume ratio, would you agree with the science in this fiction? Explain.
2. A scientist discovers and carefully describes a new species of single-celled organism. What kind of microscope would she use to do her research? Explain your answer.

FOR FURTHER READING

Bretscher, M. S. "How Animal Cells Move." *Scientific American* 257 (December 1987): 72–80.
DeDuve, Christian. *A Guided Tour of the Living Cell.* New York: Scientific American Books, 1984.
Peterson, I. "Scanning the Surface." *Science News* 135 (1 April 1989): 202–203.

CHAPTER 4

THE DYNAMIC CELL

RED BLOOD CELLS, A STEADY STATE OF ORDER AND ACTIVITY

Every second, the marrow inside your bones produces nearly 2 million new red blood cells. Each new cell has a nucleus, ribosomes, mitochondria, endoplasmic reticulum, and many of the other intricate organelles we encountered in Chapter 3. After a few days, however, the immature blood cell loses its rigid nucleus as it squeezes through a blood vessel wall into the hollow center of one of the many fine vessels that crisscross the bone marrow. Once inside the blood vessel, the cell is swept away by the moving bloodstream in the company of 25 trillion other red blood cells.

The mature cell is now bright red and disk-shaped with a dent in the center (Figure 4.1). Because it lacks a nucleus and because the mitochondria and ribosomes disintegrate within a few days, the red blood cell is one of the simpler cells in the body, and it provides a good case study for some of the more fundamental activities associated with living cells. The millions of hemoglobin molecules inside the cell operate like miniature traps, snapping open and shut, holding and then releasing oxygen. Enzyme molecules within the cell work extremely quickly, cutting and joining other molecules, releasing energy, and generating cell parts. The cell's entire water content moves back and forth across the plasma membrane 100 times a second. And glucose is constantly burned to fuel cellular activities.

The red blood cell is undeniably dynamic, and it shares this state of constant and rapid but orderly activity with all other living cells. To remain in a steady state of order and activity, cells and all living things depend on a continual flow of energy from the environment. The energy is then processed and transformed into a different state for storage or use. The special agents of this energy flow are the protein catalysts called enzymes and the energy-carrying molecules called ATP.

Cells also depend on a flow of materials—mostly water, ions, nutrients, and wastes—to help maintain their order and activity. Such materials pass into and out of the cell by a variety of finely balanced transport systems, each one a beautiful integration of form and function.

The goal of this chapter is to reconstruct the living cell we examined part by part in Chapter 3 and to look at the kinds of tasks cells perform as they take in from the environment the energy that fuels their life-support activities. Our consideration of the dynamic cell will answer several questions:

- What are the basic laws of energy flow in the universe, and how do they affect living cells?
- What are the chemical reactions that take place within cells, and how do energy carriers and enzymes propel and speed them?
- What is metabolism, and why is it so crucial to life?
- How does a cell's transport of materials allow it to maintain the appropriate chemical composition?
- How does a cell mechanically move itself or its internal organelles?

FIGURE 4.1 *Red Blood Cells Look Like Life Savers—And They Are.* Red blood cells carry oxygen to other body cells and cart away carbon dioxide.

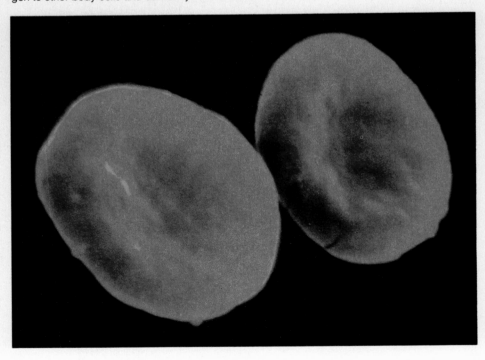

CELLS AND THE BASIC ENERGY LAWS OF THE UNIVERSE

A dynamic cell—which means any living cell—is like a bustling metropolis that runs on energy. Materials stream in and out through the "city walls"—the plasma membrane. Instructions move outward from the "city government"—the nucleus or nucleoid region—for cell maintenance and the manufacture of materials. Ribosomes, like microscopic factories, churn out proteins. Significantly, the dynamic cell needs a constant flow of energy to carry out its tasks, and that flow proceeds according to certain universal laws.

THE LAWS OF ENERGY FLOW

We use the term *energy* in many ways: We call an active person energetic, and we also consider energy a major political issue. But what *is* energy? In its most general definition, **energy** means the capacity to do work. There are two major states of energy in the universe: potential and kinetic. *Potential energy* is stored and ready to do work; a huge boulder poised at the edge of a cliff or the water pent up behind Grand Coulee Dam contains stored energy, capable of being released and of accomplishing work like smashing a hole in the road at the bottom of the cliff or turning the wheels of a hydroelectric generator (Figure 4.2a). *Kinetic energy* is the energy of motion—of a hurtling rock or of tumbling water, for example (Figure 4.2b).

Clearly, energy can be converted from the potential to the kinetic state. Energy can also be converted from one *form* to another. We are all familiar with these forms: light energy, heat energy, electrical energy, chemical energy, and mechanical energy. And we have all seen their interconversions: Electrical energy can be converted to light and heat in a light bulb; and the energy of reacting chemicals inside a battery can be converted to electricity to run a radio or flashlight. The first of the energy laws, the **first law of thermodynamics,** describes such conversions. It says that energy can be changed from one form to another, but during these conversions, it is *conserved;* that is, it is neither created anew nor destroyed. Thus, there is a constant amount of energy in the universe, but it can change form. Thermodynamics is the formal study of how energy changes form.

The **second law of thermodynamics** explains the fact that energy conversions are never 100 percent efficient. It states that systems always tend toward greater states of disorder. Systems are units that are separated in some way from their surroundings; planets and cells are systems. What the second law means is that given enough time, the energy in every system will undergo sponta-

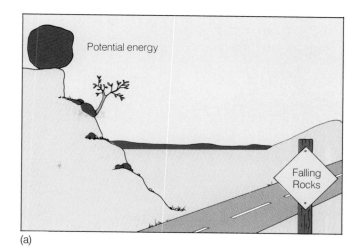

(a)

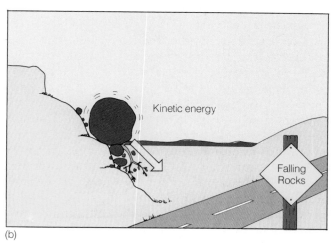

(b)

FIGURE 4.2 *Potential and Kinetic Energy.* (a) A boulder poised at the top of a cliff has potential energy; it can do work, such as thundering down the hill, uprooting a tree, and smashing a hole in the pavement. (b) Potential energy is converted to kinetic energy as the rock rolls.

neous conversions from one form to another, but these conversions are inefficient, and some energy will inevitably be lost to the surrounding environment as heat. **Heat** is nothing more than the random commotion of atoms and molecules. Since randomness is the opposite of order, the inefficient conversion of energy from one form to another (with some energy lost as unusable heat) results in increasing disorder in the system. Scientists use the term **entropy** as a measure of the disorder or randomness in a system. Entropy accounts for the energy that escapes from the system during transformations and is no longer usable. Anyone who has ever tried to keep a room tidy is already quite familiar with the concept of entropy: Without a constant input of energy to put clothes and books away, clean up dust and dirt, and keep fixtures

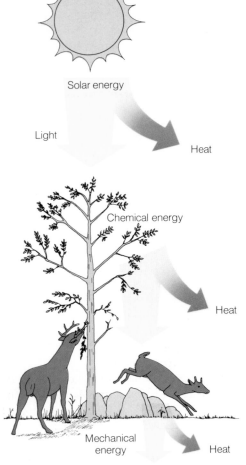

Solar energy

Light

Heat

Chemical energy

Heat

Mechanical energy

Heat

FIGURE 4.3 *Energy Interconversions and the Laws of Thermodynamics.* The first law of thermodynamics states that energy can be changed from one form to another but is neither created nor destroyed. Here, nuclear energy from atomic fusions taking place in the sun is converted to light, to chemical energy in the plant's tissues, and to mechanical energy in the animal's tissues. The second law of thermodynamics states that all such interconversions are inefficient to some degree. Thus, with each interconversion in this chain, some energy is lost as heat, diffuses away, and is no longer usable.

repaired, entropy, or disorder, tends to increase, and things go downhill fast.

Consider one chain of energy conversions and the steadily increasing disorder it represents. The nuclear reactions taking place in the sun release light but also massive amounts of heat, which are forever lost in space (Figure 4.3). Sunlight enters the earth's atmosphere and strikes the leaves of a bush. Most of the light is reflected as heat and once again lost, simply radiating away into the atmosphere in a low-quality, unusable form. Some of the light is reflected as the green light we see as the

plant's color. But a small amount is absorbed as light and converted by the plant to chemical energy. During this conversion, still more heat is lost. If a gazelle grazes on the leaves, the animal's cells release the stored chemical energy to fuel ongoing life processes; but as always, the conversion is inefficient, and energy radiates into the air as body heat. Only a tiny fraction of the original solar energy is saved in a chemical form that can be expended to "purchase" order—in this case, precise movements by the gazelle's muscles as it bounds across the African savanna. At the same time, a great deal of energy has been irreversibly lost as the random jostling of molecules accomplishing no work. Physicists predict that eventually the energy from all sources in the universe will be lost as shimmering heat, but fortunately, this "heat death of the universe" is billions of years off.

ENERGY FLOW AND DISORDER IN CELLS

The significant question for us here is how do the energy laws apply to the cells of living organisms—islands of incredible order in a universe that is slowly losing energy and winding down owing to increasing disorder? And the answer is that cells, too, are subject to the same tendency to wind down and become disordered. They only remain alive and healthy—orderly and active—at the high price of a constant flow of energy. This flow must provide enough energy to fuel all of the cell's activities but still leave enough leftover energy to satisfy the second law—the inevitable loss of heat to the surrounding environment.

A living cell, then, is a temporary repository of order purchased at the cost of a constant flow of energy. Within the cell, there is a delicate balance between order-producing activities, such as maintenance, repair, and protein synthesis, and activities that mainly generate energy and, as a by-product, heat. If the energy flow that fuels a cell's activities is impeded, order quickly fades, disorder reigns, and the cell dies. The impediment can be lack of food or an injury or aging of the cellular constituents that maintain order (such as the nucleus and ribosomes).

The red blood cell is an interesting case study for energy flow, disorder, and death, because as we saw earlier, most of the machinery for maintaining cellular order is discarded early in the cell's life cycle (Figure 4.4). The mature cell obtains its energy source—glucose molecules—through its plasma membrane from the yellowish liquid portion of blood. Enzymes in the cell then break these sugars down to fuel the cell's activities. However, over time, the precise ordering of molecules within the cell begins to decline, and since there are no organelles for synthesis and repair, the cell begins to break down. After about 120 days, the withered cell is removed from circulation and dismantled. In contrast, a typical human brain cell can live for 75 years or more by expending much of its energy to fuel maintenance and repair activities. Surprisingly, that long-lived brain cell is, in a sense, younger than the red blood cell. So long as its machinery for repair remains intact, the brain cell continually replaces aging organelles, and this constant refurbishing amounts to a cellular fountain of youth; at any given time, most of the cell parts will be less than one month old, and during the cell's lifetime, the entire contents (and the cell membrane, too) will be replaced hundreds of times.

Clearly, the nonstop organizational activity of a living cell bears a steep price tag in energy requirements. Exactly *how* cells use that energy to defer entropy, or disorder, is our next topic.

≋ CHEMICAL REACTIONS AND THE AGENTS OF ENERGY FLOW IN LIVING THINGS

We have seen that plants convert light energy to chemical energy and funnel it into nearly all living things, directly or indirectly. Once inside a cell, this energy in chemical

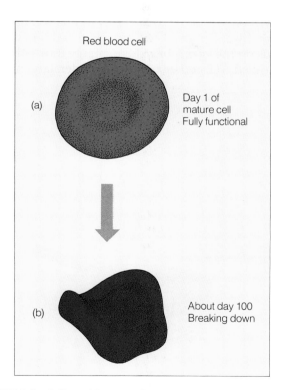

FIGURE 4.4 *Cells and Disorder.* Cells remain alive and healthy at the high price of a constant flow of energy, used for order-producing activities such as continual repair and maintenance. The red blood cell is a prime example of what happens in the absence of such activities. Since it squeezes out its nucleus, mitochondria, and many other organelles of synthesis and repair, the red blood cell's internal order steadily declines over its 120-day life span.

form is used to fuel the maintenance of order through **chemical reactions:** transformations of sets of molecules into other kinds of molecules. The transformations include a shifting of energy content from the bonds of one set of molecules to the bonds of the other. Thus, just as energy must flow into the cell, it must also flow from one set of compounds to another within the cell to power cellular activities. Let's see how.

CHEMICAL REACTIONS: MOLECULAR TRANSFORMATIONS

During a chemical reaction, starting substances, the **reactants,** interact with each other to form new substances, the **products.** There are often two reactants and two products, and this general relationship can be shown with the equation

$$\underbrace{A + B}_{\text{Reactants}} \rightarrow \underbrace{C + D}_{\text{Products}}$$

In variations on this theme, two reactants can be transformed into one product, or one reactant can become one or two products.

As a child, you may have carried out some simple kitchen chemistry, mixing vinegar and baking soda and watching the vigorous bubbling that results (Figure 4.5, page 74). A chemical reaction takes place between the reactants, forming two new products. One of these products is an unstable substance called bicarbonate ion that immediately breaks down in a second reaction into water and carbon dioxide gas, the cause of the bubbling. This reaction takes place spontaneously, as soon as the ingredients are combined. There is no need to add heat, electricity, or any other form of energy to make it "go." A spontaneously occurring reaction like this is energy-releasing (*exergonic*). The reaction between vinegar and baking soda can proceed spontaneously because there is enough energy in the chemical bonds of the reactants for the reaction to take place and still leave some energy left over to satisfy the second law of thermodynamics. This means that the products contain less energy and are more disordered than the reactants.

Some energy-releasing reactions are like the rock poised at the top of a hill: They need a little push, a small energy input before they will go. Nevertheless, even though a little bit of energy is needed to get the reaction started, the overall result is a release of energy.

A great number of cellular activities involve energy-releasing reactions, and most of these give off heat. A striking example is the contraction of muscle cells. The movement of internal cell fibers is based on numerous energy-releasing reactions. So if you've ever wondered

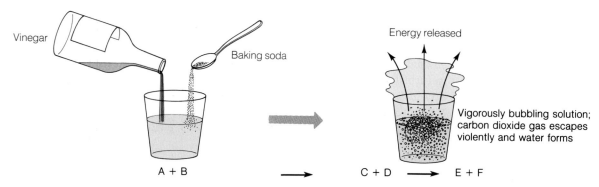

FIGURE 4.5 *Energy-Releasing Reactions.* Vinegar and baking soda react with each other in an energy-yielding reaction, manifested by vigorous bubbling. During the reaction, the reactants are transformed into products.

why running, aerobic dancing, weight lifting, or any other muscular activity makes you heat up and start to sweat, it's because of energy-releasing chemical reactions!

There is another category of chemical reactions that do not proceed spontaneously and do not usually give off heat. These are energy-requiring (*endergonic*) reactions. A familiar energy-requiring reaction takes place when you cook an egg (Figure 4.6); the added heat energy causes the egg white proteins to change configuration and solidify. Reactions that require energy are very important to living things because they include many of the molecular transformations that bring about order in the cell, such as the building of proteins, the replacement of worn sections of membrane, and the generation of new ribosomes. All energy-requiring reactions, including those that bring about these construction and maintenance activities, fail to "go" unless energy is supplied in some form, because there is less energy in the chemical bonds of the reactants than in the bonds of the products.

If the order and activity of the dynamic cell depend on an appropriate source of energy for each and every energy-requiring reaction, where does that energy come from? The answer demonstrates the beautiful economy of nature: The energy for the cell's energy-requiring reactions comes from the cell's energy-releasing reactions, although some of the released energy is lost as heat. The two kinds of reactions are energetically *coupled* so that the energy yielded by one reaction powers the other, just as a spinning treadwheel could power a blender (Figure 4.7). In nature, the coupling usually involves not just two reactions, but a long series of them in which the products of one reaction become the reactants of the next, and the overall flow of energy throughout the chain goes spon-

taneously and generates heat. These energy-converting reaction chains in cells are called *metabolic pathways* (Figure 4.8). They allow the living cell to harvest the chemical energy in food and to use it to "purchase" order by temporarily overcoming the universal tendency toward disorder. *Metabolism* is the name given to the energy changes within a living thing that arise from interrelated chemical reactions and allow the organism to utilize energy from the environment for its own survival. It is so important to life that we devote all of Chapters 5 and 6 to it. Here we focus on the agents of cellular energy flow—the molecules called ATP and enzymes—to understand how the

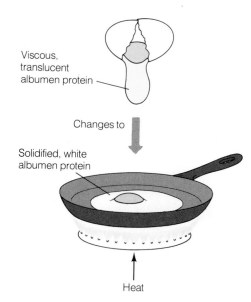

Viscous, translucent albumen protein

Changes to

Solidified, white albumen protein

Heat

FIGURE 4.6 *Energy-Requiring Reactions.* It's strange to think that frying an egg involves an endergonic, or energy-requiring, reaction—but it does. The addition of heat energy alters the hydrogen bonds in the proteins in egg white. This changes their shape, and we can see the result: A clear, viscous solution of the proteins becomes white and solid.

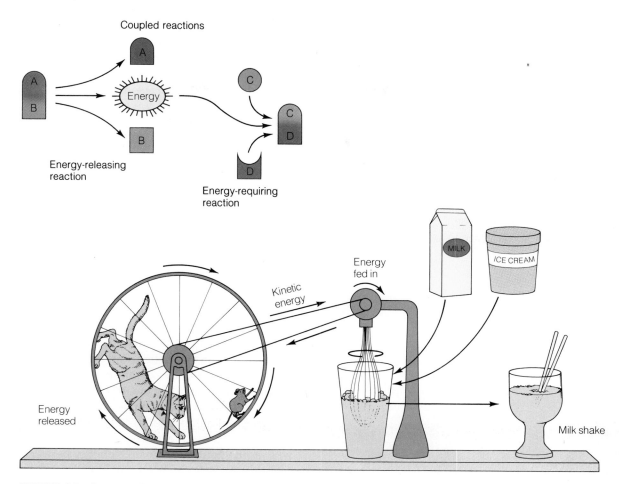

FIGURE 4.7 *Energetic Coupling.* In living things, energy-releasing reactions are coupled to energy-requiring reactions. Energy is released from the splitting of compound AB, energy is required for the combining of C and D into CD, and the energy from the first reaction can drive the second reaction. By whimsical comparison, energy from a spinning treadwheel (powered by the muscles of a hungry cat and a frightened mouse) can be mechanically coupled to a mixer that makes milk shakes out of ice cream and milk.

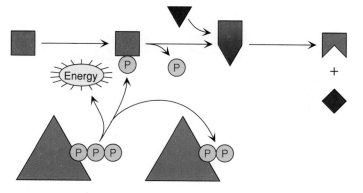

FIGURE 4.8 *A Metabolic Pathway.* In living things, coupled reactions usually occur in long series called metabolic pathways, where the products (including energy) of one reaction are used in the next reaction of the chain.

cell couples energy-releasing and energy-requiring reactions.

ATP: THE CELL'S MAIN ENERGY CARRIER

If the energy from one type of reaction in a living thing can drive another such reaction, what form does that energy take? Light? Electricity? Heat? Nuclear power? The logical assumption might be heat, since we saw that energy-releasing reactions give off heat. But the heat given off in a cell scatters, and there is no mechanism in cells for gathering it up and putting it to good use. Instead, most of the energy freed during an energy-releasing

(a) Structure of ATP

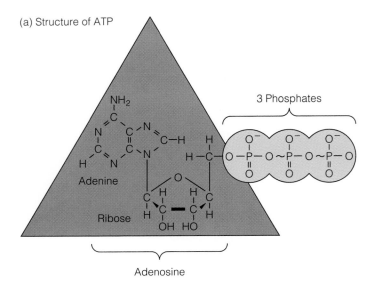

3 Phosphates

Adenine

Ribose

Adenosine

(b) ATP synthesis

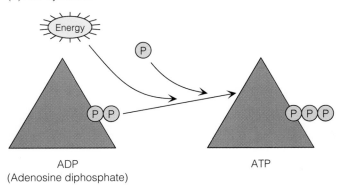

ADP
(Adenosine diphosphate)

ATP

(c) Energy release from ATP

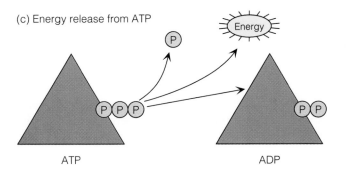

ATP

ADP

FIGURE 4.9 *Structure and Function of ATP.* ATP, or adenosine tri-phosphate, is the main energy carrier in living cells. (a) It consists of a portion called adenosine (composed of the base adenine and the sugar ribose) and, as the name *triphosphate* implies, three phosphate (PO_4) groups. (b) The synthesis of ATP is an energy-requiring reaction in which a phosphate is added to ADP. (c) In the reverse reaction, ATP releases energy and liberates a phosphate.

reaction is quickly trapped and stored in the chemical bonds of the energy carrier molecule **ATP** (Figure 4.9a). ATP can transfer energy from one molecule to another. Energy released from the reactants is used to convert a molecule called ADP (*adenosine diphosphate*), with its two phosphate groups and lower energy content, into ATP (*adenosine triphosphate*), with its three phosphate groups and higher energy content (Figure 4.9b). That stored energy can then be released by converting ATP to ADP (Figure 4.9c). The significance of ADP and ATP, as well as a few other similar energy carriers, is that they function as intermediates, or links, between all metabolic energy exchanges in the cell.

Energy intermediates are important because for energy to flow from one chemical reaction to another, the reactions must have some chemical compound in common—that is, a compound that appears among the products of one reaction and the reactants of the next. ATP often appears in the equations for both energy-releasing and energy-requiring reactions.

An analogy for ATP is money. When a farmer sells his crops and then uses the money to buy new tools, the intermediate for both exchanges is money. ATP is sometimes considered a form of currency in the cell—the chemical coin of the realm—that is "saved up" during energy-yielding reactions and "spent" during energy-costly reactions. In a sense, it *is* a form of currency, as well as an agent that transfers energy between reactions. The critical tasks of both single cells and multicellular organisms—the storage and use of hereditary informa-

FIGURE 4.10 *ATP Powers Living Reactions.* The glowing of this flashlight fish, *Photoblepharon palpebratus*, in the inky recesses of the Caribbean Sea, is fueled by ATP. Bacteria in pouches in the fish extract chemical energy from nutrients and store much of the energy in the form of ATP. Later, they break down the ATP to ADP, and the liberated energy helps convert a chemical called luciferin to a form that glows in the dark. Chemical energy is thus converted to light energy, and ATP is the intermediary.

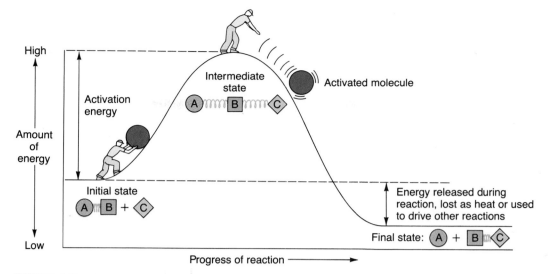

FIGURE 4.11 *Activation Energy Makes Reactions Go.* During a chemical reaction, reacting molecules need to collide with a certain minimum amount of energy before they can reach the springlike intermediate state. The amount of energy needed is called the activation energy; it resembles a hill, or barrier, when graphed. Once activated molecules achieve and then pass the intermediate state, they reach the final state where they have been transformed into new products. In this final state, the overall energy content in their bonds is lower than it was in the initial state. The excess energy has been lost as heat or used to drive cellular activities.

tion, the building of cell parts, the transport of certain materials, movement, the ability to glow in the dark, and so on—all involve ATP (Figure 4.10). In a person, an amount of ATP equal to about half the body weight cycles to ADP and back again each day, representing quadrillions of chemical reactions and energy exchanges daily.

ENZYMES: THE KEY TO WHAT MAKES A CELL'S CHEMICAL REACTIONS GO

Chemical reactions can take place in a jar or inside a living cell. However, the reactions in cells, fueled largely by ATP, could not take place continually or quickly enough without the molecular agents called enzymes. To understand the important role of enzymes in the cell, we must first consider the concept of *activation energy.*

■ **Overcoming the Energy Barrier** In a very real sense, there is a barrier, or hill, separating the reactants and products in any chemical reaction, whether in a cell or in a test tube. For reactants AB and C to be converted to products A and BC, the reactants must collide quite energetically so that the chemical bonds in the reactants can be broken and the new bonds in the products formed. For a fleeting instant during the conversion, all the bonds will be distorted like springs being stretched, and this fleeting, intermediate *transition state,* A~B~C, cannot be reached without a very energetic collision between the molecules. Because energy input is required to achieve

the intermediate state, it is often characterized as a barrier, or hill, separating reactants and products (Figure 4.11).

Now, most molecules jostling about and colliding randomly do not have enough energy of motion (kinetic energy) to achieve and overcome the energy barrier; they simply bounce off each other. (Similarly, cars colliding at a very low speed, say, 2 mph, usually bounce off and leave each other unchanged.) Some individual molecules, however, do jostle about fast enough and collide with a great enough impact to generate the springlike intermediate state A~B~C. It is only after achieving this state that reactants AB and C can be converted to products A and BC. The minimum energy needed to reach the intermediate state is called the **activation energy** (see Figure 4.11).

In a living cell, most molecules do not on their own have sufficient energy to cross the barrier and react—at least not fast enough to keep pace with the cell's needs for energy and materials. Four things, however, can speed up reactions: (1) adding heat; (2) increasing the concentration of reactants; (3) decreasing the concentration of products; and (4) adding **catalysts,** agents that speed other reactions without themselves being used up or changed. In fact, the reactions inside all living cells are made possible by a type of catalyst—the class of proteins called

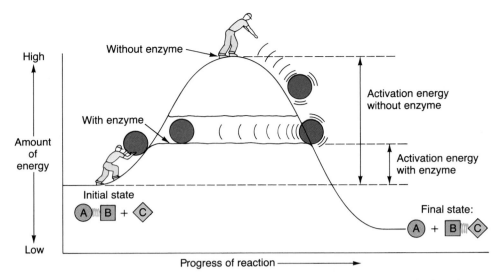

FIGURE 4.12 *Activation Energy: With and Without Enzymes.* An enzyme is a biological catalyst that can lower the activation energy—the barrier that must be crossed by reacting molecules—without itself being changed or used up. In this way, enzymes facilitate biological reactions.

enzymes (see Chapter 2). Enzymes lower the activation energy of chemical reactions, allowing the reactions to proceed more rapidly. As Figure 4.12 indicates, you can think of an enzyme as boring a tunnel through the energy barrier.

In living cells, enzymes play by far the most important role in speeding reactions. Each enzyme is specific to a given reaction. In red blood cells, for example, a specific enzyme speeds up the reaction that allows the cells to pick up or release carbon dioxide. Each single enzyme molecule facilitates the reaction more than 600,000 times per second! This enables the reaction to proceed nearly a million times faster than it would without a catalyst, allowing red blood cells to pick up and then dump the necessary quantities of carbon dioxide as they race through the blood vessels.

All enzymes share two important functions: They allow cells to carry out reactions at a high rate and low temperatures (the reactions would otherwise require high, cell-damaging temperatures to occur at a fast enough rate), and they speed the rate of a chemical reaction without changing the amounts of product and reactant. Now let's take a closer look at how enzymes carry out their important work, bearing in mind that they are like virtually all other biological molecules in one important way: Their functioning is directly related to their form.

■ **Enzyme Structure** Each enzyme has a unique three-dimensional shape, and often that shape is globular (see Chapter 2). An enzyme molecule has on its surface a deep groove, or pocket, called the **active site,** and the shape of this site fits a specific reactant or set of reactants (Figure 4.13a). Figure 4.14 (page 80) shows this deep groove in lysozyme, the bacteria-killing enzyme in egg white. The specific reactants targeted by the groove are called **substrates,** and they fit into the groove rather like a key in a lock. Just as a given key fits only a given lock, enzymes recognize and act on only specific substrates, or target molecules.

Since an enzyme's structure is so closely related to its activity, factors that can alter its three-dimensional shape can also change or destroy its ability to speed reactions. For example, a digestive enzyme in the human stomach, called pepsin, works best at an acid pH of about 3, while most enzymes need a nearly neutral pH of about 7. And when we boil an egg, heat changes the shape of the egg white proteins, making them coagulate.

■ **Enzyme Function** Molecules like the enzyme in blood cells that can speed a reaction one-millionfold are clearly remarkable. But how do such proteins with their lock-and-key grooves lower activation energy and allow specific chemical reactions to take place? Enzymes perform this clever trick with three mechanisms that help reactants more easily reach the transition state they must pass through before being rearranged into products.

First, enzymes form complexes with substrates. For a reaction to occur, reactants A and B must collide before they can be transformed (Figure 4.13b). Without a catalyst, they will meet entirely by chance (Figure 4.13c). However, the active sites of enzymes have special regions that can attract appropriate reactants (or substrates) and hold or *bind* them in place, forming an **enzyme-substrate complex** (Figure 4.13d). The enzyme-substrate complex

(a) Structure: enzyme with active sites and reactants or substrates.

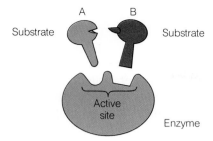

(b) For a reaction to occur, substrates A and B must collide.

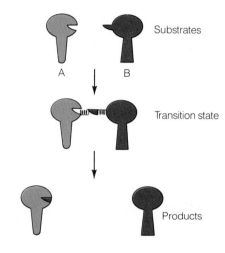

(c) Without a catalyst, substrates collide by chance.

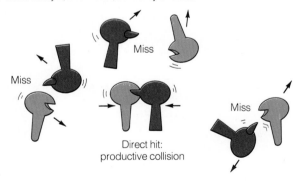

(d) With an enzyme, an enzyme-substrate complex forms.

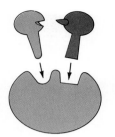

(e) Induced fit: The enzyme's active site changes shape, improving fit.

Simultaneously, the active site orients substrates.

The enzyme strains the shape of substrates, facilitating a productive collision.

The reaction finishes and the enzyme returns to normal.

FIGURE 4.13 *Enzymes: Structure and Function.* This figure shows how an enzyme, represented by the globular shape, facilitates the reaction between substrate A (without a protruding "nose") and substrate B (with the "nose"). (a) Enzyme and substrates prior to reaction. (b) The substrates, or reactants, must collide productively. (c) Without an enzyme, such collisions occur only seldom and randomly. (d) With an enzyme, the substrates form a complex. (e) This complex brings the substrates close together, orients them appropriately, strains their shape, and helps them react. When the reaction is complete, the enzyme remains unchanged and ready to catalyze a new reaction.

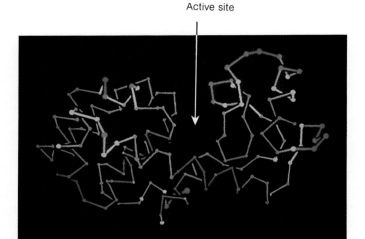

Active site

FIGURE 4.14 *Lysozyme, the Major Enzyme in Egg White.* This shows a three-dimensional computer-generated model of lysozyme. The enzyme binds substrate in its active site, the central U-shaped cleft.

has an important consequence: Since reactants are bound to the active site, they are brought close together, and this closeness effectively increases their concentrations by as much as 40,000 times—a factor that greatly speeds the reaction rate.

Second, enzymes orient the substrate. Specific regions of reacting molecules must lie close together before the molecules reach their intermediate shape, or transition state. Within the active site of an enzyme catalyst, as the fit is induced, the substrates become correctly oriented (Figure 4.13e).

Third, enzymes change the shape of substrates, helping them reach the transition state. As substrates lock into an enzyme's active site, their shape is strained or bent a little, and they are usually brought closer to the transition state through this bending. Thus, the enzyme effectively lowers the activation energy, making it easier for substrates to react and form products (see Figure 4.13e). An enzyme may have in its active site or at some other spot on its surface a nonprotein component, or *cofactor*, whose presence is necessary for the enzyme to do catalytic work, that is, help a substrate reach the transition state. Some of the B vitamins function as precursors of cofactors that help convert food to biological energy.

Because reactions catalyzed by enzymes underlie the cell's simultaneous, nonstop activity, enzymes can be seen as the effectors of the life process, enabling the cell to carry out transport, chemical, and mechanical survival tasks. Together with ATP, enzymes help ensure the flow of energy that enables the cell to resist the tendency to become disordered, fall apart, and die.

≋ METABOLISM: THE CELL'S CHEMICAL TASKS

The living cell, bustling with activity, carries out an important set of integrated chemical activities that transform energy. Called metabolism, this set of activities involves thousands of simultaneous chemical reactions catalyzed by enzymes, and accomplishes two things: (1) the harvesting of energy to fuel work and (2) the building of cell parts for maintenance, growth, and cell division. Chapters 5 and 6 detail the steps involved in harvesting chemical energy from food molecules and in converting the energy from sunlight into those food molecules in the first place. In this chapter we focus on how the cell uses energy to build and maintain its parts (Figure 4.15).

The construction of new cellular parts proceeds at a spectacular rate to keep pace with the cell's ability to function. For example, a rapidly growing bacterial cell can generate 1400 protein molecules per second at a cost of more than 2 million ATP molecules! Even in a slow-growing eukaryotic cell—say, a rat liver cell—there is a continuous turnover of cellular components. For example, in just three weeks' time, all the proteins in a rat

FIGURE 4.15 *Building Muscle Cells.* Enzymes in this muscle builder's muscle cells harvest energy, and this allows them to do the work of lifting weights. In response to the work, the muscle cells take up amino acids, fatty acids, and other small molecules and use them to build cell constituents, such as the extra protein fibers that give muscles their added bulk.

liver cell are dismantled into raw materials and replaced by new protein molecules.

Metabolism is a complex process, a web of interrelated chemical conversions in which the end products of some reactions become the reactants of others; and all must be catalyzed by enzymes so that the reactions take place quickly enough to keep the cell alive. Metabolism is ultimately responsible for the cell's ability to preserve and transmit genetic information, to develop specific functions within multicellular organisms, and to build necessary components for the maintenance of order and the staving off of disorder.

≋ THE CELL'S TRANSPORT TASKS

The cell carries out chemical tasks simultaneously with a set of transport tasks. Like any bustling metropolis, the living cell must keep enough of the right materials on hand to meet the constant demands of construction, maintenance, export, and other activities. To do this, the cell must expend energy. The flow of energy is closely tied to the flow of materials, and both are requisites for life.

Cells are minute pools of liquid surrounded by membranes (see Chapter 3). The dynamic cell is clearly like an aquatic metropolis—more like Venice than like Las Vegas. Most of a cell's internal activities take place in or very near water molecules, and the outside of the plasma membrane must remain wet as well. The pool inside each cell is a bit like seawater; it is a rich solution of ions and dissolved materials, usually with a distinctly different composition than the fluid outside the cell.

The cell's inner pool is called the **intracellular fluid** (Figure 4.16, parts 1 and 4), and the materials for maintaining that pool are taken up from the fluid outside the cell. A single-celled organism is almost always surrounded by fresh water, seawater, or the body fluids of another organism that it inhabits. In a multicellular organism, the fluid outside the cell is usually **extracellular fluid**, which fills all the spaces around and between cells. Simple animals usually have a single extracellular fluid, while some complex animals have two types: the clear extracellular tissue fluid in the spaces between cells, and the blood that flows through the animal's vessels (Figure 4.16, parts 2 and 3). Materials can move back and forth between these fluids across the membrane barriers that separate them. Such movements are governed in part by differences in the concentration, pressure, or temperature of a material from one part of a contained area to another. These differences are called *gradients*.

Materials—ions, water, amino acids, and so on—nat-

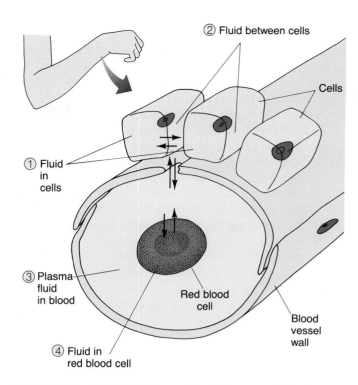

FIGURE 4.16 *Biological Fluids and Their Movements.* This diagram shows a red blood cell inside a blood vessel that is surrounded by other cells. There is fluid inside the surrounding cells (1) and inside the red blood cell (4). There is extracellular fluid between the cells (2). And there is fluid called plasma in the blood (3). Water and materials from these various fluids can move back and forth across cell membranes.

urally tend to move from regions of higher concentration to regions of lower concentration (*down* concentration gradients). When a substance enters or leaves a cell by moving from high concentration to low, the process is called **passive transport:** The material moves without energy expenditure by the cell. Cells, however, must transport certain materials in or out "uphill," that is, from regions of lower concentration to regions of higher concentration; this means that there is already more of a certain substance inside the cell than outside, for instance, but the cell needs even more. Such movement requires energy expenditure by the cell and is called **active transport.**

PASSIVE TRANSPORT: DIFFUSION

The cell needs a constant supply of materials so it can maintain itself and build new cell parts, and it needs to get rid of excess ions, organic wastes, and other materials. Most of the substances that enter or leave the cell do so through passive transport. The tendency of materials to move from areas of high concentration to areas of low concentration is called **diffusion,** and you can witness the results of diffusion if you put a spoonful of whipped cream on top of your coffee. Billions of collisions send water, cream, and coffee molecules bouncing off each other in random directions like billiard balls, until eventually, the cream molecules have spread evenly throughout the cup.

The membrane surrounding the cell is a **semipermeable membrane.** Like a permeable membrane, the semipermeable membrane contains pores that allow ions, water, and certain other small charged molecules to pass freely by passive transport. Unlike the situation seen in a fully permeable membrane, however, most large molecules (like glucose) cannot penetrate either the membrane itself or the pores and thus cannot simply diffuse in or out. Instead, substances like glucose and the waste product urea pass through a membrane with the help of a carrier molecule. As in other types of passive transport, these carriers move the molecules from an area of higher concentration to an area of lower concentration in a process that does not require energy (Figure 4.17).

PASSIVE TRANSPORT AND THE MOVEMENT OF WATER

Since a cell contains mostly water, and since its delicate outer membrane must be immersed in fluid, water itself is probably the most important substance to enter or leave the cell, and it does so by passive diffusion. The movement of water through a semipermeable membrane (such as a cell's plasma membrane) is called **osmosis.** Water moves from a region where it is present in high concentration or pressure to one where it is present in low concentration or pressure. Pure distilled water has a higher concentration of water molecules than does a solution containing both water and a solute such as salt or protein.

The red blood cell is a good example of the importance of diffusion and osmosis to living cells (Figure 4.18). As it races through the blood vessels, the red blood cell is bathed in a fluid that has a concentration of salts, sugars, and other molecules similar to that in the pool of fluid inside the cell. Because the concentration of water molecules is similar inside and outside the cell, as many water molecules move in as move out, and the cell retains its normal disk shape (Figure 4.18b; also see the box on

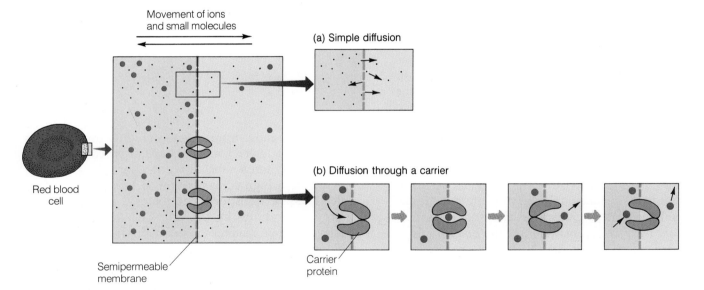

FIGURE 4.17 *Passive Transport in the Cell: Simple and Carrier-Mediated.* (a) The cell membrane is semipermeable; it allows the passage of small molecules by simple diffusion. (b) However, passage of most larger molecules requires the activity of carriers in a process called *facilitated diffusion.* Although biologists do not yet know precisely how this process works, one hypothesis suggests the involvement of *carrier* proteins. These proteins may span the membrane and change shape to allow the molecules to cross the bilayer, then change back again, ready for the next incoming molecule. Because the movement is always from regions of higher to lower concentration (passive transport), there is no energy expenditure by the cell. But it can allow substances to diffuse across a membrane that otherwise would block their passage, enabling the cell to take in or get rid of substances quickly enough to keep pace with metabolism. In a red blood cell, both simple and facilitated diffusion take place simultaneously across the plasma membrane.

FIGURE 4.18 *Osmosis and the Red Blood Cell.* Water moves across a semipermeable membrane such as the plasma membrane of a red blood cell by osmosis, that is, passively in response to concentration gradients of ions and other solutes. (a) When a red blood cell is placed in a solution that is highly concentrated in ions or other solutes (*hypertonic*), so that the concentration of water is lower outside than inside the cell, water moves out by osmosis, leaving the cell shriveled. (b) In an *isotonic* solution, with solute concentrations similar to those of the cell interior, water moves in and out at equal rates. (c) When a cell is placed in pure water, a *hypotonic* solution lacking solutes entirely, water is more concentrated outside the cell, so it rushes into the cell, and the cell expands rapidly, sometimes bursting. The red blood cells in these photos are magnified about 400 times.

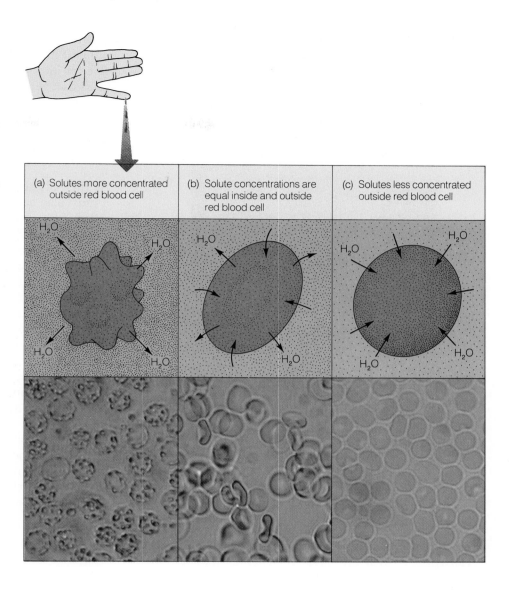

(a) Solutes more concentrated outside red blood cell

(b) Solute concentrations are equal inside and outside red blood cell

(c) Solutes less concentrated outside red blood cell

page 85). But what would happen if you placed red blood cells in a beaker of pure distilled water? The pure water would have a vastly lower concentration of ions, proteins, and other solutes than the cell's cytoplasm. At the same time, however, the concentration of water molecules in the beaker would be far greater than inside the cell, and so water would tend to move into the cell by osmosis. This would cause the cell to swell or even burst from the extra water (Figure 4.18c). One could also prepare a solution with a higher concentration of salts, sugars, proteins, and other solutes but a lower concentration of water than the fluid inside a red blood cell. Since the concentration of water molecules would be lower outside the cell than in, water would rush out of the cell by osmosis, leaving it seriously shriveled (Figure 4.18a).

You may have noticed the effects of osmosis yourself; the day after you eat a big helping of a very salty food—say, a large bag of potato chips—your weight may go up

2 or 3 pounds (Figure 4.19, page 84). The reason is that the salt concentration of your cells increases, and so you tend to retain more of the water you drink as it moves into the cells to offset the extra salt.

Inside the root cells of most plants, the solute concentration is usually higher and the water concentration lower than in the surrounding soil. Water, therefore, tends to enter the plant and its cells. The cells do not burst, however, because they are contained by their strong cellulose walls. The water in the now plump vacuoles exerts an outward force called *turgor pressure* that prevents a further net gain of water (Figure 4.20, page 84). When the water in the cells' vacuoles exerts high pressure, the plant stands erect and firm, but when the pressure falls, the plant may wilt.

FIGURE 4.19 *Osmosis, Salty Potato Chips, and Added Pounds.* Eating too many salty potato chips increases the concentration of salt in the blood and cells. As you drink to dilute the salty solution, you retain water and gain a few pounds.

ACTIVE TRANSPORT: ENERGY-ASSISTED PASSAGE

Ironically, the principle of osmosis costs the cell dearly. Cells contain such an array of materials that for the majority of cells, there is a continuous tendency for water to enter the cell and cause swelling. At the same time, cells stockpile materials that are too large to move in by passive diffusion. Thus the dilemma: To prevent swelling, the cell must move certain materials *out* (with water following osmotically), and yet to continue functioning, the cell must move certain materials into areas of higher concentration. Moving a substance "uphill" this way involves the energy-costly process called *active transport*. In many cells, active transport is accomplished by a set of proteins in the plasma membrane. These proteins break down the energy in ATP and act as pumps that push or pull substances into and out of the cell (Figure 4.21, page 86). One such pump is responsible for moving ions into and out of your nerve cells. Another active transport pump moves chloride ions across the membranes of lung and other cells; it is defective in some people and leads to the life-threatening disease called cystic fibrosis (see Figure 3.10).

Once the processes of active and passive transport have brought materials into cells, these substances are available for important chemical tasks as well as mechanical tasks that generate movement in the cells.

⁂ THE CELL'S MECHANICAL TASKS

As the cell carries out its chemical and transport tasks, it must also accomplish a set of mechanical tasks, often reflected by movements of the entire organism or its parts. Cells are capable of a tremendous range of movements. Muscle cells contract, sometimes hundreds or thousands of times per second, allowing an animal to swim, run, or fly (Figure 4.22, page 86). Single bloblike cells called amoebas creep along solid surfaces. Cells with lashing tails, or flagella, glide forward or backward. Cells with many short whips, or cilia, can sweep particles across their surfaces or dart through the water as if propelled by thousands of tiny oars. Moreover, all eukaryotic cells, whether obviously mobile and active or apparently stationary, have an array of internal movements.

Underlying the movement of cells or cell parts is the rapid assembly or breakdown of the cells' internal "skeleton" (see Chapter 3). These actions are accomplished by enzymes and fueled by ATP, and they convert chemical energy to mechanical energy. Chapter 28 discusses such conversions in your muscle cells, for example. Mechanical work is one of the cell's crucial activities,

(a) Erect

(b) Wilted

FIGURE 4.20 *Turgor Pressure.* Water in the central vacuoles of plant cells exerts an outward force called turgor pressure that keeps the plant tissues firm. In the feathery-leaved sensitive plant, the appearance of the leaf depends on the turgor pressure of cells in the junction between the leaf's stem and the plant's main stem, and on other cells at the base of each leaflet (small lobe of a compound leaf). When turgor pressure is high, the leaf stands erect (a). If you touch the plant, the cells pump ions out, and water follows by osmosis. This lowers the turgor pressure, and the leaf folds up (b).

ARTIFICIAL BLOOD: ENGINEERED ELIXIR

Blood is the elixir of human life, flowing through our vessels, minute by minute, year after year, and carrying needed oxygen to each of our cells. When a person is wounded in battle or badly injured in a car wreck, the loss of this precious fluid is often the cause of death. Medics on the scene usually transfuse the victim with saline (a saltwater solution that is osmotically similar to blood), and this keeps the victim's blood pressure from plummeting, allowing a reduced but steady flow to reach the heart, brain, and other tissues. Nevertheless, saline carries very little oxygen, and people losing blood can remain at serious risk until they begin to receive a transfusion of blood that matches their own blood type.

An obvious solution would be to keep fresh blood in ambulances and field hospitals, but the requirements for storing it are too stringent: Whole blood must be refrigerated, carefully monitored, and discarded every 21 to 35 days. What's more, even fresh blood banked in hospitals can be contaminated with AIDS or hepatitis viruses. For all these reasons, researchers have spent decades trying to create blood substitutes—chemicals that can carry oxygen in a living mammal without major side effects, risk of contamination, or the need for refrigeration and blood typing.

One early blood substitute was the class of fluorine-containing compounds called perfluorocarbons. In a dramatic demonstration of the compound's effectiveness at delivering oxygen to living tissues, an Ohio researcher in the early 1970s submerged a mouse in perfluorocarbons for several minutes (Figure 1). The fluid filled the rodent's lungs, but transported enough oxygen to sustain its metabolism. Once drained and dried off, the animal survived for two or three days, but then it died of a lung ailment. Japanese physicians are currently testing an improved version of perfluorocarbons called Fluosol-DA on humans.

In the early 1980s, a second research group collected hemoglobin molecules from the outdated, discarded red blood cells in blood banks, then chemically modified the molecules so that they tended to remain associated in clusters, as in living red cells (review Figure 4.1). Next they freeze-dried the clusters into rusty red chunks that could be reconstituted into a bloodlike slurry when mixed with water. Unfortunately, endotoxins (toxic compounds released by certain bacteria) cling to naked hemoglobin molecules and cause them to damage tissues in the body. This freeze-drying approach is still under investigation.

Finally, two researchers at an Illinois firm received a patent in late 1989 for hemoglobin encapsulated in tiny phospholipid vesicles—blood cell surrogates that contain the hemoglobin and seem to work fairly well in test animals.

Regardless of the form artificial blood may take in the future, it promises to be an invaluable commodity for rural clinics, field hospitals, and ambulances; for priming heart-lung bypass machines; and as a stockpiled synthetic elixir of life to use after earthquakes or other natural disasters.

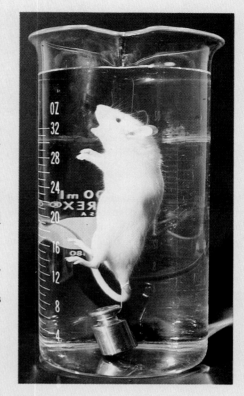

FIGURE 1 *Perfluorocarbons, an Early Blood Substitute.* This white mouse could survive several hours of submersion in the oxygen-saturated solution of fluorcarbons shown here.

FOR FURTHER READING

Andrews, E. L. "Patents Column." *New York Times*, 21 October 1989, p. 34.

Weiss, R. "Sanguine Substitutes." *Science News* 132 (26 September 1987): 200–202.

"Blood on the Shelf." *Newsweek Access* (July 1983): 27.

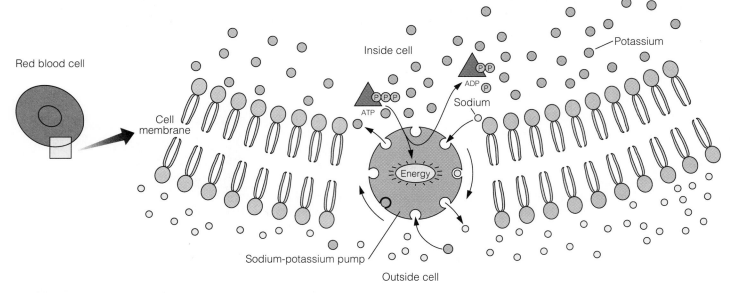

FIGURE 4.21 *Active Transport by Membrane Pumps Made Up of Proteins.* In the plasma membrane of a red blood cell, special membrane proteins actively pump in potassium ions (K$^+$) uphill from low potassium concentration to high and actively pump out sodium ions (Na$^+$) from low sodium concentration to high. This pumping requires the expenditure of much ATP energy.

FIGURE 4.22 *Dynamic Cells and Dynamic Animals.* Hummingbirds of various species can beat their wings 22 to 79 times per second, and some can fly up to 60 mph. These feats of locomotion are based on the contractile cytoskeleton inside muscle cells and require a great deal of energy. Here, the violet-capped wood nymph of Brazil sips a *Malvaviscus* blossom.

enabling it (or the organism the cell is part of) to pursue food; to divide and reproduce; and to carry out chemical work (such as the packaging and export of materials) and transport work (such as actively moving substances across the membrane).

≋ CONNECTIONS

In the time it takes you to blink or sneeze, nearly every cell in your body will have performed thousands of individual activities. Membrane pumps will have transported billions of ions in and out of your cells. Enzymes will have sped up hundreds of metabolic reactions. Some cells will have squeezed out their nuclei. Others will have divided in half. And inside them all, organelles will have shifted—vesicles moved, microtubules lengthened or shortened, and carrier proteins drifted around in the plasma membrane like icebergs.

The energy costs for cellular activity are enormous, and yet, collectively, life processes are energy-releasing. In fact, efficient energy collection and use are so critical to survival that they have been a prime cutting edge for evolutionary processes. Natural selection rewards those organisms that can grab the most energy and use it efficiently, while the less efficient life forms tend to die out.

In the next two chapters, we will trace the energy flow through cells. As we do, try to envision the bustling metropolis—the dynamic cell—that surrounds and is empowered by that flowing energy.

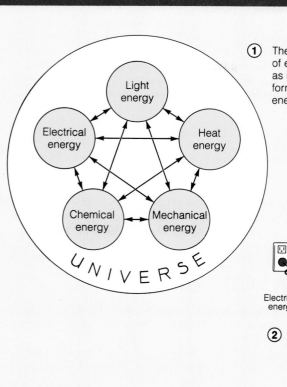

① There is a fixed amount of energy in the universe; as it converts from one form to another, no energy is lost.

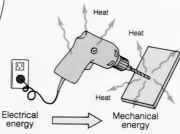

② Energy conversions are inefficient and release heat.

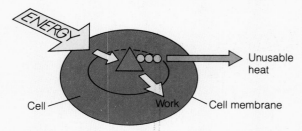

③ Within the universe, energy flows from the environment through systems such as planets or cells, and back to the environment.

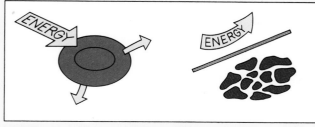

④ Each system becomes more and more disordered unless it receives a constant input of energy.

⑤ A dynamic cell takes in energy and converts it to ATP, which fuels cellular work.

⑥ There are three kinds of cellular work: chemical, transport, and mechanical. Enzymes speed and facilitate the body's chemical reactions.

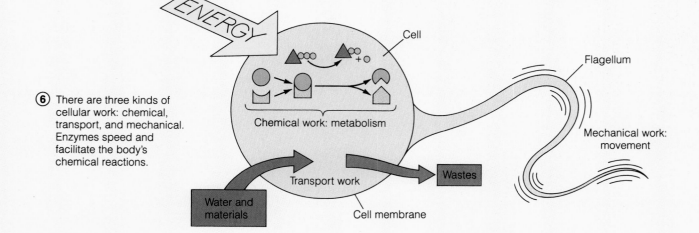

NEW TERMS

activation energy, page 77
active site, page 78
active transport, page 81
ATP, page 76
catalyst, page 77
chemical reaction, page 73
diffusion, page 82
energy, page 71
entropy, page 71
enzyme-substrate complex,
    page 78
extracellular fluid, page 81
first law of thermodynamics,
    page 71

heat, page 71
intracellular fluid, page 81
osmosis, page 82
passive transport, page 81
product, page 73
reactant, page 73
second law of thermo-
    dynamics, page 71
semipermeable membrane,
    page 82
substrate, page 78

STUDY QUESTIONS

REVIEW WHAT YOU HAVE LEARNED

1. Distinguish between potential and kinetic energy.
2. Discuss how living cells counteract the tendency toward disorder.

3. Classify the following activities as energy-releasing or energy-requiring, and explain why in each case: running; building of proteins.
4. Explain why biologists often describe ATP as energy currency.
5. What is activation energy? How do enzymes affect it?
6. Describe the structure of an enzyme, and explain why enzyme action is highly specific.
7. Compare active and passive transport.

APPLY WHAT YOU HAVE LEARNED

1. Which is younger, a mature brain cell in a 75-year-old person, or a mature red blood cell in an 18-year-old? Explain.
2. You are a jockey and need to check in at a certain weight in order to qualify for a race. Should you eat a bag of potato chips a few hours before check-in? Explain.

FOR FURTHER READING

Alberts, B., D. Bray, J. Lewis, M. Raff, K. Roberts, and J. Watson. *Molecular Biology of the Cell*. 2d ed. New York: Garland, 1989.
Prescott, D. M. *Cells*. Boston: Jones and Bartlett, 1988.

HOW LIVING THINGS HARVEST ENERGY

A MOUNTAIN HIKE: SPLITTING SUGAR FOR MUSCLE POWER

Suppose you decide one day to hike up a steep trail that winds alongside a tumbling stream—a trail, let's say, like the many crisscrossing the Green Mountains of Vermont. Fortified by a hearty break-

FIGURE 5.1 *Hiker in the Green Mountains of Vermont.* Muscles in the hiker's legs harvest energy to fuel his walking and climbing movements.

fast of pancakes and maple syrup, you start up the trail in the brisk morning air (Figure 5.1). As you ascend into the fragrant pine forest, the steepness of the terrain begins to tax you, and your heart starts to pound, your chest to heave, and your leg muscles to burn. Eventually, adopting a slower, steadier pace, you reach the summit and take in a panoramic view of blue-green forested ridges that gently fade to gray in the distance.

To carry out this simple weekend activity, you must rely on many fundamental life processes. Two of the most basic life processes are the subject of discussion here and in the next chapter. First, plants such as maple trees and wheat soak up sunlight and capture light energy in the bonds of glucose molecules; the food you ate for breakfast contained this stored chemical energy. Second, the muscles in your legs and elsewhere in your body harvest that stored chemical energy, which in turn fuels all your activities, including the movements that carried you up the mountain trail (Figure 5.2, page 90).

This chapter focuses on how living things strip energy from molecules such as sugars, and Chapter 6 describes how plants trap the sun's energy in those sugars by means of the process called photosynthesis. Together, these two chapters explain a central theme in biology: Plants, as well as certain bacteria and protists capable of photosynthesis, capture the sun's energy in sugars. The harvesting of energy from these sugar molecules then supports the life processes in virtually all living things.

The energy-harvesting processes taking place in your muscle cells are also at

work in plants, animals, fungi, protists, and monerans. In fact, the same metabolic pathways fuel all life activities. The similarity of the pathways in all organisms is a powerful reminder of life's unity, its evolutionary descent from common ancestors. In general, cells harvest energy either with or without oxygen. Muscle cells are somewhat special in that they can do it both ways, depending on environmental conditions.

Metabolic pathways are also *gradual*; that is, they occur in steps. The energy stored in a nutrient molecule is not released all at once in an incinerating burst of heat; instead, molecular bonds are broken and rearranged in numerous small steps, and the sequentially released energy is stored for future use. Since the transfer of energy is always inefficient, some heat is lost forever at each step. Despite this inevitable waste, however, living cells harvest enough energy in the form of ATP to fuel their growth, maintenance, and reproduction.

This chapter will answer several questions about how living things harvest the energy they need to sustain life:

- How do cells build and use energy carriers?
- What are the energy-harvesting reactions in a cell?
- How does glycolysis begin the energy harvest?
- What is the role of fermentation during metabolism in the absence of oxygen?
- How do cells harvest huge quantities of energy when oxygen is present?
- How does the cell control energy metabolism?

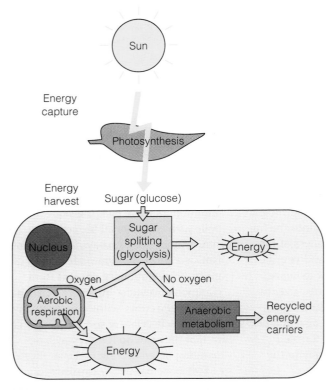

FIGURE 5.2 *Overview of Energy Capture and Harvest.* Plant cells capture the sun's light energy in the bonds of glucose molecules. Virtually all cells harvest the energy stored in sugar by breaking down glucose through a metabolic pathway called glycolysis. If oxygen is present, many cells can net a large additional energy harvest through the aerobic respiration pathway. If oxygen is absent, some cells can carry out fermentation with no further gain of ATP but a recycling of some waste products from glycolysis into necessary raw materials.

≋ ATP AND THE TRANSFER OF ENERGY FROM NUTRIENT MOLECULES

Like the bacterial cells described in Chapter 4, muscle cells contain millions of ATP molecules and expend nearly half of them every second on life-sustaining cellular activities. Obviously, the supply of this energy currency must be regenerated speedily and continuously—and so it is. This regeneration takes place through several sets of reactions that break down energy-storing compounds. Before we describe them, however, we must consider how the structure of ATP allows this small molecule to transfer energy in the cell and exactly how the energy from organic nutrient molecules is channeled into ATP for temporary storage.

ATP STRUCTURE: A POWERFUL TAIL

As Chapter 4 explained, ATP is a nucleotide with a long tail of phosphate groups (review Figure 4.9), and this structure explains its utility to the living cell. As Figure 5.3 shows, ATP, or adenosine triphosphate, contains the nucleoside adenosine and a tail of three connected phosphate groups, with each phosphate group represented by the symbol Ⓟ. ADP, or adenosine diphosphate, contains two phosphate groups; and AMP, or adenosine monophosphate, just one. When a bond between phosphates is broken, a large amount of useful energy contained in the whole molecule is readily released. This release of energy from a chemical bond is similar to pulling the plug at the bottom of a rain barrel. The energy released in the outflow was stored in the total water contents, not simply in the plug.

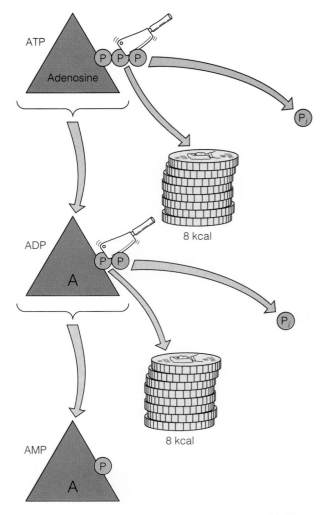

FIGURE 5.3 *Cleaving of Energy Carriers Releases Usable Energy.* Cleaving one phosphate group from ATP yields 8 kcal of energy, one inorganic phosphate ion, and ADP. Cleaving another phosphate group yields 8 kcal of energy, one inorganic phosphate ion, and AMP.

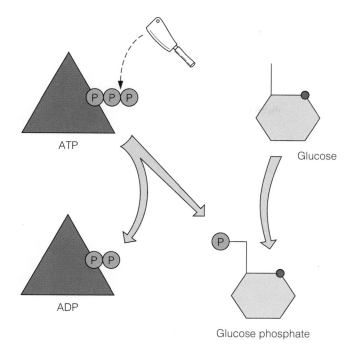

FIGURE 5.4 *Energizing Glucose.* Glucose can be "supercharged" by the cleaving of ATP and the addition of the cleaved phosphate group to the glucose carbon ring. This transferred phosphate group acts like a little battery pack, providing power that enables the glucose phosphate to participate in reactions in which a regular glucose molecule cannot take part.

The cleaving of the outer phosphate group from ATP yields 8 kcal of energy per mole of ATP and leaves ADP plus the inorganic phosphate ion (designated $\textcircled{P}_i$, in Figure 5.3). (Note that a *mole* is a common unit of measurement for molecules and atoms; it is 6×10^{23} molecules of any substance.) In most of a cell's energy transfer reactions, ATP serves as the high-energy form of the molecule, and ADP as the low-energy form.

The phosphate groups cleaved from ATP can be transferred from ATP to other kinds of molecules, thereby energizing them and allowing them to participate in reactions that could not take place otherwise. Figure 5.4 shows the transfer of a phosphate from ATP to glucose. The transferred phosphate group in effect acts like a little battery pack, providing enough power so that the sugar molecule can readily take part in later reactions that produce energy.

HOW ENERGY IS CHANNELED INTO ATP

The structure of ATP may help explain the importance of this molecule to energy flow in the cell, but other molecules with other structures can also take part in energy-harvesting reactions. The transfer of energy into ATP involves a flow of electrons, a bit like an electric current. Just as wires carry electrons from an electricity-generating plant to the outlets in your home, special *elec-*

tron carriers help move electrons around inside cells. Electron carriers can pick up electrons from one molecule and release them to another. Figure 5.5 shows an electron carrier donating a pair of electrons along with a hydrogen ion to an acceptor molecule. As the electrons move, energy is transferred to the acceptor. Electron carriers are crucial for releasing the energy you need as you hike up a steep mountain slope. Curiously, though, our cells cannot make such carriers "from scratch." Instead, we must get them nearly preformed in our diets as vitamins. We'll talk more about vitamins in Chapter 22.

The overall message in this discussion of electron transfer reactions is that as metabolic pathways proceed in the cell, energy carriers pick up electrons and/or hydrogen ions and later pass them along to other carriers. The flow of electrons through the carriers creates a current that is channeled in special ways into ATP formation. This ATP can power most types of work in the cell.

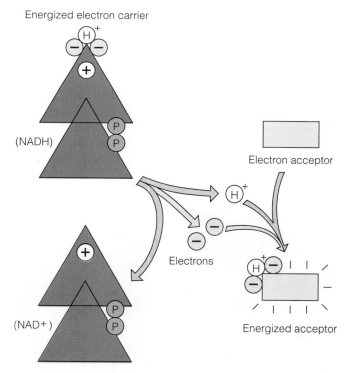

FIGURE 5.5 *Electron Transfer from Carrier to Acceptor.* Energy pathways such as glycolysis and aerobic respiration run on coupled electron transfer reactions, often linked by electron carriers. As an electron carrier releases two electrons and one hydrogen ion, these become available to an acceptor, which becomes energized, or charged, as it receives them.

Now let's begin to look at how cells harvest energy from nutrients to maintain life.

≋ AN OVERVIEW OF ENERGY-HARVESTING PATHWAYS

All cells need energy to live, whether they are budding yeast cells growing in a tank of grape juice, the billions of cooperating cells in your body as you hike in the mountains or walk to class, the muscle cells of an owl taking flight, or the cells in a willow leaf fluttering in the wind. The immediate *source* of that energy may differ: Yeasts absorb nutrients directly from their environment; humans chew and digest foods made mostly from plant and animal materials; owls swallow and digest small rodents and snakes; and willow leaves generate their own sugars via photosynthesis. Plants and the photosynthetic microbes that can generate their own nutrient molecules are considered **autotrophs** ("self-feeders"). Organisms that must take in preformed nutrient molecules from the environment are classified as **heterotrophs** ("other-feeders"), and they include the yeasts and all other fungi, most kinds of bacteria and protists, and humans and all other animals. In general, heterotrophs depend on autotrophs for their food, and autotrophs capture solar energy and build their own food. When it comes to how organisms harvest the rich source of chemical energy stored in nutrient molecules, however, it doesn't matter whether an organism is a heterotroph or an autotroph, because all organisms share the same biochemical pathways for releasing energy stored in the molecular bonds of nutrient molecules (Figure 5.6).

To get an overview of the energy-harvesting pathways in living cells, consider the three phases of energy harvest taking place in active muscle cells (see Figures 5.2 and 5.7). All during a hike, whether the muscle cells are resting or contracting, they split sugar molecules in half, a process called **glycolysis.** This splitting of sugar provides a continuous but small amount of energy. The fragments that result from sugar splitting enter one of two additional breakdown pathways. When you move rapidly up a steep stretch of trail and run out of breath, the fragments are acted on in a pathway called **anaerobic metabolism** (energy harvest without oxygen). During a slower, sustained ascent of the trail, the fragments from sugar molecules are broken down in the pathway called **aerobic respiration** (energy harvest with oxygen). Thus, energy harvest begins with the sugar-splitting process of glycolysis and then can proceed either to anaerobic metabolism if the cell has no oxygen available or to aerobic respiration if oxygen is plentiful. As we discuss these pathways, the number of ATPs each one generates will serve as an effective point of comparison for how well that particular pathway provides energy for the living cell.

Glycolysis splits the sugar glucose, with its six carbon atoms, into two molecules of a compound called pyruvate, with three carbons per molecule; this splitting process produces a small amount of energy (see Figure 5.7). The steps of glycolysis, which take place in the cell's watery cytoplasm, proceed in the presence or absence of oxygen, and they are made possible by a specific set of enzymes that occur in nearly all living cells. This uni-

FIGURE 5.6 *Metabolic Pathways Are Universal.* All organisms use the same basic molecular scheme for harvesting energy. The flapping of bird wings, for example, liberates energy a plant originally harvested from sunlight. Here, the little owl (*Athene noctua*) takes flight.

FIGURE 5.7 *An Overview of Energy-Harvesting Pathways in the Cell.* This diagram shows the energy tally for the aerobic and anaerobic pathways of glucose breakdown. The left side of the diagram shows the aerobic pathways: If oxygen is plentiful, glycolysis yields 2 ATP molecules per glucose molecule, and in the mitochondrion, aerobic respiration with its two parts, the Krebs cycle and the electron transport chain, yields an additional 2 and 32 ATPs, respectively, for a total of 36 ATPs per initial glucose molecule. The right side of the diagram shows the anaerobic pathways: If oxygen is in short supply, glycolysis still yields 2 ATPs and the energized electron carrier molecule (NADH), while fermentation yields no additional ATPs but regenerates the deenergized electron carrier (NAD $^+$) needed for the next round of glycolysis to take place.

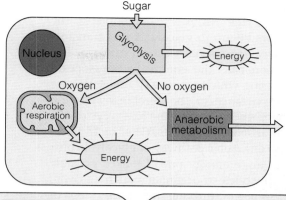

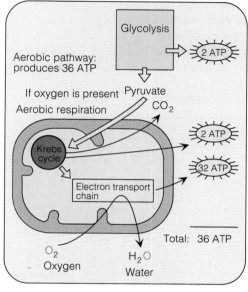

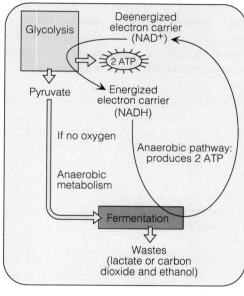

versality leads biologists to believe that glycolysis evolved in the earliest cells, perhaps 3 billion years ago, and has since been inherited by all the world's surviving organisms.

After glycolysis, energy harvest can proceed either to the set of reactions called anaerobic metabolism or to the set called aerobic respiration (see Figure 5.7). These two pathways were evolutionary add-ons to the biochemical antique of glycolysis. Anaerobic metabolism involves the process called **fermentation** (see Figure 5.7). Fermentation occurs in your leg muscles as you race up a steep slope, when your lungs and heart cannot provide oxygen to your muscles fast enough to keep pace with energy demands. Fermentation also occurs in certain yeasts and bacteria that grow in the absence of oxygen, such as in the mud at the bottom of a lake. As in glycolysis, enzymes in the cytoplasm allow the reaction steps of fermentation to take place. These steps go beyond glycolysis, however; they convert the three-carbon pyruvate molecules formed during glycolysis into potentially harmful wastes. One such waste product is the lactate that makes your muscles burn as you rush up a hill. Even though fermentation converts the pyruvate formed during glycolysis to poi-

sonous wastes, the pathway is necessary to regenerate energy carriers, which must be present in the cell for glycolysis to continue. The route from glycolysis through fermentation is crucial because it allows cells such as yeasts or your muscle cells to harvest energy and survive even when there is not much oxygen available. Nevertheless, these two phases of energy harvest—glycolysis and fermentation—are inefficient: Together they produce only two molecules of ATP per molecule of glucose.

Aerobic respiration occurs only in the cells of *aerobic organisms*—those that employ oxygen for metabolism—and these include the great majority of life forms. In aerobic cells, the respiration pathway acts on the three-carbon pyruvate generated during glycolysis to yield carbon dioxide and water plus a large number of ATP molecules (see Figure 5.7). Aerobic respiration consists of two phases, the **Krebs cycle** and the **electron transport chain**. Both the Krebs cycle and the electron transport chain are carried out in the mitochondria, which is why biologists often characterize that organelle as a cellular powerhouse. Briefly, the Krebs cycle completes the breakdown of sugars begun in glycolysis and transfers a great deal of energy to special electron carriers. The electron carriers then pass the energy to the electron transport chain, during which many ATP molecules are generated. The importance of the aerobic respiration pathway is that together, glycolysis, the Krebs cycle, and the electron transport chain make 36 ATPs, or 18 times as much ATP as the anaerobic pathway of glycolysis plus fermentation. This huge energy payoff makes possible the high-energy life-styles of organisms such as owls, people, and even spinach. Studying the specific steps of each energy-releasing pathway—our next task—is as essential to appreciating life as studying the compositional structures of symphony movements and play acts is to appreciating music and drama.

⪢ GLYCOLYSIS: THE UNIVERSAL PRELUDE

At this moment, practically every cell on earth is burning the sugar glucose via glycolysis. Glycolysis is the basis for energy metabolism in all living things, since it serves as a biochemical prelude to the pathways of fermentation and aerobic respiration. Glycolysis proceeds in the cell's cytoplasm in two basic steps (see Figure 5.8):

STEP 1: *Investment.* Glucose is "charged" with two high-energy phosphate groups at the cost of two ATP molecules. This investment step brings to mind the saying, "It takes money to make money." In this case, ATP has been spent, not harvested, to produce an activated sugar with two phosphates.

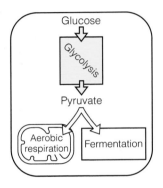

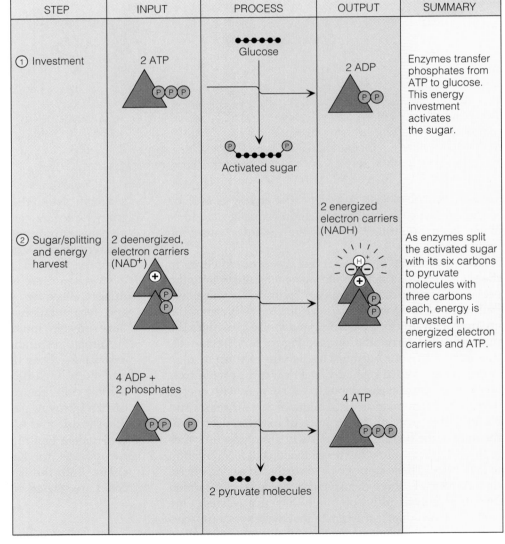

FIGURE 5.8 *Glycolysis: The Splitting of Glucose.* This diagram shows the two main steps of glycolysis, the first stage of energy metabolism in living things. The steps are described in the text. The input column shows the raw materials needed for the pathway; the output column lists the products of the reaction series; and the process column shows how the initial six-carbon glucose molecule becomes rearranged and broken down to *two* three-carbon pyruvate molecules.

STEP 2: *Sugar Splitting and Energy Harvest.* The high-energy sugar formed in step 1 is split and eventually forms two molecules of the compound pyruvate, the end product of glycolysis. During this process, energy is transferred to a deenergized electron carrier (called NAD^+) and to ADP. The result is two molecules of the energized electron carrier (NADH) and four ATPs.

Two of the four ATPs repay the initial investment made in step 1, and the other two provide the cell with a net profit of two ATPs per molecule of glucose. This is less than 5 percent of the energy stored in the bonds of glucose, and it can be merely the beginning of the much larger energy harvest that takes place during aerobic respiration in the mitochondria. But the energy harvest from glycolysis alone is enough to sustain life in anaerobic organisms that ferment, such as yeast.

≋ FERMENTATION: "LIFE WITHOUT AIR"

Louis Pasteur, the famed nineteenth-century microbiologist who discovered fermentation, called it *"la vie sans air"* ("life without air"), and indeed, *fermentation* is a major metabolic pathway for anaerobes—cells that can live where oxygen levels are very low or where oxygen is altogether absent. The specific biochemical reactions we call fermentation rest on the foundation of glycolysis but do not directly yield ATP. Fermentation enables certain kinds of cells to survive on the limited energy proceeds of glycolysis by recycling an electron carrier needed for glycolysis to continue generating ATP. Even for aerobic organisms like people, fermentation has important implications. Fermenting yeasts and bacteria generate carbon dioxide and alcohol or lactic acid, which in turn help create many of our favorite foods and drinks, including bread, cheeses, wine, and beer. And fermenting muscle cells can cause painful cramps or produce a quick burst of speed that could save one's life.

ALCOHOLIC FERMENTATION

In a few types of cells—yeast cells being the most familiar—the three-carbon compound pyruvate, produced during glycolysis, is broken down further into the two-carbon compound ethanol (ethyl alcohol); the leftover carbon is released as carbon dioxide (Figure 5.9a). For the yeast cell, the carbon dioxide and ethyl alcohol are poisonous waste products to be excreted. The significance of fermentation for the cell's continued survival lies in the regeneration of the electron carrier needed for

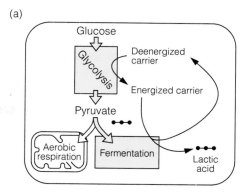

(a)

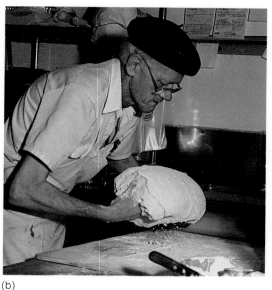

(b)

FIGURE 5.9 *Alcoholic Fermentation.* (a) During this form of fermentation, enzymes split pyruvate molecules from the glycolysis pathway into ethyl alcohol and carbon dioxide, while the energized carrier (NADH) is converted to the deenergized carrier (NAD^+). The regeneration of NAD^+ allows glycolysis to continue. (b) The carbon dioxide generated by yeast cells can cause bread dough to rise; the alcohol evaporates, giving rise to the wonderful aroma of baking bread.

glycolysis (NAD^+; review Figure 5.5). For people, however, the waste products of yeast's alcoholic fermentation are highly useful. Bakers add yeast to flour, salt, and a bit of sugar, and once inside the anaerobic environment of a big wad of dough, the yeast begins to break down the starch in flour into glucose units and to ferment the glucose. The bubbles of excreted carbon dioxide gas cause the dough to expand, and the small amount of alcohol released into the dough evaporates during baking, contributing to the homey aroma of baking bread (Figure 5.9b).

Yeasts occur naturally on grape skins, but vintners usually add special strains of the yeast *Saccharomyces cerevisiae* to grape juice, seal the "must" (as the combination is called) into airless tanks for weeks, and allow the yeast to ferment the juice's natural "grape sugar," or glucose. The alcohol level eventually reaches about 12 percent of the wine's volume—a level so high that most of the yeast is killed off.

Producers of sauerkraut and pickled cucumbers, beets, tomatoes, and other such foods employ several kinds of fermenting yeasts and bacteria whose populations rise and fall in the pickling crocks as the fermentation process proceeds. Many of these organisms, however, produce lactic acid rather than alcohol as a by-product.

LACTIC ACID FERMENTATION

Some muscle cells in people and other animals, as well as the microorganisms that create yogurt, cheeses, and

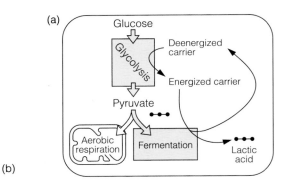

FIGURE 5.10 *Lactic Acid Fermentation.* (a) In the absence of oxygen, certain bacteria and other microorganisms, as well as certain types of muscle cells, break down pyruvate from glycolysis into lactic acid. The reaction also regenerates NAD^+ from NADH. (b) In a *fromagerie* (cheese factory), lactic-acid-generating bacteria help produce the cheddar cheese shown here.

soy sauce, carry out lactic acid fermentation under anaerobic conditions. In these cells, the pyruvate generated during glycolysis is converted to the three-carbon compound lactic acid (Figure 5.10a). As in alcoholic fermentation, there is no additional ATP harvest, but the regeneration of the electron carrier (NAD^+) allows glycolysis to proceed with its net energy gain.

Like alcohol and carbon dioxide, lactic acid is a useless waste product to the anaerobic cell, but people in the dairy industry deliberately culture lactic-acid-generating bacteria in milk or cream to give these bland fluids the sour taste of yogurt or the sharp taste of cheese (Figure 5.10b). Soy sauce—the salt and pepper of oriental cooking—is made by inoculating soybeans with a fungus, then fermenting them with three kinds of lactic acid yeasts and bacteria for nearly a year.

Ironically, lactic acid can be a bane as well as a boon, because it sometimes forms inside animal muscle cells and leads to cramping. A vigorously exercising mammal—let's say a hiker pushing quickly up a steep slope, a sprinter straining to reach the tape, or a cyclist pedaling a human-propelled airplane (Figure 5.11)—may develop a leg cramp because the muscle cells are not getting sufficient oxygen. With oxygen in short supply, the simple compound pyruvate, formed during glycolysis, cannot be further broken down via aerobic respiration, and as a result, fermentation begins in the muscle cells. The waste product lactic acid is formed, and if it builds up, it can cause a painful leg cramp or a "stitch in the side" that goes away only when the exercise lessens and the flow of oxygen to the tissues is once again sufficient to sustain aerobic respiration. One reason that exercise enthusiasts now emphasize *aerobic* exercise—exercise at a mild enough pace so that oxygen delivery outstrips oxygen use—is that it prevents muscle cramping.

FINAL ENERGY TALLY

There are just two important points to remember about fermentation. First, it yields no ATP; and second, it allows cells to function with little or no oxygen by recycling an electron carrier needed for glycolysis. Figure 5.12 shows the final energy tally for glycolysis plus fermentation. Together the two pathways fulfill the energy needs of anaerobic cells.

≋ AEROBIC RESPIRATION: THE BIG ENERGY HARVEST

While muscle cells and yeasts have the ability to metabolize sugars whether oxygen is available or not (at least for a while), the vast majority of living things are made

(a)

(b)

FIGURE 5.11 *Lactic Acid Can Build Up in Muscles.* (a) A fast sprinter can run 100 m in close to 9 seconds. Without proper training, however, the runner may develop leg cramps from lactic acid accumulation in the muscles and be unable to complete the race. (b) Amazing as it may sound, a person can generate enough muscle energy to propel a light aircraft. However, if a person pedaling a self-propelled airplane (like the Gossamer Albatross pictured here) developed a leg cramp from pedaling too fast, he or she could be forced to crash-land. Insufficient oxygen supply to the active muscles causes a buildup of lactic acid and leg cramps, serving as a warning to slow down. But lactic acid fermentation does provide a small amount of energy, which allows an oxygen-deprived muscle to continue working for a short time—in this case, perhaps long enough to speed up pedaling just a bit and prevent a crash.

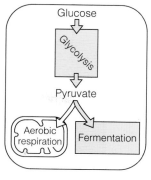

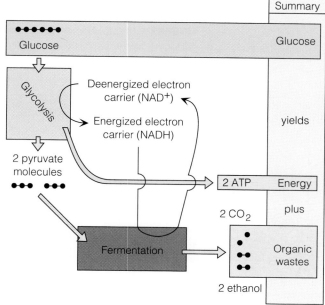

FIGURE 5.12 *Anaerobic Metabolism: A Final Energy Tally.* Some cells can survive where oxygen is scarce. In those cells, the pyruvate molecules produced by glycolysis are shunted through the fermentation pathway, where they are broken down to organic wastes; in the process, the energy carrier NADH is recycled, but no new ATP forms. Thus, in anaerobic metabolism, fermentation enables glycolysis to continue its harvest of two ATP molecules per molecule of glucose.

up of cells that require oxygen for their major metabolic pathway. This oxygen-requiring pathway is known as *aerobic respiration.* Aerobic respiration shunts the products of glycolysis through the Krebs cycle, which harvests some energy, and then through the electron transport chain, which harvests substantial quantities of energy. Being 18 times more efficient than glycolysis alone, this extended pathway of aerobic respiration can provide the huge amounts of ATP an active cell needs, and its superiority is clearly demonstrated: Mutant yeast cells that must get by solely on glycolysis and fermentation grow two to three times more slowly than their aerobic counterparts. And people with defective aerobic pathways suffer from chronic fatigue and extremely painful muscles. A case in point is the story of a young woman called

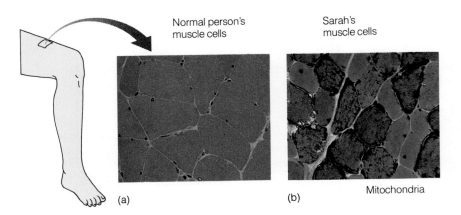

Normal person's muscle cells   Sarah's muscle cells

(a)   (b)   Mitochondria

FIGURE 5.13   (a) Normal muscle fibers. (b) Sarah's muscle fibers with masses of aberrant mitochondria (red) circling the fibers.

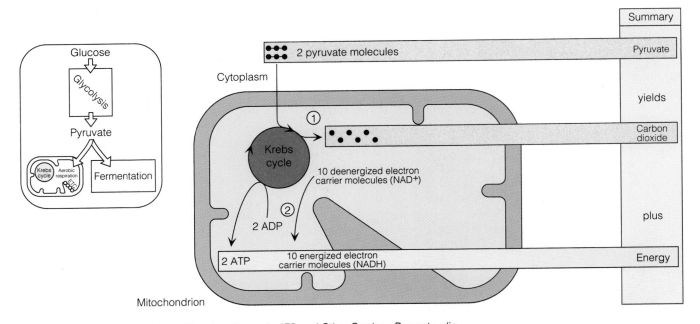

FIGURE 5.14   *The Krebs Cycle: Trapping Energy in ATP and Other Carriers.* Pyruvate, dissolved in the cytoplasm as a product of glycolysis, enters the mitochondrion. During the Krebs cycle and a reaction that precedes it, carbons originally in pyruvate molecules are split off to form carbon dioxide (1). As the Krebs cycle turns, energy is trapped in electron carriers such as NADH, and ATP is generated (2). Thus, the Krebs cycle completes the degradation of glucose to carbon dioxide initiated during glycolysis, and energy becomes stored in electron carriers.

Sarah, who was born with an altered protein in her mitochondria. Because of the mutant protein, her cells' electron transport chain—the muscles' major energy provider—could not make ATP. Instead, Sarah's muscles generated huge numbers of mitochondria but still had to rely solely on the small ATP harvest provided by glycolysis and the fermentation that accompanies it (Figure 5.13). As a result, she was confined to a wheelchair. Sarah's doctors, once they understood her problem, were able to develop a vitamin therapy that corrected it.

The reason for the greater efficiency of aerobic respiration is that the initial carbon-containing glucose molecule is broken down completely to carbon dioxide, which is not rich in energy. Thus, most of the energy residing in the molecular bonds of the sugar is released, and a large proportion of it is stored as ATP. In other words:

Glucose + Oxygen →
    Carbon dioxide + Water + Lots of ATP energy

THE KREBS CYCLE: METABOLIC CLEARINGHOUSE

As in fermentation, the two pyruvate molecules produced from each glucose during glycolysis provide the raw material for the Krebs cycle. Unlike fermentation, however, the important events of the Krebs cycle take place not in the cytoplasm but in the mitochondria of eukaryotic cells (see Chapter 3, page 62) or in the plasma membranes of bacteria. The pyruvate molecules enter the mitochondrion and are acted upon during the reaction steps of the Krebs cycle. Here, the three carbons are cleaved apart, yielding carbon dioxide (Figure 5.14). When you exhale, you are breathing out the carbon dioxide gas that originates in this dismantling of pyruvate. In a plant's cells, the carbon dioxide released during the Krebs cycle can immediately be trapped again by photosynthesis, and more glucose can be generated. The energy released during the dismantling of pyruvate is trapped in ten energized electron carriers and two ATP molecules for each molecule of glucose that initially enters glycolysis. This yield of two ATPs from the Krebs cycle provides no greater energy harvest than during glycolysis. However, the ten charged electron carriers represent a treasure trove of stored energy that will be released, bit by bit, by the components of the electron transport chain.

The Krebs cycle is an important intermediate phase in aerobic respiration for dismantling carbons and storing energy. But what makes it a "metabolic clearinghouse"? The answer is that organic nutrients other than glucose can enter the aerobic breakdown pathway at the Krebs cycle (Figure 5.15). If the supply of sugar falls, an animal or plant cell can begin to break lipids or proteins down into component parts. Some of these subunits can then be converted into pyruvate or other substances in the Krebs cycle and eventually dismantled for energy harvest. For example, after about 30 minutes of continuous vigorous exercise, your muscles and liver become low in glycogen (see Figure 2.20), the storage form of glucose, and you begin to "burn fat." This means that your cells are breaking down lipids and converting them to pyruvate and the two-carbon compound that enters the Krebs cycle (see Figure 5.15). The cycle then fuels the production of ATP by aerobic respiration. The large supply of energy stored in fat can generate enough ATP to keep your muscle cells well supplied for hours of exercise.

The clearinghouse concept extends still further, because raw materials from the Krebs cycle can also be shunted *out* of the pathway, to be converted and then used as building blocks for the biosynthesis of new fats, proteins, and carbohydrates. The Krebs cycle is, in a sense, like a foreign exchange bank: Just as one can change dollars into francs, francs into pesos, or pesos into pounds, so can the Krebs cycle turn fats into ATP energy or turn the components of sugar metabolism into proteins.

An important exception to this clearinghouse principle is the human brain cell. For complex reasons, it can use only glucose as fuel, a fact that has major implications. First, it explains why sugary foods are mood elevators. Soon after a person consumes a candy bar or soft drink, glucose subunits are cleaved from the sucrose and enter the bloodstream and brain. Infused with their basic fuel, the brain cells can function at peak efficiency, leading one to feel happier, smarter, and livelier—at least for a time. Second, this quirk of the brain cell explains why dieters are warned to consume at the very least 500 calories per day and to avoid liquid or powdered protein diets: The brain alone needs at least that many calories of glucose for normal functioning, and without it, a person can grow faint and even lapse into unconsciousness.

THE ELECTRON TRANSPORT CHAIN: AN ENERGY BUCKET BRIGADE

The final stage of aerobic respiration is really the crux: Here is where most of the energy of glucose is turned into ATP; where oxygen comes into play, making this respiration *aerobic*; and where we can clearly see why mitochondrial structure is so crucial to the survival of complex organisms.

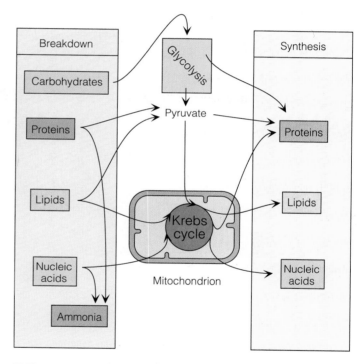

FIGURE 5.15 *The Krebs Cycle: A Metabolic Clearinghouse.* While only certain molecules can feed directly into the Krebs cycle (such as a specific derivative of pyruvate), biological polymers—proteins, nucleic acids, lipids, and polysaccharides—can themselves be broken down and their parts modified into intermediates that feed into the cycle. In this way, an organism can harvest energy not just from glucose but from any of the biological polymers. In addition, Krebs cycle intermediates can be removed from the cycle and modified into new materials for the cell, including amino acids for proteins and fatty acids for lipids.

The electron transport chain is both a series of structures and a series of events that are intertwined. To follow these structures and events, we must focus on a close-up view of the mitochondrion (Figure 5.16). Notice that the inner mitochondrial membrane separates two

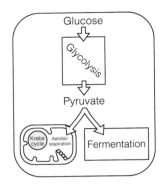

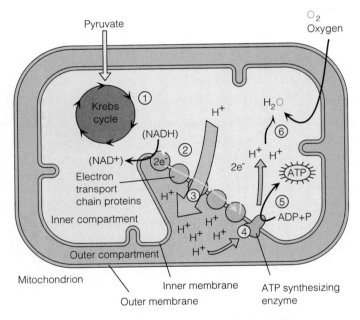

FIGURE 5.16 *Mitochondrion: Site of Krebs Cycle and Electron Transport Chain.* This diagram shows where the phases of aerobic respiration take place in the mitochondrion. Pyruvate generated during glycolysis in the cytoplasm is transported through the outer and inner membranes to the inner mitochondrial compartment, where the Krebs cycle takes place, producing energized carriers (1). Electrons from the energized carriers are passed down the electron transport chain (2), and proteins use this released energy to pump hydrogen ions across the inner membrane to the outer compartment (3). Then the hydrogen ions flow through a special channel protein back across the inner mitochondrial membrane (4), and this flow drives the process that traps energy in ATP much as a turbine captures the energy of water flowing over a dam and converts it to electricity (5). Finally, the last protein in the transport chain passes electrons with accompanying hydrogen ions to oxygen, forming water (6).

compartments within the organelle—the inner compartment and the outer compartment. The Krebs cycle occurs within the inner compartment (Figure 5.16, step 1). Embedded within the inner membrane, however, are sets of complex proteins that protrude from and sometimes span the entire membrane (step 2). These proteins are capable of extracting energy from the electron carriers created during the preceding phase, the Krebs cycle.

Electrons are passed from the energized electron carriers (step 1) down an electron bucket brigade (step 2). During this brigade, small amounts of energy are released with each downward step. The proteins use the small amounts of energy to pump hydrogen ions from the inner to the outer compartment of the mitochondrion (step 3). The accumulation of hydrogen ions in the mitochondrion's outer compartment represents potential energy, like water stored behind a dam. As the hydrogen ions flow back across the membrane into the inner compartment through a special channel protein (step 4), they drive the formation of ATP (step 5), much as the flow of water over a dam turns a turbine that collects energy in the form of electricity. Finally, the last protein in the brigade passes electrons and hydrogen ions to oxygen, forming water (step 6). Oxygen thus serves as the final acceptor of the electrons and hydrogen ions for the entire aerobic respiration pathway, explaining why it is called "aerobic." If oxygen is not available to accept the electrons and hydrogen ions once they have passed down the electron transport chain, the entire process quickly stops, and the organism or cell, if strictly aerobic, dies. That's how cyanide can kill a person—say, a captured spy in an old grade B movie: It binds tightly to a protein in the electron transport chain and prevents it from accepting the final electron transfer. The result is that ATP formation halts, and brain cells quickly starve for energy and die.

In summary, the electron transport chain takes electrons from the electron carriers generated in glycolysis and the Krebs cycle, uses the electrons' energy to produce ATP molecules, and finally transfers the electrons to oxygen to produce water. For each set of electron carriers that comes from a single glucose molecule, 32 ATPs can form during the electron transport phase—a whopping harvest of nearly 90 percent of the ATP currency "earned" during the breakdown of a glucose molecule via aerobic respiration.

The final tally for the entire three-part aerobic pathway is 36 ATPs per glucose molecule, as itemized in Figure 5.17. Despite the uncertainties of detail that remain, biologists are quite sure that the unique internal structure of the mitochondrion makes possible the high efficiency of aerobic respiration in eukaryotes. Figure 5.18 (page 102) shows a plant with a recently discovered and highly unusual level of aerobic respiration—a "fever" involved in reproduction and survival. In contrast, any impair-

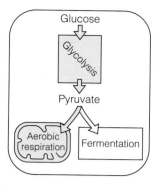

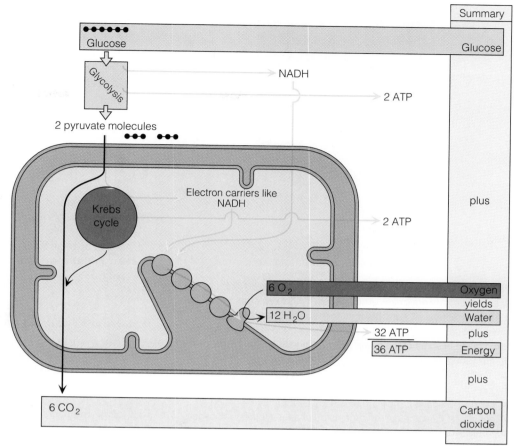

FIGURE 5.17 *Aerobic Respiration: A Final Energy Tally.* Aerobic respiration breaks down the products of glycolysis via the Krebs cycle and the electron transport chain. For every glucose molecule broken down through aerobic respiration, two ATP molecules are harvested during the glycolysis phase, and two more ATPs are gained during the Krebs cycle. The Krebs cycle also produces a set of electron carrier molecules, and these pass their electrons down the electron transport chain to generate 32 more ATP molecules. Altogether, during aerobic respiration, 36 ATPs are harvested per glucose molecule, and 6 carbon dioxide and 6 water molecules are released as inorganic waste products.

ment of mitochondrial function can have severe consequences for health and vitality (see Figure 5.13).

≋ THE CONTROL OF METABOLISM

We have seen that cells possess marvelous metabolic mechanisms—biochemical pathways that accomplish three major processes in the cell: breaking down organic nutrients; harvesting ATP energy; and building needed proteins, lipids, carbohydrates, or other biological molecules by shunting substances into and out of the Krebs cycle clearinghouse. But the intricacies of metabolism raise new questions: When glucose runs out, what makes a cell begin burning its stores of lipids or proteins? When the supplies of glucose return, what then stops the cell from burning all its lipids or proteins—literally eating itself up from the inside and thereby destroying its cellular structures? And finally, what triggers a cell to build molecules when, and only when, they are needed? Clearly, there must be a great deal of internal coordination and control over the cell's harvesting of energy and building of needed materials. But what form does it take?

Research has shown that metabolic processes inside cells can be controlled through the regulation of enzyme activity. The enzymes, in this case, are ones that perform certain steps in metabolism. For example, when a cell has plenty of ATP, a key enzyme in glycolysis stops working, and when the enzyme stops, the whole process stops. But when the level of ATP falls, the key enzyme resumes its function, glycolysis starts again, and more ATP is produced within the cell. Through these kinds of control, your body's metabolic enzymes are turned on or off so that your muscles can burn glucose when it is available (early in a hike or during a 100 m dash), but burn lipids or even proteins when glucose is lacking (after several hours of climbing or running a marathon). The control systems within cells are truly elegant—efficient and highly conserved through eons of evolution—and they occur in the smallest bacteria, the largest muscle cells, and even the baker's yeast sealed in packets on the grocer's shelf. The similarity of control mechanisms reminds us that all types of cells are related in a fundamental way by their metabolism. Glycolysis, the Krebs cycle, and the electron transport chain proceed in the same way in a yeast cell and in the baker using that yeast cell to make bread. The

FIGURE 5.18 *A Burst of Mitochondrial Activity May Increase a Flower's Chances of Fertilization.* The voodoo lily (*Sauromatum guttatum*), a native of Southeast Asia, belongs to the same plant family as the skunk cabbage. The day the flower opens, a burst of metabolism takes place in some of the flower cells—a burst as high as that of a hummingbird and that can raise the temperature inside the flower by 22° C (72° F). This heat causes odorous molecules to evaporate, waft through the air, and attract insects that may pollinate the flower. This is an unusual example of how the efficiency of aerobic respiration in mitochondria can increase an organism's chances for reproduction and thus promote its long-term genetic survival.

control and regulation of these identical processes simply ensure that order rather than metabolic chaos reigns within the organism.

≋ CONNECTIONS

Energy metabolism lies at the heart of biology because without energy for maintenance, growth, and reproduction, an organism would quickly fall prey to disorganization and death. Knowledge of how our cells harvest energy through either the anaerobic pathway of glycolysis and fermentation or the oxygen-dependent pathway of glycolysis and aerobic respiration helps us understand why we feel a burning sensation in our muscles after a short period of intense exercise, why we can sustain a moderate amount of exercise for many hours, and why only that sustained aerobic exercise can burn away fat. The concepts of energy metabolism also help us understand other subjects we study in biology. For example,

our lungs act like bellows to draw in oxygen and expel carbon dioxide, and our stomach and intestines extract raw materials from foods, making them available for metabolic reactions in our cells. Finally, an understanding of energy metabolism helps explain why our lives ultimately depend on green plants: Plants are able to convert solar energy to chemical energy and store it in nutrient molecules like glucose. The next chapter explores how plants perform this crucial function.

NEW TERMS

aerobic respiration, page 92
anaerobic metabolism, page 92
autotroph, page 92
electron transport chain, page 93

fermentation, page 93
glycolysis, page 92
heterotroph, page 92
Krebs cycle, page 93

STUDY QUESTIONS

REVIEW WHAT YOU HAVE LEARNED

1. Prepare a table that compares fermentation with aerobic respiration. Use these headings: presence or absence of oxygen; amount of ATP produced; waste products.
2. State the basic difference between a heterotroph and an autotroph, and give an example of each.
3. What are two results of glycolysis?
4. Why is aerobic respiration much more efficient than fermentation?
5. If you eat a meal that contains only protein and no glucose or other carbohydrate, how can you make ATP? Which of the following would your body use: glycolysis; fermentation; the Krebs cycle; the electron transport chain?
6. The electron transport chain provides the final step in aerobic respiration. What exactly is the electron transport chain?
7. What role does each of the following play in the electron transport chain: a mitochondrion; transport proteins; oxygen; electron carriers; hydrogen ions?
8. Give an example of how cells control metabolism.

APPLY WHAT YOU HAVE LEARNED

1. Some bacteria are harmful, causing diseases, but others are useful. How?
2. Explain how daily swimming helps some people lose weight.
3. Explain why a marathon runner may develop leg cramps.

FOR FURTHER READING

Newsholme, E., and T. Leech. "Fatigue Stops Play." *New Scientist*, 22 September 1988, pp. 39–43.

1. All organisms get the energy they need to maintain life from glucose or other nutrient molecules.

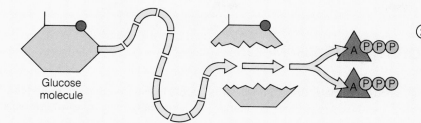

Glucose
molecule

2. An organism's cells release the energy stored in nutrient molecules through stepwise metabolic pathways. Chemical reactions in the pathways trap the released energy in ATP and other energy carriers.

3. Glycolysis is the first stage of all cellular energy harvest. For every molecule of glucose that is split, two molecules of ATP form.

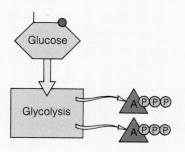

Glucose

Glycolysis

4. In low-oxygen environments, certain cells can survive on glycolysis plus fermentation. Fermentation recycles an energy carrier needed for glycolysis to continue.

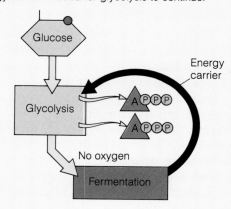

Glucose

Energy carrier

Glycolysis

No oxygen

Fermentation

5. In an oxygen-rich environment, most cells use glycolysis plus aerobic respiration. Such cells can harvest 36 molecules of ATP per molecule of glucose.

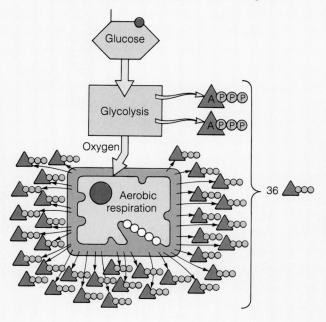

Glucose

Glycolysis

Oxygen

Aerobic respiration

36

6. Metabolism is controlled so that cells build and break down compounds when they are needed.

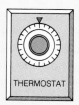

THERMOSTAT

7. Active muscle cells can harvest energy through aerobic respiration or, when oxygen becomes scarce, through fermentation.

TRAPPING SUNLIGHT AND BUILDING NUTRIENTS

THE PHOTOSYNTHETIC CHAMPION

If you've ever traveled through the midwestern United States during the summer, you know why it is called the Corn belt. And you will probably not be surprised to learn that this region's gently rolling inland sea of corn plants represents our nation's largest crop. American farmers harvest 7 billion bushels of corn each year from 70 million planted acres (Figure 6.1). Globally, only rice and wheat are bigger food sources for humans and farm animals.

The virtues of corn are well known: It is the most efficient of all the major grain crops at converting carbon dioxide and sunlight into the organic nutrients that support virtually all life on earth. Photosynthesis is the name given to this conversion, and corn can carry it out even on a hot, dry day when in many other plants, photosynthesis stops. Corn is, in short, a photosynthetic champion.

Photosynthesis is the metabolic process by which plants trap solar energy, convert it to chemical energy, and store it in the bonds of organic nutrient molecules such as glucose. Chapter 5 discussed the metabolic pathways that break down glucose to release energy, and here we find out where the glucose comes from in the first place. Nearly all types of plants and algae, as well as some protists and bacteria, are capable of photosynthesis.

Photosynthesis may be the most important metabolic pathway to familiar life forms, since it generates the nutrients that virtually all cells break down via the energy pathways we discussed in Chapter 5. Once the photosynthetic organisms we call autotrophs use light energy to generate sugars and other organic nutrients, they then break them down again for their own cellular

energy needs. Animals, fungi, and most microbes (the three groups we call heterotrophs), of course, use autotrophs (or organisms that eat autotrophs) as food.

Photosynthesis is more or less the chemical opposite of the energy breakdown pathways (see Chapter 5): Photosynthesis uses carbon dioxide, water, and energy to build glucose, instead of breaking down glucose to carbon dioxide and water and in the process releasing energy. Nevertheless, photosynthesis resembles the other energy pathways in that it involves a series of reaction steps. Thus, all the metabolic pathways that capture and release energy are related, and many of the con-

cepts in this chapter will seem familiar.

Our study of photosynthesis asks and answers a series of questions:

- What are the major chemical events of photosynthesis, and where do they occur?
- What is light energy, and how do special pigments trap it?
- What are the specific light-trapping and sugar-building steps of photosynthesis?
- What process gives plants like corn particularly efficient photosynthesis?
- How do carbon atoms cycle from autotrophs to heterotrophs in a massive global exchange?

FIGURE 6.1 *A Photosynthetic Champion.* Corn provides one-quarter of all the calories people and farm animals consume. The U.S. Corn belt includes rural Illinois, Indiana, Iowa, Minnesota, Missouri, Nebraska, Ohio, and South Dakota.

AN OVERVIEW OF PHOTOSYNTHESIS

There is a beautiful symmetry to the cellular processes of respiration and photosynthesis. This symmetry is revealed by their nearly opposite overall equations (Figure 6.2). In Chapter 5, we saw that when oxygen is present, aerobic respiration allows cells to break down glucose into carbon dioxide and water and release chemical energy.

$$\text{Glucose} + \text{Oxygen} \rightarrow \text{Carbon dioxide} + \text{Water} + \text{Energy}$$

In photosynthesis, nearly the reverse takes place in the chloroplast. Light energy is trapped, transformed, and then used to convert carbon dioxide and water into glucose and oxygen:

$$\text{Carbon dioxide} + \text{Water} + \text{Light energy} \rightarrow \text{Glucose} + \text{Oxygen} + H_2O$$

Clearly, living things must have both a source of energy as well as a means of releasing it, and for green plants and most other autotrophs, the direct energy source is sunlight.

How does a plant trap sunlight and use it in building sugar? Recall that in our discussion of harvesting energy from glucose, we followed the flow of electrons. To understand how photosynthesis captures energy and

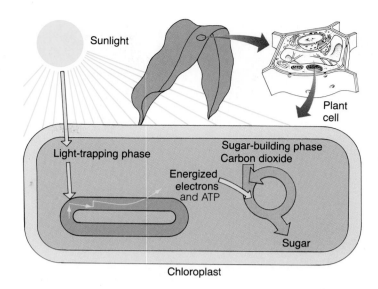

FIGURE 6.3 *Overview of the Light-Trapping and Sugar-Building Phases of Photosynthesis.* Photosynthesis takes place in green leaves, stems, and other structures within cells containing chloroplasts. Both light-trapping and sugar-building phases occur in the chloroplast. During the light-trapping phase, light energy boosts electrons to higher energy levels, and this energy is stored in chemical form in energized electrons. During the sugar-building phase, the electrons supply energy, and carbon dioxide is converted to sugar.

stores it in glucose, we must again follow the flow of electrons, this time through the two phases of photosynthesis: the "light-trapping" phase, when light energy boosts the energy level of electrons, and the "sugar-building" phase, when the energy from the electrons powers the manufacture of glucose (Figure 6.3). Biologists call the light-trapping phase of photosynthesis the **light-dependent reactions** because they require sunlight, and the sugar-building phase the **light-independent reactions** because they can proceed whether or not light is present. Since both of these phases of photosynthesis occur in the chloroplasts of plant cells, we next consider the architecture of the chloroplast and how this structure allows its unique functioning.

THE CHLOROPLAST: SOLAR CELL AND SUGAR FACTORY

In a plant such as corn, both the light-trapping and sugar-building reactions of photosynthesis occur in the *chloroplasts*. Each corn leaf cell may contain 40 to 50 chloroplasts, and each square millimeter of leaf surface more than 500,000 of the organelles. Chloroplasts are analogous to mitochondria in structure and function, but while mitochondria are "powerhouses" that burn sugar and generate ATP, chloroplasts are more a combination of

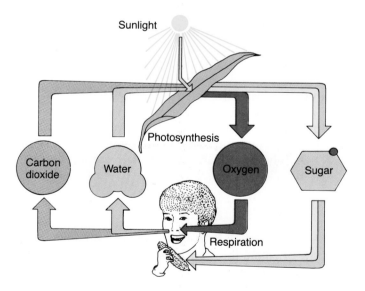

FIGURE 6.2 *Metabolic Symmetry: Photosynthesis and Aerobic Respiration Are Opposites.* Photosynthesis takes place in the cells of green leaves; it uses carbon dioxide and water, generates oxygen, and traps and stores solar energy in the chemical bonds of sugar molecules. In contrast, aerobic respiration occurs in many types of cells; it uses oxygen, expels carbon dioxide and water, and liberates energy from the bonds of organic molecules.

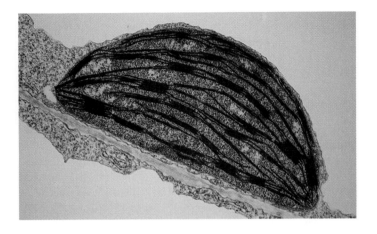

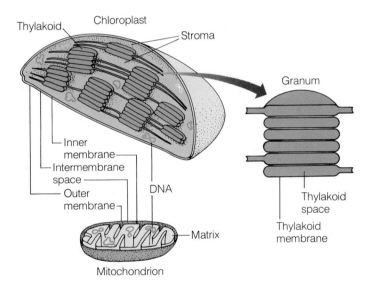

Thylakoid · Chloroplast

Stroma

Granum

Inner membrane

Intermembrane space

Outer membrane

DNA

Thylakoid space

Thylakoid membrane

Matrix

Mitochondrion

FIGURE 6.4 *Chloroplast Structure.* Chloroplasts (here magnified 25,000 times) are membrane-bounded organelles with outer and inner membranes and an intermembrane space between them, as well as a third set of membranes forming stacks of disklike sacs called thylakoids. Each thylakoid has its own membrane and internal space, plus a stroma, or matrix, surrounding the thylakoid stacks. The structure of a chloroplast resembles that of a mitochondrion, as the labels suggest.

solar cell and sugar factory that captures sunlight and generates glucose and other carbohydrates.

Chloroplasts have smooth outer and inner membranes lying side by side and collectively enclosing a space filled with watery solution, the *stroma* (Figure 6.4). A third, very thin membrane lies inside the stroma and forms a system of stacked, disklike sacs, the *thylakoids*. Chlorophyll and other colored pigments are embedded in this thylakoid membrane. The light-trapping reactions of photosynthesis take place in the thylakoid and provide the chloroplast's solar cell activity. The sugar-building reactions begin in the stroma and continue outside the chloroplast in the cytoplasm, and these reactions collectively act as the sugar factory. After sugars are produced through photosynthesis, they can be linked into starch molecules, or used as subunits in the cellulose of the plant cell wall, or broken down in the plant cell's own cytoplasm and mitochondria to power cellular activity. Ultimately, light energy is the power source for the photosynthesis in plants that builds the sugars that in turn empower your cells and those of nearly all organisms on earth. So before proceeding further, let's take a look at the nature of light and solar energy.

☷ COLORED PIGMENTS IN LIVING CELLS TRAP LIGHT

Light travels so quickly and the photosynthetic process takes place so rapidly that you can practically eat sunlight in a fresh-picked leaf: If you pluck and immediately chew a growing lettuce leaf, you are consuming energy that left the sun just 8 minutes earlier and was converted to the chemical energy in carbohydrate molecules almost instantly upon striking the plant. Photosynthesis is not only rapid, but also amazingly efficient. Researchers recently discovered a red alga growing some 884 ft below the ocean's surface and carrying out photosynthesis with only five-millionths of the peak surface sunlight.

LIGHT, CHLOROPHYLL, AND OTHER PIGMENTS

Visible light—the white sunlight one can separate into a rainbow of colors through a prism—is just a small part of the *electromagnetic spectrum,* which is the full range of electromagnetic radiation in the universe, from highly energetic gamma rays to very low-energy radio waves (Figure 6.5). All such radiation travels through space behaving both as vibrating particles (called **photons**) and as waves. The photons of visible light have wavelengths in the narrow range of 380 nm (violet light) to 750 nm (red light). These lengths are about $\frac{1}{40}$ the thickness of

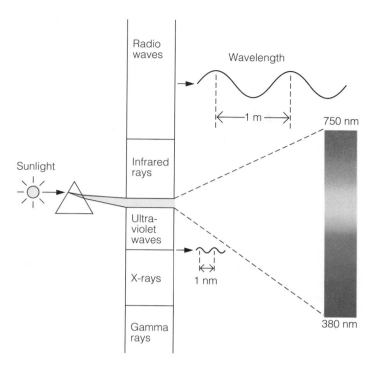

FIGURE 6.5 *Visible Light and the Electromagnetic Spectrum.* We are constantly bathed by electromagnetic energy, from radio waves to infrared rays (heat), X-rays, and gamma rays. Each category has a range of energies measured in oscillating waves of specific lengths (wavelengths). Only the small portion of the entire electromagnetic spectrum with wavelengths in the range of 380–750 nm is visible to us as white light that can be broken into colored light, each color with different wavelengths.

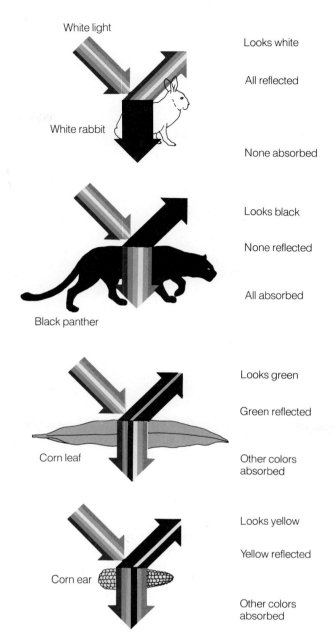

FIGURE 6.6 *Pigments Absorb and Reflect Light.* The colors we see in a given object depend on which wavelengths of light the pigments in that object absorb and which are reflected back to our eyes. A white rabbit looks white because all wavelengths are reflected. Black pigment in a panther's coat absorbs light throughout the whole color spectrum and reflects nothing; thus, the panther looks black to us. A corn leaf contains chlorophyll, which absorbs most strongly around 450 and 670 nm, while green light in the 500 nm range is reflected, and thus the leaf looks green. The peelings of fruit and corn ears often contain carotenoids, which absorb wavelengths from 400 to about 550 nm and reflect back red, yellow, and orange light. Because a leaf contains both chlorophyll and carotenoids, its pigments can absorb and use for photosynthesis most of the wavelengths of visible light that strike it.

the paper in this book. It is no coincidence that living things can absorb and use light within this restricted wavelength range. Gamma rays, with their shorter wavelengths, are so energetic that they disrupt and destroy the biological molecules they strike, while longer rays, like radio waves, are so low in energy that they cannot excite biological molecules.

Objects like red apples, black panthers, or green leaves look colored because they absorb and reflect light. Specialized cells in our eyes absorb the reflected light and send signals to the brain that enable us to "see" it. An apple looks red because light-absorbing *pigment* molecules in its skin cells absorb light from various parts of the visible spectrum and reflect only red light. A panther looks black because dark pigments in the animal's coat absorb all the wavelengths of light that strike them, reflecting none. A corn leaf looks green because its green **chlorophyll** pigments absorb violet, blue, yellow, and red light and reflect only green (Figure 6.6). But there is an essential difference between chlorophyll molecules and the pigments in a panther's coat: Chlorophyll participates in photosynthesis and gives a green leaf color.

In all photosynthetic organisms, chlorophyll is accompanied by *carotenoid pigments,* which absorb green, blue,

FIGURE 6.7 *Autumn Glory.* When chlorophyll breaks down in autumn, carotenoids can shine through, painting the forest with orange, gold, and scarlet hues. This shot was taken in northern Pennsylvania.

and violet wavelengths and reflect red, yellow, and orange light. Carotenoids are generally masked by chlorophyll and thus tend to be unnoticed in green leaves. However, they give bright and obvious color to many nonphotosynthetic plant structures, such as roots (carrots), flowers (daffodils), fruits (tomatoes), and seeds (corn kernels) (see Figure 6.6). And carotenoids are behind the glorious colors of autumn. As summer ends and the nights grow cool, chlorophyll begins to break down, allowing the gold and red carotenoids to show through and emblazon maple, oak, sumac, and other deciduous trees and vines (Figure 6.7). Since chlorophyll absorbs all but the green wavelengths, and carotenoids all but the red and yellow wavelengths, the sets of colored pigments can together absorb most of the available energy in visible light. Because they are the keys to photosynthesis, the light-trapping pigments are arguably the most important biological molecules.

PIGMENT COMPLEXES AND ENERGY CAPTURE

Leaves literally have light-receiving antennae—assemblies of pigments embedded in the thylakoid membrane of each chloroplast in every photosynthetic cell. These clusters of about 200 chlorophyll and other pigment molecules act together to harvest light energy. When the pigments are struck by light, they pass the energy they absorb (in the form of energized electrons) from one molecule to the next; you can imagine the process as Ping-Pong balls bouncing off a hard surface. Eventually the energized electrons reach a special chlorophyll molecule at the center of the antenna (the so-called *reaction center chlorophyll*) (Figure 6.8). The complex of light-trapping pigments, associated with specific proteins and electron acceptors in the thylakoid membrane, is called a **photosystem.** The actual conversion of light energy to chemical energy takes place within a photosystem, and once light energy has successfully been converted to chemical energy, the remaining steps of photosynthesis merely shuffle about the chemical energy trapped in the electron acceptors, eventually storing it in the bonds of carbohydrate molecules.

≋ THE LIGHT-TRAPPING PHASE OF PHOTOSYNTHESIS

When sunlight shines on a corn leaf, structures and processes inside the plant convert solar energy to biological energy. More specifically, during the light-trapping phase of photosynthesis, energy from sunlight is captured and used in reactions that split water, release oxygen, and form energized electron carriers. These reactions take place in the thylakoid membrane and in the compartment inside the thylakoid disc, and they involve two types of photosystems (called I and II) with their chlorophyll molecules, antenna complexes, and associated electron acceptors. The two types of photosystems lie embedded near each other in the membrane, and electrons are passed from one to the other.

Figure 6.9 outlines the steps of the light-trapping phase. When sunlight strikes green chlorophyll in the chloroplast of a corn leaf, some of the solar energy becomes trapped as it boosts electrons in the pigment molecules to higher energy levels (Figure 6.9, step 1). Then, as the electrons drop back toward their original energy levels, their trapped solar energy is stored in the chemical bonds of high-energy carriers such as ATP (step 2). Interestingly, this step in photosynthesis involves electron transport and hydrogen ion flow just like the ATP-generating system in mitochondria (review Figure 5.16). This similarity suggests an ancient evolution of a single ATP-generating system. During this process, a water molecule is split, releasing oxygen (step 3). Plant scientists usually draw the major pathway traveled by the electrons in photosynthesis as a zig-zag shape, and for this reason, they call it the Z scheme. This water-splitting process is

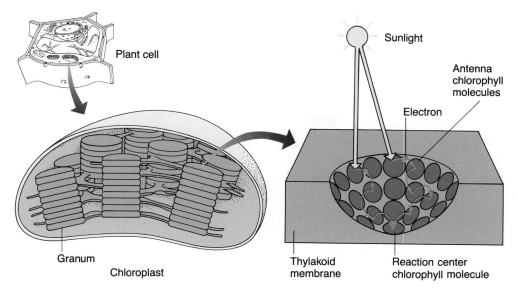

FIGURE 6.8 *Antenna Complex: Pigment Molecules Absorb Light Energy.* In the chloroplasts inside a plant's photosynthetic cells, the thylakoid membranes contain *antenna complexes*— clusters of light-absorbing pigment molecules. These clusters channel the energy from impinging light to a central chlorophyll molecule, the reaction center chlorophyll.

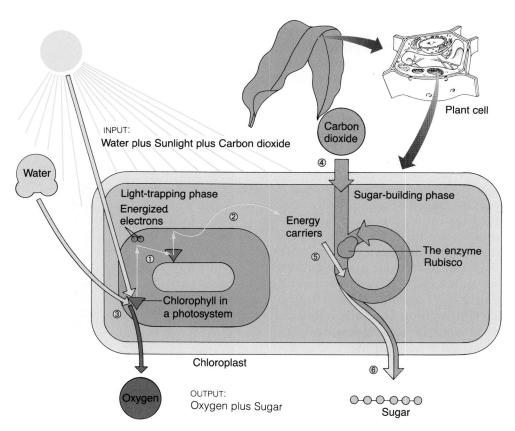

FIGURE 6.9 *In Photosynthesizing Cells, Water Plus Sunlight Plus Carbon Dioxide Yield Sugar and Oxygen.* In the light-trapping phase of photosynthesis, light energy absorbed by chlorophyll pigment molecules in a chloroplast boosts electrons to a higher energy state (1). Energy from these electrons is eventually trapped in energy carriers (2). The electrons boosted from the chlorophyll are replaced by electrons donated by the splitting of water, and oxygen is released (3). During the sugar-building phase of photosynthesis, carbon dioxide (4) is fixed into biological molecules. Hydrogen and energy from the energy carriers (5) allow sugar to form (6).

the reason plants give off oxygen to the environment and one of the reasons they need water to carry out photosynthesis and grow.

The oxygen plants produce in the light-driven, water-splitting process just described is the source of all the oxygen people, other animals, plants, and all other aerobic organisms utilize in aerobic respiration and the harvesting of energy from sugar molecules. Nevertheless, this oxygen is simply a by-product of the light-trapping phase of photosynthesis. The main product is energized electron carriers. It is these substances that fuel the crucial second phase of photosynthesis, the construction of carbohydrates such as glucose.

≋ THE SUGAR-BUILDING PHASE OF PHOTOSYNTHESIS

A corn leaf is a far better solar collector than any solar device yet built by people. It can track the sun's path without wheels and gears; it is biodegradable; it can sprout from a small seed and grow to large size in just a few weeks; and it uses only water and sunlight as it builds energy carriers during the light-trapping phase of photosynthesis. Moreover, the corn leaf does something no solar device can do: It builds organic nutrients such as glucose.

Figure 6.9 shows how the sugar-building phase of photosynthesis makes glucose from carbon dioxide (it also shows the steps of the light-trapping reactions, 1–3). Carbon dioxide, CO_2 (Figure 6.9, step 4), contains only carbon and oxygen, while sugars, like glucose, contain hydrogen as well (CHO). It is the energized electron carriers (step 5) produced by the light-trapping phase that supply the hydrogens for carbohydrates. As the six-carbon glucose is formed (step 6), these energy carriers provide energy and hydrogens that are combined with the carbon atoms from six carbon dioxide molecules. This process of removing carbon dioxide from the air and "fixing," or combining, it with hydrogen in a biological molecule is called **carbon fixation.** The carbon fixation that takes place during the sugar-building phase of photosynthesis is accomplished by an enzyme that is the most abundant protein on earth (see Figure 6.9). Called Rubisco for short, this enzyme initiates a cycle of reactions, the *Calvin-Benson cycle,* named for the scientists who first worked out its steps. Recall that another cycle of reaction steps, the Krebs cycle, is involved in aerobic respiration (see Chapter 5). The end product of the Calvin-Benson cycle is glucose, and in a very real sense, glucose represents energy from sunlight trapped in chemical bonds.

In corn plants and, in fact, all other plant types, glucose molecules generated during the sugar-building

reactions can be broken down via aerobic respiration to fuel cellular growth, division, and other activities throughout the plant. Glucose can be incorporated into starch and transported to storage areas, such as seeds or roots. Or the newly formed sugar can become subunits of cellulose for cell wall synthesis and repair.

Regardless of the specific end products, however, the significant point is this: The sugar-building phase of photosynthesis incorporates carbon dioxide from the air into carbohydrates, fats, and proteins, using solar energy

FIGURE 6.10 *Photosynthesis in All Seasons: C_4 Plants Have the Edge Over C_3 Plants.* Corn is a champion photosynthesizer because it can photosynthesize in hot, dry conditions when other plants—such as lawn grasses, beans, and rice—cannot. These plants (called C_3 plants) remove carbon dioxide from the air and fix it into a three-carbon compound before it enters the Calvin-Benson cycle and is further modified into glucose. However, corn, crabgrass, and sugarcane (so-called C_4 plants) initially fix carbon dioxide into a four-carbon compound. Because this special C_4 mechanism can fix carbon under hot, dry conditions, it circumvents photorespiration and allows C_4 plants to carry out photosynthesis on days when C_3 plants cannot.

stored during the light-trapping reactions. This harvest of biologically useful molecules that occurs during photosynthesis provides the plant, alga, or photosynthetic bacterium with raw materials for repair and growth, as well as supplying compact carbohydrate fuel (glucose) for the metabolic release of energy via aerobic respiration. Organisms like yourself cannot perform the biochemical trick of photosynthesis. But by eating the bounty from photosynthesizers (corn, wheat, spinach plants, and so on), you reap the benefits of this carbon fixation. Thus, photosynthesis lies at the very center of the web of life, fixing the carbon compounds that virtually all living things must have.

≋ A SPECIAL TYPE OF CARBON FIXATION

During very hot, dry weather, a process called **photorespiration** can interfere with normal photosynthesis, and carbon dioxide can be lost from the plant instead of fixed. This is one of the reasons lawns turn brown in summer. Perhaps you noticed, however, that in the middle of a brown lawn, annoyingly luxuriant green tufts of crabgrass may flourish, and a field of corn can be green next to a dying field of beans (Figure 6.10). This is because corn, crabgrass, and many other species that probably evolved in the tropics are so-called **C$_4$ plants**—plants that have special cells containing a "carbon dioxide pump." This pumping mechanism delivers carbon dioxide (within a compound containing four carbons) to the main enzyme of the Calvin-Benson cycle (Rubisco), thereby circumventing photorespiration and allowing photosynthesis to proceed, even in a hot, dry environment. Normal lawn grasses, beans, and most other plants lack the carbon dioxide pump and are called **C$_3$ plants** (because the first stable compound of the sugar-building phase contains three carbon atoms); they experience a slowing or shutdown of productive photosynthesis when weather conditions are hot and arid. The ability of corn to survive when many other plants cannot will make it an even more valuable food source if our climate becomes hotter and drier as a result of the "greenhouse effect" (see page 112).

≋ THE GLOBAL CARBON CYCLE

It might seem hard to relate metabolic processes like photosynthesis and aerobic respiration—which take place within microscopic cells and organelles—to ourselves and our daily lives. But consider these facts: The cells in your body contain carbon and oxygen atoms that once formed

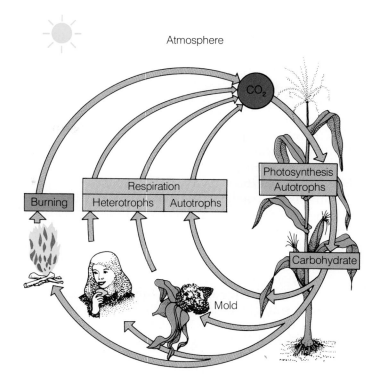

FIGURE 6.11 *The Global Carbon Cycle.* Vast amounts of carbon move through the air, soil, and water as photosynthesizing autotrophs fix carbon dioxide into organic compounds, and heterotrophs (along with nonliving combustion processes such as burning) break down those compounds and once again release carbon dioxide.

chloroplasts in a plant growing in a field near your house; and your houseplants are now built of carbon and oxygen atoms that you once exhaled. Aerobic respiration and photosynthesis are not simply chemical processes with reciprocal equations and opposite raw materials and products. They are metabolic engines that drive a global **carbon cycle.** This cycle is the flow of organic compounds (with their carbon, hydrogen, and oxygen atoms) from autotrophs (plants and other organisms that make their nutrients from nonbiological molecules) to heterotrophs (animals and other organisms that must consume the products of living cells). Organic molecules then flow to the atmosphere, soil, and water, and back to autotrophs once again (Figure 6.11). Most autotrophs fix carbon dioxide into organic compounds and release oxygen. Most heterotrophs break down those compounds and in the process consume oxygen and release carbon dioxide. Autotrophs fix that released carbon dioxide once again, release new oxygen, and the cycle continues.

These events ultimately involve all of the trillions of tons of carbon on our planet, whether it is currently tied up in wood, leaves, animal tissues, microorganisms, or rotting soil particles, and whether it is floating in the atmosphere as carbon dioxide, dissolved in seawater, or tied up as carbonates in rocks.

Biologists estimate that if all photosynthetic organisms suddenly stopped functioning today, all organic compounds would be broken down to carbon dioxide within 300 years, and no free oxygen would remain in the oceans and atmosphere after just 2000 years. This is an astonishingly short time, considering it took many millions of years for oxygen to accumulate to its current levels in air.

Figure 6.11 shows the global carbon cycle and the relationships between photosynthesizers (autotrophs) and the heterotrophs that depend on them and between photosynthesis and aerobic respiration. Notice that the burning of fossil fuels (a form of fixed carbon) makes a significant contribution to the pool of carbon dioxide available for fixation by autotrophs. The burning of fossil fuels and of our rain forests may also be causing a buildup of atmospheric carbon dioxide, leading to a phenomenon called the **greenhouse effect,** in which the gas traps heat from sunlight near the earth, just the way glass windows trap heat inside a greenhouse. This trapped heat may in turn be leading to increased global temperatures, changing climate patterns, additional photosynthesis by autotrophs, and a slight melting of the polar ice caps.

We'll return to this sobering topic in Chapter 35. The point to remember here is this: Just as the metabolic processes in individual cells have a massive collective effect—the global carbon cycle—worldwide disturbances of that cycle caused by human activity will have their own pervasive effects. Those effects, if we allow them to occur, could ultimately extend down to the basic biochemical levels and the mechanisms of energy exchange that fuel life processes on our planet and could disrupt them to the point of exterminating complex organisms.

≋ CONNECTIONS

While nearly all living things break down organic nutrients for energy, only about half a million species can build those nutrients through the solar-powered carbon fixation we call photosynthesis; those half million support themselves and all other living species. Now that we have considered the details of photosynthesis, we can see how photosynthetic efficiency underlies the bounty of our nation's largest crop—corn, a C_4 plant that can grow even under hot, dry weather conditions. We can also see that tampering with the environment could lead to disastrous changes in the fundamental interdependency of autotrophs and heterotrophs, aerobic respiration and photosynthesis.

This chapter ends the unit on cells and cellular chemistry. We have come a long way—from atoms, molecules, and cell structure to the flow of energy from the sun through living cells. Our next chapter deals with another of life's characteristics: reproduction. You will see, however, that cell structure and biochemistry are encountered again and again throughout the rest of the book. The cell is the fundamental living unit, and its properties are closely reflected in all larger and more complex living things.

NEW TERMS

C_3 plant, page 111
C_4 plant, page 111
carbon cycle, page 111
carbon fixation, page 110
chlorophyll, page 107
greenhouse effect, page 112
light-dependent (light-trapping) reactions, page 105

light-independent (sugar-building) reactions, page 105
photon, page 106
photorespiration, page 111
photosynthesis, page 104
photosystem, page 108

STUDY QUESTIONS

REVIEW WHAT YOU HAVE LEARNED

1. Create a chart that compares aerobic respiration and photosynthesis, and use these headings: location in cell; starting materials; end products; energy released or stored.
2. What are the energy sources in the light-trapping phase and sugar-building phase of photosynthesis?
3. Discuss the ways plant cells use the sugars produced during photosynthesis.
4. List the various types of photosynthetic pigments. Where is each found?
5. What is the significant product of the light-trapping reactions?
6. What happens during the Calvin-Benson cycle?
7. What is your role in the carbon cycle?

APPLY WHAT YOU HAVE LEARNED

1. Oak leaves, green all summer, begin to turn red in the fall. Why?
2. A water plant, *Elodea*, carries on photosynthesis in the light and releases bubbles of oxygen that an experimenter can collect in a test tube. Suggest one way of increasing the rate of oxygen production.

FOR FURTHER READING

Govindjee, and W. J. Coleman. "How Plants Make Oxygen." *Scientific American* 262 (February 1990): 42–51.

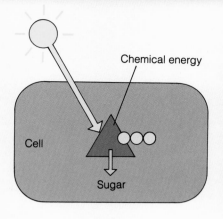

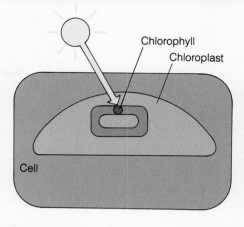

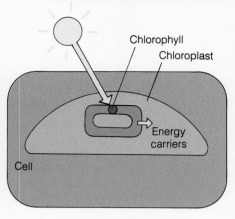

① During the metabolic process called photosynthesis, some cells trap solar energy, convert it to chemical energy, and store it in sugars.

② Photosynthesis takes place in chloroplasts, where chlorophyll and other pigment molecules trap energy from visible light.

③ The light-trapping reactions of photosynthesis convert light energy to chemical energy and store it in energy carriers.

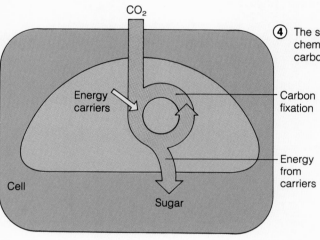

④ The sugar-building reactions transfer chemical energy from energy carriers to carbon compounds such as sugars.

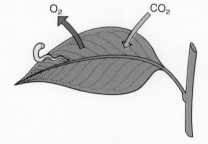

⑤ Overall, photosynthesis uses carbon dioxide from the atmosphere and releases oxygen into the atmosphere.

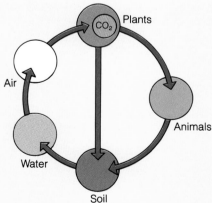

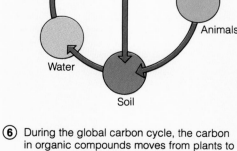

⑥ During the global carbon cycle, the carbon in organic compounds moves from plants to animals and soil, then to water, air, and back to plants.

⑦ The greenhouse effect can have serious consequences for earth and life.

PART TWO

PERPETUATION OF LIFE

CELL CYCLES AND LIFE CYCLES

CALLUSES, SCARS, AND CELL DIVISION

A ranch hand lives a hard life, taxed by the elements and by demanding physical work. Years of exposure to sun and wind, blowing sand, snapping tree branches, and slipping hand tools carve a record of honest labor into the skin of hands and face—a record that is spelled out in fine lines, deep wrinkles, callused areas, pigmented blotches, and small scars (Figure 7.1).

Calluses and scars are evidence of the skin's self-protective mechanisms, and each of us has these hardened zones on elbows, knees, and other areas. They are based on an increase in the number of skin cells. This increase produces tough, hardened regions at places of continual abrasion and very strong bands of tissue where the skin was torn or gashed. The addition of new skin cells takes place in a deep layer below our body's dry surface zones. Individual cells in this deep layer increase in size and then split, forming two new cells called *daughter cells*. The daughter cells can themselves increase in size and then split. This repeating cycle of **cell growth** (an increase in cell size) followed by **cell division** (the splitting of one cell into two) is called the **cell cycle** (Figure 7.2).

This chapter examines both the mechanisms of the cell cycle and how it propels a larger cycle called the **life cycle.** In many-celled organisms, this larger cycle moves from fertilized egg to embryo to infant to adult to the production of new eggs or sperm, and it is based on cell cycles—individual cell divisions. As we discuss the cell cycle and the life cycle, we will detect two recurrent themes. First, in both the cell cycle and the life cycle, precise replicas of cell components are made and then passed to the offspring. Our skin cells, for example, divide into new skin cells, and our lung cells into new lung cells. Said simply, like begets like.

Second, the dividing of cells and passing on of information is usually precisely controlled. If a ranch hand suffers a cut on the cheek, then the measured pace of cell divisions that normally replace worn-out skin is stepped up temporarily, and the additional skin cells fill in the wound and perhaps form a scar. After years of exposure to harsh sunlight, however, a rancher might develop a wartlike lump or a white crusty circle of renegade cells on the forehead or cheek—a skin cancer. If left untreated, these altered cells would continue to pass through the cell cycle, dividing and redividing into a larger and larger lump or crusty patch. Eventually, renegade malignant cells could leave the skin cancer site, invade other parts of the body, divide there in an uncontrolled way, and threaten the rancher's life. This same general sequence claims the lives of millions of Americans each year.

Our study of cell cycles and life cycles will answer several questions:

- Where, inside a cell, is the hereditary information that directs what our cells look like and how they grow?
- What events make up a cell cycle?
- How do the events of cell division, called *mitosis*, distribute hereditary information from the parent cell to two new daughter cells?
- How is the cell cycle regulated?
- What events make up the life cycles of people and other many-celled organisms?
- What are the events of *meiosis*, a special type of cell division necessary for life cycles involving sexual reproduction?
- What are the roles of the two types of cell division in evolution?

FIGURE 7.1 *Portrait of a Rancher.* The face and hands of a hardworking ranch hand are a living record of exposure to the elements, including calluses and a small skin cancer.

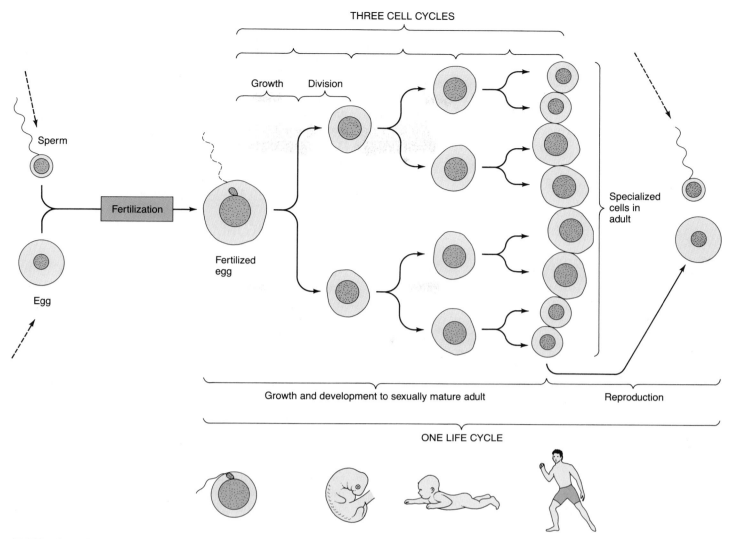

FIGURE 7.2 *Cell Cycles and the Life Cycle.* During the cell cycle, a cell grows, then divides in two; the progeny grow, then divide in two; and so on. Here we show several cell cycles. During the life cycle of a many-celled organism, a fertilized egg develops into an adult organism that produces eggs and/or sperm. These unite with other eggs and sperm to form new fertilized eggs, which can grow to new adults, which can in turn produce new fertilized eggs, and so on. Eventually, each adult dies, but its progeny may live on.

≋ CHROMOSOMES CONTAIN INFORMATION THAT DIRECTS CELL GROWTH AND REPRODUCTION

The skin cells in a rancher's face (or in your own face) share a general strategy of cell reproduction with all other living cells: They take in nutrients, increase in size, and then divide in two. (The two resulting cells are often called *daughter cells*.) But what information directs the way a cell grows or divides? What prevents a dividing human skin cell from giving rise to two new earthworm cells, or an oak cell from dividing into two clover cells? The answer is that the cell nucleus contains genetic information that governs what a cell looks like and how it performs.

INFORMATION FOR DIRECTING CELL GROWTH AND THE REPRODUCTIVE CYCLE IS STORED IN THE NUCLEUS

How did biologists learn that the nucleus directs the way a cell grows? In the 1930s, a German biologist transferred a nucleus from one cell to another cell of a slightly different type and found that the transplanted nucleus directed the growth of the second cell type (Figure 7.3). The conclusion was clear: The cell nucleus is the central repository of information for constructing new cell parts.

Other observations extended this idea, suggesting that a cell's nucleus contains information not only for constructing a new cell but also for building an entirely new many-celled individual. This is easily visualized in the embryo of a roundworm. During fertilization, only the nucleus of the sperm cell actually enters the egg cell; it then moves through the egg cytoplasm and fuses with the egg nucleus (Figure 7.4a and b). This penetration and fusion initiate the life of a new individual, with hereditary information donated in equal amounts by the mother and father. The development of the whole organism then proceeds using both sets of information. Since the nucleus is the only thing a new individual receives equally from both parents, we can conclude that the nucleus directs events not only in one cell and its cell cycle, but in the whole organism and its life cycle as well.

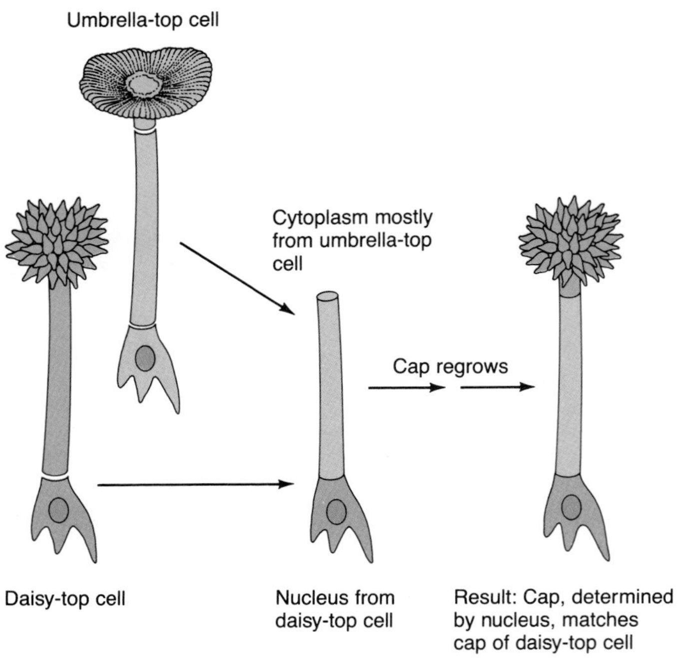

GENETIC INFORMATION IS STORED IN THE CHROMOSOMES

About a hundred years ago, researchers using simple stains and microscopes found that the nuclei of egg and sperm cells contain rodlike structures they called *chromosomes* ("colored bodies"). By watching sperm fertilize eggs under a microscope, biologists could see that chromosomes are the only structures that egg and sperm cells contribute in equal amounts to the fertilized egg and to the new individual that grows from that egg. They concluded that hereditary information lies in the chromosomes. Biologists know today that all organisms contain chromosomes, although the size, shape, and number of these structures differ from species to species. They also know that chromosomes contain DNA (see Chapter 2). Prokaryotes (bacteria) have a single long DNA molecule; the ends are joined into a circle, and the circle lies tangled in the cell. Eukaryotes (including all plants and animals) usually have several chromosomes, each containing just one long DNA molecule, but the molecule is linear, not circular. The cell nuclei in one kind of roundworm contain 12 chromosomes; those of a giant sequoia tree, 22. A goldfish nucleus contains 104 chromosomes; and a human nucleus, 46 (Figure 7.5).

FIGURE 7.3 *An Experiment Shows That Genetic Information Lies in the Nucleus.* Species of a certain green algae have large single cells with either a daisy-shaped cap or an umbrella-shaped cap that will regenerate if removed. The nucleus of each cell is in the foot of the stalk. Transplanting the upper stalk of the umbrella species to the foot of the daisy species eventually results in the regrowth of a daisy-shaped cap, proving that genetic instructions for cap shape lie in the cell nucleus.

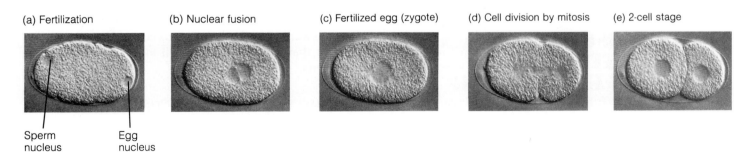

FIGURE 7.4 *The Nucleus Contains Information for the Construction of a New Individual.*
(a) The sperm and egg nuclei are the only structures both male and female donate equally to their offspring. (b, c) They fuse to form a single nucleus in the fertilized egg. (d, e) When the fertilized egg divides, each daughter cell receives a copy of this nucleus with its unique set of genes. The sperm nucleus and egg shown here are from a roundworm.

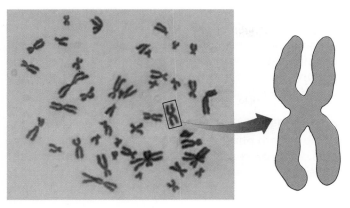

(a) The 46 chromosomes in a human cell   (b) A chromosome

FIGURE 7.5 *Chromosomes: "Colored Bodies" in the Nucleus of a Human Cell.* Just before cell division, DNA molecules are densely packaged into dark-staining structures called chromosomes. (a) A person has 46 chromosomes in each body cell nucleus. (b) A blowup of one chromosome. Since DNA molecules are of different lengths, chromosomes have different sizes and shapes, but all many-celled organisms have chromosomes in the nucleus.

The fact that chromosomes carry DNA, the genetic information needed to build a new cell, is important to our understanding of the cell cycle with its alternating phases of growth and cell division. Each new daughter cell must receive its own set of chromosomes containing a copy of the hereditary material. For that reason, the duplication and distribution of the chromosomes are central activities in the cell cycle, as our next section explains.

≋ THE CELL CYCLE

Just as it takes two very different sets of blueprints to build a suburban ranch house and an urban apartment building, each living species has its own unique genetic blueprint. This is reflected in the numbers and sizes of chromosomes, which differ from species to species. Within a given organism, however, each body cell has the same hereditary material and a set of chromosomes of the same number, size, and shape. Thus, every one of your normal body cells has 46 chromosomes in its nucleus. When a given cell—say, a skin cell—divides, the chromosomes must first double in number so that each of the new daughter cells can receive a copy of all 46 chromosomes (Figure 7.6). If this did not happen, each cell division would cut the number of chromosomes in half, and soon the cells would die out. Another "corrective mechanism" is at work, too: cell growth. After a cell divides in two, both new daughter cells grow, or enlarge in size, before

another division takes place. This keeps the cells from getting smaller and smaller with each round of division.

PHASES OF THE CELL CYCLE: GROWTH AND DIVISION

You grow because your cells grow and divide, and so do the cells of other organisms. Cells in plants and animals grow and divide as a result of a two-phase eukaryotic *cell cycle.* One phase, called *interphase,* brings about cell growth; the other phase, called the *M phase,* results in cell division (Figure 7.7, page 120).

■ **The Active Growth Phase**   The growth period of the eukaryotic cell cycle is called **interphase** because it intervenes between two division periods (see Figure 7.7). During interphase, new proteins, ribosomes, mitochondria, and other cell parts are built in preparation for eventual cell division. Without such growth, the cell would be halved in size and contents with each division, until it became too small to contain all the vital parts it needs to function.

Interphase can be quite short or very long, depending on the type of cell, its role in the organism, and conditions in the environment. The skin cells on your hand, for example, normally have a long cell cycle and a long

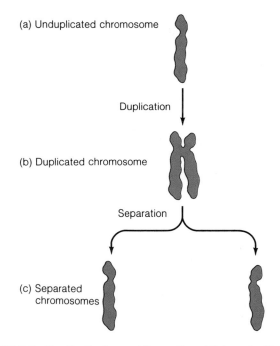

(a) Unduplicated chromosome

Duplication

(b) Duplicated chromosome

Separation

(c) Separated chromosomes

FIGURE 7.6   *The Duplication and Separation of Eukaryotic Chromosomes During Cell Division.* Before a cell divides, each of its chromosomes duplicates (a, b). During cell division, the identical halves of each duplicated chromosome separate and are apportioned into individual daughter cells (c).

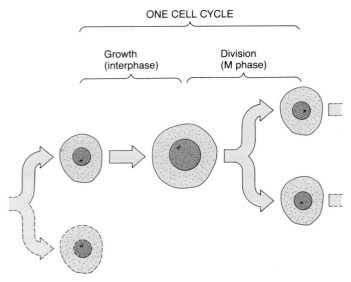

Growth
(interphase)

Division
(M phase)

FIGURE 7.7 *The Cell Cycle Generally Has Two Phases: Growth and Division.* In a eukaryotic cell, growth (interphase) results in an increase in size and preparation for cell division. The rapid division (M) phase lasts about 1 hour and includes the nuclear division known as mitosis.

interphase (up to a week or more in length). But that can change: If you tear your skin on a blackberry thorn, the interphase in those skin cells may shorten, speeding up growth and division and enabling the wound to heal rapidly. And if the bark of an aspen tree is nibbled away by a deer, then its bark-forming cells can also enter a shortened interphase, rapidly producing bark to protect the damaged area.

During the middle of interphase, the single DNA molecule in each chromosome duplicates exactly. When DNA synthesis ends, each chromosome is made up of two identical and parallel DNA molecules, held together at one point (see Figure 7.6b). Toward the end of inter-

phase, the cell prepares to divide by manufacturing the specific proteins necessary for mitosis, the phase of actual division. If, for example, skin cells near a cut in a rancher's cheek did not make these proteins, then they could not divide; and if they could not divide, then the wound would never heal.

■ **The M Phase and Cell Division** With the completion of the growth phase, a cell is prepared to divide into two cells. Division consists of two main events: **mitosis,** the division of the nucleus, and **cytokinesis,** the division of the cytoplasm (Figure 7.8). During mitosis, each duplicated chromosome splits apart, and the copies separate to opposite ends of the cell, where they are enclosed in two new daughter nuclei. During cytokinesis, the daughter nuclei are separated into new daughter cells. Other cell parts, such as mitochondria, chloroplasts, and ribosomes, duplicated during interphase, are evenly distributed to the two daughter cells. The splitting and separating of genetic material and other cell parts are so crucial to cellular reproduction that we devote the next section to discussing them fully.

⋙ MITOSIS AND CYTOKINESIS: ONE CELL BECOMES TWO

Through a microscope, interphase does not seem very active. By contrast, the two main events of cell division are spectacular: Mitosis, the division of the cell nucleus, is accompanied by a dramatic "dance of the chromosomes," while cytokinesis, the division of the cytoplasm, entails a rapid pinching of the cell in two. In the next few sections, we describe and illustrate the dynamic beauty of these cell movements.

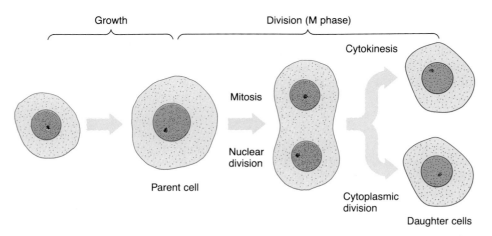

Growth

Division (M phase)

Cytokinesis

Mitosis

Nuclear division

Parent cell

Cytoplasmic division

Daughter cells

FIGURE 7.8 *Events of the Division Phase: Mitosis and Cytokinesis.* During the period of actual cell division, the parent cell undergoes a division of its nucleus (mitosis), and then the cytoplasm divides (cytokinesis), forming two daughter cells.

One chromosome with two identical halves (two *chromatids*)

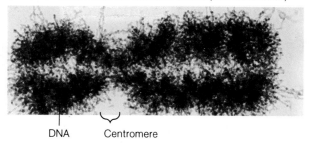

DNA   Centromere

FIGURE 7.9 *Anatomy of a Chromosome.* During the division phase, the chromosomes become clearly visible. The two rodlike halves of the duplicated chromosome are made up of DNA threads densely coiled around proteins and are held together at the centromere.

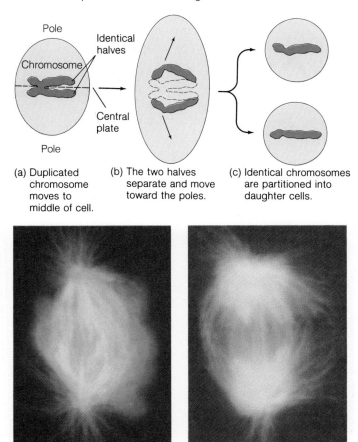

(a) Duplicated chromosome moves to middle of cell.

(b) The two halves separate and move toward the poles.

(c) Identical chromosomes are partitioned into daughter cells.

(d) Chromosomes aligning along central plate

(e) Chromosomes separating

FIGURE 7.10 *Dance of the Chromosomes During Cell Division.* During the rapid division phase, each chromosome carries out a series of dancelike movements. (a) The duplicated chromosome lines up along the cell's midline or central plate; (b) then the two halves separate toward opposite poles and (c) become partitioned into separate daughter cell nuclei. These light micrographs show chromosomes (stained reddish orange) surrounded by protein fibers (yellowish green) in the nucleus of kangaroo rat cells. In (d), the chromosomes are aligned in the middle of the cell, as in (a). In (e), the chromosomes are being pulled toward the poles, as in (b).

MITOSIS: CHROMOSOME CHOREOGRAPHY

Picture a cell watcher equipped with a light microscope and viewing the events in a set of dividing cells. This person cannot see distinct chromosomes during interphase because the long DNA molecules are unraveled and tangled in the nucleus like wadded-up string. It would be a nightmare to separate such a tangled mess; fortunately for the cell (and for the person observing mitosis under a microscope), just before mitosis, the DNA becomes tightly wound up, and distinct chromosomes suddenly appear, allowing the observer to witness the chromosome choreography as the division phase proceeds.

At first appearance, the chromosomes have already doubled (review Figure 7.6) and consist of two identical halves held together at a single point, the *centromere* (Figure 7.9). Let's follow just one chromosome to see what happens to the halves. The duplicated chromosome lines up on a central cell plate (Figure 7.10a), and the two halves split apart and move to opposite ends, or *poles,* of the cell, where they now function as separate, identical chromosomes (Figure 7.10b). These two identical chromosomes are then partitioned into two daughter nuclei (Figure 7.10c). Each of the daughters will receive one copy of every chromosome in the original parent cell and hence will be genetically identical to the other daughter and to the original parent cell. The complete "dance of the chromosomes" has four parts, summarized here and described in detail in Figure 7.11 (pages 122 and 123):

1. In *prophase,* the nucleus prepares for division. During prophase, the chromosomes condense, and the nucleolus (dark-staining region) and nuclear envelope disappear. Also during prophase, a *mitotic spindle* forms—a football-shaped bundle of microtubule fibers that suspends and moves the chromosomes and can be stained brilliant green, as in Figure 7.10d.

2. In *metaphase,* the chromosomes become aligned in the middle of the cell, like football players at the line of scrimmage (see Figure 7.10d).

3. In *anaphase,* the centromeres of each chromosome split, and the spindle fibers pull the two halves of each chromosome apart and toward opposite poles (Figure 7.10e).

4. In *telophase,* the chromosomes arrive at opposite poles, and the preparatory events reverse (the nucleolus and nuclear envelope reappear, the spindle dissolves, and the chromosomes decondense).

Once telophase is over, the division of the cell nucleus (mitosis) is complete, and the cell now has two nuclei carrying identical sets of chromosomes. The division

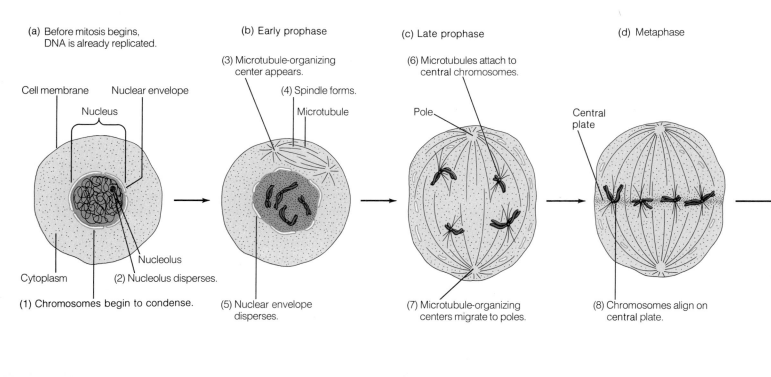

(a) Before mitosis begins, DNA is already replicated.

Cell membrane   Nuclear envelope

Nucleus

Cytoplasm   (2) Nucleolus disperses.

Nucleolus

(1) Chromosomes begin to condense.

(b) Early prophase

(3) Microtubule-organizing center appears.

(4) Spindle forms.

Microtubule

(5) Nuclear envelope disperses.

(c) Late prophase

(6) Microtubules attach to central chromosomes.

Pole

(7) Microtubule-organizing centers migrate to poles.

(d) Metaphase

Central plate

(8) Chromosomes align on central plate.

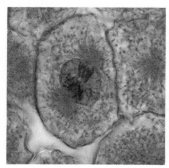

Early prophase

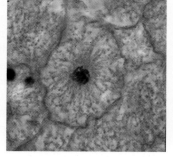

Mid-prophase

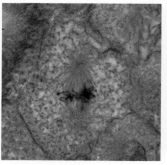

Late prophase

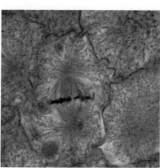

Metaphase

FIGURE 7.11 *The Phases of Mitosis.* (a) The cell's DNA has already replicated during interphase. As the cell enters the first part of mitosis, called prophase, the DNA reverts from its diffuse, tangled state during interphase and becomes more tightly packaged (1). Also, the nucleolus disperses (2), and the cell's cytoskeleton is dismantled, providing raw materials for the spindle that will suspend and move the chromosomes. (b) A pair of microtubule-organizing centers develops, from which microtubules radiate (3). Animal cells have centrioles, or short cylinders of microtubules, at the centers of their microtubule-organizing centers, but plant cells lack centrioles. The microtubule-organizing centers, shaped like a star, or aster, in animal cells, separate toward opposite ends, or poles, of the cell, spinning out a gossamer spindle that suspends and moves the chromosomes (4). During early prophase, the spindle invades the nuclear region after the nuclear envelope disperses (5). (c) Microtubules attach to chromosomes by special fibers located at the centromere of each chromosome (6). The chromosomes then jostle back and forth as the pole microtubules interact with the attaching microtubules (7). (d) During metaphase, the chromosomes become aligned in a plane on the central plate (8), a plane lying halfway between each pole, like the plane a knife makes when cutting a cantaloupe in half.

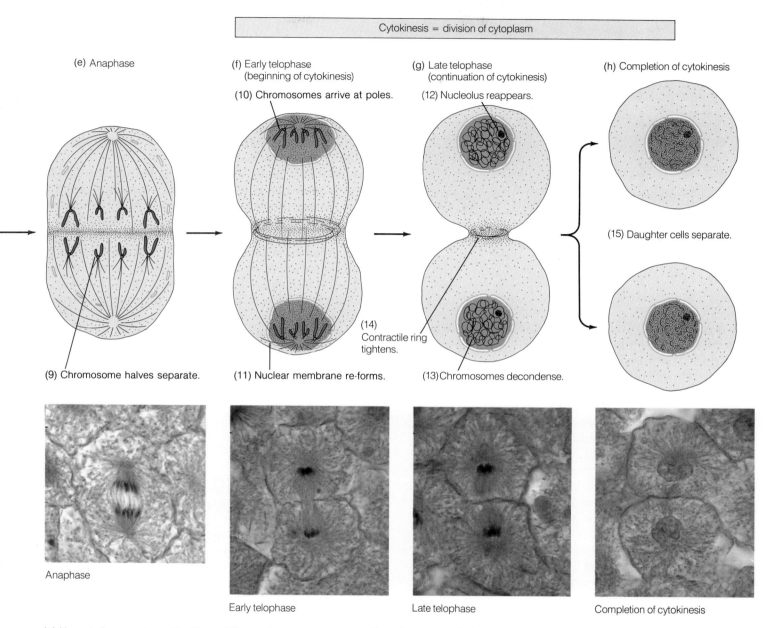

Division of nucleus

Cytokinesis = division of cytoplasm

(e) Anaphase

(f) Early telophase
(beginning of cytokinesis)

(10) Chromosomes arrive at poles.

(g) Late telophase
(continuation of cytokinesis)

(12) Nucleolus reappears.

(h) Completion of cytokinesis

(15) Daughter cells separate.

(14)
Contractile ring
tightens.

(9) Chromosome halves separate.

(11) Nuclear membrane re-forms.

(13) Chromosomes decondense.

Anaphase

Early telophase

Late telophase

Completion of cytokinesis

(e) Next, during anaphase, the fibers at the centromeres separate, pulling chromosome halves apart, toward opposite poles (9). (f) Early telophase marks the beginning of cytokinesis. The daughter chromosomes arrive at each pole (10), after which the nuclear membrane re-forms around the chromosomes (11). (g) Then the nucleolus reappears (12), the spindle dissolves, and the chromosomes reel out again into a tangled mass of DNA (13). In addition, in late telophase in animal cells, a contractile ring tightens around the cell's midline where the central plate had been, creating a furrow (14). (h) Separation of the daughter cells completes cytokinesis (15). The micrographs show cells of the white fish, magnified 450 times.

FIGURE 7.12 *Radiation Tragedy.* Marie Curie's experiments with radium exposed her to high doses of radiation from which she eventually developed cancer and died.

phase continues, however, as the cytoplasm divides and separates the nuclei into two new cells. Anything that disrupts that precise distribution can be lethal to dividing cells and eventually to the whole organism. For example, overexposure to X-rays, gamma rays, and other forms of radiation causes damage by altering the dramatic "dance of the chromosomes." The result can be cancer (Figure 7.12) or radiation sickness. If a nuclear war were ever to take place, vast amounts of radiation would be released into the environment, energizing the nuclei of organisms' cells and breaking chromosomes. Where radiation causes a single break in one of the cell's many chromosomes, there will usually be no adverse effect on the cell's functioning, since cells usually have two copies of each chromosome. Damage to one at a single point will usually be compensated for by the intact copy. When an irradiated cell starts to divide, however, broken off pieces of a chromosome will not be distributed normally to the daughter cells. As a result, chromosomal information will be unbalanced, perhaps enough to cause the daughter cells to die. One consequence of excess irradiation is to kill dividing cells. In plants, this can lead to the cessation of growth and the halting of development of new leaves, bark, roots, and flowers. The plant continues to live, but only so long as the preexisting leaves, roots, and bark can support the plant's needs. In many kinds of irradiated animals, wounds do not heal, because skin lost from a scrape or burn does not regrow, and the

body cannot produce new white blood cells to fight infections. Infections are, in fact, the earliest cause of death for radiation victims. Clearly, we depend on the restorative powers of mitosis and its orderly distribution of hereditary information from one cell generation to the next.

CYTOKINESIS: THE CYTOPLASM DIVIDES

Soon after mitosis, the cytoplasm of most animal and plant cells divides by means of cytokinesis (literally, "cell movement"). In both animal and plant cells, new cell membranes form at or near the midline once occupied by the central plate (Figure 7.13) and separate the two nuclei into the new cells.

Animal cells, with their pliable outer surface, divide *from the outside in* as a circle of microfilaments called a *contractile ring* pinches each cell in two, much as a purse string tightens around the neck of a purse (Figure 7.13a). Plant cells, with their rigid cell walls, retain their shape throughout the cell cycle and divide *from the inside out* as a central partition, or *cell plate,* made of cell membrane and cell wall material forms (Figure 7.13b).

Viewed collectively, the events of the cell cycle ensure that the cell grows and that the chromosomes and other cell parts are copied and apportioned equally into two daughter cells. Cells such as those in a rancher's skin or in your own face, however, must sometimes cycle rapidly, sometimes slowly, and sometimes not at all. Obviously, the timing of such events is controlled. But how?

≋ REGULATING THE CELL CYCLE

If you get a cut on your finger, skin cells divide and repair the damaged tissue; but once the wound is healed, the skin cells stop dividing, lest they give rise to a mass of scar tissue. Let's look briefly at the external and internal factors regulating the cell cycle.

EXTERNAL AND INTERNAL CONTROLS OVER CELL GROWTH

A major external factor controlling cell division is actual physical contact with other cells. Mammalian cells growing in a Petri dish can serve as a good model for such contact. The cells behave like people loading into an elevator. Just as the people form a layer one person high and do not climb onto each other's shoulders, the cells grow and divide only until they form a layer that is one cell thick. If you remove a swath of cells from the dish,

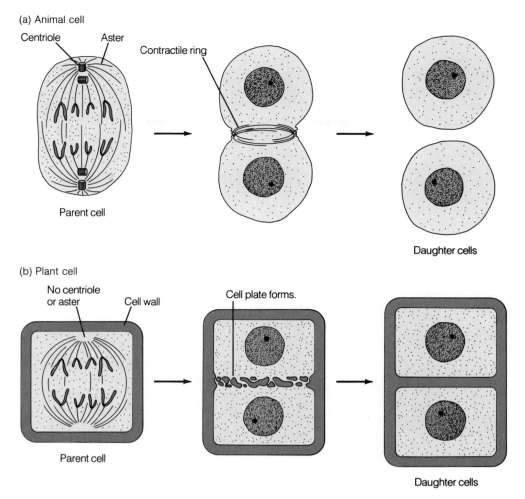

(a) Animal cell

Centriole   Aster

Contractile ring

Parent cell

Daughter cells

(b) Plant cell

No centriole
or aster   Cell wall

Cell plate forms.

Parent cell

Daughter cells

FIGURE 7.13 *Cytokinesis in Animal and Plant Cells.* (a) Animal cells are pinched in two from near the cell surface by a contractile ring that creates a furrow and eventually separates the daughter cells. (b) A cell plate partitions a plant cell from the inside out, forming two daughters.

the cells at the edge of the empty path divide until the space is once again filled, and then they stop (Figure 7.14a, page 126). The same thing happens when you cut your finger; the cells at the edge of the wound divide rapidly, fill the gap, and stop dividing when they meet.

A second external source of control is a group of proteins called **growth factors.** These are produced by one group of cells, but diffuse through the body and speed or retard the cell cycle in another group of cells. Damaged cells near a wound, for example, release growth factors that stimulate their intact neighbors to divide. Although growth factors are normally present in very small amounts, genetic engineering makes it possible to produce larger quantities that doctors can use to treat injured patients (Figure 7.14b).

Researchers have also discovered a protein inside the cell called MPF (maturation promoting factor) that appears to control some of the events of mitosis, such as chromosome condensation. The fact that MPF has a nearly identical shape and action in all cells examined, from

yeasts to people, suggests that it is a truly ancient and important internal regulator of cell division. While we do not yet know exactly how MPF controls the cell cycle, we do know that the protein itself must be controlled by genetic information in the nucleus. Thus, a basset hound has short legs and long ears and a Great Dane the reverse because hereditary information controls the number of cell divisions in the legs and ears of the two kinds of dogs as they develop. Unfortunately, sometimes this critical genetic control goes awry, and the result can be cancer (see the box on page 127).

CANCER: CELL CYCLE REGULATION GONE AWRY

Recall that the rancher in our opening example runs the risk of skin cancer from too much exposure to the sun. Since World War II, the life-threatening group of diseases called **cancer** has been on the rise, and today, cancer causes about one out of every three deaths each year in the United States. The biological basis of cancer is a loss of the normal external or internal regulation of the cell cycle we have been discussing. It isn't that cancer cells grow and divide more quickly than normal cells; the problem is that cancer cells do not stop growing or dividing on contact with other cells (Figure 7.15a). Hence,

(a) Cancer cells do not stop growing and dividing when they contact other cells.

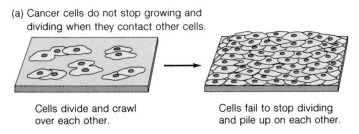

Cells divide and crawl over each other.

Cells fail to stop dividing and pile up on each other.

(b) Some cancer cells can make their own growth factors.

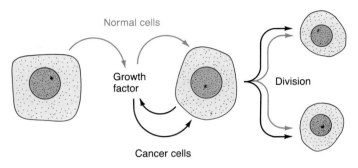

FIGURE 7.15 *Cancer Cells Do Not Obey the Rules.* In cancer cells, the contact of neighboring cells does not inhibit growth (a), and some cancer cells can produce their own growth factors (b); hence, cancer cells can form masses (tumors) that invade healthy tissue.

(a) Contact with other cells: analogy for a wound

(b) Growth factors help wounds heal faster

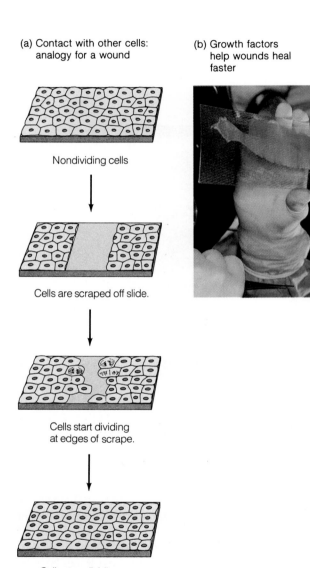

Nondividing cells

Cells are scraped off slide.

Cells start dividing at edges of scrape.

Cells stop dividing when in contact again.

FIGURE 7.14 *Regulating the Cell Cycle: External Factors.* (a) Cells stop growing when they contact other cells. For example, if a swath is cut through a layer of nondividing cells (as in a wound), cells along the edge will grow and divide until they are once more in contact with their neighbors. (b) Growth factors released from damaged cells stimulate division in specific cell types. For example, epidermal growth factor, when applied to a skin graft such as the one shown here, as well as to the wound site, helps burn wounds heal faster.

cancer cells crawl over other cells, invade healthy tissues, and multiply into masses called malignant *tumors*.

Recent evidence indicates that in a cancer called small cell carcinoma of the lung, caused by smoking, the cells, unlike normal cells, produce their own growth factor, so that they continuously stimulate themselves to divide (Figure 7.15b). Other evidence shows that in some cancer cells, genetic changes alter the action of MPF and as a result may alter how the protein regulates cell division. Chapter 13 describes other aspects of our rapidly growing understanding of cancer cells. One thing has indeed become clear: The secrets of cancer will be found in the regulation of the cell cycle.

≋ LIFE CYCLES: ONE GENERATION TO THE NEXT IN MANY-CELLED ORGANISMS

For single-celled organisms, like bacteria or *Euglena,* the cell cycle just described is the same as the *life cycle:* Each cell division reproduces the entire organism and produces the next generation. Most many-celled organisms, however, like clams and humans, are too complex to simply split down the middle and form two half-size ver-

SHOULD YOU "CATCH SOME RAYS"?

For many light-skinned people, a suntan is such an important symbol of health, attractiveness, and leisure time that they sun themselves to a blistered crimson several times a year. With every sunbath, however, they are risking premature aging and skin cancer. Like smoking, fatty diets, and so many other life-style choices, suntanning has a delayed cost: Burn now, pay later. But it is hard for some people to take the downside seriously. Are you one of them?

By definition, a suntan equals skin damage. Light-skinned people lack naturally high levels of the pigment melanin, which is more abundant in darker-skinned people. Melanin absorbs sunlight and helps protect the DNA in the nuclei of skin cells from light damage. Pale skin can become tan only after sun exposure and light damage "turn on" melanin production.

The most damaging part of sunlight is UV-B, the second-most energetic wavelengths in the ultraviolet portion of the electromagnetic spectrum (review Figure 6.5). Exposure to UV-B has been positively linked to the DNA damage that underlies three types of skin cancer: basal cell carcinoma, squamous cell carcinoma, and malignant melanoma (the most dangerous). Scientists used to call UV-B "burning rays" and the slightly more energetic UV-A "tanning rays," and they designed sunscreens to block out the former but let the latter pass through. Newer studies, however, show that UV-A itself is linked to skin cancer and also damages the collagen and elastin fibers that keep skin resilient. UV-A thus accelerates the wrinkles, sags, and droops we associate with aging. Even before this mileage begins to show on your face and hands, light damage is accumulating in the skin. The woman

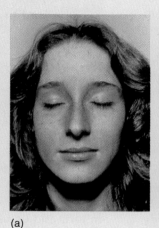
(a)

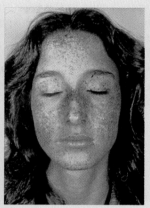
(b)

FIGURE 1 *Light Damage to Skin is Cumulative.* (a) Under normal light, this person's skin appears undamaged. (b) Under special black light, damage to the same skin's deeper layer shows up.

in Figure 1a, for example, appears to have smooth, undamaged skin under normal light. Under black light, however, permanently damaged areas (the blotches and spots) show through from the deeper layers beneath the surface skin (Figure 1b). At least one commercial sunscreen (called Photoplex) protects against both UV-B and UV-A to help prevent cancer and premature aging.

Physicians diagnose nearly half a million new cases of skin cancer each year, 27,000 of which are the potentially fatal variety called malignant melanoma (often visible as irregular, inky black, crusty patches). These numbers have increased rapidly since 1970, and the frightening upswing seems to be based in large part on the growing populations in the southwestern and southeastern United States, as well as on the rise of outdoor leisure activities and on skimpier clothing styles. Some people also frequent "tanning salons" to keep up their bronzed appearance during fall and winter. Finally, a slight but detectable decrease in the earth's protective ozone layer (details in Chapter 36) has begun to allow more ultraviolet light to reach the surface. Because there tends to be a lag time between DNA damage and the appearance of

cancers, the life-style and environmental abuses of the 1960s and 1970s are only showing up today, and the exposure of the 1990s will tell in the twenty-first century. Dermatologists predict that by the year 2000, 1 in 90 Caucasian Americans will develop malignant melanomas, while the other types of skin cancer will strike approximately 1 in 3. Skin cancer is about seven times more common in whites than in blacks.

Preventive measures range from hats and protective clothing to the judicious use of sunscreens and sunblocks, to legislation to protect the ozone layer.

Is sun worship worth the risk? You decide.

FOR FURTHER INFORMATION

Kripke, M. "Impact of Ozone Depletion on Skin Cancers." *Skin Cancer Foundation Journal* 8 (1988).

Rigel, D. S., et al. "Incidence and Mortality Rates of Malignant Melanoma in the U.S." *The Melanoma Letter* 7, 1 (1989).

"How to Ageproof Your Skin." *Woman's Day Magazine*, 16 January 1990, pp. 92–95.

Pamphlets on skin cancer from The Skin Cancer Foundation, 245 Fifth Avenue, Suite 2402, New York, N.Y. 10016.

Pamphlets on the choice and application of sunscreens from your local dermatologist or pharmacy.

sions of the original. Instead, new multicellular organisms must usually start from one or a few reproductive cells that divide repeatedly. The cells in the dividing cluster begin to take on various specialized roles and then finally develop into a new individual. At each stage in an organism's *life* cycle, individual cells in the organism pass through the *cell* cycle again and again until a new mature individual itself produces reproductive cells.

Life cycles are clearly more complicated than cell cycles (review Figure 7.2), and the reproductive stage can actually follow one of two very different strategies: asexual reproduction or sexual reproduction.

ASEXUAL REPRODUCTION: ONE INDIVIDUAL PRODUCES IDENTICAL OFFSPRING

For many types of plants and a few kinds of animals, reproduction can be **asexual,** where a new but genetically identical offspring grows directly from a few body cells of a single parent. The group of cells divides through

(a)　　　　　　　　　(b)

(c)

FIGURE 7.16 *A Gardener Exploits Asexual Reproduction in Plants: Artificial and Natural.* To make a cutting from a prize fuchsia, a gardener will cut off the new growth at the end of a young branch (a), dip the cut stem in a plant hormone that encourages rapid mitosis of cells at the cut surface (b), and stick the stem into potting soil (c). In a few weeks, roots will develop from the stem and establish the new plant. Since the cutting is genetically identical to the original plant, the eventual bush will have flowers the same showy size and color as the favorite original fuchsia.

FIGURE 7.17 *Asexual Reproduction in Animals.* The large mottled arm of this sea star with part of the body attached regenerated an entire new individual, including the four smaller mottled arms.

mitosis and usually remains attached to the parent for a while. Eventually, the cell grouping detaches from the parent and begins to function as an independent organism.

You have observed asexual reproduction if you've ever left a potato in the pantry too long. Cells in the potato's "eyes" (buds of tan tissue on the surface) begin to increase in number by mitosis, draw energy from the potato starch, and develop into stems and leaves. If cut from the potato and transferred to soil, these new individuals begin to photosynthesize, grow, and function as plants genetically identical to the one that produced the neglected kitchen potato. Farmers, in fact, usually plant sprouted eyes instead of potato seeds. Strawberry plants also reproduce asexually. The plants send out runners—special arching stems that reach the ground about 1 ft from the parent plant, take root, and develop stems and leaves.

Farmers and gardeners sometimes induce asexual reproduction artificially to duplicate a prize-winning specimen such as a fuchsia with stunning flowers. Using the technique described in Figure 7.16, one can make a cutting that will grow into a new plant with showy flowers identical to the original's.

Certain animals also display asexual reproduction. A relative of the jellyfish, the hydra, can reproduce by *budding:* Cells in special regions undergo rapid mitosis and become organized into new hydras. Another means of asexual reproduction in animals is *regeneration:* If cut in half, a sea star, such as the crown of thorns, can regenerate the missing portion (Figure 7.17). So can many other simple creatures. Obviously, though, a tail docked from a boxer puppy cannot regenerate the legs, torso, and head of a new dog. Puppies and most other animals rely on sexual reproduction. And many animals and plants that usually reproduce asexually can switch to sexual modes when conditions dictate (as we'll see later).

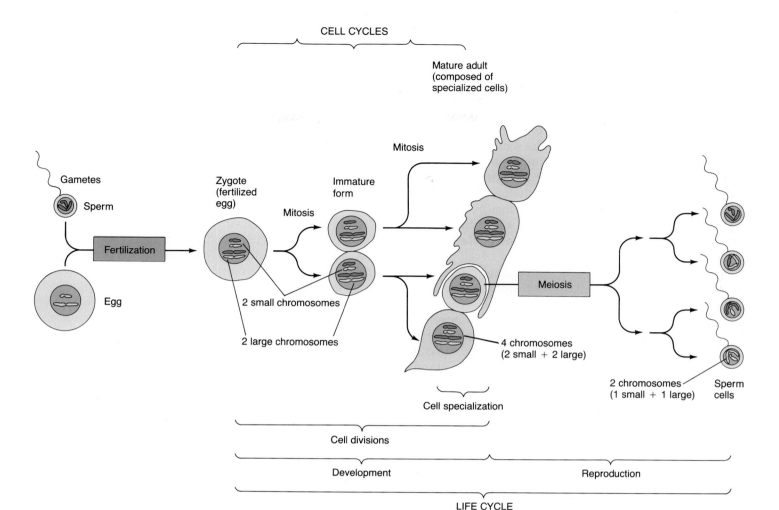

FIGURE 7.18 *A Generalized View of Sexual Reproduction in Many-Celled Organisms: Changes in Chromosome Number During Fertilization and Meiosis.* Fertilization, the union of egg and sperm, creates a zygote with a new genetic combination and two sets of chromosomes. The zygote and all other cells with two chromosome sets are diploid cells. Meiosis takes place in gamete-forming organs and halves the chromosome number, resulting in the generation of eggs or sperm with one set of chromosomes. A set here is one long and one short chromosome. Cells with one chromosome set are haploid. The production of haploid gametes via meiosis and the fusion of gametes that creates a diploid zygote during fertilization lead to a new generation and the start of a new life cycle.

SEXUAL REPRODUCTION: GAMETES FUSE AND GIVE RISE TO NEW INDIVIDUALS

Although asexual reproduction can lead to new individuals, the life cycle of most multicellular species involves **sexual reproduction.** Parents (usually two, but sometimes one) generate specialized cells called **gametes,** and when the gametes from opposite mating types (usually male and female) fuse, the life of a new individual begins. Gametes are usually large immobile cells called *eggs* or small cells called *sperm* that can move or be carried to the egg. In many organisms, such as people and ginkgo trees, each individual produces just one kind of gamete—egg or sperm. However, in pear trees, some worms, and many other plant and animal species, each adult individual can produce both eggs and sperm.

The fusion of egg and sperm is called **fertilization,** and the result is a single cell, the **zygote,** or fertilized egg, in which hereditary information from both parents unites to create a new combination that is genetically unique (Figure 7.18). The fertilized egg may then undergo *development*, a period of rapid cell division by mitosis and cell specialization, during which an immature form emerges, continues to grow, and changes into a mature adult. Within the mature adult, new eggs or sperm may form and give rise to the next generation.

Two kinds of cell division characterize the life cycle of a multicellular organism: mitosis and meiosis. Mitosis and meiosis differ in how the genetic material becomes divided into the newly forming daughter cells. After cell division by mitosis, both daughter cells contain chromosomes identical to those of the parents. For example, in Figure 7.18, the zygote has two large and two small chromosomes, and after mitosis, the daughter cells also have two large and two small chromosomes. After cell division by meiosis, however, the daughter cells have only half as many chromosomes as the parent cell. Look again at Figure 7.18. Each cell in the mature adult has four chromosomes, but after meiosis, each cell has just two chromosomes, one large and one small. **Meiosis** (literally, "to make smaller") is a special type of cell division that produces gametes or spores, specialized cells whose chromosome number is half that of the other body cells. If this chromosome reduction did not take place, eggs and sperm would have the full number of chromosomes (46 in humans, for example), and after fertilization, when a doubling of the chromosome number occurs, the children would have 92 chromosomes per cell, the grandchildren 184, and so on. But meiosis prevents this accumulation of chromosomes.

Biologists use special terms to describe a cell's chromosome count. A cell that is **haploid** (literally, "single vessel") has just one chromosome set in its nucleus. A human egg or sperm is haploid and has 23 chromosomes. A cell that is **diploid** ("double vessel") has two sets of chromosomes in the nucleus. A human body cell is diploid and contains 46 chromosomes. Figure 7.18 shows a simplified diagram of the haploid and diploid states. In the figure, a chromosome set is one long and one short chromosome. Thus, a haploid cell has two chromosomes (one long and one short), while a diploid cell has four chromosomes (two long and two short).

Meiosis and fertilization play essential but opposite roles in a life cycle involving sexual reproduction: Meiosis divides the chromosome number in half, while fertilization doubles the chromosome number. (see Figure 7.18). As in mitosis, a dramatic dance of the chromosomes during meiosis helps ensure that each daughter cell will receive the appropriate chromosome number.

≈ MEIOSIS: A REDUCTION OF CHROMOSOME SETS AND RESHUFFLING OF CHROMOSOME MATERIAL

Children never look or act exactly like either of their parents, nor do brothers and sisters look or act alike, except for identical twins. The reason for this wonderful diversity is in the chromosomes: No child is genetically identical to its parents or siblings, because meiosis reshuffles chromosomes and thereby produces **genetic variation.** Meiosis, then, achieves two basic effects: It reduces the chromosome sets from two to one, and it generates genetic variation. Let's take a look at how it accomplishes these results.

HOW MEIOSIS HALVES THE NUMBER OF CHROMOSOME SETS

Meiosis occurs in cells in your ovaries or testes, where it changes diploid cells with their two sets of chromosomes to haploid cells with just one chromosome set (Figure 7.19). It does this through two consecutive cell divisions called meiosis I and meiosis II (Figure 7.20). In meiosis I, a cell with two chromosome sets divides to form two daughter cells that each contain just one chromosome set (Figure 7.20, steps 1–4). In meiosis II, those two daughter cells divide and produce four cells, each containing just one chromosome set (steps 5–7).

Two key events account for the reduction in chromosome sets during meiosis. First, in meiosis I, when the chromosomes gather in the middle of the cell as the division phase begins, *corresponding chromosomes pair up, with one duplicated chromosome resting on either side of the central plate* (e.g., note the two long chromosomes aligned side by side across an imaginary line in Figure 7.20, step 3). As a result of this lineup, when the chromosomes separate, each daughter cell receives just one member of each pair of chromosomes, or one complete chromosome set. To understand how the reduction in chromosome sets occurs, contrast the alignment and separation of chromosomes during meiosis with what happens during mitosis: When the chromosomes gather in the middle of the cell in mitosis, they line up *individually in a row, with one-half of each duplicated chromosome on either side of the central plate* (see Figure 7.11, step 8); when the duplicated chromosomes split apart, each daughter cell receives a copy of every chromosome—that is, a copy of both members of each pair, or a double set (see Figure 7.11, step 9).

The second key reduction event in meiosis is really a nonevent: The chromosomes do not duplicate between the first and second cell divisions (see Figure 7.20, steps 4 and 5). As a result, when the chromosome halves split apart in meiosis II, each daughter cell receives just a single complete chromosome of each type, or one complete set. In contrast, the chromosomes duplicate before every mitosis (see Figure 7.6).

In short, during mitosis, the chromosomes are copied and apportioned equally to two daughter cells that are identical to the parent cell—one cell becomes two. During meiosis, however, one cell generates four daughter

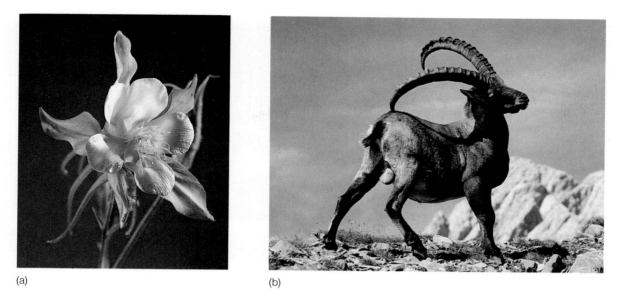

(a)   (b)

FIGURE 7.19 *Meiosis Occurs in Specialized Organs of Plants and Animals.* Meiosis takes place in (a) the anthers (here, yellow ovals) or ovaries of flowers such as the columbine and (b) the sex organs of a male ibex.

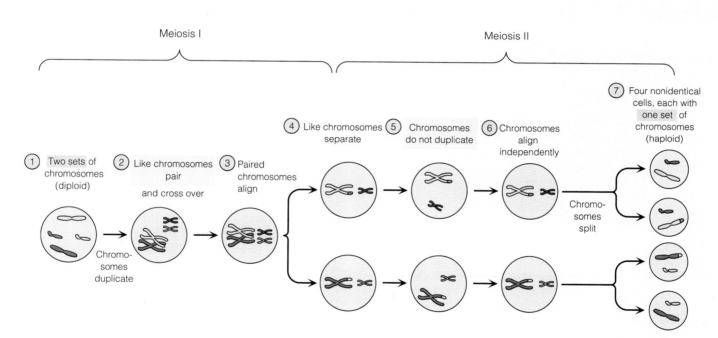

FIGURE 7.20 *Meiosis (In Reproductive Organs Only): An Overview.* An overview of meiosis shows that chromosome reduction occurs during meiosis I (1–4) as one diploid cell becomes two haploid daughter cells. A key event in this reduction division is the pairing of like chromosomes at the cell's center line (3). There is no chromosome duplication between meiosis I and meiosis II. As a result, during meiosis II, two haploid cells, containing previously duplicated chromosomes, divide to form four haploid daughter cells, each with one set of unduplicated chromosomes (5–7). Notice that the two members of the long chromosome pair exchange parts in 2–4. Because of this exchange, the four haploid cells produced by meiosis each contain a slightly different set of genetic material.

cells, each with half the original number of chromosomes. This reduction of chromosome sets from two to one is a necessary preparation for fertilization, when two haploid cells will fuse and create a new diploid individual. Errors in meiosis can lead to problems in development. If either egg or sperm lacks the correct number of chromosomes, the fertilized egg will have an incorrect number of chromosomes and perhaps develop into an individual with serious defects. **Down syndrome** is a fairly common set of genetic defects based on such errors in meiosis (Figure 7.21).

HOW GENETIC VARIATION ARISES DURING MEIOSIS

As we have already mentioned, meiosis is also responsible for the genetic variation found in a single family. A litter of puppies, all with different fur color, spot patterns, and body size, is visible evidence of the reshuffling of genetic material that occurred in the gonads of the parent dogs (Figure 7.22). The puppies resemble their parents in many ways, but differ in at least a few, and two aspects of meiosis—crossing over and independent assortment—gave rise to these differences. In *crossing over*, two like chromosomes swap corresponding parts.

FIGURE 7.21 *Down Syndrome: An Error in Meiosis.* Like one in every 1000 children born in the United States, this child has Down syndrome. Symptoms of this condition include mental retardation, malformations of the heart, a special type of eye fold, short stature, stubby hands and feet, and abnormal palm prints. A woman over 45 years old is 100 times more likely to have a baby with Down syndrome than a woman of 19. This is because the older a woman gets, the more likely it is that one of her eggs will sustain slight damage, and that damage can prevent meiosis from occurring properly. Abnormal meiosis and failure of paired chromosomes to separate can cause children to inherit an extra chromosome 21 and develop Down syndrome.

FIGURE 7.22 *Chromosome Reshuffling: A Source of Genetic Variation.* These puppy siblings of different sizes, colors, and fur patterns are visible evidence of the genetic variation produced by meiosis.

For example, if you follow the long chromosomes in Figure 7.20, you will note that the tips of the long chromosomes exchange places. Because the genetic information on the swapped parts is similar but not identical, crossing over constructs new combinations of hereditary material. In Figure 7.20, step 7, you can see that each of the four daughter cells has a distinctive long chromosome, indicated by the different combinations of dark and light blue. Since each of these four cells could develop into a different sperm or egg, each could become part of a different puppy with slightly different traits.

With *independent assortment*, genetic variability increases even further. In an organism with two long and two short chromosomes, the chromosomes could align during meiosis in one of two ways. One arrangement would yield eggs or sperm with long and short chromosomes, both of the same color. The other arrangement, in contrast, would yield eggs or sperm with a long chromosome of one color and a short chromosome of the other color (see Figure 7.20, step 7). If the genetic information on a dark and light long chromosome were slightly different, then four genetically different kinds of egg or sperm could result. Thus, independent assortment makes a great contribution to the genetic variability seen in individual families of people or puppies.

≋ MEIOSIS, MITOSIS, SEXUAL REPRODUCTION, AND EVOLUTION

As already described, mitosis and meiosis play very different but equally important roles in the life cycles of sexually reproducing, many-celled organisms (Figure

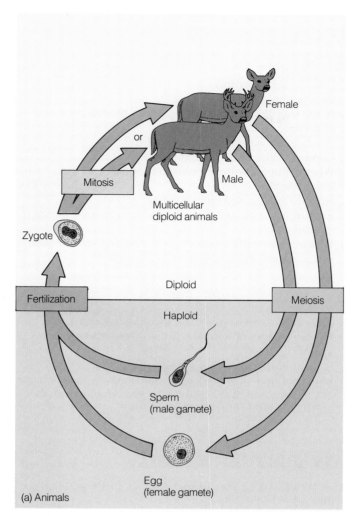

(a) Animals

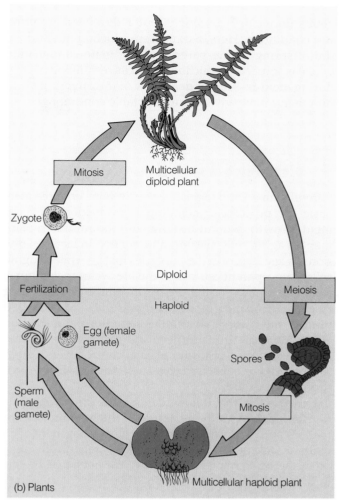

(b) Plants

FIGURE 7.23 *The Roles of Meiosis and Mitosis in the Life Cycles of Animals and Plants.* (a) In most animals, each adult produces either egg or sperm via meiosis, and the union of egg and sperm at fertilization yields a fertilized egg cell that divides by mitosis to form a new adult. (b) In many plants, meiosis produces spores with one chromosome set that grow by mitosis into multicellular plants with one set of chromosomes. Sometimes the plants are free-living, as in this fern, sometimes dependent on the adult plant, as in pines or tulips. Each plant with one chromosome set then produces eggs or sperm through mitotic divisions, and they join to make a new fertilized egg, restoring the two chromosome sets in the future adult.

7.23). In preparation for fertilization and sexual reproduction, meiosis reduces the chromosome number in the gametes and brings about a critical genetic reshuffling. Once fertilization takes place, mitotic divisions of the zygote produce the growing individual, whether it has 959 cells like a roundworm or 3 billion cells like a newborn human baby. In asexually reproducing organisms, such as potatoes and hydras, mitosis allows a few parent cells to generate identical offspring that will grow and once again reproduce, asexually or sexually.

It may seem odd that nature would invent two means of reproduction. But there are distinct advantages to each mode, depending on conditions at the time of reproduction. In an unchanging environment—say, the cold, dark mud at the bottom of the deepest oceans—asexual reproduction may be superior to the sexual mode. If a parent has a favorable genetic combination that allows it to survive these hostile conditions, then reproducing asexually would preserve the parent's time-tested genetic combination and automatically result in offspring suited to the dark, the cold, and the mud. In a changing environment, however, such as salt flats near the ocean shore, which are subject to periodic drying, flooding, and changing salt levels, one genetic combination might be advantageous under dry conditions but a disadvantage under wet conditions. Here, asexual reproduction might doom the offspring to the parent's losing gene combination, whereas sexual reproduction—with its reshuffling of genes during meiosis—might deal a few of the offspring a winning hand in surviving unpredictable conditions.

≋ CONNECTIONS

Carefully timed and regulated cell divisions underlie many basic life processes, including reproduction, development, growth, wound healing, and the routine replacement of aging cells, tissues, and organs. In fact, cell division is a key feature in the cell cycles that generate daughter cells from a parent cell and in the life cycles that generate a complex, many-celled organism from a fertilized egg. A graceful "dance of the chromosomes" during both mitosis and meiosis distributes hereditary information, either identical or reshuffled, and mistakes in chromosome distribution or loss of normal control over cell division can have consequences as serious as Down syndrome and cancer.

The chapters that follow build on the principles of cell division, delving into the nature of the hereditary information, sexual reproduction, and the development of the embryo. All these concepts, however, rely on the gene. So in the next chapter, we explore the gene and the historic experiments that led to its discovery.

NEW TERMS

asexual reproduction, page 128
cancer, page 125
cell cycle, page 116
cell division, page 116
cell growth, page 116
cytokinesis, page 120
diploid, page 130
Down syndrome, page 132
fertilization, page 129
gamete, page 129

genetic variation, page 130
growth factor, page 125
haploid, page 130
interphase, page 119
life cycle, page 116
meiosis, page 130
mitosis, page 120
sexual reproduction, page 129
zygote, page 129

STUDY QUESTIONS

REVIEW WHAT YOU HAVE LEARNED

1. How do we know that the chromosomes contain genetic information?
2. Name and describe the two phases of the cell cycle in a eukaryotic cell.
3. How do mitosis and cytokinesis differ?
4. Compare cytokinesis in animal and plant cells.
5. What might allow cancer cells to continue growing indefinitely?
6. How do mitosis and meiosis differ?
7. Explain how new genetic combinations can arise during meiosis.
8. Which is more advantageous in a changing environment, sexual reproduction or asexual reproduction? Why?

APPLY WHAT YOU HAVE LEARNED

1. How many chromosomes are there in each of the following types of human cells? skin; egg; muscle; nerve; sperm.
2. A plant breeder who has developed an attractive-looking pink geranium flower wishes to obtain many more flowers of the same variety. Should the breeder employ sexual or asexual methods of reproduction? Why?

FOR FURTHER READING

Blow, J. "Mitosis Comes Apart." *Trends in Genetics* 5 (1989): 166–167.

Chandley, A. C. "Meiosis in Man." *Trends in Genetics* 4 (March 1988): 79–84.

McIntosh, J. R., and K. L. McDonald. "The Mitotic Spindle." *Scientific American* (October 1989): 26–34.

Marx, J. L. "The Cell Cycle Coming Under Control." *Science* 245 (July 1989): 252–255.

Murray, A. W., and M. W. Kirschner. "Dominoes and Clocks: The Union of Two Views of the Cell Cycle." *Science* 246 (November 1989): 614–621.

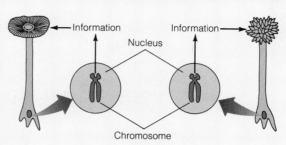

1. Chromosomes contain information that governs what a cell looks like and how it performs.

Information — Nucleus — Information

Chromosome

CELL CYCLE

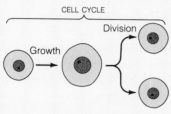

Growth — Division

2. The cell cycle has two phases: growth, when chromosomes and other cell parts duplicate, and division, when the duplicated parts are distributed to daughter cells.

Mitosis   Cytokinesis

3. During the division phase, the nucleus divides (mitosis), and then the cytoplasm divides (cytokinesis).

ON

OFF

4. Controls over the cell cycle start and stop cell division where and when needed.

ONE LIFE CYCLE

Development   Reproduction

Many cell cycles

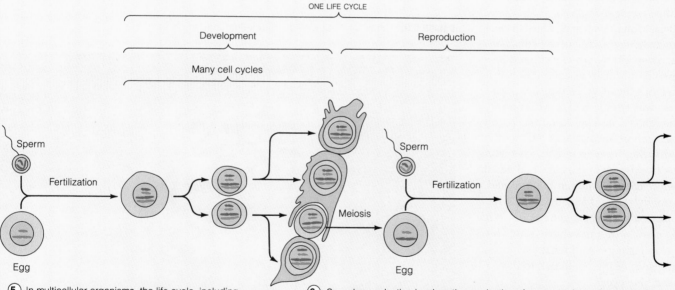

Sperm

Fertilization

Egg

Meiosis

Sperm

Fertilization

Egg

5. In multicellular organisms, the life cycle, including development and reproduction, includes many cell cycles.

6. Sexual reproduction involves the production of sperm and egg through meiosis (which halves the chromosome number) and then the fertilization of the egg (which doubles the chromosome number).

7. Asexual reproduction produces identical offspring.

THE PATTERNS OF HEREDITY

WHITE TIGERS AND HIT-AND-SKIP INHERITANCE

In the dry season of 1951, hunters found four tiger cubs playing in the hunting preserve of the maharajah of Rewa, in central India. Although the hunters were looking for bigger game, one of the cubs caused great excitement: Instead of the normal orange-and-black-striped coat, it had a pure white pelt! What's more, its nose and paw pads were pink instead of black, and its eyes were ice blue with a tendency to cross. The young tiger was brought to the maharajah's palace and named Mohan, meaning "Enchanter." Mohan grew into a magnificent creature, larger than most tigers, strong, and healthy (Figure 8.1).

When Mohan was mated with a normal female tiger, none of the cubs had Mohan's sparkling white coat, and it seemed that Mohan was a once-only occurrence, never to be repeated. But Mohan was kept in a cage with one of his daughters, and you can imagine the surprise of his keepers when they saw the offspring from an incestuous mating of the two: Some of the cubs had beautiful white coats. Although the trait had apparently skipped a generation, it had reappeared in the next generation. What could cause a trait to behave in this hit-and-skip fashion? And what was the relationship of the white pelt to the crossed blue eyes?

We will explore the answers to these questions as we investigate **heredity,** the transmission of physical, biochemical, and behavioral traits from parent to child. Observable traits, such as fur and eye color, are controlled by units called *genes*, which are specific, discrete portions of the DNA molecule in a chromosome. Your genes determine the shape of your hairline; whether you will bald as you age; whether you might become allergic to cat dander or pollen; and whether you are susceptible to certain types of heart disease and cancer. The rules of heredity describe how genes and the traits they determine move from generation to generation. Simple and straightforward, the rules hold in principle for all sexually reproducing organisms, from the single-celled *Paramecium* to peas and tigers and people.

Knowledge of genetics also helps reveal the workings of evolution by natural selection. By studying the action of genes and the ways they interact with one another, we can begin to understand how individuals come to have different shapes and functions that in turn may affect their ability to survive.

This chapter explores the rules of heredity and how they govern the transmission of traits from parents to offspring. It answers several questions:

- How did the monk Gregor Mendel discover and analyze genes?
- How do we know that genes are located on chromosomes?
- How do genes interact with one another and with the environment?

FIGURE 8.1 *A White Tiger with Blue Eyes That Tend to Cross.*

FIGURE 8.2 *Casual Observations of Traits Such as Flower Color or Shape Led to the False Blending Theory of Inheritance.* Petal color in hibiscus appears to be a blended trait: Red (right) and yellow (left) parents give rise to orange (center) offspring. However, most traits, even in hibiscus flowers, do not appear to blend. A parent with long petals and a parent with short petals, for example, give rise to offspring with long petals.

⚠ GENETICS IN THE ABBEY: HOW GENES WERE DISCOVERED AND ANALYZED

Since ancient times, farmers and gardeners have realized that many characteristics of domesticated plants and animals pass from parent to offspring. But it was not clear exactly how the traits were transmitted. Casual observations of the offspring resulting from the matings of various organisms gave rise to the notion that during reproduction, the hereditary "stuff" of the mother and father *blended* to produce the characteristics found in the offspring (Figure 8.2), just as cream mixes with black coffee to produce the tan-colored café au lait.

New ideas about the nature of matter, however, helped scientists formulate new ideas about inheritance. Gregor Mendel, an Augustinian monk who had studied physics and biology for two years at the University of Vienna, had learned there that each physical object is made up of particles called atoms and molecules. In the 1850s, he wondered if heredity could also be governed by particles that retain their identity from generation to generation, instead of blending like coffee and cream. He put this new hypothesis to the test in a carefully devised study that cataloged the number and types of pea plants in successive generations.

THE CRITICAL TEST: EXPERIMENTS WITH PEAS

To test whether hereditary traits blend over the generations or are passed from parent to offspring as if they were particles, Mendel performed mating experiments with garden peas. He chose the garden pea as his test subject for several reasons (Figure 8.3a). First, he could purchase pea strains that demonstrated clear alternatives for single traits, such as short stem versus long stem or white flowers versus purple flowers. By selecting strains that differed in only one trait, he could study inheritance of one feature unconfused by all other variations. Second, Mendel could easily control which pea plants mated with which. Pea plants normally *self-fertilize*. But Mendel could get around this simply by taking pollen-containing sperm from one plant and adding it to the egg-containing female flower parts of another plant. Mating

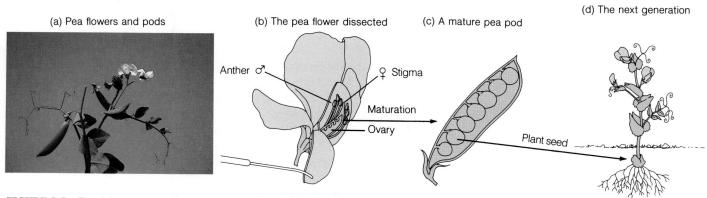

(a) Pea flowers and pods   (b) The pea flower dissected   (c) A mature pea pod   (d) The next generation

Anther ♂   ♀ Stigma   Maturation   Ovary   Plant seed

FIGURE 8.3 *The Advantages of Experimenting on Pea Plants.* Pea plants (a) not only have attractive flowers, but their petal arrangement (b) protects one pea flower from fertilization by the pollen of another pea flower. As a result, each flower fertilizes itself in nature. Mendel, however, was able to gently tease back the petals from the flower he chose to be the female parent and transfer to it pollen from the plant he wished to be the male parent. After fertilization and the maturation of the seed pod (c), Mendel could collect the seeds, plant them, and observe the characteristics of the offspring (d).

one individual plant with another individual plant is called *cross-fertilization*.

MENDEL DISCOVERS THE UNITS OF HEREDITY

Mendel analyzed the inheritance of clear-cut alternatives for seven pea traits, including stem length (long versus short), flower color (purple versus white), and seed shape (wrinkled versus smooth) (Figure 8.4). He began by demonstrating that he had strains of **pure-breeding** peas. When self-fertilized, the parental plants always produced offspring like themselves; plants with short stems had offspring with short stems, and plants with long stems had offspring with long stems.

Next, Mendel carried out matings between individuals that differed in only one trait, such as stem length. In one such cross, he planted long-stem and short-stem seeds and let them grow into the **parental (P) generation** (Figure 8.5a). When the parental plants had flowered, Mendel mated long-stem plants with short-stem plants; and when the pods became swollen with plump peas, he collected the seeds. These seeds would produce the next generation, called the **first filial (F_1) generation,** meaning the first generation in the line of descent. The seeds and the plants that grew from them are called *hybrids* because they result from the mating of parent plants with different characteristics. Planted the next season, the F_1 hybrid seeds of the long-stem/short-stem cross all grew

	Stem length	Flower color	Seed shape
Dominant characteristic (dominant allele)	Long	Purple	Round
Recessive characteristic (recessive allele)	Short	White	Wrinkled

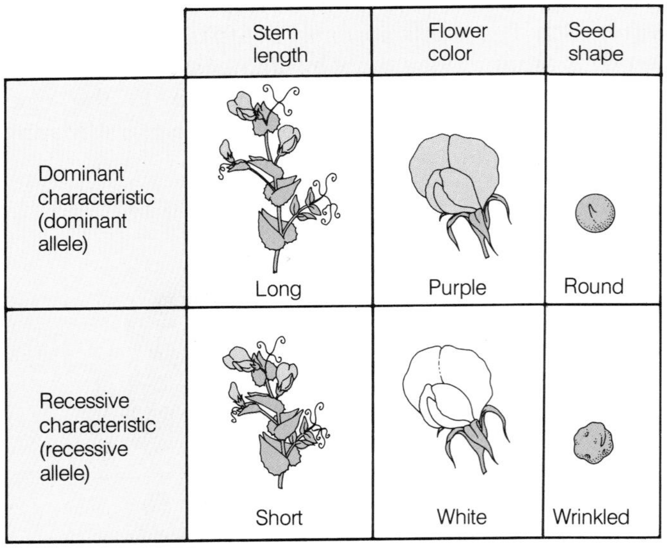

FIGURE 8.4 *Mendel Studied Several Pairs of Traits in Pea Plants.* Each of the traits (stem length, flower color, seed shape, and others) can appear in two forms: a dominant form, which shows up in the hybrid offspring of mixed parents, and a recessive form, which is hidden in the hybrid.

into long-stem plants (Figure 8.5b). The characteristic that appears in the F_1 hybrid is said to be **dominant,** while the one that does not appear is referred to as **recessive.** Long stems in peas and orange fur in tigers are dominant characteristics; short stems and white fur are recessive.

What happens to the recessive characteristic? Does it disappear completely? Does it blend with the dominant characteristic? Or does it remain intact but hidden in the first (F_1) generation? Although he would have to wait a year to find out, Mendel knew exactly how to learn the answers to these questions. He allowed the long-stem F_1 hybrid plants to self-fertilize, and the next spring he planted the seeds of the **second filial (F_2) generation.** When the second generation of pea plants matured, most of them had long stems, but significantly, there were some plants with short stems. Again, no stems of intermediate length appeared (Figure 8.5c). The reappearance of pure short stems among the offspring of long-stem hybrids proved dramatically that the short-stem trait had not blended with the long-stem trait; this was concrete evidence that hereditary traits behave like particles. This idea is known as the particulate theory of heredity.

MENDEL'S FIRST LAW: THE ALTERNATIVE FORMS OF A GENE SEGREGATE

Mendel was not satisfied with saying that "some" of the F_2 plants had short stems. He counted the number and found that 787 of the F_2 plants had long stems, while 277 had short stems (see Figure 8.5c). These numbers showed about a 3:1 ratio (2.84:1) of long-stem to short-stem plants in the F_2 generation. When Mendel examined separate crosses for all of the traits he studied, he obtained similar results. Clearly, the unit of heredity that produced short-stem flowers in the parental generation did not blend, but had been passed along to the first (F_1) generation, where it remained hidden until the second (F_2) generation.

■ **Genes and Alleles** The significance of these findings was clear to Mendel. Since short stems reappeared in the F_2 plants, the hereditary factor that causes short stems had to be an individual unit, like a particle, and not like a liquid that could be mixed. Although Mendel did not use the term himself, we now call this unit of inheritance a *gene*. A gene controls a specific trait in an organism, such as the length of a pea stem or the color of a tiger's coat or whether or not your earlobes are attached to your head. The gene is not the trait itself, but a factor that causes the organism to form the trait. We now know that genes are particular stretches of DNA along the length of a chromosome. Mendel also showed that genes come in alternative forms we now called **alleles.** In pea plants, the gene for the stem length trait has two alternative

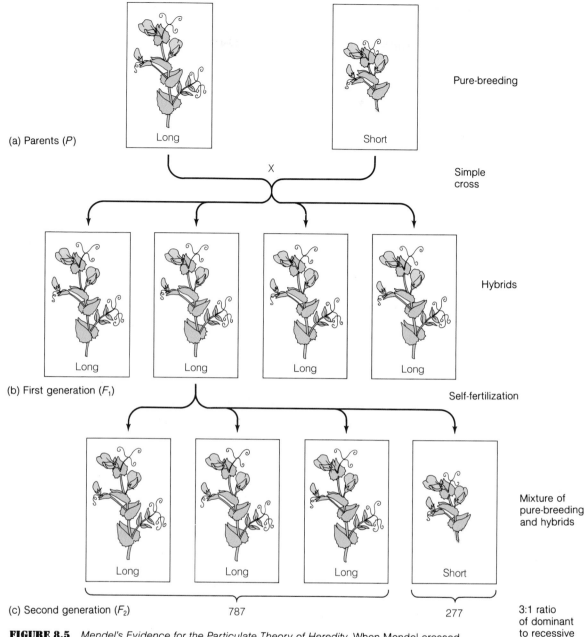

(a) Parents (*P*)

Long

Short

Pure-breeding

X

Simple cross

(b) First generation (*F₁*)

Long   Long   Long   Long

Hybrids

Self-fertilization

(c) Second generation (*F₂*)

Long   Long   Long   Short

787   277

Mixture of pure-breeding and hybrids

3:1 ratio of dominant to recessive

FIGURE 8.5 *Mendel's Evidence for the Particulate Theory of Heredity.* When Mendel crossed long- and short-stem pea plants (a), he got all long-stem progeny in the *F₁* generation (b), but short stems reappeared in one out of every four *F₂* plants (c).

alleles, one causing long stems and one causing short stems (Figure 8.6, page 140). Likewise, the gene for coat color in tigers has two alternative alleles, one for orange fur and one for white fur.

■ **Dominant and Recessive Alleles** Mendel realized that a hybrid has two alleles, one "visible" and one "invisible." The allele generating a visible trait is said to be *dominant;* a tiger with an allele for an orange coat and an allele for a white coat will always "look" just like a pure-breeding orange tiger with two alleles for orange coat.

The alternative allele that is "invisible," overshadowed each time it is paired with a dominant allele, is said to be *recessive* (like the trait it determines). The alleles for white coats in tigers and short stems in pea plants and attached earlobes in people are recessive.

By studying the hybrid plants, Mendel reached an even more general conclusion: Each individual has two alleles for each gene. An individual with two different alternative alleles for a particular trait is called a **heterozygote** (*hetero* = different). A hybrid carrying one short-stem alternative and one long-stem alternative is a heterozygote. In contrast, an individual with two identical alternative alleles for a particular trait is called a **homozygote** (*homo* = same). A pure-breeding short-stem pea plant carrying two identical short-stem alleles is a homozygote.

■ **Genetic Makeup Versus Physical Constitution** The existence of dominant and recessive alleles means that you cannot always tell what the genetic makeup is just by looking at an individual. Pea plants with long stems, for example, may have two identical copies of the long-stem allele, or they may be hybrids showing only the dominant form of the two unlike alleles. In short, two plants or animals with a different genetic makeup, or **genotype,** for a certain trait may have the same visible physical makeup, or **phenotype** (Figure 8.7). By looking in the mirror, you can easily see your physical makeup for the earlobe gene. But what is your genetic makeup? If your earlobes are attached, you have two copies of the

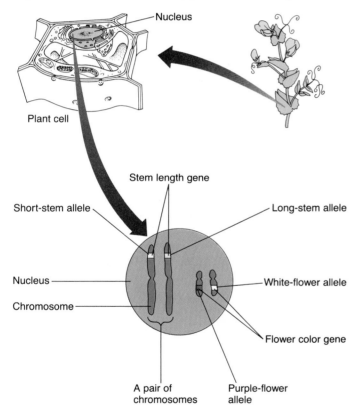

FIGURE 8.6 *Genes Are Part of a Chromosome and Come in Alternative Forms, or Alleles.* This diagram of a plant cell nucleus shows only two of the several chromosome pairs in the plant's cells. Each long chromosome depicted here carries an allele of the gene for stem length. Each short chromosome carries an allele of the gene for flower color. In reality, each chromosome carries alleles of many genes.

recessive "attached" gene form. But what if your earlobes hang free, unattached to your head? You may have two copies of the free-hanging allele, or you may have one copy of the free-hanging allele and one copy of the attached allele.

■ **Separation of Alleles in Egg and Sperm Formation** Mendel suggested that the two alleles in a parent separate from each other to produce eggs or sperm that have only one allele apiece; then, when egg and sperm fuse at fertilization, the offspring gets an allele of each gene from each parent (Figure 8.8, page 142). Generalizing from Mendel's pea experiments, we can define Mendel's first law, the **principle of segregation:** Sexually reproducing diploid organisms have two alternative alleles of each gene, and these two alleles separate (segregate) from each other as sperm and eggs form. As a result, each egg or sperm contains only one allele of each gene (review Figure 8.8a). Since we now know about chromosomes and their behavior during the special cell divisions preceding reproduction (meiosis; review Chapter 7), we can understand Mendel's principle in concrete terms. The two alleles of a gene separate (segregate) because they are part of different chromosomes in a pair, and during meiosis, those two chromosomes separate into different eggs or sperm (see Figure 8.8).

INSIGHTS FROM MENDEL'S FIRST LAW

Mendel's segregation principle can explain one of the mysteries concerning white-coated Mohan. Mohan's orange first-generation children were hybrids bearing one white-coat allele of the coat-color gene received from pure-breeding (homozygous recessive) white Mohan and one orange-coat allele of the coat-color gene from their normal, pure-breeding (homozygous dominant) orange mother. The fact that Mohan's first-generation children had orange fur shows that the white-coat allele is recessive and the orange-coat allele is dominant.

■ **Peas and Probabilities** Mendel's segregation principle can also explain the *types* of first (F_1) and second (F_2) generation pea plants, but can it account for the *numbers* Mendel observed in the F_2 generation: three long-stem plants for every plant with a short stem? Mendel applied his analytical skills to show that it could.

Geneticists usually use letters to represent genes, with an uppercase (capital) letter for the dominant allele of a gene and a lowercase letter for the recessive allele of the same gene. For the *length* gene, we will use the letter L to designate the dominant long-stem alternative allele and the letter l to designate the short-stem allele. With this system, the hybrid F_1 generation is Ll; the order of alleles is not important, but it is customary to place the uppercase letter first. Follow Figure 8.9 (page 143) and

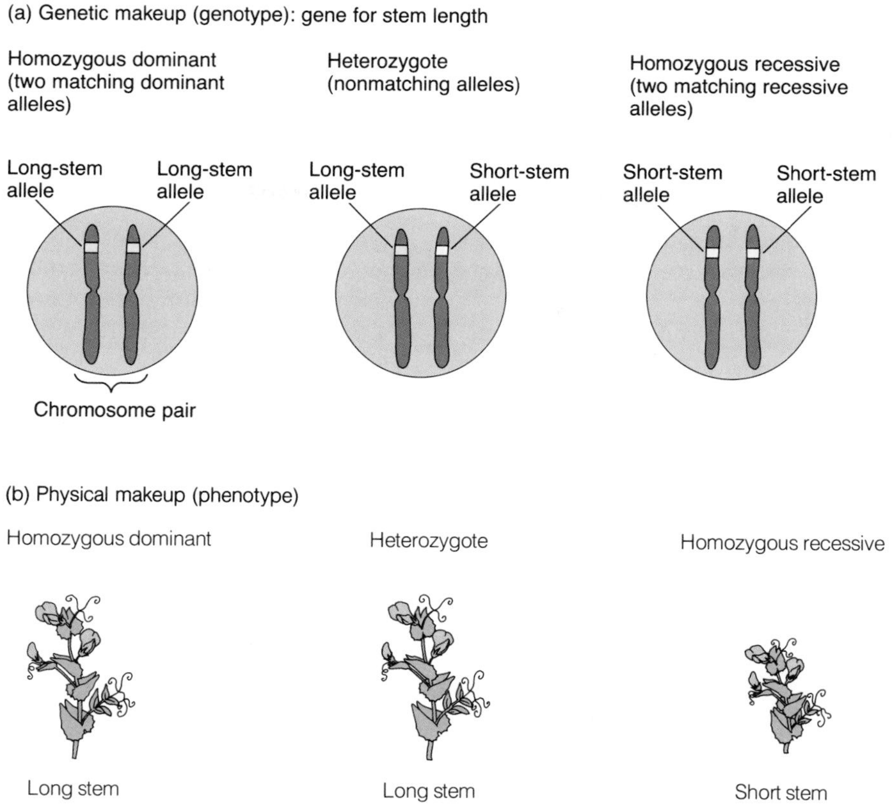

Homozygous dominant (two matching dominant alleles)

Heterozygote (nonmatching alleles)

Homozygous recessive (two matching recessive alleles)

Long-stem allele   Long-stem allele

Long-stem allele   Short-stem allele

Short-stem allele   Short-stem allele

Chromosome pair

(b) Physical makeup (phenotype)

Homozygous dominant

Heterozygote

Homozygous recessive

Long stem

Long stem

Short stem

FIGURE 8.7 *Genetic Makeup Versus Physical Makeup: How Do They Relate?* Underlying an organism's visible, physical makeup (its phenotype) is a set of two alleles (the genetic makeup, or genotype). Plants that look the same (have long stems, for instance) may have different genetic makeups: two matching dominant alleles or one dominant and one recessive allele.

you will see how Mendel used this lettering system to explain the results of a mating between two heterozygous first-generation hybrid plants.

The hybrid plants shown at the top of Figure 8.9 start out with two alternative alleles of the gene for stem length, designated by *L* and *l*. When parents make eggs or sperm, these two alleles separate (or *segregate*, to use Mendel's term) as the two chromosomes that carry them separate into different egg or sperm cells (Figure 8.9, steps 1 and 2). As a result, half the eggs or sperm carry *L*, and the other half carry *l*, but no eggs or sperm carry both. At fertilization, when egg and sperm fuse, the resulting zygote, or fertilized egg cell, randomly receives one member of each gene pair from its mother and one member of each gene pair from its father. The *l* pollen grain is just as likely to pair up with an *l* egg (step 3) as with an *L* egg (step 4), and vice versa. By continuing to fill in the boxes of the square, you can see that there are four possible combinations of egg and sperm: *LL, Ll, lL,* and *ll*. The homozygous *LL* combination produces long stems, as do the heterozygous *Ll* and *lL* combinations. The homozygous *ll* combination, on the other hand, generates short stems. Adding up these possibilities, we get

three long-stem plants for every one short-stem plant; and that, if you recall, is exactly what Mendel observed.

To recap Mendel's first law: Alleles separate as eggs or sperm (gametes) form, and alleles pair up at random during fertilization as fertilized eggs (zygotes) form. It's as simple as that. Like the toss of a coin, the combination of pollen and egg at fertilization is governed by the laws of chance. In a small number of trials, the results may differ substantially from those predicted for random tossing, but as the number of trials increases, we will obtain results closer to the mathematically predicted values. Mendel's genius was to deduce the existence of entities we now call genes from the ratios he found in hundreds of garden plants.

MENDEL'S SECOND LAW: DIFFERENT GENES ARE INHERITED INDEPENDENTLY

The two concepts we just discussed, the segregation principle and the idea of random fertilization, explain how the alleles of a single gene are inherited. But what happens if we want to follow two genes in the same mating? To answer this question, Mendel mated two pea plants that differed in two characteristics, stem length and flower color. (With the pigment gene for flower color, purple is dominant over white, so we represent the purple allele by P and the white allele by p.) For his parental generation, Mendel chose double-dominant long-stem plants with purple flowers and double-recessive short-stem plants with white flowers.

In the first (F_1) generation produced by mating long purple to short white, all the offspring were long purple. In other words, the dominant forms for both stem length and flower color appeared in the F_1 generation (Figure 8.10, page 144). Mendel then allowed the F_1 plants to self-fertilize to produce the second (F_2) generation. The outcome of this mating was not so easy to predict. Would the two dominant traits (purple flowers and long stems) always stay together and the two recessive traits (white flowers and short stems) do likewise? Or would new combinations of dominant and recessive traits appear? Look at Mendel's field of second-generation peas and discover the answer for yourself.

Most of the second-generation plants were either double dominants (long-stem plants with purple flowers) or double recessives (short-stem plants with white flowers) like the original parents; but some of them displayed new combinations—long white or short purple (see Figure 8.10). These new combinations were visible evidence that different genes (such as those for stem length and flower

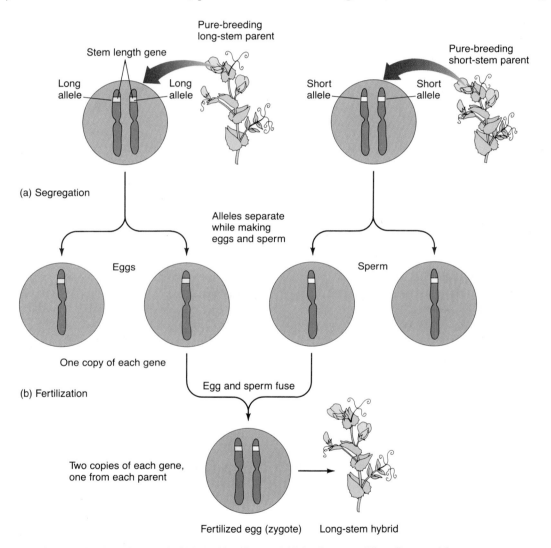

FIGURE 8.8 *The Making of a Hybrid: How Parental Alleles Separate When Eggs and Sperm Form, and How They Pair Up at Fertilization.* This figure shows the making of a hybrid from two pure-breeding parents of different physical makeups. (a) The pair of alleles for stem length separate as eggs and sperm form, so that each egg and each sperm carries only one copy of each gene. (b) At fertilization, the egg provides one stem length allele, and the sperm provides the alternative allele. As a result, the hybrid carries two different alleles, one from each parent.

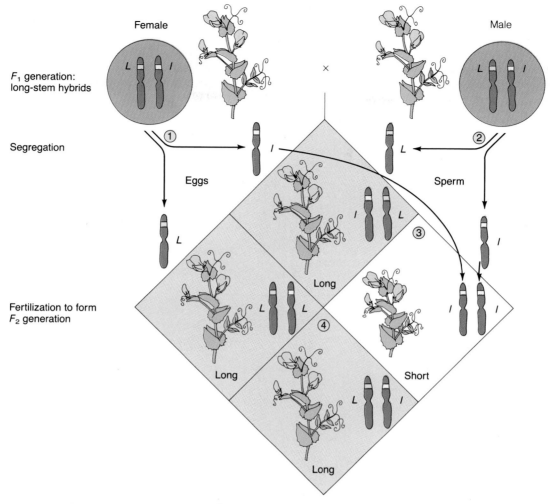

Female   Male

F_1 generation: long-stem hybrids

Segregation ① ②

Eggs   Sperm

③

Fertilization to form F_2 generation

④

Long

Long   Short

Long

Genetic and physical makeup of F_2 generation

FIGURE 8.9 *The Segregation Principle and Random Fertilization: Using a Punnett Square to Predict Genetic and Physical Makeup.* Mendel proposed that alleles separate during the formation of gametes (1, 2) and randomly pair up at fertilization (3, 4). As a result, each offspring inherits one allele from each parent for a given trait (here, stem length). The idea that alleles separate during gamete formation became known as the segregation principle. By writing the two kinds of alleles from the female parent along one side of a square and the two kinds of alleles from the male parent along the other side, we can pair the alleles and determine the genetic types that could result in the offspring. Here, parents heterozygous for stem length produce one homozygous long, two heterozygous long, and one homozygous short offspring, or a ratio of three long-stem plants to one short. This kind of square table is called a *Punnett square* after the British mathematician who first used it.

color) do not always stay together when eggs and sperm form. From his observations of the F_2 plants, Mendel arrived at his second law of heredity, the **principle of independent assortment.** Expressed in modern terms, this principle states that genes on different chromosomes assort independently of each other during the formation of eggs and sperm. In other words, the dominant alleles of two genes on different chromosomes do not have to stay together just because they were together in an ancestor. Instead, genes on different chromosomes are uncon-

nected and free to assort independently of their original partners. Figure 8.11 (page 145) shows how this actually happens when cells divide and produce eggs and sperm. Because independent assortment produces different gene combinations, it is a source of genetic diversity.

MENDEL'S RESULTS IGNORED

Biologists of Mendel's time turned a deaf ear to the monk's analysis of inherited characteristics, and although he published his results in 1865 in a journal that was sent to about 100 scientific libraries, the few plant breeders who cited them indicate by their remarks that they did not understand the importance of Mendel's three main concepts: (1) the gene as the unit of heredity; (2) the segregation of alleles and random meeting of egg and sperm at fertilization; and (3) the independent assortment of two different genes. For more than 30 years, his ideas lay dormant, until 1900, when botanists in Austria,

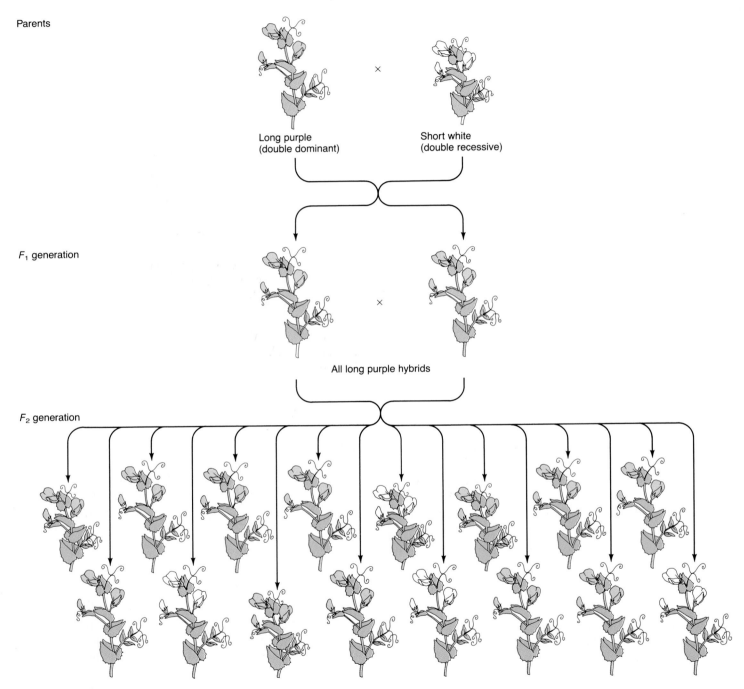

FIGURE 8.10 *A Two-Trait Cross: Alleles of Different Genes Assort Independently.* The first-generation offspring of a long purple plant and a short white plant will be all long purple hybrids. But in the second-generation offspring of the long purple hybrids, a few of the plants will have long stems and white flowers or short stems and purple flowers. These new combinations show that alleles of different genes assort independently.

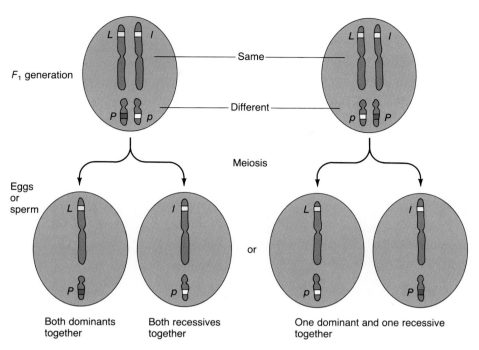

Sometimes, chromosomes line up in meiosis like this:

And other times, like this:

F_1 generation

Same

Different

Meiosis

Eggs or sperm

Both dominants together

Both recessives together

or

One dominant and one recessive together

FIGURE 8.11 *How Genes on Different Chromosomes Assort Independently.* The way chromosome pairs line up in meiosis determines how the genes they carry assort into eggs or sperm. Here we show how two different lineups result in two different assortments.

Germany, and the Netherlands duplicated Mendel's experiments in other plants and recognized the importance of his work.

☷ LOCATING AND STUDYING GENES ON CHROMOSOMES

Mendel, as we have mentioned, did not know that genes ride on chromosomes; but while the report of his discoveries was gathering dust in European libraries, scientists were making new discoveries about how chromosomes move during mitosis and meiosis (see Chapter 7). Then, shortly after the rediscovery of Mendel's principles, biologists realized that there are several parallels between the inheritance of genes and the distribution of chromosomes during meiosis: For example, each cell contains two copies of each gene and two copies of each chromosome; a pair of alleles and a chromosome pair both segregate into different gametes; and unrelated genes and separate chromosomes both assort independently when eggs and sperm form. These correlations suggested that genes are physically linked to chromosomes. In confirming that idea, investigators showed that a specific trait is always transmitted along with a specific chromosome. They started by studying sex and eye color in

tiny fruit flies, but their results hold for genes and chromosomes in all sexually reproducing organisms, including humans.

DIFFERENT SEXES—DIFFERENT CHROMOSOMES

In fruit flies, tigers, and people, females have two identical **X chromosomes,** while males have only one X chromosome and another unpaired, smaller chromosome called a **Y chromosome** (Figure 8.12, page 146). The X and Y chromosomes are known as **sex chromosomes.** They determine gender—whether we are male or female—and so we can say that this specific trait is always transmitted by these chromosomes. Although sex chromosomes are common in animals, they are rarely found in plants. In contrast to sex chromosomes, those chromosome pairs with two identical members in both sexes are called **autosomes.**

Figure 8.13 (page 146) shows how sex chromosomes are inherited. In a woman, the two X chromosomes segregate into different eggs. Similarly, a man's single X and his Y pass into different sperm. Look at the sons in Figure 8.13: They all have their mother's X chromosome and their father's Y. Abraham Lincoln once said, "All I am and have I owe to my mother." This certainly was true for the characteristics related to his *X* chromosome. The inheritance of *X* chromosomes is sometimes called crisscross inheritance because sons get their single *X* from their mother and daughters get one of their *X*'s from their father.

SEX, WHITE EYES, AND YELLOW BODIES

Since males and females have different chromosomes, we know that at least one trait—gender—is regulated by chromosomes. But are there any others? To answer that question, geneticists who did not have Mendel's monastic patience turned to an organism that breeds much more rapidly than peas—the prolific fruit fly called *Drosophila melanogaster*. Although pests in the kitchen, fruit flies are beautiful under the microscope, with brick-red eyes, tan abdominal stripes, and glistening black bristles (see Figure 8.12). They are also easy to raise and to breed, and they develop quickly: In just 12 days, an egg will become a reproductive adult ready to produce hundreds of offspring.

One day around 1910, while looking in the microscope, an American geneticist saw an intriguingly different kind of fly—a male with eyes of snow white instead of the usual brick red (Figure 8.14). A *mutation*—a permanent change in the genetic material—had altered a

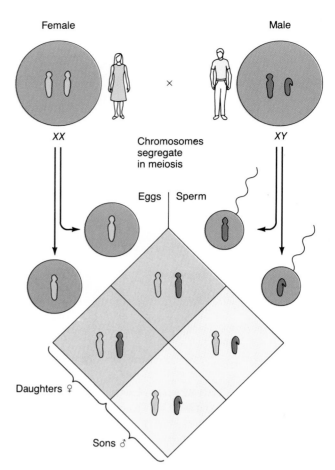

FIGURE 8.13 *Crisscross Inheritance of the X Chromosome.* Daughters inherit one of their *X* chromosomes from their father, while sons inherit their only *X* chromosome from their mother.

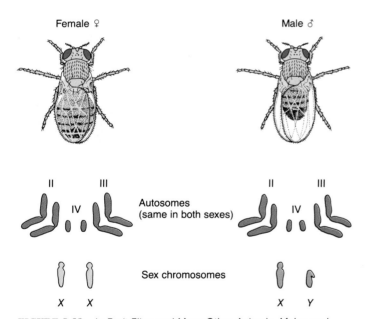

FIGURE 8.12 *In Fruit Flies and Many Other Animals, Males and Females Have Identical Sets of Autosomes But Differently Shaped Sex Chromosomes.* Fruit flies are striking creatures that have four chromosome pairs, three identical sets of autosomes and one pair of sex chromosomes. As in people, males have one *X* and one *Y* sex chromosome, while females have two *X* chromosomes.

gene for eye color from the normal red-eye allele to the white-eye allele. In the next few years, investigators found other mutations in fruit flies. One changed the fly's body color from brown to yellow, and follow-up experiments showed that the body color gene, like the eye color gene, is on the *X* chromosome.

Having identified two genes on the same chromosome, geneticists now wondered whether the two genes assort independently, as predicted by Mendel's second law. They found, to the contrary, that genes on the same chromosome tend to be inherited together. Because of this fact, geneticists say they are *linked*. Notice that we say "tend to be inherited together"; in biology, there are few absolutes, and sometimes even genes linked on the same chromosome can become separated in the next generation.

WHAT CAUSES NEW COMBINATIONS OF GENES ON THE SAME CHROMOSOME?

Rare new combinations of genes on the same chromosome can arise if crossing over occurs during meiosis

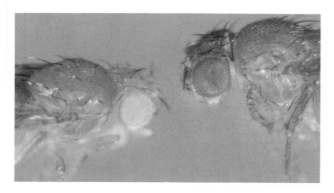

FIGURE 8.14 *The White-Eye Mutation in* Drosophila. The fly on the right with red eyes is normal, or wild-type, in color, while the fly on the left is a white-eyed mutant.

identify a gene's location has enabled medical geneticists to isolate genes causing various human diseases (such as cystic fibrosis or muscular dystrophy or certain types of cancer), and the isolation of these genes may one day lead to effective therapies.

The series of experiments leading from Mendel to mapping established fundamental principles about the gene, an entity so basic that its name reflects its role in *gene*rating our bodies, our cells, ourselves. The study of

(review Figure 7.20). Remember that, in crossing over, the ends of a pair of like chromosomes simply exchange places as if they were two pieces of string that had been cut, swapped, and retied to the opposite piece. We now know that the chromosomes carry genes, and these, too, are swapped. Figure 8.15 shows how crossing over could recombine the alleles of different genes on the same chromosome so that an original double-mutant and double-normal chromosome would give rise to a few recombined chromosomes with one mutant and one normal allele.

The occasional recombination of linked genes during meiosis has two important consequences. First, a new combination of gene forms may produce a new physical makeup—such as keener eyesight or stronger muscles—and offspring with the new physical makeup may be able to survive longer or reproduce more efficiently than the parents. Like recombination by independent assortment (see Figure 8.8), recombination by crossing over increases genetic diversity and the opportunities for evolution.

Second, the occasional recombination of linked genes allows researchers to make a **genetic map,** a drawing that locates the relative positions of genes on chromosomes (Figure 8.16, page 148). Genetic mapping is possible because genes that frequently recombine are farther apart on the chromosome than genes that almost never recombine. Comparing the recombination rates of different sets of genes gives a measure of how far apart they are, just as timing a trip between two cities on a freeway can give an idea of the distance between them on the road map. From genetic maps prepared using this logic, geneticists drew an important conclusion: Genes are located on chromosomes at specific places. In other words, the gene for white versus red eyes is in the same place on the same chromosome in any fruit fly, and the gene for attached versus free earlobes is in the same chromosomal position in all people. In our discussion of human genetic diseases (see Chapter 11), we will see that the ability to

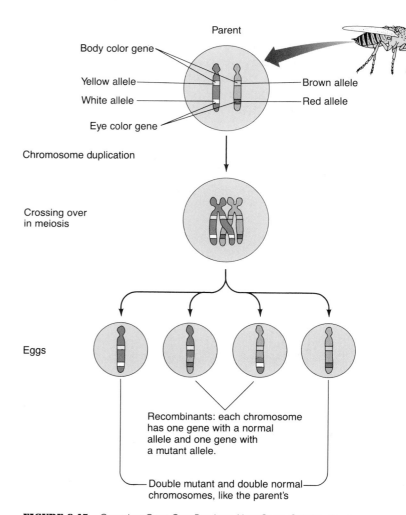

FIGURE 8.15 *Crossing Over Can Produce New Gene Combinations.* Although alleles of genes linked on the same chromosomes are usually inherited together, rare new allele combinations can arise from crossing over during meiosis. Most of the eggs produced by this female fruit fly will carry a chromosome with yellow-body and white-eye alleles *or* a chromosome with brown-body and red-eye alleles. These are the same combinations found in the parent. A few eggs, however, will carry new combinations (yellow body/red eyes or brown body/white eyes).

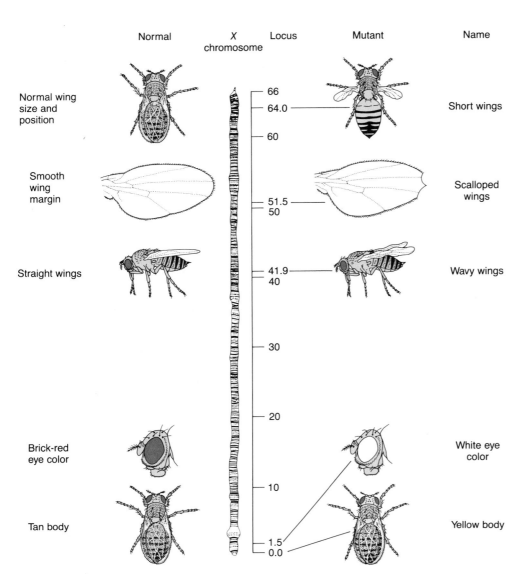

| Normal | X chromosome | Locus | Mutant | Name |

Normal wing size and position

Smooth wing margin

Straight wings

Brick-red eye color

Tan body

66
64.0
60

51.5
50

41.9
40

30

20

10

1.5
0.0

Short wings

Scalloped wings

Wavy wings

White eye color

Yellow body

FIGURE 8.16 *Map of the* Drosophila X *Chromosome.* The X chromosome of the fruit fly has dark and light bands when stained, and geneticists can locate individual genes in specific bands. This shows that genes are located at specific sites on chromosomes.

simple organisms, such as peas and flies, led to great concepts, namely, the location of genes at identified places on specific chromosomes and the rules governing the inheritance and recombination of chromosomes. As we continue, we will find that a gene does not perform its function in a vacuum, but interacts with other genes and the environment to specify how an organism develops, operates, and lives.

≋ GENE INTERACTIONS: EXTENSIONS OF MENDEL'S PRINCIPLES

In biology, every rule is challenged by phenomena that seem to contradict it, and Mendel's principles of heredity are no exception. An early geneticist admonished his students, "Treasure your exceptions." An unexpected result, a fly that is not predicted by the "rules," makes a statement about how genes work or how an organism functions. As biologists studied more and more genes, they discovered that interactions between alleles of the same gene, between different genes, and between genes and the environment often generated unexpected observations. On closer inspection, however, the unforeseen genetic behavior could be explained by slight extensions of Mendelian principles.

INTERACTIONS BETWEEN ALLELES

According to Mendel, an allele is either dominant or recessive, but later geneticists found that some alleles fail to fall clearly into either category (*incomplete dominance*). Pure-breeding snapdragons, for example, can be white or red (Figure 8.17). But a hybrid has pink flowers, not white or red as Mendel's concept of dominance would have predicted. We can explain the pink flowers if we assume that two doses of the red-flower allele make enough red pigment in snapdragon petals to produce red petals, and white-flower alleles make no pigment at

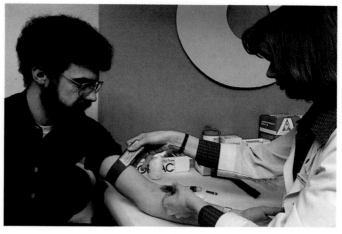

(b) Blood types

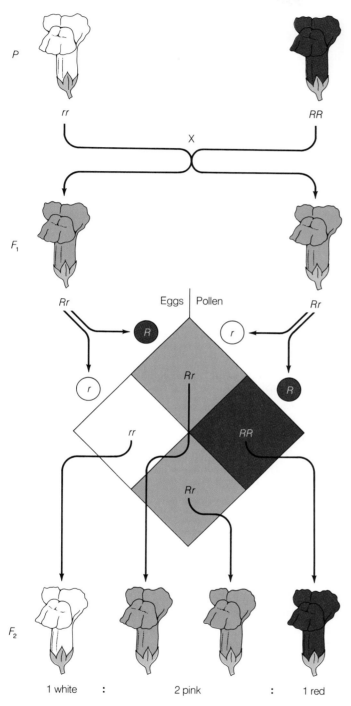

Physical makeup		Genetic makeup
Blood type	Cell surface molecule	
A	Red blood cell	AA or Ao
B		BB or Bo
AB		AB
O	(Neither A nor B)	oo

FIGURE 8.18 *Blood Types Illustrate Codominance and Multiple Alleles.* (a) Person having blood drawn from vein in arm. (b) Blood type genes control molecules on a person's red blood cells. One blood group gene has three alleles: *A*, *B*, and *o*. In an *AB* hybrid, both *A* and *B* alleles are expressed fully, leading to the AB blood type. From family studies, we know that matings between a person of A blood type and a person of B blood type can give very different progeny, depending on the genetic makeup of each parent.

all, leaving the petals snowy white. Thus, a hybrid with one red-flower and one white-flower allele would make an intermediate amount of pigment, and hence the petals would look pink.

In another type of hybrid, two alternative alleles are fully apparent; thus, *both* physical types show up in the organism. Since both alleles "appear," they are called *codominant*. A familiar example is blood type (Figure 8.18).

1 white   :   2 pink   :   1 red

FIGURE 8.17 *Flower Color in Snapdragons.* A hybrid plant with one red and one white allele (*Rr*) has pink flowers because a single red allele in an *Rr* hybrid makes less pigment than the two red alleles in an *RR* plant. With less pigment, the hybrid appears pink. This phenomenon is called *incomplete dominance.*

An allele we'll call *A* causes a marker—a certain sugar of type A—to appear on the surface of red blood cells, and the person then has blood type A. Allele *B* causes a different marker—a sugar of type B—on the blood cell surface, giving blood type B. Someone who has two *A* alleles has only the A sugar, and a person who has two *B* alleles has only the B sugar. But a heterozygote with one *A* and one *B* allele will have both A and B sugars on the surface of their red blood cells and hence will have blood type AB, a codominant physical makeup.

■ **Many Alleles of One Gene**   Although Mendel studied only two alleles for each of his seven genes, some genes have more than two alleles. Human blood groups illustrate this concept. In addition to the two codominant alleles *A* and *B*, this gene has a third allele that is fully recessive to both *A* and *B*. This recessive allele is called *o*, and a person with two doses of *o* has neither the A nor the B sugar marker and thus has blood type O. Since *o* is recessive, a person with alleles *A* and *o* will have blood type A, and a person with alleles *B* and *o* will have blood type B. Although there are three alleles of the blood type gene in the human population, a single individual can never have more than two, because as Chapter 7 explained, each child gets only one form of a gene from each parent.

The blood type gene has three alleles, but some genes have even more. The genes controlling tissue rejection in heart, liver, or kidney transplantations all have many alternative alleles. Such genes cause certain proteins to appear on a cell's surface, and these proteins serve as identification markers on each individual's tissues and organs. The protein markers help our bodies distinguish our own cells from organisms like bacteria, viruses, or parasites that might invade and cause disease.

Since there are several genes that control tissue rejection, each with multiple alleles, it is very unlikely that two unrelated (or even related) persons will have precisely the same constellation of alleles at all chromosome sites. That is why a person with kidney or liver disease must often wait a long time before becoming matched to a suitable donor. If there are too many allelic differences between the tissues of donor and recipient, the cells of the recipient will kill the cells of the donated organ. Even with drugs that quell these reactions, multiple alleles for tissue types constitute a major obstacle to life-saving transplantations for a large number of people. New drugs and therapies may change that outlook by the year 2000.

INTERACTIONS BETWEEN GENES

Until now, we have talked about genes whose alleles determine clear alternatives: white versus orange fur in

FIGURE 8.19   *Human Skin Color Is a Result of Several Genes.* Skin color differences between Africans and Europeans are probably due to differences in only four genes. Matings between blacks and whites usually give intermediate skin colors in the children, and matings between F_1 individuals show skin color variations, with some black and some white individuals resulting, but mostly brown-skinned people of many shades from dark to light.

tigers, purple versus white flowers in peas, attached versus free-hanging earlobes in people. Some traits, however, are not distinct like these, but vary continuously over a range of values and can be measured: for example, the length of a tobacco flower in millimeters, the amount of milk produced by a cow per day in liters, and the height of a person in meters. Since these traits vary in a measurable quantity, they are called **quantitative traits.** Quantitative traits are controlled by several interacting genes rather than by a single gene's pair of alleles. This makes the traits complicated for the geneticist to trace.

A good example of a quantitative trait is skin pigmentation in people (Figure 8.19). The differences in skin color between African blacks and Northern European whites stem from interactions between several genes. Matings between blacks and whites produce children with intermediate skin color, and matings between first-generation individuals produce children with a wide range of skin pigmentations—a few as light as the original white parent and a few as dark as the original black parent, but most in a middle range. This result can be explained if four genes inherited in a normal fashion act together to control the difference in skin color between African blacks and European whites. If each African allele adds color and each European allele takes away color, then the observed distribution makes sense.

Understanding the inheritance of quantitative traits is especially important for agricultural geneticists, who through cross-breeding and selecting recombinants, try to develop new gene combinations that will, for instance, grow more bushels of wheat per acre or more pounds of pig per pound of feed. However, because genes interact with each other and the environment, it is necessary to use very careful experiments and sophisticated statistical procedures to distinguish the contribution of each genetic factor. We should therefore be grateful for the perseverance of quantitative geneticists, for without them, modern farmers would produce only a fraction of the corn, wheat, and pork they now provide for our tables.

ONE GENE, MANY EFFECTS

In the preceding section, we saw that a single characteristic, such as human skin color, can be affected by more than one gene. The reverse is also true. A single gene may determine several different characteristics. The white tiger Mohan shows how one gene can have many effects. The primary effect of the white-coat allele is an absence of dark pigment in the hair and skin. However, the same allele also seems to give rise to crossed eyes, not only in tigers, but in other species that have a similar mutation— for instance, in some albino people (Figure 8.20). We do not yet know why this one allele produces two apparently unrelated characteristics. For some genes with multiple effects, however, researchers have been able to pin down a single molecular or cellular cause for the different characteristics. A recessive allele of a gene found in the Maori, the aboriginal people of New Zealand, causes both breathing problems and sterility. These seemingly unre-

FIGURE 8.20 *Albinos Show That One Gene Can Affect Several Traits.* A mutant allele of a single gene for melanin production in humans can affect several physical traits, including the pigmentation of eyes, hair, and skin and the occurrence of crossed eyes.

FIGURE 8.21 *Siamese Cats and Environmental Effects on Gene-Determined Traits.* Siamese cats have dark ears, nose, paws, and tail because those extremities are cooler than the rest of the body, and the enzyme involved in producing dark coat pigment can only function at this lower temperature. Thus, even though a gene determines whether the pigment can be produced, the environment influences whether or not it *is* produced.

lated characteristics were traced to defects in a single protein necessary for the action of cilia and flagella (see Figure 3.23). Cilia normally clear the respiratory tract of dust and debris, so if cilia work inefficiently, because of either mutation or smoking, respiratory disease may ensue. And if flagella fail to propel a man's sperm, he will be sterile.

ENVIRONMENTAL EFFECTS ON GENE EXPRESSION

The expression of a gene can be altered not only by other genes, but also by the environment. One striking example is found in a small relative of the white tiger—the Siamese cat (Figure 8.21). Siamese house cats have little color except on their ears, face, tail, and feet. The mutant allele responsible for this hair color pattern in the cat is a different form of the same gene responsible for the white coat of Mohan; but the expression of the special allele in Siamese cats is temperature-sensitive. The enzyme that catalyzes the production of dark pigment in these cats is unable to work at the normal body temperature. At the lower temperatures found in the cat's extremities, however, the enzyme is active and can produce pigment that darkens the ears, paws, and tail.

It is often difficult to distinguish the effects of environment from those of genes on an individual's appearance, because the interactions are complex, and controlled experiments are often hard to perform.

≋ CONNECTIONS

Casual observation of genetic events can be deceiving, a point the maharajah discovered through Mohan and his descendants. But as Mendel and the pioneering geneticists who followed him revealed, systematic study permits an observer to detect biological entities and patterns that he or she cannot directly see or measure. The story of Mendel's experiments is a remarkable illustration of a scientist's ability to draw concrete conclusions from indirect data. From an analysis of stem length and flower color in pea plants, he inferred the existence and behavior of genes. His insights had tremendous theoretical impact, firmly establishing the particulate theory of inheritance and giving evolutionary theorists the kind of mechanisms they needed to explain biological diversity. In the long run, Mendel's ideas, coupled with the demonstration that genes occupy specific locations on chromosomes, paved the way for understanding and treating human genetic diseases.

An amazing array of gene interactions sometimes makes it hard to penetrate the complex relationship between genes and an organism's physical makeup. But research has shown that many of the observable traits we have discussed are caused by specific chemicals in the organism's body. Color in the hair or skin of tigers, dogs, cats, and people is caused by a specific dark pigment; and cell surface substances that thwart organ transplants differ because they are composed of different sugars. Somehow, genes control these chemical differences. Exactly how they achieve that control is the subject of the next two chapters.

NEW TERMS

allele, page 138
autosome, page 145
dominant, page 138
first filial (F_1) generation, page 138
genetic map, page 147
genotype, page 140
heredity, page 136
heterozygote, page 139
homozygote, page 140
parental (P) generation, page 138
phenotype, page 140

principle of independent assortment, page 143
principle of segregation, page 140
pure-breeding, page 138
quantitative trait, page 150
recessive, page 138
second filial (F_2) generation, page 138
sex chromosome, page 145
X chromosome, page 145
Y chromosome, page 145

STUDY QUESTIONS

REVIEW WHAT YOU HAVE LEARNED

1. Why did Mendel use garden peas in his experiments?
2. Trace the results of a cross between pure round-seeded pea plants and pure wrinkled-seeded pea plants through the F_1 generation; through the F_2 generation (see Figure 8.4).
3. State Mendel's principle of segregation.
4. Show the genotype of all possible gametes made by a hybrid pea plant with long stems and purple flowers, $LlPp$.
5. How are the sex chromosomes of sperm and egg alike? How are they different?
6. If two like chromosomes contain genes AD and ad, respectively, what genotypes could occur after crossing over?
7. How is the inheritance of flower color in snapdragons an exception to Mendel's principle of dominance?
8. If two people—one with attached earlobes, one with free earlobes—mate, what will their children's earlobes look like?

APPLY WHAT YOU HAVE LEARNED

1. To observe the laws of chance, toss a dime and a quarter simultaneously 10 times and tally the number of head-head, head-tail, and tail-tail combinations. How close are your results to a $1:2:1$ ratio? Now toss the coin 90 times more for a total of 100 trials. How close are your new results to a $1:2:1$ ratio? Explain.
2. In guinea pigs, black fur is dominant over white fur. How could an animal breeder test whether a black guinea pig is homozygous or heterozygous?
3. In humans, color blindness is a recessive X-linked trait. Use a Punnett square to show the possible genotypes of offspring of a color-blind man (XY) and a normal homozygous woman (XX), where X contains the gene for normal vision, and X contains the gene for color blindness.

FOR FURTHER READING

Leyhausen, P., and T. H. Reed. "The White Tiger: Care and Breeding of a Genetic Freak." *Smithsonian* (April 1971): 24–31.

Mendel, G. *Experiments in Plant-Hybridization.* Cambridge, MA: Harvard University Press, 1965.

Orel, V. *Mendel.* Oxford: Oxford University Press, 1984.

2. Genes are the basic units of heredity. Each gene occupies a specific location on a chromosome and controls a specific hereditary trait.

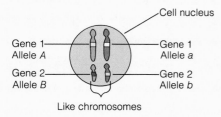

Genes

Chromosome

3. Most higher organisms have two forms (alleles) of each gene in every body cell, on paired (like) chromosomes.

Cell nucleus

Gene 1 — Allele *A*

Gene 1 — Allele *a*

Gene 2 — Allele *B*

Gene 2 — Allele *b*

Like chromosomes

1. The rules of heredity explain how traits are transmitted from parents to offspring.

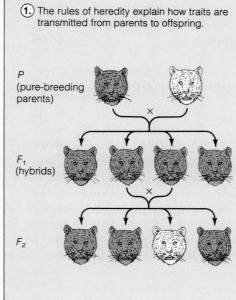

P (pure-breeding parents)

×

*F*₁ (hybrids)

×

*F*₂

4. Of the two alleles in each cell, one may be dominant and the other recessive.

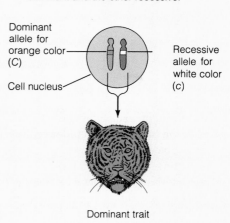

Dominant allele for orange color (*C*)

Recessive allele for white color (*c*)

Cell nucleus

Dominant trait

5. Mendel described two basic principles of heredity:

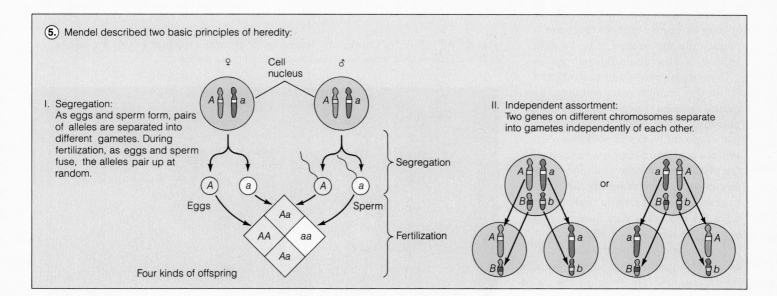

♀ Cell nucleus ♂

I. Segregation:
As eggs and sperm form, pairs of alleles are separated into different gametes. During fertilization, as eggs and sperm fuse, the alleles pair up at random.

A *a* *A* *a*

Eggs Sperm

Segregation

Fertilization

Four kinds of offspring

II. Independent assortment:
Two genes on different chromosomes separate into gametes independently of each other.

or

6. Alleles of genes linked on the same chromosome are usually inherited together, but they may become separated by crossing over, which occurs during meiosis and the formation of eggs and sperm.

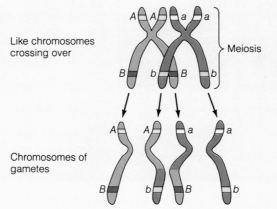

Like chromosomes crossing over

Meiosis

Chromosomes of gametes

7. Because genes interact with one another and with the environment, it is sometimes hard to detect the principles of heredity.

CHAPTER 9

DNA: THE THREAD OF LIFE

AN UNSTOPPABLE EPIDEMIC

In 1955, a deadly form of dysentery raced through a small Japanese town, and at the height of the epidemic, the four antibiotics being prescribed to treat the victims suddenly became useless. The bacterium responsible for the outbreak had somehow acquired resistance to the antibiotics, and people were dying in frightening numbers.

The local dysentery germs had become resistant to several drugs at once, and even worse, the drug resistance was spreading from one kind of bacterium to another. For example, the species that cause typhoid fever and pneumonia also became impervious to antibiotics. How could this happen?

Geneticists searched intensively and found that multiple drug resistance was caused by minute circles of DNA called **plasmids**. Many plasmids carry genes that cause antibiotics to be destroyed. When a bacterium happens to release the plasmids that have multiplied within its cell walls, these plasmids can spread to other bacteria, bringing to their new hosts the valuable ability to survive even in a person who takes antibiotics.

A plasmid represents the simplest hereditary system and displays two hallmarks of genes: It is a single molecule of DNA that can store *information*, and it can be copied, or *replicated*. From the simple genetic systems of plasmids and bacteria, biologists have learned that the structure of genes, as well as the way genetic information is encoded and replicated, are fundamentally the same in all living organisms, from bacteria to people. They also discovered that DNA's striking double helix structure helps explain its activity (Figure 9.1). This is one more example of how form dictates function in living things.

Finally, molecular biologists have developed powerful new ways to manipulate DNA. They can remove a piece of your DNA, splice it into a plasmid DNA ring, and get duplicates to form inside a bacterium. The result: a human gene multiplying in a bacterial cell! Today's experiments in genetic engineering will shape the world for generations to come, allowing people to generate new kinds of proteins, including some of the foods and drugs you will use in the future. But you and the rest of the informed public must help decide how to regulate this growing field.

As we move from Mendel's abstract idea of a gene to our modern concrete knowledge, we discuss these questions:

- How do genes direct cell function?
- What are genes made of?
- How does DNA's structure allow it to carry information and be copied?
- How do researchers alter DNA molecules in the laboratory?
- What rewards and problems lie in manipulating DNA for our needs?

FIGURE 9.1 *DNA: The Code of Life.* This computer-generated image of a cross section of a DNA molecule rivals the beauty of a medieval stained glass window.

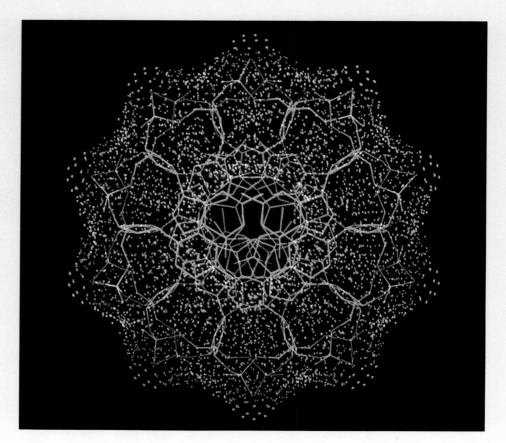

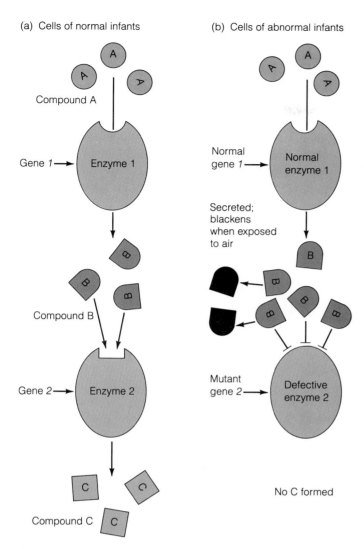

(a) Cells of normal infants

(b) Cells of abnormal infants

Compound A

Gene 1 → Enzyme 1

Compound B

Gene 2 → Enzyme 2

Compound C

Normal gene 1 → Normal enzyme 1

Secreted; blackens when exposed to air

Mutant gene 2 → Defective enzyme 2

No C formed

FIGURE 9.2 *A Defective Gene and a Defective Enzyme: Causes of a Strange Condition.* A single genetic change causes some infants' urine to stain their diapers black. (a) In the cells of normal infants, compound A is converted to B by enzyme 1, and B is converted to C by enzyme 2. (b) If enzyme 2 is defective, no C forms, and the infants excrete accumulations of compound B, which turns black on contact with air.

≋ WHAT GENES DO

Mendel's experiments showed that hereditary factors determine a pea plant's physical appearance: the color of its flowers, the length of its stem, and the shape of its seeds. But Mendel himself had no idea what genes were or how they worked. How could the information in an invisible gene control something as visible as a flower's pink color? The beginnings of an answer to this question emerged from a harmless but startling genetic defect in human babies.

WRONG GENES, WRONG ENZYMES

Much to the consternation of their parents, some otherwise normal babies produce urine that stains their diapers jet black. This rare condition, called alkaptonuria, occurs when a defective gene causes the babies to excrete a substance that turns black on contact with air. This black substance is an ordinary product of metabolism that in most people is broken down by an enzyme. There is a simple metabolic pathway in which compound A is converted to compound B by one enzyme, and B to C by another enzyme (Figure 9.2a). But in people with blackened urine, the second enzyme does not work, so B never changes to C but instead accumulates and is excreted (Figure 9.2b).

In 1909, a British physician studying the families of affected children noted that this condition is inherited as a simple recessive trait. He reasoned that if a mutant allele causes the *absence* of enzyme function, then the normal allele is responsible for the *presence* of enzyme function. In other words, a gene must function by allowing a specific enzyme reaction to occur (review Figure 9.2). Eighty years later, geneticists worked with a pink mold that you may have found at one time growing on a forgotten heel of bread in the back of your kitchen cabinet, and showed that (1) each step in a biochemical pathway is controlled by a specific gene (such as gene *1* or gene *2* in Figure 9.2a); (2) each step in a biochemical pathway is facilitated by a specific enzyme (such as enzyme 1 or 2 in Figure 9.2a); and therefore (3) a gene causes a specific enzyme to be formed. Biologists summarized this work with the phrase *one gene–one enzyme*, which implies that each gene regulates the production of a unique enzyme. According to this hypothesis, the plasmid involved in the Japanese dysentery epidemic must have carried a gene for an enzyme that breaks down certain antibiotics. Similarly, a person with brown eyes must have genes that control enzymes that help make brown pigment in the eye; by inference, a person with blue eyes lacks these genes, the enzymes, and the brown pigment.

HYPOTHESIS REVISED: ONE GENE–ONE POLYPEPTIDE

Although researchers had now shown that genes somehow determine the presence of an enzyme, a crucial question remained: How? The answer came from studies of young black Americans suffering from an inherited blood disease. Symptoms of the disease include pain in the joints and abdomen, chronic fatigue, and shortness of breath. Tests show that the blood of an affected individual has too few red blood cells and that many of the cells are shaped like crescents, or sickles, instead of nor-

FIGURE 9.3 *Normal and Sickled Blood Cells.* This scanning electron micrograph of blood from a child with sickle-cell anemia shows a sickled cell (left), then a normal cell, and then three more abnormally shaped red blood cells (right).

mal flat discs (Figure 9.3). The sickle-shaped blood cells tend to clump together and lodge in very small blood vessels—especially in the joints and abdomen—blocking blood flow and causing pain. The sickle-shaped cells are also fragile and easily destroyed, causing a shortage of red blood cells, or anemia. Because red cells carry oxygen from the lungs to the rest of the body, a deficiency of these cells results in a chronically inadequate oxygen supply and a feeling of breathlessness and fatigue. The disease, called *sickle-cell anemia*, is inherited as a recessive mutation according to Mendel's rules: Only a child who inherits a sickle-cell allele (Figure 9.4a) from both parents suffers from the anemia.

Scientists studying sickle-cell anemia knew that red blood cells are packed with the protein hemoglobin, which carries oxygen from the lungs to the rest of the body. Hemoglobin is a single protein composed of four polypeptide chains, and each chain can be visualized as a series of about 150 beads on a string, where each bead is an amino acid. When the researchers compared one of the hemoglobin polypeptide chains taken from normal people with the same hemoglobin polypeptide chain from sickle-cell patients, they found a subtle difference. Of the approximately 150 amino acids in each string, just one was changed (Figure 9.4b). This small substitution was enough to cause groups of hemoglobin molecules to stack together to form long fibers instead of remaining as individual globs (Figure 9.4c). These fibers in turn alter the shape of the red blood cell and bring about the suffering and early death of sickle-cell anemia victims (Figure 9.4d). The clear connection in sickle-cell anemia between a change in a single gene and the substitution of a single amino acid had two results: It enabled doctors

to diagnose the disease more accurately, and it showed geneticists that a gene specifies the order of amino acids in a single polypeptide chain. It also illustrates how sickness at the level of the whole organism (fatigue, pain, and breathlessness) may result from one small, invisible change in a gene.

Nevertheless, knowing that a gene specifies the amino acid sequence of a protein's polypeptides does not tell us how the gene does this encoding. To understand how

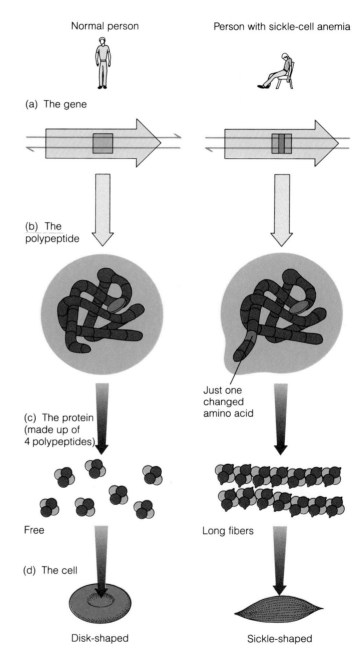

FIGURE 9.4 *Mutation in a Single Gene Causes Sickle-Cell Anemia.* As this chart shows and the text describes, a mutant gene sequence (a) leads to an altered hemoglobin polypeptide chain (b), formation of hemoglobin fibers (c), altered blood cell shape (d), and widespread effects on the body. The blue arrows in (a) represent a stretch of DNA that is a gene.

genes work, we first have to know what genes are and what in their makeup allows them to carry information and replicate.

⚅ WHAT ARE GENES?

For 50 years after genes were first discovered, it was not clear that these genetic units were molecules. But then a series of experiments between 1910 and 1920 demonstrated that genes in higher organisms are specific parts of chromosomes (see Chapter 8), and it followed that some substance in the chromosomes must constitute the genes. Experiments had also shown that chromosomes contain protein, RNA, and DNA, and so biologists presumed that one of these was the genetic material. But which one?

Geneticists at first thought that only proteins could carry the complex information in genes. Proteins have 20 different subunits (kinds of amino acids), while DNA has only 4—the bases adenine (A), thymine (T), guanine (G), and cytosine (C) (see Chapter 2). Clearly, 20 kinds of amino acid subunits can form many more combinations (and carry much more information) than can 4 bases, just as you can form more 3-letter words from an alphabet of 20 letters than from an alphabet of 4 letters.

BACTERIAL INVADERS: CONCLUSIVE PROOF FROM DNA IN VIRUSES

Definitive experimental proof that genes are made of DNA, instead of protein or RNA, came from experiments with tiny viruses called bacteriophages (or phages, for short) that infect bacteria. Shaped like lunar landing modules (Figure 9.5a), these viruses infect bacterial cells by attach-

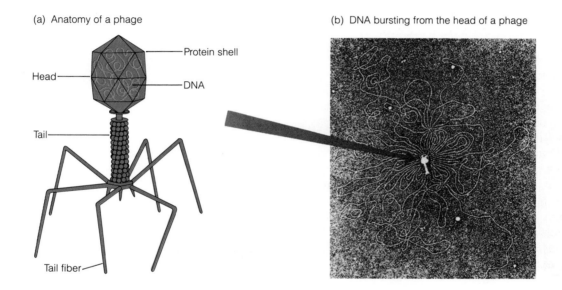

(a) Anatomy of a phage

Head — Protein shell

DNA

Tail

Tail fiber

(b) DNA bursting from the head of a phage

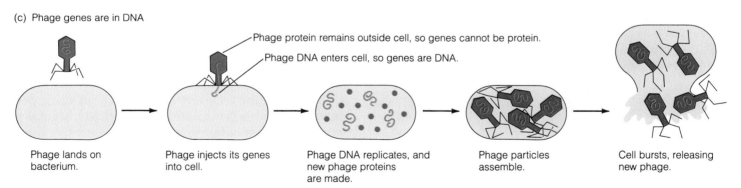

(c) Phage genes are in DNA

Phage protein remains outside cell, so genes cannot be protein.

Phage DNA enters cell, so genes are DNA.

Phage lands on bacterium.

Phage injects its genes into cell.

Phage DNA replicates, and new phage proteins are made.

Phage particles assemble.

Cell bursts, releasing new phage.

FIGURE 9.5 *Structure and Life Cycle of Bacteriophages.* (a) A bacteriophage looks like a crystal on legs. The crystalline head of a phage consists of a protein shell surrounding DNA; a columnar tail and leglike tail fibers are also made of pure protein. A relatively large amount of DNA is packed into the head of a phage. (b) The DNA from a phage has spewed out. (c) In the phage life cycle, the phage lands on a bacterial cell and injects its DNA; this directs the cell to make new phage particles, and these burst out and infect additional cells.

ing themselves to the cell surface and injecting their genes into the cell. Once inside the bacterium, the phage genes take over the cell's protein-making machinery so that it makes almost nothing but new phages. Finally, the cell bursts open and releases about a hundred new phages (Figure 9.5c), each of which can infect another bacterium and repeat the cycle. In a classic series of experiments, researchers found that phage DNA enters the bacterial cells, but phage protein does not (see Figure 9.5c). Since the genes enter the cells along with the DNA, the geneticists concluded that genes are made of DNA.

But is a simple bacteriophage representative of more complex living organisms? The answer is yes. Subsequent research has confirmed that DNA is the genetic material in all cells and many viruses. (In some viruses, such as the virus that causes acquired immune deficiency syndrome, or AIDS, the genes are made of RNA.) The experiments described here transformed the gene from an abstraction, a vague something that determines heredity, into a tangible chemical that biologists can see and manipulate. Nowadays, molecular biologists break open cells, separate the DNA from the proteins, add ice-cold ethanol to the DNA, and put it in the freezer. By

FIGURE 9.7 *DNA: The Universal Hereditary Material.* The same kind of biological molecule can encode genetic information for these light and dark species of swallowtail butterflies.

the next morning, resting at the bottom of the test tube will be a stringy white material a bit like powdered sugar—but it is pure DNA, pure genes (Figure 9.6).

Just knowing that genes are made of DNA, however, still does not explain how a gene works. The way a gene stores information and replicates is revealed by the form of the DNA molecule itself.

FIGURE 9.6 *DNA Is a Stringy White Material.* In a laboratory, researchers isolate DNA from the cells that carried it. Here, a biochemist is separating DNA from the solution it is dissolved in.

⚒ DNA: THE TWISTED LADDER OF INHERITANCE

Delicate butterflies and the plants on which they rest both have genes encoded in molecules of DNA. Such DNA can replicate so perfectly that the offspring of two swallowtail butterflies look just like the parents, and yet the genetic material can still contain enough variation so that some swallowtails turn out black rather than yellow (Figure 9.7). The variability inherent in DNA is the basis of natural selection, the key to evolution, and the explanation for life's stunning diversity.

To see how the relatively simple DNA molecule perpetuates life, we will examine it systematically, beginning with its smallest subunits and ending with the entire chromosome in which the DNA is packaged.

DNA: A LINEAR MOLECULE

One fact underlies DNA's simplicity of structure and function: DNA is a linear molecule. The structure of a single strand of DNA can be imagined as a long chain with links of four different colors, each representing a different nucleotide subunit. As Chapter 2 explained, a nucleotide consists of three parts: a phosphate, a sugar, and a base. The four kinds of nucleotides in DNA are identical except for the bases they contain—either adenine (A), cytosine (C), guanine (G), or thymine (T) (Figure 9.8)—and we can use a shorthand system that identifies these bases to show the sequence of nucleotides in a stretch of DNA. For a short stretch, it might read ACCCGTCCGTGTTAG, with each letter standing for a base.

In 1953, James D. Watson and Francis Crick, two young researchers at Britain's Cambridge University who were familiar with the biochemical evidence about DNA and were determined to unravel the molecule's structure, chose an unusual approach. They arranged and rearranged the pieces of Tinker Toy–like molecular models to see which combination of basic components might best fit the evidence about DNA's biochemical activity (Figure 9.9). From previous experiments, they knew that DNA is a linear molecule with a sugar-phosphate backbone. They also knew that hanging off each sugar of the chain, like a charm on a charm bracelet, is one of the four bases (see Figure 9.8). Finally, they were aware that the bases can lie in any order along the molecule. This is a key point. For example, if you looked at part of the sequence of 4300 bases of a certain antibiotic-destroying plasmid, you would find GAATTCTCACGTTTG. In general, after an A can come another A, a C, a T, or a G. The variability of base

FIGURE 9.9 *Watson and Crick: A Tinker Toy–like Model of DNA.* The two biologists won a Nobel Prize for their double helix model of DNA. This model helped explain some of the mysteries of heredity.

order along the molecule's enormous length is what allows DNA to store so much information. The full base-pair sequence for this plasmid with just two genes takes up 2½ typewritten single-spaced pages; the base sequence of the DNA in one of your cells would fill more than 800 books the size of this one! (Watson is currently overseeing a concerted research program to determine the entire base sequence of human DNA; see Chapter 10.)

Crick and Watson received significant help in their model-building efforts from two British biophysicists, Rosalind Franklin and Maurice Wilkins. Franklin and Wilkins took some particularly good pictures of DNA. They made these pictures by means of X-ray diffraction, a process in which a beam of X-rays shines through a fiber made of many strands of pure DNA aligned in parallel. By analyzing the pattern in the pictures, it is possible to map the relative positions of the atoms in a molecule.

When Watson went to London and saw Franklin's X-ray data, he discovered a certain symmetry that suggested that the molecule in some way consists of two paired strands of DNA. Back in Cambridge, he and Crick continued to work with their model until they hit upon an intriguing arrangement of the two DNA strands. Their final model suggested the following features:

1. A DNA molecule is composed of two nucleotide chains

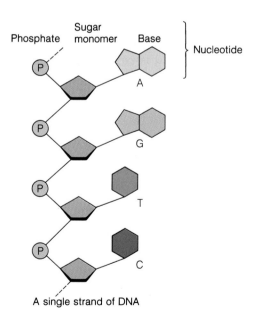

FIGURE 9.8 *Structure of DNA.* The nucleotide bases in DNA come in four varieties: A, T, G, C. Each is joined to a deoxyribose sugar monomer, which is attached to a phosphate group. The three subunits together form a nucleotide. Here we show a chain of nucleotides, forming a strand of DNA.

oriented in opposite directions, like the north- and southbound lanes of a freeway (Figure 9.10, part 1). Biologists often call the double-stranded DNA molecule a duplex.

2. The bases hanging on the two strands face inward,

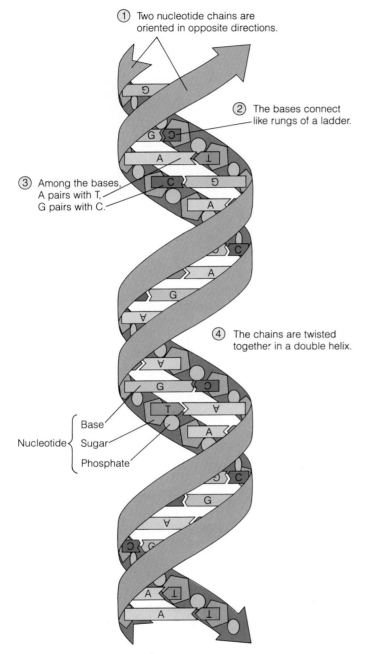

① Two nucleotide chains are oriented in opposite directions.

② The bases connect like rungs of a ladder.

③ Among the bases, A pairs with T, G pairs with C.

④ The chains are twisted together in a double helix.

Nucleotide { Base
Sugar
Phosphate

FIGURE 9.10 *Double-Stranded DNA.* DNA consists of two sugar-phosphate ribbons and internal base pairs connecting A with T and G with C like rungs on a twisted ladder.

while the two sugar-phosphate backbones face the outside of the molecule. The bases thus connect in the middle of the molecule like the rungs of a ladder (Figure 9.10, part 2). The bases of the two strands are actually connected by weak hydrogen bonds that keep the two strands bound tightly together.

3. Because of the shapes of the four bases, A pairs with T and C pairs with G (Figure 9.10, part 3). To phrase it another way, A *complements* T: The two fit together like a lock and key, readily forming hydrogen bonds with each other. But neither fits well with C or G. Likewise, C complements—pairs and forms bonds with—G. We will see that complementary base pairing between the two strands of DNA is fundamental to gene replication and function as well as to the gene splicing done by molecular geneticists.

4. Finally, the two DNA strands are twisted together to form intertwined helices, or one **double helix** (Figure 9.10, part 4). In short, DNA looks like a ladder that has been twisted along its long axis.

In 1953, Watson and Crick published a description of their double helix model in a short paper in the scientific journal *Nature*, observing, with considerable understatement, that it had not "escaped our notice" that the model immediately suggests ways in which DNA could fulfill its two major functions: making copies of itself and specifying the amino acid sequence (and thus the structure and function) of proteins. The model's simplicity plus its enormous power to explain bits of data—a combination of traits that makes a model "elegant" to a scientist—led the rest of the scientific world to rapidly accept it. In 1962, Watson and Crick shared the Nobel Prize with Wilkins for their discovery. Before exploring how DNA stores information and is copied, let us look briefly at how cells package these colossal molecules.

PACKAGING DNA

Like a belt, DNA in nature can take one of two forms: a line with two free ends, or a circle with the two ends joined. In simple systems, DNA loops back on itself to form a circle. DNA in plasmids (Figure 9.11), bacterial

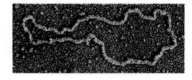

FIGURE 9.11 *A Plasmid: A Minute Circle of DNA.* This electron micrograph shows a plasmid enlarged 30,000 times. The plasmid bears genes for antibiotic resistance. It can survive inside a bacterium and make the cell resistant to antibiotics.

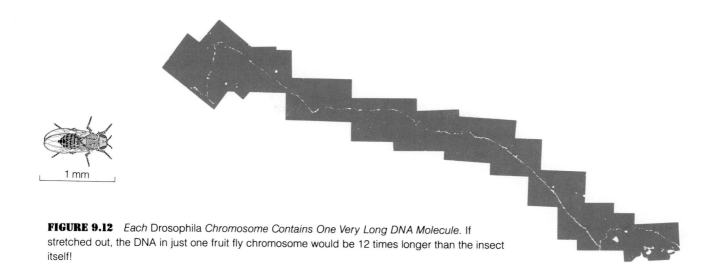

FIGURE 9.12 *Each* Drosophila *Chromosome Contains One Very Long DNA Molecule.* If stretched out, the DNA in just one fruit fly chromosome would be 12 times longer than the insect itself!

cells, some viruses, and in the chloroplasts and mitochondria of complex cells is circular in form.

In contrast, the nuclear DNA of higher organisms and the DNA of many viruses is in separate linear pieces (Figure 9.12). In higher organisms, single DNA molecules are extraordinarily long and are organized into chromosomes, with each chromosome consisting of a tightly wound DNA molecule. The DNA molecule from the largest of the four chromosomes of the fruit fly, for example, is longer than the word *genetics* printed here and carries a few thousand genes (see Figure 9.12). But the fly itself is no bigger than the letter *i*. In spite of the molecule's length, each of the fly's millions of cells contains two copies of this large chromosome as well as pairs of the other three chromosomes. Packaging such long chromosomes in small cells is like taking a piece of string the length of two football fields and stuffing it into a sausage 8 inches long. How *do* cells package huge DNA molecules into structures as small as a chromosome?

The enormous length of DNA in a eukaryotic cell cannot be wadded up haphazardly, because if it were, the separation of DNA molecules during cell division would be as difficult as untangling two kite strings. What happens instead is that DNA, like a proper kite string, is wound around small protein spools; a single spool wrapped with two loops of DNA is called a *nucleosome*. Adjacent spools pack closely together to form a larger coil, somewhat like a coiled telephone cord (Figure 9.13, page 162). This coil, in turn, is looped and packaged with other proteins into a chromosome. It is this orderly packing of DNA that prevents massive tangles during cell division.

From base pair to double helix to chromosome, the structure of DNA is a boundless biological resource. How a single molecule can create the diversity of life is the true marvel of biology.

≋ MYSTERIES OF HEREDITY UNVEILED IN DNA STRUCTURE

When Watson and Crick first publicly presented their model of the double helix at a meeting of geneticists in the spring of 1953, they were startled by the enthusiastic reception their peers gave the model. In the eyes of a scientist, a hypothesis takes on power and beauty in proportion to how well it explains diverse facts simply and coherently; and the scientific community agreed that the double helix model was a powerful tool.

In the laboratory, geneticists quickly found that complementary base pairing and linear variation in base order do indeed account for a gene's ability to replicate itself and store information. This section explores how genes can do this.

HOW DNA REPLICATES

■ **DNA Copies Itself from a Molecular Template, or Pattern** People have always recognized that like begets like. It is obvious at some family reunions, for instance, that red hair, a big nose, or short stature has been passed along for several generations. For traits to be conserved this way, genes must produce copies of themselves, and those copies must be transmitted from parent to child. Geneticists discovered that DNA is passed from generation to generation with extraordinary accuracy because the molecule itself contains a *template*, or pattern, for its own replication.

A DNA molecule makes copies of itself in three main steps: unwinding, base pairing, and joining. In *unwinding*, the two strands of the parent DNA duplex separate, and the bases on each of the separated strands are left unpaired (Figure 9.14a). Next, in *base pairing*, free nucleotides drift into the area and pair with complementary bases on the separated original strands (Figure 9.14b). A free A base that happens to float in will pair only with a T base on an original strand, and a free G will pair only with an attached C. In this way, the order of bases in an old strand specifies the base order in a complementary new strand. For the final step, *joining*, a special enzyme links together a row of newly arrived bases into a continuous new strand, which is, of course, paired with one of the original ("old") strands (Figure 9.14c). The process of replication has now produced two double helix DNA

molecules from one original parent molecule, and each of the two new molecules is identical to the original parent. Because each new duplex contains one strand of nucleotides conserved from the original molecule and one completely new strand, DNA replication is called *semiconservative*.

Now, let's take a moment to integrate this discussion of DNA replication into our knowledge of the cell cycle (review Figure 7.7). A cell's DNA replicates during the growth phase of the cell cycle, before cell division. In fact, the doubling of DNA shown in Figure 9.14 is responsible for the chromosome doubling shown in Figure 7.6, and the accuracy and precision of DNA copying, chromosome doubling, and cell division are the biological mechanisms behind the familiar phenomenon of "like begets like."

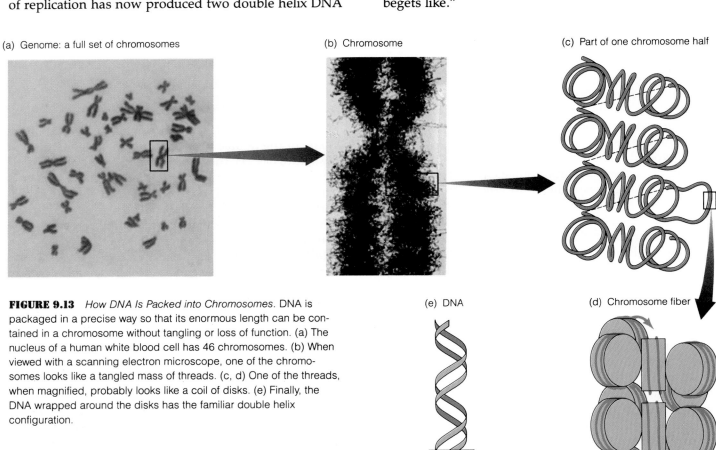

(a) Genome: a full set of chromosomes

(b) Chromosome

(c) Part of one chromosome half

FIGURE 9.13 *How DNA Is Packed into Chromosomes.* DNA is packaged in a precise way so that its enormous length can be contained in a chromosome without tangling or loss of function. (a) The nucleus of a human white blood cell has 46 chromosomes. (b) When viewed with a scanning electron microscope, one of the chromosomes looks like a tangled mass of threads. (c, d) One of the threads, when magnified, probably looks like a coil of disks. (e) Finally, the DNA wrapped around the disks has the familiar double helix configuration.

(e) DNA

(d) Chromosome fiber

DNA   A spool (nucleosome)

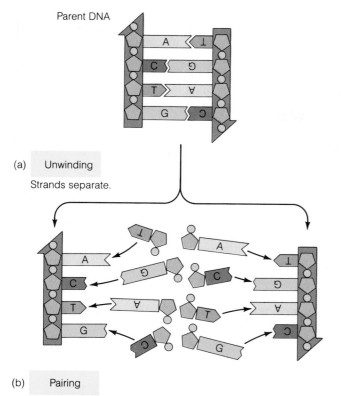

(a) Unwinding

Strands separate.

(b) Pairing

Free nucleotides diffuse in and pair up with bases on the separated strands.

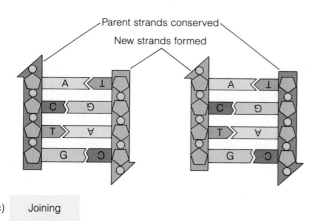

Parent strands conserved

New strands formed

(c) Joining

Each new row of bases is linked into a continuous strand.

FIGURE 9.14 *DNA Replication.* When a parent DNA strand replicates, (a) the strands separate and (b) previously unattached nucleotides (that have accumulated in the nucleus) float in and pair up with appropriate unpaired bases; (c) the new bases then link together to form new strands. The two new molecules are identical to the parent DNA molecule, and in each, one strand is conserved from the parent and one is newly formed.

■ **The Amazing Accuracy of Replication** Considering the complexities of unwinding, base pairing, and joining required for DNA replication, it is a wonder that DNA molecules are ever copied correctly. Yet, DNA synthesis is incredibly accurate: An error creeps in only once every

10^9 bases. To approach this level of accuracy, you would have to type this entire book over a thousand times and make only one typing error!

The very low rate of errors during replication depends in part on a process similar to the one used to catch errors in the typesetting of this book: proofreading. The enzyme responsible for joining the DNA bases also corrects any mistakes that occur during base pairing, rather like a correcting typewriter that lifts a wrong letter, then replaces it with a correct one.

The necessity for such an amazing level of accuracy in replication derives from DNA's other main function: information storage. If an error in replication causes the sequence of bases to change, then the genetic information will also change. Such changes in genetic information, called **mutations**, can alter protein structure, since genes specify protein structure. And as we saw earlier in this chapter, an altered protein, such as sickle-cell hemoglobin, can cause an individual to function abnormally. Fortunately, the accuracy of complementary base pairing and the proofreading action of enzymes prevent nearly all such information changes from taking place during DNA replication.

HOW DNA STORES INFORMATION

Geneticists were able to learn a great deal about genes and their activities by discovering that they specify the sequence of amino acids in polypeptides; that they are made of DNA; and that the DNA molecule's unique structure enables it to self-replicate. However, this still left the important question of how information—the information for red hair or brown eyes or attached earlobes—is stored in DNA.

The answer is that the order of the bases contains the genetic message. We have seen that DNA is a chain of bases and that genes are discrete segments along that chain. The bases relate to the polypeptides they specify by means of a code—the **genetic code**. Just as it takes several letters to specify a single word, it takes three adjacent bases to specify an amino acid in a protein. For example, the base triplet GAG specifies the amino acid glutamic acid, while GTG specifies the amino acid valine. Since the simple switch from glutamic acid to valine in hemoglobin causes sickle-cell anemia, you can see how a change in one base (in this case, A to T) can have broad repercussions, causing a base triplet to be changed (GAG to GTG), the wrong amino acid to be placed in a polypeptide chain, the protein that includes that chain to malfunction, and a person's life to be threatened. This example shows how important DNA's function of storing and retrieving information really is.

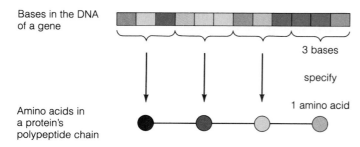

Bases in the DNA of a gene

3 bases

specify

1 amino acid

Amino acids in a protein's polypeptide chain

FIGURE 9.15 *The Genetic Code: Three Bases Specify One Amino Acid.* A gene is a stretch of DNA composed of three-letter words, three adjacent bases. Because of these sets of bases, a gene always contains at least three times as many bases as the number of amino acids it specifies.

Like the information in written words, the information in genes flows linearly in a single direction. The first base triplet in a gene specifies the first amino acid; the second triplet specifies the second amino acid; and so on, to the end of the polypeptide (Figure 9.15).

The information-bearing capacity of DNA explains how one disease-causing bacterium can rapidly transfer drug resistance to another via the simple DNA circles called plasmids (see Figure 9.11). A plasmid replicates inside a bacterial cell, and in the process of having its DNA copied provides a long section of base triplets—a gene—specifying the construction of a protein that inactivates antibiotics. Any bacterium carrying the plasmid can therefore survive in an environment poisoned with antibiotics. What's more, the plasmid can escape from one bacterial cell and be picked up by another, which then acquires the same antibiotic resistance. This is how resistance to drugs like penicillin and tetracycline can rapidly spread from one set of bacteria to another and fan an epidemic like the one we followed at the beginning of the chapter.

HOW DNA GENERATES DIVERSITY

Now that we have seen how the DNA molecule carries genetic information and how it replicates with great enough accuracy to ensure that like begets like, yet another question arises. How does this same DNA molecule generate the diversity of life we see around us—round and wrinkled peas, normal and defective hemoglobin, black and yellow butterflies, short and tall basketball players (Figure 9.16)?

The answer is *genetic variation*. In nature, two kinds of events produce new DNA sequences: *mutation*, which changes base sequences by the insertion, deletion, or rearrangement of base pairs; and *recombination* (see Chapter 7, page 130), which reshuffles normal and mutated genes to create new combinations of alleles. Of the two events, mutation is much rarer. Thanks, in part, to DNA repair enzymes, the chance of a gene mutating is only about one in a million. In contrast, the chance of recombination between genes on like chromosomes is as high as one in two (review Figure 7.20). Genetic recombination has occurred spontaneously and naturally for millions of years in the sex organs of sexually reproducing organisms, sometimes in surprising ways (Figure 9.17). This natural recombination provides much of the variation that, through natural selection, is the basis of evolution and diversity (see Chapter 1). In fact, species in which genes recombine are more adaptable and more likely to persist and evolve than species that cannot undergo recombination. Bear in mind, though, that their variation is completely random. It is purely a matter of chance whether new gene combinations help, hinder, or make no difference to an individual organism.

FIGURE 9.16 *Genes, Growth Hormone, and Body Size.* These short and tall basketball players illustrate the effects that differing amounts of growth hormone, determined by different gene combinations, can have on human height.

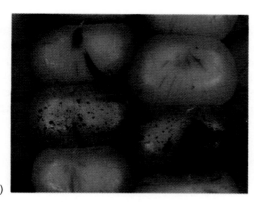

FIGURE 9.17 *Jumping Genes and Mottled Corn Kernels.* Nobel-Prize-winning geneticist Barbara McClintock, pictured here in 1947, (a) made a startling discovery that explains why Indian corn can have brilliantly decorated kernels of different colors (b). Small segments of DNA, called *transposons*, can leap around the chromosomes and alter the action of genes they enter or leave. In the cells of Indian corn kernels, transposons can turn purple pigment genes on and off, sometimes producing mostly light kernels with a few dark spots, sometimes the opposite. Transposons have been found in bacteria, fruit flies, and even people. Although some of the rearrangements they generate can be disruptive and cause cancer, others provide the genetic diversity that is the foundation of natural selection.

Not so, however, with artificial recombination in the laboratory, known as **recombinant DNA technology**. Genetic engineers with their recombinant technology have removed the element of chance and can now create specific new gene combinations to serve specific human purposes.

⩬ ALTERING THE THREAD OF LIFE: THE POWER OF RECOMBINANT DNA RESEARCH

Think back to the tiny plasmids that were so genetically active during the 1955 dysentery epidemic in Japan. These minute circles of DNA carried a specific gene—a specific sequence of bases—that bestowed drug resistance upon their bacterial hosts. Since the genetic material in all living cells is essentially the same, researchers wondered whether a plasmid could carry a gene from a person—for instance, the gene for human growth hormone—and transport it into a bacterium. They knew that a plasmid could only be copied inside a bacterium, but could a human gene be copied in such a simple cell? And if so, could the bacterium decode the gene to make a human protein? As we will see in the remainder of this chapter, the answer to all of these questions turned out to be a resounding yes.

Today, thanks to a detailed understanding of DNA structure and function, genetic engineers can design and construct brand new genes in just a few weeks—genes that might have taken thousands of years to appear naturally in organisms inhabiting streams, oceans, fields, or forests. These revolutionary microtechniques include the capacity to cut DNA molecules in specific places and to paste selected DNA fragments together into a specially designed recombined DNA molecule. Organisms that carry the recombined genes are called *recombinants*.

To make a recombinant DNA molecule, molecular biologists simply cut two different DNA molecules in specified places and glue them together as desired. They are aided in their work by enzymes that cut DNA in a precise location (within a particular sequence of bases) and by other enzymes that glue DNA molecules together. (Figure 9.18, page 166).

⩬ PROMISES AND PROBLEMS IN RECOMBINANT DNA RESEARCH

Genetic engineering holds tremendous promise for humankind. Imagine a drug that can kill the viruses that

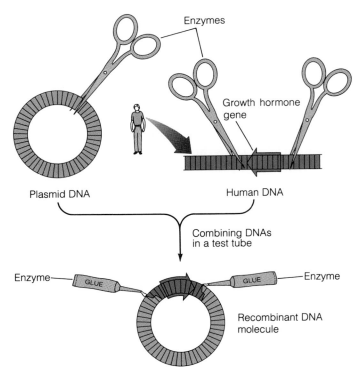

FIGURE 9.18 *How to Construct a Recombinant DNA Molecule.* In one study, genetic engineers obtained bits of human DNA carrying the gene for human growth hormone and inserted each bit into a plasmid, using the simple cutting and gluing techniques shown here. When they then inserted the recombinant plasmids into bacteria, the microbes made copies of the plasmids as they replicated their own larger chromosome. As a result, the researchers could harvest many copies of the gene for human growth hormone.

cause the common cold, available in great supply and for a few dollars; or an enzyme that can arrest a heart attack already in progress; or bacteria that can make automobile fuel from discarded corn stalks; or corn that contains proteins with the nutritional value of beef; or gene replacements to cure people and animals of crippling genetic diseases.

But now consider the potentially less favorable side to genetic engineering. While bacteria may someday be engineered to break down oil and then used to disperse oil slicks, what if these hypothetical bacteria moved beyond oil slicks and began to devour the world's dwindling oil supplies? Or what if it became possible for people to genetically engineer children? Many feel that recombinant DNA research has the potential to make dreams come true, but others worry about the social and ecological impacts that might result. Let's look at a few examples of the promise and then at the nagging questions that arise from recombinant DNA research.

RESHAPING LIFE: THE PROMISE OF GENETIC ENGINEERING

■ **Mass-Producing Proteins**   People have been exploiting microorganisms for thousands of years: Cheese, yogurt, beer, and bread are all manufactured with the aid of microorganisms. Now, through recombinant DNA technology, we can use microorganisms to manufacture virtually any protein we want by inserting into them a gene for the desired protein. The ten-year-old girl in Figure 9.19 was born unable to produce growth hormone and seemed destined to live out her life as an abnormally small individual. But in just one year of treatment with human growth hormone—mass-produced by genetically engineered bacteria containing the human growth hormone gene—the girl grew 5 inches. The precious substance is now available in pharmacies (by prescription), and physicians hope someday to use this synthetic growth hormone to treat burns, slow-healing fractures, and bone loss diseases, in addition to abnormally short stature in the thousands of children in the United States who make too little of the hormone to grow to normal height.

Growth hormone is just one of the human proteins now mass-produced through engineered bacteria. Bacteria have also been made to carry genes for insulin hormone that is chemically identical to the insulin made in the human pancreas. Until recently, diabetics had to rely on pig insulin to prevent them from suffering wide fluc-

FIGURE 9.19 *A Growth Spurt Thanks to Genetic Engineering.* As the marks on the doorframe show, this child grew several inches during treatment with growth hormone made through genetic engineering techniques.

FIGURE 9.20 *Genetically Engineered Plants.* The top two petunia plants were transformed with genes from a microorganism that is resistant to the common plant-killing substance called glyphosate. The bottom two plants are normal, nonengineered, garden-variety petunias. When experimenters sprayed all four with the plant-killing substance, the genetically transformed plants survived, while the normal plants withered and died.

tuations in blood sugar, fluctuations that can lead to blindness, coma, and death. Unfortunately, the pig protein differs slightly from human insulin, and some people are allergic to it. Genetically altered cells are also producing two other medically useful proteins: an enzyme called *tissue plasminogen activator,* which can dissolve the blood clots that block arteries during a heart attack, and a protein called *gamma interferon,* which can protect human cells from infection by the viruses that cause hepatitis, herpes, and the common cold and which can also stimulate the growth of tumor-killing cells. This partial list of early successes only hints at the many useful products with which genetically engineered microbes will ultimately provide us.

■ **Improving Plant and Animal Stocks** Recombinant DNA will accelerate a genetics project farmers began 10,000 years ago: to improve crops and livestock. In the past, breeders selected individuals with desirable traits—traits that arose by mutation and naturally occurring DNA recombination—to be parents of the next generation. Today, geneticists can select livestock such as cows, sheep, or pigs and insert into them specific genes that increase bone and tissue growth or milk production. The greatest impact on human lives, however, will be made through advances in the genetic engineering of plants. With the current capacity to insert genes directly into individual plants, geneticists can begin to use genetic engineering techniques to improve the efficiency of photosynthesis; decrease crop dependence on fertilizers; improve the nutritional quality of seeds, grains, and vegetables; and

increase plant resistance to pests, salt, drought, and extreme temperatures. Already, researchers have engineered an herbicide-resistant plant (Figure 9.20). When sprayed on a field, the herbicide will kill weeds but not the desired plant, and thus the crop can grow luxuriantly without competition from the weeds.

A natural question arises from such achievements: If beneficial genes can be inserted into plants, lab animals, and livestock, why not into people?

■ **Human Gene Therapy** Altering genes to combat an inherited disease is **gene therapy** and is, as yet, more a matter of theory than of practice. In theory, it could be applied in either of two ways: Genes could be inserted into the somatic (body) cells or into the germ cells (the cells that give rise to the sperm and eggs). The first procedure, *somatic cell gene therapy,* is the straightforward treatment of an individual's disease—a simple extension of current medical practice. But the second procedure, *germ line gene therapy,* would also affect the genetic makeup of the treated individual's offspring. Each type of therapy has important implications.

The first illness likely to be conquered in this way is a mutation-caused disease called severe combined immunodeficiency disease (SCID). Patients with SCID cannot make an enzyme called ADA, needed by bone marrow cells to mount an effective immune response and thus protect the body from bacteria, viruses, or other invaders. Lacking ADA, such children cannot resist infection, and they die of infectious diseases like pneumonia or influenza. They can survive only if raised in a sterile chamber, fed sterilized food, and denied direct physical contact with other people, even their parents. Recent studies may lead to effective gene therapy for these children.

In such therapy, medical researchers will remove defective bone marrow cells from a patient, insert the normal gene for the ADA enzyme into those cells, and return the cells to the patient's bone marrow. There, the genetically altered body cells should produce their own ADA and provide the child with a normal immune response (Figure 9.21, page 168). Since this procedure will genetically alter bone marrow cells, but not sex cells, the patient could survive to reproduce but would then pass the defective gene, not the normal gene, on to his or her children. For most people, this sort of gene therapy presents no new ethical problems: As in conventional surgery or drug therapy, a single individual is treated, a single individual benefits, and the defective genes remain to be perpetuated in the human population.

Gene therapy of germ line cells is another matter, since it constitutes a totally new approach to medicine and raises complex issues of safety and ethics. In this type of

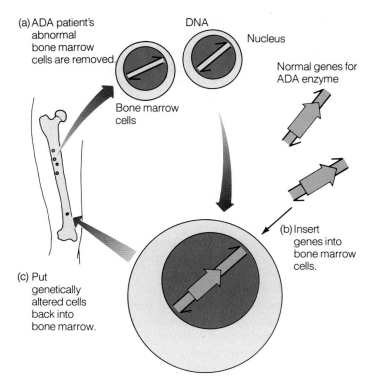

(a) ADA patient's abnormal bone marrow cells are removed.

DNA

Nucleus

Bone marrow cells

Normal genes for ADA enzyme

(b) Insert genes into bone marrow cells.

(c) Put genetically altered cells back into bone marrow.

FIGURE 9.21 *Gene Therapy in Somatic Cells.* In 1990, medical researchers received permission to start clinical trials to help children who were born unable to resist infection. In the trials, doctors would remove bone marrow cells from children born without the ability to make the enzyme ADA (a) and insert healthy genes in place of defective ones (b)–(c). The inserted genes would help the patients resist infection and lead a normal life.

gene therapy, recombinant DNA would be inserted into human sex cells. Not only would the treated individual be affected, but so would all the individual's descendants. While researchers have had some success with techniques for this kind of gene transfer in mice and sheep, they must still solve many technical and ethical problems before treating human sex cells. One problem is that the success rate is low—only 6 successes out of 300 injected eggs in a typical experiment with mice. Another problem is that the inserted DNA sometimes causes a new mutation, thus creating a new defect as serious as the disease the therapy was meant to cure. Finally, the inserted normal gene usually does not replace the native defective gene, but is simply added to the genome, so that the defective gene can still be inherited by the children of the treated individual.

The availability of mass-produced proteins, genetically improved plants and animals for agriculture, and genetic therapies for human diseases could improve our lives.

Nonetheless, the enormous potential of the new gene technology also raises some complex questions.

RECOMBINANT DNA: NOVEL PROBLEMS OF SAFETY AND ETHICS

Disagreements over both the safety and morality of recombinant DNA research have cast a shadow over its promise of a better future for the human race. Consider, for example, human gene therapy. Most people would agree that inserting a normal gene into mutant body cells containing a faulty gene is merely an extension of current medical practices and ethics. But many observers disagree with the idea of using gene therapy to "improve" an already healthy child. For instance, prospective parents might ask doctors to insert additional growth hormone genes into their newly conceived embryo so they could produce a taller basketball player or a beefier football player than might otherwise be produced. Other parents may want a child with greater intelligence or a more winning personality. From a scientific standpoint, such applications of recombinant DNA research now seem only a remote possibility. Untold numbers of genes control each of these complex traits in unknown ways. Geneticists may never be able to discover and clone (reproduce many copies of) all the relevant genes and introduce them into a person with any degree of control. However, society should consider these issues before technology achieves the applications. And if history repeats itself, the potential for such gene therapy will arrive more quickly than we expect.

Germ line gene therapy raises another ethical dilemma. Since its application would alter the genetic makeup of children not yet born, it could change the course of human evolution. Would a treatment that causes a heritable change infringe on the rights of future generations? Are the expected benefits of germ line gene therapy worth the unknown risks? Can a problem be solved in more traditional ways, whose risks are understood and whose benefits are clear? These and other questions have led scientists and lay observers to debate—at times heatedly—the potential misuse, accidental or deliberate, of recombinant DNA techniques.

Only through free and open discussion will potential problems be solved and appropriate guidelines be devised to ensure that recombinant DNA technology achieves its promise of high-yield crops, better livestock, and healthier humans with a minimum of risk. "Sunshine," as U.S. Supreme Court Justice William O. Douglas once said, "is the best disinfectant." With proper safeguards developed in an open forum, recombinant DNA technology may one day alleviate human suffering, as did such revolutionary medical advances as anesthesiology and antibiotics, and renovate agriculture to make food available for our planet's burgeoning population.

FIGURE 9.22 *Antibiotics in the Feedlot.* The routine application of antibiotics to cattle feed lies behind a dangerous increase of antibiotic-resistant bacteria. This is because the antibiotics selectively promote the growth and spread of bacteria containing plasmids with antibiotic-resistant genes. The resistant bacteria pose a serious health hazard to anyone who eats the meat of cattle contaminated with them. What's more, many doctors unnecessarily prescribe antibiotic drugs, which kill off nonresistant bacteria in patients and allow bacteria carrying plasmids with genes for drug resistance to multiply rapidly. Thus, the indiscriminate use of antibiotics by livestock farmers, doctors, patients, and others threatens human health.

≋ CONNECTIONS

The experiments discussed in this chapter revolutionized the geneticist's view of the gene, giving Mendel's abstract theory a physical reality—DNA, the stringy white matter at the bottom of a test tube or in the nucleus of a cell. As genes became tangible, geneticists learned to manipulate them, take them apart, and see how they work. Modern knowledge of gene structure and function in turn allowed geneticists to understand how a single DNA base-pair change can result in a mutated gene, an altered protein, and perhaps a lethal case of sickle-cell anemia. Likewise, knowledge of DNA replication revealed how resistance to antibiotics could quickly spread through several species of bacteria and opened our eyes to the fact that drugs are no longer the reliable cure-alls they once were for serious infections such as dysentery, pneumonia, and salmonella (Figure 9.22).

The conceptually simple DNA molecule—just a couple of strings of bases twisted around each other—unifies all life but is also the source of life's diversity. Spinach leaves are green and cardinal feathers red because the DNA in the pigment-forming cells differs in base sequence. Gene mutation and recombination are two sources of the variation that generates life's diversity. Recombination, whether it occurs in the test tube or in the testes, creates a DNA molecule with a new arrangement of genes. Today, biologists recombine DNA in the laboratory to make drugs such as insulin and growth hormone, to improve crops and livestock, and to cure human hereditary diseases. How DNA blueprints are actually decoded to form the proteins that carry out cell function is the subject of Chapter 10.

NEW TERMS

double helix, page 160
gene therapy, page 167
genetic code, page 163
mutation, page 163

plasmid, page 154
recombinant DNA technology, page 165

STUDY QUESTIONS

REVIEW WHAT YOU HAVE LEARNED

1. What is the relationship of genes to enzymes?
2. Explain the one gene–one polypeptide hypothesis.
3. How did research on viruses prove that genes are made of DNA?
4. How are the four kinds of nucleotides similar? How are they different?
5. Which of these base pairs is unlikely to be found in a DNA molecule, and why? AT; CG; TA; GA; GC.
6. What mechanism helps prevent errors in DNA replication?
7. What would be a result of a change in DNA base sequence?
8. True or false: Recombination occurs in very few organisms. Explain your answer.
9. Both mutation and recombination produce new DNA sequences. How do the two processes differ?
10. Give specific examples of the improvements biologists could bring about in animals and plants by inserting beneficial genes.
11. Describe how somatic cell gene therapy may help a child born with a genetic disease.
12. Why is research on recombinant DNA subject to strict guidelines?

APPLY WHAT YOU HAVE LEARNED

1. Because of a mutation in a single gene, albino people do not form melanin. Using the one gene–one polypeptide principle, propose a model to explain an albino's pale skin, hair, and eyes.

(1.) The genetic material in all living cells is DNA.

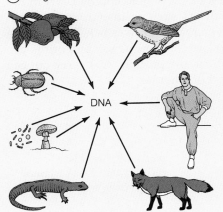

(2.) Called the double helix, the DNA molecule is shaped like a twisted ladder.

(3.) DNA stores information in the order of its bases.

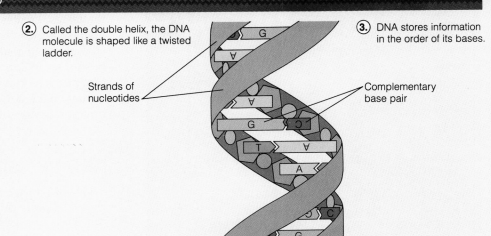

Strands of nucleotides

Complementary base pair

(4.) A gene is a segment of DNA that specifies the order of amino acids in a single polypeptide chain.

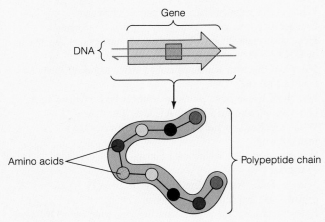

Gene

DNA

Amino acids

Polypeptide chain

(5.) DNA replicates by using each of its strands as a template for the construction of a complementary strand; the accuracy of DNA replication is the basis of like begets like.

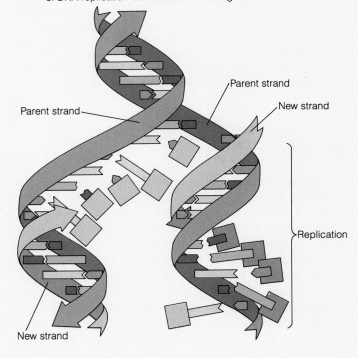

Parent strand

Parent strand

New strand

Replication

New strand

(6.) Mutation and recombination change DNA, providing the genetic variation that is the basis of evolution and diversity.

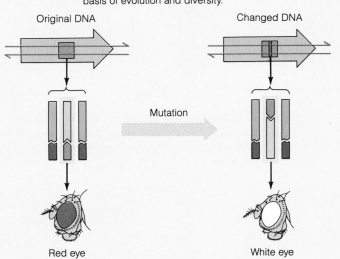

Original DNA

Changed DNA

Mutation

Red eye

White eye

(7.) Recombinant DNA technology promises to help cure genetic diseases and improve plant and animal stock; but it needs proper safeguards and open discussion.

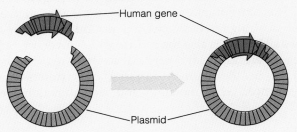

Human gene

Plasmid

2. A botanist discovers a previously undescribed plant with burgundy flowers in the Amazonian rain forest. A biologist finds that this plant's DNA has 18 percent of base A and knows that A pairs with T and that G pairs with C. What percentage of T, G, and C does the plant have?

3. An old superstition holds that each of us has a double somewhere in the world. How would you answer such a claim?

4. In the laboratory, genetic engineers have produced a strain of bacteria that protects potato crops from freezing. Genetically similar bacteria could have been made, although more laboriously, with traditional genetic methods. What arguments might some environmentalists use to oppose the release of such engineered bacteria in potato fields?

FOR FURTHER READING

Dawkins, R. *The Selfish Gene.* Rev. ed. Oxford: Oxford University Press, 1989.

Felsenfeld, G. "DNA." *Scientific American* 253 (May 1985): 58–67.

Joyce, C. "Genes Reach the Medical Market." *New Scientist* 16 (July 1987): 45–51.

Ledley, F. D. "Somatic Gene Therapy for Human Disease: A Problem of Eugenics?" *Trends in Genetics* (1987): 112–115.

Novick, R. "Plasmids." *Scientific American* 243 (December 1980): 102–127.

Radman, M., and R. Wagner. "The High Fidelity of DNA Duplication." *Scientific American* 259 (August 1988): 40–47.

Sahl, F. W. "Genetic Recombination." *Scientific American* 256 (February 1987): 90–96.

Schmitz, A. "Murder on Black Pad. The Rapist-Killer Haunting an English Village Left One Telltale Clue: A Piece of His DNA." *Hippocrates* (January/February 1988): 48–58.

Suzuki, D., and P. Knudtson. *Genethics: The Clash Between the New Genetics and Human Values.* Cambridge: Harvard University Press, 1989.

Watson, J. D. *The Double Helix.* New York: Atheneum, 1968.

Watson, J. D., and F. H. C. Crick. "Molecular Structure of Nucleic Acids: A Structure for Deoxyribosenucleic Acid." *Nature* 171 (1953): 737–738.

Zuckerman, S. "Antibiotics: Squandering a Medical Miracle." *Nutrition in Action* (January/February 1985): 9–11.

HOW GENES WORK

A CASE STUDY IN GENE ACTION

During the 1920s, Detroit physician Thomas Cooley detected a pattern of disease among certain infants brought in by their worried parents. Cooley conducted tests, and these revealed that the babies' blood had less than the normal amount of red blood cells, the cells that transport oxygen, and the infants were thus severely anemic. As they approached their first birthdays, their spleens began to enlarge, and their bones expanded monstrously, especially in the legs, arms, and head (Figure 10.1). All the infants died in childhood.

Physicians began to search for the cause of "Cooley's disease"—this peculiar group of defects consisting of anemia, enlarged spleen, and abnormal bone growth—in the hope of developing effective treatments. The disease turned out not to be contagious but instead to run in families. Children become sick if they inherit genes for the disease from both parents. In other words, Cooley's disease—today known as *thalassemia*—is inherited as a simple recessive mutation, like short stems in Mendel's peas. Although people with one mutant allele and one normal allele in their cells have no symptoms of the disease, they are carriers who can pass the defective allele on to their children. The percentage of carriers in certain populations is quite high—up to 30 percent in some villages of the Mediterranean, Middle East, India, and Southeast Asia. In fact, thalassemia is the most common single gene disorder in the entire human species.

Two unifying themes will emerge as we follow the processes by which a mutation in just one gene results in a range of potentially lethal defects in a person. The first theme is that information flows from DNA to RNA to protein

molecules. For example, the gene whose mutation causes the blood disease we have been discussing is normally transcribed into an RNA molecule, which is then translated into a major protein in the blood, called the hemoglobin protein; this protein carries oxygen to our cells. A second theme is that basic genetic mechanisms are essentially universal. All creatures, from bacteria to field mice to flowering dogwood to you, share the same pattern of gene action: DNA → RNA → protein.

In this chapter, as we explore how genes work and, ultimately, how they govern the activities of living things, we answer several important questions:

■ How does information flow from DNA to RNA to proteins?
■ What are the consequences for our health and well-being of even slight alterations in DNA?
■ How do cells determine when and where the information in a gene will be used?

FIGURE 10.1 *A Mutant Gene Can Lead to a Massive Disorder.* The skull of a child with thalassemia is enlarged and has a peculiar "hair on end" appearance because a defective gene leads to proliferation of the blood-producing bone marrow and excess growth of the skull bones.

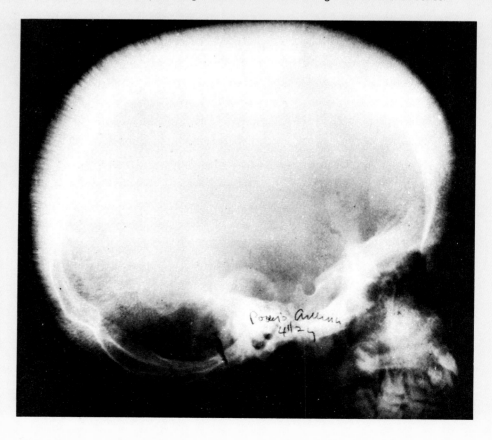

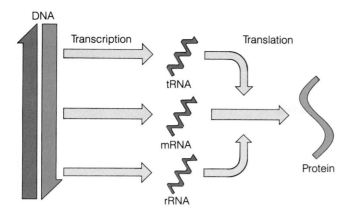

FIGURE 10.2 *Path of Information Flow in a Cell: DNA to RNA to Protein.* DNA is transcribed into a messenger RNA (mRNA) molecule; ribosomal RNA (rRNA) molecules and transfer RNA (tRNA) molecules then help translate the mRNA into the polypeptide chain of a protein.

≋ FROM DNA TO RNA TO PROTEIN

Much of the genetic research in the mid-twentieth century went to reveal that a gene is made of DNA and that it acts by specifying the amino acid sequence in a protein chain. The next question arising from this work was: Exactly how does the information in DNA become decoded and translated into protein structure?

The answer unfolded over decades of research and can be summarized this way: Genetic information within each cell flows from DNA to RNA to protein in a two-step process known as *gene expression* (Figure 10.2). In the first step of gene expression, information in a portion of the DNA molecule is copied, or *transcribed*, into an RNA molecule. This first step of making RNA from DNA is called **transcription**—a word that implies that information in one dialect (DNA) is copied into a different dialect (RNA) in the language of nucleic acids. In the second step, the information in one type of RNA molecule is *translated* by two other types of RNA molecules into a protein. This second step is called **translation** because genetic information is translated from the language of RNA into the language of proteins. Both transcription and translation are essential to every living cell; without them, proteins—the building blocks of cells—could not be manufactured. And RNA is the agent that allows both of these steps of gene expression to occur.

RNA: FUNCTION AND STRUCTURE

The three kinds of RNA involved in the two-step process of protein synthesis are messenger RNA, ribosomal RNA, and transfer RNA. **Messenger RNA (mRNA)** carries from DNA the information, or message, for manufacturing proteins during the translation phase, while both **ribosomal RNA (rRNA)** and **transfer RNA (tRNA)** assist in the translation of mRNAs to protein. All three types of RNA have a similar chemical composition and are themselves transcribed from DNA.

Like DNA, each RNA molecule consists of a string of subunits called nucleotides. RNA, however, differs from DNA in four respects (Figure 10.3). First, in each RNA molecule, the base uracil replaces the thymine found in

(a) An RNA nucleotide

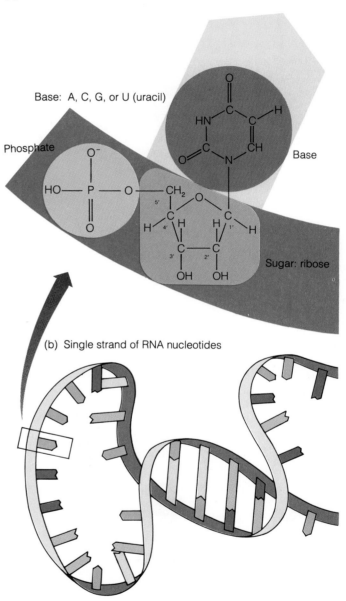

(b) Single strand of RNA nucleotides

FIGURE 10.3 *The Structure of RNA.* (a) A base (A, C, G, or U) plus a ribose sugar and a phosphate together form a single nucleotide. (b) Many RNA nucleotides join together to form an RNA molecule that can loop back and form base pairs with itself.

(a) The steps of transcription

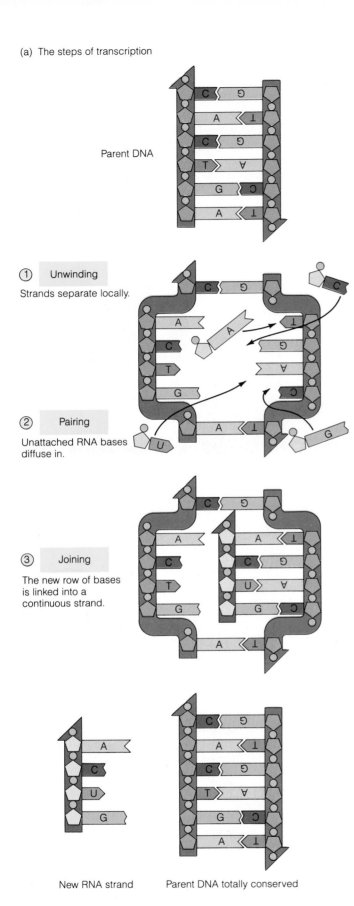

Parent DNA

① Unwinding

Strands separate locally.

② Pairing

Unattached RNA bases diffuse in.

③ Joining

The new row of bases is linked into a continuous strand.

New RNA strand   Parent DNA totally conserved

(b) Replication and transcription compared

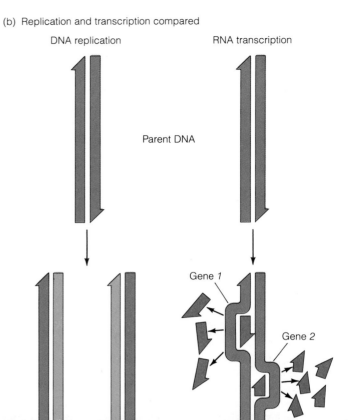

DNA replication   RNA transcription

Parent DNA

Gene 1

Gene 2

RNA

	DNA replication	RNA transcription
Parent strands copied	All of both	Small parts of each
Number of copies	Exactly one	Many, variable
Location of copies	Stay attached	Diffuse away

FIGURE 10.4 *Transcription of DNA into mRNA: The Process and Its Results.* (a) The steps of transcription in progress for any one gene. (1) Local unwinding: The two strands of the DNA separate in a local region of the DNA molecule. (2) Base pairing: The four types of RNA nucleotides float in and pair with their complementary bases on only one strand of the DNA duplex. Note that the U base of RNA, which replaces the T of DNA, pairs with an A base. (3) Joining the nucleotide to the growing RNA: Once the nucleotide units are in place, an enzyme joins the sugar-phosphate backbone together, connecting the new subunits. The result is a new strand of RNA and a totally conserved (just the way it was) parent DNA. (b) This diagram shows how DNA replication (left) results in the formation of two new strands, while RNA transcription of two genes (right) results in lots of RNA molecules (red) from each of the DNA templates (blue). The chart at the bottom summarizes the differences between the two processes.

DNA, giving us the bases A, C, G, and U in RNA (as compared to A, C, G, and T in DNA). Second, the sugar in the nucleotides of RNA is **r**ibose instead of the **d**eoxyribose in DNA (Figure 10.3a). Third, RNA usually consists of a single strand (Figure 10.3b), whereas DNA usually consists of two. Finally, RNA molecules are much shorter than the DNA molecules that make up chromosomes. Each DNA molecule carries many genes, but an RNA molecule usually contains information from only one gene. Now let's see how information passes from the DNA of a gene to an RNA molecule in the process called transcription.

TRANSCRIPTION: DNA INTO RNA

The production of new proteins—proteins to transport oxygen in red blood cells, proteins to trap sunlight in a leaf, or proteins to replace skin cells that have peeled away after a sunburn—begins with gene transcription. In transcription, a strand of DNA is copied (transcribed) into a new strand of RNA. During the process, as sections of double-stranded DNA unwind, RNA nucleotides float in and pair with the bases on one DNA strand. A special enzyme then joins the RNA subunits together to form a single strand of RNA. Figure 10.4 shows and summarizes the steps of transcription.

Accurate transcription is essential to life and good health, as Figure 10.5 illustrates. For example, many children suffering from thalassemia develop anemia because their cells cannot transcribe the normal amount of mRNA, and as a result, the cells cannot produce enough of the oxygen-carrying hemoglobin protein. Even though the child's cells transcribe every other gene normally, the failure to make enough message for hemoglobin leads to severe illness and an early death. Despite its crucial role in living cells, however, transcription alone does not guarantee gene function: mRNA must be translated into protein before it can benefit the cell.

TRANSLATION: RNA INTO PROTEIN

During translation, the information in an RNA message is converted into a specific amino acid sequence in a polypeptide chain. This translation requires some equipment. When a student sits down at a desk to translate a paragraph from French into English, a French/English dictionary acts as a sort of *code*. The student, the *translator*, interprets and applies this code; and the *desk* holds in place the passage in French and the blank writing paper ready to receive the translation into English. In a similar way, cells need a code for translating words from the language of nucleic acids (bases) into the language of proteins (amino acids). This code is called the **genetic code**. Moreover, cells need a translator to interpret the code. During protein synthesis, the translator is transfer

RNA. Finally, cells need a "desk" on which the translation can take place, and that desk is the ribosome.

■ **The Genetic Code: Dictionary of Life** The genetic code is a dictionary specifying which amino acid corresponds to which base sequence. As Chapter 9 described, the language of ribonucleic acids has an alphabet of only 4 letters (the 4 bases A, C, G, and U), while the language of proteins has 20 letters (the 20 amino acids). Obviously, one base alone cannot encode an amino acid; there just aren't enough bases. But if three bases encode an amino acid, then there are more than enough combinations to specify 20 amino acids. For example, AAC is one amino acid, ACA another, and CAA a third.

(a)

(b)

FIGURE 10.5 *Caesar Experiments with RNA Synthesis.* How long can a person live without transcribing DNA? The Roman emperor Claudius Caesar unwittingly provided an answer in A.D. 54. Caesar's wife Agrippina mixed into her husband's favorite dish of edible *Amanita caesarea* (a) a few of the poisonous species *A. phalloides* (b). These poisonous mushrooms contain a substance that inhibits transcription of DNA to RNA. For the first 10 hours after Caesar ate this delicacy, all seemed well. But as the poison was absorbed by his liver and kidneys, it began to block transcription. About 15 hours after his repast, with no new messenger RNA to make new proteins, Caesar's liver cells stopped functioning, and nausea, diarrhea, and delirium began to hit him. Two days later, he died of liver failure. It is highly doubtful that Caesar learned to appreciate the valuable role of transcription; but perhaps, in a general way, Agrippina did.

Codon			Amino acid		Codon			Amino acid
U	U	U	Phe		A	U	U	Ile
U	U	C	Phe		A	U	C	Ile
U	U	A	Leu		A	U	A	Ile
U	U	G	Leu		A	U	G	Met (START)
U	C	U	Ser		A	C	U	Thr
U	C	C	Ser		A	C	C	Thr
U	C	A	Ser		A	C	A	Thr
U	C	G	Ser		A	C	G	Thr
U	A	U	Tyr		A	A	U	Asn
U	A	C	Tyr		A	A	C	Asn
U	A	A	STOP		A	A	A	Lys
U	A	G	STOP		A	A	G	Lys
U	G	U	Cys		A	G	U	Ser
U	G	C	Cys		A	G	C	Ser
U	G	A	STOP		A	G	A	Arg
U	G	G	Trp		A	G	G	Arg
C	U	U	Leu		G	U	U	Val
C	U	C	Leu		G	U	C	Val
C	U	A	Leu		G	U	A	Val
C	U	G	Leu		G	U	G	Val
C	C	U	Pro		G	C	U	Ala
C	C	C	Pro		G	C	C	Ala
C	C	A	Pro		G	C	A	Ala
C	C	G	Pro		G	C	G	Ala
C	A	U	His		G	A	U	Asp
C	A	C	His		G	A	C	Asp
C	A	A	Gln		G	A	A	Glu
C	A	G	Gln		G	A	G	Glu
C	G	U	Arg		G	G	U	Gly
C	G	C	Arg		G	G	C	Gly
C	G	A	Arg		G	G	A	Gly
C	G	G	Arg		G	G	G	Gly

FIGURE 10.6 *The Genetic Code.* This dictionary shows how just four RNA nucleotide bases (U, C, A, and G) combined into triplets can form 20 different amino acids plus the stop codons. For example, U in the first position, A in the second position, and C in the third position (UAC) equals Tyr, or tyrosine. Note that the code contains synonyms; UUU and UUC can both mean phenylalanine; CAU and CAC can both mean histidine; and so on.

A set of three adjacent mRNA bases that encode an amino acid is called a **codon**. Early researchers found that the message UUUUUUUUUUUU, for example, translated into the polypeptide Phe-Phe-Phe-Phe, indicating that the codon UUU specifies the amino acid phenylalanine. Similar experiments revealed the meanings of

all 64 codons. The codon dictionary, or genetic code, is shown in Figure 10.6. Notice that several different codons can specify the same amino acid. For example, both AAA and AAG specify the amino acid lysine. In other words, some codons are synonyms: they mean the same thing. On the other hand, each individual codon specifies only one amino acid.

Codons, or groups of three bases along a strand of messenger RNA, are translated in sequence. Translation starts at one point along the messenger RNA, and the bases are read in groups of three; codons do not overlap, and no bases are skipped. Let's look at a simple messenger RNA and consult the genetic code in Figure 10.6 to see how it would be translated into a chain of amino acids. Our sample messenger RNA is UCGUCAUGAGUAAUUAA. The first rule is that a special *start codon* (AUG) always signifies where translation should start. This start codon also encodes the amino acid methionine. Thus, our sample messenger RNA would always be translated beginning with the *first* AUG. After the AUG start codon specifying methionine (Met), the remaining messenger RNA is read off in groups of three bases:

UCGUC AUG AGU AAU UAA

             Met   Ser   Asn   Stop

Eventually, translation encounters one of three codons (UAA, UAG, and UGA) that encode no amino acid at all. Wherever these codons appear in a message, the translation stops. Hence, they are called *stop codons*. Our sample messenger RNA would translate to the amino acid chain Met-Ser-Asn (Figure 10.7).

Numerous experiments have proved that nearly all organisms use the same genetic code. The only exceptions are found in some protists and in the mitochondria of some many-celled organisms, in which a few codons differ from the norm. It is an amazing testament to the unity of all living things that human cells and bacterial cells speak exactly the same genetic language. A gene from a human cell translates into exactly the same amino acid sequence, whether the gene resides in the nucleus of that human cell or is artificially placed in a bacterial cell.

The near universality of the genetic code is strong evidence that all living things derive from ancestral cells with a singular or similar origin early in our planet's history (see Chapter 14). Originating at the dawn of life, this ancient code has apparently been passed down intact from generation to generation. While breaking the uni-

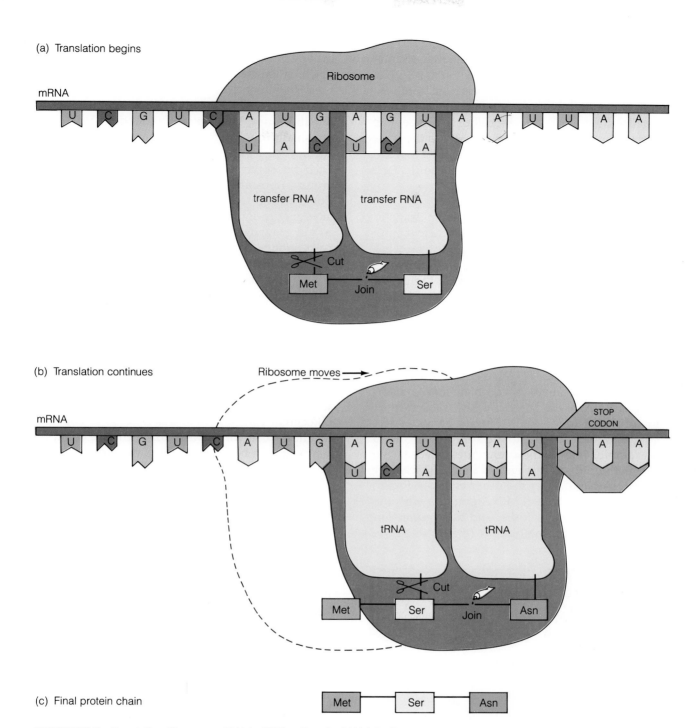

(a) Translation begins

(b) Translation continues

(c) Final protein chain

FIGURE 10.7 *Translation: Messenger RNA (mRNA) to Transfer RNA (tRNA) to Protein.*
(a) Translation begins: On the ribosome, the information in the messenger RNA (red) is translated into a polypeptide chain by transfer RNAs. One transfer RNA attached to the amino acid
methionine (Met) pairs with the AUG start codon in the messenger RNA. A second transfer RNA
carrying the amino acid serine (Ser) pairs with the messenger RNA's AGU codon. The ribosome
cuts the Met from its transfer RNA and joins it to the Ser. (b) Translation continues as the ribosome slides along the messenger RNA like a pulley on a rope: A new transfer RNA carrying
asparagine (Asn) comes in and pairs with the third messenger RNA codon, AAU. The ribosome
cuts Ser from its transfer RNA and adds it, along with Met, to Asn. (c) When the ribosome
reaches a stop codon, translation ceases, and the final protein chain is released.

versal genetic code was an important advance, it did not show us exactly how cells translate three-letter codons into amino acids.

The information in a messenger RNA is translated into polypeptide chains that will become part of a cell's proteins. To make this translation, the cell uses the genetic code plus transfer RNA as the "translator" and the ribosome as the "desk" that holds things in place so that the translation can proceed smoothly and efficiently (see Figure 10.7).

■ **The Translator Molecule: Transfer RNA** Transfer RNAs are single strands of RNA that attach to amino acids and then transfer those amino acids to the growing protein chain. Transfer RNAs loop back on themselves to form a characteristic shape rather like the lumpy ski boots depicted in Figure 10.7. Each of the 20 different amino acids has a different transfer RNA. The transfer RNAs work rather like human translators; they match a nucleic acid word, or codon, with a protein word, or amino acid. How does transfer RNA accomplish this?

At one end of each transfer RNA molecule is a sequence of three bases that recognizes its complementary codon in a messenger RNA; at the other end, the transfer RNA carries one of the 20 amino acids. Specific enzymes recognize each tRNA and attach it to the appropriate amino acid. For example, one of these enzymes attaches the amino acid methionine to one end of a specific transfer RNA. This tRNA has UAC at the other end (see Figure 10.7a). The UAC on the transfer RNA pairs up with the AUG codon on a messenger RNA, and this connection makes AUG the codon for methionine (Figure 10.7a).

(a)　　　　　　　　　　(b)

FIGURE 10.8 *Protein Synthesis Made Manifest.* The results of abundant protein synthesis are visible in (a) a spider's silk web and (b) a bear's thick coat and curving claws.

The order of codons in messenger RNA specifies the order of transfer RNAs, each attached to its own amino acid. To join the ordered amino acids to each other is then the job of the ribosome.

■ **Ribosomes: The Protein Production Desk** *Ribosomes* are small, roughly ball-shaped particles that link amino acids to each other on the growing polypeptide chain. Each ribosome is a conglomerate of proteins wrapped around ribosomal RNA molecules. The ribosome slides down the messenger RNA like a pulley on a rope, and as it goes, it joins the ordered amino acids, one by one, to the growing polypeptide chain (Figure 10.7b). Every cell contains tens of thousands of ribosomes. Working together, then, the ribosomal "desk," the tRNA "translator," and the messenger RNA bring about the life-sustaining sequence of events called **protein synthesis**.

PROTEIN SYNTHESIS: TRANSLATING GENETIC MESSAGES INTO THE STUFF OF LIFE

The synthesis of proteins is one of the cell's most important tasks, since the activities of proteins underlie nearly all cellular functions. For muscle cells to flap the wings of a honeybee, for seaweed to synthesize chlorophyll, or for oxygen to be transported in the blood of a kangaroo or child, a protein (often an enzyme) must be constructed that will then carry out a special role. Some cells make a tremendous amount of a single protein, and the result can be a dramatic mane or tail, a gossamer spider's web, or long, sharp nails (Figure 10.8).

Each protein's role is determined by its structure, and this, in turn, is dictated by its amino acid sequence. As we have seen, the order of amino acids in a protein derives ultimately from information that is stored in the base sequence of a gene.

During the translation of genetic information into protein molecules, ribosomes assist transfer RNA molecules. As the transfer RNAs line up in an order specified by messenger RNA codons (review Figure 10.7a and b), enzymes in the ribosome cut the amino acid from each transfer RNA and join the individual amino acids to each other. When a ribosome reaches a stop codon, the protein chain is completed (Figure 10.7b and c). In making a hemoglobin chain, for example, 146 amino acids are lined up and joined together before a stop codon is reached.

As in other complex biological processes, a minor change at the beginning of protein synthesis can have major effects further down the line. The fatal effects of thalassemia stem from any one of a number of simple changes in the DNA of a hemoglobin gene—changes that block the synthesis of some of the hemoglobin protein so essential to the delivery of oxygen to our brains

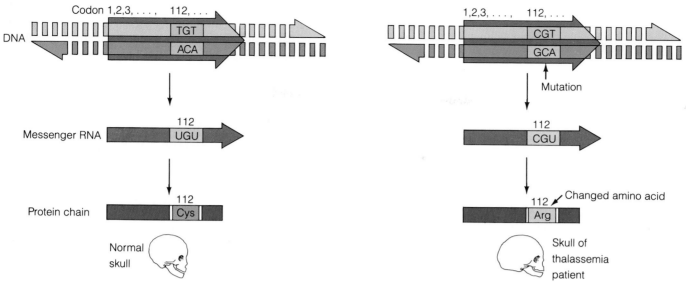

FIGURE 10.9 *A Single Gene Mutation and Its Potential Effects.* A mutation is a change in the base sequence of DNA. A single base substitution in the 112th codon of a hemoglobin gene can change a TGT sequence to a CGT. This causes the amino acid arginine to join the polypeptide chain instead of the amino acid cysteine. The hemoglobin protein containing the substituted arginine is defective and leads to a deformed skull and other symptoms of thalassemia.

and muscles. The next section explores how such changes in genes can alter their function.

GENE MUTATION: A CHANGE IN BASE SEQUENCE

A *mutation* is a change in the base sequence of an organism's DNA. This genetic alteration often has no effect on the organism's appearance or other characteristics, but sometimes it can cause a massive—often harmful—shift in the way an organism looks or acts. Geneticists recognize two general categories of mutations: **chromosomal mutations**, which affect large regions of chromosomes or even entire chromosomes and hence many genes; and **gene mutations**, which alter individual genes. Chromosomal mutations include changes in chromosome number, as in Down syndrome (see Figure 7.21), and changes in chromosome shape (detailed in Chapter 11). Here we focus on one kind of change in the base sequence of a single gene.

A mutation within a gene can result when one base pair is replaced by another (Figure 10.9). Some mutations affect the amount of protein formed, while others alter protein structure. Let us look at how a mutation in a gene for hemoglobin can affect the structure and function of this oxygen-transporting protein and cause the disease thalassemia. In most people, the gene for one chain of the hemoglobin molecule has as its 112th codon TGT, specifying the amino acid cysteine. But in some people, a mutation in the 112th codon has changed it to CGT, which specifies arginine. The altered hemoglobin molecule produced by such a mutation is very unstable, and most of it disintegrates rapidly, resulting in thalassemia (see Figure 10.9).

Although the causes of gene mutations are the same in all organisms, their effects are much more complex in a child than in a bacterium. The next section shows how a mutation in a person can precipitate a cascade of effects in many seemingly unrelated physiological processes.

HOW A GENE MUTATION CHANGES PHYSICAL APPEARANCE

Mutations can have diverse effects. The crossed eyes and white pelt of Mohan the tiger, for example, are the result of a single mutation (see Chapter 8). Thalassemia is also due to a mutation in a single gene—one that profoundly affects blood cells, spleen, and bones.

Several different mutations can lead to thalassemia, but each ultimately blocks the production of normal hemoglobin chains in red blood cells. The mutant hemoglobin proteins that accumulate in little clumps inside the

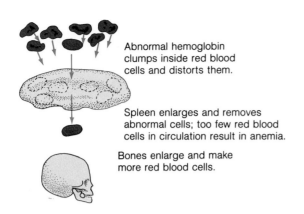

Abnormal hemoglobin clumps inside red blood cells and distorts them.

Spleen enlarges and removes abnormal cells; too few red blood cells in circulation result in anemia.

Bones enlarge and make more red blood cells.

FIGURE 10.10 *How Mutant Proteins Cause Thalassemia.* A child with the blood disease thalassemia makes defective hemoglobin which can clump inside the red blood cells. This clumping changes the cell shape, and the spleen enlarges as it removes the abnormal cells, causing severe anemia. Bones enlarge and make more red blood cells, but these also contain defective hemoglobin, so the cycle continues.

red cells distort the entire cell shape (Figure 10.10). The spleen recognizes abnormally shaped red cells and destroys them. As a result, the number of circulating red cells decreases, and the victim becomes anemic. In the course of removing the huge number of abnormal red cells, the spleen enlarges. Because the enlarged spleen so effectively removes red cells from circulation, the bone marrow must expand to produce enough cells to keep up with the body's huge demand for oxygen-carrying blood cells. As the bone marrow grows, the size of the skull and other bones that produce red cells increases grotesquely (review Figure 10.1). In this way, a simple mutation produces the multiple trademarks of thalassemia: anemia and enlarged spleen and bones.

Mutations clearly have a major, usually negative impact. So where do they come from and how do organisms protect themselves from such genetic changes?

THE ORIGIN OF MUTATIONS

Mutations arise either through spontaneous errors during DNA replication or later, through the effects of physical or chemical agents on the DNA. During replication, the wrong base can be inserted into a DNA molecule. Usually, this kind of replication error is edited out by the "proofreading" enzymes, so such mutations are very rare. More commonly, physical and chemical agents called *mutagens* change DNA structure. Mutagens are everywhere and include ultraviolet rays from the sun, chemicals in cigarette smoke, and even many natural substances from plants or fungi.

Despite the presence of mutagens throughout our environment, a permanent mutation is a relatively rare event, because *DNA repair enzymes* constantly patrol the DNA, moving up and down the long strands, detecting and fixing damaged or unpaired bases. For example, when ultraviolet sunlight strikes the DNA in the nucleus of one of a person's skin cells, a few bases can be altered. In most people, DNA repair enzymes usually correct the changes. People with the inherited skin disease called xeroderma pigmentosum, however, lack the enzymes needed to repair the DNA, and thus mutations from what would be a normal amount of exposure to the sun accumulate in their cells and cause numerous skin cancers (Figure 10.11).

Cancer-causing substances are known as *carcinogens*, and they act by generating mutations. Anything that causes mutations has the potential to cause cancer if it happens to affect a gene that regulates cell growth. Since the agents that cause mutations in bacteria are the same as those that damage human DNA, most agents that increase the rate of mutation in bacteria also increase the mutation rate—and potentially the cancer rate—in people. Owing to this fact, researchers can test the cancer-causing ability of chemical and physical agents by exposing certain bacteria to suspected cancer-causing agents and measuring mutation rates in the bacteria. With this test (called the Ames test after its originator), researchers have generated an extensive catalog of mutation-causing substances that are also potential causes of cancer. Individuals who expose themselves to known agents of mutation, such as excess sunlight or cigarette smoke, place a heavy bur-

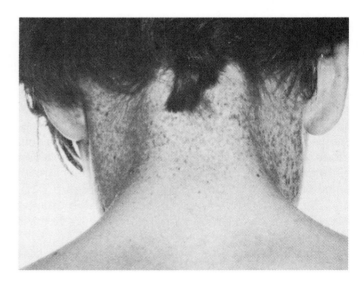

FIGURE 10.11 *DNA Repair and Cancer.* When ultraviolet light from the sun strikes and damages DNA, enzymes normally find the damage and repair it so that the DNA is usually as good as new. This sufferer of the skin disease xeroderma pigmentosum has developed hundreds of skin cancers. Deficient DNA repair leaves the skin cells vulnerable to the mutations and cancer caused by ultraviolet light.

den on their DNA repair enzymes, needlessly increasing their likelihood of getting cancer.

■ Mutation and the Structure of Individual Genes

Recent findings on gene structure help explain why some genes mutate much more frequently than others. In complex organisms, genes are split: They possess **introns**, stretches of DNA that *intr*ude into the gene but do not appear in the final messenger RNA (and hence are not translated into protein). The remaining portions of a gene, which do appear in the messenger RNA, are called **exons**, since they are *ex*pressed (Figure 10.12). In many genes, the noncoding introns are far longer than the exons. For example, in one gene that encodes a protein necessary for normal muscle function, there are 60 introns. Together, these introns are more than 100 times as long as the exons, the parts of the gene that encode the protein. Geneticists do not yet thoroughly understand the function of introns, but some think that introns might be a spin-off of the way new genes arise during evolution (see Chapter 32).

What is the significance of introns to you? Just this: A gene with many long introns is highly likely to become mutated. The human genes that mutate most frequently, like the gene behind the defect causing muscular dystrophy, tend to have more and longer introns than the genes that mutate less frequently. This insight helps establish a basic relationship between gene structure and disease and may, in the future, provide a means of treating or preventing some genetic disorders.

Some mutations do not affect the way we look or function, but others as minor as a single base-pair change among the 3 billion base pairs in a mammal's DNA can have profound effects if that mutation alters an important protein. Imagine typing a term paper 1000 times as long as this book, only to get a failing grade because your paper has one misspelled word. That is very much like what has happened to a person with thalassemia. Through transcription and translation, red blood cells build hemoglobin, and through mutations, this basic activity is disrupted and a deficient form of hemoglobin is built instead. But why is hemoglobin found *only* in red cells, and not in muscle cells or nerve cells? Why, in other words, do cells make certain proteins and not others? The next section addresses these questions.

≈ REGULATION OF GENE ACTIVITY

Cells have their particular kinds of proteins because DNA is transcribed to RNA, and RNA is translated to protein. But while all of your cell types produce small amounts of the enzymes that repair DNA, only young red blood

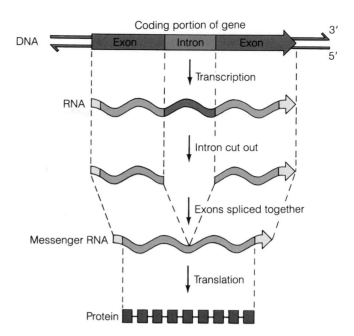

FIGURE 10.12 *Structure of a Eukaryotic Gene.* In eukaryotes, only certain parts of a gene, known as exons, actively encode proteins. An entire gene is transcribed into RNA, but introns, or intrusive stretches of DNA, are spliced out before the messenger RNA forms. This messenger RNA is then translated into a polypeptide. The figure shows one intron, but genes can have 60 or more.

cells make hemoglobin, only gut cells make some digestive enzymes, and so forth. Clearly, each cell must regulate the transcription of DNA into messenger RNA and the translation of RNA into protein so that it produces the right amounts of the right types of protein where and when they are needed. The process that controls each cell's gene activity is called **gene regulation**.

IN BACTERIA

Bacteria have streamlined and highly sophisticated mechanisms of gene control. Consider, for example, the bacteria called *Escherichia coli* growing in your intestine. They need a steady supply of tryptophan and can make their own if none is available in the environment. But just after you have eaten a prime rib dinner and your digestive enzymes have reduced the protein in the meat to amino acids, the liberated amino acid tryptophan becomes abundantly available to the resident bacteria. What a waste of energy it would be if the bacteria continued to make tryptophan. Instead, they stop making it. Later, after your vanilla ice cream dessert, the bacteria are floating in a sea of lactose, the main sugar in milk. Within seconds, the cells respond by making enzymes (proteins) that break down the milk sugar into subunits used for growth.

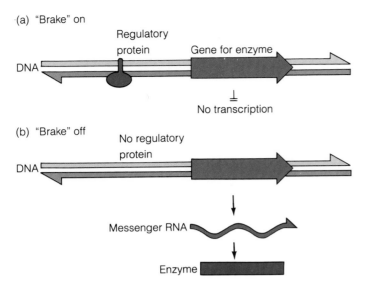

(a) "Brake" on

Regulatory protein

Gene for enzyme

DNA

No transcription

(b) "Brake" off

No regulatory protein

DNA

Messenger RNA

Enzyme

FIGURE 10.13 *A Model of Gene Regulation.* (a) If a regulatory protein binds to a special DNA site near a gene, the gene will not be transcribed into RNA. The regulatory protein thus applies a "brake" to the gene's expression. (b) If the regulatory protein is absent or nonfunctional, the gene is free to be transcribed into RNA, and a specific protein may be manufactured; the "brake" has been released.

A bacterial cell that makes unnecessary proteins expends energy and materials that could be used to prepare the cell for reproduction; thus, it grows and divides more slowly than a bacterium using its resources more efficiently. Most bacteria do not produce unnecessary proteins, because they have special protein molecules that regulate when a gene will be used. Although the details are complex, the concept of this type of gene control is simple. Basically, a *regulatory protein* can stick to a special spot on the DNA near a gene and block the gene from being transcribed rapidly (Figure 10.13a). But if the regulatory protein is absent or nonfunctional, then the gene is transcribed (Figure 10.13b). Thus, the regulatory protein acts like a brake. In your digestive tract, the milk sugar from the ice cream renders a particular regulatory protein in the bacteria nonfunctional, and a certain gene is transcribed to messenger RNA; from that messenger RNA, an enzyme (a protein) is made that digests the milk sugar. It is as if the milk sugar caused the brake to be released. Although in this case the regulatory protein acts as a brake, in other cases, regulatory proteins act as starters, igniting gene activity.

IN PEOPLE

Similar mechanisms are at work in human cells; for example, human hormones, such as estrogen and testosterone, interact with regulatory proteins that turn genes off or on. In another example, baby mammals produce an enzyme in their intestinal cells that allows them to digest the milk sugar lactose. But as the babies mature, they lose the ability to make the enzyme and along with it the means of digesting milk, probably because of some regulatory protein like the one in Figure 10.13a. Like other mammals, most of the world's people, as they mature, lose the ability to make the enzyme that digests milk sugar (Figure 10.14). This is why drinking milk gives most adults a stomachache. Most Caucasians and certain dairying tribes of Africa are exceptions. Unfortunately, many adults without the special enzyme that allows them to digest milk sugar have a taste for ice cream and other milk products. To circumvent this problem, they can enlist the aid of *Lactobacillus*—the bacterium used to make yogurt. If they take a pill containing this bacterium just before

(a)

(b)

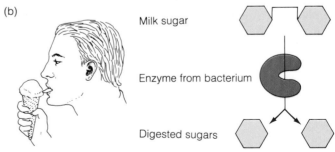

Milk sugar

Enzyme from bacterium

Digested sugars

FIGURE 10.14 *People Lose the Ability to Digest Milk Sugar.* (a) The digestive tract cells of this nursing infant make an enzyme that digests milk sugar (lactose). (b) In most adults, the gene for the enzyme that digests milk sugar is turned off, probably by a regulatory protein (as in Figure 10.14a). As a result, most adults cannot enjoy milk products without the help of bacteria that make their own lactose-digesting enzyme.

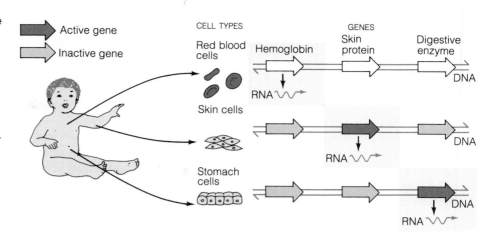

FIGURE 10.15 *People and Other Mammals Use Different Genes in Different Cells at Different Times During Development.* Each cell in a mammal has a full set of genes—genes for hemoglobin (to carry oxygen), for keratin protein (for hair and claws), for digestive enzymes (for breaking down food molecules), and so on. But different cells use only a small subset of these genes. For example, in developing red blood cells, hemoglobin genes are active, while keratin and digestive enzyme genes are inactive. In skin cells, keratin genes are on, while the others are off. And in gut cells, the genes for digestive enzymes are on, while the others are off.

eating a giant bowl of mocha almond fudge ice cream, a bacterial enzyme digests the milk sugar before it can reach the intestine and cause discomfort (Figure 10.14b).

■ **Gene Regulation and Cancer** Mutations in genes for regulatory proteins are the cause of many cancers. The mutations change the proteins, and the changed proteins bind inappropriately to DNA, turning on genes that lead to cell division. Uncontrolled cell division produces cancer. Molecular geneticists are trying to find ways to block the action of the mutant regulatory proteins as a means of reversing some types of cancer. Although mutant regulatory proteins can trigger abnormal cell division and cancer, the appropriate binding of regulatory proteins is a key to normal embryonic development, in which one fertilized egg cell divides to form an integrated complex of many different kinds of cells.

■ **Gene Regulation in Developing Organisms** Each cell in a multicellular eukaryotic individual has a full complement of genes, but different types of cells express (transcribe and translate) different sets of genes, since different cells require different proteins, and different proteins come from different genes. Both young red blood cells and intestinal cells, for example, have genes for making hemoglobin and genes for making digestive enzymes (Figure 10.15). But developing red blood cells express only the hemoglobin genes, while the gut cells express only the digestive enzyme genes. These cell types also contain the genes for making eye pigments, tooth enamel, and toenail proteins, but these and many other genes are never used by these cells. Since regulatory proteins turn genes on and off, each cell type must have a different set of regulatory proteins. A great unsolved problem of biology is how the different cells in a developing corn plant or human baby come to have different regulatory proteins. Clearly, this is an important question because if some of your blood cells began making digestive enzymes, you could literally eat your heart out!

As we mentioned earlier, biologists do not yet know how different genes are selected for expression in different cells or how the selection, once made, is stabilized. But they have already made much progress in understanding how related genes are grouped together, how individual genes are built, and how they work (see the box on page 184). This knowledge of how genes are organized and how that organization affects function may well provide the basis for understanding how mutant genes, such as those that cause thalassemia and cancer, can derail normal growth and development.

■ **Gene Families on a Chromosome** While eukaryotic genes are not clustered along the chromosome according to function, they sometimes cluster together on the same chromosome in *multigene families* according to structural similarities. The genes in these families are structurally similar because they all derive from a common ancestor gene. In humans, for example, there are five closely linked genes on chromosome 11, and each of these genes encodes a polypeptide similar in structure to a hemoglobin chain found in adults (Figure 10.16a, page 185). Each member of this gene family is active at a different time in a person's life, one in the embryo, another in the baby just before birth, and so on. This makes sense, because an unborn baby and an adult get their oxygen in different ways. Thus, although thalassemia victims have a mutant hemoglobin gene of the adult type, hemoglobin genes for the other life stages are usually all normal and active in the embryo and fetus. That is why future thalassemia sufferers survive until after birth (Figure 10.16b). The importance of gene families now becomes apparent: If medical geneticists could learn how to reactivate in the adult the same hemoglobin gene that is active in the unborn fetus, they might be able to keep people from suffering from thalassemia.

THE ULTIMATE ROADMAP OF OUR HUMAN GENES

Manic depression; Alzheimer's dementia; schizophrenia; alcoholism; thalassemia; diabetes; certain forms of cancer; obesity; and heart disease. These debilitating diseases have something in common besides human suffering and potential fatality: The tendency to develop them is inscribed in the genes. Physicians and researchers would like to be able to read those "inscriptions," to know the precise location of mutant genes along our chromosomes and how they encode defective proteins and lead to many of the chronic diseases that plague our species. With that knowledge, they could design drugs and treatments that specifically block or correct the underlying causes of a disease, not just relieve a patient's symptoms.

How can researchers locate those defective genes? Geneticists estimate that human DNA contains 50,000 to 100,000 genes, relatively short stretches of A, T, G, and C base sequences nestled within a sea of repetitive and apparently nonfunctional stretches. Our genes are truly needles in a genetic haystack: Together, they represent only about 3 percent of the estimated 3 billion total bases lying sequentially along the DNA of the 46 chromosomes.

A thorough mapping of all human genes would be an enormous boon to medical research but an equally enormous task. Luckily, biologists have developed tools in recent years to get the job done. With their emerging capability and the work's great promise for saving lives, researchers have initiated an immense effort to create the ultimate roadmap of human DNA.

Called the Human Genome Project, this tremendous undertaking will span 15 years or more, cost at least $3 billion, and require the dedication of thousands of U.S. researchers and technicians. Other countries are devoting smaller but still substantial funds and personnel for the same purpose. Different research groups will tackle different parts of the chromosomes, attempting to create low-resolution maps that place known genes and gene markers along the chromosomes. Eventually, they will generate high-resolution maps that show the exact base sequences within each of the genes as well as the nonfunctional stretches in between—all 3 billion bases in precise order (Figure 1).

Early obstacles included finding a common language for mapping the genes, since researchers can use a wide array of different mapping techniques, and developing sophisticated computer programs for storing data in a meaningful, retrievable way. Researchers are well on their way to solving these problems, and they are now optimistic that given enough time and the monetary support of Congress, they will be able to map the entire human genome and store the information in massive data banks. Some biologists, however, are concerned that increasingly scarce government monies will be diverted to this enormous undertaking, funding that might otherwise support equally valuable smaller research projects. Other scientists worry that extensive information about human genes might be used to discriminate against people with certain genetic makeups. These are issues to consider as we contemplate the next 15 years in biomedical

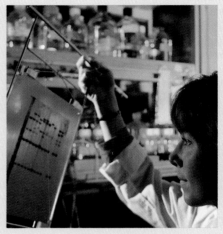

FIGURE 1 *Researcher Examining the Sequence of a Human Gene.* When researchers move an electric current through a jellylike medium to which they have applied fragments of DNA, fragments of different sizes will show up as spots in different places. By reading the spots, a researcher can tell the sequence of bases in the gene.

history, a time when, thanks to basic research, many of the most serious human diseases can be addressed in new and more effective ways.

FOR FURTHER READING

Cantor, C. R. "Orchestrating the Human Genome Project." *Science* 248 (6 April 1990): 49–51.

Koshland, D. E., Jr. "Sequences and Consequences of the Human Genome." *Science* 246 (13 October 1989): 189.

"Letters: The Genome Project: Pro and Con." *Science* 247 (19 January 1990): 270.

Roberts, L. "New Game Plan for Genome Mapping." *Science* 245 (29 September 1989): 1438–1440.

Watson, J. D. "The Human Genome Project: Past, Present, and Future." *Science* 248 (6 April 1990): 44–48.

(a) Multigene family of hemoglobin

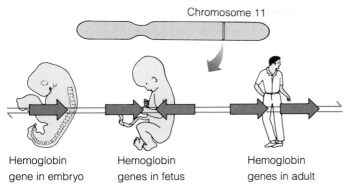

Chromosome 11

Hemoglobin gene in embryo | Hemoglobin genes in fetus | Hemoglobin genes in adult

(b)

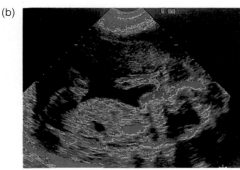

FIGURE 10.16 *Human Hemoglobin Genes: A Multigene Family.* (a) This diagram shows five different hemoglobin genes located on chromosome 11; they are turned on and off at different times in a person's development. (b) Sonogram of a human fetus. The fetal hemoglobin of future thalassemia victims is normal, which is why they show no signs of the disease until after birth.

≋ CONNECTIONS

What do the principles of transcription, translation, and gene regulation mean for a child with thalassemia? The child is a helpless spectator in the race between its bones, which are making red cells as fast as possible, and its spleen, which is destroying red cells just as fast. The starting gun can be a single base-pair substitution in any of a number of positions along a hemoglobin gene. With some mutations, the gene isn't properly transcribed; or translation may be blocked prematurely. Each mutation provokes the same result: the destruction of red blood cells containing defective hemoglobin.

Hope comes from the ability of geneticists to work with genes. Since Watson and Crick's work in the 1950s, geneticists have learned a great deal about the structure of DNA and how genes are built, regulated, and occasionally mutated. On the basis of this knowledge, biologists have developed methods for manipulating genes and have begun to use these methods to improve crops and livestock and to cure human hereditary diseases like thalassemia. The next chapter looks at the progress being made in detecting and preventing human genetic defects.

NEW TERMS

STUDY QUESTIONS

REVIEW WHAT YOU HAVE LEARNED

1. What is the difference between transcription and translation?
2. How do RNA and DNA differ?
3. During transcription, what will the order of the bases in messenger RNA be if the base sequence in the coding strand of DNA is CTAGCT?
4. Refer to the codon dictionary in Figure 10.6 and give all possible codons for the following amino acids: lysine (Lys); phenylalanine (Phe); tryptophan (Trp).
5. How do biologists explain the fact that the genetic code is identical in humans, bacteria, and other organisms?
6. What is a gene mutation?
7. Outline the model of gene regulation in prokaryotes and eukaryotes.

APPLY WHAT YOU HAVE LEARNED

1. What might be the possible consequences to individuals and society if physicians and geneticists could replace the malfunctioning gene(s) in the cells of people with genetic diseases?
2. How can a scientist use the Ames test to determine if the chemicals in a new hair dye may cause cancer?
3. Some people cannot digest milk sugar. How can bacteria be enlisted to help them enjoy a dish of ice cream?

FOR FURTHER READING

Hirsch, M. S., and J. C. Kaplan. "Antiviral Therapy." *Scientific American* 256 (April 1987): 76–85.

Schulman, L. H., and J. Abelson. "Recent Excitement in Understanding Transfer RNA Identity." *Science* 240 (1988): 1591–1592.

Watson, J. D., N. H. Hopkins, J. W. Roberts, J. A. Steitz, and A. M. Weiner. *Molecular Biology of the Gene*, 4th ed. Menlo Park, CA: Benjamin/Cummings, 1987.

Weatherall, D. J. "Other Anemias Resulting from Defective Red Cell Maturation." *Oxford Textbook of Medicine.* Edited by D. J. Weatherall, J. G. G. Ledingham, and D. A. Warrell. New York: Oxford University Press, 1983.

① A gene carries the information for making a protein.

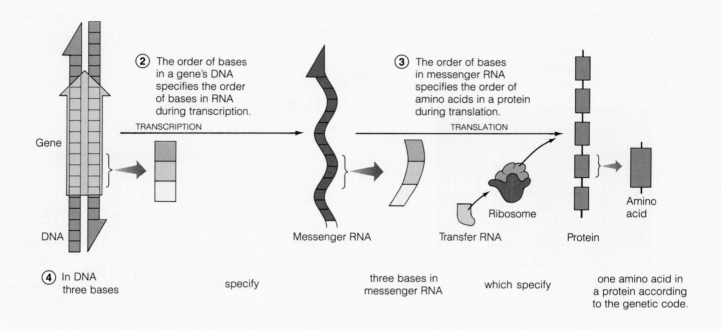

② The order of bases in a gene's DNA specifies the order of bases in RNA during transcription.

TRANSCRIPTION

③ The order of bases in messenger RNA specifies the order of amino acids in a protein during translation.

TRANSLATION

Gene

DNA

Messenger RNA

Transfer RNA

Ribosome

Protein

Amino acid

④ In DNA three bases

specify

three bases in messenger RNA

which specify

one amino acid in a protein according to the genetic code.

Mutation

Changed amino acid

⑤ A mutation is a change in DNA base sequence that may change the amino acid sequence of a protein.

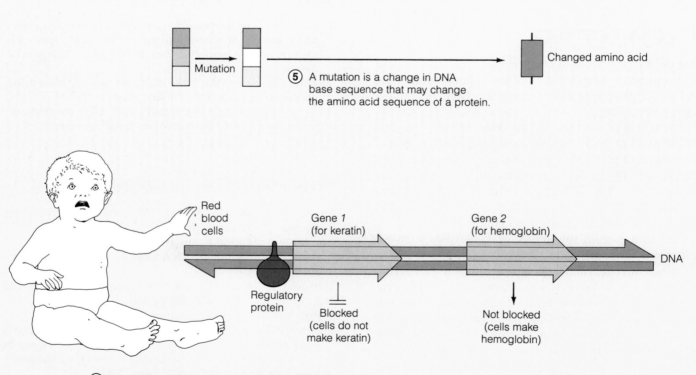

Red blood cells

Gene 1 (for keratin)

Gene 2 (for hemoglobin)

DNA

Regulatory protein

Blocked (cells do not make keratin)

Not blocked (cells make hemoglobin)

⑥ Gene regulation allows cells to use the right gene at the right time.

HUMAN GENETICS

CHAPTER 11

DIET SOFT DRINKS AND HUMAN GENETICS

Have you ever noticed warning labels on cans of diet soda that read "Phenylketonurics: Contains phenylalanine" and wondered what they mean? The answer to this question is a good case study in human genetics.

Most children (and adults) can digest the low-calorie sweetener called aspartame now used in many diet soft drinks and other foods. It is nontoxic and quite safe in moderate doses. But it does contain large amounts of the amino acid phenylalanine, a component of virtually all proteins. Most people have an enzyme that breaks down phenylalanine, whether the amino acid occurs in meat or in artificial sweeteners. Unfortunately, a small percentage of people inherit two recessive alleles for a certain

gene and as a result do not produce the enzyme. They are called *phenylketonurics*, and their disease is *phenylketonuria*, or *PKU*, because any phenylalanine they ingest in foods builds up in their bodies and is transformed into compounds known as phenylketones. These substances, which have a telltale odor and can be detected in the victim's urine, cause nerve cells in the brain to develop abnormally. If a child born with the disease PKU drinks mother's milk or diet soda or ingests any other food containing phenylalanine, its brain development will be stunted, it will suffer seizures starting about the age of six months, and its mental age will probably not advance beyond that of a two-year-old.

In the past, 1 out of every 100 men-

tally retarded children suffered from PKU. But today, thanks to modern genetic techniques, the recessive alleles that lead to PKU can be detected before the child is born; then, after birth, the parents can control the child's diet to exclude phenylalanine and thus prevent mental retardation (Figure 11.1).

As we move through this chapter, we will encounter the theme that people are unique genetic subjects, and because of this, they are hard to study. For instance, humans don't choose mates and produce offspring to satisfy a geneticist's curiosity; an investigator must search for subjects and matings that just happen to express traits of interest. We will also discover a second theme: Gene mapping and other modern techniques applicable to human genetics grew out of research on more classic genetic subjects, such as fruit flies and bacteria. Some of the same molecular-level research that fueled the recombinant DNA revolution is leading us into an era of dramatic advances in human genetics. Finally, we will see that the small genetic changes that alter a person's physiology, such as the recessive alleles that lead to PKU, usually have a negative impact on health.

As we examine how you and every other person inherit family resemblances, metabolic defects, and body characteristics, we will answer these questions:

- What standard genetic techniques do researchers use to study the inheritance of human traits?
- What molecular techniques allow scientists to map human genes?
- How are physicians currently treating human genetic diseases?
- How are geneticists helping to prevent new cases of genetic disease?

FIGURE 11.1 *Genetic Diseases Like PKU Can Be Controlled.* In this brother-sister pair, the disease phenylketonuria is under control, thanks to early diagnosis and carefully controlled diet.

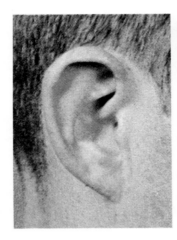
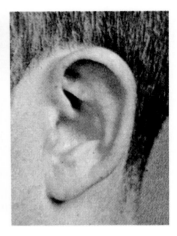

FIGURE 11.3 *A Human Genetic Trait Caused by a Single Gene.* The alleles of a single gene determine whether your earlobes are attached (left) or unattached (right).

≋ FAMILY STUDIES, CHROMOSOMES, AND HUMAN GENETICS

People are wonderfully complex organisms: We come in a nearly infinite variety of shapes, sizes, and colors and have a huge range of individual traits (Figure 11.2). Many of our traits are *polygenic,* meaning that many genes work together to determine individual characteristics such as how well our livers function, how long our limb bones are, what our facial features look like, how intelligent we are, and whether we are susceptible to various types of cancer, diabetes, allergies, heart attacks, or early senility (Alzheimer's disease).

At the same time, geneticists have discovered more than 3000 human conditions determined in simple Mendelian fashion by single genes. These include such traits as hair on the middle finger joint or elbow; attached or unattached earlobes (Figure 11.3); widow's peak; ear pits; extra fingers or toes (*polydactyly*); baldness; hemophilia; color blindness; counterclockwise cowlicks in the hair; dry, brittle ear wax; albino skin and hair coloring; thalassemia (see Chapter 10); and PKU. Many of these traits do not require medical treatment, and some, like baldness and color blindness, cannot be treated at this time. Others, however, like PKU, can be corrected if detected early enough. Table 11.1 lists a number of human traits that are inherited in simple Mendelian fashion and that are discussed in this chapter. To study and treat a genetic condition, a geneticist must know whether it is caused by a single gene or by many genes or perhaps by an alteration in chromosome shape or number. To determine these things, human geneticists have borrowed two traditional tools from geneticists studying other organisms: the analysis of family histories and the study of chromosome variations. Let's look at each tool and the kinds of conditions it can uncover.

HUMAN PEDIGREES: ANALYSIS OF FAMILY GENETIC HISTORIES

Because the human life cycle is so long and the number of offspring in the average family so small, a geneticist must follow a trait through all the members of an extended family to discover its pattern of inheritance. Geneticists can organize family data in **pedigrees,** or orderly diagrams of a family's relevant genetic features. Figure 11.4a diagrams a family with PKU expressed in certain members. A pedigree shows the family relationships, sex, and physical makeup of each member. Using the rules in Figure 11.4b, a geneticist can infer the genetic makeup of family members and determine whether a given trait is dominant or recessive and whether a gene lies on special chromosomes associated with sex. Before we look at pedigrees, let's review what we have learned about the kinds of chromosomes people have.

■ Sex Chromosomes and Autosomal Chromosomes

The chromosomes of human males and females are almost identical—but not quite. Both men and women have 23 pairs of chromosomes, for a total of 46, ranging in size from quite long (chromosome 1) to very short (chromosome 22) (Figure 11.5, page 190). In a female, the two members of each chromosome pair look identical under a microscope. In a male, however, one pair consists of chromosomes that are not identical (see Figure 11.5). The two members of this unequal pair in males are called the

FIGURE 11.2 *People: An Infinite Variety.* Genes—alone and in groups—determine the many traits that make each person (except identical twins!) wonderfully unique.

X chromosome and the Y chromosome. Females have two X chromosomes and no Y chromosome. The X chromosome is larger than the Y. Because these chromosomes differ in the two genders, they are called *sex chromosomes*. The other 22 pairs of chromosomes are identical in men and women and are called *autosomes*. There are many genes on each autosome as well as on the X chromosome (which carries so-called **X-linked** genes). The Y chromosome, however, has few genes besides those necessary for causing a fertilized egg to develop as a male embryo and fetus. Using the rules in Figure 11.4b while examining a family pedigree, it is relatively easy to tell if a gene lies along a sex chromosome or an autosome.

TABLE 11.1 ≋ SOME HUMAN GENETIC CONDITIONS	
DISEASE	EFFECT
Recessive Allele on Autosomal Chromosome	
Thalassemia (chromosome 16 or 11)	Abnormal hemoglobin; anemia, bone and spleen enlargement
Sickle-cell disease (chromosome 11)	Abnormal hemoglobin; sickle-shaped red cells, anemia, blocked circulation
Cystic fibrosis (chromosome 7)	Defective membrane protein; excessive mucus production, digestive and respiratory failure
Tay-Sachs disease (chromosome 15)	Missing enzyme; buildup of fatty deposit in brain; no mental development
Phenylketonuria (PKU) (chromosome 12)	Missing enzyme; mental deficiency
Albinism (chromosome 11)	Missing enzyme; unpigmented skin, hair, and eyes
Recessive Allele on X Chromosome	
Duchenne muscular dystrophy	Muscle cells die; weakness, respiratory failure
Hemophilia	Uncontrolled bleeding
Fragile X syndrome	Mental retardation
Dominant Allele on Autosomal Chromosome	
Hypercholesterolemia (chromosome 5)	Missing protein that takes cholesterol from the blood; heart attack by age 50
Huntington's disease (chromosome 4)	Progressive mental and neurological damage
Conditions Due to Chromosome Variations	
Down syndrome (an extra chromosome 21)	Mental retardation, heart abnormalities
Klinefelter syndrome (XXY)	Defect in sexual differentiation
Turner syndrome (XO)	Skin folds in neck; sterility

(a) Pedigree of a family with PKU

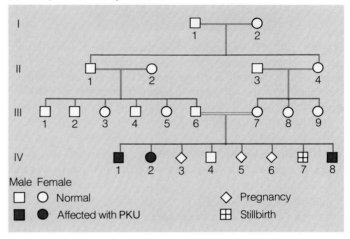

(b) Rules for applying Mendel's principles to the inheritance of human traits

Is a trait dominant or recessive?
1. If a child has a rare trait and one parent is affected, the trait is usually dominant.
2. If a child has a rare trait but neither parent is affected, the trait is usually recessive.

Is a trait X-linked or autosomal?
3. If a son inherits the trait from his father, it cannot be X-linked and must be autosomal.
4. If all or nearly all the affected people are males, the trait is probably X-linked.
5. If both males and females are affected about equally, the trait is probably autosomal.

FIGURE 11.4 *How to Interpret a Family Pedigree.* (a) In a typical pedigree, a single horizontal line (here, dark blue) connecting a male and a female represents a mating; a double line indicates a mating between relatives. A horizontal line above a series of symbols (here, red) designates brothers and sisters of one family arranged in birth order from left to right. Generations are labeled with roman numerals, and each individual in the pedigree is numbered. In this family, children with PKU (dark symbols) had undiseased carrier parents who were first cousins. The parents apparently inherited a recessive PKU allele from a grandparent. The children with PKU inherited two doses of the PKU allele, one from each parent. (b) By using the rules for applying Mendel's principles to the inheritance of human traits, you can tell whether PKU is dominant or recessive and whether it is X-linked or carried by an autosome (nonsex chromosome).

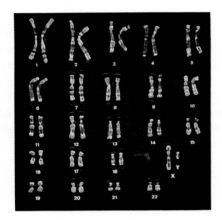

FIGURE 11.5 *Human Chromosomes.* Geneticists remove human cells, stain and photograph the cells during mitosis, then cut the chromosomes from the photo and arrange them in pairs, as shown here. This chromosome set comes from a male with sex chromosomes X and Y and 22 pairs of autosomes numbered 1 to 22.

■ Duchenne Muscular Dystrophy: A Recessive Allele on the X Chromosome The most common X-linked genetic disease is Duchenne muscular dystrophy (DMD), a degenerative muscle condition that strikes mainly boys. Figure 11.6a shows why primarily boys show the disease. A woman with one normal X chromosome and one X chromosome bearing the disease allele does not show the disease because the disease allele is recessive to the normal one. But she passes the mutant allele to half of her sons, and since they have no normal X, they show the muscular dystrophy trait. The muscle cells of these affected boys slowly die, beginning in the legs, and by age five, the child with DMD cannot stand up easily and must rise in a characteristic way (Figure 11.7). Eventually, the muscles of his diaphragm degenerate, and the child can no longer breathe. There is at present no cure for DMD, but since geneticists have recently isolated the gene and identified the protein it encodes, we might

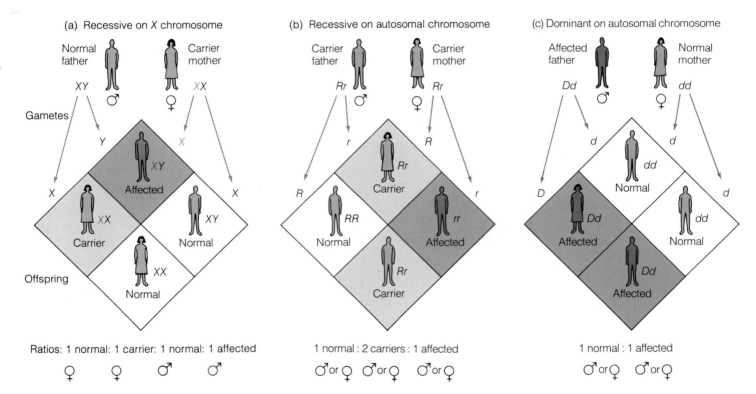

FIGURE 11.6 *Human Pedigrees and Mendel's Principles: The Inheritance of Traits Determined by a Single Gene.* A simple Punnett square (see Chapter 8) depicting information from a pedigree can reveal whether a trait is dominant or recessive, X-linked or autosomal. (a) This square traces the inheritance of a recessive gene on an X chromosome. Males are affected more often than females, and inheritance is from mother to son, not father to son. With a normal father and a carrier mother, half the daughters will be normal and half will be carriers, while half the sons will be normal and half will be affected. (b) Children affected by a trait carried by a recessive gene on an autosomal (nonsex) chromosome usually have two unaffected heterozygous parents, and the ratio of offspring will be one normal to two carriers (who don't show the condition) to one affected. (c) With dominant genes on an autosomal chromosome, all affected offspring have at least one affected parent. With one normal and one affected parent, half the offspring will be affected and half will be normal.

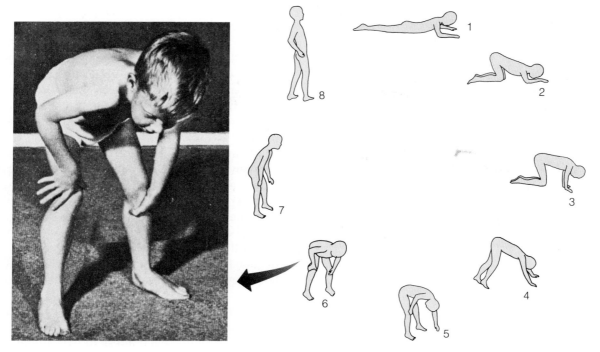

FIGURE 11.7 *Duchenne Muscular Dystrophy, Caused by a Recessive Allele on the X Chromosome.* To rise from the floor, boys with this disease must "climb up themselves," as this five-year-old is doing, because their lower limbs grow weak before their upper bodies do.

soon understand the specific mechanism of muscle degeneration and develop a treatment. Other *X*-linked genes cause hemophilia (known as bleeder's disease) and color blindness.

■ **Genes on Autosomes** Most genetic defects are caused by recessive alleles on autosomal (nonsex) chromosomes (Figure 11.6b), and PKU is a classic example. Figure 11.4 shows that two brothers and one sister in the same family inherited PKU, but neither parent showed the trait. A quick look at the rules in Figure 11.4b reveals that the trait is recessive (rule 2) and autosomal (rule 5). Since the parents pass on the mutant allele but do not show its effects, they must be **carriers,** with a dominant normal allele masking a recessive mutant one. The genetic condition called albinism (see Figure 8.20), in which a person's skin cells and hair follicles fail to make dark pigment, is another autosomal recessive condition.

Huntington's disease is the result of a dominant allele on an autosomal (nonsex) chromosome. In contrast to PKU and muscular dystrophy, which appear in infants or young children, Huntington's disease does not manifest itself until middle age. Researchers have chronicled a set of distressing symptoms in Huntington's patients that include progressive death of nerve cells, irregular and jerky movements, intellectual deterioration, and often severe depression. Pedigrees have revealed that the con-

dition is inherited as a dominant allele on an autosomal chromosome (Figure 11.6c). There is as yet no effective treatment for Huntington's disease, and the patient dies a lingering death. Tragically, the victim's children suffer whether they inherit the mutation or not. They have a 50/50 chance of inheriting the disease, but until recently, they had to wait until middle age—usually after they had reproduced—to learn whether they had the mutation and hence might pass it on to their own children (Figure 11.8, page 192). In the mid-1980s, however, molecular geneticists developed a way to detect the gene earlier in a potential victim's life (see page 195).

CHROMOSOME VARIATIONS REVEAL OTHER GENETIC CONDITIONS

While pedigrees provide the human geneticist's first tool, *karyotyping*—the preparation and study of chromosomes using special dyes and labeling techniques—enables geneticists to probe conditions caused not by single gene mutations, but rather by changes in chromosome number or structure (see Figure 11.5).

■ **Changes in Chromosome Number** As we saw earlier, people normally have 46 chromosomes: 22 pairs of

(a)

(b)

FIGURE 11.8 *Genetic Blues: Huntington's Disease, a Dominant Gene on an Autosomal Chromosome.* (a) Woody Guthrie, singer and writer of "This Land Is Your Land," "So Long, It's Been Good to Know Ya," and other popular songs, died of Huntington's disease. (b) His son, Arlo Guthrie, a popular singer in the 1970s ("You Can Get Anything You Want at Alice's Restaurant" and "The Train They Call the City of New Orleans") had a 50/50 chance of inheriting the condition.

autosomes and 2 sex chromosomes, *XX* in females and *XY* in males. If a chromosome set deviates from that pattern, development is generally abnormal. Recall that having three copies of chromosome 21 causes Down syndrome (see Figure 7.21). People can also inherit too many or too few sex chromosomes, with dramatic consequences. A person with one *X* and no *Y* chromosome (*XO*) is a sterile female with *Turner syndrome*. This is characterized by folds of skin along the neck, a low hairline

at the nape of the neck, a shield-shaped chest, and failure to develop adult sexual characteristics at puberty. A person with two *X* chromosomes and one *Y* chromosome (*XXY*) has *Klinefelter syndrome*. The recipient of this pattern develops as a male until puberty, but then develops enlarged breasts. Klinefelter males are often tall, have lower than average intelligence, and are sterile.

Many boys who inherit only one *X* but two *Y* chromosomes (*XYY*) grow to be over 6 feet tall and have below-average intelligence. Some geneticists and lawyers have suggested that an extra *Y* chromosome causes innate antisocial tendencies and point to the fact that men in prison are 20 times more likely to have the pattern than men in the general population. This argument has proved a successful legal defense in a few cases. However, others point out that 96 percent of *XYY* males lead normal lives and that perhaps society treats tall but mentally slow boys in a way that promotes aggression.

■ **Exchanges of Chromosome Parts** While extra chromosomes can cause conditions such as Down and Klinefelter syndromes, even a less drastic alteration of chromosome pattern can have serious consequences. Researchers have found that over 30 types of cancer are caused if a part of one chromosome moves spontaneously to a new location on a different chromosome. This kind of change is called a **chromosome translocation.** In one such translocation, chromosomes 8 and 14 break in specific places, and the free ends exchange places (Figure 11.9). If this translocation occurs in even just one white blood cell at any time during a person's life, that cell will divide rapidly and grow into a cancer called *Burkitt's lymphoma,* which can cause a huge tumor in lymph nodes of the face.

When this chromosome exchange occurs, a cancer gene, or **oncogene,** may be relocated. In its normal spot on chromosome 8, the oncogene regulates normal cell growth. But in its new position on chromosome 14, it is overactive and somehow causes the white blood cell to reproduce out of control, which leads to the cancerous facial tumor.

■ **Fragile *X* Chromosome and Mental Retardation** In the mid-1980s, researchers discovered a chromosome alteration that is second only to Down syndrome (trisomy 21) as a cause of mental retardation. The condition is called *fragile X syndrome.* A person with this condition has *X* chromosomes whose tips break off easily at a specific place. Geneticists have found that 15 percent of males with pure retardation (no accompanying physical symptoms) have fragile *X* chromosomes. So far, researchers haven't determined the link between a broken *X* chromosome tip and brain changes, but this syndrome underscores the importance of regular chromosome shape to normal embryonic development.

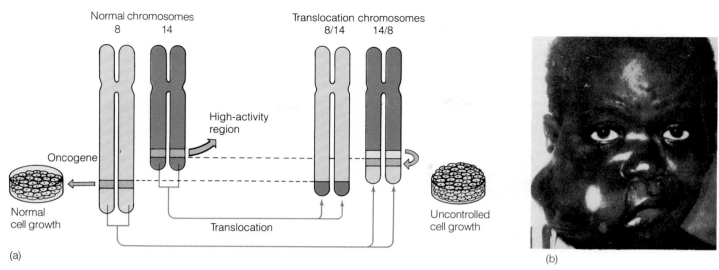

FIGURE 11.9 *Chromosome Translocation Can Turn On a Cancer Gene.* (a) A potentially cancer-causing gene (an oncogene) on a normal chromosome 8 is expressed at a level that causes white blood cells to grow at a proper rate. A specific site on a normal chromosome 14 is usually very active in those white blood cells, but not at all in other body cells. However, if the long arms of chromosomes 8 and 14 break at the positions indicated here, the tip of 8 may rejoin to the stub of 14 and vice versa. When this happens, the cancer-causing gene may end up near the high-activity site on chromosome 14 and become very active itself. The result is uncontrolled growth or cancer of the white blood cells known as Burkitt's lymphoma. (b) A jaw tumor like this boy's is a common symptom of Burkitt's, although such a tumor can also be triggered by a viral infection.

Translocations and broken chromosomes rearrange and delete genes. And significantly, a gene's specific location along a chromosome is crucial to its functioning, because adjacent DNA sequences often serve to turn neighboring genes on and off at appropriate times (review Figure 10.13). Because a gene's location is so important, geneticists have tried to map as many human genes as possible and assign them to precise locations along the 23 human chromosomes. To do this, the researchers have applied new tools first developed on nonhuman subjects, and these tools are allowing them to detect and treat a much broader range of serious human genetic diseases than ever before.

THE NEW GENETICS: RECENT REVOLUTION IN THE MAPPING OF HUMAN GENES

In 1911, geneticists noticed the striking form of color blindness called *green weakness*, a condition in which males inherit a visual defect that prevents them from seeing greens but allows them to see other colors normally (Figure 11.10, page 194). By studying pedigrees and finding that a man passes color blindness through his daughters to his grandsons, geneticists were able to assign green color blindness to the human *X* chromosome. The traditional tools of pedigrees and karyotyping, however, are so imprecise when applied to our 23 pairs of chromosomes that another half century elapsed before geneticists could use these techniques to assign a gene to any specific autosomal chromosome.

Since 1970, thanks largely to two new tools, geneticists have traced more than 700 human genes to specific places on sex and autosomal chromosomes. One of the tools uses fused somatic cells, and the other uses DNA fragments called RFLPs. Both would have amazed Gregor Mendel, and they have been called the "New Genetics." Let's look at these exciting new tools and see how the gene mapping they make possible is revolutionizing the study and treatment of genetic diseases.

GENETIC STUDIES THAT BYPASS SEX

Geneticists can mate any two flies or pea plants they choose, but they do not conduct controlled mating of people. Nevertheless, researchers can "mate" normal human body cells in a way that bypasses sex and yet gives information for mapping genes. In the 1960s, researchers discovered a way to combine a human cell

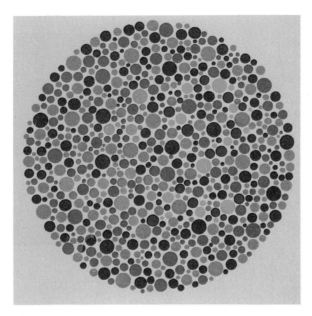

FIGURE 11.10 *Color Blindness Test: What Number Do You See?* The first human gene that geneticists ever located was the gene for green weakness, which lies on the *X* chromosome and causes a partial color blindness. About 5 percent of males of Northern European descent have this type of color blindness. The gene causes a decrease in the amount of a light-sensitive pigment in the eyes, and affected people see reds as reddish brown, bright greens as tan, and olive greens as brown. If you look at this plate and see only the number 4, you are green-blind. If you see only the number 2, you are red-blind. If you see the number 42, you have normal color vision. If the 2 is not very clear, then you are an intermediate called "green-weak," and if the 4 is not clear, then you are a "red-weak" intermediate.

with a mouse tumor cell and treat them so that they fuse into a single cell in a process analogous to fertilization (Figure 11.11). The newly created giant nucleus has 46 chromosomes from the human and 40 from the mouse. This fused cell with its huge nucleus is called a *somatic cell hybrid*.

As the hybrid cell divides, human but not mouse chromosomes are gradually lost in an irregular way. Eventually, a researcher can establish a number of cell lines, all derived from the original hybrid and each containing 40 mouse chromosomes but a different small subset of human chromosomes. A cell bank like this of somatic cell hybrids can be used to map human genes to a specific chromosome. Let's say a researcher is interested in localizing the gene for PAH, the enzyme PKU victims cannot produce. If all hybrid cells that contain a complete chromosome 12 produce the enzyme, while all hybrids that lack chromosome 12 also lack the ability to make the enzyme, then the PAH gene must lie on chromosome 12. Comparisons like this have revealed the locations of more than 700 genes since 1970, including those that

encode the proteins collagen, insulin, growth hormone, and interferon. This number is expected to rise even more dramatically, however, with the advent of a new tool for probing DNA: RFLPs.

RFLPs: NEW TOOL OF THE NEW GENETICS

In the early 1980s, about a decade after somatic cell genetics came along, molecular geneticists discovered an even better tool for mapping genes, based on a type of genetic variation that reveals a unique genetic pattern.

Geneticists discovered something in human DNA that they had already observed in bacterial and fruit fly DNA: Certain DNA-cutting enzymes can snip in two a specific DNA fragment from one person but not cut it from another

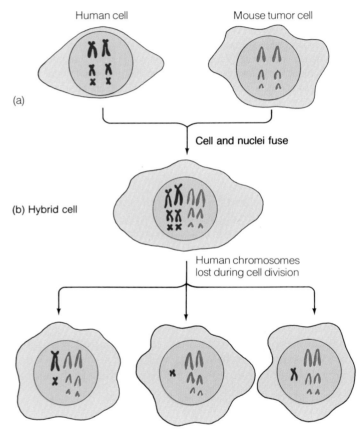

(c) Each cell line has a different subset of human chromosomes.

FIGURE 11.11 *Mating Body Cells to Map Human Genes.* (a) Geneticists bring together a human cell and a mouse tumor cell and cause them to fuse together using special treatments. (b) The fused hybrid cell contains one giant nucleus with complete sets of both human and mouse chromosomes. (c) Over several cell generations, the hybrids lose human, but not mouse, chromosomes. Since human chromosomes are lost by chance, different cell lines come to contain different human chromosomes. Any human trait that is retained by a hybrid cell must be contained on one of the human chromosomes that remain in the cell, allowing the trait to be mapped to a specific chromosome.

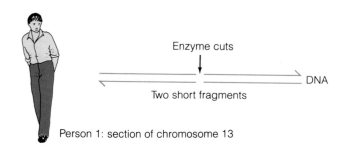

Enzyme cuts

DNA

Two short fragments

Person 1: section of chromosome 13

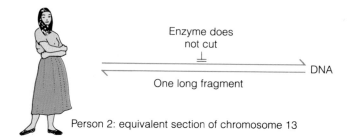

Enzyme does
not cut

DNA

One long fragment

Person 2: equivalent section of chromosome 13

FIGURE 11.12 *The Principle of RFLPs.* RFLPs (restriction fragment length polymorphisms) get their name in the following way: Enzymes called *restriction* enzymes cleave long strands of DNA into *fragments* by cutting at specific sites in the DNA's base sequence. Because of variations in DNA sequence, different people's fragments are of different sizes, and geneticists call these size differences *length polymorphisms*. RFLPs are due to mutations in the DNA sites that are cut by restriction enzymes. A restriction enzyme would cut DNA from person 1 into two fragments, but a mutation in the cutting site of the DNA from person 2 could cause the enzyme to no longer cut at that DNA position. Since a cutting site has been removed, the DNA remains joined in one very long fragment. These different fragment sizes can be used to help detect or predict diseases caused by mutated genes that contain cutting sites or that lie near cutting sites.

person (Figure 11.12). The presence or absence of a particular cutting site is inherited according to Mendel's rules. Such cutting sites (called *restriction fragment length polymorphisms*, or **RFLPs**) can be compared between members of a family with a genetic disease. If all family members with the disease share one specific cutting site, this site is almost certainly very near the defective gene. Medical researchers and counselors can use this detection system to identify newly tested family members, or even unborn fetuses, as either carriers of the gene or victims of the disease. Researchers recently were able to find a cutting site that is correlated with the expression of Huntington's disease. Geneticists can now search a suspected carrier's DNA for the marker and predict whether that person will develop the disease in middle age.

The RFLP technique is extraordinarily powerful, and hopes are high that many mysterious disease-causing genes can now be mapped, probed, and predicted with far greater accuracy and speed. A map of the X chromosome pinpointing the location of various RFLPs is

shown in Figure 11.13. A large research group in Massachusetts has localized about 400 RFLPs on the 23 human chromosomes. At least 95 percent of human genetic diseases will be linked to one of these markers. A similar technique has also been used as a genuine DNA "fingerprinting" method to solve crimes and cases of disputed paternity (see the box on page 196).

※ HELPING PEOPLE WITH GENETIC DISEASES

Gene mapping and sophisticated tools of the New Genetics are allowing modern physicians to do something doctors only dreamed of a few decades ago: to understand the entire pathway of a genetic disease—from mutant genes to defective proteins to physiological effects on the patient—and to intervene successfully so that the patient can lead a nearly normal life. The doctor must first identify the victim of a genetic disease, diagnose the condition, and then, armed with detailed knowledge about

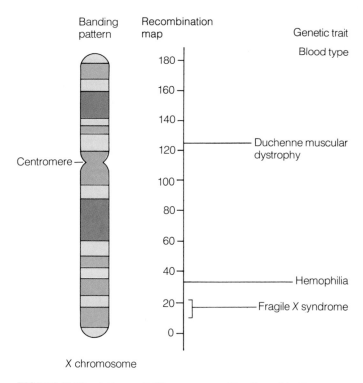

FIGURE 11.13 *A Human X Chromosome with a Few of Its Genes Mapped.* This map of a human X chromosome shows the locations of several genes that code for genetic diseases. A combination of experimental techniques has led to the precise locations of genes coding for Duchenne muscular dystrophy, hemophilia, fragile X syndrome, and other conditions.

DNA FINGERPRINTING CAN SOLVE CRIMES

Law enforcers—from Sherlock Holmes to the FBI—have collected and analyzed fingerprints as a way to identify crime suspects. But to the crime fighter's consternation, the whorls and swirls on the fingertips can be altered surgically or covered with gloves, and in some violent crimes like rape or assault, fingerprints may not be left. Nevertheless, criminals usually leave behind unique and far more revealing bits of identity—their DNA—in drops of blood, in cells clinging to bases of fallen hairs, or in semen. And geneticists are learning to read these clues.

In 1985, British molecular geneticists discovered segments of the human genome that are especially variable. Patterns of these segments are so utterly unique within each individual that they can serve as a DNA fingerprint. So-called *hypervariable regions* are short stretches of DNA, 20 to 60 bases long, repeated a variable number of times. The lengths of DNA fragments carrying the repeated sequence can reveal different alleles for the hypervariable region. And since there are

several different hypervariable regions in a complete set of human chromosomes, the number of potential length combinations is enormous.

Family studies have proved that hypervariable sequences are inherited in a normal Mendelian fashion: All fragments can be traced to one or the other parent and in turn to the grandparents. By studying 54 members of a family with four living generations, geneticists found that each individual had different fragment lengths—a unique genetic fingerprint. This pedigree potential has allowed geneticists to resolve cases of disputed paternity. In one case, British immigration authorities claimed that a boy was an illegal alien, but the technique revealed his true mother—a British citizen.

In crime cases where fingerprints are unavailable, DNA can be collected from the crime scene instead. For example, scientists have taken DNA from dried blood stains on cloth that had remained in a normal room for four years, and from the cells at the base of a single hair. The DNA in sperm collected from old semen stains is also retrievable in sufficient quantities for unambiguous identification. Some defense attorneys are calling for

certification of the forensic laboratories that work with law enforcement agencies; this would help ensure the careful use of correct procedures and limit the possibility of a false negative or false positive.

Not long ago, a dangerous rapist might have gone free if he wore a disguise and if the dazed victim was unable to pick out the assailant in a police lineup. Today, however, a rape case like this can be quickly solved if the assailant is among the suspects. In one such case, technicians took a vaginal swab from the woman 6 hours after sexual intercourse, and geneticists separated cells from the woman's reproductive tract and sorted them from the male's sperm cells. Then they purified the sperm DNA, treated it with restriction enzymes, and compared the fingerprints from semen DNA, from DNA in the blood of the prime suspect, and from DNA in the victim's blood. The DNA fingerprints of the semen and man's blood sample matched perfectly, and the probability of a match occurring by chance is less than 3 in 10^{11}. Since there are only 5×10^9 people alive on earth, he had to be the rapist.

---

the disease, treat it at one of the three levels at which a mutant gene causes its damage: physiological repercussions (the effect on the whole patient), mutant gene product (the protein that is abnormal), or the altered gene itself. Physicians treat different genetic diseases at different levels, depending on which method is most effective or, in some cases, which is currently available to them.

IDENTIFYING VICTIMS OF GENETIC DISEASE

Genetic diseases manifest themselves at various points in the life cycle, and time of onset is important, since detection must precede treatment. Several genetic conditions appear only in adulthood: Huntington's disease of middle age is an obvious one, but the list also includes breast cancer and adult-onset diabetes based on inherited tendencies, as well as less harmful traits, such as acne or hairy ears, which show up only after puberty.

The majority of genetic traits, however, become noticeable in early childhood. As the body develops, the child begins to walk and talk, and physical or behavioral abnormalities begin to show. Duchenne muscular dystrophy is an example, as are certain types of mental retardation.

Finally, some genetic traits are detectable at birth. Albinism, for example, or extra fingers and toes (polydactyly) are easily observed. And certain less obvious conditions, such as PKU, can be detected in the hospital nursery. (If you were born after 1962, you were probably tested for PKU.) Blood from a baby with PKU, taken from the infant's heel, has 20 times the normal amount of phenylalanine. Early detection is crucial in a condition like PKU, where brain damage accumulates with each ingestion of more phenylalanine. This simple blood test has spared thousands of children from mental retardation.

The earliest possible diagnosis for other conditions—whether their onset is at birth, during childhood, or in adulthood—is critical in enabling the physician to prevent as much damage to the body as possible from the effects of the defective gene. After detection, the next step, of course, is treatment.

PHYSIOLOGICAL THERAPY

A mutant gene can adversely affect the patient's entire physical health, and thus, many treatments intervene at the level of physiology. Once a PKU sufferer, for example, is identified, the doctor prescribes a diet that will protect the infant's brain from phenylketones. The diet is a monotonous but nutritious mush of dried milk powder (treated to remove phenylalanine), cornstarch, butter, and fresh fruits and vegetables (except beans and peas, which are high in phenylalanine). The baby and its parents walk a dietary tightrope, because the infant's caloric intake must be constantly high enough to prevent its body from breaking down its own proteins and thereby releasing phenylalanine from within. This high-calorie diet must be initiated within the first two or three months of life if the baby with PKU is to develop nearly normal intelligence. If treatment is delayed past six months, it is ineffective at preventing brain damage. The child must avoid phenylalanine completely until age six and to a lesser degree throughout life; hence the warning notice on cans of diet soft drinks that contain phenylalanine.

Other physiological treatments for genetic disease include dark sunglasses and chemical sunscreens for albinos and surgical removal of extra fingers and toes from polydactyl children.

PROTEIN THERAPY

Some genetic diseases can be treated by giving patients the gene product—the protein—they cannot make. In hemophilia, for example, the gene mutation blocks the synthesis of a clotting protein; thus, the blood fails to clot normally after a cut or bruise. However, after receiving blood transfusions that contain this clotting factor, a hemophiliac will form blood clots and is relatively safe from bleeding to death. Some diabetics lack the gene for normal insulin production but can respond to injections of insulin. Protein therapies such as these can work as long as the needed protein is carried by the blood, but unfortunately, that is true for only a few diseases. Most proteins must be positioned in a specific organ—for example, the enzyme that PKU victims lack is normally found in the liver—and so injecting the missing enzyme into a PKU victim serves no purpose. Ultimately, gene therapy may be the answer.

GENE THERAPY

Treating a genetic disease at the physiological or protein level may help the patient's symptoms, but it leaves the defective gene unchanged. Geneticists are trying to locate and study defective genes (through techniques such as somatic cell genetics and RFLPs) so that someday they can replace such genes with normal DNA sequences. The first successful test of genetically engineered cells in people was not to cure a disease but to follow the cells' migration in the body. Physicians removed certain white blood cells from leukemia patients and inserted a marker gene into the cells, using recombinant DNA techniques. The marker gene would allow the medical researchers to track the cells inside a person's body. They reinserted the genetically engineered cells into the patients, followed the altered white blood cells as they traveled through the body, and found that genetically engineered cells can be placed in a person without causing any harm. Because of this success, tests of genetically engineered cells in people could continue, and in one study, specially treated recombinant cells helped shrink cancerous tumors (Figure 11.14, page 198). These early efforts may eventually lead to effective gene therapy to correct many more genetic disorders at a fundamental level, including severe combined immunodeficiency (see Figure 9.21).

≋ PREVENTING NEW CASES OF GENETIC DISEASE

The recent revolution in genetic techniques has allowed scientists and physicians to carry detection and treatment one step further: Not only can they help those with

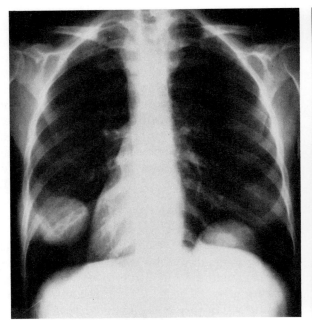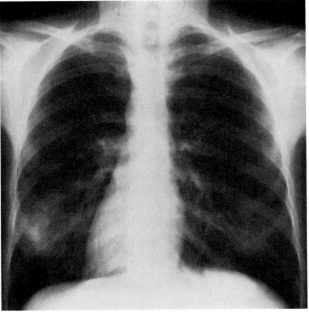

FIGURE 11.14 *Gene Therapy Helps Cancer Patients.* White blood cells treated in the laboratory caused large tumors (seen in the photo on the left) to shrink to practically nothing (photo on the right).

existing genetic problems, but they can also help prevent the conception and birth of children doomed to a lifetime of suffering. They do this by detecting harmful alleles in potential parents and in the unborn and by counseling the parents about their options (for a discussion of those options, see the box on page 200).

PRENATAL DIAGNOSIS

Tremendous progress has been made in the area of diagnosing genetic conditions before birth. In the mid-1960s, geneticists pioneered a method of collecting fetal cells called **amniocentesis,** and today, this procedure is used routinely for most pregnant women over 35 or for those with family histories of genetic disease (Figure 11.15a). The physician inserts a needle through the mother's abdominal wall and removes a few milliliters of the fluid that surrounds and protects the fetus. This fluid (called the *amniotic fluid*) contains sloughed-off fetal skin cells that are collected and grown in the laboratory, then tested for defective chromosomes or genes. In these cells, certain defective genes can be detected by the absence of normal enzymes. Nearly 100 genetic abnormalities can be identified through amniocentesis, but the technique

does have a drawback: It cannot be accomplished before about 14 to 20 weeks of pregnancy, and by that time, the mother may have already felt the fetus move and may find it emotionally hard to consider an abortion, even if test results reveal massive genetic defects.

In a more recent procedure called **chorionic villus sampling,** the physician removes fetal cells from the developing placenta, the organ that nourishes the fetus (Figure 11.15b). The fetal chromosomes can then be examined and the RFLPs studied. The mutation that causes sickle-cell disease, for example, eliminates a restriction enzyme cutting site present in normal DNA. Thus, the sickle-cell allele can be detected through RFLP analysis with 100 percent accuracy. This technique can also detect fetuses with PKU or Huntington's genes, but with somewhat less accuracy. Geneticists predict that soon, most common genetic diseases will be detectable in the unborn on the basis of RFLPs. Unfortunately, treating fetuses for these gene defects is a more distant prospect. It may be possible someday to replace defective fetal genes with normal DNA sequences and thus to cure the baby's condition before birth. For now, however, the options are limited mainly to detecting carriers and providing genetic counseling.

DETECTING CARRIERS

For decades, geneticists have been able to identify carriers of defective alleles through pedigrees: Children homozygous recessive for a trait like PKU have parents who are heterozygous carriers. This approach, however, only has predictive value for the additional children of those same specific sets of parents. But sensitive biochemical analyses can now reveal the presence of a defective enzyme in unaffected carriers. A heterozygote for PKU, for example, has just one rather than two alleles making the normal enzyme and can therefore be distinguished from a person with two normal alleles. In this way, the person is shown to be a carrier even if he or she has not yet produced children. These techniques will not work for genetic ailments whose defective protein has yet to be identified. However, RFLPs can sometimes be used instead; carriers for Huntington's, for example, are found this way. Once the potential parents have this information, they face hard choices. So do the parents of a fetus revealed as defective through prenatal diagnostic techniques. Genetic counselors try to help people with these difficult dilemmas.

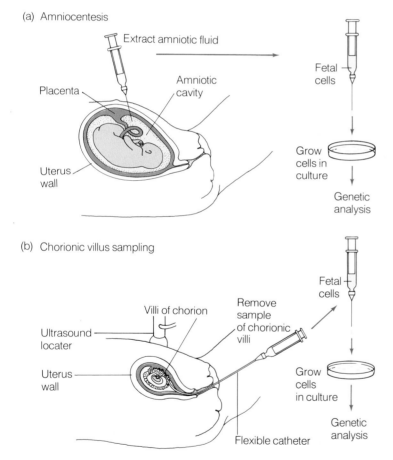

(a) Amniocentesis

Extract amniotic fluid

Placenta

Amniotic cavity

Uterus wall

Fetal cells

Grow cells in culture

Genetic analysis

(b) Chorionic villus sampling

Villi of chorion

Ultrasound locater

Uterus wall

Remove sample of chorionic villi

Flexible catheter

Fetal cells

Grow cells in culture

Genetic analysis

FIGURE 11.15 *Prenatal Diagnosis of Defective Genes and Chromosomes.* As the text explains, the differences between (a) amniocentesis and (b) chorionic villus sampling lie in when and how the fetal cells are removed. In both techniques, the cells are cultured and the chromosomes are analyzed for genetic abnormalities.

GENETIC COUNSELING

Genetic counselors provide information to families at risk for transmitting genetic defects. They acquaint the prospective parents with the manifestations of the disease they may pass along, with the potential suffering of an affected child, and with the emotional and financial impact on parents and siblings. Counselors can assess the couple's probability of having an afflicted child, help them consider the available options, and assist in their acceptance of the difficult reproductive decisions they must make.

■ **Decision Making: Who Shall Be Born?** Decisions about genetics are momentous ones. A family may elect to forgo reproduction and adopt an unrelated child. Or they may try to beat the odds, and then, if they don't, prepare themselves to love and care for a genetically defective child. Or they may conceive, undergo prenatal testing, and then terminate any pregnancy involving a defective fetus. But each such decision entails tremendous emotional costs and pits possible objections to abortion and birth control against the realities of suffering children and families.

As geneticists learn to detect more and more human genetic conditions, these difficult decisions are apt to become commonplace—and yet, no less onerous. What if someday RFLPs can reveal a couple's tendency toward cardiovascular disease, schizophrenia, diabetes, alcoholism, cancer, or Alzheimer's senility? Should they choose not to conceive or to abort a fetus found to carry an RFLP linked to one of these conditions? Or are these acceptable risks of being alive? How genetically perfect, in other words, would a child have to be if the parents had enough information to choose their offspring's physiological traits in advance? Society will face such choices in the not-too-distant future. But no matter how much complexity the New Genetics may add to our reproductive lives, the exciting tools of this field will almost certainly solve more problems than they create, allowing society to begin eliminating the burden of genetic diseases for the first time in history.

≋ CONNECTIONS

Despite the unique challenges involved in studying a person's genes and chromosomes, understanding human genetics in general and a condition like PKU in particular

THE RISKS OF GENETIC SCREENING

In the early 1970s, it became common practice for American blacks to undergo genetic screening tests for sickle-cell anemia. While this enabled physicians to detect and treat many previously overlooked cases of the blood disease (see Chapter 9), it also brought with it a subtle social cost. The tests revealed carriers (with no disease symptoms) as well as those with full-blown anemia. This gave patients new information for seeking treatment and making reproductive decisions, but on the basis of the test results, some employers and insurance companies denied people jobs and cheap premiums.

Consider, then, the recent revolution in genetic techniques and our new-found ability to map human genes, find markers for specific diseases, and screen unborn fetuses, children, and adults for various genetic conditions. There is great enthusiasm in many quarters over the potential these techniques present for studying the genetic roots of many diseases. Geneticists can already locate genetic markers for more than 400 diseases, and with the Human Genome Project (see the box in Chapter 10, page 184), that list will grow quickly. For some people, however, the social costs of "reading our genes" look high indeed.

Drawing parallels with sickle-cell anemia, some bioethicists fear that insurance companies and employers will require in-depth genetic screening for new customers or job seekers and will turn away people if they have genes that merely raise the possibility of someday developing heart disease, cancer, or other conditions. Some activists worry further that industrial firms will select for (and one day engineer people for) the ability to withstand occupational exposures rather than cleaning up their factories. And some social and religious groups are concerned that detailed fetal screening will result in more abortions, since most genetic defects that can be detected cannot presently be corrected or, in some cases, even treated effectively.

Some observers of the revolution in genetic screening have proposed the creation of new panels, laws, and rules. These, they say, should govern and probably restrict the speed and extent to which insurance companies, employers, and individuals could make use of detailed genetic information.

Some geneticists, however, believe that existing bodies, such as the Recombinant DNA Advisory Committee of the National Institutes of Health, are already sufficient to meet the ethical and legal concerns posed by the rapid expansion of genetic screening.

People on all sides of the issue seem to agree that widespread public awareness of the issues is essential. Here are places where you can get more information:

Friedman, T. "Progress Toward Human Gene Therapy." *Science* 244 (16 June 1980): 1275–1280.

The Hastings Center, 255 Elm Road, Briarcliff Manor, NY 10510.

The Kennedy Institute of Ethics National Reference Center for Bioethics Literature, Georgetown University, Washington, DC 20057.

National Maternal and Child Health Clearinghouse, 38th and R Streets N.W., Washington, DC 20057; especially the pamphlet *The New Human Genetics*.

Roberts, L. "Ethical Questions Haunt New Genetic Technologies." *Science* 243 (3 March 1989): 1134–1136.

Stanford University Center for Biomedical Ethics, Stanford, CA 94305.

Weiss, R. "Predisposition and Prejudice." *Science News* 135 (21 January 1989): 40–42.

requires that we apply the same genetic principles governing all other living things. In fact, the principles and techniques derived from research on peas, bacteria, fruit flies, and other organisms, have modified the study of our own species and have opened a new era in genetics. They have allowed scientists to detect several genetic disorders early enough to prevent organic damage, and to predict several others so that if parents wish, they can prevent the birth of an infant doomed to suffer and perhaps die prematurely. The ideal solution would be to eliminate eggs and sperm bearing defective genes by screening and removing them so that they cannot participate in fertilization. Perhaps before such manipulation becomes possible, however, geneticists may learn how to alter the genes of a defective fetus in the womb or to influence the way its characteristics unfold so as to prevent the mutant allele from causing disease. To understand how something like this could be done, we must, in the next two chapters, cross over to the realm of developmental biology. This area is the logical extension of genetics, and it explains how the genetic information in a single cell, the fertilized egg, directs it to divide, grow, and take on the shape and characteristics of a complex organism such as a human baby.

(1) People are unique genetic subjects.

Goya, *The family of Charles IV*

(2) Genetic changes usually have a negative impact on physical makeup.

Child with Down syndrome

X   X   Y

(4) Analysis of chromosomes can reveal variations that lead to certain genetic conditions.

(3) A family history can reveal whether a genetic condition is dominant or recessive and whether it is transmitted by the X or a nonsex chromosome. (Baldness is dominant in men, recessive in women, and transmitted by a nonsex chromosome.)

(5) New genetic tools help locate genes on chromosomes, detect abnormal genes, and identify DNA fingerprints.

(6) Physicians and genetic counselors use the new and traditional tools to help people with genetic diseases.

NEW TERMS

amniocentesis, page 198
carrier, page 191
chorionic villus sampling,
 page 198
chromosome translocation,
 page 192

oncogene, page 192
pedigree, page 188
RFLP, page 195
X-linked, page 189

STUDY QUESTIONS

REVIEW WHAT YOU HAVE LEARNED

1. What are the special challenges in studying human genetics?
2. Explain why boys inherit X-linked recessive traits from their mothers.
3. Give an example of a genetic defect caused by each of the following: change in chromosome number; chromosome translocation; a fragile chromosome.
4. What is an oncogene? How does it cause its effects?
5. Compare amniocentesis with chorionic villus sampling.
6. What are the purposes of genetic counseling?

APPLY WHAT YOU HAVE LEARNED

1. A couple's first child has the disease PKU. What is the likelihood that their next child will suffer from PKU?
2. A woman claims that a certain man fathered her baby. How can a molecular geneticist confirm or disprove her claim?
3. A couple has a son with hemophilia. What is the probability that the next daughter will have the disease? What is the probability that the next son will have the disease?
4. Consider the following three pedigrees and determine if the trait is inherited as a sex-linked recessive, an autosomal recessive, or an autosomal dominant. Filled-in symbols indicate affected people.

(a)

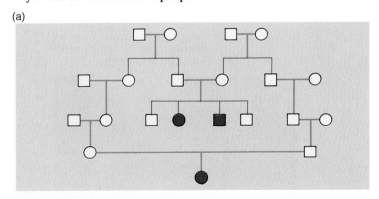

(b)

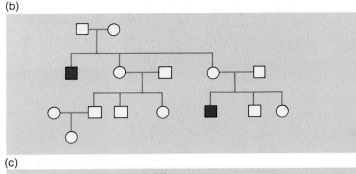

(c)

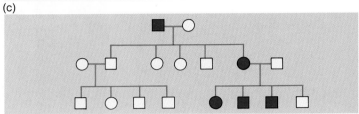

FOR FURTHER READING

Caskey, C. T. "Disease Diagnosis by Recombinant DNA Methods." *Science* 236 (1987): 1223–1228.

Hartl, D. L. *Our Uncertain Heritage: Genetics and Human Diversity.* 2d ed. New York: Harper & Row, 1985.

Lawn, R. M., and G. A. Vehar. "The Molecular Genetics of Hemophilia." *Scientific American* 254 (March 1986): 48–56.

McKusick, V. A. *Mendelian Inheritance in Man: Catalogs of Autosomal Dominant, Autosomal Recessive, and X-linked Phenotypes.* 7th ed. Baltimore: Johns Hopkins University Press, 1986.

White, R., and J. M. Lalouel. "Chromosome Mapping with DNA Markers." *Scientific American* 258 (February 1988): 40–52.

REPRODUCTION AND DEVELOPMENT: A NEW GENERATION

CHAPTER 12

THE OMNIPOTENT EGG

As dawn breaks over India's Ganges River, life begins to stir. People come down to the river to wash and draw water, birds begin to call, and crocodiles haul themselves out on the muddy bank to soak up the sun. The most frenzied morning activity, however, goes on among the small, striped zebra fish, common denizens of the slow-moving river.

As light penetrates the greenish water, the fish converge into a mass of flashing silver bodies. Darting first one way then another, pairs of zebra fish skim the submerged rocks along the river bottom. Within each pair, the male excitedly chases the larger, rounder female and frequently rams his head into her swollen abdomen. With each bump, she lays a few tiny eggs, and the male releases a cloud of sperm around them. Soon, the chasing and bumping subside, and the fish rise to feed at the water's surface.

Meanwhile, after drifting downward and settling between the rocks, the translucent, grayish eggs slowly begin to change. Within each one, fertilization by sperm has triggered a series of events that will generate a new individual. In just 40 minutes, a partition forms and cleaves the fertilized egg in two. Over the next three days, hundreds of such cleavages will generate thousands of cells inside each translucent globe. A recognizable embryo will take form within this mass, soon to develop into a tiny fish. The independent life of a new individual has begun (Figure 12.1).

When you were an embryo, you were curiously similar to a fish, right down to gill slits and tail. Such similarities suggest our close evolutionary ties to other animals with backbones (Figure 12.2, page 204) as well as the universal principles governing growth from a single fertilized egg to the many diverse cells of a mature plant or animal.

The fusion of egg and sperm is called *fertilization*, and it triggers an orderly increase in complexity called *development*. During development, the fertilized egg divides from one cell into thousands or millions of cells, each in a prescribed place and with a defined shape and function. One of the most profound mysteries in biology is how genes direct the fertilized egg to divide the correct number of times and how they designate the location, shape, and function of each of millions of cells so that the developing organism has working limbs, heart, eyes, and other body structures.

While development occurs in both plants and animals, the details are sufficiently different between the two kingdoms that we devote this chapter to animal development only and Chapters 29 and 30 to aspects of plant development. As we explore the marvels of animal development, a central theme will recur: Development is an ordered sequence of irreversible steps, with each step setting up the necessary conditions for the next step. For example, as the fertilized egg undergoes many rounds of division, the resulting cells become increasingly specialized. This specialization occurs because each cell expresses different genes at specific times, and regulatory substances control the patterns of that gene expression.

Our coverage of animal development will describe how animals reproduce and how they develop from an egg to an adult that can itself reproduce. This chapter will answer these questions:

- What events ensure that animals mate and that egg and sperm unite at fertilization?
- How do fertilized egg cells transform into embryos?
- What mechanisms cause organs to form in an embryo?
- What normal and abnormal growth patterns can occur after birth?
- How do eggs and sperm form?

FIGURE 12.1 *Embryonic Zebra Fish.* Just three days old, this embryo was photographed with polarized light, revealing developing muscles (blue), eye (dark sphere), and yolk (central red sac), which provides nutrients for the developing fish embryo.

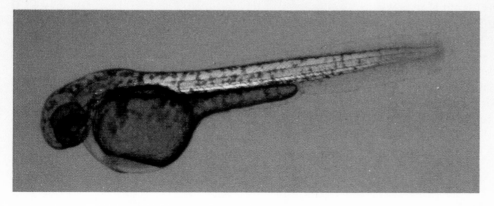

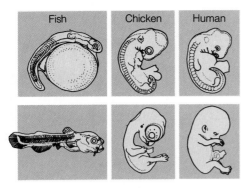

FIGURE 12.2 *When You Were an Embryo: Similarities with Fish and Other Animals with a Backbone.* In their early stages, the embryos of fish, chicken, and human look amazingly similar. Later, the differences become more obvious.

≈ MATING AND FERTILIZATION: GETTING EGG AND SPERM TOGETHER

Reproduction is essential to life: Each living thing exists solely because its ancestors succeeded in producing progeny that could themselves develop, survive, and reach reproductive age. In animals as different as zebra fish and zebras, the basic elements of reproduction are the same: A tiny mobile sperm cell fuses with a huge, immobile ovum, or egg cell, in the process of *fertilization*. The ovum contributes a set of chromosomes, a storehouse of nutrients, and crucial developmental information, while the sperm furnishes a second set of chromosomes and acts as a trigger for development. The meeting of egg and sperm requires that organisms *mate*, that is, behave in a way that brings egg and sperm together (Figure 12.3).

MATING STRATEGIES: GETTING ORGANISMS TOGETHER

Cooperation is the key to mating. Even in species like earthworms and sea slugs, in which each animal makes both eggs and sperm, individuals often cooperate by pairing up and reciprocally fertilizing each other's eggs. Such animals are called *hermaphrodites*. In most animal species, however, an individual is either male (sperm-producing) or female (egg-producing), and for mating to occur, individuals of both sexes must first recognize the opposite sex and then synchronize the release of egg and sperm.

■ **Mate Recognition** Animal mates recognize each other mainly by sight, sound, and smell. Male birds' splendid colors visually attract females: The male peacock's tail and the male frigate bird's expandable red throat sac both touch off sexual interest in females at appropriate times of the year. Such visual cues abound throughout the animal kingdom: the lion's mane, the elephant seal's proboscis, a woman's breasts and buttocks, a man's beard and chest hair, and the stickleback fish's red stripe are all examples. Crickets respond to whirring sounds made by rubbing bristled back legs against each other, and

(a)

(b)

FIGURE 12.3 *Bringing Egg and Sperm Together: The Marvels of Mating in the Animal Kingdom.* Programmed mating behavior brings animals together so that egg and sperm can unite. (a) Praying mantises will continue to mate even if the female eats the male's head, which sometimes occurs. (b) In these tree frogs, the male clasps the larger female, and the physical proximity causes both males and females to release gametes simultaneously.

these auditory cues are common noises on a summer's night. Odor cues include the *pheromones* that mammals and insects generate—chemicals given off by one individual that affect the behavior of others of the same species. Male pigs, for example, exude a strong-smelling chemical called androstenone (also known as "boar taint"). A sow that catches even a faint whiff of boar taint will assume a sexual stance and wait to be mated.

■ **The Production and Release of Egg and Sperm Together** Within a given species, eggs and sperm must reach maturity at the same time for fertilization to succeed. Some species mate only at specific times of year (often in spring), during which all adult members reach physical readiness and mate in a frenzy of activity—like the zebra fish in the Ganges. Environmental cues, such as gradual changes in day length, often trigger the synchronized maturation of eggs and sperm. Hens' eggs are more plentiful—and hence cheaper to buy—in springtime for this reason. The release of eggs and sperm is often synchronized as well and is usually stimulated by behavior. For example, when the male zebra fish rams the female's abdomen, they are stimulated to release eggs and sperm simultaneously.

This kind of synchronized release into the environment is a prelude to *external fertilization,* a process that takes place in many water-dwelling species whereby eggs and sperm are deposited directly into the surrounding water, meet by chance, and fuse. External fertilization can occur as in zebra fish or sea urchins, with little or no contact between the mating adults. In other species, such as frogs and salamanders, the males clasp the females firmly, their bodies touch for prolonged periods, and this touching stimulates the simultaneous release of gametes (see Figure 12.3b).

Animals that inhabit land, including most mammals, birds, reptiles, insects, and snails, employ a second major mechanism to ensure gamete union: *internal fertilization,* wherein the male deposits sperm directly into the female's body and the gametes meet in a chamber or tube. Their synchronized meeting is based on internal chemical signals, or hormones, as well as on behavioral cues. This type of male deposition, also called *copulation,* has a distinct advantage: Sperm can be concentrated and protected within the female's body until eggs are available for internal fertilization, thus helping ensure that viable gametes meet at appropriate times.

UNION OF EGG AND SPERM

While mating brings the eggs and sperm into close contact at appropriate times, fertilization—the actual fusion—depends on a series of events that stem from the structures and activities of the gametes themselves.

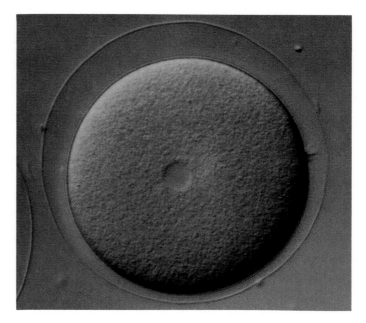

(a)

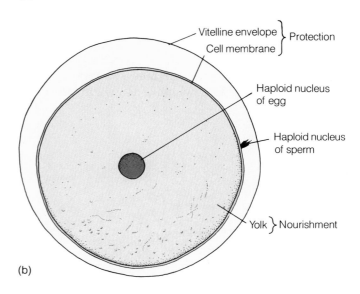

(b)

FIGURE 12.4 *The Egg.* This single large cell protects the new individual, instructs its development, and contributes half the chromosomes. Here, we show a sea urchin egg with sperm entering at the right.

■ **Cells Specialized for Reproduction** Eggs and sperm are strikingly different—and *vive la différence!* An egg is usually the largest single cell in an animal's body; the yolk of an ostrich egg, which is actually a baseball-sized *ovum,* is the largest animal cell on earth. An ovum (the name for an egg cell after it leaves the ovary) is enormous and performs many jobs (Figure 12.4): It donates a nucleus carrying just one full set of chromosomes (haploid); it protects the developing embryo inside jellylike protein coatings, strong membranes, sacs of fluid, and sometimes hard or leathery shells; it nourishes the embryo with **yolk,** which contains rich stores of lipids, carbohydrates, and special proteins; it provides the machinery for protein synthesis so that the fertilized egg can undergo very rapid cell division without slowing down for gene action; and finally, it directs development of the early embryo by special substances in the egg's cytoplasm that control the expression of its genes.

Unlike the egg, the sperm is one of the smallest cells in the body, stripped down to just a compact nucleus, mitochondria to provide energy, a long flagellum for propulsion, and a sac of enzymes to digest a path through the egg's protective coatings (Figure 12.5). The sperm's streamlined size and shape reflect its narrow objective: to reach the egg, penetrate its coatings, and deliver a haploid nucleus with just one complete set of chromosomes into the egg's cytoplasm. This penetration and delivery, together with the activation of the resting egg, make up the actual fertilization event (Figure 12.6).

■ **Fertilization: Fusion Triggers Development** Fertilization unleashes dramatic and irreversible events that permanently transform the activities and appearance of the egg. As the lashing, whiplike flagellum propels a

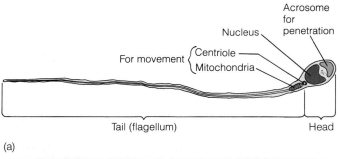

FIGURE 12.5 *The Sperm.* This mobile cell triggers development and contributes half the chromosomes. (a) The streamlined sperm cell has certain structures that allow it to penetrate and others that power the lashing tail and help it swim. (b) Electron micrograph of a human sperm.

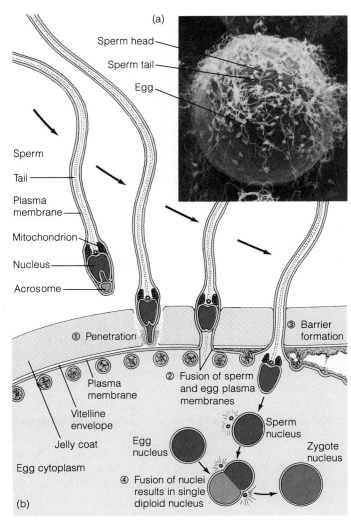

FIGURE 12.6 *Fertilization: Egg and Sperm Fuse to Form a Genetically Unique Zygote.* (a) A flotilla of sperm cells surround a sea urchin egg. (b) Fertilization requires several steps, from (1) the sperm's penetration of the egg to (2) the fusion of sperm and egg plasma membranes, (3) the sweeping wave of chemical reactions that raise a barrier to other sperm, and (4) the actual fusion of egg and sperm nuclei.

sperm headlong into an egg's jelly coating, the sperm's tip literally detonates like a microscopic bomb: The sac of digestive enzymes (the *acrosome*) at the sperm's tip explodes, releasing the contents and forming a harpoon-like protein needle that pierces through the egg's coatings (Figure 12.6b, step 1). The sperm's plasma membrane fuses with the ovum's, and the sperm nucleus plunges into the egg's cytoplasm (step 2). With this event, the egg, previously one of the body's most inactive cells, springs into a flurry of biochemical activity.

Within seconds, a wave of chemical reactions sweeps across the surface of the newly aroused egg, causing that surface to harden and present a barrier to the entry of any additional sperm (step 3). The egg's oxygen consumption skyrockets, as does its rate of protein synthesis. Male and female nuclei converge and fuse to form a single diploid nucleus with two sets of chromosomes (step 4).

While fertilization touches off dramatic events, the new individual is still no more than a single cell. To become a young zebra fish, roundworm, or tiger, the initial cell must subdivide again and again and grow into millions of cells that form an **embryo.** That subdivision begins in early embryonic development.

TABLE 12.1 ≋ KEY EVENTS OF DEVELOPMENT

EVENT	WHAT HAPPENS	CONSEQUENCES
Fertilization	Egg and sperm fuse	A unique genetic combination is created
Cleavage	Fertilized egg subdivides into a ball with many cells	Developmental determinants are partitioned into different cells
Gastrulation	Cells rearrange to produce an embryo with three cell layers	Precursors of internal organs migrate to inside of embryo; cells interact with new neighbors
Neurulation	Neural tube forms	Conditions are established for other body organs to form
Organogenesis	Body organs form	Cells interact, differentiate, change shape, proliferate, die, and migrate
Growth	Organs increase in size	Adult body form is attained
Gametogenesis	Eggs and sperm develop	Reproduction becomes possible

≋ PATTERNS OF EARLY EMBRYONIC DEVELOPMENT

Just a dozen hours or so after the remarkable events of zebra fish fertilization, a multicelled embryo has begun to take visible shape, with distinct head and tail regions, definite right and left sides, and the first signs of a backbone along the upper surface. The three stages of early development—called cleavage, gastrulation, and neurulation—cause this transformation. (Table 12.1 summarizes the key events of development.) Let's consider each stage in turn.

STAGE 1. CLEAVAGE: CREATING A MULTICELLED EMBRYO

The early developmental process called **cleavage** consists of a series of rapid cell divisions that transform the single large fertilized egg into a hollow ball of many small cells (called the **blastula,** Figure 12.7, page 208). Cleavage divisions are not interrupted by growth periods, in contrast to cell division in later life (see Chapter 7). Thus, as the huge egg is cleaved in half, then quarters, eighths, sixteenths, and so on, the cells get smaller and smaller until the embryo comes to have several hundred cells, each about the size of a normal body cell. The ball, or blastula, however, is about the same size as the original ovum.

■ **Cleavage Patterns in a Mouse Embryo** The amount of yolk originally stuffed into the egg cell affects the pattern of cleavage divisions: The eggs of mice and people have little yolk and divide one way, while frog eggs, with large amounts of yolk, divide another. After cleavage, developing mammals like mice and people obtain nourishment from the mother's body via a spongy, blood-rich organ called the **placenta,** rather than from prepackaged yolk. For this reason, the eggs of mammals (including humans) are among the smallest in the animal kingdom, even though they are far larger than the other cell types in a mammal's body. Each cleavage division can pass completely through the tiny mammalian egg, creating equal-sized cells (Figure 12.8, page 208). At first, these cells are organized into the little, bumpy, solid cluster of cells called a *morula.* As additional cells accumulate in the morula, some cells move away from the center, forming the hollow blastula stage, called the **blastocyst** in mammals. Throughout all this cell movement, a single layer of cells, the **trophoblast,** remains as the outside layer of the blastocyst, while inside the ball, the cells are piled several layers thick in an **inner cell mass** (see Figure

12.8). The trophoblast burrows into the mother's uterine wall and gives rise to the nutrient-providing placenta, while the inner cell mass becomes the embryo itself.

No matter what the cleavage pattern is, the cells of the embryo are no longer alike after cleavage: Different subsets will develop into different cell types.

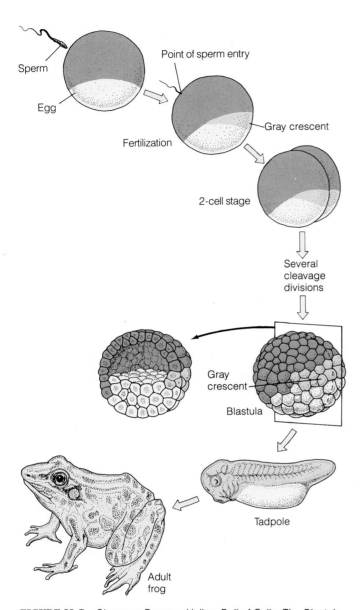

FIGURE 12.7 *Cleavage Forms a Hollow Ball of Cells: The Blastula.* A sequence of cell divisions at regular intervals transforms a single fertilized egg cell into a hollow ball of many smaller cells. Further cell divisions generate the emerging embryo. Cleavage also partitions the gray crescent, which forms opposite the point of sperm entry, into a few cells. Those cells eventually become the embryo's back (dorsal) region.

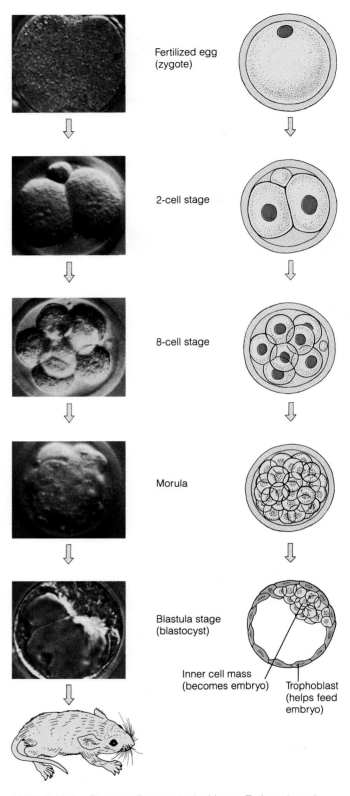

FIGURE 12.8 *Cleavage Patterns in the Mouse.* Early embryonic divisions create a solid bumpy cluster, or morula; then cell migrations create a hollow blastula-stage ball (a blastocyst) with a trophoblast on the outside (destined to become the placenta) and the inner cell mass inside (destined to become the mouse embryo).

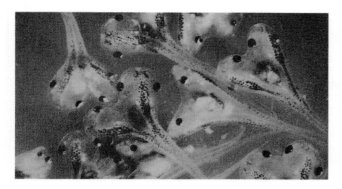

FIGURE 12.9 *The Gray Crescent Determines the Embryo's Dorsal Region: Experimental Evidence.* These two-headed tadpoles formed after researchers manipulated eggs so that gray crescent material lay on two sides of each egg. The two resulting dorsal regions produced two heads and two spinal columns.

■ **Cleavage Partitions Developmental Information** We said earlier that the egg cell contains not just nutrients but also developmental information. But what does that mean? Egg cytoplasm contains substances called **developmental determinants** that will act as instructions for the developing embryo. These instructions become localized in particular regions of the egg, and after cleavage, certain instructions end up only in certain cells. These determinants then activate the expression of specific genes in each cell, causing different cell groups to develop differently.

For example, a crescent of gray cytoplasm develops in a frog egg opposite the point of sperm entry. As cleavage proceeds, this zone of gray cytoplasm, called the *gray crescent*, is partitioned into only certain cells (review Figure 12.7). Cells containing this gray crescent material eventually become the embryo's dorsal (back) region. If a frog egg is treated to create two gray crescents, the result is a two-headed tadpole (Figure 12.9). Biologists concluded from this result that the gray crescent contains a developmental determinant that instructs cells to become dorsal parts of the embryo. Interestingly, while such determinants may help direct some developmental events in certain animals, such as frogs, other events in these animals and most events in mammals rely on a different key to early development: the position of cells relative to one another.

■ **Cleavage Can Assign a Cell's Position—And Its Fate** A mouse or human embryo, you will recall, cleaves into a solid cell cluster. At the four- and eight-cell stages, all the cells are developmentally equal. With further development, however, some cells end up inside and become the living embryo, while the outer cells become tissues that nourish the embryo but are sloughed off at birth.

The momentous decision whether to become embryo or feeding cells is made simply by a cell being at the right place at the right time.

Thus, the cleavage process accomplishes two things: First, it divides one large fertilized egg cell into many small cells; and second, it establishes differences between the small cells (via the partitioning of developmental determinants or the assignment of cell position). In the next stage of development, embryonic cells begin a dramatic series of migrations called gastrulation, which establishes the animal's body plan.

STAGE 2. GASTRULATION: ESTABLISHING THE BODY PLAN

The well-known embryologist Louis Wolpert likes to say that the most crucial event in your life is not birth, marriage, or death, but gastrulation. While this process is neither celebrated nor even perceived by embryo or mother, **gastrulation** transforms the hollow ball of cells generated during cleavage into the rudiments of an animal, with three distinct layers of cells, front and back ends, right and left sides, a future skin outside, and a future gut inside (gastrulation literally means "gut formation").

The body plan of an adult animal helps us understand how such transformations occur and what they accomplish. Think of an animal's body as three concentric tubes or layers of cells: an inner tube forming the gut; a middle layer surrounding the gut and including muscle and blood; and an outer tube of skin with its sense organs, such as eyes and ears (Figure 12.10, page 210). Gastrulation establishes these three cell layers, or *germ layers*, in the three tubes. As Figure 12.11 (page 211) shows, the embryo's inner layer, or *endoderm* ("inner skin"), gives rise to the inner linings of the gut and the organs that branch off from it, like the lungs, liver, and salivary glands. The middle layer, or *mesoderm* ("middle skin"), gives rise to muscles, bones, connective tissue, blood, and reproductive and excretory organs. The outer layer, or *ectoderm* ("outer skin"), gives rise to the skin and nervous system.

While the details of gastrulation are different in different animals, the principles are the same: The mesoderm and endoderm move inside, making a three-layered embryo. And in accordance with our theme that each step in development sets up the next, in frog embryos, it is the cells that contain the gray crescent—whose own position was determined by the sperm's point of entry—that define where mesoderm first enters the

FIGURE 12.10 *Simplified Body Plan of an Adult Animal: A Tube Within a Tube Within a Tube.* Snakes, like virtually all animals, have distinct body layers established during early development. The embryonic endoderm gives rise to the gut and most internal organs; the mesoderm generates muscles, bones, and certain other organs; and the ectoderm leads to skin and nerves. As these layers form, the animal's head-to-tail and back-to-front axes also become established.

embryo and hence where the animal's head and tail and back will form.

STAGE 3. NEURULATION: FORMING THE NERVOUS SYSTEM

Gastrulation sets up the conditions for the next developmental stage: **neurulation.** This is the formation of the *neural tube,* cells that differentiate into brain and spinal cord.

As gastrulation ends in a vertebrate embryo, an arched "sandwich" of ectoderm, mesoderm, and endoderm lies above the yolk at what will be the embryo's back. Along the embryo's midline, the mesoderm is organized into a rod, the **notochord,** which marks the future site of the backbone (the structure that distinguishes vertebrates from other animals). During neurulation, the cells of the notochord interact with the ectoderm cells lying just above them, and this interaction directs the ectoderm cells to fold over and form the neural tube. This tube will eventually develop into the nervous system (Figure 12.12, page 212).

The formation of the tube begins, once again, with the gray crescent, the cytoplasm involved in orienting the embryo's axes during gastrulation. Cells containing gray crescent cytoplasm cause nearby cells to form a neural tube, notochord, muscles, and eventually the rest of the embryo. This process by which one set of cells causes another group of cells to develop in a specific way is

called **induction.** Induction has a most important function—to organize the entire body plan. Once the body plan is "mapped out," additional inductions cause individual organs to form. The developing eye, for example, induces the overlying embryonic skin to become a lens, a crystalline oval structure that focuses light in the eye (see the box on page 213).

Unfortunately, sometimes mistakes occur during neurulation. Embryologists do not understand why, but in about 1 out of every 200 human births, the neural tube does not roll up completely and fails to close at the end, producing a birth defect called *spina bifida.* This anomaly leaves the spinal cord open to the outside, causing pain and paralysis to the infant. While it is not preventable at this time, the condition can be detected early in the pregnancy.

With the normal completion of early development—cleavage, gastrulation, and neurulation—the embryo's major axes have been established; but it is far from a finished animal. The next developmental priority is to establish the body organs that will allow the animal to move about, to search for and eat food, to escape predators, to find a mate, to reproduce—perchance to dream. That is the task of organogenesis.

≋ DEVELOPMENT OF BODY ORGANS

Let's say you were to travel to the banks of the Ganges, carefully collect some zebra fish embryos from the river

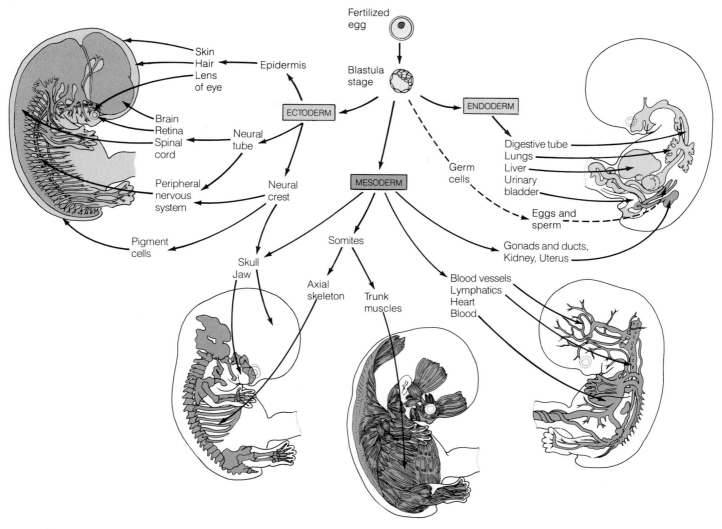

FIGURE 12.11 *Origin of an Animal's Body Parts.* This diagram shows the origin of the liver, brain, kidney, and other organs and tissues from the three embryonic cell layers: the endoderm (yellow), mesoderm (red), and ectoderm (blue). For example, mesoderm blocks off into somites, cubes of tissue that form the vertebrae of the spine and the muscles of the trunk.

bottom, then observe their continued development under a microscope at 6-hour intervals for three days. As you watched each tiny fish develop still enclosed in its clear, jellylike globe, you would see various organs take shape in the translucent animal: eyes, tail fins, liver, kidneys, muscles, and others. This **organogenesis** (literally, "origin of organs") involves two processes of change: **morphogenesis,** which causes cells to cluster in organs, and **differentiation,** which causes cells to accumulate specific proteins and hence perform specific functions.

DEVELOPMENT OF ORGAN SHAPE

The three germ layers created during gastrulation form the rudiments of the eyes, liver, muscles, and other organs.

But the cells of those layers must undergo a procession of changes for an organ to emerge: During organ development, some cells change shape; some divide rapidly; others die away; and still others move to new places. This cavalcade of proliferating, disappearing, and migrating cells goes on continuously, and the result can be endoderm bulging out from the gut to become lungs or liver or the cell death that sculpts a developing hand (Figure 12.13, page 212).

As some organs form, cells migrate long distances in the embryo. For example, as the neural tube arises from folds that fuse together, a ridge called the *neural crest* is created (review Figure 12.12). Neural crest cells scatter

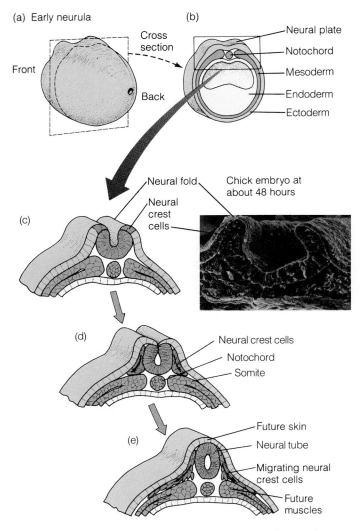

FIGURE 12.12 *Neurulation: Formation of the Future Nervous System.* (a, b) The events of neurulation, or neural tube formation, are easier to envision by considering the upper surface of an embryo cut in half. (c–e) The neural plate on the embryo's dorsal surface rolls inward forming first neural folds and then a fused neural tube. The electron micrograph in (c) shows the neural folds of a 48-hour-old chick embryo.

around the embryo like exploding fireworks, giving rise to cells and organs in distant places (Figure 12.14). Some neural crest cells stay near the epithelium, migrate into the skin, and form pigment cells (see Figure 12.12). If, as you look at your own arm, you can see brown moles or freckles or overall dark skin pigment, then you are observing the results of embryonic pigment cells that crawled from the neural crest near your spinal cord across your back and down your arm. Other neural crest cells migrated from the back of your head and around both sides of your developing mouth to form your teeth, as

well as some of the muscles, bones, and cartilage of your lower face (see Figure 12.14). If the cells fail to migrate far enough, they leave a space between the nose and mouth, or they leave a gap in the roof of the mouth. These birth defects, called *cleft lip* and *cleft palate*, respectively, can now be corrected by surgery.

While the migration of neural crest cells and the death of other cells cause organs to take shape, another set of changes—this time at the protein level—must occur before each organ can carry out its specific functions.

DEVELOPMENT OF ORGAN FUNCTION

Within an animal's body, each cell type performs a specific service: Intestinal cells digest food; muscle cells create movement; eye cells detect light. A cell can carry out its particular functions in large part because of the distinctive proteins it contains—proteins that break chemical bonds in food molecules; or proteins that cause a cell to contract; or proteins that change form when struck by light. Each cell type that looks different and contains a different set of proteins is called a *differentiated cell* and has become specialized by the process of differentiation.

Cell specialization begins when the process called **determination** identifies those genes that each cell will express as it differentiates. Determination commits cells

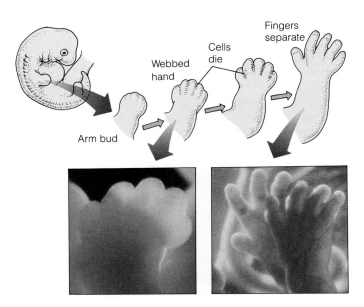

FIGURE 12.13 *Morphogenesis: How Cell Death Generates the Shape of Your Hand.* Your hand begins as a paddle, roughly like the fins on a zebra fish or the webbed feet of a duck. The fingers appear when four rows of cells radiating from the base of the paddle die away. This pattern of cell death is genetically programmed according to species, because in ducks and fish, for example, most of the cells don't die and webbing remains between the fin rays or toes. What's more, genetic mistakes can happen, leaving a duckling without its webs or a human baby with them.

DISCOVERY

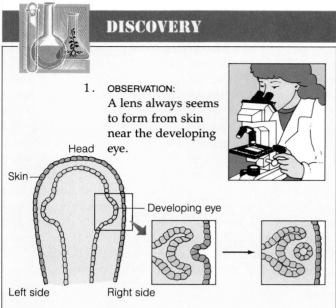

1. **OBSERVATION:** A lens always seems to form from skin near the developing eye.

Head

Skin

Developing eye

Left side　　　Right side

Chick embryo as seen from above

2. **QUESTION:** What causes a lens to form in the correct place?

3. **HYPOTHESIS:** Maybe the developing eye itself causes whatever skin is nearby to become a lens.

4. **EXPERIMENT AND PREDICTION:** If this is so, then if I move the developing eye to another place on the embryo, a lens will form there.

5. **RESULTS:** A lens did form near the relocated eye.

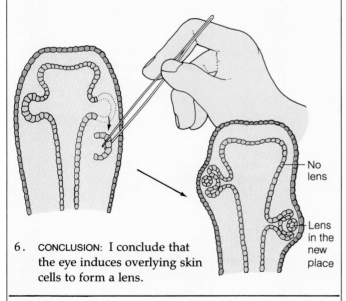

No lens

Lens in the new place

6. **CONCLUSION:** I conclude that the eye induces overlying skin cells to form a lens.

Adapted from M. S. McKeehan, *J. Exper. Zool.* **117** (1951): 31–64.

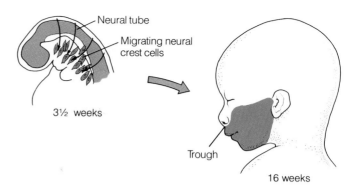

Neural tube

Migrating neural crest cells

3½ weeks

Trough

16 weeks

FIGURE 12.14 *Migrating Neural Crest Cells Help Form the Face.* Cells move from the neural tube at the site of the future spinal column around the sides of the embryonic head and help form the lower part of the face and jaw.

to a specific developmental fate, and regulatory proteins in determined cells select genes for use that are appropriate to the cell type—for example, genes for digestive enzymes in an intestinal cell or for hemoglobin in a red blood cell. Once a gene is active, the specialized proteins it encodes (such as digestive enzymes) allow the differentiated cell to do its job for the organism.

Differentiation is the last in a series of five processes that enable a fertilized egg to divide, enlarge, and grow into an animal with fully functioning organs. Table 12.2 (page 214) summarizes and reviews the five processes. Within an embryo, these five processes perform a developmental rondo: They are repeated over and over in different cell groups until the embryo has organs with proper shapes and specialized functions. The embryo is then ready to face the rigors of the world outside its transparent jelly globe, its egg shell, or its mother's uterus. But development is far from over at hatching or birth; it continues throughout the individual's lifetime.

≋ DEVELOPMENT CONTINUES THROUGHOUT LIFE

The development that continues after hatching or birth helps ensure survival of the individual and the species by allowing the organism to grow larger, to acquire adult characteristics, and to mature sexually.

CONTINUAL GROWTH AND CHANGE

If you've ever adopted a young pet, then you know that *growth* is the most apparent aspect of postembryonic development. The animal's weight doubles four or five times before stabilizing in adulthood. The same is true for a newborn child; the weight of a 7-pound infant can double to 14, 28, 56, and 112 pounds or more as it develops throughout childhood and adolescence (Figure 12.15). The weight gain comes largely from an increase in cell number due to mitosis throughout the body. Thus, while a newborn has billions of cells, an adult will have trillions.

Different parts of the body, however, grow at different rates. This accounts for some of the obvious differences between young and adult animals: An 80-year-old person usually has a large nose and large ears because the cartilage in these organs has never stopped growing. Most organs, however, stop enlarging after an animal reaches its adult size, and development continues mainly through the generation of new cells to replace those that age and die. The orderly replacement of worn-out cells usually involves **stem cells,** generative cells that retain the ability to divide throughout life. Special stem cells in the skin generate new skin cells; others in the bone marrow generate new blood cells; some in the intestines make new intestinal cells; and stem cells in the gonads make new gametes. (There are no stem cells in the brain, however, and adult brain cells are not replaceable.) When a stem cell divides, one daughter differentiates into the specialized cell type (e.g., blood or skin cell) while the second daughter becomes another stem cell (Figure 12.16).

Natural substances in the body called *growth factors* often regulate the division of cells so that growth occurs only where and when needed. Growth factors include *growth hormone,* which stimulates increased height, and *nerve growth factor,* which helps nerve cells survive and grow (Figure 12.17, page 216). Growth factors help maintain the optimal number of cells for each body part, and under normal circumstances, those numbers will not

---

TABLE 12.2 ≷ PROCESSES THAT BRING ABOUT DEVELOPMENT

PROCESS	WHAT HAPPENS	EXAMPLES IN DEVELOPING EMBRYO
Formation and storage of cytoplasmic determinants	Chemical factors stockpiled in the egg and partitioned into different cells during cleavage cause cells to select and express a set of genes that initiate a developmental program	Gray crescent cytoplasm in frog egg allows cells that received the gray material during cleavage to commence gastrulation
Induction	One cell group causes another group to start developing into a different cell type	Notochord induces neural tube
Commitment, or determination	Cells select one pathway of development	Certain cells activate genes for digestive enzymes and become intestinal cells
Morphogenesis	Organs form via changes in cell shape, cell proliferation, cell death, and cell migration	Neural tube rolls up; digits are carved; neural crest cells migrate
Differentiation	Specialized cell types develop; includes the production of specific proteins that allow a cell to carry out its particular role	Skin cells make fibrous proteins; intestinal cells make digestive enzymes

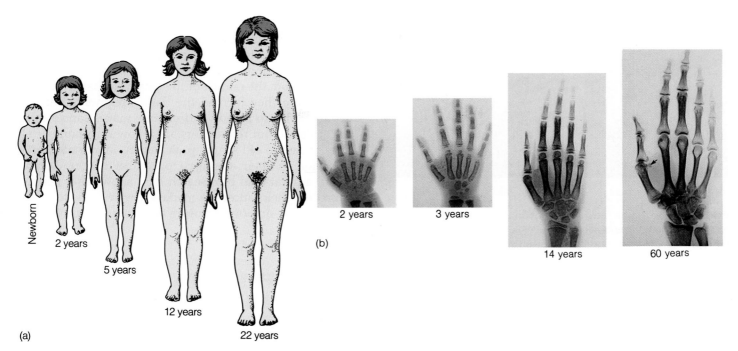

FIGURE 12.15 *Growth: Weight Quadruples and Proportions Change Between Birth and Adulthood.* (a) As this girl grows into a woman, her weight multiplies from 7 pounds to 14, 28, 56, and 112 pounds. Also, her head grows proportionately smaller, her arms and legs longer, and her trunk less dominant. (b) X-rays of human hands reveal that the finger bones lengthen and enlarge and the knuckles become more and more prominent throughout life.

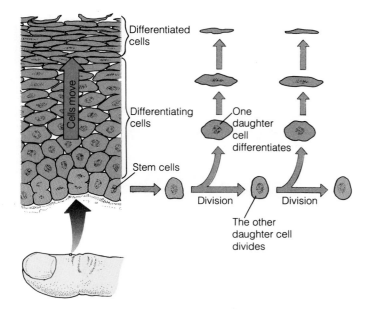

FIGURE 12.16 *Stem Cells Divide to Renew Body Parts.* In the skin on your index finger, stem cells in the dividing layer continuously reproduce as time goes by. As each stem cell divides, one daughter remains an undifferentiated stem cell with the potential for future divisions, while the other daughter becomes differentiated into an epidermal cell and is pushed outward to become part of the skin's protective layer. The differentiated skin cell can no longer divide and eventually dies and flakes off as dry skin while younger cells push up from below.

change very much as the animal ages. Sometimes, however, something goes terribly wrong: Growth regulation fails, tissues grow unchecked, and the result can be a massive tumor or invasive cancer.

CANCER: DEVELOPMENT RUNNING AMOK

A tumor is, in a sense, an inappropriate continuation of development. A *benign* tumor is a clump of cells that continues to grow unchecked, but stays localized in a tightly packed group and thus can usually be removed surgically. A cancerous, or **malignant,** tumor, however, grows without ceasing and also spreads throughout the body (**metastasizes**), invading healthy organs and destroying them; this often makes surgical removal impossible. In insidious ways, cancer cells mimic normal embryonic cells.

■ **Differentiation** We saw that most fully differentiated cells seldom divide and instead arise from stem cells that can divide but cannot perform as mature cells. Some cancer cells, such as most skin cancers, appear to be stem

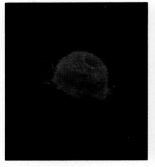 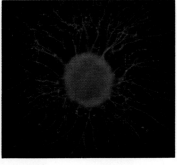

(a) Growth factor absent   (b) Growth factor present

FIGURE 12.17 *Growth Factors Regulate Cell Division and Differentiation.* With growth factors absent (a), nerve growth is inhibited, but with nerve growth factor present (b), nerves grow luxuriantly.

cell derivatives (as in Figure 12.16) that fail to completely differentiate and hence preserve their capacity to divide.

■ **Shape**   The skeleton that maintains shape in normal differentiated cells is oriented incorrectly in cancer cells. Thus, cancer cells "round up" like immature cells when grown in a laboratory dish, instead of flattening out like normal cells (Figure 12.18).

■ **Proliferation**   Cancer cells proliferate as if they were still trying to build an embryo, and they do this because they contain mutant proteins encoded by *oncogenes* (literally, "cancer genes"). The proteins mimic normal growth-controlling substances such as nerve growth factor, and they continually trigger cell division, the result being a mass, or tumor. Certain cancer treatments in the future may try to turn off the activity of oncogenes.

■ **Programmed Cell Death**   Cancer cells don't seem to age normally. Normal cells taken from an animal and grown in culture will divide between 20 and 50 times, then stop dividing and die. The number of cell divisions seems to be genetically programmed for each cell type in each species, as in the cells that die between our embryonic fingers (see Figure 12.13). Biologists do not understand this programmed aging and death, but they do know that cancer cells are immortal: They can continue to divide indefinitely.

■ **Cell Migration**   Although cell division goes on unchecked in both benign and malignant tumors, usually only the latter become life threatening, because some

of their cells migrate out and invade neighboring tissues, disrupting the tissue organization. One of the most rapidly spreading life-threatening cancers, for instance, is *melanoma*, an abnormal growth of cells that originally migrated from the middle of the embryonic back into the skin, where they formed the darkly pigmented areas. The tendency to invade and migrate makes cancer cells hard to treat, but researchers are following a number of leads related to cell shape, proliferation, immortality, and migration in their attempts to combat cancer. Meanwhile, a great deal of research data shows that people can reduce their cancer risk markedly by avoiding tobacco, high-fat diets, and overexposure to the sun (Figure 12.19).

☷ THE FORMATION OF GAMETES: THE DEVELOPMENTAL CYCLE BEGINS ANEW

We have seen that physical changes continue to accumulate after hatching or birth. As the young animal develops toward maturity, the body enlarges, the chest grows hair or swells to form breasts, the head sprouts antlers, and so on, depending on the species. Most of the changes in physical form, physiology, and behavior are under the control of hormones (see Chapters 13 and 25), and many help prepare the organism for mating and reproduction—the start of a new developmental cycle.

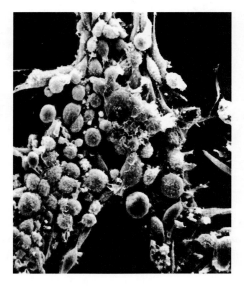

FIGURE 12.18 *Cancer Cells: Shape and Consequences.* Cancer cells lack a normal cytoskeleton; instead of remaining flattened, they round up into globular shapes.

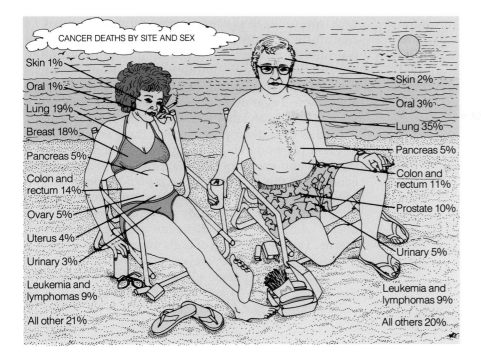

FIGURE 12.19 *Cancer Risks and Life-Style Choices.* Each year during the 1980s, doctors diagnosed cancer in about 1.3 million Americans. Of those people, about 400,000 had skin cancer, while about 930,000 had cancer at other sites. Today, one out of every five deaths in the United States is from cancer—a figure that has risen steadily since 1930 and is largely due, researchers believe, to the steep climb in lung cancers from cigarette smoking. Since 1965, lung cancer has claimed more men than has any other type of cancer. From 1930 to 1986, breast cancer was the biggest cancer killer of women. Beginning in 1986, however, lung cancer began to claim more women's lives each year than breast cancer, probably because of increased smoking among women. You can help reduce your own risk of cancer. Cigarette smoking accounts for 30 percent of *all* cancer deaths (not just lung cancer) and should be avoided. Researchers believe that a high-fat, low-fiber diet increases the risk of breast, colon, and prostate cancers; that eating large amounts of salt-cured, nitrate-treated, or smoked foods can lead to esophageal and stomach cancers; and that the heavy use of alcohol is linked to cancers of the mouth, larynx, throat, esophagus, and liver. Finally, almost all of the 400,000 cases of skin cancer per year are due to overexposure to sunlight. A few changes in diet and life-style could add years to your life by reducing your cancer risk.

Whether the external changes are dramatic or subtle, sexual maturation always involves **gametogenesis:** the construction of eggs and sperm, which are the living link between generations.

SPECIAL EGG CYTOPLASM LEADS TO REPRODUCTIVE CELLS IN THE NEW INDIVIDUAL

In many species, the egg cell contains the seeds of its own rebirth in the form of *germ plasm,* a region of distinctive granular cytoplasm within the egg cell. As cleavage divides the fertilized egg, germ plasm is partitioned into only a few cells—the future eggs or sperm (Figure 12.20, page 218). The germ plasm contains special sub-

stances that in some unknown way instruct a cell to become an egg or sperm. Thus, at this very early stage in development, the embryo contains two broad classes of cells: **germ cells,** or reproductive cells that can help form a new individual, and **somatic cells,** the body cells that will form your nose, arms, legs, liver, and so on. Somatic cells die when your body dies, but germ cells have the potential of living on in children.

Once germ cells have been determined in an embryo, they migrate to a spot that differentiates into a **gonad,** an organ that produces eggs or sperm. The male gonad is called a *testis*, and the female gonad is called an *ovary*. Inside the gonad, the germ cells mature into sperm or eggs. In a man, it takes 74 days for a sperm cell to mature, but the cells that give rise to sperm are so numerous that every hour, 100 million sperm cells develop per testis—

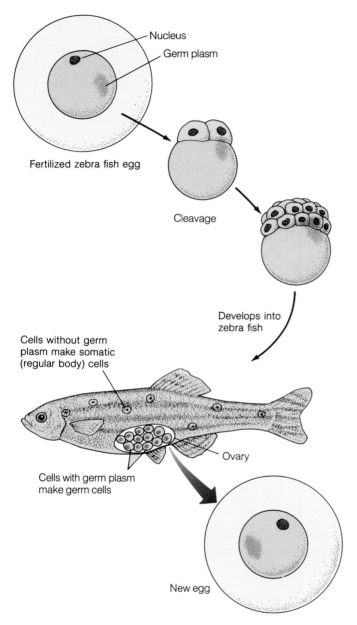

Nucleus
Germ plasm
Fertilized zebra fish egg
Cleavage
Develops into zebra fish
Cells without germ plasm make somatic (regular body) cells
Ovary
Cells with germ plasm make germ cells
New egg

FIGURE 12.20 *Continuity of the Germ Plasm: Egg to Adult to Egg.* Germ plasm is passed along as the egg gives rise to an adult, which gives rise to new eggs. The cells that will eventually form eggs or sperm are permanently segregated from the rest of an animal's cells—the body cells—very early in development.

each sperm a self-propelled missile ready to begin its race to the egg cell. Eggs are far less numerous than sperm. The production and differentiation of each egg requires a substantial investment of time and materials from several cooperating tissues. In a woman, all the eggs she will ever make—several hundred—are already present at birth, and they will take 13 to 50 years to mature.

As this brief description shows, there is a continuity of germ plasm moving directly from a mother's egg to the eggs and sperm of her offspring. If you applied the concepts of this chapter to your own development, you could trace the germ plasm in your own gonads back to an immature egg cell already present in your mother's ovary while she was still inside your grandmother's uterus. The transmission of traits via germ plasm formed in the embryo is what makes it impossible for a trait acquired during a person's lifetime—such as large muscles or the ability to speak a certain language—to be passed from parent to offspring.

≋ CONNECTIONS

The story of animal development carries us full circle from a single fertilized egg to an embryo, young animal, adult, and finally to the production of new eggs or sperm. Development helps show the unbroken continuity of living things, because none of an organism's physical structures and no event in its life cycle is made from scratch. Instead, each has a precursor in a previous part of the cycle, is constrained by prior structures and events, and develops as a result of chemical instructions or interactions among preexisting cells.

The actions of genes and proteins cause cells to become committed and differentiate, to divide and specialize into the organism's body plan and organs. These embryonic organs become the functioning anatomical parts of organisms that we will study in later chapters of the book. The embryo is also the vehicle for evolutionary change. New species arise because new traits are selected that alter development. The gradual modification of a dinosaur's forelimb into the first bird's wing, for example, occurred because mutations were selected that altered the way cells changed shape, divided, differentiated, migrated, or died away during the formation of embryonic limbs inside huge, shelled dinosaur eggs hundreds of millions of years ago.

While this chapter has laid out the general principles of animal reproduction and development, there is a great deal more to say about how these generalities apply to our own species and how we progress through the life cycle from fertilized egg to fertilized egg. That is the subject of the next chapter.

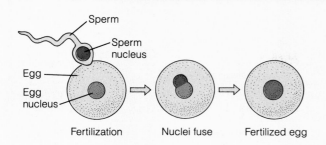

Sperm
Sperm nucleus
Egg
Egg nucleus

Fertilization   Nuclei fuse   Fertilized egg

(1) Animal mating brings egg and sperm together, and the gametes fuse in the process called fertilization.

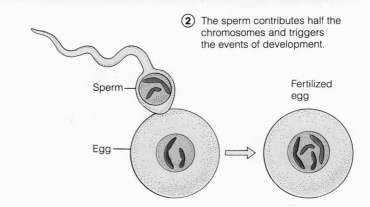

(2) The sperm contributes half the chromosomes and triggers the events of development.

Sperm

Egg

Fertilized egg

(3) The egg contributes half the chromosomes; it protects and nourishes the embryo and directs its development.

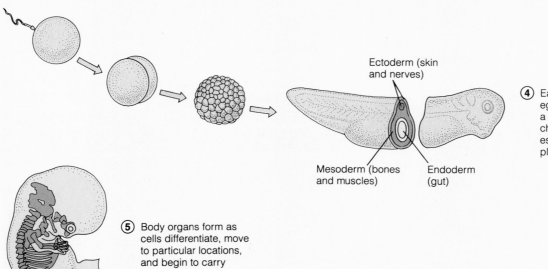

Ectoderm (skin and nerves)

Mesoderm (bones and muscles)   Endoderm (gut)

(4) Early in development, the fertilized egg divides many times, forming a hollow ball of many cells. Cell changes and migrations help establish the three-layered body plan and the nervous system.

(5) Body organs form as cells differentiate, move to particular locations, and begin to carry out specific tasks.

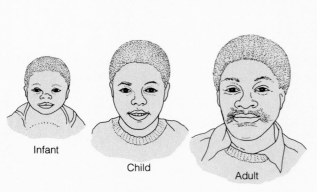

Infant

Child

Adult

(6) An individual continues to develop throughout life, growing from a newborn infant to a sexually mature adult.

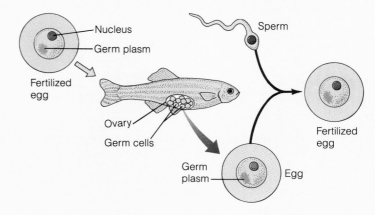

Nucleus
Germ plasm

Fertilized egg

Sperm

Ovary
Germ cells

Germ plasm   **Egg**

Fertilized egg

(7) As egg and sperm originate and mature from germ cells, the life cycle continues in the next generation.

NEW TERMS

blastocyst, page 207
blastula, page 207
cleavage, page 207
determination, page 212
developmental determinant, page 209
differentiation, page 211
embryo, page 207
gametogenesis, page 217
gastrulation, page 209
germ cell, page 217
gonad, page 217
induction, page 210

inner cell mass, page 207
malignant, page 215
metastasize, page 215
morphogenesis, page 211
neurulation, page 210
notochord, page 210
organogenesis, page 211
placenta, page 207
somatic cell, page 217
stem cell, page 214
trophoblast, page 207
yolk, page 206

STUDY QUESTIONS

REVIEW WHAT YOU HAVE LEARNED

1. Cues detected by sight and smell can lead to mating. What are some examples?

2. List the events of fertilization that follow the penetration of an egg by a sperm nucleus.
3. Is a whale egg larger than a chicken egg? Explain.
4. What role do special developmental substances play in the egg cytoplasm?
5. Trace the steps of neurulation.
6. How is a malignant tumor like a benign tumor? How is it different?

APPLY WHAT YOU HAVE LEARNED

1. How can a person reduce his or her own risk of cancer?
2. An embryologist transplanted the posterior part of a fertile fruit fly egg to the anterior part of another egg. Why did the blastula-stage embryo develop germ cells at both ends?

FOR FURTHER READING

Augier, N. *Natural Obsessions: The Search for the Oncogene.* Boston: Houghton Mifflin, 1988.
Kimmel, C., and R. Warga. "Cell Lineage and Developmental Potential of Cells in the Zebrafish Embryo." *Trends in Genetics* 4 (1988): 68–74.

THE HUMAN LIFE CYCLE

CHAPTER 13

CONCEPTION IN A LABORATORY DISH

The birth of a baby girl in Oldham, England, in 1978 created a sensation around the world. Louise Joy Brown was a healthy and normal newborn in every respect save one: She was the first baby in human history to be conceived in a laboratory dish. Her extraordinary beginning is a fitting introduction to this chapter because it symbolizes our sophisticated current knowledge of human reproduction and the subsequent stages of the human life cycle.

For nine years, Louise's parents, Lesley and John Brown, had failed to conceive a baby because a "roadblock" prevented Lesley's eggs from reaching her

FIGURE 13.1 *Louise Brown, the First "Test-Tube Baby."* After several years of trying to conceive without success, the Browns, with the help of laboratory techniques, produced a healthy baby girl.

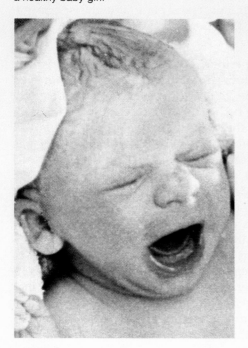

uterus, the organ that receives and protects the developing embryo. To circumvent this blockage, a team of pioneering physicians and researchers removed eggs from her ovary, mixed them with John's sperm, and reimplanted an early-stage (eight-celled) embryo into her uterus. There it burrowed into the lining of the uterine wall and developed into a larger and larger organism. Nine months later, Louise was born (Figure 13.1), the first person conceived by **in vitro fertilization** (*IVF*; literally, "fertilization in glass").

Such a technological feat is possible because biologists understand so many details of human fertilization, embryonic implantation, and fetal development. At certain stages, human embryos are amazingly similar to the fish and frog embryos we studied in Chapter 12 (review Figure 12.2). And human life cycles resemble those of most other animals in another way: For a conception to occur and for a new generation to be perpetuated, males and females must reach sexual maturity, develop sexual characteristics, attract each other, and mate.

As we discuss the physical structures and mechanisms that allow a couple to produce a baby and thus to begin a new human life cycle, several unifying themes will become apparent. First, the male and female reproductive systems are generally parallel in structure and function; they are two variations on a common motif: perpetuating the species. Second, for eggs and sperm to be manufactured and released, and for a person to develop and mature, intricate networks of communications are needed between body organs. These communi-

cations involve nerve signals as well as hormones, which are regulatory molecules produced in one part of the body that cause cellular activities to change in other parts. Third, the pregnancy that commences once egg and sperm unite is a biological partnership between mother and young. While the mother's body provides the embryo with nutrition and protection for nine months, biochemicals from the developing offspring orchestrate the duration of pregnancy, stimulate birth, and initiate a milk supply. Finally, birth is the start of independent life, but it is not the end of development. People continue to grow and mature for about two decades before they reach peak physical performance and reproductive capacity and begin to age.

Our discussion of the human life cycle will describe how humans reproduce and develop. Along the way, it will answer the following questions:

- What is the anatomy of human reproductive systems, and how do they develop in the embryo and child and function in the adult?
- How does human sexual activity unite egg and sperm, and how do birth control methods prevent reproduction?
- When conception does occur, what remarkable steps transform a single fertilized egg cell into a multicellular human baby?
- How can a mother modify her lifestyle to ensure the best chances for the developing fetus?
- How does growth continue after birth and throughout childhood, puberty, adulthood, and old age?

⚡ MALE AND FEMALE REPRODUCTIVE SYSTEMS

Each human being has a complex system of sexual characteristics, reproductive organs perfected during millions of years of evolutionary history and capable of passing along his or her genes via sexual reproduction (Figure 13.2). Male and female reproductive organs are different but parallel in structure and function. A set of *primary reproductive organs* includes the testes in males and the ovaries in females. The primary reproductive organs produce sperm or eggs, which arise through meiotic divisions, and they also produce sex hormones. Both sexes also have *accessory reproductive organs:* various ducts, chambers, and glands that act as organic plumbing to transport and store the eggs or sperm or, in the woman, to nurture the developing embryo. Let's see how human reproductive organs are built and regulated, starting with the simpler male system.

THE MALE REPRODUCTIVE SYSTEM

Male reproductive organs make sperm by the millions. They also transfer those sperm into the female's reproductive tract, where the sperm can encounter and fertil-

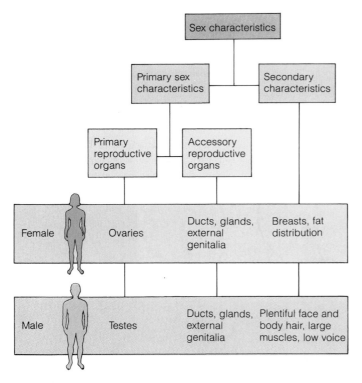

FIGURE 13.2 *Human Sex Characteristics.*

ize eggs. The two jobs involve different subsets of organs and processes.

■ **Organs for Sperm Production** The primary male reproductive organs are the **testes** (singular, testis): smooth, oval structures about 4 cm long, held in a bag-like *scrotum* (Figure 13.3a). Each testis is subdivided into about 250 compartments, and each compartment contains several highly coiled, hollow, sperm-bearing tubes called *seminiferous tubules,* which are 70 cm long (Figure 13.3b). If placed end to end, all the tubules in a man's testes would make a pipeline several hundred meters long but only about half the thickness of a sheet of paper. Sperm-forming cells (*spermatogenic cells*) in the tubule walls develop into sperm, while *supporting cells* (*Sertoli cells*) in those same walls surround and nourish the developing sperm (Figure 13.3c). Around the seminiferous tubules lie the *interstitial cells* (*Leydig cells*), which produce testosterone and other male hormones.

To develop properly, sperm must have an environment several degrees cooler than the normal internal body temperature of about 37°C (98.6°F). For this reason, the testes hang outside the body in a pouch of skin, the scrotum. If a man wears tight pants, exercises too hard, or sits in a hot tub, the temperature of the scrotum and testes can increase to a point where sperm development temporarily stops. When he cools down, however, the spermatogenic cells can once again develop into sperm.

After attaining their streamlined shape (see Figure 12.5), sperm pass down the cavity of the sperm-bearing tubules and collect in the *epididymis,* a very narrow tube that is 7 m long and sits coiled tightly on top of the testis. Here, final maturation takes place and the sperm are stored, ready for transport.

■ **Organs for Sperm Transport** When a male is sexually stimulated during intercourse, masturbation, or even sleep, sperm are rapidly transported and forcefully released from the body. The journey begins when smooth muscle cells in the walls of the epididymis contract repeatedly and propel the sperm into the 45 cm long **vas deferens,** a connecting tube that also has muscular walls that continue to propel the sperm (see Figure 13.3a and b). The vas deferens leads back into the body from the scrotum; near the urinary bladder, it merges with the *ejaculatory duct,* a duct leading from the *seminal vesicle.* Secretions from this small gland bathe the sperm in a fluid that contains sugars and other nutrients as well as substances that regulate pH and stimulate muscular contractions in the female reproductive tract. The combination of sperm and fluid empties into the ejaculatory duct, which passes through the middle of a very important secretory organ, the **prostate gland.** This chestnut-shaped structure secretes a milky alkaline fluid that mingles with

(a) Cross section of pelvic area

FIGURE 13.3 *Male Reproductive System.*
(a) Reproductive tract as it would look if the body were cut in half lengthwise and viewed from the side. (b) A testis, also partially cut in half. (c) Sperm-bearing (seminiferous) tubules, when enlarged, reveal sperm-producing (spermatogenic) cells, supporting cells, interstitial cells, and sperm.

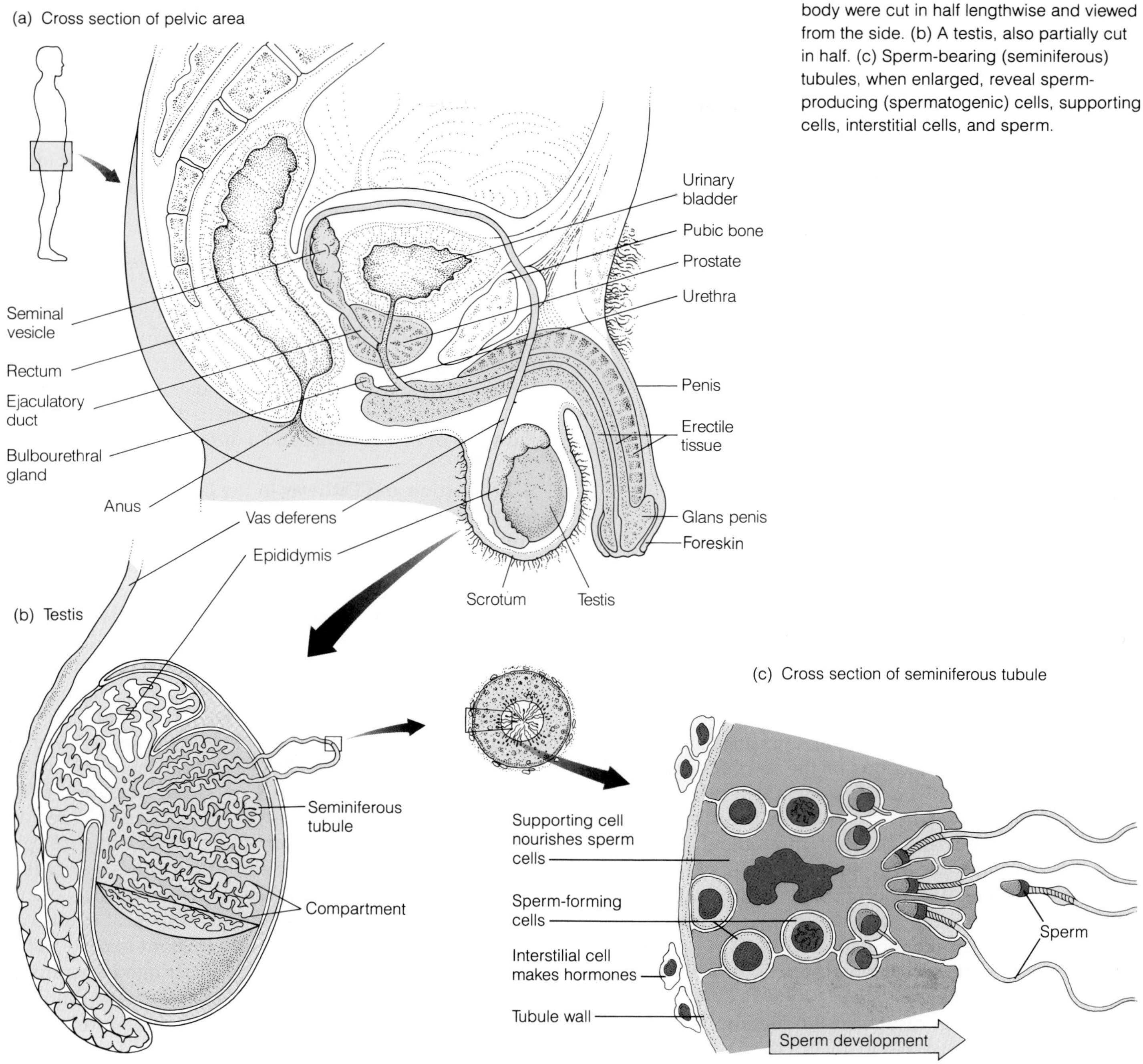

Seminal vesicle

Rectum

Ejaculatory duct

Bulbourethral gland

Anus

Vas deferens

Epididymis

Urinary bladder

Pubic bone

Prostate

Urethra

Penis

Erectile tissue

Glans penis

Foreskin

Scrotum   Testis

(b) Testis

Seminiferous tubule

Compartment

(c) Cross section of seminiferous tubule

Supporting cell nourishes sperm cells

Sperm-forming cells

Interstilial cell makes hormones

Tubule wall

Sperm

Sperm development

fluid of the seminal vesicle. The added secretion enlarges the volume of sperm and fluid and brings the mixture's pH almost to neutral. This pH will help neutralize the natural acidity of the female reproductive tract, an acidity that protects her delicate tissues from microorganisms but also tends to inhibit sperm motility.

Once surrounded by milky fluid, the sperm passes from the ejaculatory duct to the **urethra.** This is a dual-purpose tube that can also carry urine from the bladder

out through the **penis,** a cylindrical organ that transfers sperm to the female. To facilitate entry of the penis into the female's vagina for intercourse or to clear the urethra, a gland at the base of the penis, the *bulbourethral gland,* secretes a mucuslike lubricating substance, and spongy *erectile tissue* within the penis fills with blood and stiffens the entire organ.

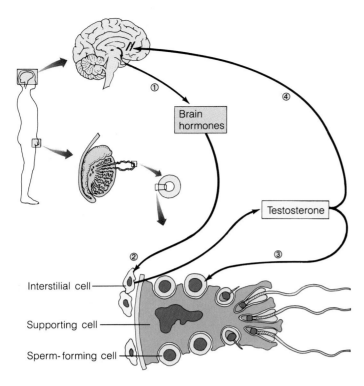

FIGURE 13.4 *How Hormones Control Sperm Production.* Several hormones operating in interlocking feedback loops control the timing of sperm production. Brain hormones (1) act on the testis, causing it to release testosterone (2), which, along with a brain hormone, stimulates sperm production (3) and feeds back to the brain, blocking the release of the brain hormones (4).

At the peak of sexual excitement, *ejaculation* takes place: Strong muscular contractions of the urethral walls and other muscles forcibly expel **semen** (sperm plus surrounding *seminal fluid,* the secretions of the seminal vesicle, prostate gland, and bulbourethral gland) from the penis. While only about 3 milliliters (ml) of semen is ejaculated (a mere teaspoonful), it normally contains 400 million sperm. So many sperm hurtling toward a single egg maximizes the chance for fertilization, and hormones help guarantee a continual supply of viable sperm.

■ **Hormonal Control of Sperm Production** Two types of hormones work together to govern the timing of sperm production. The first hormones are proteins released from two specific parts of the brain (Figure 13.4, step 1). These brain hormones act on cells in the testis, causing one cell type to produce **testosterone** (step 2), a steroid hormone that is related to cholesterol (see Chapter 2). Testosterone and the other type of brain hormone then act on the testis and stimulate sperm development (step 3). Testerone also travels in the blood back to the brain and blocks further secretion of the brain hormones (step 4).

This simple circuit keeps the amount of testosterone—and hence the rate of sperm production—constant. When the testosterone level rises too high, it blocks the release of the brain hormones, and so testosterone is no longer made. But if the testosterone level drops too low, secretion of the brain hormones is *not* blocked, and they can stimulate the testis to make more testosterone. This kind of regulatory circuit, where a substance like testosterone controls its own concentration, is called a **feedback loop.** Feedback loops are also important for controlling a woman's monthly menstrual cycle.

THE FEMALE REPRODUCTIVE SYSTEM

The female's reproductive organs not only produce and transport eggs, but also receive and nourish developing embryos and give birth to babies. Like males, females have sex organs that produce gametes (eggs) and tubes that transport them, but they also have specialized modifications of the tube wall and a different hormonal cycle.

■ **Production and Pathway of the Egg** Women produce female gametes, or eggs, within two solid, almond-shaped organs called **ovaries,** which lie inside the body cavity just below the waistline (Figure 13.5a and b). Each ovary is made up of cells called *oocytes*, which develop into eggs, and cells that surround and support the immature egg cells, called *follicular cells* (Figure 13.5c and d). An immature egg cell surrounded by follicular cells constitutes a unit called a **follicle,** and at birth, each ovary contains hundreds of follicles. After birth, no new prospective egg cells are generated. A man, in contrast, makes millions of new sperm cells every day.

Every 28 days or so, a process called **ovulation** takes place: A single follicle in one of the ovaries enlarges, its oocyte matures, the follicle ruptures, and a mature egg, now called an *ovum,* is released into the body cavity (see Figure 13.5c). British researchers took the egg that became Louise Brown from her mother's ovary just at ovulation. An ovum, the largest human cell, is about the size of the dot on this "i." It moves toward the fringed opening of one of two 10-cm-long tubes called *oviducts* (or Fallopian tubes) (see Figure 13.5b and c). Both right and left oviducts are lined with millions of hairlike cilia that act like paddles that sweep the ovum along. If sperm are present, they can encounter the egg and fertilize it—usually as it moves down the oviduct. The oviducts connect with the thick-walled, pear-shaped **uterus** (see Figure 13.5a and b). Eventually, the ovum is swept into the narrow chamber within the uterus, where, if already fertilized, it can be nourished and develop into an embryo and fetus.

Most months, the ovum is not fertilized and degenerates, or exits the uterus through the **cervix,** a muscle-lined opening with walls that secrete mucus. Depending on its consistency (which is regulated by hormones), this mucus can plug the opening during pregnancy or aid the movement of sperm. Outside the cervix lies the **vagina,** a hollow, muscular tube that receives the penis during intercourse, conveys uterine secretions to the outside, and can stretch into a birth canal that allows passage of the fetus. External reproductive organs surround and protect the vaginal opening; these include tissues sensitive to sexual stimulation, such as the *clitoris*, and the lubricating *Bartholin's glands*. The functioning of these

internal and external organs requires a hormonal control that is parallel to, but more complex than, the male's.

■ **Hormonal Control of Egg Production and Uterine Preparation** Most women have a roughly 28-day cycle coordinated by hormones and called the **menstrual cycle.** During each menstrual cycle, the uterine wall lining, or *endometrium*, builds up and prepares for pregnancy, and then the ovary releases an egg. If the egg is not fertilized, the lining sloughs off, and menstrual bleeding occurs,

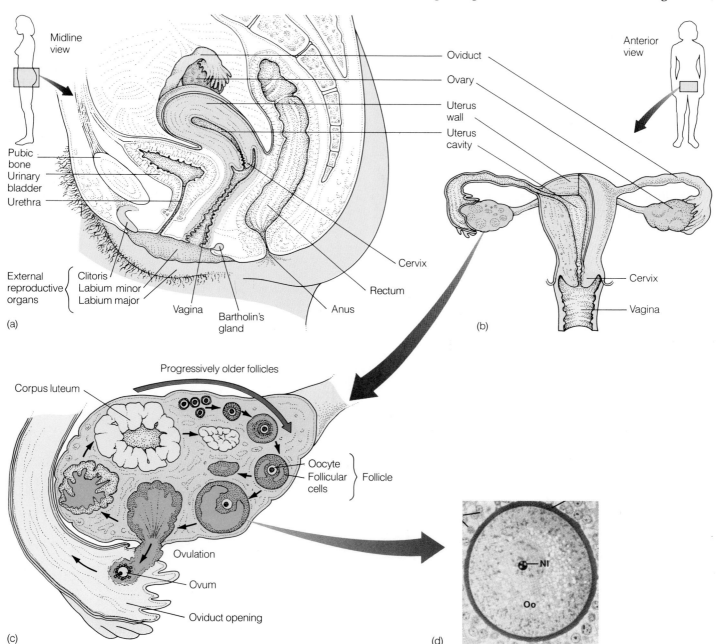

FIGURE 13.5 *Female Reproductive System.* (a) The female reproductive system as it would look if it were cut in half lengthwise and viewed from the side. (b) The reproductive system viewed from the front looks a bit like a pear with arms. (c) An enlarged view of one ovary and fringed oviduct opening. (d) A magnified view of a human egg cell, surrounded by follicle cells.

beginning a new 28-day cycle. In the menstrual cycle, parts of the brain make protein hormones (identical to the male's), the ovaries make the female steroid hormones **estrogen** and **progesterone,** and these hormones interact in a feedback loop, as in the male. Figure 13.6 depicts what happens at each stage of the cycle.

The menstrual cycle is usually counted from day 1, when menstrual flow begins. Parts of the brain secrete protein hormones (see Figure 13.6, step 1), which travel in the bloodstream to the ovary. One brain hormone stimulates follicles to grow (step 2), but usually only one follicle with its oocyte matures each month. The follicle grows rapidly and secretes increasing amounts of estrogen (step 3). This hormone causes the uterine lining to grow thicker and more heavily supplied with blood. On about the fourteenth day of a 28-day cycle, a part of the brain secretes a large pulse of hormones (step 4), and these trigger the oocyte to complete the first meiotic division, which it began before the woman was born. The developing follicle then ruptures and releases the egg culminating the process of ovulation (step 5).

The doctors who assisted in the in vitro fertilization of Louise Brown carefully monitored the levels of one of the brain hormones in Lesley Brown's body so that they could predict the exact time of ovulation and collect the egg (Figure 13.7). This monitoring is relatively easy, since excess hormone is secreted in the urine. Women who want to become pregnant can synchronize intercourse with ovulation by buying "dipsticks" coated with chemicals that turn bright blue when dipped in urine that contains peak levels of the key hormone.

Once the egg leaves the ovary and begins its passage down the oviduct, the follicle cells left behind in the ovary enlarge to form a new gland, the *corpus luteum* (literally, "yellow body") (Figure 13.6, step 6). Corpus luteum cells continue to secrete estrogen, but they now begin producing large quantities of progesterone as well (step 7). Together, estrogen and progesterone promote the continual buildup of the uterine lining and inhibit the brain from releasing reproductive hormones (step 8). This inhibition is another example of a feedback loop.

If the egg does not encounter sperm on its downward journey and is therefore not fertilized, diminishing levels of the brain hormones allow the corpus luteum to degenerate on day 24 of the cycle and thus to release less and less estrogen and progesterone. As these hormones diminish, the uterus lining begins to slough off, and menstrual flow starts again (step 10) and continues for

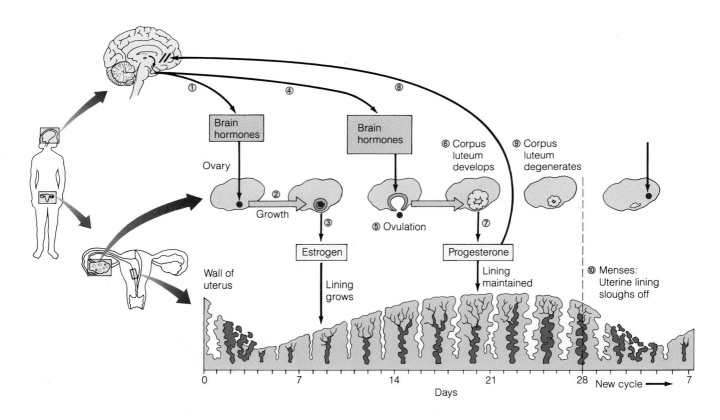

FIGURE 13.6 *How Hormones Control Egg Release: What Makes the Menstrual Cycle Cycle?* Hormones produced in a region of the brain operate on the ovary and control the ripening and release of an egg every 28 days. Shown here are the steps that regulate the cycle, as described in the text.

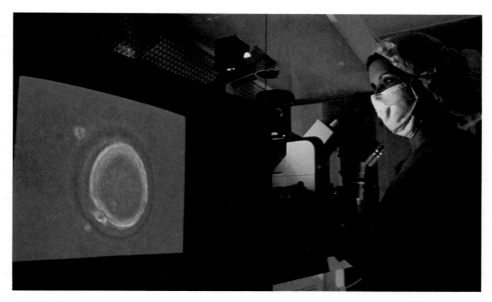

FIGURE 13.7 *Test-Tube Fertilization.* A technician examines the projected image of a ripe egg removed from a woman's ovary in preparation for in vitro fertilization—fertilization "in glass."

five days or so, marking the end of one cycle and the beginning of the next. Without estrogen and progesterone to serve as blocks, brain hormones are again produced, and the uterus lining begins to grow once again.

The female's hormonal interactions are similar to the male's in several ways. First, and most obviously, they involve the same brain hormones. Second, hormone concentrations are self-regulating in that they feed back to turn their own production off. Third, reproductive hormones ensure the continuation of the species by making reproductive cells available—continuously in the male, cyclically in the female. Finally, estrogen has developmental effects in the adolescent female that correspond to those of testosterone in the male and lead to the appearance of secondary sex characteristics (see Figure 13.2).

≋ SPERM MEETS EGG— OR DOESN'T: FERTILIZATION, BIRTH CONTROL, AND INFERTILITY

Once a female's hormones have done their job, an egg, released from the ovary, passes down the oviduct and is ready to be intercepted by a sperm. As in other animals, the rendezvous between egg and sperm depends on behavioral factors: Male and female are usually sexually attracted to each other and engage in sexual intercourse near the time of ovulation.

Human bodies are admirably suited to help sperm and egg converge. Nerve impulses generated during sexual stimulation increase blood flow to the penis, causing the organ's spongy tissue to collect blood, swell, and stiffen. The clitoris, the tissues around the vagina, and the nipples of the breasts also swell and grow sensitive to stimulation. Lubricants flow from male and female glands and the vaginal wall, easing penetration and making the vagina more hospitable to sperm. Further, clitoral stimulation and pelvic thrusts build sexual excitation in both partners, and the excitement usually peaks with *orgasm,* involuntary muscular contractions of the vaginal and uterine walls or of the muscles lining the seminal vesicles, urethra, and other male structures. These contractions are highly pleasurable and in the male cause the high-pressure ejaculation of millions of sperm.

Of the 400 million living sperm simultaneously racing toward the ovum at 1 cm per minute, only a few tens of thousands navigate the cervix and enter the uterus, and fewer still encounter the ovum in the oviduct, where fertilization usually takes place. Nevertheless, as a hundred or more sperm collide with the ovum, the tip of each sperm will detonate and release enzymes, and these will chew through the ovum's protective coats (review Figure 12.6). The race ends when one sperm enters (Figure 13.8, page 228). This union triggers the ovum to complete meiosis and to erect barriers to further sperm penetration. With this, the sperm nucleus moves toward the egg nucleus and fuses with it in the egg cytoplasm to form a unique diploid genetic constitution. The

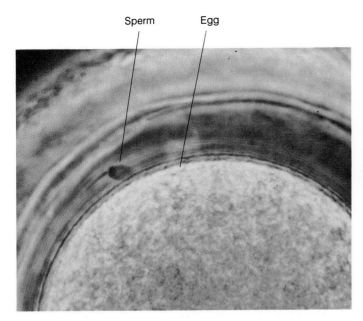

Sperm   Egg

FIGURE 13.8 *Human Fertilization.* This photo documents the brief moment of fertilization, during which the speediest sperm enters the egg (here, near center of photo), the sperm nucleus moves into the egg cytoplasm, and the two nuclei fuse to form the unique genotype of a new individual.

implantation of the developing embryo in the uterine wall marks the beginning of *pregnancy,* a series of developmental events that involve close cooperation between mother and embryo and that transform a fertilized egg into a baby.

THE MANAGEMENT OF FERTILITY

The drama of human fertilization happens quite naturally—and, many think, all too often. Every minute, 230 babies are born in the world, but only 90 people die, leaving a net increase of 140 people per minute. This is more than 1,400,000 new people every week, or the equivalent of 52 new Philadelphias every year! For many people, however, life without children is unthinkable, and infertility can seem a personal tragedy. Two active research areas involve controlling unwanted pregnancies and overcoming infertility.

■ **Birth Control** The prevention of conception or pregnancy (**contraception**) can be achieved by blocking fertilization or by inhibiting the fertilized ovum from implanting into the wall of the uterus. Table 13.1 provides details for these two types of intervention. Under the heading "Hormonal Strategies," the table describes birth control pills for women. These pills are a mixture of synthetic hormones that block production of the brain hormones and hence the maturation and ovulation of

eggs. Research toward a male pill has been slow, but the Chinese have experimented with a yellow pigment from cotton seeds known as gossypol, which blocks sperm growth. And researchers in the United States are testing injections of the synthetic hormone testosterone enanthate as a possible forerunner to an oral male contraceptive. Both male and female pills have side effects some people find unacceptable; blood clots are a particular risk to women who are over 35 or who smoke, and gossypol permanently sterilizes some men.

Even if fertilization occurs, the embryo's implantation into the uterine wall can be inhibited by an intrauterine device (IUD). This is a small plastic or metal insert worn in the uterus that can interfere mechanically with implantation of the embryo. While very effective, these devices are associated with increased risk of pelvic infections, potential scarring, and future infertility.

Even if implantation has occurred, two approaches—one controversial and the other experimental—can prevent a pregnancy. Abortion is the expulsion of a fetus, either spontaneously or deliberately induced. The Supreme Court ruled more than a decade ago that the government cannot forbid deliberate abortions early in pregnancy; many people, however, have strong moral objections to abortion. There are several medically supervised abortion procedures currently available in the United States and Western Europe, including RU486, an abortion-inducing pill developed in France. In the experimental realm, scientists in India have developed a birth control vaccine that causes the mother's immune system to destroy a pregnancy-sustaining hormone and allow any embryo that might have formed to be washed out with the menstrual flow.

■ **Overcoming Infertility** The parents of Louise Brown had a common and distressing problem: infertility. One in six couples cannot conceive without medical treatment, and that number is growing, partly because of the increase in venereal diseases (see the box on page 233) and the use of IUDs and partly because of the common practice of deferring parenthood to the 30s, when fertility naturally declines. About half the time the woman is the infertile partner, and in most of these cases, her oviducts or uterus is blocked or scarred. When the male is the infertile partner, he usually has a low sperm count or his sperm lack normal motility.

Today, physicians can successfully treat about 70 percent of infertile men and women through therapy with synthetic hormones, corrective surgery to unblock passages, or techniques like in vitro fertilization. Even with medical help, however, some women cannot produce eggs normally or carry a baby, and some men cannot make healthy sperm. These people are increasingly turning to donors of eggs or sperm and to surrogate mothers, who carry someone else's baby in utero, usually for a fee.

TABLE 13.1 ≋ BIRTH CONTROL METHODS

METHOD	HOW IT WORKS	FAILURE RATE (PREGNANCIES PER 100 WOMEN PER YEAR)	SIDE EFFECTS OR OTHER PROBLEMS
Behavioral Techniques			
Abstinence	No intercourse	0	Absolute abstinence results in zero pregnancies; in fact, however, up to 80 out of 100 women with no other method will get pregnant each year
Rhythm method	No intercourse during woman's fertile period (days 10–20)	2–30 or more	Often ineffective because eggs can be released before or after this period
Withdrawal (coitus interruptus)	Penis is withdrawn before ejaculation	16–23 or more	Depends on timing; semen may be released before ejaculation
Chemical Techniques			
Spermicides (foams, creams, vaginal rinses)	Chemicals introduced to vagina before or after intercourse kill sperm	3–15 or more	Applications often too little, too late, or not done at all, and sperm can survive
Physical Barriers			
Condom alone (some condoms are spermicidal)	Rubber sheath on erect penis catches (and kills) ejaculated semen	2–17	Effectiveness depends on how carefully used; condom can slip and allow semen to leak; allergies to condoms
Condom plus spermicidal foam	Same as above plus sperm-killing action of foam	0–2	Effectiveness depends on how carefully method is used
Diaphragm plus spermicidal jelly or cream	Round rubber dome covers cervix and holds sperm-killing jelly or cream, blocking sperm entry	2–10 or more	Effectiveness depends on how carefully both are used; some people are allergic to diaphragm or chemicals
Contraceptive sponge	Disposable sponge containing spermicide blocks cervix, absorbs and kills sperm	5–28	Effectiveness depends on how carefully used; some people are allergic to the sponge or chemicals
Sterilization			
Tubal ligation	Doctor ties off woman's Fallopian tubes, permanently blocking sperm passage	0–0.04	Slight chance operation will not completely block tubes; requires surgery
Vasectomy	Doctor severs and ties off man's vasa deferentia, permanently blocking sperm release	0–0.15	Slight chance procedure will not block vasa deferentia; procedure often performed in doctor's office; possible immune system reaction
Hormonal Strategies			
Birth control pills	Synthetic estrogens and progesterones prevent normal menstrual cycle from occurring, so eggs are not released	0.25–2	Effectiveness depends on not skipping pills; some women experience nausea, missed periods; increase risk of blood clots and strokes
Implantation Blockers			
Intrauterine devices (IUDs)	Prevent fertilized egg from implanting in uterine wall	1.5–4	Most no longer sold in U.S. because companies fear being sued for infections, scarring, and sterility risk

In addition, researchers have devised ways to freeze embryos. If the first attempt to artificially implant an embryo fails, physicians can retrieve another from cold storage and try again without having to repeat all the prior steps of in vitro fertilization. The first frozen-embryo baby was born in the United States in June 1986 and joins Louise Brown and more than a thousand other "test-tube babies" in a remarkable and rapidly growing club.

≋ PREGNANCY, HUMAN DEVELOPMENT, AND BIRTH

Whether an embryo is conceived the old-fashioned way or with the help of laboratory technicians and Petri dishes, it follows the same course: It resides in the womb for approximately nine months and grows from a single fertilized egg cell to a bouncing bundle of 200 billion cells. The unique set of genes established at conception programs the shape, distribution, and functioning of the billions of developing cells.

IMPLANTATION: THE EMBRYO SIGNALS ITS PRESENCE

The embryo begins to cleave as cilia in the oviduct sweep it toward the uterus (Figure 13.9a); by day 5, it is a spherical blastocyst with two cell groups: the embryo itself (the *inner cell mass*), and the feeding cells (the *trophoblast*). About six days after fertilization, when the blastocyst consists of 100 cells or so, it attaches to the uterine wall, secretes enzymes that break down a small portion of the lining, burrows in (*implants*), and at day 7 establishes the first physical bond between mother and young.

Trophoblast cells develop into the *chorion*, a fluid-filled sac that surrounds the embryo. The chorion is the embryo's three-way ticket to survival: It absorbs nutrients from the mother's blood and passes them on to the rapidly dividing embryo; it develops into the larger placenta that will sustain the embryo throughout the nine months of gestation; and it produces a hormone called *hCG* (*human chorionic gonadotropin*), which prevents a new menstrual cycle, the onset of which would flush the embryo from the uterus (see Figure 13.9a).

The production of hCG, the first major biochemical event of pregnancy, signals the embryo's presence and positively confirms pregnancy. Home pregnancy tests use a simple but ultrasensitive system for detecting hCG, and 97 percent of the time, they accurately reveal a pregnancy just days old.

The chorion grows and enmeshes with maternal tissue to form the dark red spongy *placenta* (see Figure 13.9b and c), which enlarges as the pregnancy continues and serves as the vital link between mother and embryo. The placenta is an exchange site where a thick tangle of embryonic blood vessels encounter blood-filled spaces in the uterine lining. Embryonic and maternal bloods do not mix, but nutrients and oxygen pass from the mother's blood across embryonic vessel walls and into the embryo's blood, and carbon dioxide and other wastes pass back in the reverse direction. After a few weeks, the placenta begins to make enough estrogen and progesterone to maintain the uterine lining and prevent menstruation.

DEVELOPMENTAL STAGES IN THE HUMAN EMBRYO

Even before a woman suspects she may be pregnant, the embryo has embarked on the early stages of development, which are similar to those of other vertebrates (see Chapter 12). Eight days after fertilization, some cells in the inner cell mass differentiate into ectoderm and endoderm layers and form cells that will enclose the *amniotic cavity*, a salty fluid-filled space that keeps the embryo moist and cushions it from blows. Later in development, the fetus sloughs off cells into this fluid, and a physician can collect these cells and analyze them for genetic disease (see Figure 11.15a).

In the third week of pregnancy, when the embryo is smaller than the length of this "l," its basic body plan has been established through gastrulation. The process is similar to that in other animals and creates the three main body layers we described in Chapter 12. By the fourth week, the *umbilical cord* has formed—a lifeline connecting the offspring to the placenta and maternal blood supplies. During the next eight weeks, the nerve cord forms and organs develop, transforming the tiny beanlike mass into an organism with characteristic human shape (Figure 13.10, page 232).

The neural tube rolls up midway through the third week of pregnancy, and a few days later, blocks of middle-layer tissue called *somites* begin to pinch off on both sides of the neural tube (Figure 13.10a); these blocks will differentiate into muscles, the backbone, and parts of the skeleton. Another tube—the developing heart—begins pumping blood early in the fourth week. About the same time, pouches poke out from the primitive gut, grow, and branch to form lungs, liver, and pancreas. By the end of the fourth week, the enlarging head bends toward the heart, and arms, legs, ears, and eyes have begun to form (see Figure 13.10a). During the fifth week, nostrils develop, paddle-shaped hands are shaped from buds, and elbows, upper arms, and shoulders take form. By the start of the sixth week, the 9 mm long embryo is

FIGURE 13.9 *Early Development and Implantation of the Human Embryo.* (a) In the human, the early stages of development take place as the fertilized egg travels down the oviduct toward the uterus. The egg starts to implant itself in the uterine wall about the sixth day after fertilization. By the tenth day, the chorion forms and gives off the hormone human chorionic gonadotropin (hCG), which maintains the corpus luteum's production of estrogen and progesterone and thus prevents the uterine lining from sloughing off in a new menstrual cycle. (b, c) After 60 days, the placenta is well established, and produces the hormones that are necessary to prevent menstruation for the remainder of the embryo's nine-month gestation. Narrow fingers of tissue, or chorionic villi, project from the chorion, and each eventually houses a tiny blood vessel. The maternal blood fills the spaces around the villi, and exchange of gases and materials takes place across the delicate layer separating the maternal and fetal blood supplies. Thus, the bloods never commingle.

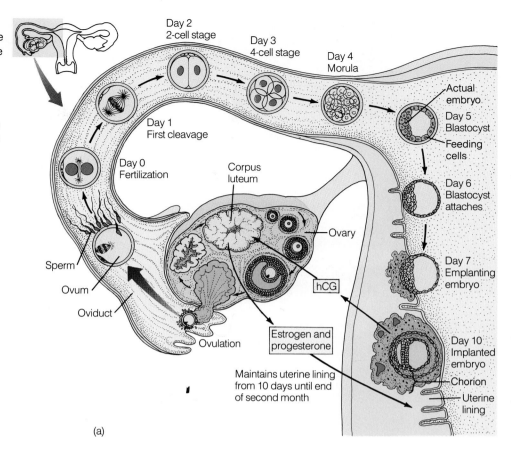

(a)

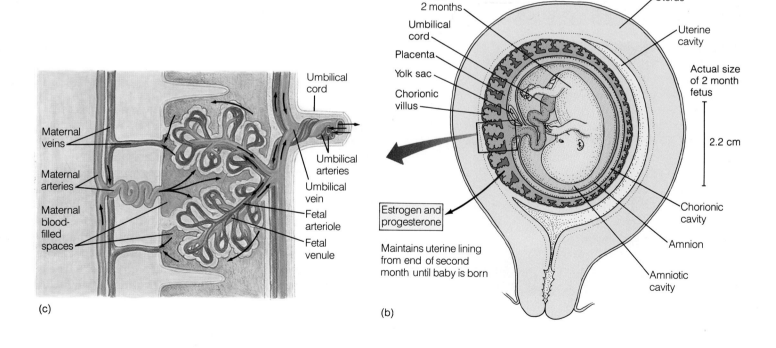

(c)

(b)

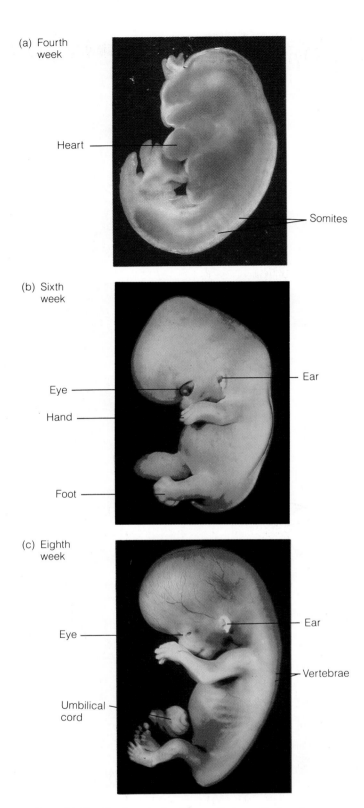

(a) Fourth week

Heart

Somites

(b) Sixth week

Eye

Hand

Ear

Foot

(c) Eighth week

Eye

Ear

Vertebrae

Umbilical cord

FIGURE 13.10 *The Developing Human Embryo.* Photos show recognizable details in the stages of the developing human embryo. (a) By the fourth week, the embryo has a future backbone and brain. (b) In the sixth week, eyes, ears, hands, and feet are beginning to form. (c) By the eighth week, the embryo has limbs, vertebrae, eyes, ears, and a jaw.

half head, with a vaguely alien face (see Figure 13.10b). As the first eight weeks end (see Figure 13.10c), the embryo, now called a fetus, has the rudiments of all its organs, including the start of its sex organs.

SEX DIFFERENTIATION: VARIATIONS ON ONE DEVELOPMENTAL THEME

The primary sex organs have an interesting—if somewhat startling—course of development. Before the eight-week mark, you cannot distinguish the genitals of a male embryo from a female one because sexual structures are *indifferent:* They have yet to complete differentiation and establish the embryo's gender.

The fascinating thing about sex differentiation is that in the absence of specific signals, these indifferent sexual structures become female. The core of each early gonad recedes, its outer shell develops follicle cells, and the organs become ovaries. Early sex ducts develop into oviducts and uterus, and the external genitalia follow suit (review Figure 13.5). In a sense, the female, with its automatic development, is the basic human gender.

If the embryo is male, however, its indifferent gonads will be made up of cells containing a Y chromosome, the core of the gonads will develop into seminiferous tubules, the shell will recede, and the gonads will become testes. Once differentiated, the testes begin to produce testosterone and another sexually significant hormone. The testosterone maintains the male duct system, while the other hormone kills the duct cells that could develop into oviducts and uterus.

FETAL LIFE: A TIME OF GROWTH

The pace of fetal change slows as the weeks pass from the first *trimester* (three-month period), with its rapid-fire developmental activity, through the second and third trimesters. A few finishing touches are added, such as hair, eyelashes, nails, and more recognizable facial features. But the main activity is growth. The fetus becomes 600 times heavier between the eighth week and birth at about the thirty-sixth week. The mother may feel flutters, rolls, kicks, and punches from within as the fetus shifts position and moves its limbs, starting at about three months. Subtler movements usually go unfelt. These include swallowing (four months), thumb sucking (five months), heartbeat detectable by a doctor's stethoscope (five months), handclasping (six months), and silent crying (eight months). The development of the fetus's body and behavior are programmed in the offspring's genes and usually proceed in a normal and healthy way. But factors that perturb the fetal environment can cause disruption.

SEXUALLY TRANSMITTED DISEASE: A GROWING CONCERN

One consequence of America's sexual revolution during the 1970s and 1980s has been a dramatic rise in the incidence of *venereal diseases*—diseases of the genital tract and reproductive organs caused by bacteria and viruses. These diseases are part of the broader category of *sexually transmitted diseases (STDs)*—infectious diseases of any body region that can be passed to a partner through sexual contact. A person with multiple sex partners has a substantial risk of contracting a sexually transmitted disease. Over 10 million cases are treated each year in the United States. And if undetected or untreated, such diseases can lead to severe complications.

The most common STD is *chlamydia*, an infection by the bacterium *Chlamydia trachomatis* picked up through sexual contact with an already infected person. A woman may have no symptoms of chlamydia at all, or she may experience pelvic pain, painful urination, vaginal discharge, fever, and swollen glands near the groin; a man may have a discharge from the penis or painful urination. The infection can be simply and effectively treated with antibiotics, but if undetected or untreated, it can lead to severe infection of the reproductive organs and even sterility. Because a person with chlamydia may have no symptoms, pregnant women can unknowingly pass the infection to

their newborns. This STD, in fact, is the most common infection in newborns, with 100,000 cases per year.

The second most common STD is *herpes genitalis*, caused by the herpes simplex virus type 2 (HSV type 1 causes cold sores and fever blisters). At least 20 million Americans have herpes, and 300,000 to 500,000 new cases arise annually. The virus causes watery blisters to form around the genitalia; these break and form painful open sores that eventually heal. The virus can lie dormant for weeks, months, or years. Then, stimulated by sunlight, emotional stress, or sexual intercourse, the virus can break out once again and cause a new cycle of pustules and sores. Right now, there is no cure for herpes, and only partially effective antiviral drugs are available. If birth coincides with an active herpes phase in the mother, the newborn can suffer death or damage to the brain, liver, or other organs.

The fastest-spreading, and in many ways most worrisome, STD is *venereal warts*, caused by the papilloma virus. These small, painless, cauliflower-like bumps grow around the sex organs, rectum, or mouth and are passed through skin-to-skin contact. Getting rid of the warts is usually no problem; they can be burned or frozen off or surgically removed. More ominously, however, researchers are finding that the papilloma virus (which can remain in the body even after the warts are removed) has been present in 90 per-

cent of the cervical cancer tissues studied so far.

Perhaps the best-known STDs are *gonorrhea* and *syphilis*. Both are caused by microorganisms; both are contracted through sexual activity with an infected person; both can have mild initial symptoms (a discharge or painless sore) or no symptoms at all; and both can be successfully treated with antibiotics. As in chlamydia, failure to find and treat gonorrhea at an early stage can lead to severe infection of reproductive organs or sterility. Failure to treat syphilis can also result in widespread damage to the heart, eyes, and brain and can result in severe damage or death to an unborn child.

Other STDs include pubic lice (crabs), scabies (parasites that burrow under the skin), certain types of vaginal yeast infections, trichomonas (flagellated protists that infect the vagina or penis), and acquired immune deficiency syndrome (AIDS; see details in Chapter 22).

Clearly, sexually transmitted diseases have become a serious threat. Planned Parenthood recommends that sexually active adults (particularly those with numerous sex partners) use condoms to prevent the passing of infections; be alert to sores, bumps, discharges, painful urination, or pelvic pain; and get tested regularly for STDs. This is especially important before or during pregnancy.

MOTHER'S CONTRIBUTION TO THE FETAL ENVIRONMENT

The development of a human fetus requires a real partnership. Hormones from the embryo help establish and maintain the pregnancy; the mother's body changes continuously to accommodate the growing young; and the mutually produced placenta acts as a support system as well as a barrier to many harmful substances. Because the foods, drugs, and other chemicals the mother takes into her body profoundly affect the baby, she must be especially careful during pregnancy.

For proper growth of muscles and bones, the fetus requires a constant supply of protein and calcium, as well as fatty acids, carbohydrates for energy, vitamins, and minerals. Because the mother's diet must provide these, obstetricians usually advise pregnant women to drink lots of milk, eat plenty of protein, and take vitamin and mineral supplements. Protein intake is especially important in the final trimester, when the fetus experiences the greatest expansion in brain size. A maternal weight gain of 25–35 lb is now believed appropriate for most women to help prevent premature or underweight infants, with their greater susceptibility to infections and breathing problems.

A mother's life-style habits, such as smoking, drinking, or drug use, can have severe repercussions for the fetus. Babies born to women who drink substantial amounts of alcohol during pregnancy suffer from fetal alcohol syndrome. They show greater incidence of mental retardation, emotional abnormalities, cleft palates, underdeveloped hearts, and facial anomalies. Many obstetricians suggest that their patients avoid alcohol altogether during pregnancy. Among mothers who smoke, miscarriage (premature expulsion of the fetus) is much more likely, and the babies, suffering from fetal tobacco syndrome, are more likely to have a low birth weight and thus greater susceptibility to respiratory disease and sudden infant death syndrome (suffocation during sleep). Overall, infants with fetal tobacco syndrome have a death rate up to twice as high as children from nonsmoking mothers. Mothers who take amphetamines or cocaine risk infants with neurological defects, and mothers who take heroin or other narcotics often give birth to addicted infants.

Some of the saddest chapters in medicine revealed that prescription drugs also carry a risk of fetal damage. In the fall of 1960, doctors in Europe and America witnessed the sudden appearance of a new birth defect: Affected infants were mentally and emotionally normal, but they had shortened, twisted legs and hands growing directly from the shoulders (Figure 13.11). After a two-year epidemic of the defects, researchers traced the cause to a new and popular sedative and antinausea drug called thalidomide. While this drug calms adult nerves, it alters the embryonic nerves that help direct proper limb development and growth. Ironically, thalidomide does not cause these effects on mice or rats, and so its devastating effects were not discovered during premarketing laboratory tests.

Premarket drug testing has become much more stringent in recent years, but pregnant women must still be wary of exposure to drugs, other chemicals, and certain viral diseases such as German measles (rubella). Pregnancy is clearly a time when a woman must take special care of her general health and nutrition—and, indirectly, her baby's.

FIGURE 13.11 *Thalidomide and the Fetal Environment.* This child is learning to compensate for his developmental alteration.

CROSSING THE THRESHOLD: THE MAGIC OF BIRTH

As a pregnant woman waits, hormones from the fetus prepare it and the mother for the impending labor and delivery. The fetal lungs grow, and special valves develop in fetal blood vessels, readying them for the newborn's first breath. Brown fat is stored around the neck and down the fetus's back, and this will produce heat after the baby is expelled from the warm uterus. Special reserves of carbohydrates are laid down in the heart and liver to tide the baby over until it can suckle milk, and the placenta makes and secretes a hormone that prepares the mother's breasts to produce milk. And like an exercise coach, the fetus generates an unknown type of signal that causes the uterus to contract periodically to build strength for the coming expulsion process.

Sometime around 260 days after conception, the fetus produces another substance that directs the mother's body to release hormones that induce strong contractions in uterine muscles (Figure 13.12a). During *labor*, the cervix

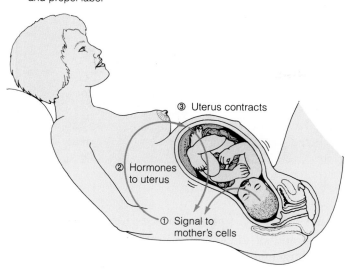

(a) Hormones initiate and propel labor

③ Uterus contracts

② Hormones to uterus

① Signal to mother's cells

(b) Shortly after birth

FIGURE 13.12 *Birth: Fetal Hormones Trigger Contractions and Delivery.* (a) Hormonal signals trigger the events of labor and delivery. The fetus sends a signal to placental cells of the mother (1), which release hormones (2) that stimulate muscle contractions in the uterus (3). These contractions send more signals to the mother's cells, causing her to release more hormone. The cycle escalates and continues until the baby is expelled through the birth canal. (b) Finally, the mother and newborn can begin to recover from the arduous process.

widens as the contractions grow longer, stronger, and more regular. Each contraction starts at the upper end of the uterus and moves toward the cervix, pulling that opening wider and pushing the baby toward the vagina, or birth canal. The squeezing eventually causes the amniotic sac to burst (the "water breaks"), and as the cervical opening reaches a width of about 10 cm, delivery is usually only minutes away. With considerable pushing of the abdominal muscles, the mother is able to force the baby's head past the pelvic opening, and then the shoulders and hips emerge. The baby, still attached to the uterus via the umbilical cord, takes its first gasp of air. Soon the blood vessels in the cord cease pulsing, and attendants sever it (Figure 13.12b). Maternal blood vessels that supply the uterus clamp down to prevent exces-

sive blood loss, and with a few more uterine contractions, the placenta is expelled.

Recent studies show that the heavy stress of being born is actually beneficial in that it causes stress hormones (adrenaline and noradrenaline) to surge through the infant's body, clearing its lungs, promoting normal breathing, mobilizing stored energy, and sending extra blood to the heart and brain. Although the transition from being a fetus immersed in fluid to an air-breathing land animal is difficult, most babies make it quite well. At the same time, the mother goes through a physical transition to nonpregnancy, and both she and the father must adjust psychologically to being parents.

GROWTH, MATURATION, AND AGING: DEVELOPMENT CONTINUES

While birth signals the end of fetal development, it is not the end of growth and change (Figure 13.13, page 236). People continue to traverse a series of more gradual thresholds as they move from infancy to childhood, puberty, adulthood, and old age.

INFANCY AND CHILDHOOD: GROWING UP FAST

Within hours after a horse gives birth to a foal, the newborn animal is standing, walking, nuzzling, and following its mother. A newborn human infant, however, is helpless—unable to feed, clean, or defend itself, sit up, talk, or walk about. *Infancy,* the period from birth to about age 2, and *childhood,* from age 2 to about 12, are times of profound physical and mental change and development, during which the child can grow to 70 percent of his or her adult height and weight, quickly amass a large vocabulary of spoken and written words, and develop coordination, perception, strength, agility, personality, and social skills.

Growth in a human follows a regular pattern that is quite evident in the infant and child. Before birth, growth is fastest in the head and body center and slowest toward the periphery. At birth, the brain is the most developed organ and is 25 percent of its final adult weight. By 6 months after birth, the brain has reached 50 percent of adult weight; by 30 months, 75 percent; and by age ten, 90 percent. Body weight lags far behind brain weight, advancing from 5 percent of adult weight at birth to only 50 percent at age ten. In keeping with the center-outward growth pattern, the child's trunk grows faster than its

FIGURE 13.13 *Development Continues Long After Birth: The Many Faces of George Burns.* The actor at ages 37, 58, 77, and 90.

arms and legs, and it gains coordination over gross limb movements before finger and toe movements.

The details of childhood development are too numerous to consider here, but we summarize a few of the major milestones of physical and mental growth in Figure 13.14. By age 12, a girl or boy has arrived at the threshold of adolescence.

PUBERTY: SEXUAL MATURITY

Adolescence is marked by the most dramatic physical changes since fetal life: *puberty*, the maturation of the reproductive system and the development of secondary sexual characteristics. In a teenage girl between about 11 and 13 years of age, estrogen produced in the ovaries enters the bloodstream and causes target tissues (around the nipples, on the hips) to differentiate into their sexually mature state. The visible results are the deposition of fat in hips and other areas (creating body contours) and the enlargement of the breasts. Normally accompanying these changes are a growth spurt that brings the girl close to her adult height, the onset of the menstrual cycle, and a heightened interest in sex.

In the male between ages 13 and 15, testosterone produced in the testes enters the bloodstream, reaches target tissues in the genitalia, face, and other parts of the body, and stimulates the development of body and facial hair, enlarged muscles and external genitalia, and a lower voice. This is usually also accompanied by a growth spurt and an interest in sex.

ADULTHOOD: THE LONGEST STAGE IN THE LIFE CYCLE

With modern nutrition, sanitation, and health care, physical maturity can begin at the end of adolescence and last for 50 to 75 years or more. Some additional development of adult features—more shifting of fat, further enlargement of sex organs, and continued growth of body hair—extend into the early 20s. But after that, all the body's organ systems are set and perform at their peak efficiency for about a decade. Sometime after age 30, the process of *aging* begins, a progressive decline in the maximum functional level of individual cells and whole organs. Very gradually—almost imperceptibly from year to year—the muscles lose strength, breathing and circulation become less efficient, hair follicles lose their ability to make pigment, the skin becomes less elastic, and reproductive functions decline. About age 50, females experience *menopause*, a cessation of the menstrual cycle, while males may lose some *potency*, or ability to maintain an erection. Sometime after age 60, people reach **senescence,** or old age, when the decline in cell and organ function is less gradual and more profound. Many researchers have studied aging and proposed intriguing theories about its causes.

WHAT CAUSES AGING?

Researchers group most hypotheses about aging into two opposing categories: genetic clock hypotheses versus wear-and-tear hypotheses. Many biologists support the idea of a genetic clock, arguing that there is a timetable for aging and death specified by the genes. Just as the genes regulate the timing of organ formation, cell death in the formation of fingers or toes, and sexual maturation, they may also regulate when our various organs and systems will wear out.

There is tantalizing evidence for these genetic clock hypotheses. First, cells seem to have preset limits to the number of times they can divide. Skin cells from a human infant will divide about 50 times and then stop, while similar cells from a 90-year-old will divide only a few times. Recent experiments show that genes on chromosome 1, the largest human chromosome, may be responsible for this programmed aging.

Second, organs seem to age in a preprogrammed way. Each year, starting at about age 30, we experience a 1 percent decline in the maximum function of various organs. This includes lung capacity, the amount of blood the heart pumps, the strength and coordination of muscles, and the highest sounds we can hear.

Third, certain genetic conditions can bring about symptoms of aging. Children with *progeria* (early aging) begin to lose their hair, show wrinkles, and experience arthritis and heart attacks by age five or six (Figure 13.15, page 238). These children, however, do not suffer the whole spectrum of age-related illnesses, which includes cataracts, diabetes, and cancer. Certain other people inherit a dominant mutation that appears to cause Alzheimer's disease, a kind of senility. The brains of Alzheimer's patients shrink and contain far greater numbers of tan-

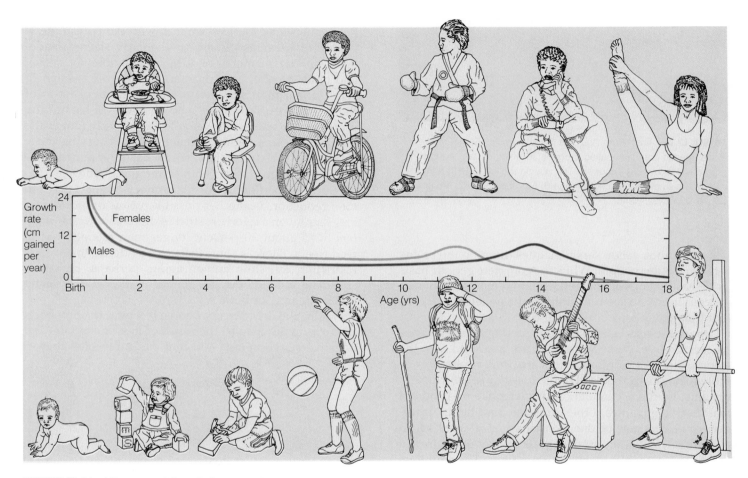

FIGURE 13.14 *Milestones of Growth.* Between birth and age two, the child experiences the major stages of brain growth. Height increases 7–8 in. per year. The child learns to sit up, stand, walk, maintain eye contact, babble, and imitate. As powers of mental reasoning begin to emerge, the child learns the elements of grammar, acquires a small vocabulary of spoken words, develops some degree of eye-hand coordination, and starts to feed itself. Between ages two and five, height increases 3–4 in. per year, and the child reaches half its adult height. The child also attains control over elimination and develops many fine motor skills (painting, using scissors, building puzzles) and gross motor skills (running, climbing, riding a tricycle, playing ball). The child completes the basic process of language acquisition at this time and has a vocabulary of several thousand words. The child also acquires more reasoning and problem-solving skills and becomes more social. Between ages 5 and 12, the child's height increases 1 in. per year, the limbs elongate, the muscles grow, and coordination improves, allowing the child to roller skate, swim, dance, play baseball, and so on. The child adds to its vocabulary, reasoning skills, and problem-solving ability; a sense of causality and morality begins to emerge; and by age 12, this individual lies at the threshold of adolescence, sexual maturity, and attainment of adult size.

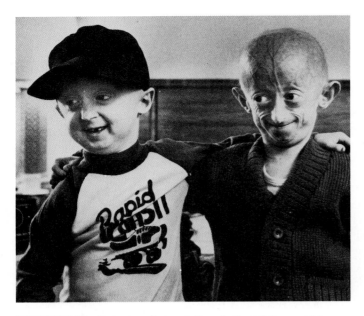

FIGURE 13.15 *Premature Aging: A Genetic Basis?* Some children have a genetic condition called progeria, in which certain signs of aging develop by age five or six. The boy on the left (age 9) from Texas and his new acquaintance (age 8) from South Africa are just about to tour Disneyland.

gles and knots, called senile plaques, than do the brains of normal older people. While the complete genetic basis for diseases like progeria and senility remains unsolved, the conditions do seem to support the genetic clock concept that genes control when and how we age.

Other biologists believe that our preprogrammed genes are less likely to cause aging than are wear and tear: the accumulation of random errors in DNA replication or protein synthesis that lead to the disarray of information systems; or the accumulation of metabolic by-products that disable enzymes, other proteins, and lipids. Informational errors may be due to environmental insults such as sunlight, radiation, or chemicals or may result from a buildup of internal metabolic by-products. The relentless piling up of metabolic garbage probably interferes with normal cell function, including the flow of information from DNA to RNA to protein, and perhaps contributes to the 1 percent per year decline in organ function we all experience.

AGING GRACEFULLY

Whatever the major causes of aging may be, there are several good reasons to think that most of us can and will live long, healthy lives despite the inevitabilities of decline and mortality.

First, compared to most animals, humans have very long *life spans*, or maximum potential ages. People can live upward of 100 years. Larger mammals, in general, tend to have longer life spans, perhaps owing to slower metabolisms and less wear and tear.

Second, human *life expectancy*, the maximum probable age a person will reach, has never been higher (Figure 13.16). Spurred by research in basic biology, advances in agriculture, emergency medicine, obstetrics, public sanitation, and immunization programs have steadily increased average life expectancy in North America to its current levels of 78 years for women and 71 for men.

Third, studies show that adhering to certain health practices can increase life expectancy significantly. Researchers have found that a man who eats regular meals (including breakfast), exercises regularly, sleeps an adequate amount, maintains his ideal weight, does not smoke, and limits alcohol consumption can live an average 11 years longer than a man who follows three or fewer of those practices. A woman who observes all six healthy practices can add seven years to her already longer life expectancy. As an added bonus, those observing these six positive life-style habits remain healthier during their extra years than cohorts with fewer good habits.

Fourth, experiments with laboratory animals point to food reduction as another possible means of prolonging life. Rats given a calorie-restricted diet *with adequate nutrition* lived about 30 percent longer than rats allowed to feed freely. There is some collateral evidence that caloric restriction can slow human aging, perhaps, biologists speculate, by slowing the genetic clock, by decreasing metabolic rate, or both.

Finally, the dread many people feel over the prospects of growing old may be the result of misconceptions and negative stereotyping. A recent study revealed that 95

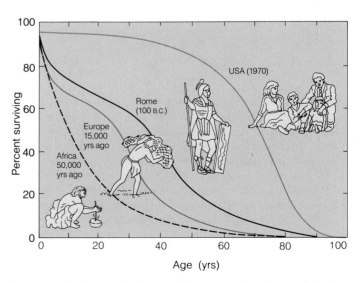

FIGURE 13.16 *Expect a Long Life.* Life expectancy in the United States is high for both females and males. This is a relatively new development; historically, human life expectancy trailed off rapidly after 40.

FIGURE 13.17 *Aging Gracefully.* Good habits, good health, and good times can help a person live a long and happy life.

percent of the elderly live independently, not in institutions; most are in regular contact with their families, not isolated or lonely; most are vigorous and active, not frail and sedentary; and most are financially secure, well educated, and integrated into their communities. Clearly, the last decades of life can hold great satisfactions rather than dependence and illness (Figure 13.17). It all seems to depend on the habits and support systems one establishes much earlier in life.

≋ CONNECTIONS

In this chapter, we followed the human life cycle from sperm and egg production in the male and female reproductive systems, through mating, fertilization, fetal development, childhood, sexual maturation, and adulthood. We saw that the developing embryo undergoes remarkable changes during the first few weeks of life, including the establishment of all major body organs and the differentiation of gender. We also saw that hormones play a key role, carrying signals from one cell group to another and thereby regulating the timing of ovulation or sperm production, as well as menstruation or maintenance of the uterine lining during pregnancy.

Reproduction solves a central problem for the human species. It is our only means of continued existence as

well as our anchor to parents, grandparents, great grandparents, ancestors, and progenitors. How far back does the unbroken chain extend? And where did it all start? We may never know the answers for certain, but the laws of nature, operating today as they did on earth billions of years ago, allow us to hypothesize how life could have started and how lineages could have begun. Chapter 14 bridges the gap from genes to the diversity of life forms by examining modern explanations for the origins of life.

NEW TERMS

cervix, page 224
contraception, page 228
estrogen, page 226
feedback loop, page 224
follicle, page 224
in vitro fertilization, page 221
menstrual cycle, page 225
ovary, page 224
ovulation, page 224
penis, page 223

progesterone, page 226
prostate gland, page 222
semen, page 224
senescence, page 236
testis, page 222
testosterone, page 224
urethra, page 223
uterus, page 224
vagina, page 225
vas deferens, page 222

STUDY QUESTIONS

REVIEW WHAT YOU HAVE LEARNED

1. Outline the feedback loop that maintains the sperm supply.
2. Where is an egg produced? What pathway does the egg follow from site of production to the outside world?
3. Of the 400 million human sperm released during sexual intercourse, how many will fertilize the egg? Explain.
4. What are some causes of infertility in women? In men?
5. Why is the chorion important to the developing embryo?
6. Does an embryo's blood mix with its mother's in the placenta? Explain.

APPLY WHAT YOU HAVE LEARNED

1. Why might a pregnant woman suffer a miscarriage if her body does not produce enough progesterone?

FOR FURTHER READING

Keeton, K., and Y. Baskin. "Birthtech." *Omni* 8 (December 1985): 91–98.

Moore, K. L. *Essentials of Human Embryology.* Toronto: Becker, 1988.

Ulmann, A., G. Teutsch, and D. Philibert. "RU486." *Scientific American* 262 (June 1990): 18–24.

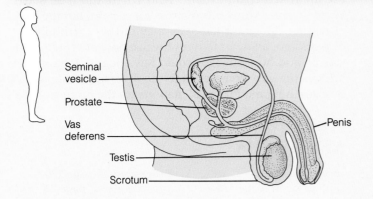

Seminal vesicle

Prostate

Vas deferens

Testis

Scrotum

Penis

① The male reproductive system makes and transports sperm.

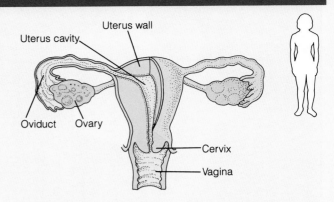

Uterus cavity

Uterus wall

Oviduct

Ovary

Cervix

Vagina

② The female reproductive system produces and transports eggs; it also supports the development of the embryo and gives birth to babies.

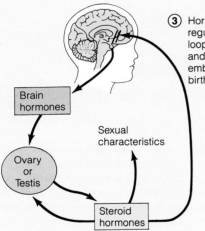

③ Hormones propel self-regulating feedback loops that control egg and sperm production, embryonic development, birth, and milk production.

Brain hormones

Sexual characteristics

Ovary or Testis

Steroid hormones

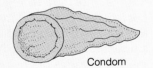

Birth control pills

④ Manipulating hormone release or other key events can make it possible for couples to block fertility or overcome infertility.

Condom

⑤ Pregnancy is a partnership between mother and fetus.

⑥ Birth is the start of an independent life and of development outside the womb, which includes growth, maturation, and aging.

PART THREE

LIFE'S VARIETY

CHAPTER 14

ORIGINS OF LIFE AND ITS DIVERSITY

THE EARLIEST MICROBES

What do sewage sludge, swamp mud, rotting manure, and cow breath have in common? The most obvious shared characteristic is an awful smell. The smell is caused by a particular kind of bacterium that lives in the cows, the mud, and the sludge and that gives off methane gas and hydrogen sulfide, with their unpleasant odor of rotten eggs and decay. The methane-producing bacteria are of special interest to us in this chapter because they may be the closest living relatives to some of the very first life forms on earth, providing clues to where we come from and to the possibility of life on other planets.

More than a dozen species of bacteria produce methane as a by-product of their metabolism (Figure 14.1), growing and thriving only where there is little or no free oxygen. Without the activity of these methane-generating bacteria, people who operate sewage treatment plants could not process sewage as efficiently or collect methane gas to burn for heat and electricity.

The methane generators are a comparatively small and restricted group today because oxygen-free environments are now relatively rare on our planet. But many biologists believe that early in the earth's history, when the entire planet was devoid of free oxygen, methane bacteria and similar oxygen-independent species were the dominant—and probably the only—life forms to exist.

Two themes will emerge as we consider the origins of life and its diversity. First, life arose as a direct result of the physical conditions on our planet, the molecules that were present, and the ways those molecules act and interact. Although we will probably never know exactly how life originated 3.5 billion

years ago, we can try to reconstruct the conditions, simulate the processes, and gather fossil and chemical evidence.

Our second theme relates to what happened once life arose. Life and earth evolved together. The biological activities of the early organisms caused changes in the earth's atmosphere and land; and geological activities that shaped the earth strongly affected living things. The result of this evolving together, or *coevolution*, was a multiplicity of habitats and the diversification of living organisms to more than 10 million species.

As we examine the current scientific theories for how life evolved and diversified on the planet earth, we will answer the following questions:

- How did our planet form, and what made it hospitable to life's origin?
- What steps might have led to the emergence of living cells?
- How did life and earth coevolve?
- How do scientists use their classifications of living species to understand evolutionary history?

FIGURE 14.1 *Methane-Producing Bacteria: Similarities to the Earliest Microbes.* Rod-shaped cells of the bacterium *Methanobacterium ruminantium* inhabit a cow's intestines, live without oxygen, and produce methane gas. These bacterial cells are magnified about 30,000 times.

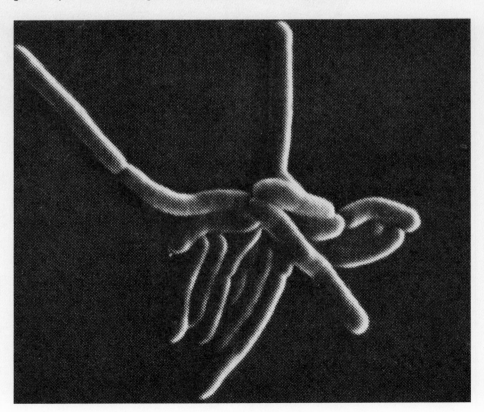

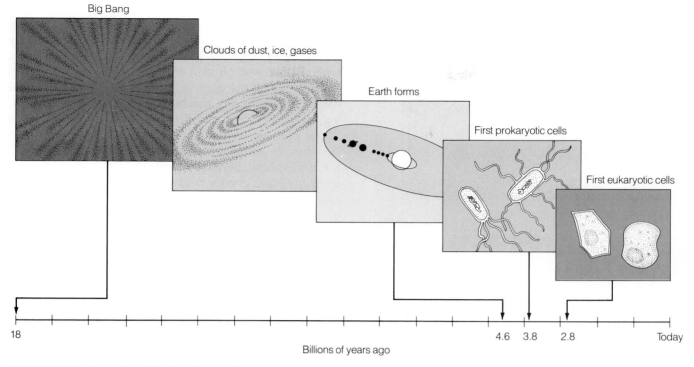

FIGURE 14.2 *Formation of the Solar System.* Billions of years after the Big Bang, a monstrous cloud of gases and dust condensed to form the sun and planets of our solar system, including the earth. This figure shows a cosmic time frame, from the Big Bang to the first cells.

≋ EARTH AS A STAGE FOR LIFE

One of the twentieth century's most dramatic achievements, the exploration of space, has provided tremendous new knowledge about our solar system's central star, the sun, and the nine planets that orbit it. The more we explore and learn, however, the lonelier we feel, for many astronomers believe that we earthly organisms inhabit the solar system alone. Why did living things arise and flourish here and yet not originate (or at least not survive) on neighboring planets?

HOW THE EARTH FORMED: FROM BIG BANG TO BIG ROCK

Most astronomers believe that the universe began with the so-called Big Bang, an event about 18 billion years ago during which an enormous explosion created all the matter in the universe, 99 percent of which is hydrogen and helium atoms. Radio astronomers have detected cosmic radiation throughout space that appears to be left over, even now, from the Big Bang. In our own solar system, clouds of hydrogen and helium orbiting the sun cooled and combined into the planets, moons, and countless smaller bodies. Geologists have set the time of earth's formation at 4.6 billion years ago (Figure 14.2).

A BLACK AND BLUE WORLD

After its formation, the earth is thought to have been a huge barren ball of ice and rock, devoid of oceans and atmosphere. Deep within the planet, however, radioactive elements released energy that combined with the crush of gravity to heat the rock slowly into a molten mass that did not cool for several hundred million years (Figure 14.3a, page 244). As the cooling took place, a dense core of iron, nickel, and other heavy elements formed, surrounded by a liquid outer core, a thick mantle of hot rock, and, at the surface, a hardened crust of cool black rock. Molten rock frequently erupted through this surface, building cinder cones that punctuated the flat black landscape, spewing lava, and issuing clouds of gases from the planet's interior. The earth's first atmosphere began to form from these vapor clouds, and gravity held a blanket of gases including carbon dioxide, water vapor, nitrogen, hydrogen sulfide, and even traces of methane as the first air; there was no free oxygen. Moisture in this air fell to the earth's steaming surface as rain, then evaporated and fell again, eventually cooling the surface and creating a shallow ocean. The ocean covered the entire globe and was broken only by the sharply rising black cones spewing forth more lava and gases.

(a)

(b)

FIGURE 14.3 *The Primordial Earth.* (a) A molten scene about 4 billion years ago, as the forces of gravity and radioactive decay melt the initial ball of rock and ice and create a red inferno of lava that will boil and smoke for millions of years. (b) By about 3.5 billion years ago, the planet has cooled, and the landscape is nothing but somber shades of black and blue. (Artist's conceptions.)

Because of the volcanic ash in the air, the sky would have looked pale blue, as it sometimes does today (Figure 14.3b); it would also have been dotted with clouds and frequently streaked by flaming meteors and asteroids—small hunks of primordial rock hurtling in from space. This planetary study in black and blue, with its oxygen-free air, was the stage on which life made its first appearance.

EARTH'S ADVANTAGEOUS PLACE IN THE SUN

Life as we know it—based on liquid water and carbon compounds—could have arisen only on a planet with sufficient amounts of both. Mercury and Venus, the sun's closest satellites, are blisteringly hot, with daytime surface temperatures higher than a self-cleaning oven's.

Under such conditions, all water occurs as vapor and all carbon compounds as inorganic gases. The earth orbits between Venus and Mars—a planet that may once have had flowing water and living things, but is now barren, with all of its water frozen and most of its carbon trapped in rocks. Of all the planets in our solar system, only the earth had a winning combination of composition, geological activity, size, and distance from the sun. Because of our planet's molten core and volcanic eruptions, carbon dioxide and water vapor were (and are) thrown into the atmosphere. Because of our planet's size and distance from the sun, gravity held the gases like an encircling blanket, which, in turn, trapped enough of the sun's heat energy to keep the surface temperate and most of the water liquid.

ANCIENT EARTH AND THE RAW MATERIALS OF LIFE

One of the biggest puzzles in the study of life's origins concerns the availability of organic building blocks for the earliest cells. Did organic subunits exist on the black and blue world we have described, and if so, where did they come from?

Astronomers have recently discovered enormous clouds of organic molecules in the outer reaches of our galaxy and think that such clouds could take part in the formation of planets (Figure 14.4). They also believe that some organic compounds were very likely present from the earth's beginning.

A second source of organic raw materials might have been a class of meteorites composed of nondescript dark gray stones; they formed when the planets arose, and they continue to hurtle through space, occasionally falling to earth even now. Chemical analysis of a single such meteorite has revealed all of the nucleotide bases found

FIGURE 14.4 *Organic Raw Materials of Life.* Clouds of organic molecules aglow in the light of Orion nebula M42 take part in a planet's formation.

FIGURE 14.5 *The Urey-Miller Experiments: Simulations of the Earth's Primordial Conditions Produce Biological Subunits.* Stanley L. Miller (pictured) and Harold C. Urey, working at the University of Chicago in 1955, designed a simple system to re-create the conditions on the early earth. They filled a flask with gases they thought might have been present in the atmosphere 4 billion years ago, then shot bolts of electric current through the gaseous mixture to simulate primordial lightning storms. They continued the experiment for a week and watched as the water in the "shallow sea" (a connected flask below the gas chamber) grew pink and then dark red with organic material. They analyzed the contents of this liquid and found a high concentration of amino acids and sugars. This experiment and dozens of others like it altering only the composition of the "atmosphere" and the source of energy (ultraviolet light, heat, radioactivity, and so on) led to the "primordial soup" theory. This is the idea that life arose in a rich broth of biological subunits in some warm, shallow ocean basin on the primitive planet. As the text explains, this is just one of the theories for how life began on earth.

in DNA and RNA. Some scientists hypothesize that millions of tons of such meteorites struck the earth during its early history, perhaps providing organic compounds.

Finally, many scientists point out that natural energy sources—lightning, the ultraviolet component of sunlight, heat from volcanic activity—could have driven the energy-requiring reactions that turned atmospheric gases into the subunits of biological molecules. Laboratory experiments have shown that organic compounds form easily under conditions probably close to those found on earth more than 4 billion years ago (Figure 14.5). These

organic molecules may have accumulated in pools and formed what some investigators think of as a "primordial soup." Many biologists conclude that the raw materials of living things did exist on the early earth and could have formed from molecules originally present in the atmosphere and from the energy sources and physical conditions present at the time. The stage was then ready for the drama of life to begin.

≋ THE UNSEEN DRAMA: FROM MOLECULES TO CELLS

A record of the chemical events leading to life's emergence may well have been laid down in sediments that solidified into rocks, starting about 4 billion years ago. However, we will almost certainly never find that fossilized sequence, because the earth was heavily bombarded by giant meteors for the first 800 million years of its history. Every square kilometer of the earth's initial crust probably melted during these impact events, and as a result, the oldest earth rocks that can ever be discovered are probably no more than 3.8 billion years old.

Now here is the fascinating—and frustrating—thing: Fossils of the earliest cells, similar in many ways to the methane-generating bacteria mentioned earlier, have been uncovered in rocks nearly 3.5 billion years old, and cellular traces have been found in rocks 3.8 billion years old. Clearly, life emerged sometime before that period, but the evidence melted eons ago! Biologists, therefore, must be content to simulate, experiment, and speculate on the steps in that emergence with little hope of ever finding irrefutable proof of how it actually happened. Here we outline one possible scenario for how life arose: five key steps that could have led to the emergence of life (Figure 14.6, page 246).

FIVE POSSIBLE STAGES IN THE EMERGENCE OF LIFE

■ **1. Polymers Form as Molecules Combine into Long Chains** Biologists believe that seawater containing organic precursors like amino acids or nucleotides may have collected in ancient tide pools, become concentrated via solar evaporation, and finally dried in the hot sun or become frozen at night, with polypeptides and RNA molecules resulting. Experiments show that these harsh conditions can cause subunits to join together in chains and that clay or other minerals can sometimes speed up the reactions.

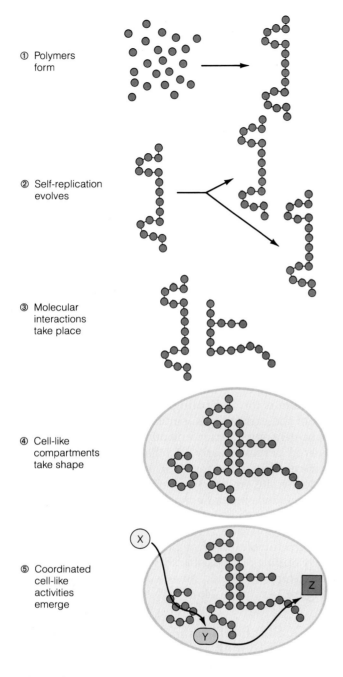

① Polymers form

② Self-replication evolves

③ Molecular interactions take place

④ Cell-like compartments take shape

⑤ Coordinated cell-like activities emerge

X

Y

Z

FIGURE 14.6 *Five Theoretical Steps in the Emergence of Living Cells.* No one knows exactly how living cells arose, but biologists propose that polymers such as polypeptides and polynucleotides must have formed from monomers and that the polynucleotides (primitive RNAs) began to self-replicate. Interactions between these RNAs and polypeptides (simple enzymes) took place, so that the information encoded in the base sequences of the RNA molecules came to specify an amino acid sequence in the polypeptide chains. Eventually, cell-like compartments formed around the interacting molecules, and finally, coordinated cell-like activities emerged inside the "protocells," including simple metabolic pathways that would allow the compartments to take up and use energy from the environment.

■ **2. Molecules Begin to Copy Themselves** Most biologists agree that the first nucleotide chains would have included simple, single-stranded RNA, and experiments show that strands of RNA can make copies of themselves. If RNA is added to a solution containing the bases A, U, G, and C, new complementary RNAs sometimes form. Next come complementary copies of these new RNAs, and this second generation of copies consists of exact copies of the original RNAs. What's more, certain RNAs (referred to as *ribozymes*) can splice out pieces of themselves that in turn act as primitive enzymes that facilitate the breaking and rejoining of other RNA molecules. Perhaps in this way, simple genes made of RNA formed and copied themselves (replicated) in quiet tide pools or in soggy clays. Biologists sometimes refer to this stage as "the RNA world."

■ **3. Molecules Evolve Interactions That Transmit Information** We know today that genetic information flows from genes to proteins, but how did the crucial step from random RNA structure to information-containing RNA structure to association with proteins come about? Biologists do not really know, even though this is *the* crucial step, the coevolution of chicken and egg, as it were. Researchers have observed that amino acids and nucleotides in watery solutions do tend to associate with each other; there is even a plan to search deep-sea vents for evidence of such associations taking place today (Figure 14.7). Over the course of millions of years of such associations, certain nucleotide sequences must have come to signify specific amino acid sequences, and both types of molecules somehow took on evolutionary "meaning": Their shapes and interlinked activities were selected (perpetuated) by natural chemical processes. As modern-day organisms confirm, a stabler storage form for genes—DNA—eventually emerged, and the storage and use of genetic information came to include RNA, DNA, and proteins.

■ **4. Cell-Like Compartments Take Shape** There is very clear evidence that polypeptides and phospholipids spontaneously form tiny spheres under certain conditions of heat, dryness, and pH (Figure 14.8). Some biologists have suggested that such spheres may have occasionally enclosed self-replicating associations of genes and proteins. Some of these compartmentalized systems, or "protocells," may have been selected for—and thus perpetuated through time—on the basis of total performance: faithful replication, absorbance of raw materials, and so on. All of today's organisms show similarities in the sequences of certain RNA molecules. For this reason, some biologists suggest that all living organisms originated from one type of these early cells designated the *progenote*—the earliest common ancestor of all life.

FIGURE 14.7 *The Search Continues at Deep-Sea Vents.* In 1979, two oceanographers in a tiny submarine called *Alvin* descended 2500 m (more than 1½ miles) into the deep ocean abyss off the southern tip of Baja California and made a spectacular discovery: they found a field of tall chimneys spewing great plumes of hot, mineral-laden water. The chimneys, called *hydrothermal vents,* lie above rock beds crisscrossed by tiny cracks and warmed by underlying zones of hot, melted rock. Seawater percolates down through the cracks, is heated, then seeps back upward through the chimney, bearing hydrogen, methane, and other compounds that can serve as precursors to the organic building blocks of life. Strange communities of marine animals surround the vents, including clams, mussels, and large tube worms whose bodies are packed with bacteria such as methane generators. One researcher has suggested that submarine hot springs were the site of life's origins and that the bacteria found there recently might be descendants of the earliest cells. Furthermore, he suggests that a sequence of chemical events leading from inorganic precursors to living cells could still be going on today in the dark, hot recesses of the abyssal vents, and some type of self-replicating RNA molecules might even be found there. The answer awaits some future deep-sea mission.

■ **5. Coordinated Cell-Like Activities Emerge** Somehow, the protocells came to have metabolic pathways, perhaps as a result of competition between compartments for the limited raw materials of the surrounding environment. With the competition came selection for series of enzymes that could modify more abundant materials into less available ones. In this way, metabolic pathways would have evolved with the ability to harvest energy from chemical bonds and synthesize new materials. At some point, the coordinated activities—replication, protein synthesis, energy harvest, new synthesis, and repair—became so closely integrated and interdependent that the "compartments" were indistinguishable from true biological units—living cells—with all the fundamental characteristics outlined in Chapter 1, including order, adaptation, and reproduction.

We can't say when or where this happened. But all available evidence suggests that given the conditions on the early earth, plus a span of several hundred million years and the basic physical and chemical properties of matter, life units could have emerged—and did. Thus, sometime after the earth's crust ceased melting from meteoric bombardment, living cells evolved and multiplied, and biologists believe that these cells emerged through an orderly and natural sequence of events: origin of the universe → formation of stars → formation of planets → accumulation of organic molecules → formation of polymers → interactions between polymers → compartmentalization → life.

≋ EARTH AND LIFE EVOLVE TOGETHER

The earth has probably been inhabited for more than 80 percent of its history, or about 3.8 billion years. Fossil evidence suggests, however, that organisms with many cells—animals, plants, and some fungi—first appeared only about 700 million years ago. Biologists conclude that single-celled organisms ruled the earth for the vast majority of its history.

FIGURE 14.8 *Cell-Like Lipid-Coated Droplets Can Form Spontaneously in the Laboratory.* Under specific conditions of temperature, moisture, and acidity, tiny spheres will form in solutions that contain amino acids and phospholipids. These lipid-coated droplets (magnified about 10,000 times) form as the solution of phospholipids cools. Perhaps the earliest cell membranes formed spontaneously under similar circumstances.

TABLE 14.1 ≋ MAJOR STEPS IN THE EARLY EVOLUTION OF LIFE

EVENT	BILLIONS OF YEARS AGO	MANIFESTATION	OXYGEN (%)	EVENTS AND CONSEQUENCES
8 Full diversity of life forms present	0.4	Large fishes	20	Complete ozone screen; atmosphere same as today; large, active fishes, land plants
7 Shelled animals and early land plants appear	0.55	Hard-shelled animals, early land plants	10	Diversity evident in fossil record
6 Many-celled organisms appear	0.67	Soft-bodied multicellular animals	7	Fossils and tracks made; oxygen and ozone accumulate
5 First eukaryotic cells appear	1.4	Large nucleated cells	2	Mitosis, meiosis, genetic recombination, and aerobic respiration occur
4 Oxygen-tolerating blue-green algae appear	2.0	Enlarged, thick-walled cells at intervals on algal filaments	1	Ozone screen thickens; iron deposits appear
3 Photosynthetic organisms appear	2.8	Stromatolites, filaments, blue-green algae	less than 1	Oxygen given off into atmosphere
2 Autotrophs, methane-generating bacteria, and sulfur bacteria appear	3.5	Stromatolites and precursors to blue-green algae	0	Atmosphere still contains very little oxygen
1 Origin of life	3.8	The first cells evolve	0	Atmosphere lacks oxygen

EARLY LIFE FORMS EVOLVE AND CHANGE THE EARTH

Researchers have amassed considerable evidence that during the 3-billion-year Age of Microbes, at least five major developments modified the planet in dramatic and permanent ways. An additional three evolutionary milestones involved many-celled organisms. Table 14.1 summarizes these eight events.

■ **First Life** In ancient rocks 3.8 billion years old, found in southwestern Greenland, scientists have observed layers that suggest different periods of microbial activity. The evidence is too scanty to speculate much on these traces of life, but, as already mentioned, the very earliest cells must have taken in organic materials already present in the "primordial soup" and harvested energy without the use of oxygen, since the earth's primordial atmosphere contained no oxygen at this time (see Table 14.1, event 1).

■ **Autotrophs (Self-Feeders)** Fossilized remains of structures called stromatolites have been found near the desolate Australian town of North Pole and dated at 3.5 billion years old. Figure 14.9 shows modern stromatolites, which are pillow-shaped mounds built layer upon layer by massive colonies of anaerobic blue-green algae that can make their own food through photosynthesis. Although conclusive evidence is lacking, the stromatolite fossils suggest that ancient organisms, probably capable of photosynthesis, may have been among the earliest cells and may have begun releasing oxygen into the atmosphere as long ago as 3.5 billion years (event 2).

FIGURE 14.9 *Modern Stromatolites: Evidence of Photosynthetic Cells.* These pillow-shaped mounds were built by large colonies of photosynthetic cells (blue-green algae) in Hamlin Pool, along Australia's west coast. Fossils of similar structures built 3.5 billion years ago may be evidence of some of the earliest cells on earth.

Other fossils from this period reveal a trail of biochemical evidence left by anaerobes that were probably similar to modern methane bacteria, which use carbon dioxide, hydrogen, and hydrogen sulfide as raw materials for making their own food. Some form of autotrophy—self-feeding—clearly existed by this time.

■ **Photosynthesis** Australian rocks dating from 2.8 billion years ago show definite filaments that strongly resemble today's blue-green algae. From these, biologists have concluded that while photosynthesis may have evolved earlier, it was *definitely* present by 2.8 billion years ago, and oxygen was being given off steadily into the seas and atmosphere (event 3). Ironically, oxygen is poisonous to anaerobic cells and thus would have been harmful to all early cells, self-feeders and other-feeders alike.

■ **Tolerance to Oxygen** Fossils 2 billion years old from the Lake Superior area show thick-walled cells that were probably resistant to the harmful effects of oxygen. Such fossils suggest that oxygen given off by blue-green algae all over the earth had begun to accumulate in the atmosphere and that cells began to evolve mechanisms for avoiding oxygen poisoning (event 4). Beyond its effects on evolution, the buildup of oxygen would have caused the formation of an ozone screen. Sunlight energizes oxygen in the air, creating ozone molecules, which in turn absorb ultraviolet light. The accumulating ozone would have screened living things from some of the damage that ultraviolet light inflicts on DNA. Collectively, living things were starting to have dramatic effects on the planet.

■ **Eukaryotic Cells** Fossils from 1.4 billion years ago show clear evidence of large cells containing a true nucleus surrounded by a membrane (event 5). Since most eukaryotes use oxygen when they break down sugars for energy, aerobic respiration must have evolved by that time. Aerobic respiration would have helped the cells dispose of oxygen and also would have provided the added energy to support their larger size. Eukaryotic cells are characterized by their true nucleus, as well as by the other membrane-enclosed organelles they contain. Figure 14.10 (page 250) illustrates a widely accepted theory for how they got these organelles.

■ **Multicellular Organisms** The Age of Microbes ended about 670 million years ago when many-celled animals (called *metazoans*) appeared. Fossils from five continents show impressions in sediment left by soft-bodied marine animals that crawled in the sand or mud (event 6).

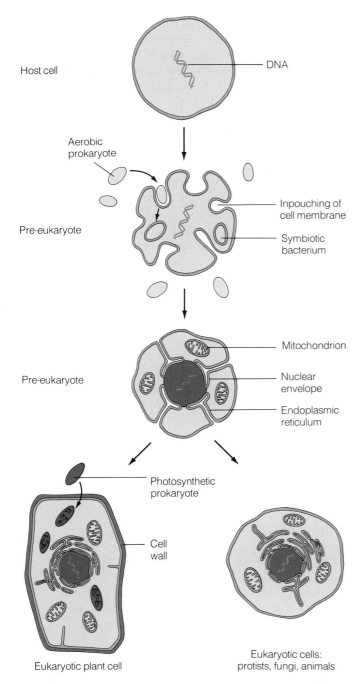

Host cell

DNA

Aerobic prokaryote

Pre-eukaryote

Inpouching of cell membrane

Symbiotic bacterium

Pre-eukaryote

Mitochondrion

Nuclear envelope

Endoplasmic reticulum

Photosynthetic prokaryote

Cell wall

Eukaryotic plant cell

Eukaryotic cells: protists, fungi, animals

FIGURE 14.10 *The Origin of Complex Cells.* Many biologists believe that in some early prokaryotic cells (so-called pre-eukaryotes, lacking specific internal organelles) an infolding of the cell membrane may have given rise to the nuclear membrane and the endoplasmic reticulum. As this infolding occurred, smaller bacteria may have entered the larger host cell and then become *symbionts*—living together in a mutually beneficial situation. The nonphotosynthetic symbionts could have evolved into mitochondria, and the photosynthetic "invaders" could have given rise to chloroplasts. Biologists call this the *endosymbiotic theory.*

■ **Hard-Shelled Animals** About 550 million years ago, animals with hard, protective outer shells (such as clams and horseshoe crabs) left fossil imprints (event 7). The size and activity of these organisms were increasing, as was the atmospheric oxygen. Besides oxygenating the air, living things had begun to affect the earth in other dramatic ways. Methane from methane-generating bacteria and carbon dioxide from oxygen-breathing organisms were accumulating in the atmosphere. The remains of living things were creating massive deposits of some minerals and building thick layers of soil and organic sediments.

■ **Large Plants and Animals** By 400 million years ago, the atmosphere was essentially like today's. The ozone screen was fully formed, and very large, complex life forms were appearing in profusion. Large fishes swam in the ancient seas, and the first primitive land plants grew along moist shores (event 8). Within another 200 million years, amphibians, reptiles, birds, and mammals would move about the continents, and large stands of conifer trees and flowering plants would grow abundantly. Today's plants and animals are the current representatives of a long history of life forms. It is only because the collective activities of millions of microbes over billions of years altered the earth in permanent ways that larger, later life forms—ourselves included—were able to evolve.

THE EARTH EVOLVES AND ALTERS LIFE

Just as life's evolution has affected the earth, the earth, too, has changed geologically and influenced living things. Geologists, who study the earth, believe that for hundreds of millions of years after our planet formed, black volcanic islands jutted upward through blue ocean, and as the lava from nearby cones accumulated, some areas of high, flat terrain—the first continents—emerged from the seas. The early landmasses were probably flat areas of lighter rock upon huge *plates* of heavier rock. The crustal plates drift slowly as heat-driven currents in the mantle cause the formation of new crust between the great plates, a process called *plate tectonics.* Stresses created by the slow movements of plates cause the earthquakes, volcanoes, and uplifting of mountain ranges that mold the face of the earth.

Although the movement of crustal plates probably began before life originated, stresses did not build up quickly, and the face of the earth did not change much for the first 3 billion years or so. This time span is usually called the *Proterozoic* ("early life") era and is one of four time spans, or *geological eras,* into which scientists divide the earth's history. Each era is further subdivided into *geological periods.* Figure 14.11 groups the geological periods

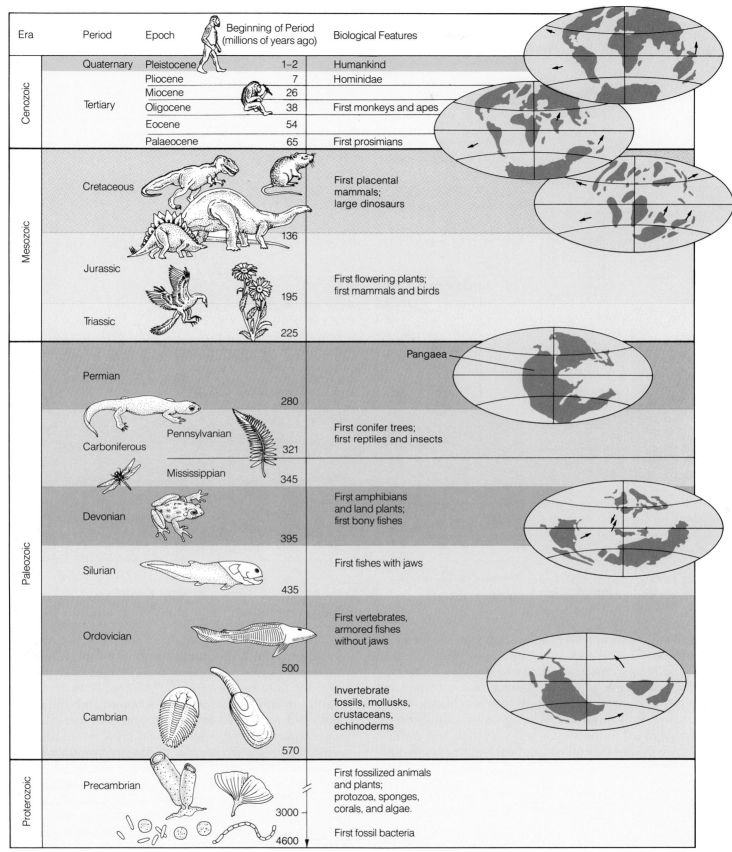

Era	Period	Epoch		Beginning of Period (millions of years ago)	Biological Features
Cenozoic	Quaternary	Pleistocene		1–2	Humankind
	Tertiary	Pliocene		7	Hominidae
		Miocene		26	
		Oligocene		38	First monkeys and apes
		Eocene		54	
		Palaeocene		65	First prosimians
Mesozoic	Cretaceous				First placental mammals; large dinosaurs
	Jurassic			136	First flowering plants; first mammals and birds
	Triassic			195	
				225	
Paleozoic	Permian				
	Carboniferous	Pennsylvanian		280	First conifer trees; first reptiles and insects
		Mississippian		321	
	Devonian			345	First amphibians and land plants; first bony fishes
	Silurian			395	First fishes with jaws
	Ordovician			435	First vertebrates, armored fishes without jaws
	Cambrian			500	Invertebrate fossils, mollusks, crustaceans, echinoderms
				570	
Proterozoic	Precambrian				First fossilized animals and plants; protozoa, sponges, corals, and algae.
				3000	
				4600	First fossil bacteria

FIGURE 14.11 *Earth's History: Geological Eras and Periods.* This chart relates the emergence of life forms to changes in the earth's landmasses. Starting with the Paleozoic era, it elaborates on what took place during events 6, 7, and 8 of Table 14.1.

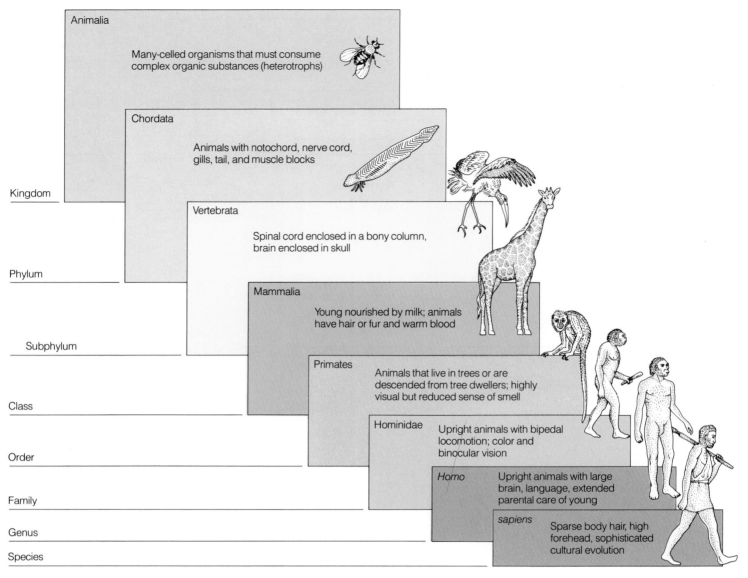

FIGURE 14.12 *Taxonomic Classification of Human Beings.* Note that subphylum, the fourth level above genus, is really a subdivision of phylum.

into the four eras and shows the relation of life forms to landmasses during each era. The figure illustrates this main point: As crustal movements remodeled the face of the earth, the physical changes, accompanied by changes in climate, created new habitats (places where organisms live) or altered existing ones. Living things either adjusted to these changes, survived, and continued to evolve, or they failed to adjust and as a result died out.

The early Proterozoic era had just one period, the Precambrian, which was dominated by the evolution of bacteria and single-celled eukaryotes. By the end of the next era, the *Paleozoic* ("ancient life"; 225 to 570 million years ago), a million species or more of aquatic and land organisms had descended from the progenitors that existed at the start of the era, while crustal movements had consolidated several landmasses to one giant continent called Pangaea ("all earth"). The *Mesozoic* ("middle life"; 65 to 225 million years ago) then saw the giant Pangaea break apart and a wave of extinctions of early aquatic and land animals; these events were accompanied by the coming of the giant reptiles, the dinosaurs, as well as the growth of great swampy forests of tropical plants and trees and the beginning of the birds, mammals, and flowering plants. Finally, during the current era, the *Cenozoic* ("recent life"; 65 million years ago to the present), extensive crustal plate movement brought the continents to their present places, carrying many groups of organisms to entirely new locations and conditions and further encouraging the divergence of lineages. There was another wave of extinctions that ended the dominion of the reptiles, and after this time, birds, mammals, and flowering plants diverged into large and varied groups of species.

In the next four chapters, we will consider each major group—the single-celled organisms, the fungi, the plants, and the animals—in detail. Here it is important simply to understand the overall effects of geological activity on life's evolution: The continually changing earth put pressures on living things that led to the appearance and disappearance of millions of species over time. Earth and life truly coevolved.

≋ THE SCIENCE OF TAXONOMY: CATALOGING LIFE'S DIVERSITY

Since before the time of Aristotle, naturalists have delighted in searching out, describing, naming, and grouping the myriad kinds of living things. The science of assigning living things to categories of like organisms is called **taxonomy.** In the mid-eighteenth century, a Swedish biologist named Carolus Linnaeus classified life's diversity by assigning every organism then known to science to a series of increasingly specific groups, depending on the number of structural traits shared by group members. In this system, each organism is assigned a two-word name: a **genus** name followed by a **species** name. A species, as we saw in Chapter 1, is a unique group whose members share the same set of structural traits and can successfully interbreed only with members of the same species. Taxonomists group several related species in a genus (plural, genera), a collection of very similar organisms related by common descent from a recent ancestor and sharing similar physical traits. We humans, for example, belong to the genus *Homo* and the species *sapiens* ("wise"). Note that the genus name begins with a capital letter, and both names are written in italics.

Linnaeus placed each organism, with its genus and species name, in a series of categories, or *taxonomic groups,*

based on shared characteristics. Biologists recognize five major levels beyond species and genus as shown in Figure 14.12 for our own group of *Homo sapiens.* Members of similar genera belong to the same family, members of similar families to the same order, similar orders to the same class, similar classes to the same phylum, and similar phyla to the same **kingdom.**

In recent years, most biologists have generally recognized five kingdoms (Figure 14.13, page 254). These include the **Monera,** or single-celled prokaryotes, such as various kinds of bacteria; the **Protista,** or single-celled eukaryotes, such as *Euglena;* the **Fungi,** or multicellular decomposers, such as mushrooms or molds that break down other biological tissues for food; the **Plantae,** or multicellular photosynthesizing organisms, such as algae, mosses, ferns, and flowering plants; and the **Animalia,** or multicellular life forms that feed on other organisms and usually exhibit movement, such as insects, clams, birds, and reptiles.

While the five-kingdom system has logical appeal, a growing body of genetic and biochemical data seems to separate organisms into just three related groups (Figure 14.14, page 255). One group includes all organisms whose cells have a nucleus surrounded by an envelope (the *eukaryotes*), including the plants, animals, fungi, and protists. The second group is the *eubacteria,* the common bacteria that cause many diseases, turn milk to yogurt, and decompose leaf litter in the soil. These lack a membrane-enclosed nucleus and are prokaryotes. The third group is the *archaebacteria,* including the methane generators with which we began this chapter. These are also prokaryotes that lack true nuclei, but they tend to inhabit extreme environments, such as a cow's digestive tract, salt flats, alkaline lakes, volcanic sulfur fields, and the ocean depths. These bizarre cells are genetically closer to humans than they are to the eubacteria!

So far, taxonomists have classified more than 2 million species, currently alive or extinct, into the kingdoms of life and the other taxonomic subdivisions below the kingdom level. But they still have a leviathan task ahead, because as many as 8 million additional species, still undiscovered and unnamed, are believed to inhabit the oceans, rain forests, and other hard-to-explore regions. The grouping of these organisms into kingdoms and other taxonomic groups brings out their evolutionary relationships to each other. Taxonomy, in general, also helps us appreciate the enormous diversity of species that arose as earth and life coevolved and changed each other over 4 billion years of history.

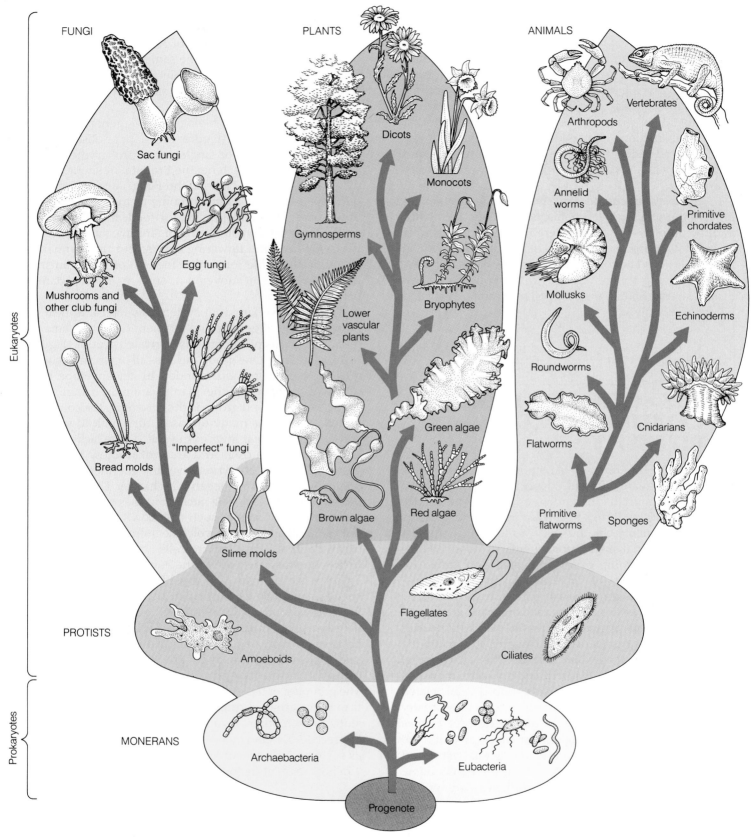

FIGURE 14.13 *Stunning Diversity in Life's Kingdoms.* Living things can take extremely different forms. Many evolutionary biologists believe that some common ancestor gave rise to all the prokaryotes; that these simple cells gave rise to the more complex single-celled eukaryotes, the protists; and that branches of the protists then gave rise to the fungi, plants, and animals.

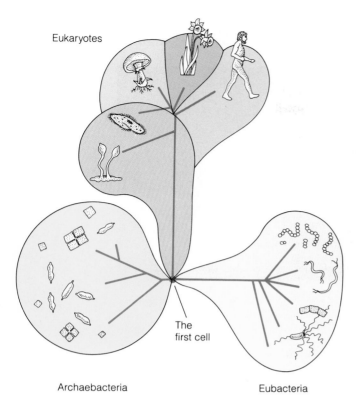

Eukaryotes

The first cell

Archaebacteria

Eubacteria

FIGURE 14.14 *The Three-Kingdom System.* Genetic evidence suggests that archaebacteria are as different from other bacteria as they are from higher life forms. This diagram, based on genetic similarities between various organisms, suggests that the original cells may have given rise to three groups—the eukaryotes, the eubacteria, and the archaebacteria—simultaneously. This view contrasts with the five-kingdom scheme illustrated in Figure 14.13.

≋ CONNECTIONS

Understanding the origins of life is a profound scientific challenge because the drama unfolded on a stage that no longer exists and under conditions that changed forever once photosynthetic organisms began to thrive and give off oxygen. Nevertheless, by knowing both the characteristic behavior of matter—particularly organic molecules—and the attributes of all living things, biologists have reconstructed a plausible sequence of the prebiotic events that could have led to the emergence of the first cells. Since these early cells may have been similar to methane-producing bacteria, the bad smells that come from sewage treatment plants and swamp mud can serve as sensory reminders of life's enormously long history and its primitive beginnings on an equally primitive planet. The evolution of life and its effects on our planet have produced the diversity of species and habitats we see today. The remaining chapters in this unit explore life's splendid diversity, kingdom by kingdom.

NEW TERMS

Animalia, page 253
Fungi, page 253
genus, page 253
kingdom, page 253
Monera, page 253

Plantae, page 253
Protista, page 253
species, page 253
taxonomy, page 253

STUDY QUESTIONS

REVIEW WHAT YOU HAVE LEARNED

1. What was the Big Bang?
2. Rearrange the following steps to reflect the probable sequence of events as living cells arose; self-replication, molecular interactions, coordinated cell-like activities, formation of polymers, and development of cell-like containers.
3. What important events in the earth's history may have occurred in the following time periods: about 4.6 billion years ago; about 2.8 billion years ago; about 1.4 billion years ago?
4. What causes most of the geological activity that alters the earth's surface?
5. What did Linnaeus contribute to taxonomy?
6. Define species.

APPLY WHAT YOU HAVE LEARNED

1. If you could design a mobile space laboratory to detect the presence of some precellular step in the origin of life on a distant planet, what would the module look for?

FOR FURTHER READING

Amato, I. "Tracing Living Signs of Ancient Life Forms." *Science News,* 7 October 1989, p. 229.

Dickerson, R. E. "Chemical Evolution and the Origin of Life." *Scientific American* 239 (September 1978): 70–86.

Kunzig, R. "Stardust Memories: Kiss of Life." *Discover* (March 1988): 68–76.

McDermott, J. "A Biologist Whose Heresy Redraws Earth's Tree of Life." *Smithsonian* (August 1989): 72–80.

North, G. "Origin of Life: Back to the RNA World—and Beyond." *Nature* 328 (1987): 18–19.

Pool, R. "Pushing the Envelope of Life." *Science* 247 (January 1990): 158–160.

Waldrop, M. M. "Did Life Really Start Out in an RNA World?" *Science* 246 (8 December 1989): 1248–1249.

Weiss, R. "Seekers of Ancestral Cell Debate New Data." *Science News,* 16 January 1988, p. 36.

(1) More than 4 billion years ago, natural processes in the solar system and on the early earth could have generated the energy sources and raw materials that gave rise to early life.

(2) Life could have emerged in five stages similar to these:

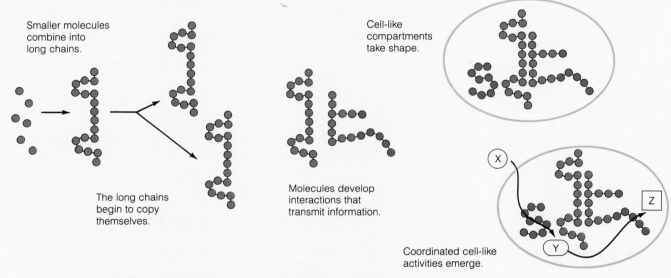

Smaller molecules combine into long chains.

The long chains begin to copy themselves.

Molecules develop interactions that transmit information.

Cell-like compartments take shape.

Coordinated cell-like activities emerge.

(3) Earth and life coevolved: A great diversity of life forms arose and changed the earth's surface and atmosphere; changes in the planet, in turn, affected the evolution of life.

Earth

Life forms

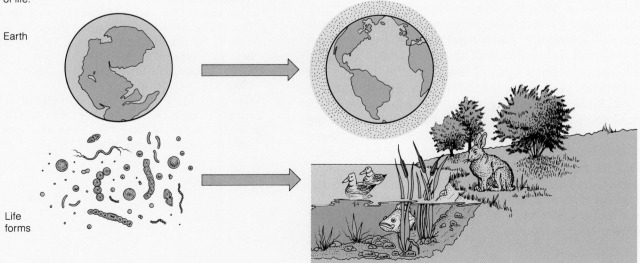

THE SINGLE-CELLED KINGDOMS

CHAPTER 15

THE COMPLEX LIFE OF A MICROSCOPIC HUNTER

Watching through a microscope as the single-celled hunter *Didinium nasutum* stalks its prey is a bit like viewing a silent horror film. *Didinium* is juglike in shape, and two rows of lashing cilia propel it through the water in a zigzag course in search of food. The food is usually slipper-shaped *Paramecium* cells, which are themselves gliding around in search of small food particles. But *Didinium* has a deadly weapon: a cone-shaped region that protrudes from one end of the cell and is ringed by poisoned darts. With this "snout," *Didinium* can stab a *Paramecium,* impaling it long enough to fire darts and paralyze the hapless victim. Then, as the tip of its cone stretches open wider and wider,

the hunter swallows alive the prey cell larger than itself (Figure 15.1). Digestive enzymes immediately begin breaking down the meal, and within 20 minutes, *Didinium* is ready to hunt again.

Didinium and *Paramecium* are both protists, that is, members of the kingdom Protista (review Figure 14.13). All protists are complex single cells that are larger than bacteria and that contain a true nucleus and other membrane-bounded organelles. Protists include cells that move and hunt like animals, cells that carry out photosynthesis like plants, and cells that break down organic matter like fungi.

By contrast, the smaller, simpler, single-celled bacteria belong to the kingdom

Monera (see Figure 14.13). The monerans, like the protists, exhibit a variety of life-styles. Some generate their own organic nutrients, while others must absorb them. Some can live without oxygen, while others need it.

A fundamental theme of this chapter is that while monerans and protists are independent single cells, they carry on all the essential life processes. *Didinium,* for example, can swim, jab, swallow, excrete, digest, shoot darts, detect the presence of prey, and reproduce sexually. In fact, practically all the behaviors we usually associate with larger, more complicated beings appeared first in the single-celled kingdoms.

Another theme underlies our discussion of monerans and protists. Single-celled organisms are so widespread on our planet that they collectively have a tremendous impact on our lives. The biggest scourges of humankind are the infectious diseases caused by bacteria, viruses (which are not cells), and protists. Every year, more people die of malaria—caused by a protist with a very complex life cycle—than of any other single cause. Microscopic organisms, however, can also be useful. For example, they release massive quantities of oxygen and decompose vast amounts of organic matter. They are also invaluable subjects in scientific research and crucial tools of genetic engineering.

This chapter will describe the basic characteristics of single-celled life forms and answer these questions:

- What are the characteristics of the monerans, and what are the nonliving particles called viruses, viroids, and prions?
- What are the characteristics of the protists?

FIGURE 15.1 *A Single-Celled Hunter and Its Prey.* The juglike *Didinium* (left) swallowing a slipper-shaped *Paramecium.*

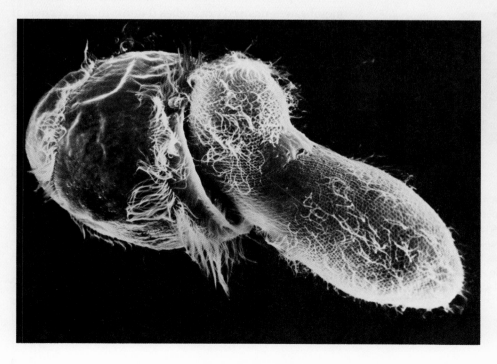

☰ MONERA: THE KINGDOM OF BACTERIA

Monera is the name of the kingdom that includes all prokaryotes, or cells lacking a true nucleus. It includes bacteria and **cyanobacteria** (formerly called blue-green algae), which live in an amazing range of habitats anywhere from arctic snow drifts to deep-sea hydrothermal vents (see Figure 15.2a).

STRUCTURE OF MONERANS

As prokaryotic cells, monerans have an outer cell wall, an inner plasma membrane, and a noncompart-

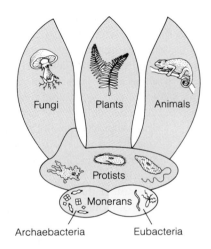

Cell wall

Plasma membrane

DNA

Cytoplasm

(b)

FIGURE 15.2 *Monera: The Prokaryotic Kingdom.* (a) The monerans fit at the base of our evolutionary diagram because they evolved earlier than the other four kingdoms and almost certainly gave rise to them. Monerans include the archaebacteria (darker yellow) and the eubacteria (paler yellow). (b) Monerans, such as this dividing *Bacillus subtilis*, are prokaryotic cells; their DNA floats in a central area without a nuclear envelope, and they have a tough cell wall, a plasma membrane, and ribosomes in the cytoplasm.

Fungi Plants Animals

Protists

Monerans

Archaebacteria   Eubacteria

(a)

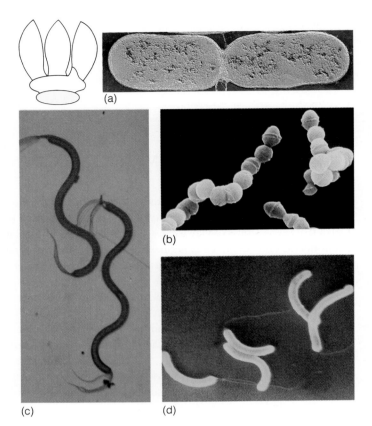

(a)

(b)

(c)   (d)

FIGURE 15.3 *Bacterial Zoo: A Diversity of Shapes Among the Microscopic.* Bacteria come in a wide range of shapes and sizes, including (a) rod-shaped bacilli such as *Shigella*, (b) round cocci such as streptococci, (c) spiral-shaped spirilla like the spirochetes shown here, and (d) comma-shaped vibrios such as *Vibrio cholerae*.

mentalized cytoplasm dotted with ribosomes; they lack membrane-enclosed organelles (Figure 15.2b). A circular strand of DNA, usually coiled into the cell's center, serves as the single chromosome. (Remember, *prokaryote* means "before nucleus.") The outer cell wall is a strong but flexible covering made primarily of sugars and proteins. The antibiotic penicillin interferes with the building of these sugar-protein coverings in some species, so that as the cell grows, the wall does not, and soon the cell bursts.

Most prokaryotes are only about one-tenth the size of a eukaryotic cell, but they come in diverse shapes, which include spheres (*cocci*), rods (*bacilli*), spirals (*spirilla*), and curved rods (*vibrios*) (Figure 15.3). The names of prokaryotes often reflect their shape. For example, *Streptococcus mutans*, the cause of strep throat, is spherical.

NUTRITION

Monerans can be heterotrophs that consume organic nutrients, or they can be autotrophs that can make their own organic nutrients. Together, these two modes of nutrition mean that monerans survive on an enormous range of energy sources and hence are everywhere—in air, soil, water, and in other life forms. Only a few species of monerans can make all the substances they need from simple inorganic compounds. They include green and purple photosynthetic bacteria, cyanobacteria (blue-green algae), and the sulfur and methane bacteria.

Most monerans are *decomposers*, living on dead or dying organisms. Without the activity of these microbial decomposers (as well as the multicellular decomposers, the fungi), the earth would quickly accumulate a thick layer of fallen leaves, dead animals, and other organic matter that would choke out living things. Some monerans, however, live on or inside other living organisms

as harmful *parasites* or as neutral or beneficial *symbionts* (literally, "organisms that live together"). Most disease-causing bacteria are parasites, while the beneficial symbionts include intestinal bacteria that generate vitamins or help digest cellulose in animals such as koala bears.

REPRODUCTION

Monerans usually reproduce into identical daughter cells by splitting in two (Figure 15.4). If the dividing cells remain connected, small clusters or long filaments can form (as in Figure 15.3b). Reproduction can be incredibly fast and efficient, occurring as often as every 20 minutes. If a single bacterium (such as the one diagrammed in Figure 15.4a) and all its progeny continued dividing every 20 minutes, after just 48 hours there would be a mass of cells 4000 times the weight of the earth. This doesn't happen, of course, because the microbes have neither an unlimited source of nutrients nor ideal conditions of temperature and humidity. Nevertheless, the capacity for rapid reproduction, together with a high mutation rate (about one mutant cell per million dividing cells) and several types of genetic recombination (see Chapter 9), means that bacteria can adapt quickly to changes in the environment.

Many monerans have yet another reproductive adaptation that enables them to survive unfavorable conditions; they form small, tough-walled resting cells called **spores** (Figure 15.5, page 260). Spores can withstand extremes of heat, cold, drought, and even radiation for long periods. When conditions improve, the spores grow into new bacterial cells. Spores are the reason why surgical instruments must be sterilized with high heat and pressure, and they are also the reason why home-canned foods must be processed so carefully. The bacteria that cause botulism poisoning form spores that can withstand several hours of boiling. If these spores aren't destroyed during canning, they can later germinate into cells that release a deadly toxin.

BEHAVIOR

Some monerans are unable to move under their own power and are subject to the currents and movements of the water, blood, plant sap, air, or other fluid medium that surrounds them. Many monerans, however, have whiplike flagella that enable the cells to move toward nutrients and light and away from predators or noxious

(a)
Chromosome attachment site

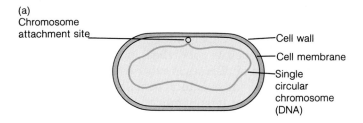

Cell wall
Cell membrane
Single circular chromosome (DNA)

(b)
Chromosome doubles and cell elongates

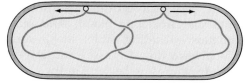

(c)
Division into two daughter cells completed

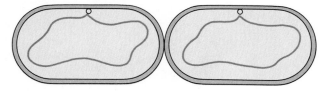

(d)

FIGURE 15.4 *Monerans Reproduce by Cell Division.* A bacterium reproduces as one cell (a) divides in two. The stages of this cell division include chromosome doubling and cell elongation (b) and the formation of a central partition that splits the parent cell into two daughter cells (c). (d) Two cells that have just formed by cell division. These are *Escherichia coli* cells, a bacterial species that inhabits the human intestines.

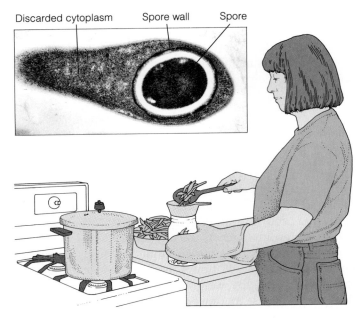

Discarded cytoplasm   Spore wall   Spore

FIGURE 15.5 *A Bacterial Spore*. Certain species of bacteria can withstand extreme environmental conditions (freezing, boiling, high pressure, radiation, and so on) by forming tough-walled spores, or resting cells. Adverse conditions may kill the remainder of the cell, but the spore will survive, and when conditions improve, it may germinate and release a new individual. If this happens in improperly canned vegetables, fruits, or meats contaminated with *Clostridum botulinum*, the result can be severe food poisoning. To be on the safe side, you must boil the food under high pressure (in a pressure cooker) for a certain period of time.

chemicals (Figure 15.6). The capacity for movement thus enables monerans to find and exploit appropriate food sources or avoid danger.

TYPES OF MONERANS

Biologists have various ways of classifying the nearly 3000 bacterial species that make up the kingdom Monera: by nutritional mode, by shape, by cell wall characteristics, and by evolutionary relationships. As we saw in Chapter 14, fairly recent evidence suggests that bacteria like the methane producers are as different from other kinds of bacteria as they are from plants and animals. For this reason, we recognize three main subdivisions of the monerans: the archaebacteria, perhaps the oldest living species on earth (Figure 15.7); the cyanobacteria (blue-green algae), the ancient and still ubiquitous self-feeders (Figure 15.8, page 262); and the Schizophyta, the remaining bacteria and related single-celled organisms (Figure 15.9, page 262). Table 15.1 describes the prominent characteristics of each group.

THE IMPORTANCE OF MONERANS

Monerans have a great ecological impact. Cyanobacteria give off oxygen and help cycle carbon, nitrogen, and other elements by fixing them into usable substances. Soil bacteria also cycle elements, and decomposing bacteria break down enormous quantities of dead animals, fungi, and plants as well as human and animal wastes, pesticides, and pollutants that would otherwise poison the environment. When the *Exxon Valdez* supertanker ran aground in March of 1989, spilling millions of gallons of oil into Alaska's Prince William Sound and endangering wildlife for miles around, populations of bacteria in the sound played an inadvertent role in the massive cleanup effort: the bacteria could break down many of the toxic compounds in the oil (Figure 15.10, page 263).

Monerans also have great medical impact, both negative and positive. Bacteria cause hundreds of human diseases, including blood poisoning, some sexually transmitted diseases (such as gonorrhea), and tooth decay, as well as thousands of diseases in other animals and plants. Bacteria do confer some medical benefits, however. *Escherichia coli* and other inhabitants of the human gut, for example, produce vitamins K and B_{12} and other cofactors that we probably absorb and use. Usually harmless residents like *E. coli* may also blanket the intestinal walls so heavily that harmful bacteria cannot gain access and pass through into the circulating blood. And many plant-eating mammals, including cattle, sheep, and rabbits, would be unable to digest grasses and leaves without the cellulose-decomposing bacteria in their intestines.

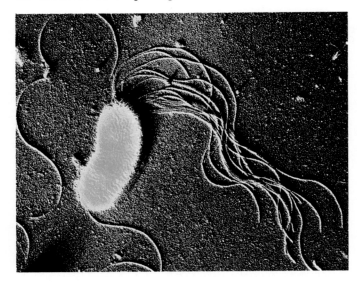

FIGURE 15.6 *Bacterial Flagella: The Only Living Wheels*. Some bacteria, such as this *Pseudomonas fluorescens*, have a flagellum that propels the bacterium toward favorable conditions the way a rotating propeller shaft pushes a boat through the water. The base of the flagellum is the only rotating wheel in a living thing. The enlargement drawing shows how the base of a flagellum is inserted in the cell wall and membranes.

TABLE 15.1 ≋ THE KINGDOM MONERA

PHYLUM/DIVISION	REPRESENTATIVE GROUPS	SIGNIFICANCE
Archaebacteria 	Methane generators, salt-tolerant bacteria, heat- and acid-resistant bacteria	Break down organic matter in extreme and anaerobic environments
Cyanobacteria	"Blue-green algae"	Contain chlorophyll; origin of chloroplasts; show division of labor (specialized cells fix free nitrogen into compounds organisms can use); fix carbon and nitrogen; release oxygen
Schizophyta	Green and purple photosynthetic bacteria	Contain one kind of chlorophyll; fix carbon under extreme conditions without producing oxygen
	Sheathed bacteria	Found in polluted streams and sewage sludge plants; break down organic matter
	Actinomycetes	Cause leprosy, tuberculosis; decomposers; make antibiotics
	Rickettsias	Rod-shaped parasites that cause typhus and Rocky Mountain spotted fever; transmitted by fleas, ticks, and lice
	Mycoplasmas	Smallest living cells; live only inside other cells; lack cell walls; cause one type of pneumonia, urinary tract and other infections
	Spirochetes	Cause syphilis, other diseases
	Eubacteria	Some killed by penicillin and sulfa drugs and can cause diarrhea; others killed by streptomycin and tetracycline and can cause skin and other infections

In the economic arena, people spend millions of dollars each year on antibiotics and disinfectants to counter the effects of bacteria, and on various medical treatments to fight infectious diseases. (*Antibiotics* are substances that inhibit or destroy bacteria and other microorganisms.) In a different part of the arena, manufacturers harness bacteria to produce foods—including cheese, yogurt, pickles, soy sauce, and chocolate—and to generate chemical reagents, such as butanol, fructose, and lysine. Researchers recently discovered a deep-sea-vent bacterium that grows best in nearly boiling water; this species of archaebacteria may someday yield heat-resistant enzymes that can be used to speed up a variety of high-temperature industrial processes.

Finally, bacteria have had an immeasurable scientific impact. The techniques of genetic engineering, for example, are largely based on bacterial chromosomes and plasmids (see Chapter 9). Without prokaryotic research subjects, biology would be a century behind its current state.

FIGURE 15.7 *Salt Tolerators: One Kind of Archaebacterium.* These salt pans in the southern part of San Francisco Bay look bright crimson in large part because of the trillions of individual bacteria that thrive under such highly saline conditions.

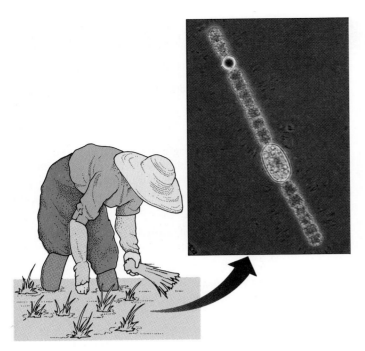

FIGURE 15.8 *Cyanobacteria: Hardy Prokaryotes in Many Environments.* Cyanobacteria appear green because they contain chlorophyll, and many biologists believe that an ancient cyanobacterium may have been enveloped to become the first chloroplast. The bacteria often grow as filaments with individual cells joined end to end, interspersed by an occasional thick-walled cell (arrows) that shuts out oxygen and allows nitrogen fixation to take place. In nitrogen fixation, simple nitrogen gas is converted to ammonia or some other form that plants can use. Some rice farmers allow cyanobacteria to grow in the rice paddies where they raise their crops to fertilize the rice plants naturally.

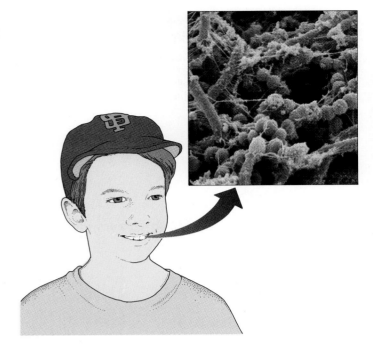

FIGURE 15.9 *Our Teeth Teem with Schizophyta Bacteria That Cause Decay.* Here we see groups of rod-shaped *Streptococcus mutans*, which, if not brushed away, can cause cavities.

VIRUSES AND OTHER NONCELLULAR AGENTS OF DISEASE

Often considered with the monerans are a set of microscopic parasites that are technically not alive because they lack one or more of life's characteristics: They cannot, for example, carry out metabolism or reproduce on their own. Instead, they depend on living cells for continued existence, and they probably evolved from them.

Best known are the **viruses,** geometrically shaped particles that are 1000 times smaller than a bacterium. A virus particle is, in effect, a minute package of DNA or RNA surrounded by a protein coat and sometimes a lipid membrane (Figure 15.11). Viruses infect both prokaryotic and eukaryotic cells by attaching to the plasma membrane and then allowing their DNA or RNA to enter the cell. Once inside, the viral genes take over the cell's protein-synthesizing machinery for the production of new virus particles. Eventually, the cell liberates thousands of viruses that infect additional cells. The speed of viral reproduction is even more astounding than the speed of cell division in bacteria. In 24 hours, one virus particle could generate enough particles to fill the universe! In nature, however, reproduction is always limited by the availability of cells, since viruses lack the machinery for self-reproduction or metabolism.

There are hundreds of kinds of viruses, many of which cause plant and animal diseases such as those listed in Table 15.2. Some, like the viruses that cause colds and the influenza viruses that cause flu, are transitory parasites: We encounter them in sneeze droplets or mucus sprayed into the air by cold or flu sufferers, or we pick them up from contaminated hands or surfaces, and they begin to multiply in our bodies. Eventually, our immune system destroys them. Other types of viruses, however, like herpes simplex viruses, which cause cold sores and genital herpes, insert their DNA or RNA into the chromosomes inside nerve or other body cells and take up permanent residence. There they lie dormant, erupting only occasionally to cause symptoms when triggered by fever, sunlight, or other environmental stimuli. One of the biggest challenges in modern medicine is to develop drugs that can fight viruses; although antibiotics kill many kinds of bacteria, they do not destroy viruses, because viruses are not cells. The box on page 267 describes recent efforts to fight the common cold, genital herpes, and AIDS, all caused by viruses.

The *viroids* are a group of intracellular parasites that lack a protein coat, consist only of small RNA molecules,

FIGURE 15.10 *Bacteria Help Clean Up the Environment.* When the *Exxon Valdez* supertanker ran aground in March 1989, spilling more than 10 million gallons of crude oil and blackening more than 368 miles of shoreline, roughly 2500 people participated in the cleanup effort, along with millions of air-breathing bacteria. The bacteria, local residents of the sound, can break down compounds that account for most of the oil's toxicity.

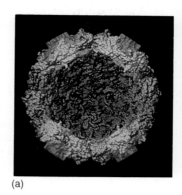
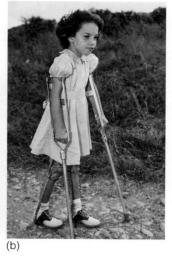

(a)   (b)

FIGURE 15.11 *The Polio Agent: An RNA Virus.* (a) Like all viruses, the poliovirus has a protein coat surrounding genetic material, but here the genes are composed of RNA, not DNA. (b) Since the virus attacks nerve cells that enable muscles to contract, muscles that power arms and legs may stop functioning and hence atrophy.

FIGURE 15.12 *Viroids: Parasites That Infect Crop Plants.* Here, viroids have infected artichoke leaves and damaged or killed rings of tissue in numerous places.

TABLE 15.2 ≋ NONCELLULAR AGENTS OF DISEASE

AGENT	CONSTITUENTS	EXAMPLE
Viruses	DNA plus protein	Hepatitis A, herpes simplex, mononucleosis, smallpox
	RNA plus protein	Measles, rubella (German measles), common cold, influenza, poliomyelitis, mumps, rabies, cancer (some forms), AIDS
Viroids	RNA only	Potato spindle tuber disease, citrus exocortis disease
Prions	Protein only	Kuru (a brain disease in Borneo and elsewhere), scrapie (a disease of sheep)

and affect only plants. Diseases of potatoes, cucumbers, citrus trees, and chrysanthemums are blamed on viroids (Figure 15.12).

Finally, the *prions* are the smallest and strangest of the intracellular parasites. They appear to lack genetic material entirely and to consist of nothing but protein, yet they are implicated in serious nerve and brain diseases, including scrapie in sheep and goats; kuru in humans; and possibly Alzheimer's disease, the most common form of senility and a leading cause of death. Biologists are not sure how prions reproduce or cause disease, but they

TABLE 15.3 ≋ THE KINGDOM PROTISTA

SECTION	PHYLUM/DIVISION	PROMINENT CHARACTERISTICS	COMMON MEMBERS	SIGNIFICANCE
Animal-like protists (protozoa)	Mastigophora	Most primitive protists; use flagella	*Giardia*	Intestinal parasite
			Trypanosomes	Common human parasite
	Sarcodina	Move like amoebas with the help of pseudopodia; some enclosed in glassy shells	Foraminiferans; radiolarians	Make limestone deposits (calcium, silicon shells)
	Sporozoa	Nonmotile (not capable of movement); live as parasites inside other organisms	*Plasmodium vivax*	Cause of malaria
	Ciliophora	Use cilia to move and to sweep food particles into mouth; have anal pores that discharge wastes; contain vacuoles filled with enzymes that digest food	*Paramecium*	Feed on bacteria in ponds
			Didinium	Hunter of other protists
Plantlike protists	Euglenoids	Photosynthetic; have flagella and eyespot that allows the cell to swim toward light	*Euglena*	Combine plant and animal characteristics
	Dinoflagellates	Have flagella in groove; photosynthetic; some wear a coat of cellulose armor	*Gymnodinium breve*	Cause of red tide
	Golden-brown algae and diatoms	Have carotenoids; photosynthetic; enclosed in glassy shell walls made of silicon	*Pinularia*	Shells used as abrasives; important component of phytoplankton in oceans
Funguslike protists	True slime molds	Contain many nuclei in one large cell; eat bacteria and other protists	*Physarum*	Break down organic matter on forest floor
	Cellular slime molds	Single cells converge to form a slug; some cells specialize for reproduction	*Dictyostelium*	Break down organic matter on forest floor

believe that like viruses and viroids, prions may be evolutionary remnants of parasitic prokaryotes that left behind all their lifelike characteristics and retained only the structures necessary to infect host cells and be reproduced by them.

THE PROTISTS: SINGLE-CELLED EUKARYOTES

We now arrive at Protista, the kingdom of free-living eukaryotic cells. This group numbers 35,000 species and includes the gluttonous *Didinium*, described at the beginning of the chapter.

GENERAL CHARACTERISTICS

Recall that the monerans are prokaryotes and that organisms in the other four kingdoms all consist of eukaryotic cells. As a eukaryotic cell, the average protist is 10 times longer than a prokaryote and has 1000 times the volume. This larger size is made possible in part by the cell's expanded membranous surface area, which enables it to carry out rapid metabolism, protein synthesis, and other cell functions. (See Table 3.1 for a review of membrane-enclosed eukaryotic cell organelles that contribute to the extensive system of membranes.) Fossil evidence shows that true eukaryotic cells had evolved from their prokaryotic ancestors by 1.4 billion years ago, and these early eukaryotes, in turn, diversified into a wide range of animal-like, plantlike, and funguslike protists that were able to exploit a number of food sources and habitats (Table 15.3). The single-celled protists can be seen as a "testing ground" for the major modes of living reflected in all higher colonial and many-celled organisms. And indeed, several of the tests were successful, because protists themselves gave rise to the early members of the plant, animal, and fungal kingdoms (Figure 15.13).

THREE TYPES OF PROTISTS

■ **The Animal-Like Protists** As the hunters of the microbial world, the **protozoa** (literally, "first animals"), are cells that usually stalk and consume other cells or food particles. As Table 15.3 shows, each of four protozoan phyla is distinguished by its means of locomotion: Members of one phylum bear long, whiplike flagella (Figure 15.14, page 266); those of another move like amoebas by extending and retracting false feet (*pseudo-podia*) (Figure 15.15; see also Figure 3.24); members of a third phylum are sporelike and do not move about at all as mature cells (Figure 15.16); and members of the phylum that includes *Didinium* use the numerous cilia on their surfaces to propel themselves (see Figure 15.1).

■ **The Plantlike Protists** Three divisions of organisms within the kingdom Protista contain chlorophyll and carry out photosynthesis: the green, gliding, spindle-shaped *euglenoids* (see Figure 3.1); the spinning *dinoflagellates*, whose two flagella set in grooves cause them to whirl around as they swim (Figure 15.17, page 268); and the *golden-brown algae and diatoms*, whose color is the result of golden pigment molecules (Figure 15.18, page 268). These protists are part of the mass of cells called **phytoplankton** ("floating plants") that grows near the surface of fresh and marine waterways. This mass of cells has great ecological significance: It releases oxygen and forms the base of the aquatic food chain, in which larger organisms graze on the phytoplankton, still larger creatures eat the grazers, and so on. Because many plantlike protists move about by means of flagella, they are thought of as hybrids of different life-styles, straddling the boundary between plantlike and animal-like cells.

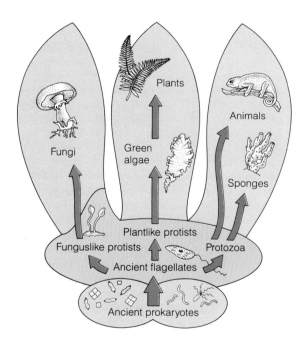

FIGURE 15.13 *Protists Gave Rise to the Multicellular Kingdoms.* As this simplified evolutionary tree shows, protists with funguslike characteristics gave rise to the fungi; plantlike protists led to the green algae and then gave rise to the plants; and protozoa preceded the sponges and other animals.

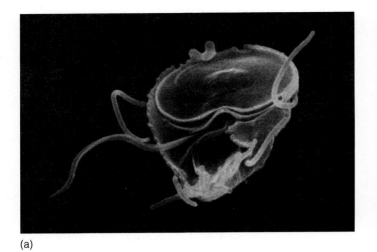

(a)

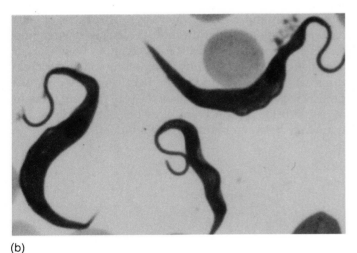

(b)

FIGURE 15.14 *Flagellum-Bearing Protozoa: Mastigophorans.* (a) Many of the once-pristine rivers, streams, and lakes in America's national parks and wilderness areas now harbor a mastigophoran called *Giardia lamblia.* This flagellated protozoan diverged from other life forms that have nuclei in their cells earlier than any other organism yet examined. It is the most common human parasite and can live and multiply in the intestines of people and certain animals, such as dogs, cattle, beavers, and deer. A person or other animal can shed millions of tiny *Giardia* cysts in the feces. If a camper infected by the parasite buries human waste too close to a lake or stream, fecal contamination can reach the waterway, and the durable cysts can infect someone else who drinks the water even two or three months later. A case of *giardiasis,* as the infection is called, usually involves severe diarrhea, painful intestinal cramps, and sulfurous belching that can last for days or weeks. To avoid giardiasis, you should treat any water you drink directly from streams or lakes. (b) Trypanosomes are agents of African sleeping sickness. Here, eel-like trypanosomes, magnified 5000 times, wriggle among a person's lozenge-shaped red blood cells. The trypanosome parasite enters the bloodstream with the bite of a tsetse fly and then invades the brain and spinal cord. Trypanosomes are very dangerous disease-causing protists because they can evade the human immune system by changing the molecules in their cell membrane.

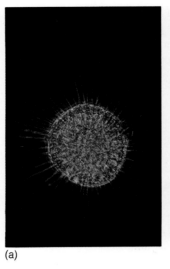

(a)   (b)

FIGURE 15.15 *Sarcodina: Pseudopodia and Glassy Shells.* Sarcodines extend narrow ribbons of cytoplasm, or pseudopodia, to capture prey or pull themselves forward. (a) An *Actinosphaerium* resembles a pincushion with hundreds of glass needles. (b) Radiolarians, such as this mixed group of species, have intricately patterned glassy shells. These protozoa and others that secrete mineral-based shells (of calcium, for example) contribute to the formation of thick limestone and other rock deposits. The White Cliffs of Dover formed in this way.

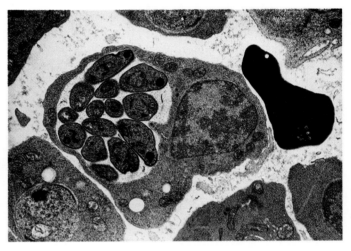

FIGURE 15.16 *Sporozoans Include the Protist That Causes Malaria. Plasmodium vivax* is the best known sporozoan and the cause of malaria, humankind's most prevalent infectious disease. Once the sporozoans infect cells as shown here, symptoms develop, including severe headaches, fever, and general discomfort. Malaria is spread by tropical mosquitoes, afflicts 300 million people each year, and kills 2 to 4 million of them. Vigorous eradication of mosquitoes with DDT and treatment of the protozoan with quinine drugs has fostered a growing natural resistance among the organisms, resulting in mosquitoes and parasites that are harder to kill. Researchers are now working on a vaccine to help protect against malaria.

ANTIVIRAL DRUGS: FIGHTING A NONLIVING ENEMY

Antibiotics that kill living monerans and protists leave virus particles untouched. Until recently, therefore, most drugs prescribed for colds, flu, herpes, and other viral infections have done little more than treat uncomfortable symptoms and control secondary bacterial or protistan infections, while the body's immune system geared up to destroy the virus on its own. But antiviral drugs are now beginning to arrive.

Recall that a virus must shed its outer protein coat in order to enter a cell and be replicated. One new drug prevents certain viruses from shedding their coats. For example, mice infected with poliovirus and then given the drug remained unparalyzed. A second drug, called interferon, is proving effective against the human cold. Interferon is a natural substance produced by the immune system, and it somehow prevents cells from replicating viruses. A few biotechnology companies have been able to produce large quantities of interferon, which drug researchers have incorporated into a very effective nasal spray. Clinical trials are now under way to get the necessary government approval to market both of these drugs.

Yet another drug called acyclovir is proving effective against herpes infections. By blocking the activity of certain herpes viral enzymes, acyclovir helps control shingles, cold sores, and genital herpes, a venereal disease affecting 10 to 20 million Americans.

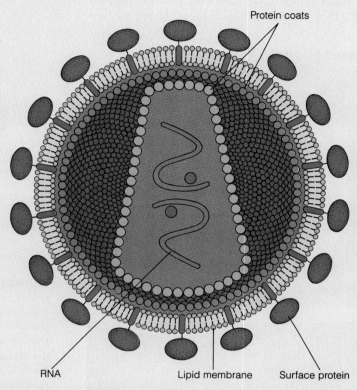

FIGURE 1 *The AIDS Virus.* The virus that causes AIDS consists of a lipid membrane surrounding protein coats; inside the coats are two molecules of RNA, which make up the genetic material of the virus.

In the desperate battle against the deadly AIDS virus (Figure 1), researchers have produced three antiviral drugs with the capacity to prolong the lives of AIDS patients and reduce their symptoms. These drugs, abbreviated AZT, DDI, and DDC, all act by blocking the reverse transcription of RNA from the AIDS virus into DNA, which would lead to immune cell destruction. The blocking thus prevents some degradation of the immune system, and with it, some of the devastating symptoms of the disease.

Many observers hope that within a decade viruses, including the lethal AIDS virus, will be at least as controllable as the bacterial and protistan parasites we have battled so long and hard with drugs.

■ **Slime Molds: The Funguslike Protists** Among the most bizarre members of the kingdom Protista are the glistening, single-celled *slime molds*. Slime molds derive food and energy as a fungus would—by secreting digestive enzymes that break down organic matter and then absorbing the digested material back into the cell. While this resembles the nutritional strategy of mushrooms and molds, the slime molds have life cycle phases that can also resemble those of animals and plants. There are two

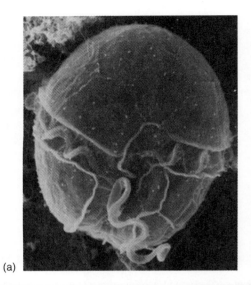

(a)

(b)

FIGURE 15.17 *Dinoflagellates: Armored Protists Behind Red Tides.* (a) The microscopic phytoplankton *Protogonyaulax catenella* has the dual flagella in grooves characteristic of so many dinoflagellates and wears a coat of cellulose armor. (b) Some dinoflagellates cause dangerous red tides—dense blooms of certain species that can tint the water blood red and produce deadly nerve toxins, as happened during this red tide along the coast of Ethiopia. The toxins kill people who eat fish and shellfish that have consumed the dangerous dinoflagellates. Many coastal areas ban the collection of shellfish from May through August, when red tides occur.

FIGURE 15.18 *Diatoms: Golden-Brown "Pillboxes."* Diatoms have a golden-brown pigment, as well as jewellike shells. Diatoms have silicon instead of cellulose in their cell wall and store oils rather than starches. Diatom shells fall to the ocean floor and accumulate in crumbly white sediments called diatomaceous earth, which people use in toothpaste and swimming pool filters. Diatoms and golden-brown algae contain chlorophyll and are so abundant that biologists estimate they contribute more oxygen to the atmosphere than all land plants combined.

FIGURE 15.19 *A True Slime Mold.* A true slime mold is a large, flat, fan-shaped mass of cytoplasm that contains many nuclei and that moves about slowly like a giant amoeba in damp soil, leaf litter, and downed wood.

types of slime molds: the *true slime molds,* which consist of a single very large, flat cell with many nuclei (Figure 15.19), and the *cellular slime molds,* which consist of individual amoeba-like cells that live independently for a time and then can join together in a beautiful reproductive structure (Figure 15.20). If you look closely at fallen leaves and branches in the woods, you are likely to see these colorful little structures, which look like miniature golf balls on tees and are so important for recycling biological materials. No one is sure how the slime molds evolved; however, they do show a hint of the cellular cooperation we will see so much of in the multicellular kingdoms we consider in the next three chapters.

≋ CONNECTIONS

The most numerous and in many ways most important organisms on earth are ironically the smallest and the oldest. All around us, in the soil, air, and water, unseen millions of single-celled organisms absorb and use a broad range of nutrients and reproduce in as little as 20 minutes. In some ways, our own survival depends on the widespread success of these monerans and protists. Golden-brown algae and cyanobacteria together produce most of the oxygen we breathe; soil microbes "clean up the dirt" by breaking down dead organisms; and phytoplankton form the first link in the oceanic food chain. Nevertheless, parasitic monerans and protists do take their toll. To a disease-producing single-celled organism, a human body or a towering plant is a place to live and reproduce, and generations of such cells live and die on our teeth and in our guts. For the remainder of this century, we humans will expend considerable effort trying to control the harmful microorganisms and trying to use the others for our own purposes. Our next chapter explores the fungal and plant hosts for some of these microbes, including their structures, modes of reproduction, and crucial ecological roles.

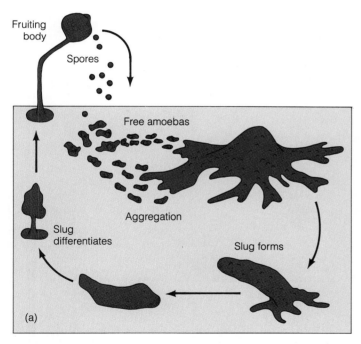

FIGURE 15.20 *Life Cycle of the Cellular Slime Mold.* (a) Cellular slime molds normally move about like amoebas—as single cells in search of bacteria. If environmental conditions worsen, however, they can aggregate, form a slug, and slither to some likely location. There they stop and differentiate into a fruiting body with a stalk and a spore chamber that eventually releases new, genetically recombined spores. These, in turn, germinate into new free amoebas that crawl off once again in search of bacteria to consume. (b) Close-up of slime mold fruiting bodies on a tree trunk.

① The trillions of free-living, single-celled organisms all around us can reproduce quickly and use a broad range of nutrients.

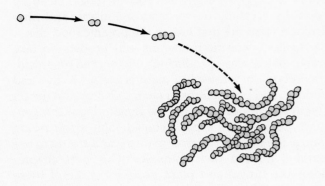

② The monerans are the simplest and oldest living things. Protists are more complex cells and gave rise to plants, animals, and fungi.

Fungi Plants Animals

Protists

Monerans

Monerans

Protists

③ Monerans are bacteria, which lack membrane-enclosed organelles. Bacteria have great medical, ecological, and scientific impact on our lives.

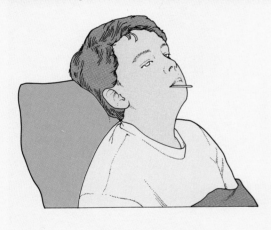

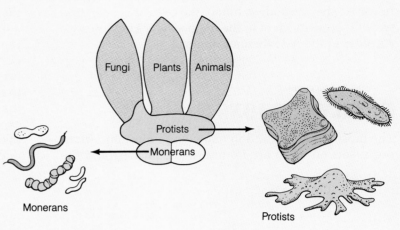

Bacteria

Oil spill

Bacteria break down oil

④ Protists have a nucleus and other membrane-enclosed organelles and are animal-like, plantlike, or funguslike in life-style. Some cause serious disease; most contribute to and help sustain our environment.

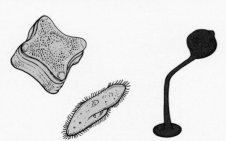

⑤ Viruses are nonliving, parasitic agents of disease, made up of protein and nucleic acid. They use a host cell's life-sustaining machinery for their own reproduction.

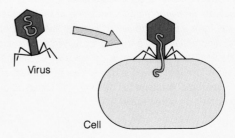

Virus

Cell

NEW TERMS

cyanobacteria, page 258
phytoplankton, page 265
protozoa, page 265

spore, page 259
virus, page 262

STUDY QUESTIONS

REVIEW WHAT YOU HAVE LEARNED

1. How do protists and monerans differ?
2. Why are microbes important organisms?
3. Describe the cell structure of a typical moneran.
4. Describe the shapes of the following bacterial species: *Bacillus subtilis, Staphylococcus aureus, Spirillum volutans,* and *Vibrio comma.*
5. How do a decomposer, a parasite, and a symbiont differ?
6. How are some types of bacteria useful to humans?
7. Trace the steps by which a virus infects a cell.
8. How do a viroid and a prion differ from a virus?
9. Name the four major groups of protozoa, and give the chief characteristics of each. How do they affect our health and our environment?
10. What is phytoplankton? Why is it important?
11. What are slime molds? How are they important?

APPLY WHAT YOU HAVE LEARNED

1. A newly hired laboratory worker believes that he can thoroughly clean surgical instruments using hot, soapy water. Is he correct? Explain.

2. You are suffering from measles, a disease caused by a virus. Why is the doctor unlikely to prescribe an antibiotic?
3. Campers in the wilderness decide that it is safe to drink from a cool, clear, backcountry stream. What illness may result from this decision? Why?

FOR FURTHER READING

Begley, S., and T. Waldrop. "Microbes to the Rescue." *Newsweek,* 19 June 1989, pp. 56–57.

Corliss, J. "The Kingdom Protista and Its 45 Phyla." *Biosystems* 17 (1984): 87–126.

Desowitz, R. S. *New Guinea Tapeworms and Jewish Grandmothers: Tales of Parasites and People.* New York: Norton, 1987.

Finlay, B., and T. Fenchel. "Everlasting Picnic for Protozoa." *New Science,* 1 July 1989, pp. 66–69.

Magasanik, B. "Research on Bacteria in the Mainstream of Biology." *Science* 240 (1988): 1435–1439.

Oldstone, M. B. A. "Viral Alteration of Cell Function." *Scientific American* (August 1989): 42–48.

Postgate, J. "Bacterial Worlds Built on Sulphur." *New Science,* 21 January 1989, pp. 40–44.

Scott, A. "Viruses Work to Improve Their Image." *New Science,* 19 May 1988, pp. 52–55.

Shapiro, J. A. "Bacteria as Multicellular Organisms." *Scientific American* (June 1988): 82–89.

PLANTS AND FUNGI: PRODUCERS AND DECOMPOSERS

REDWOODS AND ROOT FUNGI

The most massive organisms ever to exist on our planet are still alive today: the giant sequoia trees growing on the western slope of the Sierra Nevada in central California (Figure 16.1). The biggest sequoia on record, the General Sherman Tree, stands in a quiet grove in Sequoia National Park and has a trunk weighing 1300 metric tons (almost 3 million pounds). The trunk of this gargantuan plant is 11 m (37 ft) in diameter—about the size of a large classroom. The plant reached reproductive age at 10 to 14 years, but can still produce cones and seeds even at age 3000.

For all their bulk and long life, sequoias have one seeming anatomical deficiency: Their roots push outward only 15 m (50 ft) or so from the trunk and extend downward less than 1 m (3 ft) into the soil. By contrast, a corn plant can have roots twice as deep! How can such a shallow and relatively small root system anchor such a massive tree and take in enough water and nutrients to support its growth? The answer is that the roots are infected with a minute fungus that sends billions of tiny hairlike extensions into the soil around the sequoia roots. These extensions absorb far more water and other nutrients than the plant's roots could on their own.

With this chapter, we move from the single-celled kingdoms to the multicellular realms of the fungi and plants. You depend on both fungi and plants for your very survival. For example, fungi recycle fallen trees, unlocking and returning to the soil and air many compounds required for life. Flowering plants provide grains for the bread and cereal you eat as well as the fruits and vegetables so vital to a healthy diet. And

wood from many different trees shows up in your tables and chairs, beds and bookcases. Because fungi evolved earlier than plants and are for the most part smaller and simpler organisms, we will consider them first.

Several themes will emerge in our discussion of fungi and plants. Fungi are the great decomposers: They break down organic matter and in the process make nutrients available to other life forms. Fungi are also highly interdependent with plants. Most fungi survive by breaking down plant matter, and some 90 percent of all plants that live on land have the kind of root-fungi associations the giant sequoias depend on.

In contrast to the decomposers, plants are the great producers: Each year, the earth's plants collectively produce, through photosynthesis, 150 billion tons of carbohydrates, an amount that could fill the boxcars of 100 trains, each long enough to stretch from the earth to the moon! People and most other organisms live directly or indirectly on the carbohydrates that plants produce.

Finally, as plants evolved from water dwellers to land dwellers, a series of structures and functions emerged that helped them meet the challenges of life on land. New systems for internal transport, vertical support, and reproduction enabled complex land plants to diverge and greatly outnumber the kinds of simpler plants that remained in bodies of water or at their damp fringes.

This chapter describes the basic traits of fungi and plants and answers several questions:

- How do fungi live, grow, and reproduce?

- What are the major groups of fungi, and why are they important?
- How do plants live, grow, and reproduce?
- What trends help us understand plant evolution?

FIGURE 16.1 *A Giant Sequoia.* A majestic individual, a member of the largest living species, towers over neighboring trees in the Mariposa Grove, Yosemite National Park, California.

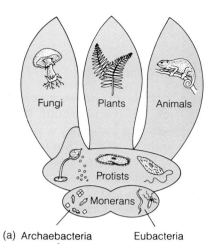

(a) Archaebacteria   Eubacteria

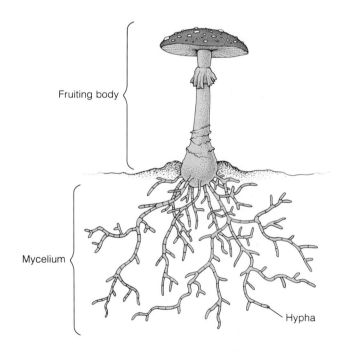

(b)

FIGURE 16.2 *Decomposers and the Interdependence of Two Kingdoms.* (a) The fungi are an entirely separate kingdom of life that arose from protistan ancestors. (b) In nature, the survival of fungi often depends on plants. Here, a *Pycnoporus* fungus grows on a decaying log, recycling its nutrients.

≋ AN OVERVIEW OF THE KINGDOM FUNGI: THE GREAT DECOMPOSERS

Biologists believe that fungi arose from single-celled decomposers as much as 900 million years ago (Figure 16.2a). Today's fungi include the familiar mushrooms, molds, and puffballs. Along with the various decomposing bacteria and protists, fungi are the earth's major decomposers (Figure 16.2b). Most fungi decompose *nonliving* organic matter, such as fallen wood and leaves and dead animals. Called *saprobes*, such fungi will consume everything from leather and cloth to paper, wood, paint, and other materials, slowly reducing old buildings, books, and shoes to crumbled ruin. Some fungi, however, are *parasites*, organisms that attack living things. Today, parasitic fungi are the main cause of plant diseases, attacking crops in fields, gardens, and orchards. Other types

of parasitic fungi attack humans and other animals and include the species that cause athlete's foot and vaginal yeast infections.

THE FUNGAL BODY

Fungi are distinguishable from the other four kingdoms by the way they obtain nutrients and by their unique physical structure. Fungi obtain organic nutrients by secreting enzymes that break down organic matter and then absorbing the released nutrients through their cell membranes. Because the actual breakdown process takes place outside the organism's body, it is called *extracellular digestion*.

Some fungi are free-living single cells; these include the yeasts that help create beer and wine. Other fungi are composed of many cells—like toadstools and mushrooms. The body of a multicellular fungus consists mostly of cells joined into filaments. In a mushroom, these filaments, called **hyphae** (singular, hypha), grow underground in a mat, or *mycelium*, that is rather like a string sculpture or a piece of steel wool (Figure 16.3). A mushroom's hyphae can penetrate the soil for many square meters and grow very quickly: Just one day's new growth of filaments, if placed end to end, can easily exceed half

Fruiting body

Mycelium

Hypha

FIGURE 16.3 *Fungi: General Structures.* A typical mushroom is just a small portion of the total organism. The mycelium makes up the rest, a tangled mat of loosely packed hyphae beneath the ground surface. The mushroom, or fruiting body, is made of tightly packed hyphae and produces spores that germinate and grow into new hyphal mats.

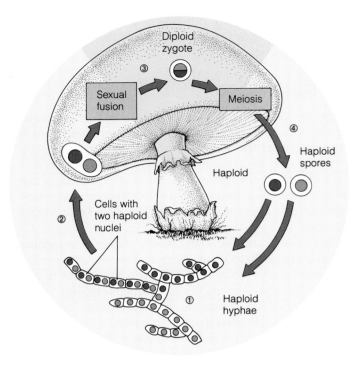

FIGURE 16.4 *Sexual Reproduction in a Fungus.* Cells from two genetically different hyphal filaments can fuse, giving rise in some fungi to single cells with two haploid nuclei (1). These cells with dual nuclei can develop into a fruiting body, such as this mushroom (2). Inside the mushroom, the two nuclei in a cell can fuse together and form a diploid zygote (3), which then undergoes meiosis and forms haploid spores (4). The spores can germinate, and new hyphae arise. Note the brief duration of the diploid phase of the life cycle and the predominance of the haploid phase.

a mile. This explains why a mushroom's aboveground reproductive structure *(fruiting body)* can literally spring up overnight on damp logs or soil. The mushroom looks like a solid mass but is actually a meshwork of hyphae.

FUNGAL REPRODUCTION

Fungi generally reproduce asexually. Filaments can break off the main body and grow into new individuals; or the fungus can produce asexual **spores,** cells that are dispersed and that divide into new, genetically identical fungi. The hyphae and the asexual spores are usually haploid (contain one set of chromosomes), and the haploid phase dominates the fungal life cycle. Spores are adaptations for survival, able to withstand extreme conditions of dryness or cold, then produce a new fungus when conditions improve. Fungi can also reproduce sexually (Figure 16.4). Sexual reproduction results in a short-lived diploid cell (with two sets of chromosomes) that produces haploid spores through meiosis. The spores grow into hap-

loid hyphae capable of either asexual or sexual reproduction.

FUNGAL INTERACTIONS WITH PLANTS

Despite the damage that they cause each year, fungi as a group are more beneficial than harmful. It is easy to imagine the earth becoming buried by a deep layer of dead wood and fallen leaves were it not for the activity of fungi, since only they can decompose cellulose and lignin, the hard substances in wood. Also, out of sight, below the soil surface, fungi and plant roots—like those of the giant sequoias—interact in important and mutually beneficial ways. These associations between roots and fungi are called "fungus roots" *(mycorrhizae).* It wasn't always clear that such fungal infections would benefit plants, but experiments showed that they do (see the box on page 275).

In many woody plants, hairlike fungi associated with cells of the plant's roots expand the surface area for uptake of water and minerals (Figure 16.5). Infected sequoia trees

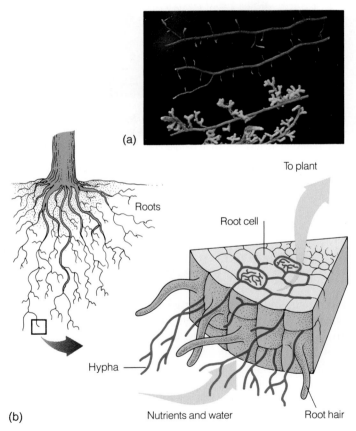

FIGURE 16.5 *Fungus Roots, or Mycorrhizae.* Fungi associated with plant roots absorb water and nutrients, and the host plant benefits. (a) Called mycorrhizae, associated fungi give pine roots a fuzzy white appearance compared to bare roots. (b) Some beneficial fungi can grow inside plant root cells, while their microscopic hyphae (red) branch into the soil amid the root hairs.

1. OBSERVATION: The roots of this tree have a fungus growing on them.

2. QUESTION: I wonder what effect the fungi have on the plant?

3. HYPOTHESIS: I'll bet the fungi are harmful, like my athlete's foot infection.

4. EXPERIMENT AND PREDICTION: To find out, I'll add fungi to one group of seedlings, but prevent fungi from growing on another group, and if my guess is right, the infected plants will grow slower.

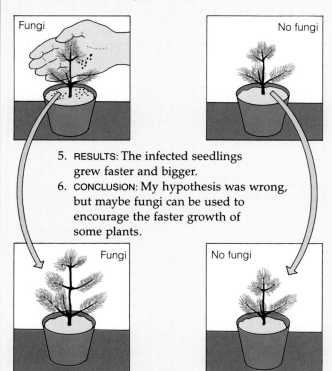

5. RESULTS: The infected seedlings grew faster and bigger.
6. CONCLUSION: My hypothesis was wrong, but maybe fungi can be used to encourage the faster growth of some plants.

Adapted from J. A. Vozzo, "Field innoculations with Mycorrhizal Fungi." In: *Mycorrhizae* editor E. Hacskaylo. US Gov. Printing office, Sept. 1971

FIGURE 16.6 *Lichens: Alga-Fungus Associations.* Lichens can be colorful or camouflaged, but in each type, a fungus is closely associated with photosynthetic algal cells. Several types of lichens grow on this boulder in Colorado.

can take in ten times more phosphorus and other nutrients than they could without the fungal symbionts. In exchange for augmenting the plant's nutrition, the fungus gets a home and probably other benefits not yet understood.

A second important type of fungus-plant interaction can be found in **lichens:** gray, orange, or greenish crusts that grow on bark, soil, or rocks (Figure 16.6). A lichen is an association between fungus and an alga, with the fungus forming a dense hyphal mat around the algal cells. The fungus lives off the alga's photosynthetic products, while the alga receives protection and a supply of water from the fungus. Lichens take few nutrients from the rock or tree to which they are attached, and they are often the first organisms to inhabit lava flows or other newly exposed rocks. While lichens can survive in hostile environments, they are particularly sensitive to some industrial pollutants. Therefore, they have recently been used as early indicators of polluted environments.

Fossils reveal that fungi grew on land about 500 million years ago—long before plants left the oceans. What's more, fossilized fungi have been identified among the roots of the oldest fossilized land plants. This evidence suggests that "fungus roots" helped plants make the transition to land. Without fungus-plant interactions, plants might have evolved in very different ways, and life's giants, the sequoias, might never have appeared.

MAJOR GROUPS OF FUNGI: A SURVEY

Like plant cells, fungal cells have a thick wall. This makes it impossible for fungi to move from place to place on their own. Spores are the only means a fungus has of dispersing to new areas. Let's look at five major classes of fungi, focusing on how they form spores (Table 16.1).

■ **The Egg Fungi, or Water Molds**   These simple inhabitants of soil and water (called the *Oomycota*, from the Greek words for "egg" and "mushroom") form large, immobile egg cells (hence the name "egg fungi"); and after fertilization, they form spores that disperse by swimming (hence the popular name "water molds"). One member of this group causes late blight in potatoes, a disease that kills potato vines. This fungus attacked Ireland's potato crop in 1845–1847 (Figure 16.7). The Irish depended so heavily on potatoes for their daily diet that more than a million people starved, and 1.5 million more emigrated, mostly to North America.

■ **Bread Molds and Related Fungi**   The typical members of this group grow in stringy, cottony masses and form dark, thick-walled spores called *zygospores* (the class is named *Zygomycota*). The common fuzzy whitish or grayish mold that grows on bread *(rhizopus)* is a zygomycote. Most of the ecologically important associations between fungi and plant roots involve species of this group.

■ **Mildews and Other Sac Fungi**   This is the largest class of fungi and includes mostly decomposers but some important plant parasites as well. During sexual reproduction, these fungi produce spores in a little sac called an *ascus* (the group is called *Ascomycota*) (Figure 16.8). The single-celled yeasts, including those used in making bread and beer, are ascomycotes. The yeasts that we employ in the brewing and baking industries are probably the most economically useful fungi. One yeast, however, causes vaginal yeast infections.

Some many-celled ascomycotes are used directly as food, including the morels and truffles so favored by gourmet cooks. Others, however, attack our food crops. Powdery mildews parasitize apple and cherry trees, as

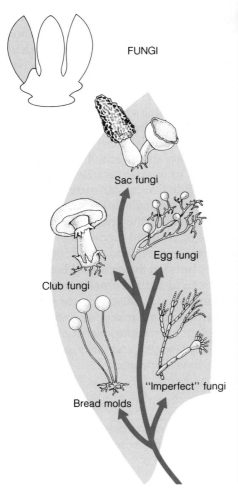

FUNGI

Sac fungi

Club fungi

Egg fungi

Bread molds

"Imperfect" fungi

TABLE 16.1 ≋ FUNGI: THE GREAT DECOMPOSERS

CLASS	EXAMPLES	CHARACTERISTICS AND SIGNIFICANCE
Oomycota (egg fungi; 400 species)	Potato blight fungus	Thrive in damp conditions and cause major mold damage to potatoes, grapes, and other crops; their motile spores disperse by swimming
Zygomycota (bread molds; 600 species)	Common black bread mold	Produce diploid spores and a cottony mat of hyphae on breads, grains, or other foods and organic materials
Ascomycota (sac fungi; 30,000 species)	Pink bread mold, brewer's yeast, morels, truffles	Produce spores in an ascus, or sac, borne in a cup-shaped body; because they include the yeasts, they are the most economically useful fungal group; powdery mildews harm fruit trees and grain crops
Basidiomycota (club fungi; 25,000 species)	Common field mushrooms, giant puffballs, bracket fungi, toadstools, smuts	Produce spores in club-shaped basidia; the fruiting body is the familiar mushroom or toadstool, which can be extremely poisonous
Deuteromycota ("imperfect" fungi; 25,000 species)	Species that produce penicillin	Lack sexual reproduction and instead produce asexual spores; various species are used in making drugs, cheeses, and soy sauce

well as grain crops. Bread baked with rye infected by one ascomycote causes ergot poisoning, characterized by hallucinations, convulsions, premature labor, and gangrene of the arms and legs. Some historians now believe that the "possessed" witches of Salem suffered from this condition. Still other ascomycotes cause Dutch elm disease and chestnut blight, diseases that have robbed us of millions of our most beautiful hardwood trees.

■ **Club Fungi** The most familiar fungi—the edible mushrooms, bracket or shelf fungi, puffballs, toadstools, and smuts—are all members of *Basidiomycota* (Figure 16.9). The spore-producing abilities of the club fungi are truly amazing. An average-sized grocery store mushroom can produce about 16 billion spores. If all these spores germinated and developed into adult mushrooms, they would collectively weigh 800 times the weight of the earth! Obviously, since we are not buried by mushrooms, the spores rarely germinate and reach maturity.

(a) (b)

FIGURE 16.9 *Mushrooms: Poisonous and Edible.* Most familiar forest fungi are basidiomycotes. (a) This deadly poisonous *Amanita muscaria* grows in a forest in New York State. (b) A giant puffball, *Calvatia gigantica*, dwarfs this boy's hands and arms.

FIGURE 16.7 *Potato Blight and Ireland's Food Staple.* This Irish potato farmer may have had relatives who lived through the great potato famine of the mid-nineteenth century, caused by a water mold.

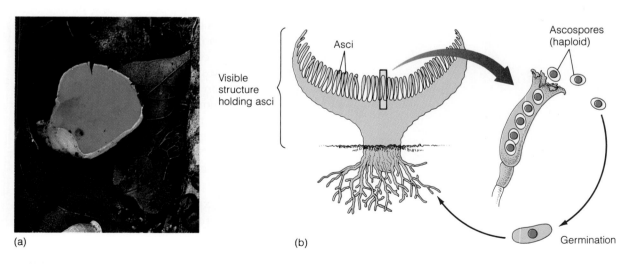
(a) (b)

FIGURE 16.8 *Sac Fungi.* These fungi house a series of little sacs, or asci; meiosis within each sac yields spores, and these blow about, germinate, and grow into new sac fungi.

The mushroom we see is just the *fruiting body,* the reproductive portion of the individual fungus made up of densely packed hyphae. Below the soil, sometimes extending for many meters, is a large mat of loosely tangled hyphae that sends up mushrooms at times appropriate for spore production and dispersal.

■ **Imperfect Fungi** One group of fungi (called *Fungi imperfecti,* or *Deuteromycota*) have one "imperfection": They lack sexual reproduction and propagate only through asexual spores. We use "imperfect" fungi for making penicillin and other drugs, Roquefort and other blue cheeses, soy sauce, and sake (Japanese rice wine). An "imperfect" fungus that grows on stored peanuts and various grains produces a compound called *aflatoxin,* one of the most potent cancer-causing substances ever identified.

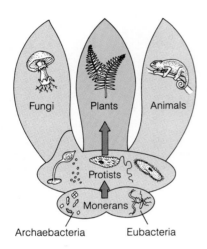
(a)

(b)

FIGURE 16.10 *Plants: The Great Producers.* (a) Plants probably evolved from photosynthetic protists that were predecessors to the green algae. (b) This beautiful carpet of bright green duckweed helps camouflage the frog.

We now shift our focus to a second kingdom of organisms, the plants, with which fungi share so many interactions and which produce the carbohydrates that directly or indirectly support most other life forms.

❊ AN OVERVIEW OF THE KINGDOM PLANTAE: THE GREAT PRODUCERS

Plants can be gigantic like the sequoia, minute like the duckweed on a pond's surface, brilliantly colored, or drably camouflaged (Figure 16.10). Regardless of differences, however, plants have certain common physical characteristics, the ability to turn solar energy into sugar, and a fascinating evolutionary history.

PLANT LIFE CYCLES

Plants are multicellular organisms that generate their own organic nutrients through photosynthesis. Most plants have leaves (or equivalent structures) that act as solar collectors, stems (or the equivalent) that support these collectors, and roots (or similar structures) that anchor the plant and absorb water and nutrients. Plant life cycles are characterized by an alternation of haploid and diploid phases, which we saw in the fungi. Unlike the fungi, however, both phases in plants are generally many-celled.

The haploid phase is the "gamete-making (that is, sex-cell-making) plant" **(gametophyte).** As the name implies, each such plant can produce male and/or female gametes. In sexual fusion, the gametes unite, forming a zygote that develops into the spore-making plant **(sporophyte),** which is the diploid phase. When most people think of a plant, they are visualizing the spore-forming part of the life cycle. Meiosis in the spore-making plant results in the production of male and/or female haploid spores, not gametes. The spores develop into the gamete-making plant and this completes the life cycle.

While the two phases alternate in the life cycles of all plants (Figure 16.11), in some species, including several algae, the gamete-making generation is a sizable individual that occupies a large part of the life cycle. In other plant species, including daisies and sequoia trees, the gamete-making phase consists of but a few short-lived cells, and the spore-making phase dominates the life cycle.

A CHANGING PLANET AND THE CHALLENGES OF LIFE ON LAND

Plant life probably originated in water as photosynthetic protists evolved into primitive green algae, which may have been the ancestors to all land plants. The first tran-

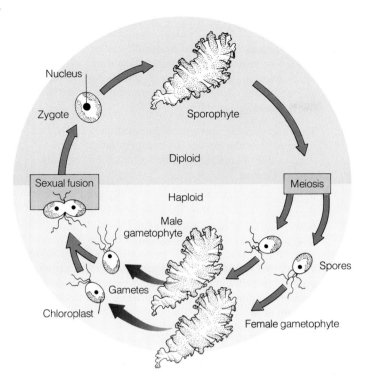

FIGURE 16.11 *A Representative Plant Life Cycle.* Plants have an alternation between a multicelled haploid generation (the gametophyte, or gamete-making plant) and a multicelled diploid generation (the sporophyte, or spore-making plant). One phase usually dominates the other, but both are present at some time in the plant's life cycle. Here we show the life cycle of a sheetlike green alga (see Figure 16.15), where neither phase dominates in the form of a larger, longer-lived plant.

sitional plants invaded the damp fringes of ancient oceans some 400 million years ago. At that time, the continental landmasses slowly buckled and relaxed dozens of times. This raising and lowering of the land allowed vast, shallow inland seas to accumulate and drain again and again. But it also stressed aquatic plants, submerging them at some times and leaving them high and dry at others. Organisms with the ability to procure energy, reproduce, and absorb water and minerals on drying land could have withstood the extreme environmental pressures, and biologists believe that plants with exactly these abilities—derived from mutations and random genetic combinations—became the earliest surviving land plants. In our survey of the plant kingdom, we will see certain evolutionary trends, including: (1) the emergence of a vascular system for transport of materials; (2) the increasing dominance of diploid plant forms; and (3) the production of seeds (Figure 16.12). Let's begin now with the algal ancestors of the land plants and trace the appearance of land plants step by step.

≋ ALGAE: ANCESTRAL PLANTS THAT REMAINED AQUATIC

If one thinks of the geological eras as four acts in a long-running play, then the plant stars of the earliest era—the Proterozoic—were the algae. But what exactly are algae? The term **algae** (singular, alga), from the Latin

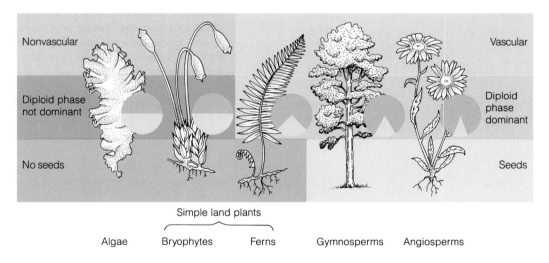

FIGURE 16.12 *Trends in Plant Evolution.* In the evolution of the five major plant groups, three main trends emerged: (1) an increase in the importance of the diploid (sporophyte) phase of the life cycle, as shown by the darker green sectors in the life cycle circles; (2) the evolution of a vascular system (shown in red in Figure 16.18a) and (3) the evolution of seeds.

TABLE 16.2 ≋ CHARACTERISTICS OF THE MAJOR PLANT DIVISIONS

DIVISION		EXAMPLES	EVOLUTIONARY TRENDS AND SIGNIFICANCE
Algae*	Red algae (Rhodophyta; 4000 species)	Many fanlike or filamentous types	Single-celled, colonial, or multicellular; reproduction asexual or sexual; haploid generation dominates; light-absorbing red or purple pigments can function at great water depths; used commercially for extracting thickening agents.
	Brown algae (Phaeophyta; 1500 species)	Giant kelps and *Sargassum*	The largest algae; have tubelike conducting cells, but not true vascular tissue; diploid generation dominant in many species; kelp harvested for chemicals and cattle feed
	Green algae (Chlorophyta; 7000 species)	*Ulva, Chlamydomonas*	Produce carotene, chlorophyll, like the land plants; many have conspicuous haploid and diploid generations; may have been ancestors to the land plants; frequent freshwater environments
Simple Land Plants	Bryophytes (Bryophyta; 20,000 species)	Mosses, liverworts, hornworts	Waterproof coatings, rigid tissues for upright growth on land, rootlike rhizoids; haploid generation (gametophyte) is dominant; often the first plants to colonize an area
	Vascular plants (Tracheophyta; 300,000 species)		Vascular tissue for transport and vertical support on land; most have roots, stem, and leaves; diploid sporophyte usually dominant; most successful and diverse group of plants
	Subdivisions of vascular plants: Lower vascular plants (12,000 species)	Ferns, horsetails, club mosses	System of vascular pipelines; rhizomes, stems, and fronds; gametophyte can be tiny, independent plant or can grow from sporophyte; usually grow on shady forest floors in low-lying damp areas
Gymnosperms	Seed plants		Produce seeds; gametophyte reduced to a few cells, dependent on sporophyte; do not depend on water for reproduction
	Classes of seed plants: Coniferophyta (600 species)	Conifers, cycads, ginkgos	Naked seeds produced in cones; usually have needlelike leaves or scales; produce pollen; well-developed vascular system; true roots, stems, and leaves; sporophyte is dominant and supports gametophyte; conifers harvested in great numbers for wood products
Flowering Plants	Anthophyta	Flowering plants	Produce flowers, seeds, and fruit; economically useful for food, drugs, landscaping
	Subclasses of Anthophyta: Monocots (50,000 species)	Lily, corn, onion, palm, daffodil	Leaves usually have parallel veins; seedlings from newly germinated seeds have just one "seed leaf" or cotyledon; flower parts usually occur in multiples of three; seed stores much endosperm
	Dicots (225,000 species)	Rose, apple, bean, daisy	Leaves usually have netlike veins; seedlings have two cotyledons; flower parts usually occur in multiples of four or five; seed stores little endosperm

*Some biologists consider red, brown, and green algae to be protists because certain of their members are single-celled.

word for "seaweed," has been used to describe aquatic photosynthetic organisms in the kingdoms Monera, Fungi, and Plantae. The algae considered plants are mostly multicellular aquatic organisms with simple reproductive structures. Algae probably arose about 600 million years ago from photosynthetic protists and were the only plants "on stage" that early. They have never exited the evolutionary play, however, and since their early stardom have diversified into 12,000 species that vary from tiny threads to giant kelps and tangled seaweeds. Algae live in virtually all bodies of water and sometimes on damp soils, rocks, trees, and even snow banks. Although they are generally simpler than the complex plants now dominating land habitats, modern algae continue to serve a critical ecological role: Aquatic habitats are so widespread (covering nearly 75 percent of the earth's surface) and algae are so abundant, biologists believe that collectively the algae capture 90 percent of all the solar energy trapped by plants, notwithstanding the vast forests and prairies on land.

The aquatic habitat helped shape the organisms that live there. Because water supports the algal plant body, algae lack rigidity and usually undulate gently with waves and water currents. Most algae are quite flat and translucent, with maximum surface area. This allows cells to absorb water, minerals, and sunlight directly from their surroundings, with no need for the specialized conducting tubes of a complex land plant. Finally, reproduction can be asexual, involving the fragmenting of cells or body parts that then develop into a whole new plant; or it can be sexual, with the production of eggs and sperm. In the sexual life cycle, the diploid phase (sporophyte) does not generally dominate (see Figure 16.11). The main secret to the algae's success is a range of photosynthetic pigments that can absorb the light of the different wavelengths that penetrate to varying water depths. Botanists use these same pigments to distinguish between red, brown, and green algae (Table 16.2).

RED ALGAE: RESIDENTS OF TROPICAL SHALLOWS OR DIM OCEAN DEPTHS

Most red algae are small, delicate plants that occur as thin filaments or flat sheets with an ornate, fanlike appearance and generally live in shallow tropical ocean waters (Figure 16.13). Some, however, exist as single cells or as cells associated in colonies, and a few species survive at depths of up to 268 m (884 ft). These deep-sea denizens are 100 times more efficient at capturing sunlight than are the red algae found in shallow waters, and they rely on reddish pigments that can absorb the dim blue-green light that penetrates to great depths. The reddish compounds give these plants their stunning range of colors—from pink to purple to reddish black—and pass light energy to chlorophyll for photosynthesis. Red

FIGURE 16.13 *Red Algae: Deep or Shallow Ocean Dwellers.* Because of their special red pigments, red algae can live beneath the ocean surface. Some live deeper than any other plants; others, like the one shown here, live in shallow seas.

algae produce a starchy substance (carrageenan) that is used as a stabilizing agent in ice cream, puddings, cosmetics, and paint.

BROWN ALGAE: GIANTS OF THE OFFSHORE REALM

Most brown algae inhabit cool, offshore waters and occur as small multicellular plants. However, the largest members of the algal world are the multicellular *kelps*, brown algae that can grow 100 m long and float vertically like tall trees (Figure 16.14, page 282). Huge floating masses of the brown alga *Sargassum* thrive in the Sargasso Sea, north of the Caribbean, sometimes entangling hapless divers and ships alike.

Brown algae range from golden brown to dark brown to black because of pigments that collect the blue and violet light penetrating medium-deep water and pass it on to chlorophyll. These pigments explain why kelps can grow so tall—the sequoias of the sea.

Many brown algae possess complex leaflike, stemlike, and rootlike structures analogous to parts of land plants; these structures collect sunlight and produce sugars, support the plant vertically, and anchor the plant to submerged rocks.

GREEN ALGAE: ANCESTORS OF THE LAND PLANTS

Most species of green algae live in shallow, freshwater environments or on moist rocks, trees, and soil, although

FIGURE 16.14 *Brown Algae: Giant Underwater Kelps.* As a diver looks up at tall, fluttering kelp through the sunlight that slants below the surface, the plants tower like forest trees and can reach as high as a 30-story building.

a few inhabit shallow ocean waters. Green algae usually occur as single cells or as multicelled threads, hollow balls, or wide, flat sheets (Figure 16.15).

Green algae are most notable for producing yellow pigments (carotenoids) and chlorophyll; together these pigments absorb the sunlight penetrating shallow water or air with maximum efficiency. Green algae share this pigment combination only with the land plants. This is a main reason why botanists consider green algae the direct ancestors of land plants.

We now move on to the simplest land plants, which still rely on standing water for reproduction, and recount how several trends allowed life to inhabit the land.

※ SIMPLE LAND PLANTS: STILL TIED TO WATER

In the long-running play of plant evolution, it wasn't until halfway through the second act—about 350 million years ago—that plants made the transition to land. There were two groups of players in this act: the **bryophytes**, or mosses and relatives, which grow low to the ground and require standing water for reproduction; and the **lower** (primitive) **vascular plants**, including ferns, horsetails, and club mosses, which grow taller than the bryophytes and have an internal transport system, but still require standing water to reproduce (Figure 16.16). The term *vascular* describes this latter group's internal transport system—a network of tubes, or vessels, that carries nutrient-filled fluid to all parts of the plant. The vascular

group played the starring role in this time period, but both kinds of simple land plants were later upstaged by the higher vascular plants, which include the cone-bearing gymnosperms and the flowering angiosperms, with their seeds and freedom from moisture-dependent reproduction. Botanists believe that a green alga made the transition to land and then later diversified into the bryophytes and lower vascular plants, the latter in turn giving rise to the higher vascular plants.

The challenge of life on land remains today, and in meeting that challenge, plants evolved (1) a way to prevent drying out while still allowing for the exchange of oxygen and carbon dioxide; (2) a way to support the plant body in the absence of buoyancy from surrounding water; (3) a way to absorb water and minerals from the soil, mud, or sand and transport them through the plant body; and (4) a way to reproduce sexually without shedding sex cells (gametes) directly into an ocean, lake, or pond. As we shall see, the relative success of each group of land plants depends, in part, on how its physical structures evolved and met these challenges.

PLANT PIONEERS ON LAND: THE BRYOPHYTES

The modern bryophytes include mosses, liverworts ("lobed plants"), and hornworts ("horn-shaped plants") (see Table 16.2). Second in number of species only to the flowering plants, bryophytes are small organisms that either lie flat or stand less than about 1–2 cm tall (Figure 16.17). They are often the first plants to colonize a new area, and in most of the species, the spore-bearing plant grows like a miniature street lamp from the dominant

FIGURE 16.15 *Sea Lettuce: A Sheetlike Green Alga.* An observer of *Ulva*, a green alga undulating in a shallow tidepool, would find it impossible to tell male or female gametophyte from sporophyte. This is because both phases of the plant's life cycle are identical, and neither the haploid nor the diploid phase dominates (see Figure 16.11).

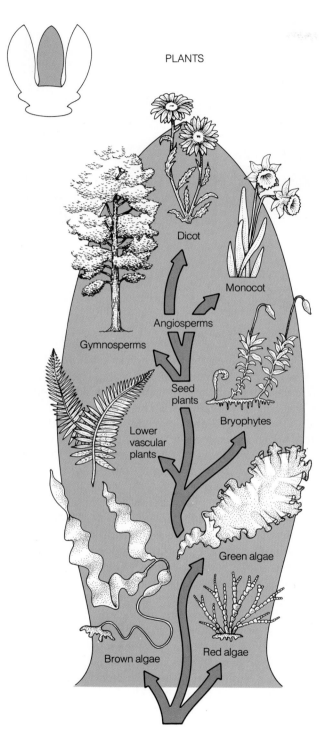

PLANTS

Dicot

Monocot

Angiosperms

Gymnosperms

Seed plants

Bryophytes

Lower vascular plants

Green algae

Brown algae

Red algae

FIGURE 16.16 *The Transition to Land.* Within the plants, ancient green algae led to the simple land plants—bryophytes (such as mosses) and lower vascular plants (such as ferns). Such early land plants, in turn, gave rise to the larger, cone-bearing gymnosperms ("naked-seed" plants) and flowering angiosperms.

(a)

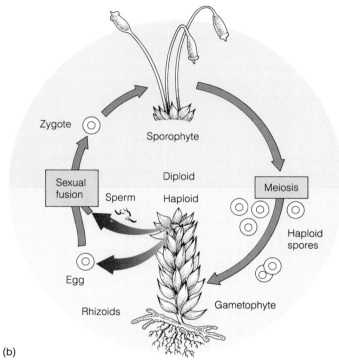

Zygote

Sporophyte

Diploid

Sexual fusion

Sperm

Haploid

Meiosis

Haploid spores

Egg

Rhizoids

Gametophyte

(b)

FIGURE 16.17 *Mosses: The Most Familiar Bryophytes.* (a) Although mosses grow in thick, velvety carpets on fallen trees or patches of soil, each individual moss is a leafy gametophyte (haploid) that periodically bears a slender, green-brown sporophyte (diploid). Hair-like projections called *rhizoids* anchor the plant. (b) The moss life cycle alternates between the conspicuous haploid and diploid phases. A spore chamber (*sporangium*) produces spores, which germinate and grow into gametophytes; and these haploid plants form egg- and sperm-producing organs that make gametes, which in turn fuse and grow into new sporophytes.

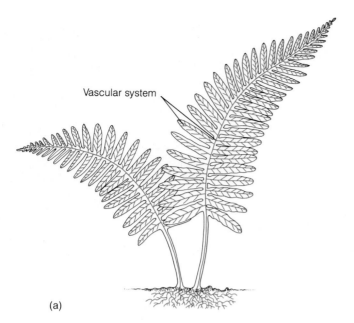

(a) (b) (c)

FIGURE 16.18 *The Lower Vascular Plants.* (a) With their vascular pipelines for support and transport of water and nutrients, the horsetails, ferns, and tree ferns can grow larger and taller than the mosses and relatives. (b) A horsetail, *Equisetum.* These curious plants are common in roadside ditches and other soggy areas. (c) A tree fern, *Dicksonia antarctica*, in a grove of eucalyptus trees in Australia.

phase, the leafy gamete producer. The small gametophyte plants have waterproof coatings that prevent drying out, as well as tiny "portholes" in their leaves that allow gas exchange through the coatings. They also have tissues that are strong enough to help keep the low plants upright, but they lack the rigid internal vascular system that transports water and supports other land plants. Instead of true roots, bryophytes have hairlike projections *(rhizoids)* that act only as anchors, absorbing neither water nor minerals. Water reaches the individual plant cells by diffusing through the entire organism, as in a simple wick system. Because of this dependence on direct diffusion, the plants are limited to very small size and to growing in shady, moist places. What's more, bryophytes must live where it is at least seasonally wet because they have retained swimming sperm that can only reach and fertilize an egg when the plant is drenched.

While many bryophyte species have survived in moist habitats, this pioneering group did not itself give rise to other more complex land plants, and the bryophyte lifestyle is considered an evolutionary dead end.

HORSETAILS AND FERNS: THE LOWER VASCULAR PLANTS

The real stars of the distant era that began about 350 million years ago were the *vascular plants*, which evolved one more of the solutions to life on land. This "solution" was a *vascular system* consisting of specialized cells that grow end to end in long internal pipelines extending from the tips of rootlike structures in the ground, up through the **stem** (the erect part of the plant), to the leafy photosynthetic surfaces (Figure 16.18a). These vascular pipelines can transport water, minerals, and the products of photosynthesis throughout the plant and also lend vertical support that allows the plants to grow taller than the bryophytes. Together with a layer of waterproofing on the outer cells of the leaves and stems, the vascular system enabled this new group to occupy drier habitats than bryophytes and thus move further inland. This group includes the horsetails (Figure 16.18b), the club mosses and relatives, and the familiar ferns and tree ferns (Figure 16.18c; see Table 16.2). In the tropical climates of 350 million years ago, these first vascular plants grew in vast primordial forests and reached sizes much larger than today's tree ferns. However, the climate gradually grew colder and drier, and these giants exited the stage, leaving behind the more diminutive horsetails and ferns that we see today on the shady forest floor. In a different sort of legacy, the bodies of the fallen giants were decomposed by fungi and bacteria, and the organic compounds carbonized into the enormous coal deposits that now provide much of our fossil fuel energy.

Horsetails, club mosses, and ferns show a continuation of the trend toward dominance of the diploid phase: In these lower vascular plants, the diploid sporophyte is the conspicuous adult, and the haploid gametophyte is a small free-living green plant (see Figure 16.12). For sexual reproduction, these plants still retain swimming sperm, like the bryophytes, and thus require standing water for reproduction. In a sense, the lower vascular plants are the botanical equivalents of the amphibians, animals such as frogs and salamanders that can live on dry land but must return to water to lay their eggs. Interestingly, giant amphibians were the dominant land animals during the Paleozoic era, the warm, swampy period dominated by the ferns and their relatives. Along with the giant club mosses, horsetails, and ferns of that era, the amphibians were replaced in later, drier times by other stars that did not require standing water for reproduction: the seed plants and the reptiles.

CONE-BEARING CONIFERS: CONQUERORS OF DRY LAND

About 230 million years ago, the supercontinent Pangaea was breaking apart, the continental seas had retreated, and new masses of high, dry land were becoming open to plant and animal colonization (see Figure 14.11). While the early land plants were tied to swampy lowlands, riverbanks, and coastal lagoons, new lines of plants appeared with reproductive modifications that broke the final barriers and completed colonization of land. These organisms were the *Coniferophyta* (conifers), sometimes referred to as **gymnosperms** ("naked-seed" plants), and they included the now extinct seed ferns as well as the cycads, ginkgos, and evergreens, such as pines, firs, sequoias, and redwoods (see Table 16.2).

The reproductive innovations of the gymnosperms include the pollen grain and the seed, as well as a further shift toward dominance of the diploid phase (review Figure 16.12). To understand these innovations, follow the life cycle of a gymnosperm, such as a giant sequoia or a pine tree, as shown in Figure 16.19. The spore-making phase (sporophytes) of gymnosperms are the conspicuous woody trees and include the most massive and the oldest living things on earth—the sequoia tree and the bristle cone pine tree. In the **cones** of these spore-making plants, meiosis occurs, forming the gamete-making phase, the gametophytes. These consist of just a few short-lived cells: pollen grains produced in male cones and an egg as part of small cell clusters in female cones. Pollen grains are produced in huge numbers and are disseminated by the wind. In effect, the pollen grain "airlifts" the sperm cells it contains to the egg cells harbored inside special chambers in female cones. Both male and female gamete-making structures are reduced to small nonphotosyn-

thetic forms housed entirely by the spore-producing tree. Thus, the diploid sporophyte tree is the dominant phase, continuing the dominance-of-the-diploid trend. One could also say that the water needed for fertilization is provided by the moist tissues of the parent plants themselves, rather than by splashing raindrops or standing water, and transportation of sperm is usually by "flying" rather than swimming.

Once sperm and egg meet and sexual fusion takes place, the gymnosperm embryo develops, exposed on the surface of a scale of the female cone. (The term *naked seed* refers to this exposed position.) Eventually, the embryo and surrounding cells from the pine cone scale develop into a **seed,** which is encased in a tough coat housing

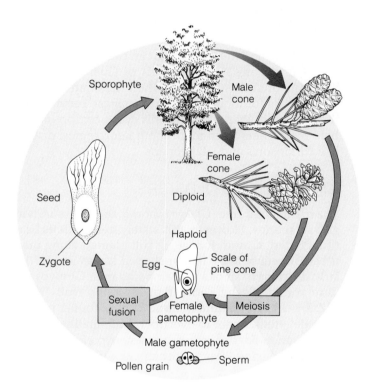

FIGURE 16.19 *Life Cycle of a Gymnosperm.* The often huge sporophyte dominates the gymnosperm life cycle. Spores formed in cones undergo meiosis and produce male and female gametophytes. These develop sperm (contained in pollen grains) and eggs (retained in ovules of the female cone). Pollen grains are carried by the wind, and the sperm cells within those grains fuse with eggs inside ovules. The result is a diploid zygote that becomes a plant embryo inside a seed. The seed eventually falls to the ground, where it may germinate and grow into another sporophyte, such as a pine or sequoia tree.

both the young plant and a supply of food for the early stages of its growth. Seeds are the most successful dispersal devices to have evolved in the plant kingdom, and they greatly facilitated the further colonization of land. Reptiles—the dominant land animals during the time when gymnosperms were the ruling land plants—evolved parallel structures: eggs with leathery shells that could be laid and hatched on land. These freed them from the dependence on standing water seen in amphibians.

CONE-BEARING RELICS: CYCADS

A small group of tropical plant species often mistaken for palm trees, the *cycads* are cone-bearing remnants of a group that flourished at the time of the dinosaurs. *Zamia pumila* is a cycad native to the sandy-soiled forests of Florida (Figure 16.20a). People sometimes gather and eat cycad seeds and roots, but these structures are mainly used as unusual ornamental plants in warm regions. A natural chemical in cycad seeds can cause weak muscles, tremors, and malaise. Residents of Guam suffered an epidemic of such symptoms when they had to rely on cycad seeds for food during the Japanese occupation of their island in World War II. This cycad compound may help researchers learn more about neurological diseases like Parkinson's and Alzheimer's.

GINKGOS: ODD OLDSTERS

The maidenhair tree *(Ginkgo biloba)* is the only surviving ginkgo species. However, this stately tree, with its beautiful golden autumn foliage, is still planted along many urban streets because its fan-shaped leaves are particularly resistant to smoke and air pollutants. Ginkgos produce round, fleshy cones that resemble fruit (Figure 16.20b). In late summer, these cones, found only on "female" trees, begin to overripen and smell strongly of rancid butter. For this reason, gardeners prefer the "male" pollen-producing trees for ornamental plantings.

CONIFERS: THE FAMILIAR EVERGREENS

The cone-bearing conifers are the most familiar and largest remaining group of gymnosperms, and they include many ecologically and economically important species. Millions of acres in mountainous regions are dominated by pine, spruce, fir, and cedar, many of which are harvested for wood, paper, and resins (Figure 16.21). Smaller conifers, such as yews, hemlocks, junipers, and larches, are often used for the graceful landscaping of buildings and parks. Finally, the largest living things, the giant sequoias, are conifers as well as living reminders of the Mesozoic era, an age of giants.

(a)

(b)

FIGURE 16.20 *Cycads and Ginkgos: Two Cone-Bearing Relics.* (a) Cycads, such as this native of Florida, *Zamia pumila*, resemble palms but are ancient gymnosperms bearing cones. (b) The graceful fan-shaped leaves of the ginkgo remind some people of a maiden's flowing hair, giving the tree its common name: the maidenhair tree. In autumn, the fleshy yellowish cones ripen and fall, and the tree gives off a strong rancid smell.

Conifers have two characteristics that distinguish them from other seed-producing plants: Their leaves are narrow and needlelike, and they produce familiar woody cones. Their needle-shaped leaves are covered by a waterproof waxy layer. The needle shape resists drying because it has little surface area for evaporation. Woody cones, such as the larger pine cone in Figure 16.19, are actually the female cones. Each leaf, or scale, of a female cone bears an *ovule*, which is a skin surrounding a spore-producing structure (the *sporangium*). Inside, meiotic divisions give rise to large spores, which produce the female gametophytes. Conifer trees also produce smaller, shorter-lived male cones made of soft tissue and containing many scales. Structures inside the scales produce microspores, which develop into pollen grains, or **pollen,** the actual male gametophytes.

Conifers were able to conquer higher and drier reaches of the continents because of their numerous evolutionary advances: drought-resistant leaves; protective seeds; greatly reduced male gametophytes that give rise to airborne pollen grains; equally reduced female gametophytes that give rise to eggs protected inside the ovule; a well-developed vascular system that produces wood to stiffen the trunk, branches, and roots; and, as we saw with the sequoias, the tendency to form fungus-root associations and the ability to survive for many centuries. Because of these modifications for life on land, conifers are still a successful group, with more than 500 modern species. Nevertheless, a final set of evolutionary novelties gave the flowering plants still greater success.

≋ THE FLOWERING PLANTS: A MODERN SUCCESS STORY

Geological processes continued to lift and tilt the continents, and the landmasses became colder and drier in general. The scene was set for the latest act in the evolutionary play—the modern era (the Cenozoic)—during which the mammals and the flowering plants (**angiosperms,** or Anthophyta) would become the dominant life forms on land. The mammals had evolved milk production to suckle the young, and a special organ, the placenta, that nourishes the developing embryo inside the protective womb. Meanwhile, the angiosperms evolved a new reproductive structure, the ovary, which houses and protects the ovules; they also evolved fruit and flowers, which allow efficient pollination and seed dispersal. These innovations occurred along with other adaptations for life on land, such as broad leaves for efficient solar collection, and enabled the flowering plants to evolve into more species than the combined numbers of all other plant groups.

Today, flowering plants are the most common and conspicuous species in the earth's tropical and temperate regions, and from this monumentally successful group come virtually all of our crop plants (wheat, rice, corn, soybeans, fruits, and vegetables) and beverages (coffee, tea, colas, and fermented drinks), as well as spices, cloth, medicines, hardwoods, ornamental plantings, and, of course, flowers—symbols of beauty, affection, and renewal throughout human history. Table 16.2 shows the two major subclasses of the flowering plants, the *monocots* and the *dicots,* and their main characteristics.

REPRODUCTION IN FLOWERING PLANTS: A KEY TO THEIR SUCCESS

As in gymnosperms, the angiosperm life cycle displays the complete dominance of the diploid (sporophyte) plant phase, as reflected in the haploid gametophytes' complete dependence on the conspicuous adult plant. The word *angiosperm* means "seed in a vessel," and indeed, seed development takes place within a structure at the base of the flower, called the *ovary,* which gives the ovules much more protection than the gymnosperm cone scale. The wall of the ovary eventually matures into a fruit that surrounds the seeds and aids in their dispersal. In some cases, animals attracted to the fruit's colorful skin or enticing flavor eat the fruit and then excrete the seeds along with feces in some new location (Figure 16.22a). Angiosperms have many other dispersal mechanisms, including "parachutes" that loft the seed in the wind and hooks for hitchhiking on a passerby (Figure 16.22b, page 288).

Flowers often consist of a ring of colorful petals surrounded by a ring of small green sepals. Female reproductive structures within the flower include a sticky top, or *stigma,* and a slender neck, or *style,* which leads to the

FIGURE 16.21 *Conifer Forest.* Cool, fragrant forests of pine, spruce, cedar, fir, or other conifers cover millions of acres at high altitudes or northern latitudes. Here a forestry worker is planting conifer seedlings as part of a reforestation project.

(a)

(b)

FIGURE 16.22 *How Plants Disperse Seeds.* (a) Cedar waxwing feeding bright red berries to its young. (b) Prickly seed of the common burdock, *Artimum munus*, clinging to a terrier's face.

ovary (Figure 16.23). Inside the ovary are one or more ovules, structures containing cells that will undergo meiosis and give rise to an egg as well as a few accessory cells. Together, these make up the embryo sac, or mature female gametophyte (Figure 16.23, step 1). The male flower part is the *stamen,* made up of the *filament,* a slender stalk, and a club-shaped structure called the **anther.** Special cells in the anther give rise via meiosis to pollen grains, or immature male gametophytes (step 2), and as they do, the anther splits open, and mature pollen grains

are released. In some species, these are dispersed by wind, while in others, they are carried by bees, birds, bats, or other animals, called **pollinators.** By chance, pollen grains will fall or be carried to a sticky stigma, and a pollen tube will begin to grow rapidly down through the style (step 3).

Once the tube has reached the opening of the ovule housing the egg and accessory cells, two sperm cells migrate down, and a *double fertilization,* an event unique to flowering plants, takes place. One sperm cell fuses with the egg to form the diploid zygote that becomes the embryo. The nucleus of the other sperm fuses with two non-egg-cell nuclei in the ovule, and the resulting nucleus (now triploid, with three sets of chromosomes) divides and forms *endosperm,* a nutritive tissue that becomes enclosed, along with the embryo, inside the developing seed.

As summer progresses, the wall of the ovary begins to enlarge into a fleshy **fruit** (step 4) or becomes modified into a winged structure such as a maple "squirt," the "parachute" of a milkweed seed, or one of the many other kinds of dispersal mechanisms (review Figure 16.22). We tend to think of fruits as round, sweet objects, but certain vegetables, such as cucumbers, tomatoes, and squash, are really fruits, and so are walnuts, peanuts, and pecans. Fruits fall off the plant or are harvested by animals. These processes disperse the seeds, which later sprout and grow into new plants (step 5).

FLOWERING PLANTS AND POLLINATORS: A COEVOLUTION

The success of the flowering plants can be only partly explained by the protective enclosures missing in the naked-seed gymnosperms. Flowers, fruits, and seeds also help ensure pollination and seed dispersal by attracting animals. Most flowers attract insect, bird, or mammal pollinators that inadvertently carry loads of pollen from one plant to another as they forage for sweet nectar or for the protein-rich pollen grains. A walking or flying animal, such as a bee, dusted with pollen and scouting for more of its favorite food will substantially increase the chances of pollination. And once fruits have developed on a plant, animals are attracted either by sight or smell to harvest and eat the fruits. The tough, highly resistant coats of many plant seeds allow them to pass through an animal's digestive tract intact. The seeds are then deposited in new locations, and this helps the plant spread.

The mutual dependence of flowering plants on pollinators and seed dispersers and of these animal species on the same plants for nutrition is no coincidence: It is the result of **coevolution,** a natural selection for increasing interdependence based on selective advantages for both parties. Plants that had genes for flowers with sweet nectar, a strong fragrance, or a bright color might have survived in greater numbers because more animals would have visited them in their constant foraging for food and carried away pollen. The animals that were attracted to the new food sources probably also survived in greater numbers, thus perpetuating the interdependence.

This coevolution has produced specialized physical structures in both plants and animals that help ensure the trade-off of nutrients for pollination and seed dispersal. The mouthparts of many types of bees, butterflies, moths, birds, and even bats are precisely the right shapes for tapping the nectar or pollen of the flowers they visit. Hummingbirds, for example, which see best in the red spectrum and have long, narrow beaks and a poor sense of smell, are attracted by bright red fuchsia flowers with their long, slender shape but little fragrance

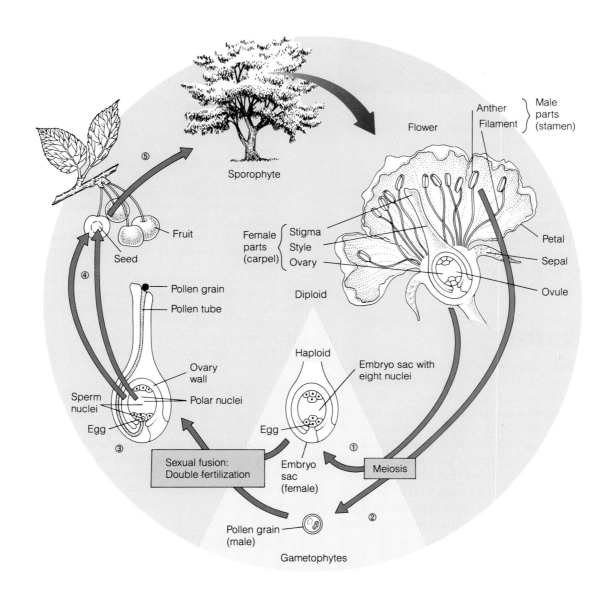

FIGURE 16.23 *Life Cycle of a Flowering Plant.* As you follow the description given in the text, you can see the many steps in the angiosperm life cycle. The ovary produces eggs, the anthers make microspores that develop into pollen, and their double fertilization gives rise to embryo and endosperm inside a seed (steps 1–4). Seeds usually form inside fruits, fleshy or dry, and when the seeds are released, they germinate into seedlings—the new generation of free-living sporophytes (step 5).

FIGURE 16.24 *Sipping Beauty: A Study in Coevolution.* An Anna's hummingbird drinking from a red and magenta fuchsia blossom. Both organisms evolved with compatible anatomy.

STUDY QUESTIONS

REVIEW WHAT YOU HAVE LEARNED

1. Describe the extracellular digestion that takes place in a fungus.
2. What are fungus roots? How do they affect a plant's nutrition?
3. What are the distinguishing features of the five major classes of fungi?
4. Name the stages in a plant's alternation of generations.
5. How did changes in the earth's surface affect the evolution of land plants?
6. How do the three types of algae differ from one another?
7. Are algae too simple to have ecological importance? Explain.
8. What adaptations enable a bryophyte to live on land?
9. Compare the life cycles of a moss and a pine.
10. Name the female and male reproductive structures of a flowering plant.
11. In angiosperms, fertilization involves two sperm nuclei. How do the roles of the two nuclei differ?

APPLY WHAT YOU HAVE LEARNED

1. When an Alaskan glacier retreats, it leaves behind bare rock. How are lichens able to grow on this barren surface?
2. A sequoia tree, which manufactures its own food by photosynthesis, may live thousands of years and grow hundreds of feet tall. Yet, the tree's survival depends on a microscopic fungus. Explain.

≋ CONNECTIONS

As producers, plants capture sunlight and use its energy in the building of carbohydrates. As decomposers, fungi dismantle carbohydrates and other organic molecules, use some of the nutrients, and release more into the soil and air where other life forms—including plants—can use them once again. It is the collective activity of the tons of fungal hyphae in each acre of topsoil and the tons of plant matter rooted in that soil that gives decomposition and production so global an impact. The life cycles of fungi and plants show what is unique about these groups; they also show the trends from simpler to more complex processes, revealing evolutionary directions, and how specific kinds of organisms reproduce—one of the major tasks of any living thing.

As we consider the animal kingdom in the next two chapters, we will learn more about the interrelatedness of the kingdoms. Just as plant roots need fungi for maximum absorption and nutrient recycling, and fungi need plant and animal matter for energy, so do animals—the great consumers—rely on all the other kingdoms.

(Figure 16.24). Conversely, flowers pollinated by bees tend to be yellow or blue and very fragrant, corresponding to the bees' vision and ability to perceive odors.

Plants pollinated by the wind, such as oak trees or sequoias, must produce more than 200 times as much pollen as a plant with its own army of animal pollinators.

FOR FURTHER READING

Bell, P. R., and C. L. F. Woodcock. *The Diversity of Green Plants*, 3d ed. London: E. Arnold Ltd., 1983.

Gensel, P. G., and H. N. Andrews. "The Evolution of Early Land Plants." *American Scientist* 75 (1987): 478–489.

Niklas, R. J. "Aerodynamics of Wind Pollination." *Scientific American* (July 1987): 90–95.

Ross, I. K. *Biology of the Fungi: Their Development, Regulation, and Associations.* New York: McGraw-Hill, 1979.

Vogel, S. "Taming the Wild Morel." *Discover* (May 1988): 58–60.

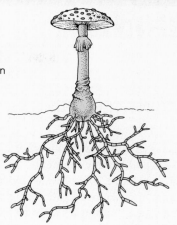

① Fungi, the great decomposers, release digestive enzymes into external organic matter and then absorb the digestive nutrients.

② Multicellular fungi exist mainly as filaments of haploid cells, either packed together in a reproductive structure like a mushroom or in loose mats like bread mold.

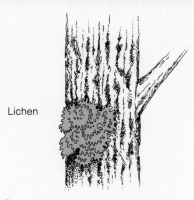

Lichen

③ Fungi are interdependent with plants.

④ Plants are the great producers; they arose from algae, and their evolution includes mosses, ferns, conifers, and flowering plants.

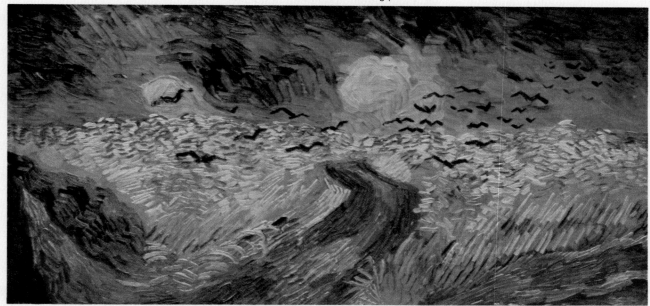

Vincent Van Gogh, *Crows Over Wheatfield*

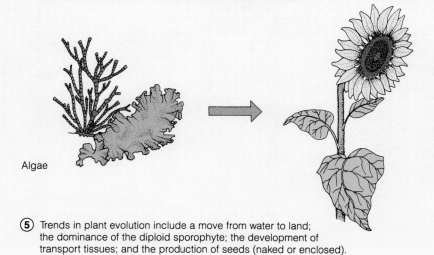

Algae

⑤ Trends in plant evolution include a move from water to land; the dominance of the diploid sporophyte; the development of transport tissues; and the production of seeds (naked or enclosed).

⑥ Insects, birds, and mammals coevolved with plants, and many of these animals aid pollination and seed dispersal.

CHAPTER 17

ANIMALS WITHOUT BACKBONES

SELF-PROTECTING TERMITES

Termites build towering, hard earthen mounds that look like very tall, furrowed monuments. Inside, millions of jostling white insects live in complex societies, safeguarded by an amazing arsenal of self-defenses that keeps most potential enemies at bay. For example, some termite species have a caste of soldiers that protect the workers, queen, and others in the colony. At the first sign of an intruder, each soldier rushes forward and squirts an irritating gluey substance from its conical head (Figure 17.1). A hungry anteater is just as likely to retreat with a painful tongue as to collect a meal.

Termites and the anteaters that devour them are **animals,** multicellular "other-feeders" (heterotrophs) that move themselves about at some point in their life cycle. When you think of the animal kingdom, you probably envision organisms such as goldfish, horses, eagles, or people. All these animals are **vertebrates:** animals with backbones; while familiar to us, vertebrates make up only 5 percent of the animal kingdom. The overwhelming majority of all animal species are **invertebrates:** animals without backbones, like termites, earthworms, and oysters. We consider the invertebrates in this chapter, and vertebrates and their relatives in the next.

As we survey the colorful, strange, and sometimes fantastic invertebrates, we will follow certain trends that led to the basic physical features most animals share. The first trend was away from a circular body plan toward a saclike body with mirror-image right and left halves and a head. The head, with centralized regions for detecting and responding to external cues such as light and sound, provided many advantages, including

more ways of interacting with other organisms and the environment.

In another trend, invertebrates moved away from a saclike body with a single opening at one end toward a longer body containing a food-digesting tube with openings at both ends; the tube eventually became suspended in a fluid-filled body cavity. This trend in the evolution of body structures led to a more complete breakdown and use of food and in turn the release of more energy that could power rapid movements.

Yet another trend was toward segmentation, the development of a series of body units. The series of vertebrae you can feel as you run your fingers up your spine is a reflection of segmentation.

While these trends conferred certain advantages, they were not prerequisites for survival; some animals survive today without one or more of them.

This chapter will discuss:

- The characteristics that animals share, and how invertebrates differ from vertebrates like ourselves.
- Sponges, the simplest invertebrates, which survive with saclike bodies.
- Radial animals, such as jellyfish.
- Flatworms and roundworms, with their heads and mirror-image right and left sides.
- Mollusks, including clams, snails, and squid—the most intelligent invertebrates on earth.
- Segmented worms, with their advances in digestion and circulation.
- Arthropods, including lobsters, spiders, and insects.
- Echinoderms, the sea stars and relatives, with the first internal skeletons—but not the last.

FIGURE 17.1 *A Soldier Termite.* This insect is a termite in one of the 500 species of the genus *Nasutitermes.* It protects the communal termite nest by squirting intruders with a sticky substance from its nozzle-shaped head.

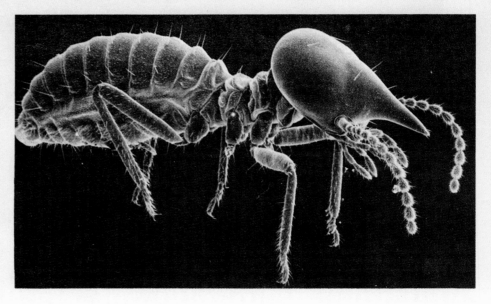

ANIMALS AND EVOLUTION: AN OVERVIEW OF THE ANIMAL KINGDOM

Animals are quite distinct from the members of the other kingdoms. Unlike the single-celled monerans and protists, animals have many cells. In fact, some of the largest animals—an elephant, say, or a giraffe—have trillions of cells. And unlike plants, animals eat other living things because they cannot manufacture their own food and instead must eat it and break it down (digest it) for its energy content. Unlike most fungi, which are also multicellular heterotrophs, animals can move about at some point in their life cycle—usually throughout the entire cycle—to search for food and mates and to avoid danger. (Fungi and plants often produce mobile spores, but these do not move about under their own power.) Finally, animals' primary mode of reproduction is sexual, and because each individual grows and changes from a single fertilized egg to a multicellular organism, it passes through various distinct stages of development.

As a group, animals have a long and interesting evolutionary history, and because they have left a more complete fossil record than most members of the other kingdoms, that history is better understood. The earliest fossils of soft-bodied animals are burrows, trails, and impressions found in 700-million-year-old rocks in southern Australia. Many zoologists (biologists who study animals) believe that animals arose from animal-like protists. Figure 17.2 shows the probable relationship of the sponges and other animals to the single-celled protists that preceded them. Soft-bodied animals existed for at least 120 million years before other species started generating hard parts (shells, scales, teeth, and bones); the earliest fossils of these hard structures were found in 580-million-year-old rocks (Figure 17.3, page 294).

The fossil record shows that the overall result of geological change and continued evolution was a great *radiation* of the animal kingdom—the fanning out, from a common ancestor, of large numbers of animal species with a variety of forms adapted for various environments. All the animals that ever existed—including invertebrates, such as the first worms to crawl on land and the giant dragonflies of primordial forests, and vertebrates, such as the leather-winged pterosaurs and the lumbering mastodons and ground sloths of the last ice age—were products of the great radiation. Dozens of phyla evolved; animals in a phylum share the same general body plan. Most of the phyla still have living members, and they include more than a million individual species, of which 95 percent are invertebrates. In this chapter, we discuss only the most important invertebrate groups, beginning with the sponges, the simplest animals (Table 17.1, page 302).

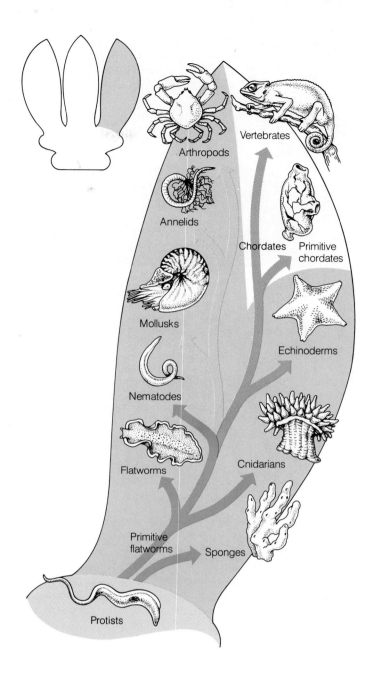

FIGURE 17.2 *Evolutionary Relationships of the Major Animal Phyla.* Animals are the fifth living kingdom, and they arose from protistan ancestors. The sponges were an early branch off the main line. The other branches, however, have diverged into large groups, including the mollusks, arthropods, and chordates. In this chapter, we examine the invertebrates, which include all but the primitive chordates and vertebrates.

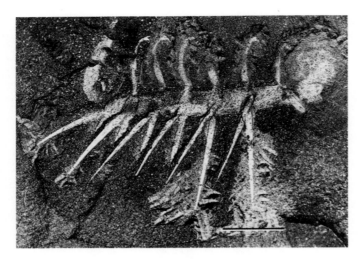

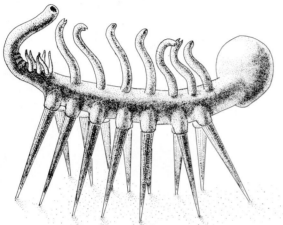

FIGURE 17.3 *Hallucigenia: A Vision from the Mists of Invertebrate Evolution.* This photo and interpretive drawing depict a small, strange animal with a bulbous head and seven sets of rigid pointed spines projecting from a tubelike body. Dubbed *Hallucigenia*, the animal lived 570 million years ago and may have been an odd, immobile consumer perched stiffly on its unjointed legs, or it may have been an appendage on some larger beast. Regardless, fossil impressions of this living nightmare were discovered in the Burgess Shale deposits high in the Canadian Rockies. Here, archaeologists found dozens of now extinct invertebrates (mostly soft-bodied and many as peculiar as *Hallucigenia*) that lived at the dawn of animal evolution.

≋ SPONGES: THE SIMPLEST ANIMALS

Sponges are simple animals in the phylum Porifera ("organisms with pores") that filter and consume fine food particles, bacteria, or protozoa from the water. The 5000 or so species of sponges live attached to rocks, pilings, sticks, plants, or other animals in oceans, rivers, and lakes.

THE SPONGE'S BODY PLAN

The simplest sponges resemble vases or irregular-shaped clusters of tubes (Figure 17.4a and b). Each "container" has a central cavity, a large central opening, and hundreds of tiny holes, or pores (incurrent pores), through the body wall (Figure 17.4c). If you place red dye in water that surrounds a living sponge, you can see the colored water being drawn through the pores along with food particles. Special cells lining the cavity remove and digest the suspended food particles; and the water and waste matter exit through the central opening. Sponges range

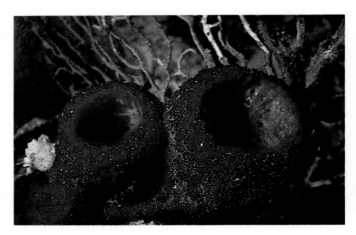

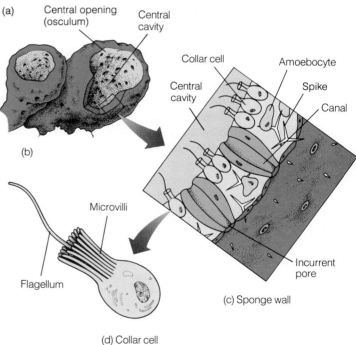

FIGURE 17.4 *The Sponges: The Simplest Animals.* (a) This red vase sponge (*Mycele* species) is a common resident of coral reefs. (b) The simple vaselike body has a central opening, the *osculum*, and a central cavity. (c, d) The body wall contains collar cells and amoebocytes and is perforated by incurrent pores and channels.

from the size of a pinhead to the size of a wine barrel. This large size is possible because canals permeate the body and provide avenues for the transport of food and wastes to and from the deepest parts.

Most sponges are not symmetrical, and have a simple organization of cells. The sponge's body wall is a thin "sandwich" with a protective layer of flattened cells outside; an inner lining containing "collar cells" (choanocytes), each with a long flagellum surrounded by a delicate collar (Figure 17.4d); and a gelatinous filling containing wandering cells called amoebocytes (see Figure 17.4c). The constant beating of the collar cells' flagella draws the water current in through the body. The embedded amoebocytes help digest food particles and differentiate into gametes. The sponge's cells are supported by a protein "scaffolding" and may contain tiny pointed spikes made of silicon or calcium compounds, which help protect the sponge from predators. The irregular, tan bath sponges sometimes sold commercially are the tough fibrous skeletons that remain after people harvest the animals, trample them, leave them to decay, then process their "scaffolding."

The different cell types in sponges act fairly independently and are much less highly organized than cell groups in other animals. For this reason, biologists believe that sponges probably split from the line leading to all other animal groups very early in the history of animal evolution (see Figure 17.2).

SPONGE ACTIVITIES AND REPRODUCTION

Although a sponge looks inert from the outside, its cells are continuously active. An average-sized sponge can move about 50 gal of water a day through its canals and pores. Food particles in this water are taken into the sponge's cells, where they are digested. Sponges can excrete wastes and gases directly through the cells of the thin body wall itself; sponges have no muscles, kidneys, or other organs. Sponges can reproduce sexually or asexually.

Sponges probably evolved from protozoa that had collar cells and flagella and that lived in colonies. The sponges have advanced beyond the colonial protozoa, however, because sponge cells, despite their degree of independence, do function together as a single individual. Sponges are not directly related to any other animal groups, nor did they give rise to more complex organisms.

≋ THE RADIAL ANIMALS: CNIDARIANS

Some of the most ephemeral and beautiful invertebrates are members of the phylum Cnidaria (pronounced "ni-

(a) (b)

(c) (d)

FIGURE 17.5 *The Radial Animals: Ephemeral Invertebrates.* Most radial animals are translucent and billowing, but they are protected by stinging weapons called nematocysts. (a) A brown hydra. (b) A pink-tipped sea anemone. (c) A coral. (d) A jellyfish.

DAIR-ee-a"; from the Greek word for "nettles" or "stinging plants"). This phylum includes the translucent hydras, the gossamer jellyfish, the colorful sea anemones, and the corals (Figure 17.5). Most live in the oceans, but a few, such as the hydras, inhabit fresh water. **Cnidarians** have well-organized tissues and a **radial body plan;** that is, they are circular, with structures that radiate outward from the center like the spokes of a wheel. They also possess fearsome weapons: tentacles armed with stinging devices that assist the animals in self-defense and in capturing food. One kind of cnidarian, the corals, are tiny, cup-shaped animals that live in huge colonies and excrete limestone skeletons. Over time, these can form giant reefs and atolls, including the only biological structure visible from space—the 1200-mile Great Barrier Reef off Australia's eastern coast—and islands such as Bermuda, the Bahamas, and Fiji.

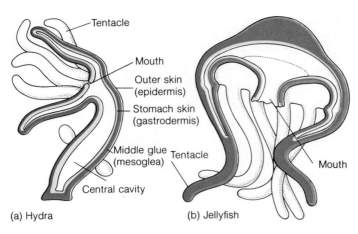

FIGURE 17.6 *Anatomy of the Radial Invertebrates.* The body layers, main structures, and vaselike polyp shapes of (a) a hydra, and (b) a jellyfish's inverted polyp shape.

"VASES" AND "UMBRELLAS": TWO BODY PLANS

The radial invertebrates grow in a range of sizes, colors, and strange appearances, but on close inspection, all have one of two basic body forms: the *polyp,* a hollow, vaselike body that stands erect on a base and has a whorl of tentacles surrounding a mouth near the top (Figure 17.6a), or the *medusa* (plural, medusae), an inverted umbrella-shaped version of the polyp, with tentacles and mouth pointing downward (Figure 17.6b). Sea anemones, corals, and most hydras are polyps, while jellyfish are medusae.

Each polyp or medusa has a three-layered body wall with an outer skin *(epidermis),* an inner "stomach skin" *(gastrodermis),* and a jellylike substance in between called *mesoglea* (literally, "middle glue"). Jellyfish resemble jelly because of the extreme thickness of this middle layer. In one monstrous North Atlantic jellyfish that is nearly 10 ft across and weighs a ton, virtually all the mass is the "middle glue."

ADAPTATIONS OF CNIDARIANS

The radial animals have three important adaptations relating to the capture and digestion of food. Embedded in the outer skin of the tentacles are remarkable stinging capsules. When triggered by the approach or contact of an enemy or potential prey, the tube of the tentacle turns inside out like a sock. But this "sock" has a sharply pointed end that can strike prey animals and release a paralytic toxin. Several human deaths have been attributed to the toxin from one kind of large tropical jellyfish. In contrast, the well-known Portuguese man-of-war can inflict nasty stings, but rarely kills people.

Once a cnidarian has immobilized a prey or simply chanced to find a food morsel, it swallows the food and digests it within the body's central cavity (the *gastrovascular cavity*), the second important adaptation (review Figure 17.6). Cells lining this cavity produce enzymes that begin to break down the food; this food breakdown within a cavity is called **extracellular digestion,** and cnidarians and all of the more complex animals make use of it. Extracellular digestion allows animals to digest food pieces larger than individual cells and thus expands their range of food sources.

A third adaptation helps radial animals like jellyfish and anemones detect prey, coordinate body movements, and capture and swallow the victim. *Nerve cells* arranged in a loose network permeate the animal's tissues. These cells have long extensions and electrochemical functions that allow them to detect stimuli and activate cells in response. Cnidarians and all more complex animals have nerve cells and nervous systems. The nerve cells and other cell types in cnidarians are organized into **tissues,** groups of cells performing the same function. However, different cnidarian tissues are not grouped together into *organs,* as tissues are in more complex animals, such as the flatworms, the next group we discuss.

≋ FLATWORMS: A HEAD AND MIRROR-IMAGE SYMMETRY EVOLVE

The simplest invertebrates to display right-left mirror-image symmetry and "headedness"—traits seen in most animal species—are the **flatworms** (phylum Platyhelminthes). Planarians are free-living flatworms that inhabit freshwater lakes, rivers, or bodies of salt water (Figure 17.7a). Other flatworms, such as flukes and tapeworms, are parasites that exist on the host's body tissues or food reserves (Figure 17.8, page 298).

BILATERAL SYMMETRY AND CEPHALIZATION

Flatworms have **bilateral symmetry;** the right and left halves are mirror images, but the *anterior* and *posterior* (front and back) ends are different, and so are the *dorsal* and *ventral* (top and bottom) surfaces (Figure 17.7b). In addition, one end of the flatworm functions as a head (the animal is *cephalized,* from the Greek word for "head"). It contains both a nerve mass that serves as a brain and specialized regions for sensing light, chemicals, and pressure. When a flatworm moves forward, the head region—with its brain and senses—encounters a new region first. Depending on the information collected by the head region, an animal can continue forward or back up and try a different direction. The evolutionary adap-

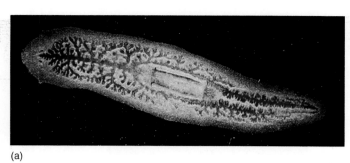

(a)

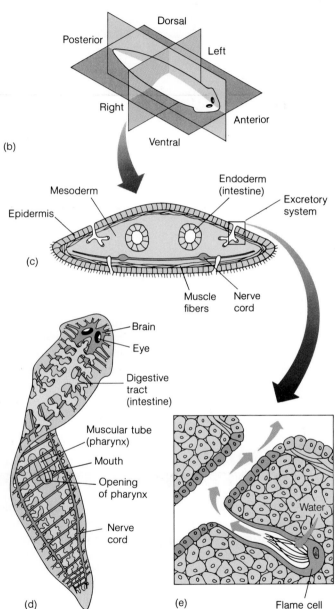

(b)

Dorsal
Posterior
Left
Right
Anterior
Ventral

(c)

Mesoderm
Endoderm (intestine)
Epidermis
Excretory system
Muscle fibers
Nerve cord

(d)

Brain
Eye
Digestive tract (intestine)
Muscular tube (pharynx)
Mouth
Opening of pharynx
Nerve cord

(e)

Water
Flame cell

FIGURE 17.7 *Flatworms: Simple Animals with Primitive Organ Systems.* (a) A planarian flatworm. (b) Its elongated body shows bilateral symmetry; the right and left sides are mirror images of each other. The animal also has a head in the front (anterior) and a tail in the rear (posterior). (c) The flatworm in cross section is characterized by distinct body layers. (d) The intestines and excretory and nervous systems play important roles. (e) The flame cells of the excretory system remove excess water from the body.

tation of a head is so successful that almost all animals more advanced than flatworms have one.

DISTINCT TISSUES, ORGANS, AND ORGAN SYSTEMS

Flatworms display two additional advances seen in all higher animals. Flatworms have three distinct tissue layers, and they develop true organs and organ systems. Recall that the radial invertebrates have a three-layered body wall, but the middle layer is a gelatinous material containing scattered cells. In flatworms, the middle layer (*mesoderm*) is made up of living cells (not jelly) and lies between an outer cell layer (epidermis) and an inner cell layer (endodermis) (Figure 17.7c). The middle layer is important because it gives rise to muscles and other organs. **Organs** are structures made up of two or more tissues that function together, and flatworms have a number of organs that carry out digestion, movement, nervous responses, excretion, and reproduction.

Organs usually occur in *organ systems*, sets of organs with related functions, and flatworms have five such systems. The *digestive system* of the planarian has a muscular tube (*pharynx*) that can thrust out of the body during feeding, as well as a branched digestive tract (*intestine*) lined with cells that absorb nutrients (Figure 17.7c and d). The intestine, with only one opening, is a blind, dead-end tube; food enters and wastes exit through the same opening, the mouth. The *excretory system* rids the body of excess water and includes a network of water-collecting tubules adjacent to saclike cells that contain clusters of cilia. Because the cilia appear to flicker like flames, the saclike cells are called *flame cells* (Figure 17.7e); as the cilia move, they drive water into the tubules and out through pores in the body wall.

Planarians move by means of cilia on their external surfaces and by layers of contractile muscle cells that lie below the epidermis in a *muscular system*. Muscles receive signals from the *nervous system*, which includes the light-detecting cells of the eye, a brain, longitudinal "trunk lines," or **nerve cords,** and a network of lateral nerves (see Figure 17.7c and d). Finally, there is a *reproductive system*. Most flatworms have both testes and ovaries; thus, they are what biologists call *hermaphrodites,* and they pair up to exchange both sperm and eggs with another individual.

Contributing to the flatworm's success is its flatness: Because it is so thin, oxygen and carbon dioxide diffuse straight through the thin layers to every cell, nutrients diffuse outward from the branching intestine to all body cells, and the animal needs no extra internal or external support.

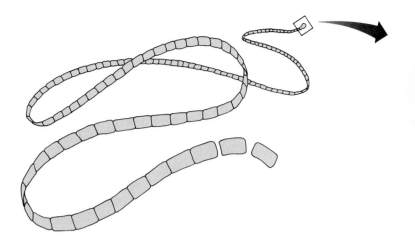
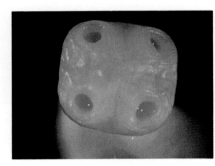

FIGURE 17.8 *Tapeworms: Appearance and Life Cycle.* The beef tapeworm infects about 60 million people worldwide. It can reach 60 ft in length inside a person's intestines and is spread by the ingestion of raw or undercooked beef. The life cycle begins when the larva is eaten. The head of a beef tapeworm has a structure for attaching the parasite securely to the host's tissues (see photo). The larva attaches itself to the host's gut and there develops into an adult. In the adult stage, tapeworms may cause the host to lose weight, suffer chronic indigestion, and have persistent diarrhea. The many sections detach and are released in the person's feces; there the segments expel their eggs. If a cow eats food contaminated by egg-bearing human feces, the egg hatches in the new host, and the newly hatched organism penetrates the gut wall, enters the bloodstream, and eventually bores into the muscles and forms a cyst.

FLATWORM LIFE-STYLES

Free-living flatworms, such as planarians, must hunt and avoid danger to survive, and their nervous systems and senses are especially crucial. Planarians have tiny *eye-spots* that sense light and movement, as well as regions on the head for detecting food (see Figure 17.7d). By contrast, parasites like flukes and tapeworms live a more sheltered life, with most of their needs provided for by a host; for them, prodigious reproduction is the major survival strategy. Parasitic flatworms are usually quite flat, with heads that are little more than knobs with hooks or adhesive suckers around the mouth for attaching to host tissues (see Figure 17.8). The tapeworms that infect cows' intestines have about 900 body units filled with reproductive organs, and they shed embryos that bore into the cow's muscles and form protective cysts. If a person consumes undercooked beef bearing these cysts, the cysts can grow into new tapeworms in the person's intestines.

In the earth's tropical zones, parasitic flatworms, with their tremendous capacity for reproduction, are major sources of debilitating diseases for people and their livestock. Their free-living cousins have little direct impact on people, but the group, overall, has had great evolutionary importance. Biologists have reason to believe that a small, bilaterally symmetrical creature similar to a flatworm probably gave rise to all the more complex animal groups.

ROUNDWORMS: ADVANCES IN DIGESTION

In sheer numbers of individuals, the most abundant animals on earth are roundworms, members of the phylum Nematoda. **Roundworms,** as the name implies, are round in cross section rather than flat, and most are very small; many, in fact, are microscopic (Figure 17.9). A cubic meter of rich soil can contain 3 billion roundworms (also called nematodes), and a biologist once said that if all our planet's lands and seas were swept away but the nematodes somehow stayed in place, a clear outline of the earth and its geological features would remain. When the environment grows harsh, roundworms can curl up, dry out, and shut down metabolically for up to 30 years. In this "latent" state, they can withstand extremes of heat, cold, radiation, and chemicals, then revive when conditions improve. While many roundworms are free-living, hundreds of the 12,000 or more nematode species are plant and animal parasites that damage crops and cause diseases.

ROUNDWORM ADVANCES: A FLUID-FILLED BODY CAVITY AND A TWO-ENDED GUT

The roundworm's success is often attributed to two new characteristics. While they share bilateral symmetry and cephalization with the flatworms, roundworms also possess a digestive tract, or gut tube, suspended in a partially lined body cavity. Because food enters the mouth and moves in just one direction through the gut tube, with wastes exiting the anus, regions of specialized function developed along the gut for grinding food into small pieces, breaking it down with enzymes, absorbing nutrients and water, and expelling wastes.

In roundworms, this one-way gut is suspended in a so-called *false body cavity*. Like cnidarians and flatworms, roundworms have three tissue layers (Figure 17.10a). In roundworms, however, there is a space between the innermost layer and the outer two (Figure 17.10b). All animals more complex than roundworms have a fluid-filled space not between two body layers but *within the middle layer* itself, the mesoderm (Figure 17.10c). This space, bounded on all sides by mesoderm, is a true **body cavity** (a *coelom*). The advantages of a true or false body cavity (which work basically the same way) are threefold. First, because of this space, the reproductive and digestive organs can evolve more complex shapes and functions. Second, in a fluid-filled chamber, the gut tube and other organs are cushioned and thus better protected; and because the gut is suspended, its activities can take place undisturbed by the activity or inactivity of the ani-

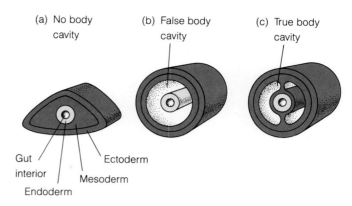

(a) No body cavity
(b) False body cavity
(c) True body cavity

Gut interior
Endoderm
Ectoderm
Mesoderm

FIGURE 17.10 *Evolution of the Body Cavity.* More complex invertebrates evolved three tissue layers with a split in the middle layer that forms the true body cavity. This mesoderm-lined, fluid-filled space cushions internal organs.

mal's outer body wall. Third, liquids cannot be compressed; therefore, the cavity can provide support and rigidity for the soft animal, just as water does when poured into a balloon. With their one-way digestive tube protected in a false body cavity, roundworms became highly efficient digesters, capable of consuming a wide variety of foods and therefore of inhabiting environments all over the world.

IMPACT OF ROUNDWORMS ON THE BIOSPHERE

Many types of free-living roundworms help consume rotting plant and animal matter and thus are ecologically important decomposers. However, the parasitic roundworms get most of the fame—or infamy. At least 1000 nematode species parasitize plants, and some observers estimate that they consume fully 10 percent of all crops annually. Nearly 50 species parasitize people, entering in food or contaminated water or through bare skin. They cause a list of diseases, including trichinosis (from eating undercooked worm-infested pork), hookworm (an infestation in the internal organs, also common in the tropics), and elephantiasis (Figure 17.11, page 300). Clearly, much of the nematode's evolutionary success comes at the expense of other organisms, ourselves included.

EVOLUTION IS NOT LINEAR: TWO ROADS DIVERGE

Our discussion so far has seemed linear—a history of increasing complexity in successive animal groups, as summarized in Table 17.1, page 302. It may seem logical, therefore, to assume that evolution is also linear, with each group giving rise to the next more complex set of

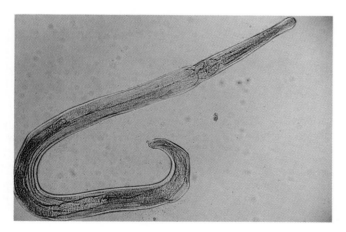

FIGURE 17.9 *The Roundworm: Phylum Nematoda.* Nematodes thrive in mud, sand, soil, running or standing water—even in acidic fruit juices—and were the first animals to have a body cavity. Here is the nearly transparent pin worm, *Enterobius vermicularis*, that infects humans.

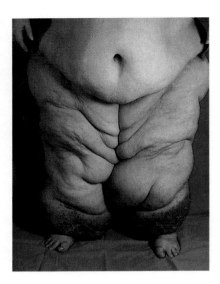

FIGURE 17.11 *Elephantiasis.* Grotesque limb enlargement can occur when roundworms block lymphatic vessels (see Chapter 20) and the tissues accumulate fluid. This woman with elephantiasis has hugely swollen legs but normal feet.

animals. Zoologists, however, think that evolution was not linear. They believe that an ancient ancestor similar to a free-living flatworm gave rise both to the roundworms and to all the higher bilateral phyla. Zoologists also think that a branching off took place very soon after bilateral organisms arose, with one road leading to the mollusks, annelids, and arthropods, and the other to the echinoderms and chordates (review Figure 17.2). Let's look at those groups now, beginning with the mollusks.

≋ MOLLUSKS: SOFT-BODIED ANIMALS OF WATER AND LAND

The phylum Mollusca is a collection of soft-bodied animals that includes snails, clams, oysters, squid, and octopuses. All but the land snails and slugs are aquatic. Although **mollusks** are soft, some members secrete a hard shell; therefore, unlike the sketchy histories left by the groups we've covered so far, a fairly good fossil record of the mollusks remains. The phylum probably arose from wormlike ancestors over 550 million years ago.

MOLLUSCAN BODY PLAN

You may have the impression from eating raw oysters that mollusks are slimy blobs with no distinct shape. However, there is an identifiable body plan to all mol-

lusks, no matter how different or soft they look inside their shells. As Figure 17.12 shows, each mollusk has a *foot,* a muscular organ used for gripping or creeping over surfaces; a *head* housing the mouth, brain, and sense organs; a group of internal organs (heart, gut, sex organs, and excretory and respiratory apparatus); and a *mantle,* a thick fold of tissue that covers the internal organs and in some mollusks secretes the calcium carbonate that makes up the hard shell. The space between the mantle and the organs is called the *mantle cavity,* and suspended in this space are *gills,* special surfaces for exchanging oxygen and carbon dioxide in water.

MOLLUSCAN ADVANCES

The mollusk's respiratory gills function in tandem with a *circulatory system* that includes a *heart* with chambers for pumping blood; *vessels,* or blood-carrying tubes, that pass through the gills; and a blood-filled cavity that bathes internal organs and tissues (see Figure 17.12). This kind of circulatory system in which the blood flows through open spaces part of the time is called an *open circulatory system.* As blood passes through the gills, it releases carbon dioxide and picks up oxygen; the blood then carries its oxygen cargo away from the gills and toward all internal cells. Together, the respiratory and circulatory systems provide each cell in the body with oxygen and nutrients in an efficient manner, and they have allowed many mollusks to reach large sizes; some, as we will see, are veritable monsters of the deep.

Some mollusks have well-developed nervous systems with large brains and acute senses. Like nematodes, mollusks have a one-way gut suspended in a body cavity, but it is a true coelom, not a false one (see Figure 17.10c).

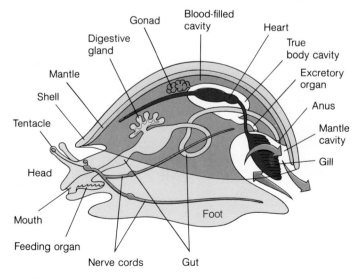

FIGURE 17.12 *The Molluscan Body Plan.* This generalized drawing shows the foot, head, and other major molluscan organs.

Mollusks rasp food off in layers with a feeding organ that works something like a cheese grater. And finally, most mollusks produce highly mobile fringed larvae; their similarity to certain larvae of the next animal group, the segmented worms, suggests that as different as they look, mollusks and annelids are related historically.

SOME CLASSES OF MOLLUSKS

Figures 17.13 to 17.15 show representatives of three classes of mollusks: gastropods (literally, "belly feet"), bivalves ("two valves"), and cephalopods ("head feet"). The gastropods include land snails and slugs—the only land-dwelling mollusks—as well as brightly colored nudibranchs, or sea slugs. The bivalves include oysters, clams, mussels, and scallops, whose shells are divided into two parts; rapid closing of the shells offers protection from predators and in scallops produces a self-propelling water jet. The most complex and fastest-swimming mollusks are the cephalopods, including squid, octopuses, and chambered nautiluses. Their "head foot" bears a circle of eight or ten arms studded with suckers and terminates in a funnel; with such modifications, it becomes an organ

FIGURE 17.14 *Oyster and Pearl: The Bivalve Jeweller.* Some mollusks line their shells with nacre—calcium carbonate plus a gluelike protein called conchiolin. Nacre is what we call mother-of-pearl. When the lined shells are open, particles can enter and irritate the soft tissue. In oysters, a region of the mantle responds to foreign particles by secreting nacre around the irritant, and the result can be a lustrous jewel. Oysters and other bivalves are *filter feeders;* they have gills that collect oxygen, release carbon dioxide, and are modified to strain out and collect tiny food particles suspended in water.

(a)

(b)

FIGURE 17.13 *Gastropods: Snails, Slugs, and Nudibranchs.* (a) Gastropods can be drab and familiar garden pests like this Hawaiian land snail, sliding across its slime trail. Since it lives on land, it has air-breathing lungs and can close off its shell to retain moisture during dry weather. (b) The soft, shell-less mollusks called nudibranchs are often dazzlingly colored and patterned, such as this elegant eolid *(Flabellina iodinea)* of California's Monterey Bay. They have no shells and breathe by gills. Nudibranchs can produce sour or bitter substances that repel attackers; their bright color serves as a warning that a bad meal lies ahead.

FIGURE 17.15 *Cephalopods: Predators of the Deep.* The most complex mollusks are squid, octopuses with eight tentacles like the lesser octopus *(Elgolone cirrhoser)* shown here, which lives in dark crevices in the deep Pacific. Another is the chambered nautilus, whose shell is coiled, coated with nacre, and divided internally by cross walls. These cephalopods have highly sensitive eyes that can form images like our own; in some giant deep-sea squid, the eyes grow larger than a car's headlights.

TABLE 17.1 ≈ KEY INVERTEBRATE ANIMALS

PHYLUM	EXAMPLES	NUMBER OF SPECIES	NOTABLE FEATURES
Porifera	Sponges	5000	Asymmetrical saclike bodies; central opening; body wall perforations; spikes; no evolutionary descendants
Cnidaria	Jellyfish, hydras, corals, sea anemones	10,000	Radial symmetry; grow as polyps or medusae; three-layered body wall; gastrovascular cavity; nerve cells
Platyhelminthes	Flatworms, including tapeworms, flukes	10,000	Bilaterally symmetrical with a head; three body layers; true organs and organ systems; life cycles often complex and include two or more hosts
Nematoda	Roundworms	12,000	Bilaterally symmetrical with a head and gut tube; false body cavity; hydroskeleton; extremely common in soils and as parasites on other animals and plants
Mollusca	Snails, clams, octopuses, squid, slugs	120,000	Bilaterally symmetrical with a head and gut tube; true body cavity; gills, open circulatory system; very common marine and freshwater organisms
Annelida	Earthworms, polychaete worms, leeches	5000	Bilaterally symmetrical with a head and gut tube; true body cavity; segmentation; hydroskeleton; move with bristles pushing against ground; crop, gizzard; closed circulatory system; earthworms occur widely and help aerate soils
Arthropoda	Spiders, mites, ticks, scorpions, millipedes, centipedes, insects, lobsters, shrimp	1 million	Bilaterally symmetrical with a head and gut tube; body cavity; segmentation; exoskeleton for support and protection; specialized segments and appendages; jointed legs; tracheae and gills; acute senses; most diverse phylum in living world
Echinodermata	Sea stars, sea urchins, sea cucumbers	6000	Gut tube and body cavity; head end in some larvae and adults; no segmentation; first endoskeleton; unique water vascular system for locomotion; separate evolutionary line from Mollusca, Annelida, and Arthropoda

specialized for hunting, swimming, and feeding. The cephalopod mantle is also modified into a muscular enclosure; this can expand and draw water into the mantle cavity or contract and force it out of the siphon, jet-propelling the mollusk backward. These explosive bursts can carry the animal to safety or bring its suckered tentacles within reach of prey. The coordination for hunting and feeding in cephalopods depends on acute senses, a large brain, and the most complex nervous system among the invertebrates. In fact, some squid can be trained to accomplish a number of tasks and are the most intelligent creatures without backbones. While members of the next phylum are far from intelligent, they and their descendants, the insects, are highly successful and show many innovations.

≈ ANNELID WORMS: SEGMENTATION AND A CLOSED CIRCULATORY SYSTEM

Marine sandworms, common earthworms, and leeches (Figure 17.16) are all members of the phylum Annelida, the **segmented worms,** whose 9000 species belong to three classes. Most are members of the class Polychaeta (meaning "many bristled")—often colorful marine worms that burrow in the mud or sand and bear common names such as fireworms, clam worms, and feather dusters. The familiar reddish earthworms are in the class Oligochaeta ("few bristles"). These widespread inhabitants of moist soils, often numbering 50,000 or more per acre, literally eat their way through dense, compacted earth, excreting the displaced material in small, dark piles. Each year, earthworms carry as much as 18 tons of rich soil per acre to the surface (Figure 17.17).

The leeches are in a separate class, Hirudinea, and live mostly in fresh water. Many prey upon worms and mollusks, but the most infamous types parasitize large animals and suck their blood. A leech can consume three times its weight in blood and go for as long as nine months between meals (see Figure 17.16).

Annelid reproduction is generally sexual. Many marine annelids shed sperm and eggs into seawater, where the gametes unite and develop into larvae that are similar to mollusk larvae. Earthworms and leeches are hermaphrodites; pairs reciprocally fertilize each other.

ANNELIDS SHOW ALL FIVE EVOLUTIONARY TRENDS

The term *annelid* means "tiny rings" and refers to the external segments visible on members of this phylum. Annelids are the first group possessing all five physical traits we have been tracing: right-left bilateral symmetry; cephalization; a gut tube; a body cavity; and segmentation. Although more than a million animal species evolved after the annelids, no new trends as basic as the emergence of these features occurred; all later animals show variations on the anatomical themes first "stated" together in the annelids.

Earthworms—typical annelids—have 100 or more body segments (see Figure 17.17a). Each segment is separated from the next by an internal partition, and each contains a set of internal structures. Most segments contain two excretory units (*nephridia*; see Figure 17.17b). Each nephridium removes excess water and wastes from the body fluids by means of a tiny funnel structure and excretes them through a pore in the segment wall. Each segment also contains a fluid-filled compartment of the body cavity (also called the *coelom*) and is surrounded by circular and longitudinal muscles in the body wall. As the circular muscles squeeze against the incompressible fluids in the cavity and gut, force is transmitted to adjacent segments, creating a *hydroskeleton*, an internal skel-

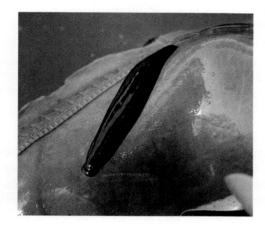

FIGURE 17.16 *Leeches: Predators and Parasites.* Segmented worms range from the delicate to the despised, including this large, black leech. When surgeons reattach severed human fingers, they occasionally use laboratory-raised leeches during the patient's recovery, to suck blood from and thereby relieve pressure around the reattachment sites.

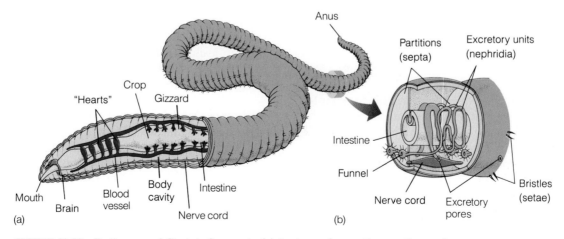

FIGURE 17.17 *Earthworms: A Study in Segments.* (a) Anatomy of an earthworm, the most familiar annelid. (b) An enlarged view of a segment.

eton made of fluid. Contractions in sequential segments produce waves of force that propel the animal forward. Pairs of bristles (setae) attached to each segment push against the ground with each contraction and help the animal move. In addition to nephridia, muscles, and cavity compartments, each segment contains clusters of nerve cells connected to the brain by nerve cords.

MORE ANNELID CHARACTERISTICS

Organ systems for digestion and circulation run the length of the annelid worm and represent further evolutionary advances. The gut tube has two specialized regions not seen in roundworms: a *crop* for storing food and a *gizzard* for grinding it. After the gizzard has ground some food, the region called the intestine absorbs nutrients released from the food by digestive enzymes (see Figure 17.17a). Annelids also have a *closed circulatory system:* blood carried entirely in tubes, or vessels, rather than partially in open spaces, as in mollusks. The contraction of several hearts, or muscular vessels, pumps the blood continuously through the closed circuit, and tiny vessel branches carry blood close to each cell in the body. The digestive and circulatory systems are interlinked in that blood passing through the gut picks up nutrients and transports them to all tissues. In turn, the blood picks up metabolic wastes from the tissue cells and transports them to the body cavity, from where they can be eliminated by the nephridia.

Advances in segmentation, digestion, and circulation allowed the annelids to grow longer and thicker than previous worms. Segmentation also had a special evolutionary significance: The descendants of the segmented worms, the arthropods, became the immensely successful group they are largely because of their segments and the appendages for chewing, sucking, flying, and running that developed from them.

≋ ARTHROPODS: THE JOINT-LEGGED MAJORITY

The termites introduced at the beginning of this chapter are members of the largest animal group on earth: the phylum Arthropoda (Figure 17.18). **Arthropods** include fossil trilobites, spiders, mites, ticks, scorpions, centipedes, millipedes, lobsters, crabs, and insects. Hard as it is to believe, the insects, with over 900,000 species, make up the great majority of all living animal species. In fact, if you took a book that listed every animal species

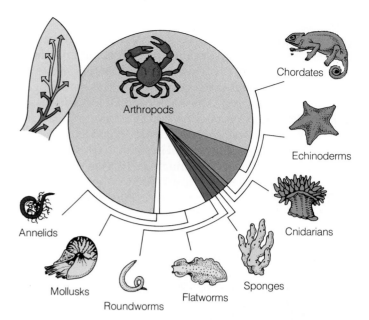

FIGURE 17.18 *Arthropods: The Vast Majority of All Animal Species.* Compared to the nearly 1 million species of insects and other arthropods, all other species make up a small slice of the animal kingdom. Our era could be called the Age of Arthropods.

on earth and pointed to any name at random, it would probably be an insect. The insects' astounding diversity and success are based on several characteristics shared by all arthropods: (1) an external skeleton; (2) modified, specialized segments; (3) rapid movement and metabolism due to special respiratory structures; and (4) acute sensory systems.

THE ARTHROPOD EXOSKELETON

As animals—particularly land animals—grew larger and thicker, they needed support for the body. The key to the success of insects and other arthropods is the **exoskeleton,** or external skeleton, a hard, waterproof shell that completely surrounds the animal and provides strong support as well as a counterforce for muscle movements. This inflexible armor prevents internal tissues from drying out, which is extremely important, since most arthropods live on land.

In all arthropods, the exoskeleton is thin and flexible only at the *joints,* the hingelike areas of the legs and body. The term *arthropod,* in fact, means "jointed foot." The presence of jointed appendages allows arthropods to move quickly and efficiently above the ground or seafloor instead of dragging the body directly along the ground on stubby legs or bristles. Muscles attached to the inside of the exoskeleton on either side of the joints help move each appendage. Exoskeletons do, however, have a major disadvantage: The animal cannot grow larger unless it *molts,* periodically shedding its constricting armor.

SPECIALIZED ARTHROPOD SEGMENTS

Unlike annelid worms, which have the same simple segments repeated throughout the body, arthropods have evolved different and highly modified segments that give them a far greater repertoire of activities. And their segments are usually fused into a few major body regions. Insects, for example, have three regions—the head, *thorax*, and *abdomen;* in spiders and crustaceans, there is an abdomen plus one fused head-thorax region. During evolution, specific appendages for walking, swimming, and flying arose, with each species having its own modifications. Some arthropods also developed pincers or feelers for hunting and feeding. And from the head region grew other appendages: *mouthparts* for chewing and sucking, *antennae* for sensing. The brightly colored shrimp in Figure 17.19a, for example, has appendages modified for eating (mouthparts), grasping (pincers), walking (legs), mating (swimmerets), and fast swimming (tail). The appendages themselves became further modified during evolution. Just compare the flapping wings of the butterfly (Figure 17.19b), the small buzzing wings of the housefly, and the hard but colorful shell-like wings covering the functional flying wings of the Harlequin beetles

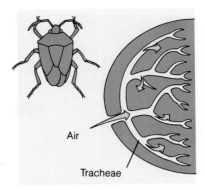

FIGURE 17.20 *Arthropod Respiration: How an Insect Breathes.* An insect's tracheae are branching hollow tubes that carry air deep into the body tissues.

on this book's cover. The many modifications enabled the arthropods to exploit a wide variety of environments.

ARTHROPOD RESPIRATORY ADVANCES

Special respiratory structures allow arthropods to generate energy and move rapidly, and in fact, some insects do both at the highest rates in the animal kingdom. Some tiny flies can generate enough energy to beat their wings 1000 times per second. Such high rates require high levels of oxygen, and the arthropods' efficient respiratory organs provide a large surface area for collecting oxygen and releasing carbon dioxide quickly. Insects, for example, have branching networks of hollow air passages called *tracheae* (Figure 17.20). High energy-producing rates and rapid movements allow arthropods to fly, run, and swim faster than any previous animal group and hence to escape from predators with agility and to disperse over wider ranges.

ACUTE SENSES

Most arthropods have antennae capable of detecting movement, sound, or chemicals with great sensitivity. The antennae of several male moth species, for instance, can detect the *pheromones*, or odor signals, given off by adult female moths at distances of over 7 miles. Arthropods also have various types of separate organs on the head and body for detecting sound and taste, and many have special compound eyes made up of 2500 or more six-sided segments called *facets* (Figure 17.21, page 306). Many compound eyes are capable of color vision and can detect the slightest movements of prey, mates, or predators.

(a)

(b)

FIGURE 17.19 *Arthropod Appendages: Modified for Different Tasks.* (a) A brilliantly colored creature like the scarlet cleaner shrimp *(Hippolysmata grabhami)* of the tropical western Atlantic has a fused head and thorax and a separate abdomen, as well as appendages modified for grasping, walking, mating, and swimming. (b) The giant swallowtail *(Papilio cresphontes).*

FIGURE 17.21 *The Many Facets of an Arthropod's Eye.* The compound eye of this fruit fly has about 800 geometric facets. Each facet detects light and movement, and the insect sees a composite image created by the many individual segments.

IMPORTANT ARTHROPOD CLASSES

The successful and highly divergent phylum Arthropoda is divided into a number of taxonomic classes, most of which will sound quite familiar.

■ **Centipedes and Millipedes** Centipedes look rather like annelid worms and have a series of flattened body segments. Centipedes kill their prey with *poison claws,* modified legs on the first body segment (Figure 17.22a). Some huge tropical centipedes are dangerous to humans, but the common varieties that lurk in damp basements are harmless and consume insects.

Millipedes are slow-moving counterparts to the centipedes. They have two pairs of much smaller legs per segment; a round, not flattened, body; and a preference for decaying vegetable matter instead of live prey.

■ **Crustaceans** Crabs, shrimp, lobsters, barnacles, crayfish, and sowbugs are so different from each other that only two generalizations apply: Almost all have an exoskeleton hardened with calcium salts that covers most of the animal as a protective shell, and all have two pairs of antennae (Figure 17.22b).

Crustaceans are so numerous and so diverse that most bodies of water contain them. There are even a few species that live on land, like the sowbugs. Many small aquatic species, such as fairy shrimp and brine shrimp, serve as the primary food for large fishes, whales, and other animals. Barnacles, an aquatic species, secrete a natural cement that is strong enough to keep the animals glued to tidal-zone rocks despite the pounding ocean surf. This feat is comparable to a person standing upright in winds

up to 400 mph. Some biologists are studying barnacle glue in hopes of developing medical cements for fastening dentures and rejoining broken bones.

■ **Spiders and Relatives** Spiders, ticks, mites, and scorpions (*arachnids*) are closely related to the extinct trilobites and the primitive-looking horseshoe crabs often seen on Atlantic coastal beaches. Arachnids lack antennae, have a fused head-thorax and abdominal regions, each divided into segments, and usually have six pairs of legs (Figure 17.22c). The first pair of legs is modified into poison fangs, used for killing prey or for self-defense. The second pair of legs holds the prey while the arachnid injects poison or enzymes. The other four pairs are walking legs. Spiders also have silk-spinning organs at the rear of the abdomen; these reel out threads that are, size for size, stronger than steel. Despite their sinister reputation, most spiders are only capable of poisoning small animals. With their potent poisons, however, the black widow and brown recluse can be dangerous to people.

Present-day scorpions have a body form very similar to the huge water scorpions that lived 500 million years ago and that probably gave rise to the land scorpions and spiders. Mites and ticks are, for the most part, merely irritating biters, but some ticks carry serious diseases, including Lyme disease with its characteristic circular rash and delayed joint swelling, and Rocky Mountain spotted fever with its fever, rash, and joint pain.

■ **Insects** It is difficult, in a few paragraphs, to do justice to **insects,** the largest—and, by that measure, most successful—class of animals on earth. Most insects are terrestrial animals living in habitats from the tropics to the poles, but a few are aquatic. Many zoologists attribute insect success to the general arthropod traits we have already considered, but also to the insect's small size. Smallness enables the insects to exploit a vast array of microhabitats—the bark of a tree, the planar landscapes on the backs of leaves, the dense thickets of an animal's fur, or the universe of midair.

Specific body parts also help account for the insects' success. Many insects have organs for smelling, touch reception, tasting, seeing, and hearing in various parts of the body. An insect's head bears one pair of antennae; its thorax bears three pairs of legs and usually one or two pairs of wings; and its abdomen is usually free of appendages. A series of modified segments, the mouthparts, enables the insect to feed efficiently. Some, like mosquitoes, have superb pointed stylets for piercing and sucking. Others, like locusts and grasshoppers, have chewing mouthparts that can quickly decimate foliage (Figure 17.22d). In 1985, a long, dark column of grasshoppers descended upon Idaho's Magic Valley, eating a swath about 2 miles wide through the bean, beet, and potato fields. In one day, seven grasshoppers can eat as

(a)

(b)

(c)

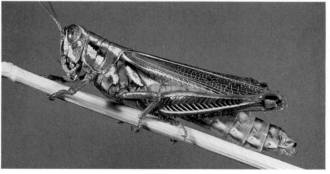

(d)

FIGURE 17.22 *Arthropods: Four Major Classes.* (a) Centipedes like this one have many segments and march along on an army of small legs; they also have poison claws. (b) Crustaceans: The blue lobster lives in the Atlantic Ocean; shown in a rare blue phase, it can grow to 1 m in length and weigh 18 kg (40 lb). Like the shrimp (see Figure 17.19a), it has a fused head and thorax, as well as several sets of specialized appendages, including mighty pincers that help in food gathering and defense. (c) Arachnids: This banded garden spider is fearsome but harmless to humans; it weaves a broad web and preys on small insects. The one shown here has ensnared a grasshopper. (d) Insects are mostly terrestrial and highly diverse. Grasshoppers like this common member of the genus *Melanoplus*, with its hearty appetite and chewing mouthparts, can denude a farmer's field in minutes.

much as a grazing cow; and during a plague, they can occur 1800 to the square meter.

Insects have evolved various ways to grow, despite the confining exoskeleton, and various ways to thrive, despite the changing seasons. In some insects, like grasshoppers and cockroaches, the embryo becomes a miniature version of the adult without wings or mature reproductive organs. This organism feeds, grows, and molts five or six times until it reaches adult size, then does not molt again. In most insects, however, including butterflies, flies, and beetles, the embryo develops into an immature form, or larva, eats voraciously, then forms a transitional stage, or *pupa*, sometimes inside a cocoon. A complete change, or *metamorphosis*, takes place in the body within the pupal exoskeleton. Finally, a nonmolting, reproductively mature adult emerges (see Figure 25.1). In insects that metamorphose, the larvae and adults are adapted to very different foods and environmental conditions—a successful evolutionary solution to surviving in a changing climate.

Perhaps the most fascinating of the arthropods are the **social insects:** the termites, ants, wasps, and bees. Most species of these insects live in large colonies with labor divided among *castes,* or subgroups, that differ in appearance and behavior. Termites have the most complex social life of all the insects. There is a sexual caste of kings and queens, a worker caste with sterile males and females, and a soldier caste. Some soldiers have huge, biting mouthparts, and others have the conical, bazooka-shaped heads for squirting defensive chemicals on intruders (see Figure 17.1). Such insect colonies are highly evolved and function, in a sense, as a single well-coordinated organism with the capacity to simultaneously build homes and cities, defend their own, harvest food, and reproduce. When you consider that the roots of the animal kingdom lie in creatures no more complicated than sponges, radial animals, and flatworms, the complex social insects like termites demonstrate vividly the great evolutionary distance the invertebrates traveled.

⚶ ECHINODERMS: THE FIRST INTERNAL SKELETONS

An expert on invertebrates once called the **echinoderms** "a noble group especially designed to puzzle the zoologist." Indeed, the members of the phylum Echinodermata, including the sea stars, brittle stars, sea urchins, sea cucumbers, and sea lilies, have an odd mixture of traits, some of which are innovations appearing in all higher animals and some of which are apparent regressions toward simpler forms. Besides their bizarre and interesting life-styles and appearances (Figure 17.23), the echinoderms display two new features: an internal skeleton and a hydraulic pressure system for locomoting that is unique in the animal kingdom. If we consider certain fundamental features of their embryos, echinoderms appear to lie on a distinct evolutionary pathway that sets them off from all other invertebrates and suggests an ancestry common to our own phylum, Chordata.

The word *echinoderm* means "spiny-skinned," and virtually all members have spines, bumps, spikes, or unappetizing projections that help to protect these slow-moving marine creatures from predators. Echinoderm larvae are bilaterally symmetrical, but adults of many species are radial, headless, and brainless. Nerve trunks run along each of the adult's arms (usually numbering five or multiples of five) and unite in a ring structure around the mouth. This simple system allows coordinated—but very slow—movement of the limbs; a sea star travels only 15 cm (6 in.) per minute.

Echinoderms reproduce sexually by shedding sperm and eggs directly into the ocean and forming larvae. But their most remarkable means of reproduction is by regeneration. A single sea star arm with a small piece of the central hub still attached can regenerate an entire body. A sea cucumber, when irritated, will sometimes violently expel its internal organs, leaving intact only small fragments—and regenerate the lost parts.

The radial body plan is not an advanced trait, nor is the absence of a brain. Echinoderms also lack excretory and respiratory systems, relying mainly on diffusion for all exchanges with the outside. However, they do show one important evolutionary development: an **endoskeleton,** or true internal support system. Just below the outer skin lie calcium-based plates that protect the internal organs and give rise to the outer spines and bumps. (We will consider other types of internal skeletons in Chapter 28.)

FIGURE 17.23 *Echinoderms: Spiny-Skinned Ocean-Goers.* Echinoderms such as this sea lily from Palau in the South Pacific can resemble flowers blooming on coral reefs.

Echinoderms have a large body cavity, and in many species it contains a unique *water vascular system,* a set of canals filled with modified seawater (Figure 17.24). Hundreds of short branches off the canals become the tube feet that dot the undersurface of a sea star, for example, and enable the animal to move across the seafloor. A sea star can also force open a tightly shut clam shell with its arms and tube feet, then push its stomach out through its mouth into the open shell and digest the prey. Capable of eating more than ten clams or oysters in a day this way, sea stars can be serious predators on commercial beds.

⚶ CONNECTIONS

We tend to think of vertebrates as physically more complex than invertebrates, and many are. But the invertebrate world, which includes 95 percent of all animal species, is full of astonishing structures and complicated processes: headlamp eyes, poison fangs, metamorphosis, caste systems. These belie any attempt to dismiss all invertebrates as "simple" or "primitive." In fact, the invertebrates, during an immense history that began more than 700 million years ago, evolved most of the body structures and systems employed throughout the animal kingdom in invertebrates and vertebrates alike. In Chapter 18, we will encounter additional anatomical innovations that are more familiar to us, including bones, teeth, scales, feathers, hair, and warm blood. But these are really just evolutionary overlays on the basic "inventions" of the invertebrates.

FIGURE 17.24 *Sea Stars and Hydraulic Tube Feet.* Sea stars have radial bodies with 5 to 25 arms or more, internally branching digestive pouches, and well-developed gonads. The sieve plate is the site where water enters and leaves the vascular system. Most remarkable are the hundreds of tube feet on the animal's lower surface; the tube feet can jut outward to an extended position under the hydraulic force of fluid from the internal water vascular system and then contract to rounded nubbins when the pressure diminishes. By sequentially jutting and contracting these tube feet in waves, the animal can slowly move along rocks or sand below.

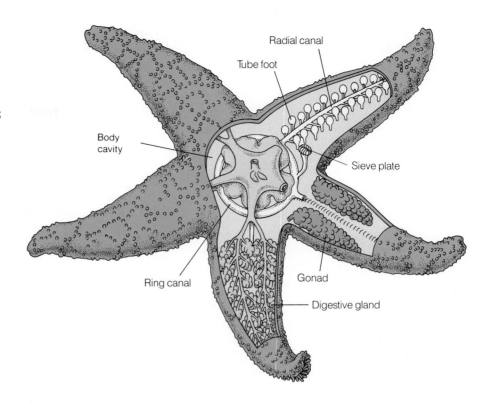

NEW TERMS

animal, page 292
arthropod, page 304
bilateral symmetry, page 296
body cavity, page 299
cnidarian, page 295
echinoderm, page 308
endoskeleton, page 308
exoskeleton, page 304
extracellular digestion, page 296
flatworm, page 296
insect, page 306

invertebrate, page 292
mollusk, page 300
nerve cord, page 297
organ, page 297
radial body plan, page 295
roundworm, page 298
segmented worm, page 302
social insect, page 307
sponge, page 294
tissue, page 296
vertebrate, page 292

STUDY QUESTIONS

REVIEW WHAT YOU HAVE LEARNED

1. List and explain five trends in invertebrate evolution.
2. How does a sponge obtain nutrients?
3. What features would you see in a typical cnidarian?
4. List five organ systems in the planarian and name a specialized structure in each system.
5. What two new features arose in roundworm evolution?
6. Describe the evolutionary adaptations of mollusks with respect to respiration, circulation, the nervous system, and larvae.
7. List the classes of mollusks, and name one member of each class.
8. Annelids show numerous evolutionary adaptations. List five.

9. What characteristics distinguish the arthropods as a group?
10. List the arthropod classes, and name at least one member of each class.
11. What are the main characteristics of echinoderms?
12. How does a sea star move?

APPLY WHAT YOU HAVE LEARNED

1. The laws of Islam and Judaism prohibit followers from eating pork. From a biological perspective, what might have been the benefit of this prohibition throughout much of history?
2. Commercial scallop farmers often kill sea stars found in their catch because the echinoderms feed on scallops. The farmers cut each sea star into several pieces and toss the bits overboard, but this strategy usually backfires. Why?

FOR FURTHER READING

Brownlee, S. "Jellyfish Aren't Out to Get Us." *Discover* (August 1987): 42–54.

Field, K. G., G. J. Olsen, D. J. Lane, S. J. Giovannoni, M. T. Ghiselin, E. C. Raff, N. R. Pace, and R. A. Raff. "Molecular Phylogeny of the Animal Kingdom." *Science* 239 (1988): 748–773.

Gould, S. J. *Wonderful Life: The Burgess Shale and the Nature of History.* New York: Norton, 1989.

McMenamin, M. A. S. "The Emergence of Animals." *Scientific American* 256 (April 1987): 94–102.

(1) Animals are many-celled organisms that can generally move about and must take in preformed organic compounds.

(2) Invertebrates are animals without backbones.

(3) Sponges, the simplest animals, are a primitive lineage, unrelated to other invertebrates.

(4) Evolutionary trends among invertebrates include:
- Bilateral symmetry (first seen in flatworms)
- A head (first in flatworms)
- A gut tube with two openings (first in roundworms)
- A true fluid-filled body cavity (first in mollusks)
- Segmentation (first in annelid worms)

(5) Of the invertebrates, only the annelid worms and arthropods reflect all five trends.

Vertebrates
Arthropods
Annelids   Primitive chordates
Mollusks
Echinoderms
Nematodes
Flatworms   Cnidarians
                  Lower invertebrates
Sponges

(6) Insects and other arthropods radiated to over 1 million species.

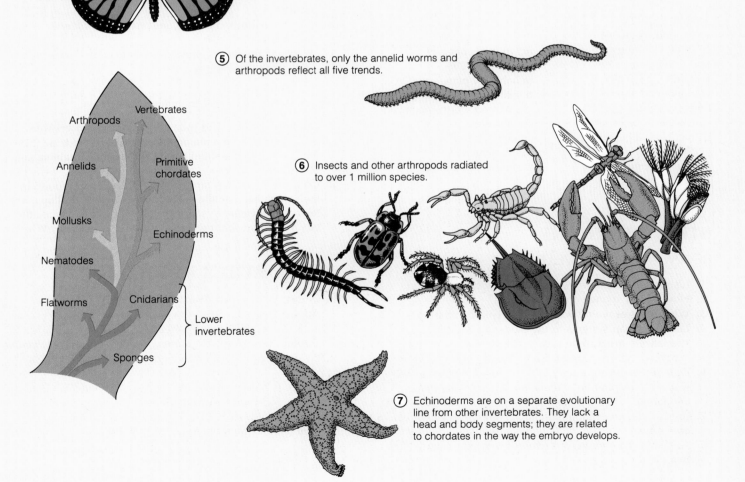

(7) Echinoderms are on a separate evolutionary line from other invertebrates. They lack a head and body segments; they are related to chordates in the way the embryo develops.

THE CHORDATES: VERTEBRATES AND THEIR RELATIVES

CHAPTER 18

BATS: BENEFICIAL VERTEBRATES WITH A BAD REPUTATION

Bats fly silently about at night, but they are not the dirty, rabid vampires many people imagine. The animals clean themselves meticulously, like cats, and rarely have rabies. Only three species rely on blood meals, flapping noiselessly up to cattle or other victims, making quick, painless incisions with razor-sharp teeth and lapping up the flowing blood. In fact, most species of bats eat fruit, pollen, insects, spiders, or other small animals. These same bats consume enormous numbers of insects, pollinate many species of flowering plants, and help reforest denuded lands by excreting millions of undigested seeds.

A bat's most obvious feature is a set of prehistoric-looking wings (Figure 18.1). It also has bizarre facial appendages that help the animal receive sonar signals and navigate in the dark. In addition,

bats are highly social: Thousands of individuals hang upside down together in caves or trees, and within these colonies, they assume specialized roles, such as guarding the cave entrance, scouting for food sources, or warning of danger.

With the bats and their relatives, we move away from the vast majority of animal species—the invertebrates—and focus on the remaining 5 percent—the **chordates,** members of the phylum Chordata. As expected from their name, chordates are distinguished by a cord—a flexible rod, or *notochord*—that runs the length of the body and supports it. Chordates also have a central nerve cord; in humans and other complex chordates, it's called the spinal cord. In some chordates, a series of interlocking bones called the *backbone* replaces the notochord, protects the nerve cord, and

provides internal support for the body. Chordates with a backbone (the great majority) are called *vertebrates* after the sets of interlocking bones, or **vertebrae,** that make up the bony column.

As we study the characteristics and evolution of the chordates, we will see four unifying themes. First, chordate evolution is a history of innovations that built upon the major invertebrate traits. These innovations enabled chordates to grow larger than their predecessors, on average, and to successfully exploit their environments in new ways.

Second, chordate evolution is marked by physical and behavioral specializations. The bat's leathery wings, use of sonar, and complex social behavior, for example, are all specialized traits that allowed the bat to hunt at night, feed on fruit, and hide from enemies in caves and trees.

Third, evolutionary innovations and specializations led to *adaptive radiations*: the development of a variety of forms from a single ancestral group. The earliest bats, for example, radiated into 900 species, some as large as eagles and some as small as bumblebees.

Finally, we will see that although we humans have some unique physical traits and capabilities, evolutionary principles common to all other organisms guided our emergence.

This chapter will examine several features of the chordates:

- The characteristics that chordates share and how they evolved.
- Chordates that lack a backbone.
- The characteristics of chordates with a backbone—the fishes, amphibians, reptiles, birds, and mammals.
- The primates, the mammalian order that gave rise to our own species.

FIGURE 18.1 *Bats: Prehistoric-Looking Vertebrates.* Bats can look a fright, but most are beneficial insect eaters and plant pollinators.

CHORDATE CHARACTERISTICS

The diverse vertebrates—including many graceful, powerful, and fleet animals of land, sea, and air—are higher chordates that trace their evolutionary roots to the small, inconspicuous, bottom-dwelling sea creatures without vertebrae called *tunicates* and *lancelets* (Figure 18.2). All chordates display bilateral symmetry, a head, a body cavity, a one-way gut tube, and body segmentation—traits that first emerged in the invertebrates; indeed, the chordates almost certainly evolved from ancient aquatic invertebrates. But chordates, both with and without vertebrae, also share a number of new structures—important evolutionary innovations not found in their invertebrate relatives (Figure 18.3).

One of these new structures is the *notochord,* a stiff but flexible rod that provides internal support and runs the length of the animal. A second new structure is the hollow *nerve cord,* or **spinal cord,** which is a tube of nerve tissue that also runs the length of the animal, just above the notochord, and which serves to integrate the body's movements and sensations.

Chordates also have **gill slits,** pairs of openings that penetrate the back of the mouth from the inside to the outside. Tunicates and lancelets employ gill slits for filtering food. Young fish and certain amphibians obtain oxygen using gill slits. In most reptiles, birds, and mammals, gill slits occur only in the embryo or develop into other structures.

All chordates have a **tail** that extends beyond the anus at some point in their development. The tail has the internal support of a notochord or vertebral column, and

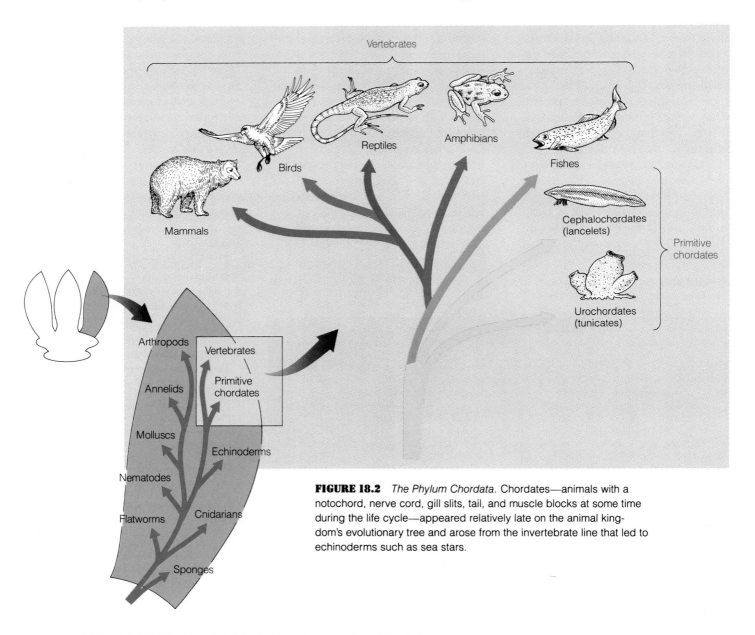

FIGURE 18.2 *The Phylum Chordata.* Chordates—animals with a notochord, nerve cord, gill slits, tail, and muscle blocks at some time during the life cycle—appeared relatively late on the animal kingdom's evolutionary tree and arose from the invertebrate line that led to echinoderms such as sea stars.

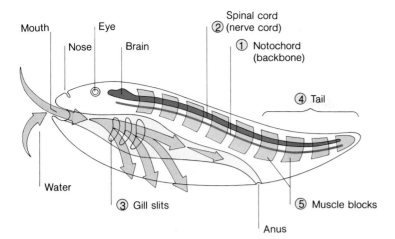

FIGURE 18.3 *Generalized Chordate Body Plan.* Chordates show the major animal characteristics that emerged in the invertebrates, but five additional traits as well: the notochord, nerve cord, gill slits, tail, and muscle blocks. In complex chordates, such as our own species, some of the five chordate innovations occur only briefly during early embryonic development, then disappear.

it moves by means of yet a fifth chordate trait, muscle blocks *(myotomes).* Muscle blocks, which occur along the length of the body behind the head, are the major signs of segmentation in chordates. When you eat fish, you can easily see these stacked muscle layers in the "flaky" white or pink meat. Most chordates have a tail throughout life, but in some chordates, including humans, the tail appears only briefly in the embryo.

The chordate innovations—notochord, spinal cord, gill slits, tail, and muscle blocks—had dramatic implications (Figure 18.4). The physical support of a notochord, and later of a vertebral column, allowed chordates to grow larger and larger—as large, in fact, as dinosaurs and whales! The centralized nerve coordination and control involves the spinal cord and allows acute senses, a wide range of behaviors, and greater intelligence than in most invertebrates. The gill slits could be used for feeding or breathing. And the tail, in its many manifestations, became a major organ of locomotion (especially in fishes), balance (in hopping animals), and even communication (in many birds and mammals).

Many biologists agree that the evolution of chordates began with some common ancestral echinoderm related to the sea stars. This single ancestral echinoderm may have given rise to the chordates without backbones and to the earliest vertebrates.

FIGURE 18.4 *Major Innovations in the Chordates.* The five distinguishing chordate characteristics evolved in chordates without backbones. The backbone emerged in the early fishes; and each of the five vertebrate classes shows a slightly different set of innovations, each modifying the adaptations of progenitor species.

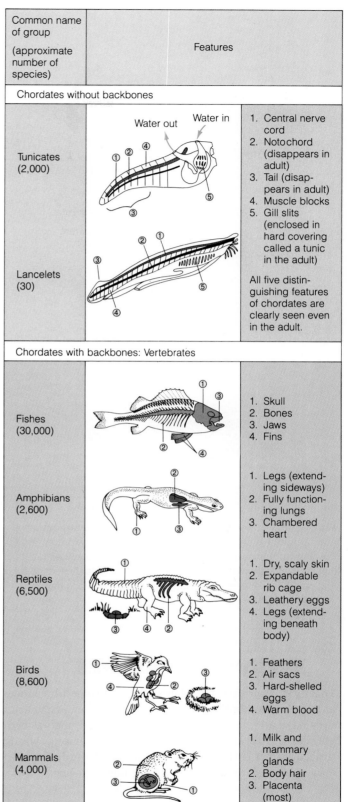

Common name of group (approximate number of species)	Features	
Chordates without backbones		
Tunicates (2,000)		1. Central nerve cord 2. Notochord (disappears in adult) 3. Tail (disappears in adult) 4. Muscle blocks 5. Gill slits (enclosed in hard covering called a tunic in the adult) All five distinguishing features of chordates are clearly seen even in the adult.
Lancelets (30)		
Chordates with backbones: Vertebrates		
Fishes (30,000)		1. Skull 2. Bones 3. Jaws 4. Fins
Amphibians (2,600)		1. Legs (extending sideways) 2. Fully functioning lungs 3. Chambered heart
Reptiles (6,500)		1. Dry, scaly skin 2. Expandable rib cage 3. Leathery eggs 4. Legs (extending beneath body)
Birds (8,600)		1. Feathers 2. Air sacs 3. Hard-shelled eggs 4. Warm blood
Mammals (4,000)		1. Milk and mammary glands 2. Body hair 3. Placenta (most)

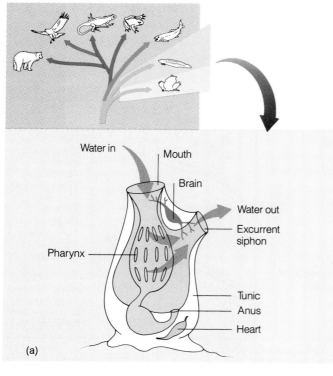

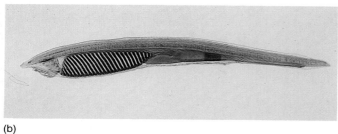

(b)

FIGURE 18.5 *Chordates Without Backbones: Tunicates and Lancelets.* (a) Adult tunicates are stationary filter feeders enclosed in a baglike tunic. (b) Lancelets, such as *Amphioxus*, are vaguely fishlike, but they rest vertically with transparent bodies planted tail first in the sand and only the anterior end exposed for capturing tiny food morsels.

≋ CHORDATES WITHOUT BACKBONES

The most primitive chordates, the tunicates and lancelets, lack a backbone. But, at least when they are young, they have all five chordate traits: notochord, nerve cord, gill slits, muscle blocks, and tail. Only 1–5 cm long, young tunicates and lancelets look like miniature tadpoles and move about by thrashing their long tails (see Figure 18.4). They eventually settle down on a rock or in the sand and use combs of cilia to filter food from water passing through their gills. Tunicates undergo a dramatic metamorphosis; the tail, with its notochord and nerve cord, disappears,

and the stationary adult develops a hard outer coat, or *tunic* (Figure 18.5a). The tunicate filters food from water drawn in through the mouth and out through a hole in the tunic. If disturbed, a tunicate will often squirt water forcibly from its siphon. Hence the common name "sea squirt." The small, streamlined lancelets also settle down in mud or sand, but if food becomes scarce, they can pull up anchor and move to more fertile grounds (Figure 18.5b). Some biologists believe that the ability to move as adults gives lancelets a distinct feeding advantage over the stationary tunicates, and for this reason, they think that lancelets more closely resemble the mobile ancestor that led to fish and other vertebrates.

≋ THE VERTEBRATES: CHORDATES WITH BACKBONES

FISHES: THE EARLIEST VERTEBRATES

Emerging 570 million years ago, at the dawn of the Paleozoic era, the first **fishes** were streamlined filter feeders, about 1 ft long, that lived in the muddy bottoms of ancient seas. They had fixed circular mouths that lacked jaws and that could suck up sediments like organic vacuum cleaners; hence, these fishes are called jawless fishes. While they retained the chordate notochord for internal support and the gill slits for feeding, each had a **skull** to protect its brain, and muscles rather than tiny cilia to powerfully draw water and suspended food into the mouth. Muscle-powered gill slits allowed the jawless fishes to consume greater quantities of food than their earlier cousins and to grow as much as 30 times larger. In addition, the gill openings were supported by the bony gill bars—the tissue between the gill slits—and bony plates beneath the skin served as armor that probably protected these animals against dangerous invertebrates, such as giant sea scorpions.

Modern jawless fishes like the *lamprey* have lost the bony plates and skulls of their ancestors and instead have very flexible internal skeletons of cartilage, the firm but flexible material that supports your ears and the tip of your nose. The lamprey is a parasite that attaches to its prey by suction, rasps through the victim's body wall with a sharp tongue, and then drinks its blood (Figure 18.6). At times, lampreys have been serious pests of commercially fished species in Lake Michigan.

■ **The First Jawed Fishes** Ancient jawless fishes gave rise to the first jawed fishes (*placoderms*, meaning "plate-skins") about 425 million years ago. These descendants had three basic innovations that were so useful that they appeared in nearly all the vertebrates that followed. First,

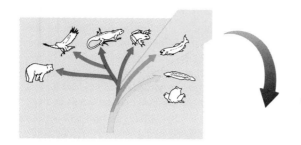

FIGURE 18.6 *Lampreys Parasitizing a Carp.* Modern relatives of the ancient jawless fishes, lampreys live by rasping off the flesh and sucking the blood of other aquatic animals. Here, three lampreys *(Lampetra fluviatilis)* parasitize a large carp.

the early jawed fishes had **hinged jaws,** derived from their predecessors' gill support bones. Jaws allowed these fishes to consume large chunks of food—kelp fronds, clams, other fish—instead of filtering small particles. As a result, some became huge; a person can stand upright in the fossilized jaws of one placoderm. Second, these jawed fishes produced the first *vertebrae:* They had a series of separate bones that fused with the notochord and arched over the spinal cord. The resulting vertebral column was incomplete compared to the full vertebral column of modern bony fishes, but it provided a site of attachment for muscles in the body wall, anchoring them as the body bent to and fro and allowing the muscles to grow larger and propel the fish more powerfully. The backbone also encased and protected the delicate spinal cord. Third, early jawed fishes had the first *fins,* appendages that provided more control over swimming direction, speed, and depth. The sockets in which the fins fit gave rise to the hips and shoulder joints of later land animals.

Although the early jawed fishes were heavily armored with bony plates beneath the skin, the sharks, skates, and rays that are their modern descendants have skulls,

vertebrae, and the rest of the skeleton made entirely of cartilage. Cartilage is lighter than bone and affords these predatory fishes the speed and agility that allows them to catch prey. In skates and rays, large fins provide lift, much like graceful underwater wings, while in sharks, stiff fins knife through the water, helping these largest and most common oceanic hunters to maneuver easily toward their victims (Figure 18.7).

■ **Bony, Jawed Fishes** The final group derived from the ancient jawless fishes evolved into the modern bony fishes, which include bass, trout, tuna, and virtually all of today's familiar fishes. Their skulls, vertebrae, and the rest of the skeleton are composed mainly of bone. There

(a)

(b)

FIGURE 18.7 *Cartilaginous Fishes: Hunters of the Sea.* (a) A skeleton of cartilage is lightweight and flexible and gives the spotted eagle manta ray of Baja California a slight resemblance to a predatory bird flapping beneath the waves. (b) The fearsome jaws of a great white shark. Sharks are earth's most common predators, and their jaws, inherited from placoderm ancestors, serve them admirably.

FIGURE 18.8 *Bony, Jawed Fishes: The Most Diverse Vertebrates.* The bony, jawed fishes have radiated into the largest vertebrate phylum, with 30,000 living species and thousands of additional species that are now extinct. The oldest descendants, the lobe-finned fishes, had muscular appendages supporting their fins, as well as lungs that allowed air breathing during migration across land from pool to pool. The first teleosts, or spiny-finned fishes, had delicate bones in their fins, as well as a swim bladder. The modern teleosts include brilliantly colored species, such as this painted coral trout in the South Pacific.

are two major groups of bony fishes, each displaying the cumulative hallmarks of vertebrate evolution—backbone, skull, jaws, and fins—but showing innovations as well. The oldest bony fishes are the *lobe-finned fishes,* which arose more than 400 million years ago. These fishes had large, muscular, lobed fins that allowed them to "walk" across the bottom of shallow bays, as well as **lungs,** or air sacs, for breathing. When water levels fell, their innovations allowed them to survive by breathing air and migrating over land to other pools or bays. Present-day lobe-finned fishes include the *lungfishes* and the *coelacanths.*

A second group of bony, jawed fishes are the *spiny-finned fishes (teleosts).* These lost their ancestors' fleshy fins, but gained much more versatile spiny fins with webs of skin over delicate rays of bone. They also lost the lungs, but developed instead a swim bladder, an internal balloon below the backbone that can change volume and allow the animal to adjust its swimming depth. Although teleosts lost the ability to survive out of water, their versatile fins and swim bladders were highly advantageous for aquatic life. These fishes radiated, early on, into the largest group of vertebrates, and they owe much of their success to specializations that evolved as the group radiated. Some bony fishes have an antifreeze substance in the blood, allowing them to live below the polar ice. Some are brilliantly colored, an aid to courtship and a warning to predators (Figure 18.8). Some have "flashlights" (luminescent organs) that enable them to find food in the cold, black reaches of the ocean abyss. Sea horses swim vertically, propelled by intricate undulations of the bony fins. However, except for a very few air breathers, fish are relegated by their anatomy to life in water. The great landmasses were to be dominated by other vertebrates—descendants of the early fishes. Several features of these other vertebrates, however, derived from important evolutionary innovations within the fishes including the skull, gill slits, bone, hinged jaws, and vertebrae.

AMPHIBIANS: FIRST VERTEBRATES TO LIVE ON LAND

Fossils show that during the Age of Fishes (roughly 400 million years ago), the vertebrate lineage that included air-breathing, lobe-finned fishes produced the **amphibians:** vertebrates that can live both on land and in water (*amphi* means "both"; *bios* means "live"). Modern amphibians include the frogs, toads, and salamanders—animals whose ancestors overcame the formidable problems of life on land and evolved means of walking, breathing air, and staying moist. Early amphibians had front and hind *legs* containing strong bones and powerful muscles. Extending sideways from the body, the legs supported the animal's weight far better than lobed fins; and with two pairs of legs, the animal could move about on land—albeit slowly and clumsily—for greater distances, even without water's buoyant support.

Laborious walking, however, would have required a great deal of energy from food, and large quantities of oxygen would have been needed for aerobic respiration. Fossil evidence reveals that early amphibians had fully functioning lungs, or air sacs, that provided a site for gas exchange. Air was probably pumped in by swallowing movements, much as it is in modern frogs and toads (Figure 18.9a). Moreover, specializations arose in the heart and circulatory system that tended to separate oxygenated blood en route to the body tissues from unoxygenated blood bound for the lungs—a separation that became complete in the birds and mammals (see Chapter 20). Finally, an amphibian's smooth, moist skin could absorb about half of the necessary oxygen and release most of the carbon dioxide.

These innovations provided sufficient oxygen but did not solve the problem of drying out. Since the amphibian skin had to remain moist for gas exchange (to supplement the lungs), the animals were restricted to life near the water's edge. In addition, amphibians lay eggs with a clear, jellylike coating that must also stay moist, lest the embryos die before the fishlike tadpoles emerge. Because their eggs require water, amphibians must return to the water to reproduce, and in this sense, they are analogous to the first land plants whose sperm required standing water to swim from one plant organ to another.

Today, frogs, toads, salamanders, and a few relatives are the only remaining members of the class Amphibia and continue to display the evolutionary innovations of the amphibian lineage: legs, fully functioning lungs, and partial separation of oxygenated and deoxygenated blood. These generally small vertebrates are very common in freshwater environments and show a range of interesting specializations. In some, the moist skin has become brightly colored, attracting potential mates and warning would-be predators of poison skin and glands (see Fig-

ure 18.9a). And in others, the skin is dry and adapted to life in dry habitats (Figure 18.9b). The nervous system of many amphibians, protected by the vertebral column and bony skull, has become well developed and accompanied by keen senses of sight and hearing. What's more, rapid-fire nervous reactions enable these animals to catch flies by flipping out their long tongues. And amphibian limbs are often specialized with webbed feet for efficient swimming and with thick muscles for hopping or running on land.

REPTILES: CONQUERORS OF THE CONTINENTS

Long after amphibians began crawling about in the early swamps, the first **reptiles** appeared among them. Like the early seed plants, the reptiles evolved important innovations for life on land that freed them from dependence on moist environments.

The earliest reptiles resembled crocodiles and had four innovations for terrestrial life (review Figure 18.4). First, these ancestral reptiles had dry, scaly skin that provided a barrier to evaporation and sealed in body moisture. This eliminated the skin surface as a major site for gas exchange, but a second reptilian innovation made up for this: respiratory modifications consisting of lungs with a larger surface area for gas exchange than in amphibians, an expandable rib cage that could draw in large quantities of air like a bellows, and a heart and circulatory system that separated oxygenated and deoxygenated

(b)

(a)

FIGURE 18.9 *Amphibians: The Transition to Land.* Some of the first land animals were large amphibious creatures that moved about on sideways-jutting legs and probably spent much of their time in shallow waters. The brilliant gold and red pigments in the skin of (a) the tropical poison arrow frog *(Dendrobates histrionicus)* and (b) the Chinese salamander *(Tylototriton verrucosus)* warn would-be predators of the amphibian's poison skin glands. The salamander, with its dry skin, is adapted to life in dry habitats.

blood more fully than the amphibian's two-chambered heart. These modifications allowed more oxygen to reach body tissues. Third, early reptiles displayed changes that eliminated reliance on open water for reproduction. Males had sex organs that could deliver sperm directly into the female's body rather than into the surrounding water, and females produced a new type of egg. These eggs essentially encased the developing embryo in a pool of water, provided a source of food, the yolk, and surrounded both embryo and food with membranes and a leathery shell to prevent drying out in open air. Finally, the early reptiles' legs extended directly beneath the body rather than out to the side, and this made walking and running easier.

Later reptiles built on the legacy of these innovations, which were, in fact, pivotal to all subsequent vertebrate history. They gave rise not only to the modern reptiles—the crocodiles, turtles, lizards, and snakes—but indirectly to the birds, mammals, and dinosaurs.

The grand radiation of reptiles into forms that swam, flew, and lumbered across the earth spanned nearly 150 million years (see Figure 14.11) and inspired the common name for the entire era: the Age of Reptiles. Most of the great and diverse reptiles died out, however, in a massive extinction event about 65 million years ago (Figure 18.10).

A few small reptiles weighing less than 44 lb survived the extinctions and radiated once again during the Cenozoic, the current geological era beginning about 90 million years ago. Today, there are about 500 species of reptiles—animals that continue to show the innovations of dry scaly skin, legs, expandable rib cages, fully separated oxygenated and unoxygenated blood supplies, sex organs for direct delivery of sperm, and eggs in a leathery shell. Alligators, caimans, and crocodiles are all streamlined meat eaters that inhabit warm climates. The tortoises and turtles have a tough protective shell. The lizards, snakes, and iguanas are elongated reptiles that inhabit wet, dry, or hot areas and sometimes reach great size. The heaviest lizards today are the Komodo dragons of Indonesia, which can weigh up to 255 lb (Figure 18.11a); and the longest snakes are the pythons of the same region, which can grow to 40 ft (Figure 18.11b).

BIRDS: AIRBORNE VERTEBRATES

One might expect that the first flying vertebrates—the giant, leathery-winged pterosaurs—gave rise to the **birds,** but they didn't; they died out when the dinosaurs did. Biologists believe instead that small, two-legged, lizard-like reptiles *(thecodonts)* were the real ancestors of birds, and they have important fossil evidence to prove it. Five skeletons of the oldest bird, the crow-sized *Archaeopteryx,* have been found at different sites in rocks 150 million years old. The fossil imprints suggest that this animal was a true intermediate: It had scaly skin, curving claws, a long, jointed tail, and sharp teeth like a reptile, but it had feathered forelimbs and tail like a bird.

Most evolutionary adaptations in birds prepared the animals for efficient flight (see Figure 18.4). Lightweight *feathers* make an aerodynamically sound "fabric" (Figure 18.12). Fluffy down feathers act as insulators.

Another set of innovations in birds involves the skeleton and muscles. Birds have lightweight, hollow bones and a breastbone enlarged into a blade-shaped anchor for the powerful muscles that raise and lower the wings (see Figure 18.12). The legs are reduced to skin, bone, and tendons and can be folded up like an airplane's landing gear to reduce drag during flight.

Flight is a strenuous activity that requires plenty of oxygen, and birds have yet another set of modifications to ensure a sufficient supply. First, they are **warm-blooded** *(homeothermic);* they maintain constant internal temper-

FIGURE 18.10 *The Mystery of the Disappearing Dinosaurs.* Based on physical evidence, including remains such as this skeleton of a *Tyrannosaurus rex*, most paleontologists agree that the dinosaurs and probably 70 percent of the other plant and animal species on earth died out about 65 million years ago. But why? One theory holds that a massive meteorite perhaps 6 miles wide struck the earth at that time and created a huge cloud of dust, smoke, and rain as corrosive as battery acid. This cloud heavily obscured the sunlight and led to a chilly year of darkness in an acidic fog polluted with toxic trace metals, during which most of the plants and animals died out. There are a handful of other hypotheses (e.g., the dinosaurs died out because mammals ate their eggs or because the reptiles' brains were so small that the animals could not adapt to environmental changes). Many scientists maintain, however, that no single event or condition could have caused the disappearance of numerous species over a geologically short period of a few million years. They suggest that decreases in global temperature, dropping sea levels, and competition from smaller warm-blooded mammals, among other factors, contributed to the great demise.

(a)

(b)

FIGURE 18.11 *Modern Reptiles.* (a) The Komodo dragon of Indonesia can be up to 10 ft long and weigh 255 lb. The lizard has such a huge appetite that it will readily kill and devour other Komodo dragons and consume the dead bodies of nearly every other animal. (b) A Burmese python, also of the Indo-Malayan region, continues to grow for 20 years before reaching more than 20 ft in length, and it can survive for upwards of 70 years.

atures despite environmental changes, and this steady temperature allows a steady production of energy that fuels the activities of flight and leg muscles. **Cold-blooded** *(poikilothermic)* animals, such as fishes, amphibians, and reptiles, grow active in warm environments but sluggish in cold ones. Second, birds have a series of connected lungs and *air sacs* for exchanging oxygen and carbon dioxide. And third, birds have a four-chambered heart that completely separates oxygenated and deoxygenated blood, so only blood with plenty of oxygen reaches body tissues.

Finally, in addition to the evolutionary innovations of feathers, warm blood, and air sacs, birds have hard-shelled eggs, which free them from dependence on water for reproduction.

So successful were these innovations that birds radiated into more than 8600 species specialized for distinct modes of life. Flamingos, wading in large flocks, look like tall-stemmed pink flowers (Figure 18.13a, page 320). Condors with wingspans of 10 ft soar on mountain updrafts, seldom flapping. Cassowaries, meanwhile, have powerful legs that enable them to run rapidly and deal deadly blows with their feet (Figure 18.13b). And champion migrators like the arctic tern have an innate sense of direction and the physical stamina to travel more than 25,000 miles each year.

MAMMALS: RULERS OF THE MODERN ERA

This chapter began with bats—the only true flying mammals—and now we arrive at the group that includes them, and our own species as well. Although mammals arose

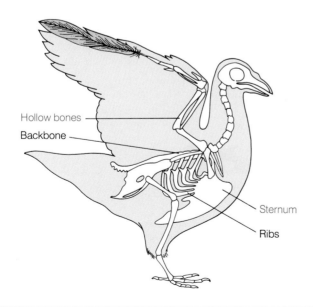

Hollow bones

Backbone

Sternum

Ribs

FIGURE 18.12 *Feathers and Lightweight Bones: Innovations for Flight.* A generalized bird skeleton, showing the bladelike sternum and the reduced bones of legs and feet, as well as wing feather.

from fierce, heavy-set reptiles, they developed their own distinct body traits at least 180 million years ago. Since that time, they have radiated into more than 4000 modern species that give the current geological era its common name: the Age of Mammals (Figure 18.14).

Like birds, **mammals** are warm-blooded and have a four-chambered heart. But they have two unique innovations as well: *milk* and *body hair* or fur (see Figure 18.4). The majority of mammals also have a special reproductive structure, the *placenta*, that supports the growth of

FIGURE 18.13 *Modern Birds.* (a) Pale pink East African flamingos cluster in a courtship display. (b) The double wattled cassowary *(Casuarius)* is the most colorful flightless bird, with its naked blue head skin, vivid red wattles, and jet black feathers.

(a)

(b)

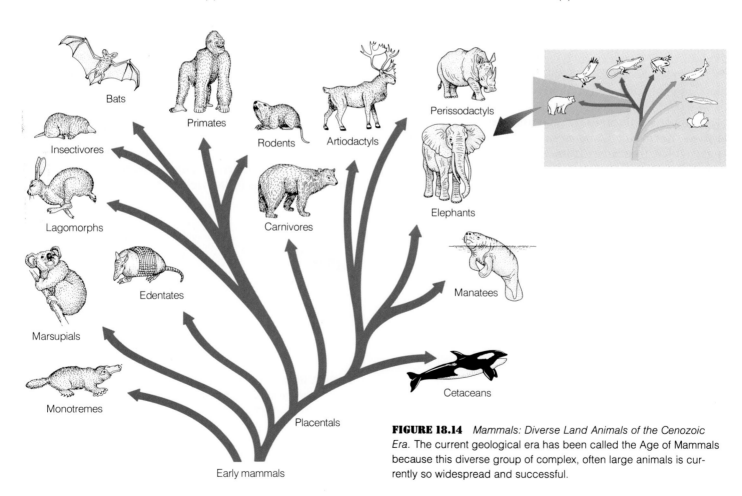

FIGURE 18.14 *Mammals: Diverse Land Animals of the Cenozoic Era.* The current geological era has been called the Age of Mammals because this diverse group of complex, often large animals is currently so widespread and successful.

(a)

FIGURE 18.15 *A Modest Collection of Modern Mammals.* All mammals produce body hair and milk for feeding the young, but mammals are remarkably diverse in body size and appearance. (a) The duck-billed platypus is an egg-laying monotreme that suckles its young. It protects itself with a venom-spurting spur, a rare poisonous weapon for a mammal. (b) The wombat, a marsupial from Australia, resembles a very large woodchuck and can live in deep burrows up to 100 ft long. (c) A North American cougar chases a snowshoe hare. (d) The Florida manatee is a mild-mannered, herbivorous "sea cow" that can weigh 1 ton.

(b)

(c)

(d)

the embryo to a fairly complete stage of development before birth. These *placental mammals* descended from one branch of the earliest mammals. Two other branches led to nonplacental mammals with different ways of harboring embryos. The **monotremes,** which include only the duck-billed platypuses and spiny anteaters of Australia and New Guinea, lay leathery eggs and warm them until the young hatch. The females then suckle the young with milk. The **marsupials,** including kangaroos, opossums, koala bears, wombats, and dozens of other animals from Australia and the Americas, give birth to immature live young no bigger than a kidney bean. When the newborn emerges from the birth canal, it crawls upward into an elastic pouch of skin on the mother's abdomen, attaches to a teat, starts to consume milk, and continues to develop inside the pouch.

With their successful reproductive strategies based on their evolutionary innovations of milk, body hair, and the placenta, mammals can be compared to the flowering plants, which underwent a parallel radiation during the same geological era. Mammals radiated into a large and diverse class that can live in more environments and with more life-styles than any other class of animals except perhaps the birds (Figure 18.15). Various specializations of structure and behavior underlie this success.

■ **Temperature Regulation** Mammals can keep a constant body temperature that is warmer than the surrounding air on cool days and cooler than the air on extremely hot days. This is partly because of their high rate of metabolism, partly because of insulating hair or blubber, and partly because they have evolved special behaviors such as hibernating and migrating. Bats, for example, eat insatiably when awake, then return to caves or similar hiding places with constant mid-range temperatures. They curl up tight and sleep in clusters to retain warmth, and in winter, when food is in short supply, they hibernate, spending weeks in a dormant state with lowered body temperature.

■ **Specialized Limbs and Teeth** Mammals display a range of limb types, with bones elongated, shortened, or broadened, depending on the animal's particular locomotor or food-gathering habits. In bats, the forelimb bones are light and strong, and the "finger" bones are greatly elongated and widely spread and support the flight membranes (see Figure 18.1). In moles and other digging mammals, the forelimbs are short and powerful, with oversized claws, while antelope forelimbs have strong, slender bones that allow swift running. Mammalian teeth are modified into chisels for gnawing wood (as in rats and beavers); flat molars for grinding grain (as in cows); and sharp points for tearing flesh (as in some bats, foxes, and other meat eaters).

■ **Parental Care** Mammals not only suckle their young for days or months, depending on the species, but usually guard the new generation fiercely and teach them survival skills. The extended care of young is especially pronounced in *primates* (monkeys, apes, and humans), as we shall see.

■ **Highly Developed Nervous Systems and Senses**
The keen senses of smell and hearing in most mammals, and of sight in a few, are instrumental in helping the animals find food, interact with mates and young, and avoid danger. The sonar of bats, which gives them the ability to navigate and locate objects in complete darkness, is but one example. Such sensory information is integrated and often stored in memory, enabling the animals to learn from experience and minimize the effort they expend to survive. The brains of primates, dolphins, porpoises, and whales represent the highest level of nervous system development among all animals.

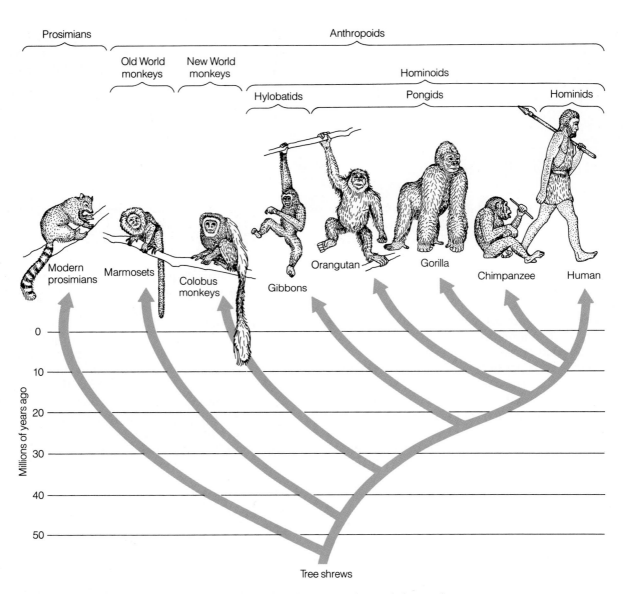

FIGURE 18.16 *Primate Family Tree.* As seen in this evolutionary tree, the prosimians and marmosets branched off quite early from the tree shrews that gave rise to all the primates. The gibbons and other apes (which include the orangutans, the gorillas, and the chimpanzees) are much more recent lineages, with chimpanzees and humans diverging and evolving less than 10 million years ago.

EVOLUTION OF THE PRIMATES, INCLUDING HUMANS

One of the most intriguing questions in modern biology is, How did human beings evolve? We have all the requisite traits for membership in the chordate phylum and the vertebrate subphylum, and with our warm blood, mammary glands, placenta to support the developing embryo, and body hair, we are obviously members of the mammalian class and the placental subclass. But our unique combination of behavioral abilities—including spoken and written language, agriculture, and extensive tool use—have allowed us to dominate the environment like no other animals before us. Despite this, there is ample evidence today—both fossil and genetic—that humans are simply one branch of the primate order and that our branch separated no more than 6 million years ago from the lineage leading to chimpanzees (Figure 18.16). One animal behaviorist has called humans "naked apes," and so, it seems, we are.

THE PRIMATE FAMILY TREE

There are several dozen species in the order **Primates,** divided into two suborders: the **prosimians** (which means "before monkeys"), and the **anthropoids** (literally meaning "humanlike"). Prosimians include small, tree-dwelling animals, such as the lemurs and tarsiers (Figure 18.17). They have two hallmarks of later primates: an **opposable thumb** (a first digit that can touch each of the others) and an acute sense of sight. Anthropoids include the Old and New World monkeys (Figures 18.18 and 18.19) and the apes (Figure 18.20, page 325), as well as humans. The apes, our closest relatives, are large, tailless animals characterized by long arms, a large brain, and complex social behavior.

PRIMATE CHARACTERISTICS

Primates evolved several specializations for living in trees (and later, on grassy savannas) that have added to their success as a group and that form the background for human physical features.

■ **Vision** Life in the trees is rich with visual information, such as fluttering leaves, moving spots of sunlight and shade, and tangled tree limbs. Thus, it is not surprising that primates evolved **stereoscopic vision,** which includes good depth perception that allows the animal to discriminate distances well. This ability is important for maintaining balance up in the trees and for catching small, mobile prey such as insects, a favorite prosimian food. Bony sockets protect the forward-facing eyes in the

FIGURE 18.17 *Tarsiers: Primitive Primates.* Tarsiers, with their huge, forward-directed eyes and ears and their knobby, adhesive toes, are tiny nocturnal prosimians that live in the tropical forests of the Philippines, Sumatra, and Borneo. There they hunt insects, lizards, and birds by leaping through the tree limbs. They also harvest fruit.

FIGURE 18.18 *An Old World Monkey.* The primates that live in forests as well as adjacent arid savannas throughout the Eastern Hemisphere (from Africa to India and Southeast Asia) are Old World monkeys characterized by closely set, down-pointing nostrils. They usually lack a prehensile tail and have a fairly large brain. Many have brightly colored buttock calluses involved in sexual attraction. The baboons and mandrills of Africa are Old World monkeys. The rhesus monkey shown here, *Macaca mulatta,* a favorite laboratory species, is widespread in the forest of Southeast Asia and northern India and lives in close-knit social groups. The adult on the right is grooming its young.

FIGURE 18.19 *A New World Monkey.* Dozens of primate species inhabit the steamy jungles and drier forests of Mexico and Central and South America. The New World monkeys are characterized by a flat nose and grasping, or prehensile, tail. The black howler monkey, *Alouatta caraya*, inhabits forests from Ecuador to Paraguay and produces a loud, sharp call that can be heard for many miles. The animal has enlarged, cavernous bones in the throat that act, in the words of one zoologist, as a sort of "bony trumpet." Shown here is the mantled howler monkey (*Alouatta villosa*) of Guatemala.

later monkeys and apes, and together, the eyes and brain create a realistic three-dimensional picture of the environment, which helps the animals to identify foods, mates, and predators.

■ **Brain**   During the evolution of life in the trees, keen vision was accompanied by the development of powerful arms that allowed climbing and swinging movements, as well as dexterous hands with opposable thumbs for grasping tree limbs. The primates also evolved complex social systems. This vision, dexterity, and ability to communicate and cooperate all depended on large brains. Prosimians have fairly large brains for animals of their size and body weight; monkeys have proportionately larger brains than prosimians; and apes and humans have the most complex brains of any mammals—both in terms of size (in relation to the rest of the body) and number of nerve cells and in terms of the intricacy of internal connections between brain regions (see Chapters 26 and 27).

■ **Infant Care**   Along with their mobile life in the trees, primates evolved new reproductive strategies. Compared with most other mammals, they give birth to smaller litters of young, with most species producing just one young at a time. The infants tend to be helpless (especially those of apes and humans) and depend on their parent(s) for complete physical care for a long period after birth. The higher a species' intelligence, the more the parent must educate the young. This great investment of parental care pays off in the high survival rates of the young.

■ **Upright Gait**   Although humans are the only fully upright primates, most monkeys and apes spend more time vertically than horizontally, clinging upright to tree trunks, sitting upright on the ground, running on two legs with the tail providing balance, and leaping with body held vertically. Even though gorillas and chimpanzees walk on all fours by touching the ground with back feet and the knuckles of front hands, their arms are so long that the body is still fairly erect (Figure 18.21, page 326). Upright posture improves visibility and leaves the hands free for other activities. As we will see, this was one of the major factors in human evolution.

■ **Teeth**   Primates have teeth modified for a diet of both plant and animal matter. The earliest mammals had 44 pointed teeth, adapted for slicing up the bodies of insects. Early primates had similar teeth, but as later primates changed to diets of leaves or fruit, their back teeth, through natural selection, began to broaden and flatten, and they were able to consume a larger variety of foods.

Each of these primate characteristics—stereoscopic vision, large brain, extended infant care, upright gait, and modified teeth—played an important role in the success of the primate lineages.

PRIMATE EVOLUTION

How did early insect eaters lead to the modern groups of prosimians, monkeys, apes, and humans? Genetic analyses and fossilized bones and footprints suggest that all mammals hark back to small animals that lived some 65 million years ago—animals that radiated into hundreds of species and prospered in the earth's then extensive tropical forests. Midway through this prosimian heyday, monkeys arose from prosimian ancestors, diverged into the Old and New World families, and began branching into the nearly 50 genera still represented today (see Figure 18.16).

By comparing the similarity of DNA from pairs of living primates, geneticists can quantify the genetic differences between them. They have found that the genetic distance between people and chimpanzees is less than the genetic distance between chimps and gorillas. This

(a)

(b)

FIGURE 18.20 *The Apes, or Hominoids.* The apes—the gibbons and orangutans of Asia and the gorillas and chimpanzees of Africa—are our closest primate relatives. Often called hominoids ("resembling or related to humans"), they are the largest primates and have long arms, large brains, and complex social behavior. The gibbons of Southeast Asia are slender animals that swing swiftly through the trees, sleep in trees during the day, and at night hunt for shoots, fruit, and occasionally eggs and small vertebrates. (a) The orangutans live in remote forests of Sumatra and Borneo. The name means "man of the woods," and indeed, the orangutan's face is capable of varied, sensitive expression. It moves about rather clumsily in search of fruits and tender shoots and constructs sleeping shelters in the trees. (b) Gorillas are the largest, strongest primates and they inhabit the mist forests of western and central Africa. Male gorillas can be nearly 2 m (6½ ft) tall and weigh over 200 kg (450 lb), and while they will occasionally climb trees, they are mainly animals of the forest floor. The males beat their breasts and roar loudly when aroused. Gorillas live as mixed-sex and mixed-age communities and migrate through the forests in search of abundant sources of fruit. Experiments have revealed that gorillas are quite intelligent, have a good sense of humor, and can comprehend many words and phrases. (c) Chimpanzees are our closest relatives, and 99 percent of their genes are similar to ours. Their brains are highly developed and resemble our own. They live in the forests of western and central Africa in family groups with a few males, several females, and numerous offspring. Chimpanzees forage during the day and huddle together at night on platforms that they construct from branches and leaves. They use tools (sticks and grass) in nature, and in captivity can be trained to carry out complex tasks, including—according to some scientists—communicating in American Sign Language. They express affection by hugging and kissing and happiness by laughing and jumping up and down; and when unhappy, they cry or sob loudly.

(c)

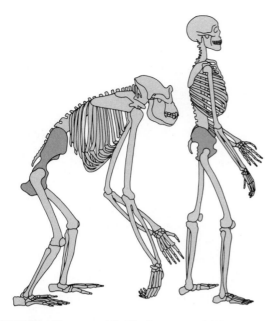

FIGURE 18.21 *Human and Gorilla Compared: A Matter of Skeletal Proportions.* The bones of the gorilla (left) and the human (right) are somewhat similar in shape and function, but the proportions are very different. The gorilla pelvis, for example, here shown in orange, tilts and cants the large rib cage and heavy neck and head forward, while the human pelvis is vertical and helps hold the entire skeleton upright.

number means that a chimp is genetically closer to *you* than it is to a gorilla! Using other techniques, biologists have interpreted the genetic distances to mean that gorillas probably branched off the evolutionary line leading to humans about 10 million years ago, while chimpanzees branched off about 6 million years ago (see Figure 18.16).

All of the final evolutionary branching of early primate species leading to *Homo sapiens* took place in less than 6 million years—an extremely short span in geological history. Using a 24-hour analogy, the earth formed at midnight, life appeared at 8:20 A.M., the first vertebrates appeared at 9:35 P.M., and the first humans arose only 38 seconds before the clock struck midnight. Let's examine that last "38 seconds."

THE RISE OF *HOMO SAPIENS*

Most anthropologists agree that at present, the primitive tree-dwelling ape known as *Proconsul africanus* represents the earliest identifiable genus on the path that eventually led to humans. And exciting fossil finds in Tanzania and Ethiopia in the mid-1970s revealed a species that virtually everyone agrees is the earliest true

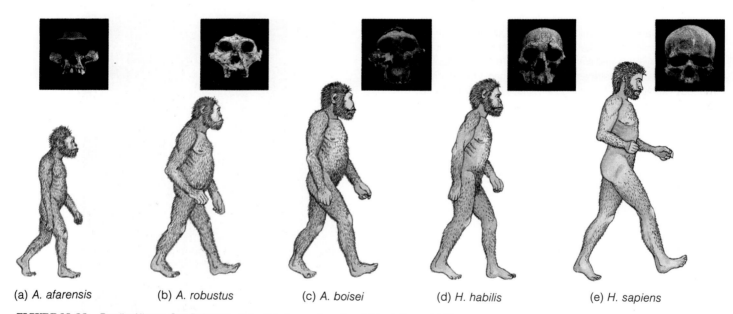

(a) *A. afarensis*   (b) *A. robustus*   (c) *A. boisei*   (d) *H. habilis*   (e) *H. sapiens*

FIGURE 18.22 *Family Album: Our Relatives.* A family album of our hominid relations. (a) The apelike *Australopithecus afarensis* lived about 3.75 million years ago and was the first fully upright hominid. (b) The large-boned *A. robustus* lived about 3 million years ago, and recent evidence shows that its hand was adapted for precision grasping and it may have used tools. (c) *A. boisei* lived nearly 2 million years ago and was probably concurrent with (d) *Homo habilis*, an active tool user. *Homo habilis* preceded *H. erectus* (see Figure 18.23), who lived about 1.5 million years ago. (e) Then *H. sapiens* emerged about 1 to 1.5 million years ago. These species did not lead to humans in an advancing ladder, but occupied ends of branches in a family tree whose precise arrangement is still under investigation.

hominid (early human) yet discovered: *Australopithecus afarensis* (Figure 18.22a).

■ ***Australopithecus afarensis* Walked Upright** *Australopithecus afarensis* had a very apelike skull and teeth, a brain only slightly larger than a chimpanzee's (450 cc, or cubic centimeters), and long upper but short lower limbs. At the same time, however, the head sat on top of the backbone like ours does rather than projecting forward like an ape's, and the hands had humanlike bones (although chimpanzee-like joints). Most importantly, footprints show that the legs, feet, and pelvis were modified for fully upright walking on two legs *(bipedalism)*. *Australopithecus afarensis* was small, standing only 1.0–1.4 m (3½–4½ ft) tall and weighing just 18–22.7 kg (40–50 lb). But its discovery proves fairly conclusively that an upright, two-legged stance was the first major human trait to evolve, long before an enlarged brain or evidence of tool use. It also suggests that these animals had hands free for carrying food or young and could see long distances in search of predator or prey.

Fossils show that australopithecines ("southern apes") radiated in Africa between about 2 and 3 million years ago and that several of the species resulting from this radiation lived simultaneously or in overlapping time periods in a multibranched human family tree. All were upright animals with large faces, small brains, heavy jaws, and large, crushing molars up to 2½ times the size of our own (Figure 18.22a–c). The australopithecines were probably all vegetarians who used their massive teeth to process coarse, abrasive foods, such as tubers, seeds, and hard nuts. No conclusive evidence has been found to suggest that these early hominids used tools or hunted; thus, their behavior was probably much simpler than that of later humans. Paleontologists vigorously debate how these early hominids were related to each other and to the humans that followed.

■ ***Homo habilis*** By 2 million years ago, one of the early australopithecines had given rise to the first human (member of the genus *Homo*): *Homo habilis*, the "handy man" (Figure 18.22d). *Homo habilis* individuals resembled their forebears, having large faces, big teeth, and trunks and limbs fully adapted for walking upright. At 1.5 m (5 ft) tall, however, they were larger than australopithecines, while their brains, at 700 cc, were half again as large as those of their predecessors. Even though this was still only half the size of a modern human's brain, it suggests a real increase in intelligence and probably a heightened degree of finger control. Significantly, the first evidence of toolmaking and the butchering of animals appears in the fossil record about the time of *Homo habilis*. Anthropologists speculate that the "handy man" hunted small animals and scavenged larger animals that lions or hyenas had killed. *Homo habilis* individuals probably made crude stone tools by cracking rocks and flaking off chips, and they probably butchered their game. Meat eating sets the species apart from all earlier hominids and hominoids (apes), and a way of life based on their newly evolved use of tools dominated the rest of human evolution.

■ ***Homo erectus*** Whereas *Homo habilis* had been restricted to the savannas and woodlands stretching from northeastern to southern Africa, a new species, *Homo erectus* ("erect man"), arose sometime before 1.5 million years ago and spread throughout northern Africa, southern Asia into Indonesia, and probably into southern Europe. A nearly complete skeleton of a *Homo erectus* male was discovered in 1985 in northern Kenya and determined to be 1.6 million years old (Figure 18.23). The stage of bone growth showed that the male was a 12-

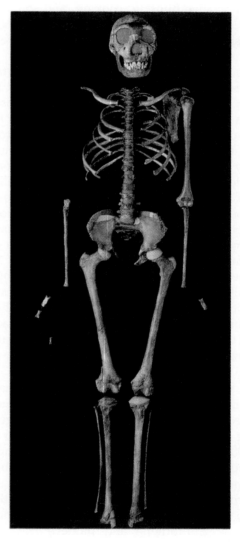

FIGURE 18.23 *Homo erectus: A 1.6-Million-Year-Old Skeleton.* This nearly complete skeleton is that of a 12-year-old boy who would have been 6 ft tall as an adult.

year-old boy who would have been 6 ft tall had he lived to adulthood. The brain volume (800 cc) was larger than that of *Homo habilis*, the teeth and skull had changed shape, and scientists speculate that the parts of the brain that govern abstract thought and reasoning may have undergone some internal reorganization. *Homo erectus* made a new kind of stone tool that required finer skill: a sharp hand ax.

Both the geographical range and the new tools suggest a greater ability to deal with varied food resources and extremes of climate. Fossil evidence indicates that *Homo erectus* hunted elephants, bears, antelope, and other large game in cooperative bands and transported, slaughtered, and cooked the meat over camp fires (which also provided warmth for long winter nights). *Homo erectus* probably first used fire about 500,000 years ago as the species spread to the temperate zones of Europe and Asia. Cooperative hunting, cooking, and sharing were probably developed and maintained through **cultural transmission;** that is, they were taught to the next generation through actions and language.

The discovery of this fairly complete *Homo erectus* skeleton revealed some very important clues about human evolution. The head of the 12-year-old boy was rather large, the pelvis was fairly narrow, and the boy had yet to reach adult size and sexual maturity. These factors suggest that the same pattern of human development we see today may have already emerged 1.6 million years ago. Modern human babies are born in a physically helpless state because the large head (and brain) could not pass through a woman's pelvis at birth if gestation were any longer and physical development in the womb any more complete. At the same time, our postbirth growth and maturation process continues for upward of 12 to 15 years. These two factors converge to ensure that our large brains will be able to store enormous amounts of survival information, while the long period of physical dependence during childhood will keep us in contact with potential teachers—parents, grandparents, siblings, and neighbors.

■ *Homo sapiens* During the million years that early humans inhabited the Old World, their faces and teeth slowly decreased in size, and their brains enlarged from a volume of 800 cc around 1.5 million years ago to 1200 cc by 500,000 years ago. These gradual changes led imperceptibly from *Homo erectus* to **Homo sapiens** by about 400,000 years ago (Figure 18.22e).

Two groups of humans have occupied the earth in the millennia since that time: archaic *Homo sapiens* (including Neandertals and others) and modern *Homo sapiens* (including the Cro-Magnons and all the current races of people). Archaic humans dominated the Old World from about 400,000 years ago until they were superseded by modern humans between 30,000 and 100,000 years ago.

Neandertal fossils, first discovered in Germany's Neander Valley in 1856, were the first hominid remains to be studied scientifically. Neandertals have been much maligned as brutish, ignorant savages, but the accumulated facts speak quite differently. Living in Europe and the Middle East, Neandertals were short, stocky, powerfully built people able to move with a strength and speed surpassing that of today's best Olympic athletes. They had large protruding faces and characteristic projecting brow ridges (Figure 18.24). Their brains were similar in organization to a modern human's, but *larger* (1400 cc) on average. The tools and other artifacts they left behind suggest an ability to deal with the environment through learned cultural behavior rather than brute physical force. They made spears and spearheads for hunting large game and scrapers for cleaning animal hides. They routinely built shelters, and their large front teeth were often worn, perhaps from chewing hides, as the Eskimo did traditionally, to make clothing. The skeletons of old and crippled Neandertals showed that the strong cared for the elderly and infirm, and they buried their dead ceremoniously with fine stone tools, servings of game meat, and even flowers—evidence of belief in an afterlife.

Modern humans did not descend from Neandertals but rather arose elsewhere (fossil evidence suggests Africa), migrated northward, and eventually outcompeted the archaics on their own turf. The two types may have coexisted for thousands of years, but by 34,000 years ago, the Neandertals had died out.

FIGURE 18.24 *At Home with the Neandertals: Cave Burial in the South of France, 100,000* B.C. This Neandertal widow and child are preparing to bury their deceased loved one. As this museum model shows, they lived in a cave (their summer camp) in southern France and wore clothing (skins), built fires, used tools, and had ritual beliefs about death.

FIGURE 18.25 *The Origins of Art.* Cro-Magnon peoples left numerous cave paintings throughout Europe and other continents beginning about 29,000 years ago. This famous cave painting from Lascaux Cave in southern France depicts a bull and running horses. The appearance of art in human evolution indicates a major increase in the quality of information used by social groups and the need to encode it for future retrieval.

The Cro-Magnons looked distinctly different from their archaic cohorts. Their faces were smaller, flatter, and less projecting than the Neandertals', the heavy brow ridges had all but disappeared, and their skulls were higher and rounder. Their limbs were more slender (but still stoutly athletic compared to our own), their teeth were smaller, and most important, their culture was vastly more complex. Their tool kits contained knives, chisels, scrapers, spearheads, axes, and tools for shaping other tools of rock, bone, and ivory. They left dozens of cave paintings, engravings, and sculptures, suggesting a major development of symbolic forms of communication, probably accompanied by increased language abilities (Figure 18.25). Their living sites and shelters became larger and more complex, and human burials became more common and elaborate, indicating the establishment of religion.

The success of these cultural developments spurred a population expansion. Modern humans followed the herds of mammoths, woolly rhinoceroses, reindeer, and other game into the arctic regions of Eurasia and across the Bering land bridge into the Americas. They also built boats to carry them across uncharted waters to New Guinea and Australia. By 15,000 to 20,000 years ago, people had occupied virtually all the inhabitable regions of the earth.

Anthropologists have debated whether *Homo sapiens* arose at a single site and later spread around the world, or whether *Homo erectus*, which was already widely distributed, transformed into *Homo sapiens* at the same time in different places. Both fossil and genetic evidence support the single-origin model. Interesting data come from analyzing the DNA of mitochondria. Since animals inherit their mitochondria only from their mother via the egg cell, the data suggest to some scientists that all living people can trace their ancestry through their mothers, grandmothers, and great-grandmothers back to a single female who lived about 200,000 years ago, most likely in Africa. The media have dubbed this woman "Mitochondrial Eve." Males of other families may have contributed genes in cell nuclei, but this one African woman may have been mother of us all. It was her descendants who left Africa and, in a series of major migrations, eventually populated the globe.

THE AGRICULTURAL REVOLUTION

Beginning in the Middle East at least 10,000 years ago, some peoples began to purposely sow the seeds of their food plants, a practice that allowed them to produce adequate amounts of food in the areas near their settlements rather than pursuing game and living as nomads (Figure 18.26). They began with wild plants, but soon developed new strains with better yields. At the same time, they gradually tamed and domesticated wild animals for their milk, blood, meat, and labor.

With agriculture came a more settled life-style and the first towns and cities in areas with abundant water and

FIGURE 18.26 *Agricultural Scene at Jericho.* By 7000 B.C., humans had developed communities based on agriculture. As portrayed in this scene outside the city walls of Jericho, men reaped wheat with sickles carved of wood or deer antler. Semidomesticated goats can be seen grazing in the background, and a hand-dug irrigation ditch waters the fields.

ELEVENTH HOUR REPRIEVE FOR ELEPHANTS?

When Columbus and Magellan were sailing the world's seas in search of trade routes, precious metals, spices, and ivory, there were nearly as many elephants as people on the African continent. In those times, the idea would have been unthinkable that the earth's biggest land mammal could someday be nearly exterminated to satisfy the human demand for ivory trinkets and piano keys. Yet, in the twentieth century, that is precisely what has happened, and the battle to save the elephants is still being fought.

As African human populations have grown, as elephant habitats have dwindled, and as growing worldwide prosperity has fueled the demand for luxury items made of ivory, the numbers of elephants have dropped sharply to just 6 percent of their historic, stable populations. Most of this plunge has occurred in just the last two decades. Between 1973 and 1989, for example, poachers reduced Kenya's elephant populations by 85 percent, shooting the animals with powerful rifles, hacking off their tusks, leaving the carcasses to predators, and shipping the tusks to Hong Kong and other handicraft centers, where they are carved into figurines, printing stamps, instrument keys, knife handles, and jewelry. The Japanese are the biggest buyers of ivory products, but Americans and Europeans purchase their fair share, as well.

Responding at the eleventh hour, several conservation groups have launched "don't buy ivory" campaigns in recent years, and in 1989, a coalition pressed international wildlife authorities to change the elephant's status from threatened species to endangered species. That change suspended the global ivory trade temporarily. Then, after considerable political and diplomatic effort, a number of European nations, the United States, Japan, Hong Kong, and nine African countries cooperated to turn the temporary suspension into a two-year outright ban as of early 1990. Three African nations, however—Zimbabwe, Botswana, and Mozambique—are pushing for a legal waiver that would allow them to sell ivory from their managed elephant herds. The waiver would pose a difficult question: How will international trade officials distinguish legally culled tusks from a poacher's illicit spoils?

Scientists can use isotopes and DNA fingerprinting techniques to help trace samples from individual ivory tusks to their animal herd and region of origin. But the process is time-consuming, expensive, and technically difficult, and the fact remains that as long as consumers continue to buy ivory products, poachers will slaughter elephants to supply their tusks to the black market. That leaves us with a choice: to support international bans on the selling and buying of ivory products (avoiding stores that sell ivory both at home and abroad) or within our lifetimes to see the largest living land animal relegated to a few fenced game parks and pens in zoos.

Each of us has a vote.

FOR MORE INFORMATION, CONSULT:

Your local zoological park.

U.S. Fish and Wildlife Service Office of Endangered Species, Washington, D.C.

World Wildlife Fund, Washington, D.C.

Booth, W. "Africa Is Becoming an Elephant Graveyard." *Science* 243, (10 February 1989): 1732.

Cherfas, J. "Science Gives Ivory a Sense of Identity." *Science* 246 (1 December 1989): 1120–1121.

Cohn, J. P. "Elephants: Remarkable and Endangered." *Bioscience* 40 (January 1990): 10–14.

Lewin, R. "Global Ban Sought on Ivory Trade." *Science* 244 (9 June 1989): 1135.

good soil. Farmers could trade surplus food for tools, baskets, and pots made by nonagricultural craftspeople, and this led to mercantilism and the development of new crafts and skills. Among these were record-keeping systems that used writing and mathematics to keep track of food reserves.

The accumulation of skills and knowledge, passed down through thousands of human generations in a *cultural evolution*, has allowed humans to control food supplies, eliminate predators and diseases, and blunt the effects of natural selection on themselves and their domesticated species. While we *Homo sapiens* arose via the same mechanisms and faced the same survival pressures as the other primates, mammals, and vertebrates before us, cultural evolution has changed our relationship to the planet. Today, we are the only creatures to control most aspects of our own existence, and with that control lies the future of all species (see the box above).

CONNECTIONS

The evolutionary distance is great from a tunicate to a tiger or a Thailand farmer, yet each is a member of the same phylum, Chordata, and each arose from the same marine ancestors that lived more than 500 million years ago. Innovations, specializations, and radiations of the vertebrates led to the emergence of amphibians, reptiles, birds, and finally mammals, including the biggest, fastest-running, and most intelligent animals ever to live.

It is interesting that only one mammal, *Homo sapiens*, is in a position to study the emergences, extinctions, and adaptations of fellow living things. Our species evolved from primitive, tree-dwelling ancestors in just 65 million years—far less time than elapsed between the origin of the lobe-finned fishes and their first labored steps on land. But the tortuous evolutionary pathway leading to the human brain, for example, was no different in principle from the roads that began at the origin of life and ended in the modern monerans, protists, fungi, plants, and animals. The following parts of the book will explore how animals and plants, once evolved, could overcome the day-to-day challenges of survival.

NEW TERMS

amphibian, page 316	marsupial, page 321
anthropoid, page 323	monotreme, page 321
bird, page 318	opposable thumb, page 323
chordate, page 311	primate, page 323
cold-blooded, page 319	prosimian, page 323
cultural transmission, page 328	reptile, page 317
	skull, page 314
fish, page 314	spinal cord, page 312
gill slit, page 312	stereoscopic vision, page 323
hinged jaw, page 314	tail, page 312
Homo sapiens, page 328	vertebra, page 311
lung, page 316	warm-blooded, page 318
mammal, page 320	

STUDY QUESTIONS

REVIEW WHAT YOU HAVE LEARNED

1. What characteristic structures do all chordates possess at some time in their life cycle?
2. What distinguishes the two broad categories of chordates?
3. How do the cartilaginous and bony fishes differ? How are they similar?
4. Describe the lobe-finned fishes, and name a living example.
5. List three typical characteristics of amphibians. Of reptiles.

6. Why do researchers consider *Archaeopteryx* evidence that birds arose from early reptiles?
7. Give the main characteristics of birds. Of mammals.
8. Biologists group mammals according to mode of reproduction. List the three groups, and name at least two members of each.
9. List five characteristics of primates.
10. How did *Homo habilis* and *Homo erectus* differ?
11. When did Neandertals live? What were the major features of Neandertal body structure and culture?
12. How did the agricultural revolution affect human society?

APPLY WHAT YOU HAVE LEARNED

1. Cartoons frequently show dinosaurs and humans together. Is this picture scientifically accurate? Explain.
2. In the 1930s, fishermen off the coast of Madagascar discovered some living coelacanths. Some biologists refer to these fish as "living fossils." Why?

FOR FURTHER READING

Alexander, R. M. *The Chordates*. Cambridge, England: Cambridge University Press, 1975.

Bakker, R. T. *The Dinosaur Heresies: New Theories Unlocking the Mystery of the Dinosaurs and Their Extinction*. New York: Zebra Books, 1990.

Brown, M. H. *The Search for Eve*. New York: Harper & Row, 1990.

Edey, M. A., and D. C. Johanson. *Blueprints: Solving the Mystery of Evolution*. Boston: Little, Brown, 1989.

Forsyth, A. "Snakes Maximize Their Success with Minimal Equipment." *Smithsonian* (February 1988): 159–165.

Griffiths, M. "The Platypus." *Scientific American* 258 (May 1988): 84–92.

Lewin, R. *Bones of Contention: Controversies in the Search for Human Origins*. New York: Touchstone, Simon & Schuster, 1988.

Lewin, R. "Myths and Methods in Ice Age Art." *Science* 234 (1986): 936–938.

Lewin, R. "The Unmasking of Mitochondrial Eve." *Science* 238 (1987): 24–26.

Stringer, C. B., and P. Andrews. "Genetic and Fossil Evidence for the Origin of Modern Humans." *Science* 239 (1988): 1263–1268.

Waldrop, M. M. "After the Fall." *Science* 239 (1988): 977.

Young, J. Z. *The Life of Vertebrates*, 3d ed. Oxford, England: Oxford University Press, 1981.

(1) All chordates share five characteristics:

Tail (at some stage of life cycle) ③

② Notochord

① Nerve cord

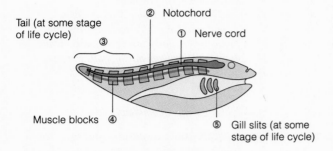

Muscle blocks ④

⑤ Gill slits (at some stage of life cycle)

(2) The chordates include two main groups: a few kinds of marine animals without a backbone and thousands of species of vertebrates.

Tunicates

(3) The vertebrates radiated to:

Reptiles: the first not needing water for reproduction

Birds: the first warm-blooded animals, well adapted for flight

Amphibians: the first to have lungs and legs

Fishes: the first vertebrates

Mammals: animals with body hair and milk production

(4) The mammals include primates, which have an opposable thumb, stereoscopic vision, a large brain, and extended parental care of their young.

(5) Humans evolved from earlier primates; they have the most highly evolved brain of any animal and have developed complex cultures, including agriculture and art.

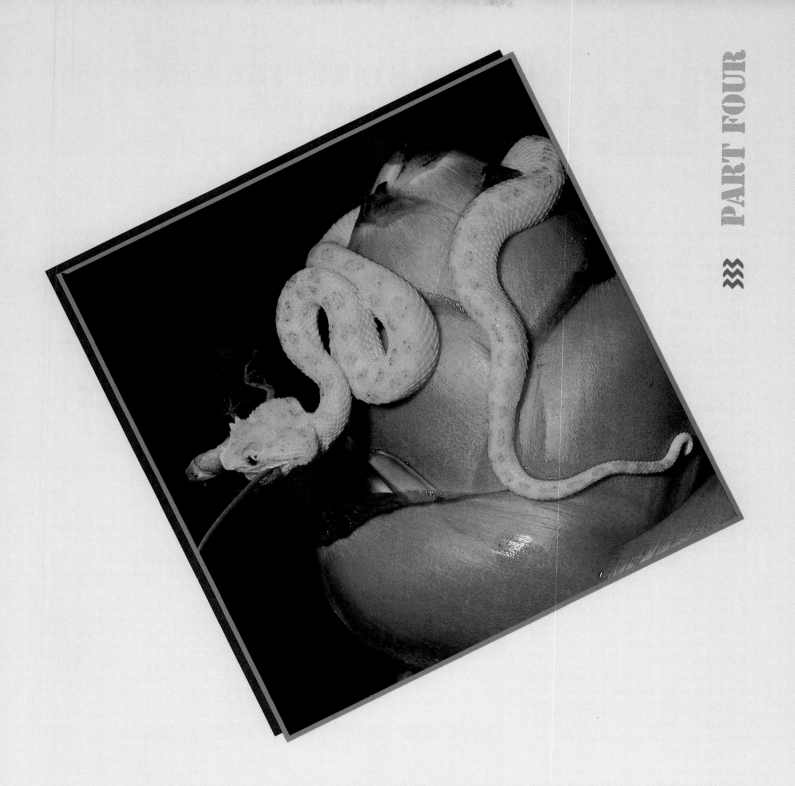

PART FOUR

HOW ANIMALS SURVIVE

CHAPTER 19

A STEADY STATE: THE KEY TO ANIMAL SURVIVAL

A WHALE OF A SURVIVAL PROBLEM

An intrepid visitor to the perpetually frozen Antarctic could stand at the coastline, raise binoculars, and witness a dramatic sight just a few hundred yards offshore: a spout as tall and straight as a telephone pole fountaining upward from the blowhole of a blue whale, then condensing into a massive cloud of steam in the frigid air. The gigantic animal beneath the water jet would be expelling stale air from its 1-ton lungs after a dive in search of food. Then, resting at the surface only long enough to take four deep breaths of fresh air, the streamlined animal would raise its broad tail, thrust mightily, and plunge into the ocean again (Figure 19.1). The observer on shore might see such a sequence only twice per hour, since the blue whale can hold its breath for 30 minutes as it glides along underwater like a submarine, at speeds up to 30 mph.

It is difficult to comprehend the immense proportions of the blue whale, the largest animal ever to inhabit our planet. At 25–30 m (80–100 ft) in length, this marine mammal is longer than three railroad boxcars and weighs more than 25 elephants. Its tongue alone is the size of a grown elephant.

Although blue whales are the largest animals that have ever lived, they share with all other animals the same fundamental physical problems of day-to-day survival: how to extract sufficient energy from the environment; how to exchange nutrients, wastes, and gases; how to distribute materials to all the cells in the body; how to support the body; and how to protect it. The whale also faces the central challenge of maintaining a constant internal environment—a steady state (of body temperature, for example)—despite fluctuations in the external

environment (in water temperature, for example). Blue whales and all other animals have evolved unique adaptations of form and function that suit them to their way of life. The evolution of structures and systems that adapt organisms of all sizes to their environments is a major theme of this book.

Although the blue whale's solutions are specific to huge size, they represent general approaches that we will encounter again and again in this part of the book. Our purpose in this chapter is to investigate the general processes that keep animals functioning on an even keel. Then, in Chapters 20 through 28, we delve into the specifics of different

structures and systems that contribute to the steady state. And while animals and plants share some basic elements of form and function, there are also important differences; for this reason, we separate the two subjects, covering animals in Part Four and plants in Part Five.

In this chapter we will answer these basic questions:

■ How do animals solve the general problem of maintaining a steady state despite a changing physical environment?
■ What general cellular mechanisms help an animal maintain a constant internal environment?

FIGURE 19.1 *Blue Whale Plunging for Krill.* The earth's largest animal, the blue whale (*Balaenoptera musculus*) must solve a classic problem faced by most animals: the maintenance of constant internal conditions.

FIGURE 19.2 *Staying Alive: Maintaining a Steady State in the Face of a Changing Environment.* Despite the cold, howling wind, this Canadian musk ox maintains a constant, warm internal temperature.

≋ MAINTAINING A STEADY STATE: PROBLEMS AND SOLUTIONS

Living organisms are subjected to wide variations in temperature, light, acidity, salt, wind speed, and availability of water, minerals, and nutrients. These and other environmental factors create a shifting external setting to which the organisms must adjust or they will die (Figure 19.2). For example, the blue whale maintains a constant body temperature whether feeding in the frigid waters of the Antarctic or giving birth in warm tropical seas. At the same time, each living thing carries out certain activities basic to the life process, such as gathering energy, reproducing, and maintaining body integrity, and these activities must continue in a smooth, steady, orderly manner or, again, the organism will die. This *steady state* usually requires a continuous supply of energy and materials from the environment, as well as the release of wastes into it. Let's look again at the blue whale: It consumes and burns 1 million kilocalories a day to stay warm and active in its icy ocean environment, and it does this by straining 8000 lb of krill daily from the ocean water on special food-gathering sieve plates; meanwhile, it releases its organic wastes back into the ocean. In short, the central problem for a living thing is to maintain a steady state internally in the face of an often harsh and fluctuating external environment.

Biologists call the maintenance of a steady state **homeostasis** (from the Greek words for "same standing"). This process keeps within a narrow range the amount of water and salt in body fluids, the levels of oxygen and carbon dioxide in blood, the amount of glucose reaching the brain, and the animal's body temperature, among other things. Let's look at some of the structures and strategies that contribute to homeostasis.

GENERAL BODY ORGANIZATION: A HIERARCHY OF TISSUES, ORGANS, AND ORGAN SYSTEMS

A multicellular organism, especially one as enormous as a blue whale, could not carry out its many survival tasks if its cells were crammed haphazardly into the body. Instead, the cells are organized into tissues, organs, and organ systems (Figure 19.3). A *tissue*, as we have seen,

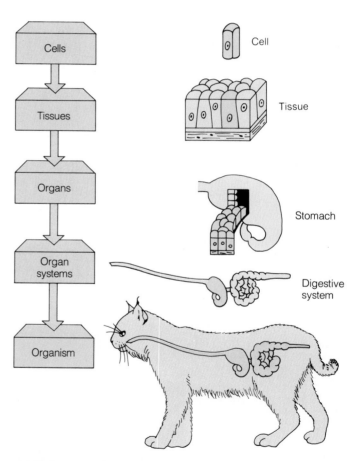

FIGURE 19.3 *Hierarchy of Organization in the Animal Body.* The billions of cells in an animal's body are neatly organized into tissues, organs, and organ systems to carry out body functions efficiently. Organ systems working together in a coordinated fashion help an organism, such as an individual animal, stay alive.

(a) Connective tissue

White collagenous fibers
Connective tissue cell
Yellow (elastic) fiber
Matrix

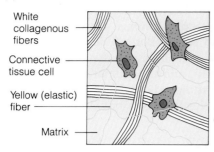
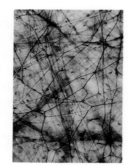

(b) Epithelial tissue

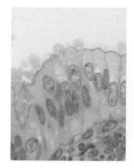
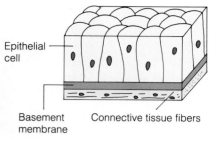

Epithelial cell

Basement membrane
Connective tissue fibers

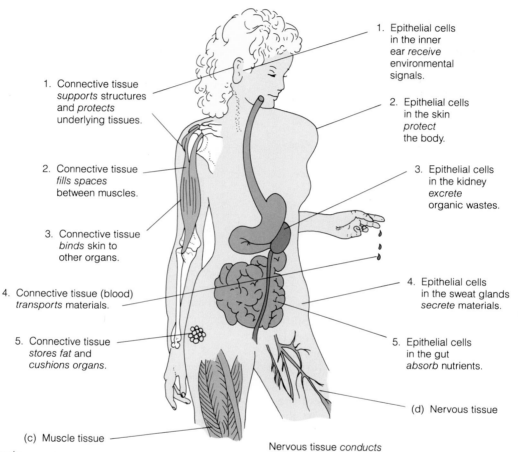

1. Connective tissue *supports* structures and *protects* underlying tissues.

2. Connective tissue *fills spaces* between muscles.

3. Connective tissue *binds* skin to other organs.

4. Connective tissue (blood) *transports* materials.

5. Connective tissue *stores fat* and *cushions organs.*

1. Epithelial cells in the inner ear *receive* environmental signals.

2. Epithelial cells in the skin *protect* the body.

3. Epithelial cells in the kidney *excrete* organic wastes.

4. Epithelial cells in the sweat glands *secrete* materials.

5. Epithelial cells in the gut *absorb* nutrients.

(d) Nervous tissue

(c) Muscle tissue

Muscle tissue *contracts* (shortens) and moves parts of the skeleton, the heart and some internal organs.

Nervous tissue *conducts* electrochemical signals, sometimes long distances, and helps coordinate body activities.

Nerve cell body

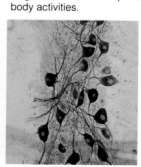
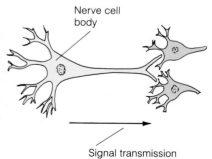

Signal transmission

FIGURE 19.4 *The Four Tissue Types and Their Roles in the Body.* (a) Connective tissue binds other tissues together and supports flexible body parts. It produces extracellular material that forms a meshwork, or framework, for other structures. This matrix includes fibers of the protein collagen and elastic proteins, and it spaces connective tissue cells much farther apart than the densely packed epithelial cells shown in (b). Weight for weight, collagen is as strong as steel. There are several types of connective tissue. The photo and interpretive drawing show loose connective tissue, which fills spaces between muscles and forms delicate networks that connect epithelial layers to underlying organs. Other connective tissues include adipose tissue, or fat, which stores fat droplets and acts as a cushion; fibrous connective tissue, which makes up tendons and ligaments; and cartilage, which provides the rigid framework of the nose and outer ears and attachment sites for muscles. Some scientists also consider blood a connective tissue, as well as bone, which is made up of a hardened matrix plus living cells. (b) Epithelial tissues are sheets of cells that cover body surfaces, lining the outer layer of the skin, the inner surface of body cavities, and the chambers of the lungs and other organs. Closely packed epithelial cells can have one surface free to face either the environment or a body cavity, while the other surface may be pressed to a dense tangled mat of proteins and carbohydrates called the basement membrane. Epithelial tissues receive environmental signals; protect the body from invaders; secrete sweat, milk, wax, or other materials; excrete wastes; and absorb nutrients, drugs, or other substances. Damaged epithelial tissue can greatly interfere with normal body functioning. For example, when cigarette smoke damages the cilium-covered epithelium lining the breathing tubes, that epithelial lining can no longer remove harmful debris from the lungs and passageways as efficiently. (c) Muscle tissue, found in limbs, the heart, and many internal organs, moves body parts. It will be discussed in detail in Chapter 28. (d) Nervous tissue transmits electrochemical signals that coordinate the body's activities. Chapter 26 discusses this tissue in greater depth.

TABLE 19.1 ≋ ESSENTIAL LIFE PROCESSES AND RESPONSIBLE SYSTEMS COMMON TO ALL ORGANISMS

PROCESS	RESPONSIBLE SYSTEMS
Obtaining energy from the environment	Digestive system
Exchanging gases (oxygen and carbon dioxide) with the environment	Respiratory system
Removing wastes	Excretory system
Regulating body fluid composition	Excretory system
Distributing materials within the body	Circulatory system
Defending the body from attack	Skin, immune system
Protecting the body from temperature extremes	Skin, circulatory system
Regulating body shape	Skeleton, muscular system
Reproducing	Reproductive system
Coordinating body activities	Nervous system, endocrine system

is a group of cells of the same kind performing the same function within the body. There are four types of tissue in the multicellular animal body: *epithelial tissue,* sheets of cells that cover body surfaces; *connective tissue,* which binds other tissues together and supports flexible body parts; *nervous tissue,* which transmits electrochemical impulses; and *muscle tissue,* which, by contracting, enables the animal to move (Figure 19.4). An *organ* is a unit composed of two or more tissues that together perform a certain function. Your stomach, for example, contains all four tissue types cooperating to briefly store and transport food. Several organs may work together to form an **organ system,** which is defined as two or more interrelated organs that serve a common function. The various systems listed in the right-hand column of Table 19.1 are organ systems; the digestive system, for example, contains the salivary glands, liver, stomach, intestines, and pancreas functioning in concert to process food. Collectively, all the organ systems make up an **organism,** in this case, an individual animal.

Many organ systems depend on a transport tube and the process of diffusion (Figure 19.5). Diffusion, which

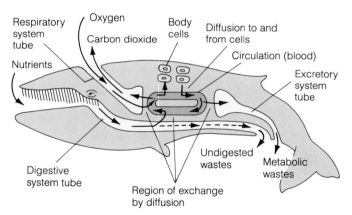

FIGURE 19.5 *Tubes and Diffusion: The Common Strategy for Exchanging Materials with the Environment.* Animals obtain nutrients, exchange gases, and rid themselves of organic wastes by tubes that penetrate deep into the body. The tubes exchange substances with the blood by diffusion. Finally, substances diffuse to and from the blood and the individual cells. The digestive, respiratory, and excretory systems all work by means of a similar strategy involving tubes and diffusion.

is the tendency of materials to move from areas of high concentration to areas of low concentration, is a means of transport into and out of cells. But diffusion only works over short distances, and cells in a large animal like the blue whale can be separated from the outside world by 2 m (6 ft) or more. This is where the transport tube plays a role; it carries materials deep into the body, where they can diffuse over short distances into an internal transport system—the circulatory system—that can service all the cells. Transport tubes include airways, the gut, and excretory tubes that carry wastes out of the body.

A large animal's organ systems not only overcome the problem of servicing billions of individual cells by carrying materials to within easy diffusion range of each cell surface. They also provide strong structural support and sophisticated avenues of coordination and communication to tie together far-flung body parts. The blue whale, for example, has a multi-ton skeletal system that serves as a scaffolding for muscle attachment and keeps its massive internal organs (including a heart the size of a Volkswagen bug) from collapsing together. And like most other animals, it has two systems that integrate the actions of all other organ systems so that the organism functions efficiently as a single entity. The nervous system acts via a network that carries electrical signals that rapidly regulate body functions. The endocrine system, in contrast, acts more slowly via chemicals that diffuse out of the blood and cause slower, longer-lasting body reactions and changes. As we will see throughout this part of the book, both of these systems are necessary for an organism to maintain homeostasis—the internal constancy it needs to survive.

ADAPTATIONS: FORM AND FUNCTION SUIT THE ENVIRONMENT

It may seem ironic that the world's largest animal eats minute krill, but the blue whale's mouth is perfectly suited to harvesting this food source—the most abundant in cold ocean waters. The blue whale's sieve plates, or *baleen* plates, are evolutionary extensions of the ridges you can feel with your tongue just behind your front teeth (Figure 19.6). The whale has several hundred such ridges, each about 1 m (about 1 yd) long, that function like a giant sieve. While gliding along underwater in search of food, the whale finds a cloudlike school of tiny krill, herds it into a tight mass, then lunges forward, trapping thousands of liters of food and water in its mouth, which expands greatly by means of special pleats in the throat. A contraction of the jaw muscles forces the water out between the baleen plates, and bushels of krill remain behind for the whale to devour with one mighty swallow. In general, specialized structural adaptations allow different organisms to exploit their environments in different ways, and the adaptation's physical form is appropriate to its function. Biologists call the study of biological structures **anatomy** and the study of how those structures work **physiology**.

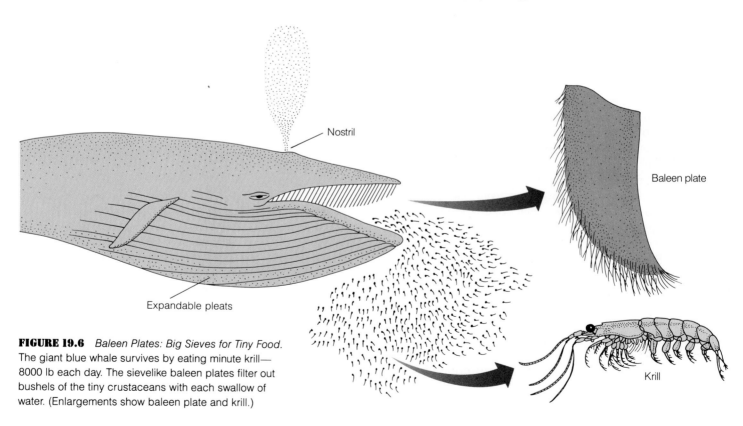

FIGURE 19.6 *Baleen Plates: Big Sieves for Tiny Food.* The giant blue whale survives by eating minute krill—8000 lb each day. The sievelike baleen plates filter out bushels of the tiny crustaceans with each swallow of water. (Enlargements show baleen plate and krill.)

Nostril

Baleen plate

Expandable pleats

Krill

The blue whale's breathing mechanism is another adaptation that is fine-tuned to the animal's life-style and environment. As the animal glides along with most of its body submerged, it can breathe like a swimmer with a snorkel. This is because its nostrils are positioned on top of its head—a perfect place for an air-breathing aquatic mammal (see Figure 19.6). When the animal dives, powerful muscles close these nostrils tight.

The whale's baleen plates, nostril position, and other adaptations for living in an icy sea arose through evolution by natural selection. Presumably, the higher the nostrils, for example, the more efficiently the animals could breathe (compared to whales lacking genes for high nostrils). And the better they could breathe, the better they could reproduce, the more genes they could pass on to surviving progeny, and the more prevalent the genes could become within the whale population. In general, the source of anatomical and physiological adaptations is natural selection, whether those adaptations are in a beetle, a sponge, a bird, a person—or a whale.

With tissues, organs, and organ systems well adapted to a particular environment, animals' bodies are able to maintain the steady internal state they need to survive. Let's turn our attention now to how that steady state is maintained at the cellular level.

≋ KEEPING THE CELLULAR ENVIRONMENT CONSTANT

During a regular checkup, your doctor can get a fairly accurate reading of how healthy you are by taking your temperature and blood pressure, listening to your heart and lungs, analyzing the contents of your blood and urine, and so on. Since medical researchers have established a normal range of values for each measurement, chances are good that if yours fall within the correct ranges, your body is maintaining homeostasis, that is, keeping the environment healthy at the cellular level. The box on page 342 describes the case of a young woman whose blood chemistry became badly disrupted because of hepatitis but was corrected through proper medical care. The message is this: Your physiological systems continuously adjust to aspects of the environment in and around your cells, maintaining the environment outside the cells within narrow limits suitable to the cells' efficient functioning.

Although each of the systems covered in the following chapters contributes in a different way to the overall stability of the internal milieu, they have some general strategies in common. After outlining two such strategies, we focus on one case history: how the blue whale stays warm in near-freezing polar waters but cool in tropical seas.

POSITIVE AND NEGATIVE FEEDBACK LOOPS

A highway driver keeping a vehicle within a narrow lane is a type of homeostatic system. This person must *sense* the situation at any given point ("Where am I on the road?"), *evaluate* it ("I'm too close to the shoulder"), then *act* (steer left). To do this requires three separate elements of the homeostatic system (Figure 19.7): a *receptor* to sense the environmental conditions (in this case, the eyes); an *integrator* to evaluate the situation and make decisions (here, the brain); and an *effector* to execute the commands (the muscles and bones of the hands and arms). The system in this example is under conscious control, but most homeostatic mechanisms are not.

Again and again in our discussion of physiological systems and their homeostatic functioning, we will see these common elements: receptors, integrators, and effectors. And we will see that the interplay of each set of elements constitutes a **feedback loop,** a circuit of sensing, evaluating, and reacting that can resist change in the internal environment to maintain homeostasis—the "same standing"—or bring about rapid change for a specific purpose.

Positive feedback loops bring about rapid change; they occur when an initial internal condition triggers change, and the change triggers more change. An example is the positive feedback loop that drives labor and delivery in

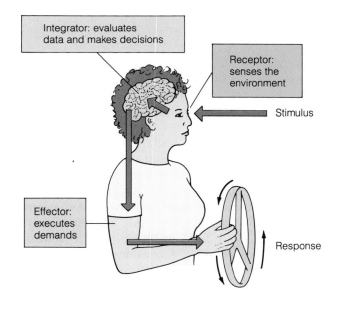

FIGURE 19.7 *Common Elements of Physiological Systems: Receptors, Integrators, and Effectors.* These elements work together in feedback loops that help maintain homeostasis.

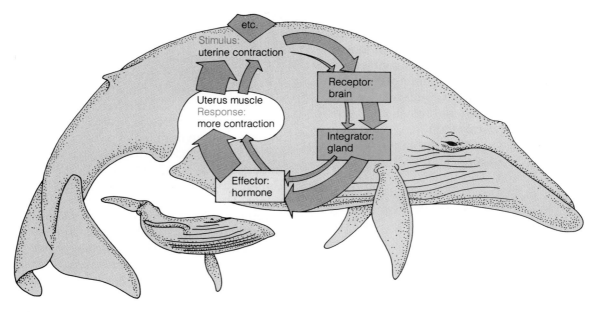

FIGURE 19.8 *Whale Labor and Birth: A Positive Feedback Loop.* In a positive feedback loop, the response to the stimulus triggers more stimulus and more response in an increasing spiral. In mammals, the hormonal loop that leads to stronger and stronger uterine contractions and eventually to birth is a positive feedback loop.

the birth of a person or whale (Figure 19.8). A hormone causes uterine contractions, and these strong muscular actions cause more hormone to be released, more contractions to occur, and so on. The feedback produces stronger and longer contractions spaced closer and closer together until at the climax of the cycle, the infant blue whale (a whopping 7.5 m, or 25 ft, long!) is expelled into the sea. This explosive action stops the loop, and homeostasis—in this case, a noncontracting uterus—is restored.

While positive feedback loops promote rapid change, they are not as common as negative feedback loops; these help keep conditions constant, which the organism generally requires.

Negative feedback loops resist change by sensing a stimulus (a change from a baseline condition), then activating mechanisms that oppose the change. The maintenance of body temperature within a narrow range serves as a good example of how such loops work. As a normal mammal, the blue whale's body has a constant warm temperature of about 37°C, or 98.6°F (i.e., it is warm-blooded; see Chapter 18). Maintaining this body temperature is an enormous problem for the whale, since it swims in icy waters, where it must resist chilling, and in tropical waters, where it must resist overheating. Keeping the whale's delicate thermal balance requires both anatomical and physiological solutions.

The whale has a physiological mechanism based on negative feedback that keeps its body temperature hovering around 37°C (Figure 19.9a); a similar mechanism helps maintain our body at about the same temperature. Like all mammals, whales have delicate sensors in the skin and in a small region of the brain lying just above the roof of the mouth (called the *hypothalamus*). These sensory receptors can detect surface and internal body temperature and relay a signal to other cells in the brain's hypothalamus. The signal of a drop in temperature, for instance, can trigger a control message from the brain to the muscles to begin contracting—doing physical work that gives off heat and warms the animal. As the internal temperature becomes normal again, sensors detect the warmth and stop signaling the brain. A similar negative feedback loop causes a person to shiver when cold, and the work of shivering warms us. Notice that in this negative feedback loop, the response—a rise in temperature—is the opposite of the stimulus—a drop in temperature. That is why this mechanism is called a *negative* feedback loop. It is a *feedback loop* because the initial change—a change in temperature—triggers several events in a certain order and creates a second change in the same parameter, in this case, temperature.

In addition to the physiological mechanism of the feedback loop, whales have anatomical features that help them stay warm in icy waters. One is the special arrangement of blood vessels in the flippers (Figure 19.9b). Another is a layer of fat called *blubber* that is 30 cm (1 ft) thick and completely encases the animal, much like a wet suit on a scuba diver. Blubber is a wonderful insulator, capable of blocking the flow of heat from the body core outward into the ocean water, which removes body heat 50 times more effectively than air. (People used to kill blue whales for their blubber, but mass slaughter nearly drove the whales to extinction. Current international whaling laws protect them.)

When winter hits Antarctica, the blue whales move to warmer tropical waters, and there, encased in their thick, insulating blubber, they must employ a different set of behavioral, anatomical, and physiological mechanisms to solve the problem of staying cool. Since they cannot find krill, their main food, in the tropics, blue whales

fast for the six months they spend in warm waters (a behavioral adaptation). While fasting, they live off the fat in their blubber layer, and the steadily thinning anatomical insulation improves the animal's ability to stay cool. If the whale is still too hot, a physiological mechanism based on blood flow can switch on, much the way an air conditioner begins to cool an overheated house. The blubber layer is crisscrossed by blood vessels, and when the animal is overheated—say, from exercising in warm water—blood can be shunted into these vessels (Figure 19.9c). This conveys heat toward the body surface, where the surrounding lower-temperature water can absorb it. Your own skin may feel flushed and look red when you exercise heavily owing to a similar shunt-

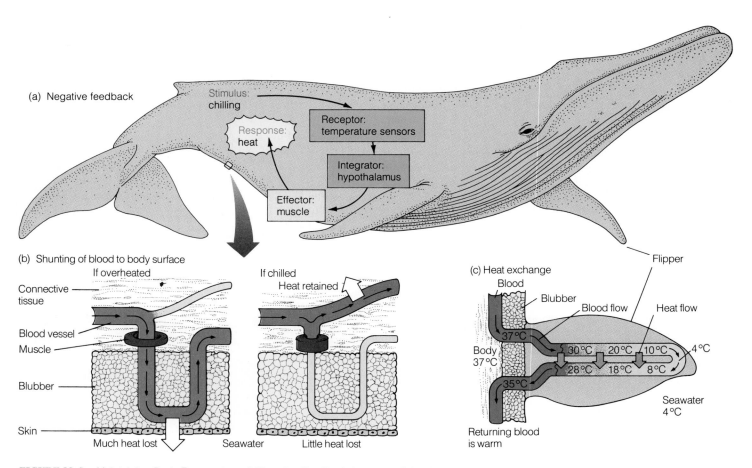

FIGURE 19.9 *Maintaining Body Temperature: A Negative Feedback Loop and Other Mechanisms.* (a) A negative feedback loop involving sensors in the whale's skin and hypothalamus region of the brain stimulates muscles to flex and warm the massive animal. (b) The blood vessels in a whale's flipper are arranged in a *rete mirabile*, or "wonderful net." The vessels lie side by side in such a way that the current of hot blood flowing away from the body core moves counter to the current of warm blood flowing back toward the body, and heat flows directly from outgoing blood to incoming blood. With such an arrangement, precious body heat is conserved rather than lost to the environment. (c) An additional homeostatic mechanism involving tiny circular muscles that open and close blood vessels near the body surface allows the animal to retain body heat when it is surrounded by cold seawater and to dump body heat when it is overheated.

THE CASE OF MARTHA'S LIVER: A BLOOD TEST FOR HOMEOSTASIS

As a floor nurse at a major hospital, Martha depended on her boundless supply of energy. When a strange fatigue set in, however, and when the whites of her eyes took on a yellowish cast, her upper abdomen began to hurt, and she grew nauseated, she was certain that she had contracted something serious and went to see her own doctor.

To the physician, the *jaundice* (yellowish pigmentation) suggested viral hepatitis, a disease in which a virus attacks liver cells. When aged red blood cells break down, they produce an orange-yellow pigment called *bilirubin*. Now, a healthy liver can pick up the bilirubin and secrete it into the intestines for elimination from the body. In a diseased liver, however, pigment breakdown is disrupted, and the yellowish substance builds up in the skin and eyes.

To test her suspected diagnosis of viral hepatitis, Martha's doctor drew some of her patient's blood (Figure 1) and sent it to a medical laboratory for a *chem screen*. This is an analysis of ions, small biological molecules, and enzymes found in the blood (Table 1, column A). If the person is healthy, homeostatic mechanisms keep the quantity of each substance within a definable range of values (column B). But if an illness interferes with homeostatic mechanisms, one or more of the substances will appear in excessive or scanty quantities (column D).

For example, a healthy person has 0.1–1.2 mg of bilirubin per 100 ml of blood, but Martha had 3.6 mg, or three times more than the normal upper limit (column D). She also had up to 70 times the normal levels of certain enzymes. These enzymes appeared in her blood because the liver cells, which normally store the enzymes, were being broken down.

This particular pattern on the chem screen indeed indicates liver disease, and in combination with the other clues—jaundice, tender abdomen, fatigue, nausea, and Martha's exposure to patients—the doctor was able to

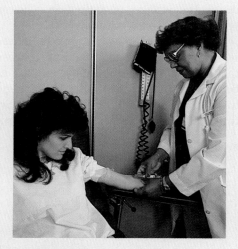

FIGURE 1 *Taking a Blood Sample.*

confirm a diagnosis of viral hepatitis. Modern medicine as yet has no effective drugs for this condition, but Martha's physician prescribed bed rest and fluids, and within a few weeks, Martha's blood chemistry—and her old energy level—returned to normal (column E).

TABLE 1 ≋ READOUT OF PATIENTS' BLOOD CHEMISTRY, WITH HEPATITIS AND WITHOUT

(A) CHEM SCREEN	(B) NORMAL RANGE	(C) UNITS*	(D) HEPATITIS PATIENT	(E) AFTER RECOVERY
SODIUM:	⟨135–145⟩	mEq/L	140	136
POTASSIUM:	⟨3.5–5.2⟩	mEq/L	4.5	4.0
CHLORIDE:	⟨95–109⟩	mEq/L	104	105
GLUCOSE:	⟨65–110⟩	mg/dl	93	86
CALCIUM:	⟨8.5–10.5⟩	mg/dl	9.4	9.4
PHOSPHORUS:	⟨2.5–4.5⟩	mg/dl	3.6	3.9
BILIRUBIN:	⟨0.1–1.2⟩	mg/dl	3.6	0.4
SGPT(ALT):	⟨0–45⟩	IU/L	2970	41
SGOT(AST):	⟨7–40⟩	IU/L	1990	40
IRON:	⟨35–200⟩	μg/dl	146	88
CHOLESTEROL:	⟨130–240⟩	mg/dl	142	143
TRIGLYCERIDES:	⟨10–150⟩	mg/dl	114	40

*mEq = milliequivalent weight; dl = deciliter (100 ml); IU = International Units; μg = microgram; mg = milligram; L = liter. These units are commonly used in medical laboratories for measuring substances in the blood.

ing mechanism, and jackrabbits likewise get rid of excess heat through many surface vessels in their large ears (Figure 19.10).

Thanks to these homeostatic mechanisms, the whale's body temperature can remain at 37°C despite icy oceans or warm seas, polar blizzards or blistering equatorial sun. Mammals, in general, employ such mechanisms, as well as others, including fever and sweating. Here's how fever works in a sick person.

A fever can start after virus particles or bacterial cells invade the body and begin multiplying. The infectious agents can turn up the brain's natural thermostat indirectly through white blood cells that protect the body from disease. Just as turning up the thermostat in your home signals the furnace to fire up, the resetting of the hypothalamus to a higher value triggers the heat-generated mechanisms of a person's body to step up their operation. The muscles start contracting, the body starts shivering, and this makes the temperature climb higher and higher. The sick person can still feel very cold, however, because it takes time for the body to reach the high setting on the thermostat. If the thermostat is set at 43.3°C (110°F) and the body is still only 39.4°C (103°F), the patient will feel cold. Elevated temperature helps to stop or slow bacterial growth, however, and eventually the thermostat is reset to normal and the fever breaks. But since the temperature is still high, sweating begins, and the evaporative cooling provided by perspiration quickly reduces the body temperature. Aspirin can help reduce fever because it interferes with the production of molecules that turn up the thermostat, and the thermostat shifts to a lower setting.

So-called cold-blooded animals (see Chapter 18) use different temperature-regulating mechanisms, many of them behavioral. An alligator, for example, will bask in the sun with its jaws gaping open to keep its body temperature between 35°C and 40°C (Figure 19.11a), and a lizard will instinctively move in or out of the sun and do

FIGURE 19.10 *Tall Ears: Adaptations for Dumping Heat.* An antelope jackrabbit radiates excess internal heat through huge, thin ears with a large amount of surface area—organic radiators.

FIGURE 19.11 *Cold-Blooded Animals Control Temperature Through Behavior.* (a) Alligator gasping, first observed by ancient Egyptians, helps reduce heat buildup in the animal's brain. Modern experiments show that a basking 'gator can keep its body temperature high but its brain temperature beneath the point of brain damage by locking its jaws open as it rests. (b) A rainbow lizard basking on a rock in Tanzania. Normally, lizards crawl into or out of the sunshine to keep their body temperature fairly warm and steady. A lizard with a bacterial infection, however, will actually induce its own artificial fever, baking itself in the sun until its body temperature soars and the bacterial growth slows. Clearly, behavioral homeostatic mechanisms are crucial for survival.

(a)

(b)

① Animals must maintain homeostasis—a constant internal environment—despite fluctuations in the external environment.

② Tubes and diffusion help maintain homeostasis.

③ Many body structures and function contribute to homeostasis, including the increasingly complex organization of tissues, organs, and organ systems.

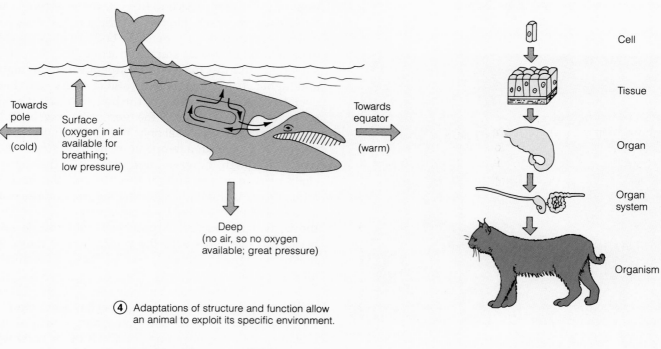

Towards pole

(cold)

Surface (oxygen in air available for breathing; low pressure)

Towards equator

(warm)

Deep (no air, so no oxygen available; great pressure)

Cell

Tissue

Organ

Organ system

Organism

④ Adaptations of structure and function allow an animal to exploit its specific environment.

⑤ Negative feedback loops help keep conditions constant; positive feedback loops promote useful changes.

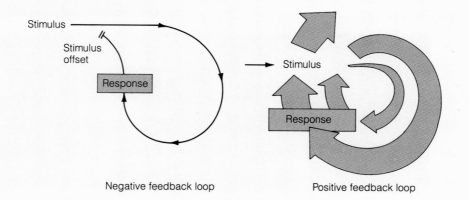

Stimulus

Stimulus offset

Response

Stimulus

Response

Negative feedback loop

Positive feedback loop

heat-generating push-ups to generate a fever (Figure 19.11b).

Whether the animal is warm-blooded or cold-blooded, body temperature is just one feature kept in check by homeostatic mechanisms. As we will see in upcoming chapters, it takes many systems working simultaneously and under elaborate coordination and control to maintain an animal's total internal milieu despite the vagaries of the external world.

⚎ CONNECTIONS

The subject of animal anatomy and physiology allows us to see how members of the animal kingdom function internally to survive in and exploit their environments. Throughout Part Four, we will focus primarily on the human body. Not only do biologists know more about people than they do about any other organism, but understanding human physiology helps us take care of our bodies in intelligent ways.

Although we separate individual organ systems into different chapters to make learning easier, keep in mind that this "dissection" is largely artificial. Each organ system depends on the others; for instance, the brain cannot function unless the blood delivers to it a constant supply of oxygen from the lungs and glucose from the digestive system. To give a sense of this dynamic interaction among organ systems, we will end Part Four by showing in detail how the human body responds to the physical stress of heavy exercise and yet continues to maintain the homeostasis it needs to survive.

NEW TERMS

anatomy, page 338
feedback loop, page 339
homeostasis, page 335

organism, page 337
organ system, page 337
physiology, page 338

STUDY QUESTIONS

REVIEW WHAT YOU HAVE LEARNED

1. Define anatomy and physiology.
2. What does homeostasis mean? Give some examples of homeostatic mechanisms.
3. The digestive system illustrates a common strategy in servicing the body's cells. What is it?

4. How are a tissue and an organ alike? How are they different? Use specific examples in your answer.
5. Does a given organ system function independently of other organ systems? Explain.
6. Name four special adaptations that suit a blue whale to its life in the sea.
7. How does negative feedback help the blue whale maintain a constant body temperature in cold water? In warm water?
8. Give an example of a positive feedback loop.
9. How does a lizard regulate its body temperature?

APPLY WHAT YOU HAVE LEARNED

1. An expedition to study the blue whale's feeding patterns in tropical waters failed to gain useful data. Why?
2. During the process of sexual arousal, does the body use a positive or a negative feedback loop? Defend your answer.

FOR FURTHER READING

Gould, J. L., and C. G. Gould. *Life at the Edge*. New York: Freeman, 1988.

Kiester, E. "A Little Fever Is Good for You." *Science* 84 (1984): 168–173.

Minasian, S. M., K. C. Balcomb III, and L. Foster. *The World's Whales: The Complete Illustrated Guide*. Washington, DC: Smithsonian Books, 1979.

Schmidt-Nielsen, K. *Animal Physiology: Adaptation and Environment*, 3d ed. New York: Cambridge University Press, 1983.

Scott, J. "Molecules That Keep You in Shape." *New Scientist* 24 (July 1986): 49–53.

Wursig, B. "The Behavior of Baleen Whales." *Scientific American* 258 (April 1988): 102–107.

CHAPTER 20

CIRCULATION: THE INTERNAL TRANSPORT SYSTEM

THE IMPROBABLE GIRAFFE

Giraffes have always aroused irresistible curiosity in people. These strange African mammals are the third heaviest land animals after the elephant and rhinoceros, and they are by far the tallest: An adult male stands 5.5 m (18 ft) high (Figure 20.1), and a female 4.5 m (15 ft). Three adult basketball players standing feet to shoulders could scarcely look a male giraffe in the eye.

From a physiologist's point of view, a giraffe's body is even more enigmatic than it looks. The giraffe, for example, has very high blood pressure. Yet, physiologists have found that the high pressure is an evolutionary adaptation. It helps distribute blood, with its cargo of oxygen, glucose, and other materials, throughout the large animal's lanky body. The high blood pressure, in fact, explains why, upon rising from a prone position, giraffes don't faint from lack of blood to the brain. You have probably experienced the vague dizziness that sometimes comes with jumping out of bed too quickly; for a split second, your heart isn't pumping hard enough to get sufficient blood to your brain. A giraffe rises slowly. When standing, its head is 2–3 m (6½–10 ft) above its heart, and its blood is forced through the vessels with high enough pressure to travel all that way with its vital cargo of oxygen and glucose. The giraffe is indeed a physiological marvel and makes a good case study for the functioning of the **circulatory system,** which consists of the heart, blood, and blood vessels.

Despite its unusual physiological problems, the giraffe's basic needs are the same as all other animals': It requires a source of energy (sugars and other organic molecules), a supply of oxygen (for aerobic respiration), and a means of disposing of carbon dioxide wastes. Like other large animals, the giraffe has a complex set of internal systems that together enable it to fulfill these needs. Take the need for oxygen, for example. The respiratory system includes tubes that bring oxygen to the lungs from outside the body. Blood circulating through the lungs then picks up this vital gas and delivers it throughout the body to brain cells, muscle cells, and all the others. Because the body's other systems are so closely tied to the circulatory system, circulation is the central activity in maintaining the steady state conducive to life (see Chapter 19).

Two unifying themes emerge from our discussion of animal circulation. First, the simple physical principles of *diffusion* and *bulk flow* underlie the circulatory system's transport of materials. Diffusion allows the conveyance of materials over short distances, between cells and their surroundings; bulk flow, which is the overall movement of a fluid from an area of higher pressure to an area of lower pressure, allows the conveyance of large amounts of blood throughout the entire body. Second, circulatory adaptations in different animals reflect the constraints of an animal's environment and way of living. Without its high blood pressure and other adaptations of its circulatory system, the giraffe, for example, could not browse from tall trees and stoop to drink from seasonal water holes.

In our discussion of how an animal's circulatory system transports materials and helps maintain homeostasis, we answer several important questions:

- How do the fluid and cellular components of blood contribute to the transport of materials?
- How do diffusion and bulk flow underlie transport in the circulatory system?
- How does the circulatory system function in humans and other mammals?
- What is the lymphatic system, and how does it return fluid to the bloodstream?

FIGURE 20.1 *Giraffes: Lofty Leaf Eaters.* The giraffe's long legs and neck pose special problems for blood circulation.

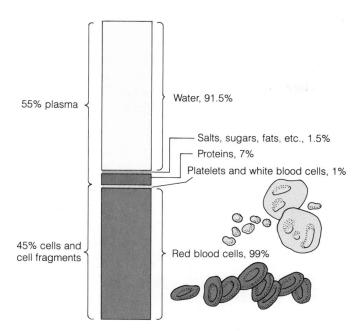

55% plasma

Water, 91.5%

Salts, sugars, fats, etc., 1.5%

Proteins, 7%

Platelets and white blood cells, 1%

45% cells and cell fragments

Red blood cells, 99%

FIGURE 20.2 *The Fractions of Human Blood.* If spun at high speed in a centrifuge, blood will separate into yellowish plasma and a denser portion containing red and white blood cells and fragments called platelets, which aid in blood clotting. Clearly, most of the blood is water and red blood cells.

≈ BLOOD: A MULTIPURPOSE LIQUID TISSUE FOR INTERNAL TRANSPORT

As a large mammal, you have about 5 liters (a little more than a gallon) of blood in your body, coursing through an amazing 80,000 km (50,000 miles) of vessels. This moving stream of liquid tissue operates like a rapid courier service, circulating oxygen from the lungs, nutrients from the digestive tract, organic wastes to the kidney, and regulatory substances such as hormones from one region of the body to another. Blood also distributes heat and assists in cooling (see Chapter 19). These functions are all based on the individual components of blood and the roles they perform.

BLOOD: LIQUID AND SOLIDS

When a medical practitioner draws a vial of blood from a vein in the inner crick of your arm, the blood sample is then sent to a laboratory for analysis. By whirling the sample in a centrifuge, a technician can separate the components of this life-giving liquid into a pale yellow liquid called **plasma, red** and **white blood cells,** and cell fragments known as **platelets** (Figure 20.2). Plasma is

more than nine-tenths water, and its main function is to transport the blood cells and platelets as well as other particles, including salts, sugars, and fats from the foods we eat. Plasma also contains dissolved proteins, which can help defend the body from invaders and aid in blood clotting. The white blood cells are important in the body's defense, and the platelets are another element important in blood clotting. The red cells (or *erythrocytes,* from the Greek for "red" and "cell"), contain hemoglobin and transport oxygen throughout the body.

RED BLOOD CELLS AND OXYGEN TRANSPORT

The human body contains 25 trillion red blood cells—fully one-third of all its 75 trillion cells. Three features help red blood cells perform their vital function of transporting oxygen to tissues: their hemoglobin content; a special enzyme that aids in carbon dioxide transport; and their biconcave disk shape that is a bit like a doughnut without a hole (Figure 20.3). Among other benefits, this disk shape allows all of the red blood cell's hemoglobin molecules to lie near the outer membrane, which oxygen must diffuse across.

Recall from Chapter 4 that mature red blood cells lose their nucleus and most other organelles during development. Biologists believe that this evolutionary adaptation may make room for more hemoglobin and increase the cell's oxygen-carrying capacity. Regardless of why it happens, it dooms the cell to die after about 120 days and to be replaced by a subset of dividing stem cells in the bone marrow. (Other stem cells generate white blood cells and the precursors of platelets.)

FIGURE 20.3 *Blood Cells Transport and Defend.* In this scanning electron micrograph, red blood cells (which transport oxygen) are clearly distinguishable as biconcave disks, while white blood cells (which help defend the body) look like fuzzy balls (3500×).

A negative feedback loop controls red blood cell replacement. When the red cell count decreases because of either normal red cell death or sudden blood loss from a serious wound, less oxygen reaches the kidneys. As a result, these paired organs make a hormone that triggers the production of more red blood cells. Vigorous exercise can also set in motion this same cycle of events. A runner's muscles, for example, will use massive amounts of oxygen, decreasing the amounts reaching the kidneys and triggering release of the hormone and a rise in red blood cell production.

Researchers have made progress recently in manufacturing this hormone, which promises to decrease the need for blood transfusions, simplify bone marrow transplants, and bolster the fight against disease-causing microorganisms, AIDS, and some cancers.

WHITE BLOOD CELLS: DEFENSE OF THE BODY

For every 1000 red blood cells, there are only 2 white blood cells (*leukocytes*, literally, "white cells"; see Figure 20.3). These spherical white cells, however, are larger in volume than the flatter red cells, they retain their nucleus, and they have a changeable, amoeba-like shape that allows them to squeeze through the walls of even the smallest blood vessels (*capillaries*) and patrol the fluid-filled spaces between cells. These characteristics allow white blood cells to defend the body against invasions by microorganisms and other foreign materials.

There are five classes of white blood cells, and together they travel in the blood like a mobile militia. Some are specialists at phagocytosis ("cell eating"): They are attracted to damaged or infected tissues and consume bacteria, viruses, and cellular debris. Others carry enzymes that break down foreign proteins and break up blood clots. Still others, called *lymphocytes*, are active in the immune responses that protect the body against infectious agents and foreign intruders (including, unfortunately, transplanted organs). There is so much to say about white blood cells and the body's immune defense that Chapter 22 is devoted entirely to the subject.

PLATELETS: PLUGGING LEAKS IN THE SYSTEM

Streaming along with the red and white blood cells are millions of cell fragments called platelets, or *thrombocytes* (review Figure 20.2). These irregularly lobed bits and pieces have broken off larger cells in the bone marrow. Platelets are crucial blood components because like certain chemicals we can add to the water in a car's radiator, platelets help to plug small leaks in the circulatory system. As we will see later in the chapter, were it not for the blood-clotting action of platelets, an animal's blood might literally drain away through even a minor wound.

Blood, with all its separate components, is capable of delivering oxygen and nutrients, carting off waste gases, and patrolling and defending the body. But static, stagnant pools of blood could not accomplish these tasks. The key to carrying out these tasks is constant movement: The blood must be pumped at high pressure on a continuous circuit through the body, and that is the function of the pump and tubing of the circulatory system.

≋ CIRCULATORY SYSTEMS: STRATEGIES FOR MATERIAL TRANSPORT

Just as delivery trucks would be useless in a town without roads, blood would be of little value to an animal unless it could circulate its cargo of gases, nutrients, and other substances throughout the body. A constant flow of blood with its storehouse of materials must pass near each body cell for diffusion of those materials to keep pace with the cells' consumption. Similarly, wastes that would accumulate and poison the cell must quickly diffuse into a nearby channel and are swept away in the bloodstream. Two basic types of circulatory systems, open and closed, perform these functions.

EVOLUTION TOWARD GREATER EFFICIENCY

In *open circulatory systems,* short blood vessels pipe a blood equivalent called hemolymph into and out of the heart; in the rest of the body, the fluid sloshes freely around and through the animal's tissues. Most arthropods and mollusks have an open circulatory system (see Chapter 17).

■ **Closed Circulatory Systems** Segmented worms and more complex animals, such as vertebrates, have *closed circulatory systems*, in which the circulating blood is completely contained within a system of vessels. In a representative vertebrate, such as the sea bass, the heart pumps blood through tiny vessels in the gills, where it comes in contact with environmental oxygen (Figure 20.4). Once the blood has picked up oxygen, it courses through smaller and smaller vessels that spread throughout the fish's tissues, delivering oxygen to every body cell. On its circuit, the blood passes by the gut, where it picks up nutrients, and the kidneys, where organic wastes are removed from circulation for excretion from the body.

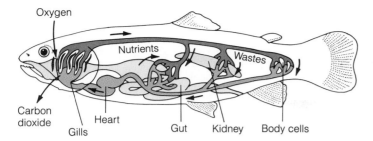

FIGURE 20.4 *The Closed Circulatory System of the Sea Bass.* In the closed circulatory system of a vertebrate (here, a sea bass), the blood remains in a system of interconnected vessels as it circulates through the body. Here, red signifies oxygen-rich blood leaving the gills and heart and traveling to the body tissues, and blue signifies oxygen-poor blood returning from the tissues in vessels leading to the heart and gills.

Eventually, the continued pumping of the heart pushes the blood around once more to the gills, where it dumps carbon dioxide and picks up new oxygen.

A closed circulatory system has several advantages over an open one. First, fluid contained within a network of closed tubes can be shunted to specific areas where it is needed, much as a farmer can dispatch the water in irrigation pipes to different fields. Second, since the fluid is completely contained within pipelines, more pressure can be exerted on it; thus, the fluid can be forcefully distributed to areas distant from the pump (like a giraffe's head or feet). You can demonstrate the results of pressurized fluid by placing your thumb over the opening of a garden hose: You can squirt water on a tomato plant—or on an unsuspecting friend—much farther away than you could by dribbling the water out unobstructed. These features allow closed circulatory systems to deliver blood throughout larger, more complex, and more active organisms in a much more efficient way than an open system could.

■ **From Single to Double Loops** In fishes, blood flows in a simple loop from the heart, to the gills (where it takes up oxygen), to body tissues, and back to the heart (review Figure 20.4 and see Figure 20.5a). A fish's heart has two chambers: a less muscular one called the *atrium* (plural, atria), which receives oxygen-depleted blood from vessels leading to the heart, and a more muscular cavity, the *ventricle*. The ventricle receives blood from the atrium and pumps it through a vessel leading away from the heart. The blood travels through increasingly smaller vessels to the smallest blood vessels, called *capillaries,* in the gills. There, oxygen is absorbed and carbon dioxide is released to the environment. Oxygenated blood then leaves the gills and collects in large vessels, called arteries, which pipe fresh blood throughout the fish's body. After the oxygen and other materials are delivered to

tissue cells, deoxygenated blood returns to the atrium through vessels called veins, and the cycle is repeated.

This single-loop system solves the problem of separating oxygenated and deoxygenated blood, but it fails to solve the problem of low blood pressure: Blood leaving the heart is under high pressure, but once it fans out through a dense network of fine vessels in the gills, pressure diminishes and remains low as the blood moves through the body. While this low pressure is adequate to meet the demands of a "cold-blooded" animal that is supported and buoyed up by water, it could not push blood quickly enough through the body of a land animal, whose cells require high levels of oxygen and nutrients. Instead, land vertebrates have a two-loop circulatory system. Blood is first pumped to the lungs to take on oxygen, after which it flows back to the heart, where it is pumped out a second time—at high pressure—to circulate to body tissues (Figure 20.5b).

■ **The Four-Chambered Heart** During the evolution of the two-loop circulatory system, the two-chambered heart of early fishes also underwent significant change, first

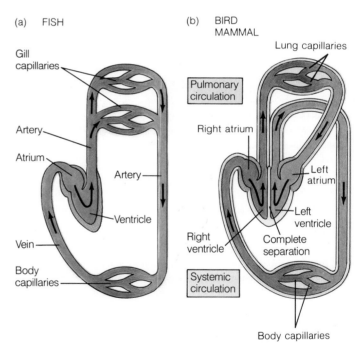

FIGURE 20.5 *Circulatory Loops in Fishes, Birds, and Mammals.* Birds and mammals have two circulatory loops, a modification of the pattern seen in fishes. (a) A fish has a two-chambered heart (one atrium, one ventricle) and a single loop of blood vessels leading through the gills and to the body tissues. (b) In birds and mammals, the heart has four chambers (two atria and two ventricles) that together prevent blood mixing, and the vessels are arranged in pulmonary (lung) and systemic (body) circulatory loops.

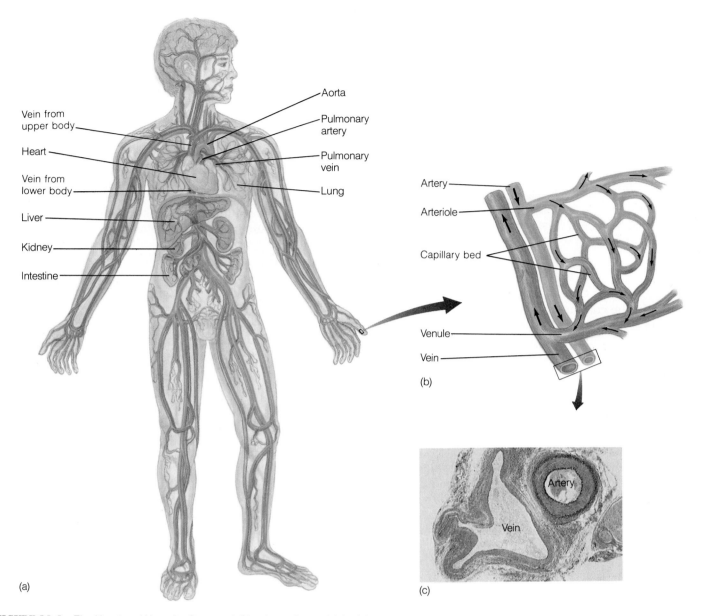

FIGURE 20.6 *The Heart and Vascular System: A Circulatory River of Life.* (a) The human circulatory system contains 80,000 km of tubing and circulates 5 liters of blood each minute. (b) Arteries leading away from the heart break into arterioles, and these divide into delicate capillaries in beds that permeate the body tissues and then coalesce into venules and veins. These return blood to the heart. (c) Arteries are narrower than veins and have muscular walls, while veins are larger and have thinner, less muscular walls.

to a three-chambered heart in amphibians and reptiles, and then to a four-chambered heart (see Figure 20.5b). All birds and mammals alive today, including giraffes and basketball players, have four-chambered hearts with an atrium and ventricle on the left side and an atrium and ventricle on the right side. In a sense, these complex animals have two hearts that pump blood in two completely separate circulatory loops (Figure 20.6). In the "lung loop," or **pulmonary circulation,** blood arrives from the tissues, enters the right side of the heart, and is pumped directly to the lungs, where it picks up oxygen. The oxygenated blood then enters the left side of the heart and is pumped out into a "body loop," the **systemic**

circulation. This double loop means that the blood flowing to the muscles, brain, and extremities will have the highest possible oxygen content and that "used," oxygen-poor blood can be shunted from an isolated right heart to the nearby lungs without ever mixing with oxygen-rich blood. It also means that the left heart chamber can send blood out through the body loop at high enough pressure to reach all the tissues quickly.

To appreciate how the four-chambered heart and separate circulatory loops operate in birds and mammals and how blood pressure and blood flow are controlled, let's examine the mammalian circulatory system in detail, using the human system as our model.

≈ CIRCULATION IN HUMANS AND OTHER MAMMALS

It takes an immense network of circulatory tubing to deliver blood close to each of your cells (Figure 20.6a). If all of the blood vessels in your body could be removed and placed end to end, the single long pipeline would encircle the earth twice. To keep the blood flowing through all these vessels, your heart must beat with a regular rhythm for eight or nine decades, and the flow must be maintained at a high enough pressure to force blood into your brain, nose, toes, and all the tissues in between. Both the anatomy and the activity of the blood vessels and heart make these tasks possible.

BLOOD VESSELS: THE VASCULAR NETWORK

Blood moves away from the heart in **arteries,** which are large, hoselike vessels with thick, multilayered, muscular walls (Figure 20.6b and c). The contraction of these wall muscles helps keep blood under pressure—the force we call blood pressure. The easily felt vessels through which blood pulses in your wrists and neck are arteries. Arteries, in turn, branch to form *arterioles.* These vessels are too small to be seen with the unaided eye and have thinner, less muscular walls. Arterioles branch again into the dense, weblike network of delicate **capillaries** that permeate the fingertips, earlobes, and all the tissues of the body. Capillaries are microscopic vessels only about 8 μm (0.0003 in.) in diameter and never more than 1 mm (0.04 in.) from any body cell. A capillary's diameter is so small that red blood cells can pass through in single file only, and so delicate that their walls are just a single cell thick. The vessels are so numerous, however, that they make up almost all of the 80,000 km (50,000 miles) of a person's blood vessels, and if all were open at the same time, they could contain the entire 5 liters of human blood.

The capillaries' ultrathin structure is one key to the efficiency of the circulatory system: Materials can readily diffuse outward or inward through the single cell layer.

Capillaries are interwoven into other tissues in networks called *capillary beds* that link the arteries and veins (see Figure 20.6b). The arterial side of each bed (shown in red) conveys fresh blood, and oxygen, nutrients, and other materials move through the capillary walls into the extracellular fluid and then into the tissue cells around the bed. The depleted blood then continues to move through the bed to the venous side (shown in blue); there the capillaries leading away from the tissues coalesce into larger vessels known as *venules,* which in turn merge to become **veins,** which carry blood toward the heart, where the cycle of circulation begins again. Because venous blood has already traveled some distance from the heart and been slowed by the tight passage through the narrow capillaries, and because veins have a larger diameter and thinner, less muscular walls than arteries (see Figure 20.6c), blood in veins is under lower pressure. Large body muscles, such as the ones that move your legs, help prevent pooling of this low-pressure fluid and keep it flowing in the proper direction. As the leg muscles contract, they flatten the veins that run through them. The blood in these flattened veins can move only one way, toward the heart's right side, because any back-tracking is ~~provided~~ by a system of *valves*—tonguelike flaps that extend into the internal space of the vein. Like a one-way door, when a valve is pushed from one direction, it opens and allows blood to pass, but when pressure is exerted from the other direction, it stays locked tight. The constant flow of blood from heart to arteries to capillary beds to veins and back to heart depends partially on the structure of arteries and veins, as we have just seen, but is perpetuated by the action of the beating heart. *prevented*

THE TIRELESS HEART

At the center of the circulatory system lies the heart (see Figure 20.6a). Roughly the size of a large lopsided apple, the human heart pumps a teacupful of blood with every three beats, 5 liters ~~of blood~~ every minute, and upward of 7200 liters (70 barrelfuls) of blood every day of your life. Over a lifetime, ~~this~~ amounts to 2.5 billion heartbeats and 18 million barrels of blood. Although medical engineers have tried valiantly, replacing the living circulatory pump has proved exceedingly difficult.

This tireless organ, like the hearts of all mammals, has four chambers: the right atrium and right ventricle, and the left atrium and left ventricle. Each atrium-ventricle pair plays a specialized role in circulating the blood. The right set receives deoxygenated blood from the upper

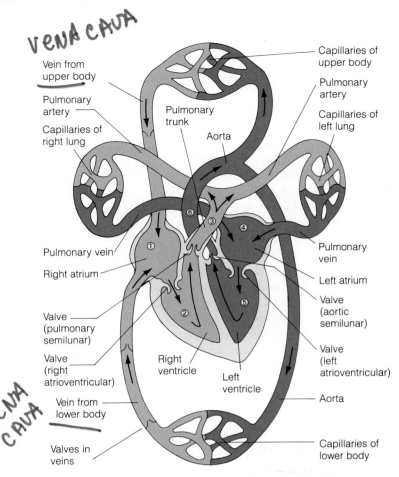

VENA CAVA

VENA CAVA

FIGURE 20.7 *The Pumping Heart.* Blood flows in a circuit from large veins to the right atrium (1), right ventricle (2), pulmonary artery (3), lungs, pulmonary veins, left atrium (4), left ventricle (5), aorta (6), and body tissues.

and lower body and pumps it through the Y-shaped **pulmonary artery** to the lungs (Figure 20.7, steps 1–3), where gases are exchanged in the capillaries.

In the next step of the circuit, blood freshly oxygenated in the lungs collects in the pulmonary veins (the only veins that carry oxygen-rich blood) and enters the heart's left atrium and then left ventricle (steps 4 and 5). The thick walls of the muscular left ventricle contract around this blood in a wringing motion until enough pressure develops to push open a valve and squirt the blood into the **aorta** (step 6)—the main artery leading to the systemic circulation. This body loop conveys blood to capillary beds in distant tissues, then returns it via venules and veins back to the right atrium once again. There are four valves in the heart, which together prevent blood from reversing its normal flow. In a person with diseased valves, the heart's pumping strength is diminished by backwashing blood, which is sometimes heard as a heart murmur.

HOW THE HEART BEATS

Like the muscles in your arm, each of the heart's four chambers is made up of specialized cells organized into fibers that can contract. The muscles in your arm, however, can contract weakly (as when you pick up a pencil) or more strongly (as when you heft a load of books), and the strength of contraction depends on what percentage of the muscle fibers contract at any given time. By contrast, all the fibers in the heart (cardiac) muscle contract with each heartbeat; thus, the healthy heart is always pumping at full strength.

Several elements account for this specialized characteristic of heart muscle. First, heart muscle cells are linked electrically. As a result, neighboring sections of the heart wall contract and relax together in a superbly coordinated pumping action that keeps blood flowing smoothly through the system.

Second, cardiac muscle contracts automatically—that is, without stimulation from the nervous system, although nerves do play a role in speeding or slowing the heart rate. Some heart muscle cells, called *pacemaker cells*, contract slightly earlier than others, and because each contraction spreads quickly throughout a region of the muscle, pacemaker cells ignite contractions in the entire heart and set the rate of the heartbeat.

People sometimes have heart conditions in which the beat is irregular, and they can often be helped with an artificial pacemaker. This small electrical stimulator, powered by a battery or by the decay of a small amount of radioactive material, is implanted beneath the skin of the shoulder or abdomen. Then, by way of electrodes threaded through veins into the heart, the pacemaker sends rhythmic electrical impulses that stimulate the cardiac muscle to contract at an appropriate time.

Each contraction/relaxation sequence of the atria and ventricles makes up a single heartbeat and is called the *cardiac cycle.* Recent evidence suggests that as the ventricles relax, they actively suck blood into their chambers from the atria. The "lub-dub" sounds you can hear through a stethoscope are generated by the closing of the valves between heart chambers and/or arteries. As the ventricles begin to contract (see Figure 20.7, steps 2 and 5), the pressure of surging blood pushes shut the valves between heart chambers, and you hear a low-pitched "lub." Then, when the ventricles begin to relax (steps 3 and 6), pressure in the aorta and pulmonary artery rapidly forces valves in both arteries to slam closed, producing the quicker, higher-pitched "dub."

BLOOD PRESSURE; THE FORCE BEHIND BLOOD FLOW

With each heartbeat, blood courses through your arteries at high speed (33 cm, or about 13 in., per second) and under high pressure. If a person severs the aorta, blood spurts out 2 m (more than 6 ft.) high! Blood pressure initiated by the heart's strong contractions serves an important purpose. It delivers blood quickly and continuously to cells far from the heart.

The "blood pressure" that a doctor evaluates with a tight cuff around the patient's arm is a measure of the force that blood exerts against artery walls as the heart alternately pumps (the *systole* phase) and rests (the *diastole* phase). The muscular, relatively inflexible walls of arteries help to maintain the blood pressure, but the farther blood travels from its central pump and the more it is dispersed through the miles and miles of capillaries, the more the pressure falls. Just as toothpaste gushes from a tube when it is squeezed, blood flows from areas of higher pressure to areas of lower pressure when the heart muscle contracts. The consumption of high-cholesterol foods can lead to a narrowing of the arteries, a consequent increase in blood pressure, and a heightened risk of heart attacks and strokes, as the box on page 354 explains.

Blood pressure is measured in millimeters of mercury (mm Hg), that is, how high the pressure could cause a column of mercury to rise. A normal reading for a person at rest is 120 mm during pumping (systole) and 80 mm during rest (diastole), or 120/80. A giraffe's blood pressure, in comparison, averages 260/160—the highest in the animal kingdom. The giraffe is not sick, though; without such high pressures, blood could never climb the animal's long neck and reach its head.

The giraffe has two special behaviors that offset its naturally high blood pressure: It rises from a prone position fairly slowly, thereby preventing fainting during the few moments required for the heart to pump blood up to the brain; and it spreads its legs wide apart when it drinks (Figure 20.8), an adaptation that biologists believe moves the animal's heart closer to the ground and reduces the rush of blood toward the brain. Just as important, however, is a "wonderful net" of capillaries in the head that can open rapidly to divert and slow down rushing blood before the surging fluid can overtax vessels in the brain or eyes and cause them to burst. The giraffe also has very thick-walled arteries that resist the pressurized blood inside.

Once blood has squeezed through narrow capillaries, the force originally exerted by the heart is nearly spent. This is why veins are often called the body's "blood reservoir," capable of holding as much as 80 percent of total blood volume at any given time. It is also why venous blood tends to pool in inactive limbs and why physical activity helps maintain good blood circulation. In the giraffe's legs, a thick hide applies pressure from the outside, acting a bit like support hose and preventing blood from pooling in the legs.

In giraffes and other animals, the movement of muscles in the arms, legs, and rib cage propels blood through individual veins, while the one-way valves in each ves-

FIGURE 20.8 *The Giraffe's Ungainly Drinking Posture Helps Protect Its Brain from a Rush of Blood.* With the heart closer to the ground, the downward rush of blood slows.

sel's central cavity prevent backflow and keep the blood moving toward the heart. If no muscular movement takes place over a long period of time, blood and other fluids accumulate in the extremities. In a rather macabre example, scholars believe that the lingering deaths of crucifixion victims in ancient times were due in large part to blood pooling in the legs.

SHUNTING BLOOD WHERE IT IS NEEDED: VASOCONSTRICTION AND VASODILATION

From moment to moment, the distribution of blood varies in an animal's arteries, veins, and capillaries, depending on local oxygen use. During exercise, for example, blood flow to your muscles increases; then later, if you eat a sandwich, blood flow to the intestines rises and the gut absorbs nutrient molecules from this increased blood supply. Variations in the amount of blood flowing through the vessels in a particular body region are generally controlled by the action of nerves or, as in the examples just cited, by the chemical messengers called hormones. Signals from the nerves or hormones cause vessels to constrict or dilate. If blood is needed in one region of the body, say, the arm and leg muscles, then contraction will take place in the walls of small arteries in other body

CHOLESTEROL AND HEART DISEASE

The next time you're in a library or movie theater, notice the person sitting next to you: Current statistics indicate that either that person or you yourself will die of *atherosclerosis*, a disease sometimes referred to as "hardening of the arteries." What causes an artery to "harden"? Decades of research have focused on the role of cholesterol, a glistening white, fatty substance that is manufactured naturally by the liver. Cholesterol is an important component of cell membranes and is present in foods, especially eggs, meat, and dairy products. Scientific evidence suggests that cholesterol can build up in an artery and lead to the formation of *plaques*, thickened regions of the artery wall that occlude the passageway and prevent blood from flowing freely (Figure 1). As blood transit becomes increasingly impeded in a given artery, a blood clot may form and block the flow completely, leading to a heart attack or, if the affected artery is in the brain, to a stroke. Today, 3400 Americans a day suffer heart attacks, and 1600 suffer strokes; this makes cardiovascular diseases (ailments of the heart and blood vessels) the leading cause of death in this country.

Is cholesterol really the main culprit? As in so many areas of human physiology and medicine, the picture is complicated. More and more evidence suggests that susceptibility to atherosclerosis is highly individual and is related both to genetic tendencies and to life-style habits. Physiologists now know that within a person's bloodstream, cholesterol travels attached to carrier molecules called lipoproteins; high-density lipoproteins (HDLs, or "good cholesterol") have very few cholesterol molecules attached, while low-density lipoproteins (LDLs) have great numbers of attached cholesterols. (Cholesterol floats; that's why LDLs are called "low density.") Evidence suggests that by exercising regularly, eating more vegetable and fish oils, and avoiding smoking, heavy drinking, and obesity, a person will have higher levels of HDLs, and these can help prevent plaques from building up in the blood vessels and decrease the risk of heart disease and strokes. People with high levels of LDLs, on the other hand, seem to be at greater risk of developing atherosclerosis, and researchers think the problem lies with the numbers of LDL receptors on the outer membranes of their cell surfaces.

If one has high numbers of LDL receptors on his or her cell surfaces, then cholesterol can be taken into cells and little will remain to circulate and build up plaques. Unfortunately, many Americans have low levels of LDL receptors. As a result, their cells remove *less* cholesterol from the bloodstream, and the fatty material continues to circulate. This can contribute to plaque buildup, which can in turn predispose the person to atherosclerosis, heart attacks, and strokes.

The important question, then, is, What leads to the low levels of LDL receptors? In some individuals, the cause is genetic, but in many people

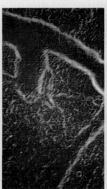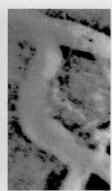

FIGURE 1 *A Plaque-Clogged Artery.* Before surgery (left) and after (right).

the characteristic may be related to a high-fat diet and a resulting chronic oversupply of cholesterol in the blood. In effect, to keep themselves from being inundated with cholesterol, cells shut down the metabolic machinery that manufactures LDL receptors, and thus the blood cannot be sufficiently cleansed of the circulating fatty sludge.

This understanding of the link between the body's handling of cholesterol and its development of atherosclerosis has generated a search for drugs that can stimulate cells to build new LDL receptors. Another approach is represented by a new drug, lovastatin, which inhibits the action of a liver enzyme that makes cholesterol. In the meantime, the most prudent course is the one that health experts have been recommending all along: Eat fewer fatty foods, get more exercise, maintain ideal weight, don't smoke, and drink moderately if at all.

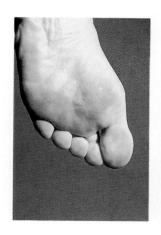

(a)   (b)

FIGURE 20.9 *White Toes, Red Face: A Sign of Blood Flow Control.*
(a) In this person with Reynaud's syndrome, whitish, mottled toes are
a sign of restricted blood flow. (b) The full flush of heavy exercise
comes from dilated vessels and increased blood flow to the skin,
where excess heat can disperse to the environment.

regions, such as the intestines. As a result, blood flow
in the intestines is reduced, and more blood can be
shunted to the arms and legs, where it will deliver oxygen to rapidly metabolizing cells. In capillaries within
body regions that require a greater blood flow, a reverse
process takes place; the vessels' internal spaces are
enlarged, and more blood can flow through.

Turning blue from the cold is a sign of restricted blood
flow. When we get extremely cold, tiny muscles close
down capillary beds near the skin surface in the face; the
remaining blood in the facial tissues becomes oxygen
depleted, and the complexion can take on a bluish cast.
Reduced blood flow to the hands or feet can turn fingers
white (Figure 20.9a). Likewise, eye drops that "get the
red out" contain chemicals that close down the tiny capillaries in the eyeball. But when we are very warm, as
during active exercise, vessel dilation allows the capillary
beds to become engorged, and the skin can feel hot and
look red (Figure 20.9b). Medical researchers are testing
ways to maintain dilating hormones in newly grafted
vessels to keep the small-diameter grafts from clogging
or collapsing.

HOW BLEEDING STOPS: FORMATION OF A CLOT

Although special muscles can shunt blood away from a
capillary bed, more drastic mechanisms are sometimes

needed to stop blood from flowing to a particular spot,
such as a finger wound made with a sharp paring knife
(Figure 20.10a). Smooth muscles in the walls of a damaged vessel contract and partly close off the vessel, and
circulating cell fragments called platelets release a hormone that keeps the muscles contracting. Then, platelets
at the injured site begin sticking to each other and to
rough surfaces at torn edges of the wound to form a
plug. The next stage is *coagulation*, the actual formation
of a blood clot. In creating the clot, tough, insoluble threads
of the protein *fibrin* become woven into a strong, wiry
mesh that traps red blood cells. The entire multistage
process is subject to rigid controls, and for good reason;
a clot that forms unnecessarily, such as in vessels of the
heart or brain, can lead to a heart attack or a stroke. The
agents of such fine control are substances that circulate
in the blood plasma, stimulating coagulation when a vessel is injured and inhibiting it most other times.

People with "bleeder's disease," or hemophilia, have
a mutant gene that codes for a defective clotting protein.

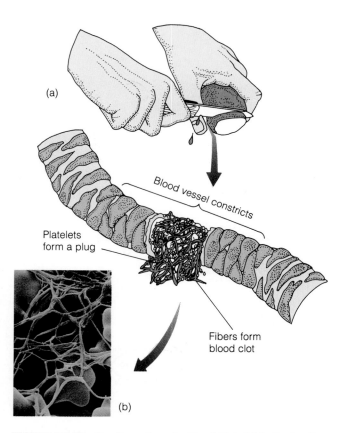

(a)

Blood vessel constricts

Platelets
form a plug

Fibers form
blood clot

(b)

FIGURE 20.10 *The Formation of a Blood Clot: A Life-Saving Series
of Events.* (a) When a wound severs a tiny blood vessel, numerous
processes are set in motion, including constriction of the vessel, the
formation of a platelet plug, and then coagulation, or blood clotting.
(b) Fibrin threads and red blood cells are clearly visible in this blood
clot.

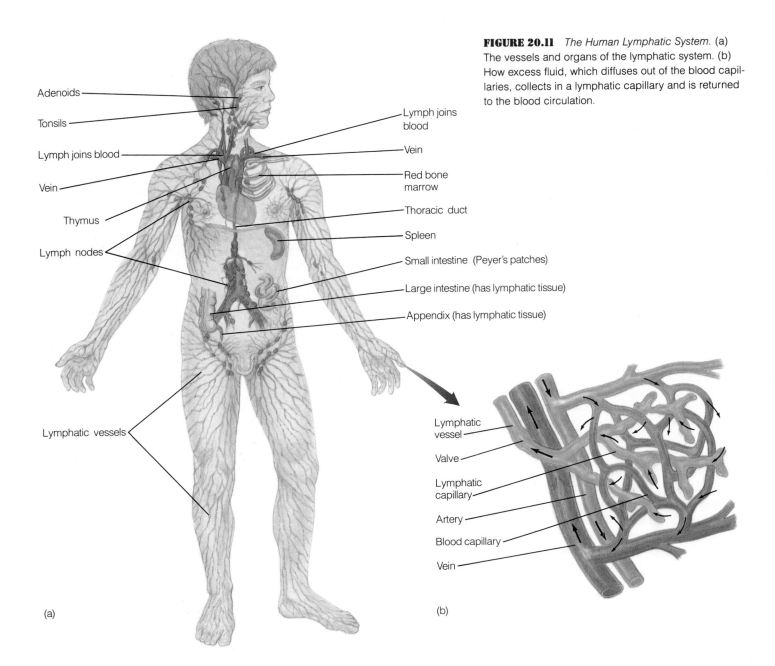

FIGURE 20.11 *The Human Lymphatic System.* (a) The vessels and organs of the lymphatic system. (b) How excess fluid, which diffuses out of the blood capillaries, collects in a lymphatic capillary and is returned to the blood circulation.

Adenoids

Tonsils

Lymph joins blood

Vein

Thymus

Lymph nodes

Lymphatic vessels

Lymph joins blood

Vein

Red bone marrow

Thoracic duct

Spleen

Small intestine (Peyer's patches)

Large intestine (has lymphatic tissue)

Appendix (has lymphatic tissue)

Lymphatic vessel

Valve

Lymphatic capillary

Artery

Blood capillary

Vein

(a)

(b)

Unless a hemophiliac receives periodic intravenous transfusions of plasma containing the normal protein, even a minor injury can lead to major blood loss and even death.

Despite the efficiency of the circulatory system, fluid tends to leak from blood vessels. Complex animals have thus developed a "second circulatory system" to drain away this leaked fluid.

THE LYMPHATIC SYSTEM: THE SECOND CIRCULATORY HIGHWAY

One consequence of a pressurized blood system is that fluids can be forced through the walls of delicate vessels and can build up in the extracellular spaces. The second system of fluid-containing vessels, the **lymphatic system,** is a network of tubes that collects fluid that has been forced out of capillaries and returns it to the bloodstream (Figure 20.11a). As Chapter 22 describes, lymphatic vessels also play a major role in the immune system.

Unlike the blood vascular system, the lymphatic system is made up of capillaries and larger vessels that do *not* form a circulatory loop. Instead, lymphatic capillaries are minute tubes with closed ends that permeate and drain the body tissues (Figure 20.11b). Like creeks merging to form a river, these capillaries coalesce to form larger lymph vessels that move fluids toward the heart. Because the lymphatic capillary walls are permeable to extracellular fluid and its contents, excess fluid in the spaces between cells seeps into this tributary system and forms the **lymph.** This yellowish fluid also contains proteins, dead cells, and sometimes invasive microorganisms, such as bacteria and viruses. Eventually, some of the smaller vessels drain their contents into two major lymphatic vessels that disgorge a steady stream of lymph into veins of the circulatory system near the heart. The lymph fluid then becomes part of the blood plasma once again.

Debris in the lymph is prevented from mixing into the blood by several bean-shaped filtering organs called **lymph nodes** located at intervals along the lymphatic vessels. As lymph moves along a vessel, it percolates through the nodes, like oil passing through an oil filter, and in the process, dead cells and other debris are filtered out. The nodes also harbor large numbers of infection-fighting white blood cells that can attack bacteria and other foreign materials. You have no doubt experienced painful swollen lymph nodes in the neck, underarms, or groin area during a bout of infection. Just as lymph vessels can carry dead cells away from an infection site, they can also transport migrating cancer cells to new sites. Doctors often remove lymph nodes and vessels along with cancerous tissue in patients with cancer of the breast or other sites, and this removes a highway for the cancer's spread.

Like veins, lymphatic vessels are designed to transport fluids under low pressure; valves keep the lymph flowing in one direction, and the squeezing force of contractions in muscle tissue surrounding the lymph vessels propels the fluid. The giraffe's very tight leg skin helps to massage lymph upward from the animal's feet. A person with inactive muscles, say, an injured skier in a hospital bed, can suffer lymph accumulation and resultant tissue swelling. In rare cases, the lymph vessels draining an extremity become blocked entirely, as by the infectious parasite that causes elephantiasis. The results can be swelling so enormous that the affected limb becomes barely recognizable (see Figure 17.11).

The lymphatic system also has two components, the *thymus* and the *spleen,* that are separate from the network of vessels. The thymus is a soft, V-shaped organ located at the base of the neck in front of the aorta. The thymus plays a role in the body's immune system (see Chapter 22).

An adult's spleen, which is about the size and shape of a banana, lies just behind the stomach and acts as a spongelike holding area for a portion of the body's blood supply. Much of the spleen consists of filtering tissue that removes impurities from the blood (including old red blood cells), and the organ harbors dense masses of white blood cells, as well.

≋ CONNECTIONS

Most large animals have complex circulatory systems for propelling blood throughout the body. Different environments, such as salty oceans and dry African savannas, and different physical constraints, such as the giraffe's long legs and long neck, have helped shape the kinds of structures and mechanisms that have evolved to make circulation possible. Although this chapter focused on the circulatory system in isolation, we must keep in mind that an organism functions as an integrated whole. That is why diseases of the heart have such catastrophic consequences. Heart rate is regulated by the nervous and hormonal systems, and these, too, depend on a steady supply of blood, with its cargo of nutrients and oxygen. And the heart muscle could not continue to contract without a steady supply of oxygen diffusing across gill or lung surfaces. It is this apparatus for gas exchange— the respiratory system—that we describe in the next chapter.

NEW TERMS

aorta, page 352
artery, page 351
capillary, page 351
circulatory system, page 346
lymph, page 357
lymphatic system, page 356
lymph node, page 357
plasma, page 347
platelet, page 347

pulmonary artery, page 352
pulmonary circulation, page 350
red blood cell, page 347
systemic circulation, page 350
vein, page 351
white blood cell, page 347

STUDY QUESTIONS

REVIEW WHAT YOU HAVE LEARNED

1. What are the special adaptations of a giraffe's circulatory system, and how do they allow the animal to successfully exploit its environment?
2. List the constituents of blood plasma.

(Continued on page 359)

① The circulatory system transports material to and from every part of the body; in higher animals, it is completely enclosed in vessels.

② The human circulatory system has a four-chambered heart that pumps blood under pressure through two separate loops; one travels through the lungs, the other to and from the rest of the body.

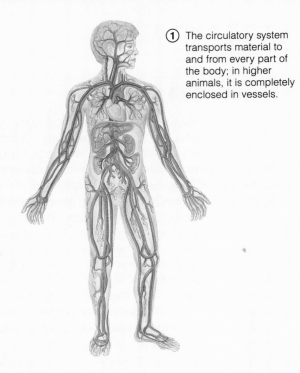

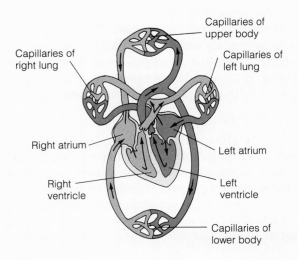

Capillaries of upper body

Capillaries of right lung

Capillaries of left lung

Right atrium

Left atrium

Right ventricle

Left ventricle

Capillaries of lower body

③ Blood is a liquid transport tissue consisting of a large fluid portion as well as red cells, white cells, and platelets.

Red blood cells carry oxygen

White blood cells fight intruders

Fluid moves cells and materials

Platelets help clot blood

Lymphatic capillaries

Lymph node

Blood capillaries

④ Various mechanisms shunt blood where it is needed most, and prevent loss of blood from the circulatory system.

Lymphatic vessels

Arteries

Veins

Heart

Blood capillaries

Lymph node

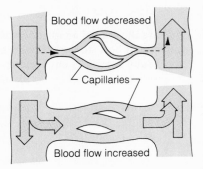

Blood flow decreased

Capillaries

Blood flow increased

One-way valves

Lymphatic capillaries

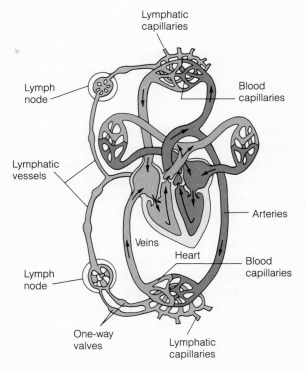

⑤ The lymphatic system returns fluid seeping from blood vessels to large veins near the heart; it includes infection-fighting lymph nodes and a blood-filtering spleen.

3. What are three special characteristics of red blood cells?
4. Describe the negative feedback loop that governs red blood cell replacement.
5. What route does blood follow in the single-loop circulatory system of a sea bass?
6. List the different types of vessels that transport blood on its circuit through a mammal's body.
7. What are the two parts of the heartbeat? How do they differ?
8. What keeps blood from flowing backward in an artery? In a vein?
9. Describe the sequence of events that produces a blood clot at a wound site.
10. Describe the function of lymph nodes.
11. Name two large lymph organs, and give the function of each.

APPLY WHAT YOU HAVE LEARNED

1. Create a chart to record your pulse at different times of the day and during active and quiet periods. What do you conclude about the rate of heart contractions?
2. A nurse takes the blood pressure of a patient who has sat in the waiting room for half an hour and finds that the pressure when the ventricles are relaxed is 90. Later, after a physical exam, the patient's pressure at this phase in the cycle is 75. What could account for the difference?
3. Should a person try to rid his or her body of all cholesterol? Explain.

FOR FURTHER READING

Brown, M. S., and J. L. Goldstein. "How LDL Receptors Influence Cholesterol and Atherosclerosis." *Scientific American* 251 (November 1984): 58–67.

Golde, D. W., and J. C. Gasson. "Hormones That Stimulate the Growth of Blood Cells." *Scientific American* 259 (July 1988): 62–70.

Lillywhite, H. B. "Snakes, Blood Circulation, and Gravity." *Scientific American* (December 1988): 92–98.

Pedley, T. J. "How Giraffes Prevent Edema." *Nature* 329 (1987): 13–14.

Robinson, T. F., S. M. Factor, and E. H. Sonnenblick. "The Heart as a Suction Pump." *Scientific American* 254 (June 1986): 84–91.

Vines, G. "Diet, Drugs, and Heart Disease." *New Scientist* 25 (February 1989): 44–49.

CHAPTER 21
RESPIRATION: GAS EXCHANGE WITH THE ENVIRONMENT

OXYGEN DEPRIVATION AND THE ASCENT OF MT. EVEREST

"I progress so slowly! How long my pauses to breathe are each time I don't know. With the ski sticks I succeed in going 15 paces, then I must rest several minutes. All strength seems to depend on the lungs. . . . And while standing I must use all my willpower to force my lungs to work. Only when they pump regularly does the pain disappear, and I experience something like energy. Now my legs have strength again. I . . . follow the north flank of Mt. Everest." [Reinhold Messner]

World-class European mountain climber Reinhold Messner recorded these sensations and events after his attempt in 1980 to reach the summit of Mt. Everest without the aid of an oxygen tank (Figure 21.1). So thin is the air surrounding the earth's tallest mountain peak (8848 m, or 29,028 ft) that Messner's ascent was both difficult and dangerous. He had to breathe about six times harder than at sea level, and still he faced the constant threat of losing consciousness and falling, as well as the risk of brain damage from lack of oxygen to that sensitive organ. Despite these tremendous odds, Messner did reach the summit, setting a world's record.

As Messner pushed himself upward, what events took place in his body that allowed him to succeed under such extreme conditions? How did his system supply enough oxygen to the cells of his brain and muscles? And why did his body—or yours, or any animal's—need oxygen in the first place? We will address these and other questions in this chapter as we study **respiration,** the process by which organisms exchange gases with the environment.

Most of us have never experienced oxygen deprivation. Nevertheless, we become keenly aware of our need for oxygen while bounding up a few flights of stairs or holding our breath underwater. Like all animals, people must obtain sufficient oxygen for the rapid conversion of sugars to usable energy; they must also dispose of carbon dioxide, a waste product of energy metabolism.

As we discuss gas exchange in animals, two unifying themes will emerge. First, the process of diffusion dictates the architecture of an animal's respiratory system. Your lungs and a fish's gills, although very different in form, both overcome a simple but limiting physical principle: diffusion of gases takes place rapidly only over very short distances. The second theme is that an animal's activities are limited by the amount of oxygen it can obtain. Thus, an animal's respiratory architecture and its way of life are closely entwined.

This chapter explains how animals exchange gases with the environment, and in so doing, it answers several questions about respiration.

- How does diffusion underlie all forms of respiration?
- How do lungs, gills, and other organs of gas exchange compare?
- How does the human respiratory system function?

FIGURE 21.1 *Scaling High Peaks Without Supplementary Oxygen.* At the top of the world, where oxygen is rare, climbers like Reinhold Messner must struggle to breathe in enough of that life-sustaining gas to keep the mind and muscles in gear.

DIFFUSION: UNDERLYING MECHANISM OF GAS EXCHANGE

All animals must steadily take in oxygen from the environment if their cells are to burn carbohydrates aerobically and expel the metabolic by-product carbon dioxide quickly and efficiently (review Figure 5.7). Inside the animal, the oxygen and carbon dioxide enter and leave cells by diffusion, the spontaneous migration of a substance from a region of higher concentration to a region of lower concentration. Two facts about diffusion in living creatures have shaped the evolution of the respiratory structures that bring oxygen into the animal and release gaseous wastes to the environment. First, to enter and leave cells, gases must diffuse through a watery medium. Second, diffusion is much less rapid—an amazing 300,000 times slower—in water than in air. In some animals, the constraints of diffusion have affected the shape and activity of the entire body. The porous nature of the sponge, for example, permeated as it is by channels, allows each body cell to lie only a fraction of a millimeter away from oxygen-bearing water. Even though this setup provides a relatively meager supply of the gas, it is enough to sustain the sponge's low metabolic rate. And as we have seen, the flat bodies of flatworms are so thin that oxygen and carbon dioxide can diffuse directly between cells and the animal's moist surroundings.

Most large animals employ a different, more complex strategy that relies not on direct diffusion from the environment, but on a circulating internal fluid (blood or its equivalent). This fluid picks up and transports oxygen across a thin moist surface—a physical boundary with the environment, such as the thin, moist skin of a frog (Figure 21.2) or the air sacs of our lungs—and distributes it throughout the body. Then it picks up carbon dioxide in the tissues and releases it back into the outside world, usually through the same moist surface. The specialized structures that have evolved to act as exchange surfaces between the body and the environment are gills, tracheae, and lungs.

ORGANS OF GAS EXCHANGE: ADAPTATIONS IN WATER AND AIR

A large tuna, using its gills, can extract enough oxygen from seawater to meet the needs of its powerful swimming muscles, and yet, a tiny mouse could not do this, even if its lungs could function in water. Some insects, such as stag beetles, are as large as mice and yet suc-

FIGURE 21.2 *Skin Folds and Gas Exchange: A Matter of Surface Area.* The Lake Titicaca frog has skin that falls into deep folds. These look strange but provide the animal with a large surface area through which oxygen can enter and carbon dioxide can escape. The surface area is so large, in fact, that this intriguing amphibian can survive without breathing through its lungs at all.

cessfully extract oxygen from the environment using neither gills nor lungs. The explanation for these apparent puzzles lies in the wide variety of breathing structures and mechanisms that have evolved to carry out gas exchange in animals, ranging from the gills of fishes to the tracheae and lungs of land animals.

EXTRACTING OXYGEN FROM WATER: THE EFFICIENT GILL

Fishes, tadpoles, and most other animals that must extract oxygen from water have *gills*: organs specialized for gas exchange that develop as outgrowths of the body surface. Water holds 1/20 or less the amount of oxygen in air, and gills have evolved to exchange gases with great efficiency. They can be external or internal, simple or complex, but each provides a large surface area through which gases from the watery environment can diffuse and enter the organism's bloodstream.

Many amphibians, such as the axolotl in Figure 21.3 (page 362), have frilly external gills that wave about in water currents. The outer layer of the gill is only a single cell thick and lies in contact with the equally thin walls of extremely fine blood vessels. As water moves past the waving gill surfaces, oxygen can diffuse in and carbon dioxide out through the gill membrane and capillary walls.

External gills like these lie dangerously exposed, making them vulnerable to predators and the elements. For this reason, most fishes have evolved more protected internal gills, which are nevertheless surrounded and supported by water to prevent them from collapsing.

and snails, and land vertebrates, such as reptiles, birds, and mammals. Not coincidentally, the arthropods and vertebrates have evolved specialized internal respiratory channels that help bring oxygen close to each body cell and carry away carbon dioxide.

■ **Tracheae: Tubes for Gas Exchange** Insects and certain other arthropods have a system of air tubes, the *tracheae*, that are physically linked to the outside world through holes (*spiracles*) in the sides of the abdomen.

FIGURE 21.3 *The Goal of a Gill: Getting Oxygen into the Blood.* This albino axolotl sports elaborate external gills that have a large surface area for exchanging gases with the environment.

Internal gills are often protected by stiff flaps that are easily visible on the side of a fish's head and that cover the openings, or gill slits, between the gill bars (Figure 21.4a and b). Beneath each flap, the **gill** is subdivided into hundreds of filaments composed of delicate, flexible plates. Embedded within each plate is a lacy meshwork of capillaries lying just one cell layer away from water that passes through the gill filament. The steady opening and closing of a fish's mouth pumps a constant stream of water through the mouth to the gills, over the gills, then out through the protective flaps (Figure 21.4b and c). This proximity of blood to oxygen-bearing water means that oxygen readily diffuses across the cells into capillaries, while carbon dioxide diffuses outward just as easily. The pumping heart then circulates the oxygen-rich blood throughout the animal's body.

In contrast to fishes, land animals have access to the higher oxygen content of air, but they face a different threat: the drying out of their respiratory surfaces, which would bring gas exchange to a lethal halt. Part of the evolutionary solution to this problem has been the moist gas exchange surface of structures such as the branching tracheae or pouchlike lungs located deep within an animal's body.

ADAPTATIONS FOR RESPIRATION IN A DRY ENVIRONMENT: TRACHEAE AND LUNGS

The only animals that live surrounded entirely by dry air are terrestrial arthropods and mollusks, such as insects

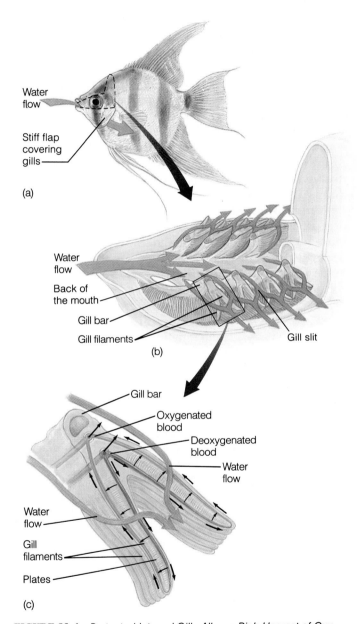

(a)

(b)

(c)

FIGURE 21.4 *Protected Internal Gills Allow a Rich Harvest of Oxygen from Oxygen-Poor Water.* (a) Water moves into a fish's mouth and passes across its gill filaments before exiting through the gill flaps. (b,c) The water current runs counter to the blood current in tiny vessels inside the gill filaments, allowing a maximal amount of oxygen to diffuse into the blood vessels.

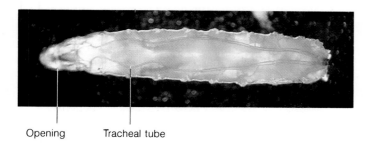

Opening   Tracheal tube

FIGURE 21.5 *Tracheal Tubes: Efficient Gas Exchange in Insects.* Two parallel tracheal tubes are clearly visible in this fruit fly larva. Air entering the ends of these tubes can fan out into the branching air capillaries, diffuse across fluid in the capillary tips, and reach internal cells.

Within the animal, the tracheae branch into ever finer tubes (far narrower than a human hair) that end in *air capillaries*, which allow air from the environment to penetrate deep within the animal's tissues (Figure 21.5). Atmospheric oxygen entering the holes diffuses rapidly through the air in the tracheal tubes until it reaches a minute droplet of fluid pooled at the inner tip of each air capillary; here, the incoming oxygen dissolves in the fluid and diffuses across adjacent cell surfaces. In highly active tissues such as the flight muscles of a honeybee, no cell is more than a few micrometers away from an air capillary.

■ **Lungs: Complex Air Bags** In all air-breathing vertebrates, the lungs are internal pouches connected to the outside by a system of hollow tubes. The lung pouches are enclosed and supported in a fluid-filled sac (analogous to the gill's watery support); the pouch walls are richly endowed with a dense lacework of blood capillaries; and the walls are lined with a thin, moist membrane through which oxygen can diffuse into the bloodstream and carbon dioxide can exit as blood moves through the capillaries. Figure 21.6 compares the lungs in amphibians and birds. In frogs and other amphibians, air flows in and out via a single, two-way path known as **tidal ventilation** (because it is rather like the ebb and flow of tides). In birds, the respiratory arrangement is a bit different. A bird must take two breaths to move air completely through the system of air sacs and lungs; the first breath draws air into the air sacs and the lungs, and the second pushes the first breath into other air sacs and then out of the body. This one-way airflow arrangement is so effective that some birds can flap actively at altitudes like those found at the top of Mt. Everest, apparently with no ill effect. Only a few humans, such as Reinhold Messner, can even stand at rest without an oxygen tank in such an oxygen-starved environment.

Unlike the respiratory structures of birds or fishes, a mammal's lungs do not have the benefit of a unidirectional air or water current. Instead, like an amphibian's lungs, they operate by way of tidal ventilation. This tidal pattern means that fresh air enters the lungs only during half of the respiratory cycle and that a quantity of unexpelled, stale *dead air* filled with carbon dioxide remains in the lungs at all times, mixing with the fresh air that enters from outside. To compensate for tidal ventilation, mammalian lungs possess special adaptations that increase the rate of gas exchange.

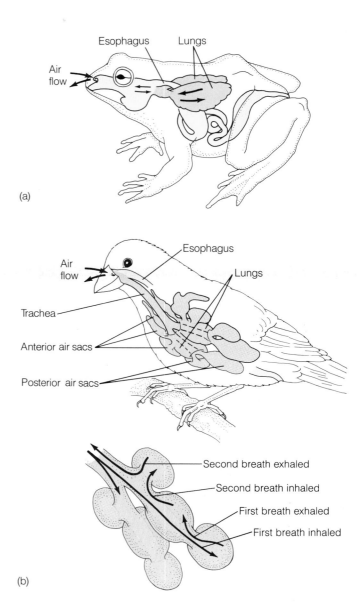

FIGURE 21.6 *Pouches for Air Exchange.* (a) Amphibians have bag-like lungs with relatively little interior surface area for gas exchange. The thin, moist skin—through which gases can readily diffuse—acts as a supplement to the lungs. (b) Birds have a system of air sacs that interconnect with the lungs, lighten the animal, and provide a higher level of oxygen for the bird's actively contracting wing muscles. To move air through the entire network, a bird must take two breaths.

≋ RESPIRATION IN HUMANS AND OTHER MAMMALS

You are a mammal, and by the time you have finished reading this chapter—an hour, let's say—your efficiently operating respiratory system, superbly adapted to exchanging gases in a dry environment, will have drawn in oxygen and expelled carbon dioxide about 800 times. In contrast, a giraffe would have breathed as many as 1200 times to overcome the huge volume of dead air that must be cleared from the animal's very long windpipe before fresh air can enter the lungs, and Messner, at Everest's summit, would have taken about 4000 breaths. Despite such differences, giraffes, humans, and all mammals share a basic set of respiratory structures and mechanisms.

RESPIRATORY PLUMBING: PASSAGEWAYS FOR AIR FLOW

When a mammal breathes in, air enters the respiratory system through the nose and sometimes through the mouth and is warmed and humidified by the moist mouth cavity or by the twin **nasal cavities,** chambers that open into the throat, or *pharynx* (Figure 21.7a). The pharynx branches into a pair of tubes; one, the *esophagus,* leads to the stomach, while the other, the windpipe, or **trachea** (not the tracheae of insects), is the major airway leading into the lungs. At the anterior end of the trachea lies the *larynx,* or "voice box," housing the vocal cords. Just above the opening to the larynx is a flap of tissue called the *epiglottis,* which normally closes off the larynx during swallowing to prevent food from accidentally entering into the lungs.

A few centimeters below the larynx, the trachea branches into two hollow passageways called **bronchi** (singular, bronchus), each of which enters a lung. Finer and finer branchings of these tubes create an "inverted tree," with thousands of narrowed airways, or **bronchioles,** that eventually lead to millions of tiny, bubble-shaped sacs called **alveoli** (singular, alveolus), where gas exchange takes place (Figure 21.7b). Each alveolus is surrounded by blood capillaries, and the inside of each tiny pouch is lined with a moist layer of epithelial cells. Human lungs contain roughly 300 million alveoli, and if the linings of all these delicate bubbles were stretched out simultaneously, they would occupy enough surface area to cover a badminton court or 20 times the body's entire skin surface. At those places where the wall of a capillary lies near the outer wall of an alveolus, oxygen can easily diffuse out of the alveolus and into red blood cells

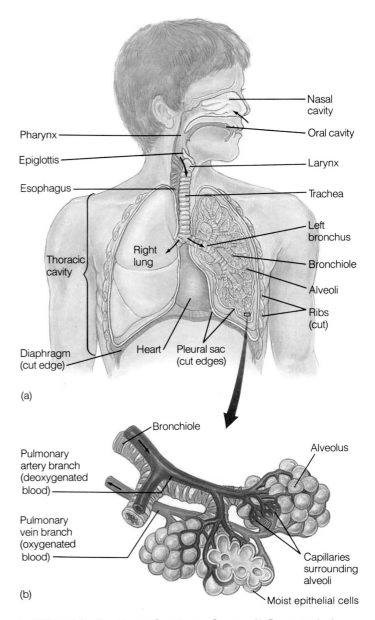

FIGURE 21.7 *The Human Respiratory System: Air Passages in the Mouth, Throat, and Chest.* (a) Anatomy of the airways and lungs. (b) Close-up view of the branched passageways, or bronchioles, and the cluster of tiny air sacs, or alveoli.

squeezing down the center of the narrow capillary. Conversely, carbon dioxide waste can leave the blood, diffuse out of the capillary, enter the alveolus, and be expelled to the outside with the next exhalation.

Many of the cells that line the larger airways produce a sticky mucus ideally suited to capturing inhaled dirt particles or microorganisms. This mucus is continuously cleared from the bronchi by the beating of cilia, tiny hair-like structures on the cell surface that sweep the mucus and any trapped debris up toward the throat, to be swallowed or expelled (Figure 21.8).

Among Caucasians, the most common lethal genetic disease is *cystic fibrosis,* an inherited condition in which the lungs produce large quantities of a heavy, sticky mucus that interferes with gas exchange. Life expectancy is only 30 years in cystic fibrosis victims, but strides are being made in the detection and treatment of the disease.

In people who smoke cigarettes, inhaled tobacco smoke damages the cells lining the lungs and paralyzes the cilia in the airways. The smoke from a single cigarette can immobilize the cilia for hours and lead fairly quickly to a hacking "smoker's cough"—the respiratory system's attempt to rid itself of airborne garbage that accumulates because the cilia no longer sweep it out. Long-term abuse of the airways can lead to near total breakdown of the system; as cilia deteriorate and other natural defenses are overwhelmed, it becomes increasingly likely that genetic changes in lung cells will go unrepaired and that the lungs will develop cancer or emphysema, a degenerative disease in which the alveoli steadily deteriorate. Also, the tars and gases in cigarette smoke damage blood vessel walls (see Figure 21.8b) and increase one's chances of heart attack or stroke. Studies show that four out of every ten smokers will eventually die as a direct result of their habit.

Recent research also suggests that passive smoking—breathing the smoke from someone else's cigarettes—can have serious health consequences. Children raised around smokers are more likely to have respiratory and other medical problems, and nonsmokers who live with smokers have a 35 percent higher risk of developing lung cancer and heart disease.

VENTILATION: MOVING AIR INTO AND OUT OF HEALTHY LUNGS

Healthy lungs look like pink, spongy balloons. These soft, elastic sacs are suspended in the **thoracic cavity,** which is the region within the rib cage directly over the heart. In humans and other mammals, the lungs rest on a domed sheet of muscle, the **diaphragm,** but they don't simply hang there; each is enclosed in a fluid-filled **pleural sac** (review Figure 21.7). The fluid surrounding the lungs is under lower pressure than the air inside the lungs, and this pressure difference enables the lungs to remain slightly expanded, even when no air is being taken in; a "collapsed lung" results when a pleural sac is punctured and the fluid drains away. Breathing in and out, or **ventilation** of the lungs, is possible because the bellowslike activity of the diaphragm and expandable rib cage draws fresh air in and allows stale air to rush back out (Figure 21.9, page 366).

At the start of each inhalation, the muscles that move the ribs contract, and the ribs are lifted and pulled apart slightly. Simultaneously, the diaphragm moves downward toward the stomach, and its "dome" flattens out. As the chest cavity enlarges, the pressure of the fluid in the pleural sacs drops, and with it drops the air pressure in the lungs. As a result, air flows in from outside, moving from an area of higher pressure to one of lower pressure, and moves through the airways to the alveoli, making the lungs expand.

Exhalation, or the passive release of air from the lungs, results when these steps are reversed. The muscles that expanded the chest cavity relax, the ribs are lowered, the diaphragm moves upward once again, and the pressure on the pleural sacs increases. This pressure causes the lungs to deflate, squeezing air from the millions of alveoli out through the bronchioles, bronchi, trachea, and mouth or nose.

■ **Control of Ventilation by Centers in the Brain** Day after day, year after year, the contraction and relaxation of the respiratory muscles continue at just the right pace so that ventilation speeds up or slows down, depending on whether one is sleeping, studying, riding a bicycle, or scaling a mountain peak. What keeps this vital rhythm going at the appropriate pace so that the changing metabolic needs of every cell are always met?

The answer is that respiratory control centers in the brain monitor gas pressure in the blood and help determine when and how we breathe. The muscles that expand

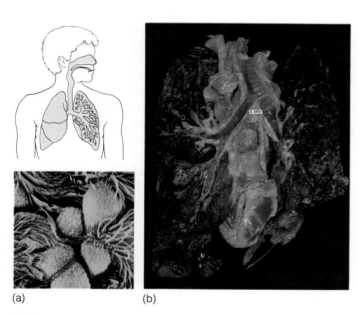

(a)   (b)

FIGURE 21.8 *Bronchial Cilia: Microscopic Custodians of the Respiratory Tract.* (a) Hairlike cilia protrude from the cells that line the trachea and bronchioles and sweep out mucus and debris unless the cilia are paralyzed by cigarette smoke. When that happens, debris can reach and accumulate in the lungs. (b) Tar has blackened and clogged the delicate tissues of a smoker's lungs.

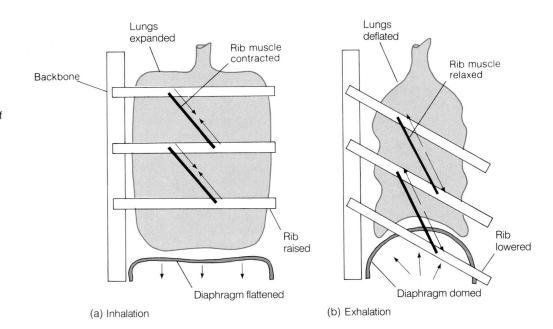

FIGURE 21.9 *Ventilation: How Muscular Movements Expand the Rib Cage and Inflate the Lungs.* (a) During inhalation, the muscles attached to the ribs contract and lift the ribs apart, and the diaphragm flattens. These two actions increase the size of the chest cavity and hence the lung capacity, and this results in a decrease in air pressure within the lungs. Atmospheric pressure then forces air into the lungs. (b) During exhalation, the rib muscles relax, the diaphragm assumes a dome shape, the ribs move closer together, the chest capacity drops, and air is expelled.

(a) Inhalation

(b) Exhalation

the chest cavity, for example, respond to nerve impulses generated by a region of the brain stem. Breathing is involuntary; we cannot consciously prevent these nerves from firing or the chest from expanding (at least not indefinitely). Other nerve cells then act to inhibit respiratory muscle stimulation so that the rib muscles relax and exhalation begins. One exception to this comes from taking an overload of chemical depressants, such as alcohol or barbiturates: These inhibit nerve firing by the brain's respiratory centers and can cause breathing to become irregular—or to cease forever.

Curiously, the human brain is much better at sensing a rise in blood carbon dioxide than it is at sensing a decrease in blood oxygen, and three French physiologists proved this fact in 1875 with a dramatic and disastrous experiment. The three ascended to a high altitude in a hot-air balloon equipped with bags of pure oxygen to be used as they perceived the need. Their balloons passed the 7000 m (24,600 ft) level, still far below Everest's summit, and one of the scientists recorded—in oddly scrawled handwriting—that they were feeling no ill effects. When the balloon finally descended, two of the three were dead. The survivor's sad conclusion: The human body is very poor at sensing its own need for oxygen. A climber like Messner, research shows, has an exaggerated hyperventilation response (very rapid, very deep breathing) that allows him to draw in sufficient oxygen even under extreme conditions.

INTERNAL OXYGEN PICKUP AND DELIVERY

If a person is deprived of oxygen for as little as 10 minutes, brain damage or death is almost certain to occur.

Clearly, ventilation must continue uninterrupted in the manner just described to bring air into the lungs, and diffusion across respiratory membranes must occur without hesitation so that gases can be rapidly picked up in the lungs and transported in the blood in sufficient quantity to service each cell throughout the body. Like ventilation, diffusion is based on gas pressures, or concentrations.

■ **Gas Concentration and Diffusion** When you inhale, the new air entering the lungs has a higher concentration of oxygen than the oxygen-depleted blood returning to the lungs from tissues in arms, legs, and elsewhere (Figure 21.10, steps 1–3). As a result, oxygen diffuses out of the tiny air sacs and into the bloodstream in nearby lung capillaries, where its concentration is low (step 4). It is then carried to distant body regions, and there moves into tissue cells where the oxygen concentration is lowest (step 5). Simultaneously, carbon dioxide moves out of the tissue cells, where its concentration is highest (step 6), across the capillary walls, and into the bloodstream (where the concentration is lower); it is then carried back to the lungs and there moves into the alveoli, where the carbon dioxide concentration is lowest (step 7). From the lungs, exhalation can carry the waste gas out of the body (step 8).

At the top of Mt. Everest, the concentration of oxygen in the air is less than one-third its concentration at sea level, and as a result, it is hard for a person to obtain enough oxygen for the body's respiratory needs. Reinhold Messner barely made it back from the summit to his camp (altitude 6500 m, or 21,125 ft) before collapsing from oxygen deprivation; and he might never have returned to the foot of Everest without the help of Nena Holguin, the American woman who waited alone at the

1. OBSERVATION: It says here that when people get used to high altitudes, they hyperventilate, that is, they breathe more rapidly and more deeply.

2. QUESTION: I wonder if there is a relationship between hyperventilation and success in mountain climbing?

3. HYPOTHESIS: My hunch is that people who naturally hyperventilate can better tolerate low oxygen levels.

4. EXPERIMENT: To find out, we could get a group of eight people to climb Mt. Everest, but before going, we should measure each person at sea level to see how fast and how deeply they breathe air with a reduced oxygen content.

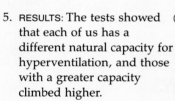

Volume of air with reduced oxygen content breathed per minute at sea level

5. RESULTS: The tests showed that each of us has a different natural capacity for hyperventilation, and those with a greater capacity climbed higher.

6. CONCLUSION: We could probably use this test to predict a person's success at tolerating high altitudes and therefore at safely conquering tall peaks.

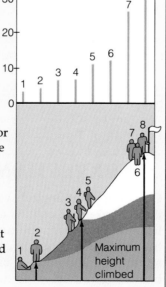

Maximum height climbed

Adapted from: J. B. West, *Science* 223 (1984): 784–788, and J. B. West, *Canadian J. Physiol. Pharmacol.* 67 (1989): 173–178.

camp, revived him, and helped him descend the mountain. Severe oxygen deprivation also has long-term effects. Members of a 1981 research expedition to Everest had trouble remembering telephone numbers several months after their climb and were still defective in some manual skills up to a year later (see the box at the left).

Even at sea level, where the concentration of oxygen in air is sufficient to drive diffusion of the gas into the blood at a rapid rate, the provision of this oxygen cannot by itself meet the respiratory needs of active animals. If your vessels were filled with tap water, that fluid simply could not take up enough oxygen to supply the needs

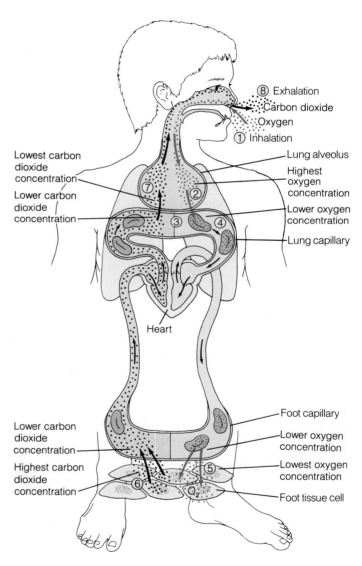

FIGURE 21.10 *The Diffusion of Gases in the Body.* Oxygen entering the respiratory system diffuses through the delicate walls of the lung and its capillaries and is carried in the bloodstream to distant body tissues (here, the foot). Carbon dioxide moves in the opposite direction.

of your active cells as you swam laps or jogged around a track. Evolution's answer to this problem is hemoglobin, an iron-containing blood component specialized to take on large amounts of oxygen quickly and to release it readily.

■ **Hemoglobin Enhances Oxygen Transport** In each cubic millimeter of blood, there are millions of red blood cells, each filled with millions of molecules of the oxygen-binding pigment hemoglobin. Multiplying this number by the 25 trillion red cells in an average adult's bloodstream, one discovers that there are twice as many hemoglobin molecules in a person's body as there are stars in the Milky Way galaxy! Hemoglobin's specialized structure enables it to act like an oxygen sponge and allows blood to soak up 70 times more oxygen than an equivalent quantity of water could absorb. In a mammalian lung at sea level, the concentration of oxygen is high enough to ensure that all the hemoglobin molecules take on a full load of oxygen. In oxygen-poor environments, such as those found at high altitudes, hemoglobin loads oxygen even more readily than at sea level, because it grips it more tightly. Oxygen loading changes hemoglobin to a brilliant red and explains the bright color of blood as it leave the lungs. In contrast, blood returning to the lungs, which carries hemoglobin depleted of oxygen, has a darker, almost bluish cast and causes veins, like those in the wrist, to look blue (see Figure 21.10).

The oxygen-saturated hemoglobin inside the red blood cells leaving the lungs is circulated to the muscles, fat, skin, and other body tissues, where it encounters low oxygen concentrations, since the active tissue cells steadily use up the available oxygen. In such an environment, the hemoglobin molecules begin to give up their oxygen, the liberated oxygen dissolves in the fluid portion of the blood, and then it diffuses into the tissue cells. Curiously, the short-legged, long-haired mammals called yaks, native beasts of burden in the Himalayas of Tibet and Nepal, have blood that contains surprisingly little hemoglobin—in fact, 25 percent less than a person's blood at sea level (Figure 21.11). But yaks have a different adaptation—three times as many red blood cells per cubic centimeter of blood as humans. Some observers speculate that the smaller cells, with their higher surface-to-volume ratio, can pick up oxygen quite readily. They also suspect that yak hemoglobin itself may have a special affinity for oxygen, much as the hemoglobin of the human fetus can pick up the gas with particularly high efficiency.

MOVING CARBON DIOXIDE OUT OF THE BODY

A second crucial part of the respiration process is releasing carbon dioxide waste to the outside environment, and once again, the bloodstream plays the major transport role. In tissue cells, carbon dioxide levels are high, since the cells have been metabolizing and giving it off; thus, the gas tends to diffuse out of the cells and into the bloodstream, where the carbon dioxide concentrations are lower. Once in the blood, the carbon dioxide dissolves in the plasma, combines with deoxygenated hemoglobin, or combines with water to form bicarbonate ions; in all of these forms, it is transported to the lungs, where it can be exhaled.

The beauty of the way hemoglobin works is clear: Hemoglobin picks up oxygen in the lungs and releases it effectively under conditions found in working muscles and other body tissues. Hemoglobin also helps transport carbon dioxide back to the lungs for release from the body.

FIGURE 21.11 *The Yak: Extra Red Blood Cells Help at High Elevation.* Yaks, the short-legged beasts of burden in the Himalayas of Nepal and Tibet, have three times more red blood cells than a person at sea level, but the yak's cells are small and contain less hemoglobin. Reinhold Messner used yaks to carry life-support equipment and food to his camp, located 6500 m (21,125 ft) above sea level.

≈ CONNECTIONS

Of all the physiological systems, respiration is perhaps the best example of how an animal's anatomical structures and functions are tuned to the environment and the physical principles of nature. The respiratory exchange surface may be folds of skin, filamentous gills, finely branching tracheal tubes, lung sacs, or bubblelike alveoli. Regardless, in each case, the architecture allows gases to diffuse with maximum efficiency from the environment, through moist cell membranes, into the animal's cells, and back out again. Your own elaborate respiratory

① Respiration, the exchange of gases with the environment, depends on diffusion across a moist membrane, and the requirements of diffusion helped shape the evolution of respiratory structures.

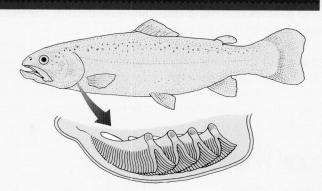

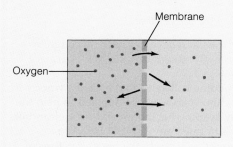

Oxygen

Membrane

Diffusion across a membrane

② Most animals that live in water have gills; a large surface area allows efficient gas exchange.

③ Land animals have a system of tubes (tracheae) or of tubes and pouches (airways and lungs) that allows respiration.

Airway

Lungs

Tracheae

Fruit fly larva

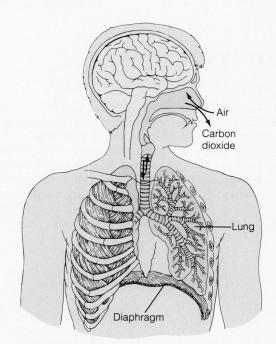

Air

Carbon dioxide

Lung

Diaphragm

④ The human brain controls ventilation, during which oxygen-containing air is drawn through branching tubes into the lungs and carbon dioxide is expelled.

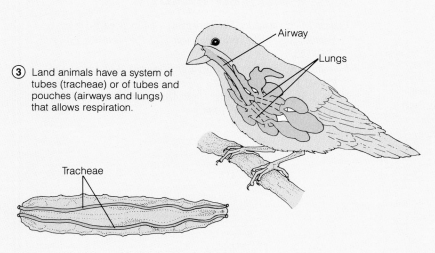

Oxygen concentration

Air in lungs

Red blood cells

Tissues

Carbon dioxide concentration

⑤ Efficient oxygen pickup in the lungs and delivery to the tissues depend on the gas's concentration in air and blood as well as on the oxygen-carrying capabilities of hemoglobin.

⑥ The removal of carbon dioxide depends on diffusion from high concentrations in the tissues to low concentrations in the lungs.

structures, those of Reinhold Messner, and even those of a yak are, at last analysis, merely a set of transit pipes and bellows that bring sufficient quantities of gas in and out of the multicellular organism.

The respiratory system is not the only body system that relies intimately on a continual circulation of blood. As we will see in the next chapter, the immune system, which protects our bodies from invasion, is also closely tied to the moving internal sea of oxygen-carrying blood.

NEW TERMS

alveolus, page 364
bronchiole, page 364
bronchus, page 364
diaphragm, page 365
gill, page 362
nasal cavity, page 364

pleural sac, page 365
respiration, page 360
thoracic cavity, page 365
tidal ventilation, page 363
trachea, page 364
ventilation, page 365

STUDY QUESTIONS

REVIEW WHAT YOU HAVE LEARNED

1. How do gills maximize oxygen intake?
2. What path does oxygen follow from the outside environment to an insect's individual cells?
3. What makes a bird's respiratory system highly efficient?
4. Place the following human respiratory structures in their proper order to show oxygen's path from the outside world to the blood: bronchioles, larynx, nasal cavities, bronchi, trachea, alveoli, capillaries.
5. List the effects of smoking on the body.

6. Do your lungs regulate your breathing rate? Explain.
7. Describe how hemoglobin enhances oxygen transport.
8. How is the yak adapted to surviving in the "thin" air of the Himalayan Mountains?
9. Trace the path of a carbon dioxide molecule from a blood capillary in your little toe to the air about to be exhaled from your lungs.

APPLY WHAT YOU HAVE LEARNED

1. A heavy smoker tries to relieve her hacking cough by buying "extra-strength" cough drops at the pharmacy. Why will her strategy fail?
2. A condemned building collapsed and buried a worker in debris up to his chin. A co-worker struggled to help the victim breathe by keeping his mouth and nose clear, but the buried man suffocated anyway. Why?
3. As a research project, you study the breathing rates of students sitting in a crowded, unventilated classroom for an hour. What is your study likely to reveal, and why?

FOR FURTHER READING

Feder, M. E., and W. W. Burggren. "Skin Breathing in Vertebrates." *Scientific American* 253 (November 1985): 126–143.

Messner, R. *The Crystal Horizon.* England: The Crowood Press, 1989.

West, J. B., and S. Lahiri. *High Altitude and Man.* Baltimore: Williams & Wilkins, 1984.

THE IMMUNE SYSTEM: DEFENSE FROM DISEASE

CHAPTER 22

AIDS, THE NEW PLAGUE

Baby Anna was, to all appearances, a normal infant, and by the end of her first year of life, she could speak a few words and stand on her own. By the time she turned two, however, she had contracted several severe respiratory infections and had regressed mentally and physically to the level of a seven-month-old. Although Anna recovered from the infections, she was prone to others and never regained her normal pattern of growth and development. Her parents later learned she was suffering from AIDS, acquired immune deficiency syndrome.

AIDS is the most devastating of the sexually transmitted diseases described in Chapter 13. AIDS is contracted mainly by young adults and is caused by a virus that passes from person to person in blood, blood products, and semen. The virus can be transmitted in the unsterilized needles of intravenous drug users and in contaminated blood transfusions, as well as through sexual intercourse without the safe sex practices described in the box on page 378. If a woman infected with the AIDS virus becomes pregnant, she can pass the virus to an unborn child like baby Anna.

When active, the AIDS virus attacks the **immune system**—the network of organs, cells, and molecules that defends the body from invaders of all kinds (Figure 22.1). Once a person's immune defenses are gone, he or she is vulnerable to certain cancers as well as to most of the viruses, bacteria, and parasites found in the environment. AIDS also leads to the destruction of brain cells and, as a result, to impaired thinking, speaking, and performance.

AIDS is the most threatening epidemic of the twentieth century and shows how crucial the immune system is to our day-to-day survival. The interacting cells and molecules of the immune system circulate in the blood and lymph or lodge in organs such as the skin and spleen. As they circulate and congregate, they carry out three essential activities, which provide the recurring themes of this chapter. First, the immune system *recognizes* foreign invaders, that is, all molecules and cells originating outside the body. Second, cells of the immune system *communicate* with each other. Finally, the immune system *eliminates* foreign invaders and thus defeats the attack.

Vertebrates have evolved finely tuned immune responses to an almost infinite number of health hazards. In our brief tour of the immune system we will draw most of our examples from the human immune system, but the principles we describe apply to all vertebrates. This chapter will provide answers to the following questions:

- How does the immune system promote lifelong protection against disease?
- How do physicians use vaccinations to enhance immune protection?
- How does AIDS destroy a healthy immune system?
- How do arthritis and pregnancy relate to immunity?

FIGURE 22.1 *T Cell Infected by AIDS Virus.* A T cell, part of the body's protective armory of immune cells, normally helps fight off infections by viruses. But the AIDS virus (shown here as particles emerging from a stricken T cell) attaches and incapacitates those immune cells. When that happens, the whole immune system suffers a breakdown. People infected by the AIDS virus die of diseases such as pneumonia, a certain skin cancer, and blood poisoning, which a healthy immune system could have resisted.

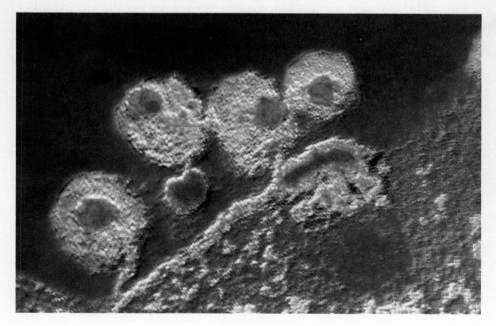

THE BODY'S DEFENSES: MAJOR PLAYERS, MAJOR ACTIVITIES

THE FIRST LINES OF PROTECTION: SKIN AND INFLAMMATION

A slip of your paring knife while peeling potatoes can open a chink in your skin—the body's first line of defense against unwanted outsiders. The skin forms a protective barrier that, along with mucous membranes of the nose, throat, and digestive passages, keeps out most minute organisms capable of multiplying within the body and causing infection. A break in this smooth, elastic barrier provides an opening through which dirt and bacteria clinging to the knife's blade can enter the bloodstream.

Eventually, a blood clot will seal the gap and scar tissue will form, but even before that, skin and blood cells release chemicals that raise the temperature around the cut, and blood flow to the area increases. Shortly thereafter, scavenging white blood cells known as *phagocytes* are attracted to the area in search of invading microorganisms (Figure 22.2). The rise in temperature slows bacterial multiplication; the increased flow of blood cleanses the wound and brings in more white blood cells; and the phagocytes

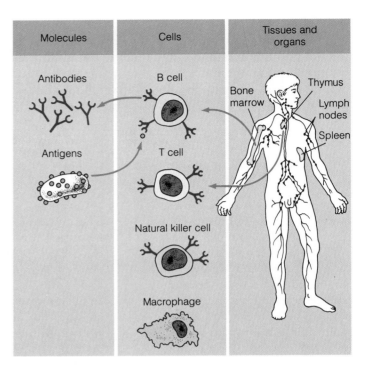

FIGURE 22.3 *Players in the Immune Response.* The functioning of the immune system is based on certain molecules, cells, and organs. The molecules include *antigens*, which stimulate an immune response, and *antibodies*, or specific immune proteins that bind to antigens. Immune system cells include *B cells*, which, after stimulation by antigens, divide and differentiate into cells that produce antibodies. In contrast, *T cells* are special white blood cells that mature in the thymus; some T cells stimulate or inhibit the action of other immune system cells. *Macrophages* are large white blood cells (a type of specialized phagocyte) that devour foreign substances, process them, and present them to other immune system cells. *Natural killer cells* can eliminate cells infected with a virus or parasite; they can also eliminate cancer cells. Immune system tissues and organs include the *bone marrow*, which produces the precursors of all four kinds of white blood cells shown here; the *thymus*, where T cells mature; and the *lymph nodes* and *spleen*, both being sites where immune system cells congregate.

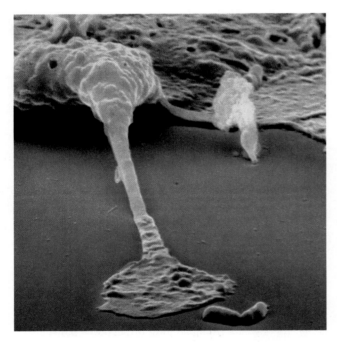

FIGURE 22.2 *White Blood Cell Stalking Bacteria: Nonspecific Internal Defenses Guard Against Invaders.* As this phagocyte (white cell) approaches the small, rodlike bacteria (foreground, tinted green), a mobile cellular extension (pseudopodium) of the macrophage reaches out to capture the invader. Both cells appear about 5000 times larger than life size in this scanning electron micrograph.

ease their way through the walls of fine blood capillaries and into tissue spaces, where they engulf bacteria and cell scraps without disturbing any of the body's own intact cells. These physiological reactions, which result in the redness, heat, and swelling we know as *inflammation*, clear the cut of debris and often keep the bacteria from spreading. You can feel inflammation in many types of infection—fungal, parasitic, bacterial, and viral—because, like the skin and mucous membranes, the inflammatory response is *nonspecific*: It is deployed against all types of microorganisms.

THE MOST POWERFUL DEFENSE: A SPECIFIC IMMUNE RESPONSE

Suppose that the bacteria that enter the wound in your finger find ample nutrients and begin to multiply faster than inflammation and scavenging phagocytes can kill them. The body then falls back on its most powerful line of defense, an all-out **specific immune response.** In this response, white blood cells incapacitate a specific invader (in this case, the bacteria in the cut) while leaving the nontargeted cells of the body untouched. Behind this specific response lies the entire arsenal of the immune system: its several kinds of white blood cells, the proteins (such as antibodies) they secrete, and the organs and fluids in which they multiply, mature, and operate. Let's survey the principal players (depicted in Figure 22.3), then see how they interact in the defense of a cut.

■ **The Cells and the Organs Behind the Immune Response** Of the several types of white blood cells that cooperate to generate immunity, the **lymphocytes** are the driving force. There are two main types of lymphocytes: **B cells,** which make and secrete **antibodies,** special proteins that coat invaders and mark them for destruction, and **T cells,** which kill foreign cells directly and help regulate the activities of the other lymphocytes. **Natural killer cells** are a third type of lymphocyte. They help guard the body against cancer cells and are probably the frontline attackers against virus-infected cells. **Macrophages** are another class of white blood cells involved in specific immunity. Macrophages (literally, "big eaters") are large, specialized phagocytes that stimulate lymphocytes to attack invaders, then help clean up by consuming debris when the lymphocytes are through.

The white blood cells of the immune system constantly circulate in the bloodstream and lymph. Like red blood cells, they arise from stem cells in the bone marrow. Unlike red blood cells, however, they mature to active immune system cells inside of the *spleen, thymus,* and *lymph nodes.* The nodes are weblike organs that trap viruses, bacteria, and other foreign particles and harbor macrophages as well as lymphocytes ready to move into action. The nodes and other *lymphoid organs*—so called because they are connected by the lymphatic system as well as by the bloodstream (see Figure 22.3)—make up a comprehensive dragnet that provides immune protection to virtually every tissue of the body.

■ **Molecules Involved in Defense** No survey of the principal players in a specific immune response is complete without mention of antibodies. Antibodies are the protective proteins made by B cells, which secrete them into the blood and lymph. Different B cells make antibodies of slightly different shape. The key to the amazing effectiveness with which antibodies guard the body and help destroy its enemies lies in their two-tiered, Y-shaped construction, with the arms of the Y geared for recognizing invaders and the stem of the Y geared for eliminating them (Figure 22.4). During an immune response, the recognition part of each Y-shaped antibody, called a *binding site,* attaches to a projecting portion of an infecting bacterium or other foreign agent. Those projecting parts are examples of **antigens**, which are defined simply

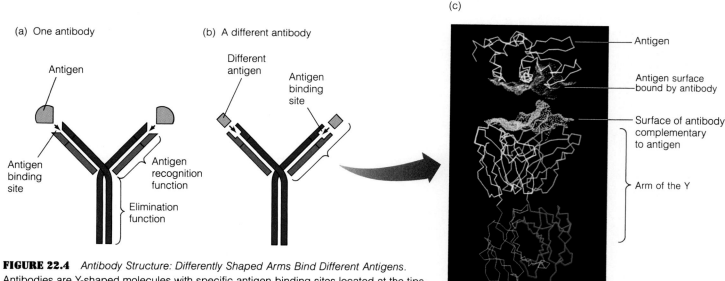

FIGURE 22.4 *Antibody Structure: Differently Shaped Arms Bind Different Antigens.* Antibodies are Y-shaped molecules with specific antigen-binding sites located at the tips of the Y's arms. Antigen molecules, wedge-shaped in (a), rectangular in (b), fit into the binding sites of antibody molecules like keys in locks. The elimination of an antigen, once it is bound to a receptor of reciprocal shape, depends on the stem of the Y. (c) A computer graphic reveals how an antigen—in this case, a small protein (top, in white)—binds to an antibody (here, just one arm of the Y is shown in dark and light blue, yellow, and red). The pink crumpled segment depicts the surface of the antigen, and the green crumpled segment shows the complementary surface of the antibody. These are the elements that fit together like lock and key.

as any molecules that cause antibody production. It is the binding of antibody and antigen that generates an immune response, including the production of more antibodies, molecules our bodies produce to fight the foreign invaders. Because antibodies bind to antigens of reciprocal shape, antibodies and antigens are often likened to locks and keys. The outer coats of most invading viruses, bacteria, fungal cells, and larger parasites contain many different antigens. Most adult humans have antibodies with a total of more than 100 million distinct antigen-binding sites. These recognize and combine with 100 million different antigens, including the outer coats of thousands of different viruses, bacteria, and molds.

The ability of antibodies to recognize and respond to millions of specific invaders enables the healthy immune system to promote resistance against measles, mumps, recurring colds, and many other infections. The diversity and specificity of antibody shape and activity also give physicians a powerful diagnostic tool and a potential therapy against sexually transmitted diseases and cancer. For example, laboratory technicians can now isolate single B cells able to make antibodies to particular antigens; the antibodies produced by a particular B cell's daughter cells are called *monoclonal antibodies* because they come from one clone, that is, from the genetically iden-

FIGURE 22.5 *Home Pregnancy Tests Are Based on Monoclonal Antibodies.* This woman is using such a test to determine whether a missed menstrual period is due to pregnancy. Antibodies in the test solution bind to a pregnancy hormone (if present) in human urine and cause an easily detected color change.

tical progeny of one cell. Physicians can use monoclonal antibodies to detect the AIDS virus in a person's blood. And that information, although it does not indicate if and when the person will develop full-blown AIDS, can be used to identify blood from exposed people and prevent transmission of the AIDS virus in blood transfusions. There are many other uses for monoclonal antibodies. Home test kits for detecting pregnancy contain monoclonal antibodies that bind to a hormone produced during pregnancy (Figure 22.5), and another laboratory test kit includes monoclonal antibodies that detect a protein present in high levels in men with prostate cancer. Doctors also use monoclonal antibodies to help fight tumor cells as well as some forms of venereal disease.

The stems of antibody molecules are not as varied as the antigen-binding sites in the arms of the Y-shaped molecules. However, slight variations in stem shape result in different classes of antibodies, able to go different places in the body and carry out different elimination activities. One class, for example, is called *IgG*. It circulates in the blood and is very efficient at eliminating viruses and bacteria. IgG antibodies are the only ones able to cross the placenta from a pregnant woman to her unborn baby (Figure 22.6a). They are also the main component of *gamma globulin*, which is prepared from blood fractions that are pooled. Gamma globulin is often injected for immediate protection into a patient who has just had surgery or into a traveler bound for a country with contaminated food or water. Another class of antibodies with a different stem shape binds to specialized cells in the skin and nasal and intestinal linings. Besides helping to fight parasites, this kind of antibody triggers the irritating condition we call allergy (Figure 22.6b and the box on page 376). A third class of antibody is secreted in saliva, tears, and milk and can help a breast-feeding infant resist infection.

■ **A Specific Response: Immune Players in Action** With this cast of cellular and molecular characters in mind, let us return to our example of the cut finger and follow the steps involved in a specific immune response. As we saw earlier, once the skin is broken, the nonspecific defense responses battle the bacteria that have invaded the wound, and the battle creates debris (broken bacterial cells and cell parts; Figure 22.7, step 1). This debris begins to accumulate near the wound and enters the bloodstream and lymphatic system (step 2). There, the debris encounters B cells (step 3), and antibodies anchored in the membrane of some of these circulating B cells recognize antigens on the bacterial debris. The antigens bind to antibodies on the surface of the B lymphocytes, and this act of recognition generates a specific response.

B cells carrying antibody-antigen complexes move to the lymph node nearest the wound—in this case, probably in the elbow—and settle in (step 4). Here, helper T

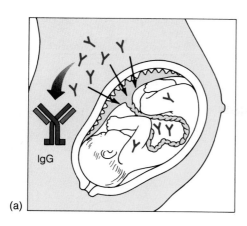

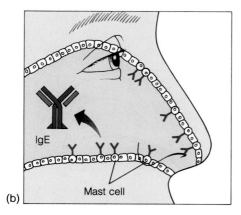

FIGURE 22.6 *Different Antibody Classes Have Different Destinations.* (a) IgG antibodies have a specially shaped stem that enables them to cross the placenta and protect the fetus from infection. (b) IgE molecules, with a different stem shape, bind to cells responsible for allergies in the skin and mucous membranes lining the nose, throat, airways, and intestines, as the box on page 376 describes.

cells (which have already received communication from macrophages about the bacterial antigens) communicate with the B cells (step 5) and stimulate them to divide into 2, 4, 8, 16, and eventually hundreds of daughter B cells that are, in fact, antibody-secreting factories (step 6). Every second, each of these cell factories secretes into the lymph fluid 1000 antibody molecules targeted to bind to the bacterial antigens (step 7).

The secreted antibodies now leave the lymph node, and when the lymph drains back into the bloodstream near the heart (step 8), the antibodies enter the blood and circle back through the heart to capillaries near the cut (step 9). Here, they pass into spaces between the cells and combine with their targets—the bacterial antigens (step 10). When millions of antibodies bind to bacterial antigens, they neutralize the bacteria by causing them to clump together. Such clumped bacteria covered with antibody can be eliminated from the body in two ways: The antibody coating can attract macrophages that devour, digest, and eliminate the antibody-bacteria complexes (step 11), or special molecules called *complement* proteins can

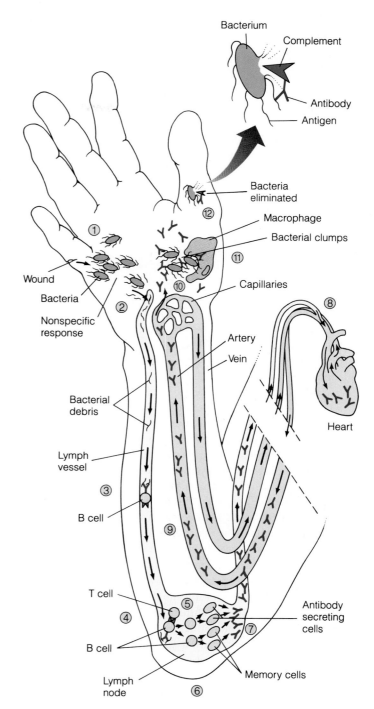

FIGURE 22.7 *How the Immune System Fights a Bacterial Infection: Events of a Specific Immune Response.* As the text explains, a finger wound touches off a multistep immune response, beginning with inflammation and other aspects of a nonspecific response (step 1) and proceeding through the recognition of antigens on the bacterial invader and formation of antibodies with reciprocal shape (steps 2–7), the clumping of antibodies and bacteria (steps 8–10), and the elimination of killed invaders (steps 11 and 12). In this figure, the bloodstream is pink, lymph vessels are yellow, bacteria are green, and antibodies are purple.

ANATOMY OF AN ALLERGY

Does pollen from grass, trees, or ragweed make you sneeze? If so, you may be one of the 35 million Americans with allergies—extreme immune reactions to foreign substances, or *allergens*. Allergens can range from pollen and house dust to pet dander, bee venom, milk, molds, and mites. Researchers are not completely certain how these substances can set off sneezing bouts, but they are studying the basic mechanisms of allergy in search of more successful treatments.

A person's outer barrier—the skin; mucous membranes of the eyes, nose, and throat; and linings of respiratory passages, lungs, and intestines—all contain *mast cells*, specialized cells packed with 1000 or more large, globular granules (Figure 1). These granules account for the runny nose, watery eyes, and sneezing of hay fever; the diarrhea and stomach cramps of food allergies; and the wheezing of asthma. Allergens somehow trigger mast cells to disgorge their granules. These contain histamines and other chemicals, and these agents cause the capillaries to leak fluid into surrounding tissues. This in turn can lead to swelling and redness and cause constricted airways, mucous secretion, lack of intestinal absorption, and other dysfunctions.

When an allergic individual first encounters an allergen, say, a noseful of ragweed pollen, he or she will generate millions of the IgE class of antibodies for that antigen. These antibodies will move through the bloodstream and attach to the surface of each mast cell (see Figure 22.6b). The next time the allergic individual encounters allergen molecules, the allergens will link to the antibodies on the mast cells and trigger those cells to explosively release their granules and the histamines and other substances they store.

Physicians often treat allergy patients with a series of desensitization shots to

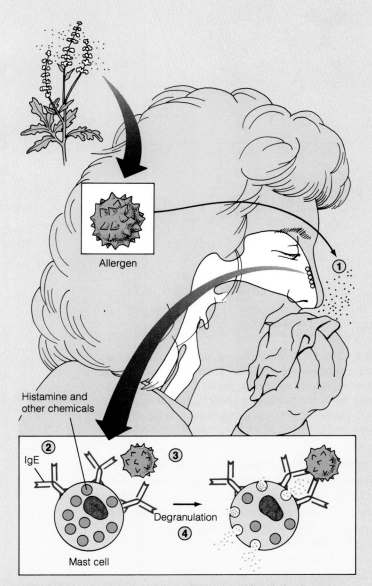

encourage antibody formation to allergens from grass, pine, or goldenrod; house dust; milk, shellfish, or other food; or some combination of these. Later, when the patient encounters naturally occurring allergens, the preformed antibodies can bind to the allergen molecules before they trigger an allergic reaction.

While mast cell granules spew out

histamines, they also release compounds called leukotrienes, which are 100 times more powerful in causing allergy symptoms. Antileukotriene drugs are now under development and may work alongside standard antihistamine drugs and other new treatments to someday stop allergy miseries once and for all.

bind to the antibody-bacteria complexes and perforate the bacterial membranes, killing the bacterial cells (step 12). There is a lag time between the start of an infection and a fully active immune response. This is because it takes four to ten days for a few B cells to divide and redivide into thousands or millions of antibody factories that can in turn secrete sufficient numbers of antibodies to stem the tide of infection. Once those antibodies are racing through the body, however, the bacteria that have grown in the cut are killed, and the infection is well on its way to being cured.

≋ VACCINATIONS: HARNESSING THE IMMUNE SYSTEM FOR ENHANCED PROTECTION

In the last 200 years, medical practitioners have learned to manipulate the immune system to promote health and prevent disease. They can quickly neutralize snake venom by injecting into a snakebite victim antibodies produced by and gathered from another animal. And doctors can prevent diseases for which there is no cure by activating a person's own immune responses with vaccines.

PASSIVE IMMUNITY: SHORT-TERM PROTECTION BY BORROWED ANTIBODIES

If you were bitten by a rattlesnake, there would be no time to spare. The venom contains toxic proteins that stop nerves from functioning, and you would need a quick-acting remedy to keep the toxin from damaging your nervous system. The best treatment would be *passive immunization:* an injection of specific antibodies against the toxin—antibodies that your own cells did not produce (Figure 22.8). Technicians prepare the commercial product by injecting a horse or rabbit with inactivated snake venom and inducing the mammal to form antibodies. They then collect these antibodies, now called antivenin, which a doctor can inject shortly after a snakebite occurs. The antivenin antibodies circulate through your bloodstream and bind to the snake venom in a typical antigen-antibody fashion before the venom can reach your nerve cells.

Passive immunization has one great advantage: It works very fast. But it also has the disadvantage of acting for only a short time. Your immune system would soon recognize as foreign and eliminate the borrowed antibody molecules. With them would go the passive protection, leaving you vulnerable to snake venom if bitten again. One day, passive immunization with antibodies that neutralize circulating AIDS virus may be part of the treatment for keeping a virus-infected person from devel-

oping full-blown AIDS. Since we do not yet have effective ways to provide passive immunization against the AIDS virus, it is a matter of life and health to observe the safe sex practices outlined in the box on page 378. Those are the only prevention techniques now available.

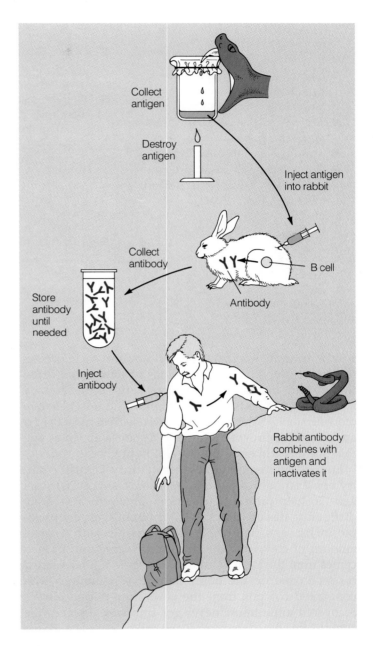

FIGURE 22.8 *Passive Immunization: Transferred Antibodies.* As the text explains, technicians prepare commercial snakebite medicine (so-called antivenin) by collecting venom, injecting a rabbit with a harmless altered form of the venom, then collecting the antibodies it produces. If a snakebite victim receives an injection of these antibodies, they will bind to and inactivate the toxic venom molecules, rendering them harmless.

PROTECT YOURSELF FROM AIDS

Risk Factors

- Sex with an individual who has had a positive AIDS test, indicating exposure to the virus
- Intravenous drug use
- Injections of any kind with used hypodermic needles and syringes
- Multiple sex partners
- Sex with anyone who has had multiple sex partners
- Transfusions with tainted blood
- Sex with an intravenous drug user or with anyone engaging in certain other high-risk activities

- Sex with any individual whose sexual history and exposure status (positive or negative) are unknown to you
- Sex that involves contact with blood, such as sex during menstruation or anal intercourse
- Other activities that involve contact with blood, including some types of work in medicine, dentistry, or undertaking

Protection

- Find out your own exposure status and that of any potential sexual partner.
- Avoid high-risk sexual activities.

- Seek help for addiction, and use only sterile needles and syringes for any necessary injection.
- Use condoms and, if possible, preparations that contain a virus-fighting compound such as nonoxynol-9.
- Avoid unnecessary contact with blood, and accept only transfusions that test free of the AIDS virus.
- Stay informed of new developments in the fight against AIDS. Seek additional information from your college or university health service, your local public health service, or your private physician.

ACTIVE IMMUNITY: LONG-TERM PROTECTION BY ALTERED ANTIGEN

Unlike an antivenin that provides the short-term protection of passive immunity, vaccines against polio, diphtheria, and measles provide long-term protection by stimulating the body's own immune system to generate *active immunity.* Safer and more effective than most drugs, such vaccines provoke a specific response aimed at one microbe or toxin and nothing else in the body.

Edward Jenner, an eighteenth-century English country physician, developed the first vaccine almost 200 years ago. To help protect people from smallpox, a severely disfiguring disease that killed one out of every four people in Jenner's time, he injected people with a small amount of cowpox virus (the word *vaccination*, in fact, comes from the Latin word *vacca* for "cow"). Jenner based his technique on the observation that milkmaids who contracted cowpox from the cows they milked always recovered and almost never got smallpox. Indeed, the cowpox virus he injected caused mild sickness and discomfort, but it nonetheless served as effective protection against smallpox. By 1978, modern vaccines based on Jenner's original ones had successfully eradicated smallpox worldwide.

Modern vaccines are made of microbes and toxins that laboratory technicians have killed or otherwise modified so that the invaders cannot themselves cause disease. In the first shot of a series of vaccines, antigens on the surfaces of the killed microbes bind to appropriately shaped antibodies on the surface membranes of a few B cells already circulating in the body, as in Figure 22.7, step 3. These B cells divide rapidly (steps 4–6), and their progeny cells are of two types: short-lived antibody-secreting factories and long-lived *memory cells* (step 7). The antibodies secreted into the bloodstream confer a small amount of immediate protection. The memory cells are similar to the original B cells and carry the same-shaped antibodies on their surface; however, now there are many of these cells instead of just a few. A booster shot, which is simply a second dose of the same vaccine, then stimulates these many memory cells to divide and produce even more antibody-secreting cells and more memory cells. When a vaccinated person later comes in contact with live bacteria, virus, or toxin, his or her body will already contain many antibodies and a large number of specific B cells and T cells that can quickly eliminate the dangerous agents and prevent disease. Vaccination is a slow-acting process, requiring a booster shot and several weeks for the development of memory cells and adequate protection. Nevertheless, the active immunity it stimulates lasts a long time, sometimes a lifetime.

New vaccines are the best hope for combating infectious diseases that cannot be treated or cured by other techniques of modern medicine. Unfortunately, many microbes have a shape or life-style that enables them to evade active immunity. The flu virus, for instance, changes its shape so readily that this year's flu shot may not protect against next year's virus. In the two centuries since Jenner's first vaccine, fewer than 20 safe, effective vaccines have been developed.

But investigators are diligently working on vaccines against the AIDS virus and all other viruses.

THE HEALTHY IMMUNE SYSTEM AND HOW AIDS DESTROYS IT

B cells and T cells are the driving force of the immune system, forming mobile protective units with the ability to recognize and eliminate an almost infinite array of antigens. AIDS undermines the well-coordinated operations of these cells. To understand what happens in AIDS, let's take a closer look at how the healthy immune system functions.

B CELLS AND IMMUNOLOGICAL MEMORY

At this moment, your body contains B cells with the ability to make antibodies shaped to match virtually any antigen you could ever encounter: a rare virus from Borneo, a pollen grain from a Mongolian steppe grass, and even molecules of an organic chemical just invented by a chemist in her laboratory. Each of these B cells carries on its cell surface membrane close to 100,000 copies of the kind of antibody it will secrete when stimulated by an intruder's antigen of reciprocal shape. But while all B cells carry antibodies on their membranes and can mature and produce thousands of the molecules, each B cell carries and secretes antibodies with binding sites of only one shape. "One cell, one antibody" is the shorthand for this phenomenon.

When an antigen enters the body and binds to a reciprocally shaped antibody on a B cell (review Figure 22.4), that binding stimulates the B cell to divide rapidly. This rapid division produces a *clone*—a group of daughter cells all descended from the same parent cell. In just ten days, a single B cell can generate a clone of 1000 daughter cells. As we mentioned earlier, some of the daughter cells stop dividing and become antibody factories, focusing all their energy on antibody production and surviving only a few days. Other cells of the clone become memory cells that can divide and redivide when stimulated by the same type of antigen that triggered the original B cell.

■ **The More Exposure, the More Protection** The first time a specific antigen stimulates an animal, only a small number of cells respond. But the next time the animal encounters the same antigen, it is better prepared. It has a large pool of memory cells, and the net result is a stronger, swifter reaction to the invader. In short, the more exposure to an antigen, the more memory cells and the faster and stronger the response. Memory cells thus serve as remnants of former encounters with antigens and constitute a kind of **immunological memory** that can make the difference between resisting disease and succumbing to it. Recall that this memory is the basis of booster shots and the active, long-term protection resulting from vaccination.

By the time a healthy newborn grows to adolescence, he or she will have developed many sets of protective memory cells, providing immunity to many kinds of infections. The infant or child with AIDS, however, loses (or never develops) the ability to generate such memory. As an infant, the boy in Figure 22.9 received a blood transfusion bearing the AIDS virus. As a consequence, his immune system could neither produce memory B cells nor provide appropriate long-term immunity; he died a few years after the photo was taken.

In people with fully functioning immune systems, memory cells can survive for several decades, and this explains why most of us rarely contract measles or chicken pox a second time. If a grandparent had measles as a second-grader and is then exposed to measles while

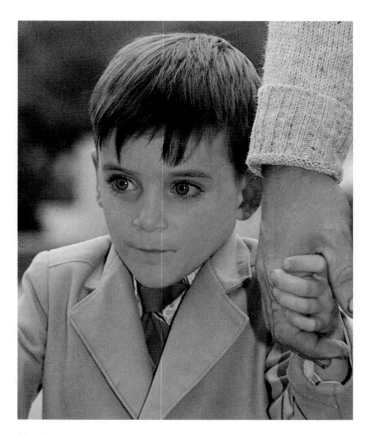

FIGURE 22.9 *AIDS: No Immunological Memory.* This child contracted AIDS from a tainted blood transfusion when he was just two months old.

babysitting a grandchild, a large pool of memory cells in the older person is stimulated to produce enough antibodies to help eliminate the viruses before they have a chance to cause disease.

T CELLS: DIRECT COMBAT AND REGULATION

Fully 50 percent of the lymphocytes in our bodies are T cells. Normally, T cells directly destroy body cells transformed by infection or cancer (Figure 22.10) and attack cells transplanted from other animals. Healthy T cells also communicate with B cells and with each other to initiate, amplify, diminish, or stop immune responses. Infection by the AIDS virus undermines the function of these important cells, as we shall see shortly.

As the "T" implies, T cells pass through the thymus for processing and maturation and lodge in the lymph nodes, spleen, and skin (whereas B cells begin to mature in the bone marrow before eventually lodging in either the lymph nodes or the spleen; review Figure 22.3). The thymus shrinks and functions less effectively as we age; this is one reason the immune system's capacity diminishes in time and why many senior citizens need flu shots at the start of flu season. T cells have other distinctive characteristics. While the surface of each B cell is studded with antibody molecules that bind antigen molecules, T cells have a different (although related) kind of protein in their outer membrane, which also binds to antigens. Ironically, the AIDS-causing virus uses this T-cell antigen receptor as a doorknob, binding to it and

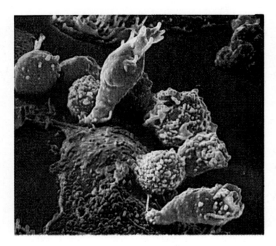

FIGURE 22.10 *T Cells Destroying a Cancer Cell.* This scanning electron micrograph captures several T cells (white) in the process of attacking a single larger cancer cell (brown). The constant vigilance of T cells generally prevents the many genetically transformed cells that arise in the body from multiplying into cancerous tumors.

opening a doorway through which the virus enters the cell. In healthy cells, this doorknob-like antigen receptor only works when associated with molecules called **histocompatibility proteins** (from the Greek *histos* for "tissue"). Each person's cells have a unique array of histocompatibility, or tissue-typing, proteins serving as cellular fingerprints that distinguish "self"—molecules that are part of an individual's body—from "nonself"—molecules that invade from the outside. These proteins were first recognized for their role in determining whether an animal accepts or rejects a tissue graft, such as skin transplanted from a donor onto a burn victim. Rejection occurs when the "fingerprints" of the graft differ from those of the recipient. The T cells recognize that the grafted tissue is "nonself" instead of "self," and attack it. T cells can also recognize and attack "altered self," as when a virus infects a cell and alters the normal pattern of proteins on the cell membrane.

■ **Eliminators** Half of all T cells eliminate cells with foreign antigens on their surface. Most do this directly by poking holes in the membranes of foreign cells as well as in self-cells altered through infection by viruses or transformed by cancer. When doctors transplant a heart or kidney from one person to another, they try to find a donated organ with closely matched histocompatibility markers. Unless the donor is an identical twin to the recipient, however, the physician must administer drugs that suppress immune responses and keep T cells from attacking the foreign organ. Instead of killing cells directly, some T cells release enzymes and hormones that wall off an infected area, cause inflammation, and remove the offending antigens. Before the first day of school, children and teachers must have a routine test for tuberculosis that depends on a reaction by this type of T cell (Figure 22.11).

■ **Regulators Help or Suppress** Half of all T cells control other T cells and B cells rather than directly or indirectly disposing of antigens. *Helper T cells* issue "directions" that cause B cells and other T cells to spring into action. In contrast, *suppressor T cells* prevent helpers and eliminators from taking action. Both helper and suppressor T cells communicate with other cells by secreting molecules such as interferons and interleukins, which alter the activity of specific targets.

The virus that causes AIDS infects and destroys a type of helper T cell before moving on to infect other immune system cells and eventually the brain. If enough helpers are destroyed, the immune system stops functioning (Figure 22.12a). Interestingly, a molecule secreted by healthy, activated T cells may provide a treatment for AIDS-infected patients. The name of the molecule is interleukin 2 (IL-2); it is secreted by helper T cells in the process of responding to antigens. The secreted IL-2 acti-

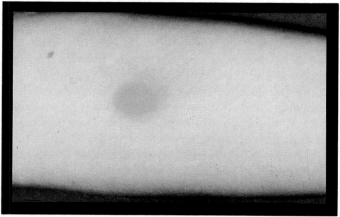

FIGURE 22.11 *T Cells and the TB Test.* This person's arm has the swollen red patch that stimulates a positive reaction to the tuberculosis skin test. A medical worker injects antigens from the tubulosis bacterium under the skin. If T cells recognize and react to them and release a protein signal that brings on immune activity, swelling, and redness, this reveals previous exposure to or possibly current infection by the tuberculosis bacterium.

vates other T cells that go after the same antigen. Researchers can produce IL-2 by recombinant DNA techniques and inject it into patients who have been infected by the AIDS virus but have not shown overt signs of disease. This may help activate more T cells to kill cells that are infected with the AIDS virus before it can spread to new, healthy cells (Figure 22.12b). As a bonus, IL-2 also activates natural killer cells to become much more effective at destroying virus-infected cells.

≋ ARTHRITIS AND PREGNANCY: DIFFERENT VIEWS OF SELF AND NONSELF

The overriding task of the immune system is to protect us from foreign antigens without harming any part of ourselves. Without the ability to accomplish this task, the body would produce antibodies and killer T cells against its own proteins and hence destroy itself. So how does the powerful protective network normally tolerate self-components and yet not tolerate any invader that is marked nonself?

ARTHRITIS: TURNING AGAINST THE SELF

In a person with rheumatoid arthritis, the joints swell painfully, the fingers and wrists become gnarled and twisted, and everyday movements, like buttoning a shirt, can be painful or even impossible (Figure 22.13, page 382). Arthritis is one of the *autoimmune diseases*—attacks on certain cells or tissues by the person's own immune system. It afflicts millions of people in the United States alone.

Many immunologists think that macrophages are a key

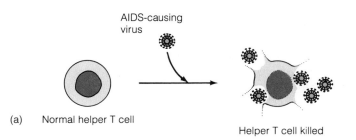

(a)   Normal helper T cell

Helper T cell killed

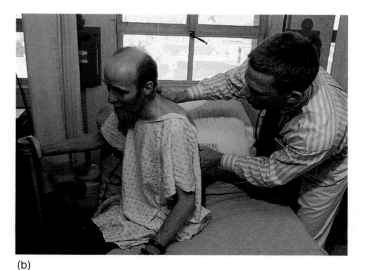

(b)

FIGURE 22.12 *AIDS and T Cells.* (a) By attacking helper T cells, the AIDS virus destroys a person's immunity to disease. Infected helper T cells can no longer induce other T cells to form, and thus these latter cells do not kill cancer cells or infections of parasitic protozoa. Without helper T cells, B cells are not stimulated to produce antibodies, so the victim's body cannot combat viral and bacterial invaders circulating in blood and lymph. Destroyed helper T cells cannot stimulate bone marrow stem cells, and thus fewer new lymphocytes and macrophages form. (b) Here, a physician examines an AIDS patient at San Francisco General Hospital. In an AIDS patient, such as this one, the AIDS virus destroys T cells, which can no longer defend the body against cancerous cells that arise. Many AIDS victims develop carcinomas, or cancerous tumors.

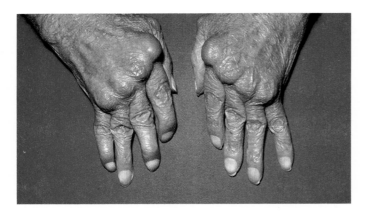

FIGURE 22.13 *Autoimmune Disease: A Breakdown of Self-Tolerance.* In a person with rheumatoid arthritis, overly active immune cells spark inflammation and joint destruction. Some immunologists think that excessive, misdirected helper T-cell activity is the major culprit. Anti-inflammatory drugs and the injection of gold salts provide some relief to many of the 7 million American victims of the disease, but researchers are still looking for more specific drugs with fewer side effects to fight autoimmune diseases.

to **self-tolerance**—the lack of responsiveness to one's own cells and molecules (self-components). It is this tolerance that somehow goes awry in arthritis. In healthy people, the large, specialized phagocytes called macrophages not only engulf antigens and cellular debris, but also present this material to helper T cells. Macrophages do not normally process and present intact substances that originate in the same body as they do. Thus, self-substances do not normally trigger T helpers. Moreover, during development in the fetus and newborn, self-substances may kill immature B cells carrying complementary antibodies in their membrane. In arthritis, the balancing act that produces immune responses to foreign antigens but tolerance of self goes awry and a person's immune system attacks his or her own cells and tissues. Other autoimmune diseases include lupus erythematosus and some forms of diabetes. On the other hand, too much suppression of the immune system can leave us susceptible to cancer or infection by bacteria and viruses. This is what happens once the AIDS virus has infected and destroyed a large number of helper T cells.

PREGNANCY: TOLERANCE OF NONSELF

Pregnancy is an interesting exception to the immune system's recognition and elimination of foreign antigens. Many immunologists consider the human fetus to be a bit like a grafted organ: Although it contains some of the mother's own genes and proteins, it also contains the father's, including the same histocompatibility proteins that would lead the mother to reject one of his organs if transplanted. The mother's immune system obviously continues to function during pregnancy to protect her from foreign invaders. So why doesn't her body reject the half-foreign fetus?

Early studies showed that the uterus is a special immunological zone during pregnancy. At this time, a woman's body will still reject a skin graft from her husband anywhere on her exterior surface and will even reject material from the fetus if it is placed anywhere but the uterus. An embryo may be protected from its mother's immune system because it lacks histocompatibility markers or because embryonic cells make protein signals that quiet killer T cells. Researchers are still studying the actual strategy of embryo protection, but it must be quite complicated, precise, and efficient to defend the tiny amount of tissues from the mother's powerful protective network.

This is not to suggest, however, that the fetus never comes under attack. In about one couple in 300, the woman's body rejects the fetus, perhaps because the man's tissue type is so similar to the woman's that the special protective strategies are not stimulated and the normal maternal defenses against foreign tissues go to work. And in about 1 couple in 15, there is the potential for a dangerous type of anemia in newborn infants, called Rh disease, based on an attack by the mother's immune system.

■ **What Happens in Rh Disease?** Most people in North America are Rh-positive (they have Rh antigens on the membranes of their red blood cells), but some people are Rh-negative and lack the antigens. When an Rh-positive man and an Rh-negative woman produce a baby, it may be Rh-positive. If the baby's blood cells mingle with the mother's during delivery and its Rh antigens enter her bloodstream, her immune system may secrete anti-Rh antibodies. These don't affect the newly delivered baby, since a few days pass before the antibodies form in the mother's system. During a subsequent pregnancy, however, the antibodies are already in the mother's blood, and these can cross the placenta and attack the new fetus's red blood cells, leading to anemia, brain damage, or even death (Figure 22.14).

A therapy has been found that prevents Rh disease. Since about 1970, it has been standard procedure to inject Rh-negative mothers with anti-Rh antibodies (called Rhogam) at the birth of the first child. These ready-made antibodies bind to the Rh antigens on any fetal blood cells that enter the mother's circulation during delivery, cover them up, and prevent her own B cells from recognizing the antigens and making antibodies to them. Thus, maternal B cells do not gear up to produce anti-Rh antibodies, and since antibodies do not lie in wait for the next embryo, its tissues will be safe in the womb—the special immunological safe haven—until the fetus is born.

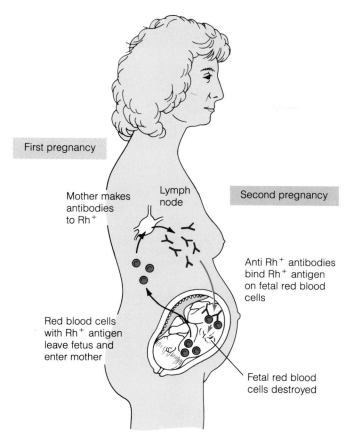

FIGURE 22.14 *Pregnancy and Rh Disease.* During the first pregnancy, a fetus's Rh-positive blood cells may mingle with the mother's Rh-negative blood, enter a lymph node, and react with B cells, which then produce anti-Rh antibodies. During a subsequent pregnancy, the mother's anti-Rh antibodies may enter the fetal bloodstream, attack red blood cells, and destroy them.

Labels on figure:
First pregnancy
Lymph node
Second pregnancy
Mother makes antibodies to Rh⁺
Anti Rh⁺ antibodies bind Rh⁺ antigen on fetal red blood cells
Red blood cells with Rh⁺ antigen leave fetus and enter mother
Fetal red blood cells destroyed

≋ CONNECTIONS

Immune protection is a numbers game that helps each one of us adapt to living in our own particular environment of plants, animals, and microbes. In this game of chance, enormous diversity of specific immune responses is essential for health and survival. Specific immunity gives us the ability to resist specific invaders when they infect us. It also generates an immunological memory that helps us resist the same invaders in the future.

The immune system is not self-contained. It relies on the circulation of blood and lymph (see Chapter 20) and on control and communication via the nervous and hormonal systems (see Chapters 25 through 27). This may be one reason why undue stress or prolonged mental depression seems to lower our resistance to infection. In the next chapter, we will see that the same tight physiological interdependence underlies the smooth functioning of the digestive system.

NEW TERMS

antibody, page 373
antigen, page 373
B cell, page 373
histocompatibility protein, page 380
immune system, page 371
immunological memory, page 379

lymphocyte, page 373
macrophage, page 373
natural killer cell, page 373
self-tolerance, page 381
specific immune response, page 373
T cell, page 373

STUDY QUESTIONS

REVIEW WHAT YOU HAVE LEARNED

1. When there is a break in the skin, what nonspecific immune responses occur?
2. What are the roles of B cells? T cells? Macrophages?
3. Why are lymphoid organs important to health?
4. Trace the sequence of events in the immune system's response to invading bacteria.
5. How is the Y-shaped molecular structure of antibodies the key to their effectiveness? Explain.
6. Describe immunological memory.
7. What is self-tolerance?
8. How is the body protected by passive immunization? By active immunization?

APPLY WHAT YOU HAVE LEARNED

1. It is unlikely that a flu vaccine will confer lasting immunity against disease. Why?
2. A boy with measles receives a visit from his grandmother, who is sure she will not catch the disease. Why might she be right?
3. Researchers believe that it will be many years before there is an effective vaccine against AIDS. Why might they be right?

FOR FURTHER READING

Gallo, R. C., and L. Montagnier. "AIDS." *Scientific American* 259 (October 1988): 47–48.
Koshland, D. E. "Frontiers in Immunology." *Science* 238 (1987): 1023.
Rosenberg, A. "Adoptive Immunotherapy for Cancer." *Scientific American* (May 1990): 62–69.
Wechsler, R. "Hostile Womb." *Discover* (March 1988): 83–87.
Young, J. D., and Z. A. Cohn. "How Killer Cells Kill." *Scientific American* (January 1988): 38–44.

① The cells and molecules of the immune system communicate with each other and defend the body by recognizing and eliminating specific invaders of all kinds.

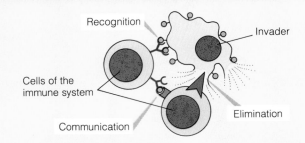

Recognition

Invader

Cells of the immune system

Elimination

Communication

② Medical workers can gather antibodies from rabbits and other animals and inject them into patients; the immunological proteins provide immediate short-term protection against snake bites and some infections.

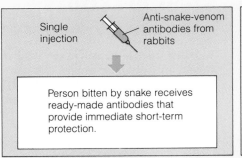

Single injection

Anti-snake-venom antibodies from rabbits

Person bitten by snake receives ready-made antibodies that provide immediate short-term protection.

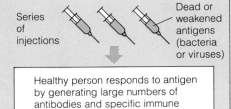

Series of injections

Dead or weakened antigens (bacteria or viruses)

Healthy person responds to antigen by generating large numbers of antibodies and specific immune cells; takes time, but provides long-term protection.

③ Vaccinations confer long-term resistance to infection based on immunological memory.

④ The normal immune system can distinguish the body's own tissue from foreign antigens. In autoimmune diseases, one's own tissue is attacked; in pregnancy, a partially foreign entity is tolerated.

NORMAL	AUTOIMMUNE DISEASE	PREGNANCY
Immune system	Immune system	Immune system
versus	versus	versus
foreign tissues (antigens)	Own tissues	foreign tissues (embryo)

⑤ The AIDS virus incapacitates the immune system by infecting some of the white blood cells that initiate and control immune responses.

ANIMAL NUTRITION AND DIGESTION

CHAPTER 23

UP A TREE: THE KOALA'S EXCLUSIVE DIET

What do you suppose would happen if, for some strange reason, you were forced to spend a week in a eucalyptus tree, sitting on its large, smooth limbs and relying on its aromatic, silvery green leaves for all your needs? Among other things, you would get very hungry, and you might eventually try eating a handful or two of the leaves to stave off the gnawing hunger. You would find eucalyptus leaves to be leathery, however, and to have a bitter, nauseating, turpentine flavor.

Ironically, one mammal—the koala—is perfectly adapted to spending a lifetime in a eucalyptus tree (Figure 23.1). This cuddly-looking Australian marsupial (mammal with a pouch) can meet all of its needs for shelter, food, and water in the very eucalyptus tree that would be so inhospitable to a stranded human. In fact, the koala's ability to live on nothing but eucalyptus leaves is most remarkable. An exclusive diet of these pungent leaves—filled, as they are, with toxic oils and containing fairly low levels of nitro-

gen and other basic nutrients—would starve and poison other mammals. Yet, a koala's digestive tract can detoxify the leaves and simultaneously extract all the nutrients and water this 9 kg (20 lb) animal needs.

The koala's ability—unique among mammals—to survive on eucalyptus leaves focuses our attention on animals' special adaptations for obtaining energy and materials. Two unifying themes emerge from our discussion of this important topic. First, animals must take in sufficient amounts of specific substances or they develop diseases and die. As heterotrophs, animals must take in foods from the environment to provide themselves—and ultimately each of their millions or trillions of individual cells—with a supply of **nutrients,** substances that organisms extract from their surroundings in order to survive and reproduce. Animals, for example, need amino acids, sugars, fatty acids, vitamins, and minerals for aerobic respiration, replacement of worn parts, and growth.

Second, an animal's anatomy and physiology are tied closely to its diet and its nutrient needs. For example, a koala's mouth is suited for crushing vegetation, and its digestive system has a special tubular sack—shaped by natural selection—that allows the koala to effectively use the food it consumes.

This chapter answers several questions about how animals obtain the nutrients they need:

- What organic and inorganic nutrients must animals have to survive?
- What are the various ways animals consume and utilize foods?
- What structures and mechanisms allow people to take in food, break it down, and use the nutrients?

FIGURE 23.1 *A Craving for Eucalyptus.* The koala (*Phascolarctos cinereus*, or "ash-gray pouched bear"), an Australian marsupial, survives on a diet of nothing but eucalyptus leaves. Its lengthy digestive tract extracts all the nutrients and water it needs from this herbaceous diet.

☰ NUTRIENTS: ENERGY AND MATERIALS TO SUSTAIN LIFE

Food is a huge industry that is central to most human economies, and yet, we often lose sight of this industry's underlying biological function: to provide living organisms with the organic and inorganic materials they require on a day-to-day basis, including a source of energy and a dependable supply of carbon and nitrogen atoms for the building of biological molecules. Plants get their energy directly from the sun, their carbon from the air, and their nitrogen from the soil. Animals, however, must take in all three as food molecules. The science of **nutrition** is concerned with the precise amounts of protein, carbohydrates, lipids, vitamins, and minerals and the total number of calories of food energy an animal must consume to stay alive and healthy. Without this knowledge, dairy farmers could not be sure their cows were producing the highest quality of milk, and parents might wonder if their children were growing and developing to fullest mental and physical potential.

An animal can have a highly specialized diet: A koala consuming eucalyptus leaves and a butterfly sipping flower nectar are both *herbivores*, eating only plant matter. In contrast, a Doberman pinscher and a crocodile are both meat-eating *carnivores*, with teeth that can tear flesh and a digestive tract well suited to utilizing large amounts of protein. And a person and a wild boar are both *omnivores* (*omni* means "all"), with teeth that can efficiently grind up both plant and animal matter and a digestive tract suitable to handling both. Because the human diet is both varied and familiar, we will use it as our main model in studying organic and inorganic nutrients, beginning with carbohydrates, then moving on to lipids, proteins, vitamins, minerals, and total calories.

CARBOHYDRATES: CARBON AND ENERGY FROM SUGARS AND STARCHES

The sugars in fruit and the starches in potatoes, rice, bread, and pasta are a rich source of energy and carbon atoms and provide the nutrients we call *carbohydrates*. Our digestive system can cleave units of glucose from the sugars in fruit or the starch in rice, wheat, or potatoes. After a meal, the glucose units pass into the bloodstream, and the circulatory system carries them to cells throughout the body, where the simple sugar serves as the main supplier of energy for cellular respiration (review Figure 5.17). Cells in the brain and nerves are particularly sensitive to fluctuations in blood glucose levels, and if starving, the body will break down first its fat stores, then its own muscle tissues and convert the subunits to glucose to provide the sensitive nervous system cells with the levels they need to stay fully active. People are often told to avoid table sugar and the desserts and processed foods containing it. Sugar is said to provide nutritionally "empty calories" and is blamed for causing, among other things, tooth decay. The fact is, sugar provides calories, our largest nutritional need. And while it doesn't provide vitamins or minerals in significant amounts, neither do other sweeteners, such as honey, brown sugar, raw sugar, corn syrup, or maple syrup. The issue is whether a person substitutes sugar-laden foods (such as a candy bar) for more nutritious ones (such as an apple). The causative role of sugar in tooth decay is a bit clearer: Sucrose readily promotes the growth of oral bacteria that produce acid and cause cavities. Whether you eat cake or a more complex carbohydrate, regular brushing is the way to prevent tooth decay.

The cellulose fibers in plant cell walls—most obvious in leaf veins, celery "strings," the tissue surrounding grapefruit sections, or the woody core of a pineapple—are also carbohydrates, but ones that we humans cannot digest or utilize. Some herbivores, however, such as termites, koalas, and horses, have cellulose-digesting microbes in their digestive tracts that cleave glucose subunits from cellulose and provide the animals with a rich source of energy. The cellulose we consume in plant foods does provide fibrous "roughage," however, which helps to stimulate the mechanical movements that propel wastes through the large intestine.

LIPIDS: HIGHLY COMPACT ENERGY STORAGE NUTRIENTS

Most people have only to visualize the solid white fat of a bacon slice or the slippery golden oil in salad dressing to know what a lipid is. And many have only to reach down and "pinch an inch" to see that the body stores lipids as an energy reserve.

Lipids, or fats, are a compact energy source and the form most plants and animals use for long-term energy storage. When needed, the stored fats can be oxidized during aerobic respiration; they provide more usable energy than an equivalent amount of carbohydrates. After a meal of eggs, oily sunflower seeds, or fatty meat, an animal's digestive system breaks down some of the lipids in these foods to fatty acids and other compounds; some of these other compounds can then enter mitochondria in cells and contribute to aerobic respiration. Without fat stores, walruses could not live in the Arctic (Figure 23.2), birds could not migrate long distances, carnivores would have more trouble surviving between irregular meals, and hibernators might not be able to survive the long winter without eating.

FIGURE 23.2 *Lipids: High-Energy Storage.* Fats are a compact way to store calories. Active animals, such as walruses living near the Arctic Circle, have thick fat layers that act as superb insulation as well as a source of stored energy for generating heat and powering activity.

Besides their energy storage function, fats help the body absorb and use certain essential nutrients. Vitamins A, D, E, and K, for example, are fat-soluble, and lipids must be present for these vitamins to be absorbed into the circulation and delivered to cells.

PROTEINS: BASIC TO THE STRUCTURE AND FUNCTION OF CELLS

The body's structure and its vital activities depend on proteins, composed of long chains of amino acids. The body's most abundant protein, collagen, is a major constituent of skin, cartilage, tendons, and bone. Muscle tissue is largely protein, and so are hair and the cornea of the eye. Enzymes, antibodies, hemoglobin, and some hormones are composed of protein molecules; without such proteins, most cellular activities would grind to a halt. What's more, there is a steady turnover of protein: Enzymes and cell constituents are continuously broken down and rebuilt, and new cells are generated to replace dying cells. This constant turnover means that animals need a continuous supply of protein in food, from which their digestive systems can extract amino acids for the building of new protein as needed.

A human is a large, active animal that requires about 1 g of protein per kilogram of body weight per day. As a rough measure, a 5'8", 150 lb college student needs approximately one-sixth of a pound of pure protein each day, about what you would obtain from a filleted chicken breast or three portions of beans and rice. But knowing how much protein one needs does not answer the question of what *kind* of protein to eat. The body can synthesize many of the 20 amino acids it needs if nitrogen is available. The body cannot, however, manufacture eight so-called **essential amino acids:** These must be obtained in the diet, and all within a few hours, since free amino acids are not stored (Figure 23.3). Children also need extra supplies of two other amino acids (histidine and arginine), since their bodies make only enough for maintenance, not growth.

If one or more of the essential amino acids are missing, the body's cells will not be able to synthesize the full spectrum of proteins necessary for replacing lost macromolecules or generating new cells. Why? First of all, the body cannot store amino acids; what it doesn't incorporate immediately into proteins, it excretes or converts to other biological molecules. Second, as Chapter 10 described, cells manufacture proteins by adding one amino acid at a time to the end of a chain. If even one necessary type of amino acid is lacking, elongation of the chain stops. It is as if the typesetter trying to set this

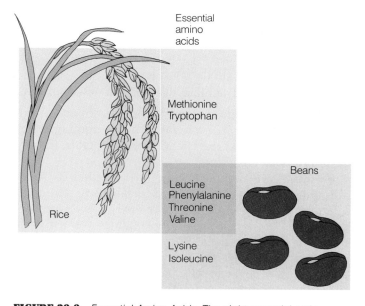

FIGURE 23.3 *Essential Amino Acids.* The eight essential amino acids on this list must be included in the daily diet; if they aren't, the body begins to break down its own tissues and use the amino acids they contain for critical functions, such as maintaining the lungs, heart, and brain. Meat, fish, eggs, and other high-protein foods contain all eight essential amino acids. Rice contains only six, and beans a slightly different set of six, but a diet that includes both rice and beans in the same meal will provide all eight.

book had run out of the letter *z* right before starting the project. He or she could begin typesetting Chapter 1, but as soon as a word like *enzyme* or *zebra* appeared, the process would have to stop. In the same way, if even one essential amino acid is absent, protein synthesis in the body comes to a screeching halt.

Most animal proteins in food like meat, cheese, eggs, and milk contain the eight essential amino acids; however, many plant proteins do not. This is why strict vegetarians (those who avoid all animal products, including eggs and dairy products) must be careful to include combinations of foods in their daily diets that ensure an adequate amino acid intake. One such combination is rice and beans, which individually lack certain amino acids but together contain all the essential ones (see Figure 23.3). The incomplete nutrition provided by individual plant species also explains why some people in poor and underdeveloped nations show the swollen belly, patchy skin, and other symptoms of protein deprivation. Their diet, consisting mostly of cereal grains such as corn, wheat, or rice, all too often lacks one or more essential amino acids. Consequently, their bodies draw proteins from their own muscles and other tissues, dismantling them to provide the missing amino acids for new protein synthesis. The rest of the amino acids present in excess are simply excreted and wasted. As medical observers have noted sadly, when a human being starves, it is usually the lack of protein, rather than the lack of food energy, that leads to death.

VITAMINS AND MINERALS: NUTRIENTS OF GREAT IMPORTANCE

While the body requires relatively large amounts of protein and carbohydrate each day, it needs another set of nutritive substances, the vitamins and minerals, in only extremely small amounts.

Vitamins are organic compounds needed in small amounts for normal growth and metabolism. Although vitamins are not a direct source of energy, most are necessary to help energy-harvesting enzymes perform their functions. Most animals, however, cannot make many of these helper molecules themselves and can store only small supplies in their cells. Thus, they must acquire a set of specific vitamins in their daily food.

Nutrition researchers have determined that humans require the 14 vitamins listed in Table 23.1. As noted earlier, some vitamins are soluble in fat, while others are water-soluble. Fat-soluble vitamins tend to be stored in the body's fat tissues; and since accumulations can produce serious side effects, nutritionists warn against taking high doses of vitamins A, D, E, or K. Water-soluble

vitamins, on the other hand, dissolve directly into the bloodstream. Because amounts beyond the cells' immediate needs are excreted, taking large amounts of vitamin C, for example, is largely pointless; any excess beyond the normal daily requirement is filtered out by the kidneys and excreted in the urine soon after it enters the bloodstream.

A prevalent myth is that vitamins from "organic" foods or natural sources are somehow healthier or more effective than vitamins from processed foods or those that are manufactured directly. An "organic" tomato has no more nutritive value than one grown on a larger, more mechanized farm, although the latter may contain more pesticide residues. And since vitamin C, for instance, is the specific biochemical ascorbic acid and has a unique molecular structure, it will be the same whether it is extracted from rose hips or synthesized in a laboratory.

The body also needs small amounts of another type of nutrient: **minerals,** or specific chemical elements (see Chapter 2). *Major minerals* are those elements we need in amounts greater than 0.1 g each day; *minor minerals* are those we need in amounts less than 0.01 g daily (Table 23.2, page 390). An adult's body contains about 2 kg (4.4 lb) of minerals, and of this, about three-quarters is the calcium and phosphorus in the bones and teeth. The minerals that give tears, blood, and sweat their salty taste are sodium, potassium, and chloride. Sulfur is found in many proteins, and magnesium is a constituent of enzymes and, along with the calcium, is held in reserve in the bones.

The human body contains less than 1 teaspoon of minor minerals, but these elements still play critical roles. Perhaps most important is iron—an essential component of the oxygen-binding pigment hemoglobin. Iodine is a constituent of thyroid hormones, zinc is a common enzyme activator, copper is needed for hemoglobin production, and fluorine is needed for healthy bones and teeth. A person's need for specific minerals can change over time. A woman, for example, needs more iron before menopause (because she loses blood periodically), more calcium after menopause (when hormonal changes alter the way her body uses that mineral), and more of both during pregnancy.

Because vitamins and minerals are required in such small quantities, most people in affluent countries get a more than adequate supply simply by eating a reasonably balanced diet. When poor diet leads to vitamin or mineral deficiencies, however, the consequences can be serious. The long-term absence of vitamin A, for instance, can lead to blindness, while the lack of B vitamins may lead to convulsions and other neurological disorders (see Table 23.1). For residents of a wealthy nation, a far greater concern than malnutrition is overeating and obesity. An estimated 25 percent of teenagers and 50 percent of adults

TABLE 23.1 ≋ VITAMINS

VITAMIN	SOURCES	FUNCTIONS IN BODY	DEFICIENCY SYMPTOMS
Water-Soluble Vitamins			
Choline	Egg yolks, liver, beans, peas, grains	Part of phospholipids; needed for nerve cell function	Not seen in humans
Vitamin C (ascorbic acid)	Dark green vegetables, citrus fruits, strawberries, brussel sprouts, other fruits and vegetables	Helps form collagen and bone; enzyme helper; blocks toxic effects of oxygen	Gum bleeding; hemorrhages under skin; rough skin; failure of wounds to heal; bone degeneration
Niacin	Milk, meats, cereals, and starchy vegetables	Part of enzyme helpers involved in electron exchange reactions	Sore skin; smooth tongue; diarrhea; mental confusion; irritability
Pantothenic acid	Widespread in foods	Central to energy metabolism	Rarely seen
Vitamin B_6 (pyridoxine)	Whole grain cereals, vegetables, meats	Enzyme helper involved in amino acid metabolism	Skin soreness; smooth tongue; abnormal brain activity
Vitamin B_2 (riboflavin)	Milk, meat, vegetables, whole grains	Helper of enzyme active in energy metabolism	Cracks at corner of mouth; sensitivity to light
Vitamin B_1 (thiamine)	Milk, dairy products, fruits, breads, vegetables	Helper of enzyme that removes carbon dioxide from nutrients	Beriberi (paralysis, swelling, heart failure); mental confusion
Folacin (folic acid)	Fruits, leafy and other vegetables, grains	Helper of enzyme involved in metabolism of amino acids and nucleic acids	Anemia; diarrhea; smooth tongue
Biotin	Widespread in foods	Constituent of many enzymes in metabolism	Not seen in humans
Vitamin B_{12}	Meat and dairy products	Helper of enzyme involved in nucleic acid metabolism	Anemia; nerve degeneration
Fat-Soluble Vitamins			
Vitamin E (tocopherol)	Vegetable oils, margarine, salad dressings	Counters toxic effects of oxygen	Breakage of red blood cells; anemia
Vitamin A (retinol)	Carrots, milk, vegetables, fruits	Part of visual pigment; helps maintain living tissues; promotes bone growth	Impaired night vision; dry eyes; diarrhea; lung infections; bone changes; tooth cracking and decay; impaired brain growth; anemia
Vitamin K	Cabbage, milk, green leafy vegetables	Essential for synthesis of certain proteins, including blood-clotting proteins	Unchecked bleeding
Vitamin D	Milk, sunshine, eggs, liver	Helps bones and teeth take up calcium for proper growth	Rickets (bowed legs, other bone deformities); tooth decay; blood changes; lax muscles

TABLE 23.2 ≋ MINERALS

MINERAL*	SOURCES	FUNCTIONS IN BODY	DEFICIENCY SYMPTOMS
Major Minerals (more than 0.1 g/day)			
Calcium	Milk and dairy products, fish bones, collard greens, spinach, broccoli	Major part of bones and teeth; cell membrane integrity; involved in nerve transmission	Fragility of bones (osteoporosis); stunted growth
Phosphorus	Widespread in foods	Major constituent of bones, blood plasma; part of DNA, RNA; needed for energy metabolism	Rare
Potassium	Bananas, orange juice, potatoes, tomatoes, other vegetables	Critical to normal heartbeat; principal positive ion in body cells	Mental confusion
Sulfur	Protein foods	Present in body proteins; helps proteins assume proper shapes	Unknown
Sodium	Salt, common in foods	Constituent of salt in body fluids; helps regulate fluid content of body	Rare; excess leads to water retention
Chlorine	Salt	Major negative ion in body fluids; part of hydrochloric acid in stomach	Rare
Some Minor Minerals (less than 0.01 g/day)			
Magnesium	Nuts, legumes, cereals, dark green vegetables, seafood, chocolate	Constituent of bones; roles in protein synthesis and energy metabolism	Tetany; prolonged muscle contraction; hallucinations
Iron	Oysters, liver, bran flakes, lean beef, spinach, greens, strawberries, raisins	Part of many major enzymes active in DNA synthesis and cellular respiration; part of hemoglobin and myoglobin	Anemia; exhaustion; headache; weakness
Iodine	Salt (iodized)	Part of thyroid hormone thyroxine; controls metabolic rate	Goiter (enlarged thyroid gland)
Zinc	Oysters, milk, egg yolks, meat, whole grains	Part of many enzymes; helps bone form; DNA and protein synthesis; wound healing	Retarded growth; small sex organs; loss of sense of taste
Copper	Grains, shellfish, organ meat, legumes, dried fruit, fresh fruit, vegetables	Helps form hemoglobin, collagen, nerves	Rare; retarded growth, sluggish metabolism
Fluoride	Fluoridated water	Normal formation of bones and teeth	Tooth decay
Chromium	Yeast, organ meats	Involved in carbohydrate metabolism	Stunted growth; adult-onset diabetes

*Arranged in order of decreasing amounts needed in daily human diet.

are either *overweight* (body weight more than 10 percent over ideal according to weight charts such as the one in Figure 23.4) or *obese* (body weight more than 20 percent over ideal; see Figure 23.5).

FOOD AS FUEL: CALORIES COUNT

Like all animal cells, human cells require energy to carry out biochemical, mechanical, and transportation tasks. As Chapter 5 explained, cells derive energy from the chemical bonds in the fats, carbohydrates, and proteins an animal eats, digests, and absorbs; any energy an animal does not immediately use can be stored as the carbohydrate glycogen in the liver or muscles or as fat in fat cells (see the box on page 396). Food energy is usually measured in kilocalories (kcal), also called Calories (Cal). A large apple contains about 70 Cal worth of energy-producing compounds; and jogging 1.6 km (1 mile) burns about 100 Cal of stored energy. The food energy in a slice of bread could bring a liter (about a quart) of water to a boil, and a pound of body fat has enough energy to bring 52 liters (13 gal, or 52 qt) to a boil!

Each person has a minimum daily energy requirement that varies with age, sex, body size, activity level, and other factors. In general, a normally active female college student needs around 1800–2000 Cal a day to fuel her total metabolic needs; a male college student needs about

FIGURE 23.5 *Obesity: Overstorage of Physiological Fuel.* Sumo wrestlers from Japan eat huge quantities of rice, as well as fish and other protein-rich food, to generate the body fat that gives them their formidable bulk. Although a Sumo wrestler, like this now-retired Hawaiian, can easily weigh 300 pounds, he is extremely strong and skilled, not simply fat.

FIGURE 23.4 *Ideal Weight.* Ideal weight depends on frame size and physical condition. At 6'0" and 195 pounds, a heavy-boned, muscular football tackle would not be overweight, while a sedentary office worker with a medium frame probably would be. Height and weight are given for people not wearing shoes or other clothing.

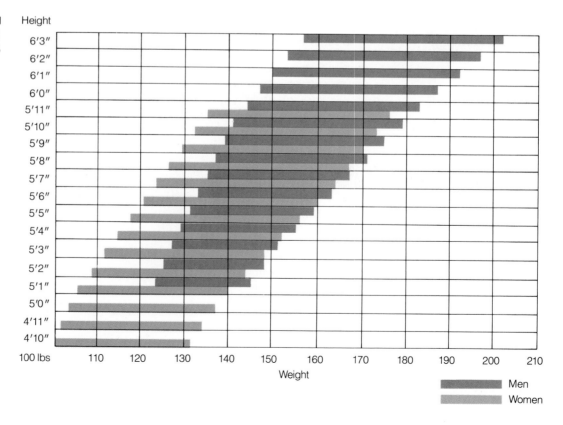

2200–2500 Cal. Carbohydrates and protein each provide about 4 Cal/g, while fat provides more than twice as much, or 9 Cal/g.

Figure 23.6 provides a revealing look at the caloric values for several common snack foods and the amount of energy that a person must expend in various physical activities to work off that food. A person would have to run for about 30 minutes, for example, to burn off the calories in a cheeseburger. When an animal's food intake exceeds its energy needs, the inevitable result is an increase in the amount of leftover energy stored as body fat. This basic biological fact means that the secret of weight control lies in taking in only as many kilocalories as the body needs for fuel. To stay trim and fit with a desirable percentage of body fat and healthy, efficient heart and lungs,

the American College of Sports Medicine recommends that a person perform continuous, rhythmic, aerobic exercise (such as walking, jogging, cross-country skiing, swimming, or cycling) three to five times per week for 15–60 minutes, depending on the intensity of the exercise. Sound weight-loss diets combine calorie reductions, primarily through diminished intake of sugar and fat, with increased physical activity. Many physiologists believe that physical exercise can cause the body's basal metabolic rate to increase, that is, to burn calories at a faster rate both during and after exercise. The result is less energy in and more energy out, with the differences made up from the body's fat reserves. Some researchers have concluded that fat cells help determine our appetites as well as our waistlines (see the box on page 396).

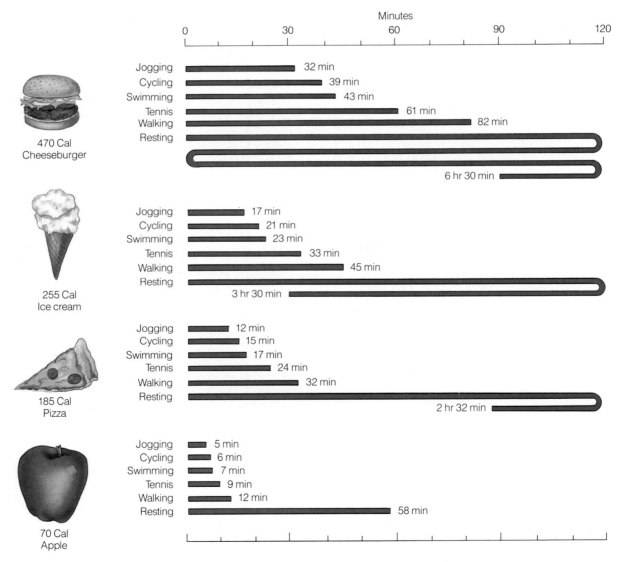

FIGURE 23.6 *Food and Exercise Equivalents.* As this chart shows, you can swim off a cheeseburger in 43 minutes, but at rest, your body will take 6 hours and 30 minutes to burn those same 470 calories!

TABLE 23.3 ≋ GUIDELINES FOR GOOD NUTRITION

These recommendations are not meant to prohibit the use of any specific food item or to prevent you from eating a variety of foods.

1. EAT A VARIETY OF FOODS DAILY
- Fresh fruits and vegetables
- Whole grain breads and cereals
- Milk, cheese, or yogurt
- Lean meat, poultry, fish, and eggs
- Legumes (dried peas and beans)

2. MAINTAIN IDEAL WEIGHT
- Increase physical activity.
- Prepare smaller portions of less sweet, less fatty food.
- Avoid "seconds."

3. AVOID TOO MUCH FAT, SATURATED FAT, AND CHOLESTEROL
- Choose lean meat, fish, poultry, tofu, or dried beans and peas.
- Limit your use of egg yolks and organ meats (such as liver).
- Limit your use of butter, cream, margarine, and other fats.

4. EAT FOODS WITH FIBER
- Choose foods that are good sources of fiber, such as whole grain breads and cereals, fruits, vegetables, dried beans and peas, nuts, and seeds.

5. AVOID TOO MUCH SUGAR
- Use less of all sugars, including white sugar, brown sugar, raw sugar, honey, and syrups.
- Eat less food containing these sugars, such as candy, soft drinks, ice cream, cakes, and cookies.
- Choose fresh fruit, or fruit canned in its own juice without sugar or in a light syrup.

6. AVOID TOO MUCH SALT (SODIUM)
- Learn to enjoy the natural flavors of food.
- Cook with only a small amount of added salt.
- Remove salt shaker from the table.
- Limit the amount of potato chips, salted nuts, cheese, and other salty foods you eat.
- Use fresh seasonings and herbs in place of salt.

7. DRINK ALCOHOL IN MODERATION, OR NONE AT ALL
- One or two alcoholic drinks a day are considered moderate; this includes beer, wine, or mixed drinks.

8. REDUCE YOUR CANCER RISK
- Limit saturated fats.
- Eat more fiber.
- Avoid foods containing the sweetener saccharin.
- Avoid foods containing nitrates or nitrites.
- Limit alcohol.
- Get plenty of vitamins A, C, and E by eating dark green vegetables, citrus fruits, whole grains, nuts, seeds, and vegetable oils.
- Avoid moldy nuts, grain, or seeds.
- Reduce consumption of smoked foods or foods fried at high temperature.

Adapted from *Nutrition and Cancer Prevention: A Guide to Food Choices*, June 1981, Northern California Cancer Program.

Some dieters find it useful to focus on the biological function of eating: to take in the nutrients the body needs for energy, maintenance, and repair (Table 23.3). As one wise saying goes, "Eat to live, don't live to eat."

≋ DIGESTION: ANIMAL STRATEGIES FOR CONSUMING AND USING FOOD

Most animals, whether they eat eucalyptus leaves, raw meat, or a chef's salad, are consuming food in a form that cells cannot use directly. The mechanical and chemical breakdown of that food into small molecules that the organism can absorb and use is called **digestion.**

In the simplest animals—the sponges—some body cells take in tiny whole food particles directly from the water and break them down with enzymes to obtain nutrients.

This strategy is called **intracellular** ("within the cell") **digestion.** Because it can act on only very tiny bits of material as food, intracellular digestion puts an upper limit on an animal's size and complexity.

With the evolution of larger animals came the evolution of the structures and mechanisms necessary for **extracellular digestion,** the enzymatic breakdown of larger pieces of food into constituent molecules, usually within a special body organ or cavity. The vast majority of animals rely on extracellular digestion.

EXTRACELLULAR DIGESTION: STRUCTURES AND MECHANISMS

In a simple animal, such as a flatworm, extracellular digestion takes place in an internal sac with a single opening through which whole particles of food enter and undigested wastes leave. In more complex animals, food enters one end of the digestive tract, the mouth,

(a)

(b)

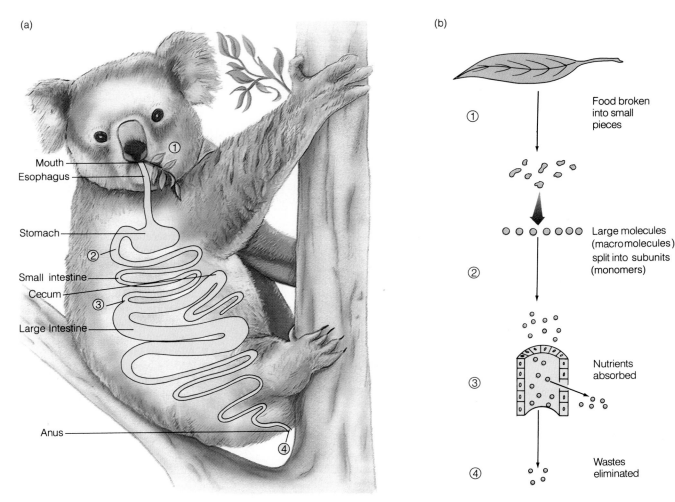

① Food broken into small pieces

Large molecules (macromolecules) split into subunits (monomers) ②

③ Nutrients absorbed

④ Wastes eliminated

Mouth
Esophagus
Stomach ①
②
Small intestine
Cecum ③
Large Intestine
Anus
④

FIGURE 23.7 *Four-Step Digestion Process: Ingestion, Digestion, Absorption, Elimination.* As the text explains, the koala ingests eucalyptus leaves, and its stomach enzymes break down the plant's macromolecules (mainly carbohydrates) into monomers (sugars). Cells in the gut wall then absorb these nutrients, and indigestible wastes eventually exit the gut tube.

and moves in a single direction through the gut tube, and wastes exit the other end, the anus; in between, a variety of specialized regions and structures perform special digestive roles that break the food first into pieces, then into small molecules (Figure 23.7).

In virtually all vertebrates, the gut is divided into five main regions: mouth, esophagus, stomach, small intestine, and large intestine (see Figure 23.7a). Together, these form the digestive tract, or **alimentary canal.** In addition, nearby *accessory organs,* such as the salivary glands, liver, gallbladder, and pancreas, produce enzymes, bile, and other materials that are funneled into the tract at appropriate times to aid digestion (Figure 23.8).

Within the alimentary canal, a four-step digestive process takes place to extract nutrients from food. First,

the animal—a koala, let's say—ingests food—in this case, eucalyptus leaves. It breaks the leaves into small pieces with its grinding teeth (Figure 23.7, step 1), and enzymes from salivary glands are mixed with food as the animal chews and swallows. In the second stage of digestion, additional enzymes in the gut break down the large macromolecules of food—in this case, cellulose and plant juices from the leaves—into smaller molecules, or monomers (step 2). Third, the monomers are absorbed; they pass across the gut wall and into the animal's lymphatic system or bloodstream, which transports the small nutrient molecules to each cell in the body (step 3). Finally, undigested residues are eventually eliminated (step 4).

Evolution has resulted in a number of intriguing variations in extracellular digestive systems that allow animals to exploit very different foods. Carnivores, such as hyenas and wolves, have short intestinal tracts because meat is relatively easy to digest, whereas cows and rab-

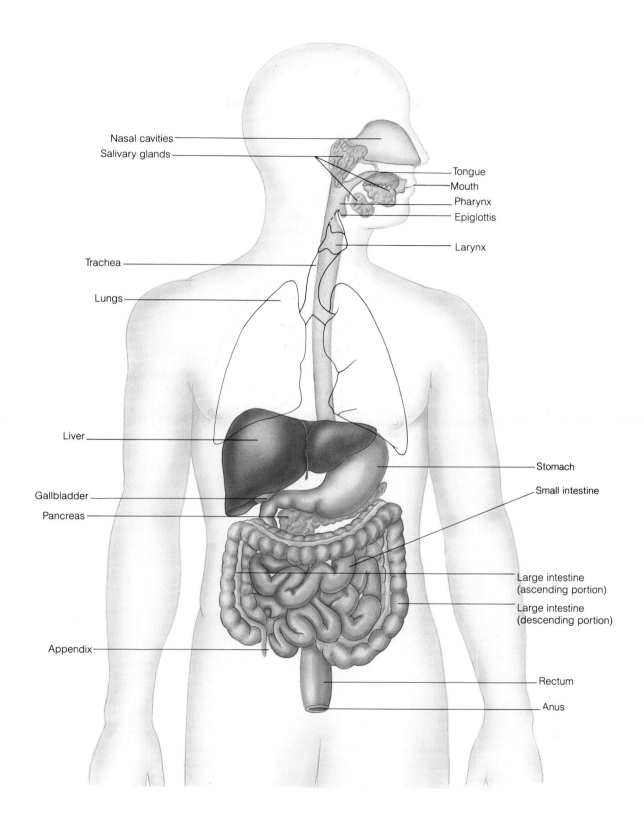

Nasal cavities

Salivary glands

Tongue

Mouth

Pharynx

Epiglottis

Larynx

Trachea

Lungs

Liver

Stomach

Small intestine

Gallbladder

Pancreas

Large intestine
(ascending portion)

Large intestine
(descending portion)

Appendix

Rectum

Anus

FIGURE 23.8 *The Human Digestive System.* The digestive system is essentially one long tube (which lines the mouth, curves around in fold upon fold, then exits at the anus) plus associated organs. See the text for details.

FAT CELLS, SET POINT, AND WEIGHT LOSS

To quote the illustrious cat Garfield, for many overweight people, "Diet is die with a T." Despite their best efforts, fully 95 percent of all people who diet regain every bit of weight they lose. Why is it so difficult to lose weight and keep it off? Researchers have begun to investigate the possibility that people have a fixed *set point*, a level of fat storage and body weight that is genetically determined and difficult—but not impossible—to alter.

The theory suggests that some people have naturally high set points (above ideal weight) while others have low ones (at or below ideal weight) and that the set point is based on the number and size of fat storage cells (Figure 1). Individuals vary in the number of fat cells their bodies contain, and fat cells in humans are much more numerous than in any other mammal save the hedgehog and the fin whale—both notoriously fat.

Once gained, fat cells appear never to be lost; they merely increase or decrease in size by storing more or less fat, depending on dietary excesses. Significantly, a person's fat cells tend to remain a given size and *to return to that original size* soon after a diet ends. Fat cells act as if they had a mind of their own, and in fact, they do appear to communicate with the brain. They

seem to signal any drop in lipid stores, trigger increased appetite and eating behavior to compensate, and initiate a change in metabolic rate so that the body uses its calories more efficiently—all as if to defend the fat cell's genetically determined size.

In obese people, fat-shuttling enzymes may be overactive and may store fat molecules that would be burned for energy in naturally thin people. Research also suggests that "yo-yo" weight loss and gain from one failed diet after another may actually train the body to cling to every calorie, making it harder and harder to lose weight each time. Finally, it appears that the set point can be raised by the smell or taste of fatty food, an evolutionary adaptation, perhaps, to allow animals to take advantage of energy-rich resources when they find them.

Given this discouraging picture, how can one lower the set point and thus reduce the size of fat cells and with it the weight of the body? Evidence suggests that a moderate reduction in total calories (especially from lipids and sugars) is a beginning, but that *dieting must be accompanied by a consistent increase in physical activity*. This seems to turn up the metabolic rate so that the body burns more calories—and not just during exercise sessions, but for all the hours at rest between bouts of exertion. Exercise decreases fat tissue and

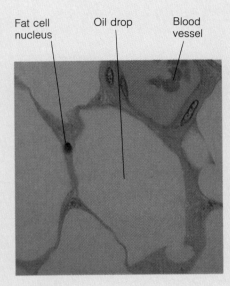

FIGURE 1 *A Fat Cell with Its Large Stored Lipid Droplet.*

increases muscle mass; thus, the body looks and feels trimmer. Finally, moderate daily exercise reduces the appetite. Considering all these health benefits, regular exercise (four to five sessions per week) is probably the major reason why people who jog, swim, bicycle, or do aerobic dance regularly find it easier to control their appetites and to maintain their weights at lower levels—that is, to lower their set points.

---

bits have elaborate guts with several stomachlike pouches containing fermenting bacteria that help break down all the cellulose in the animals' vegetarian diets. Another variation on this theme is the koala's 2.5 m fermenting chamber, a blind pouch called the *cecum* lying coiled near the animal's stomach like a huge appendix (see Figure 23.7a). The largest such structure known in the animal kingdom, the koala's cecum is 30 times bigger than a

person's appendix (which is only 8 cm, or 3 in., long) and contains bacteria that break down the cellulose in eucalyptus leaves, enabling the animal to survive on its limited diet. A person's appendix (see Figure 23.8) has no known digestive function and sometimes gets inflamed, leading to the medical emergency we call appendicitis. The human appendix does have some immune activity and may fight infections in the gut. In the next section, we will see how the human digestive tract works on its normal omnivorous diet.

≋ THE HUMAN DIGESTIVE SYSTEM

The length of the human digestive tract (Figure 23.8), from mouth to anus, is roughly 8 m, or about the height of a two-story house. All the sandwiches, brownies, apples, milk, and other foods a person eats pass through the central cavity of this tube and undergo one digestive process after another. To see in detail how the human alimentary canal liberates the nutrients in foods and transports them across cell layers of the gut tube, we will follow a food item—a turkey sandwich, let's say—throughout each step of its digestive journey.

THE MOUTH, PHARYNX, AND ESOPHAGUS: MECHANICAL BREAKDOWN OF FOOD

Your teeth are superbly adapted to the job of cutting, tearing, and grinding the plant and animal tissues you consume as an omnivorous mammal. (By contrast, a koala's teeth—especially its broad, flat molars—are well suited for grinding the fibrous, leathery leaves it eats all day and night.) Teeth can stand up to this regular wear and tear because the enamel covering them is the hardest substance in the body: The enamel on a person's permanent teeth will generate sparks if struck against steel. Working with the teeth is a muscular *tongue*, the principal organ of taste, but also, in our species, an organ for forming the sounds of spoken language. When you take a bite of a sandwich, your tongue moves some of the food toward the molars for grinding and some to the incisors for cutting, and shapes each small bite into a soft, moist lump that can be easily swallowed.

At the same time, each bite is also mixed with clear, watery **saliva,** which is secreted by three large pairs of *salivary glands* that lie in the tissues surrounding the oral cavity. Saliva contains primarily mucus and water, which moisten food particles and help them cling together in a moist lump. Saliva also contains small amounts of an enzyme that begins the breakdown of carbohydrates in the bread of a sandwich.

Swallowing often begins as a voluntary act once food reaches the back of the mouth. As the tongue pushes the lump of food up and back against the roof of the mouth, however, sensitive touch receptors trigger the reflex action of the pharynx. When this happens, the *soft palate* at the very back of the mouth rises to prevent food from entering the nasal cavities, and the flaplike epiglottis moves backward and downward to close off the opening to the trachea (windpipe). Simultaneously, the larynx (voice box) moves up, and muscle contractions in the throat push food or liquid into the **esophagus,** the "pipeline" to the stomach. By placing your fingers lightly on either side of your throat just below your chin as you swallow, you can feel the larynx move up and forward to open the esophagus to receive food.

To enter the stomach, a lump of food must pass through a smooth ring of muscle, or *sphincter,* located at the junction of the stomach and esophagus (Figure 23.9). This sphincter (the *cardiac sphincter*) usually remains tightly contracted, like the drawstring on a purse, and thus can prevent the stomach's contents from moving back up into the esophagus. The stomach's gastric juices are so acidic that they could severely damage the esophagus, which lacks the stomach's heavy mucous lining. People often experience "heartburn" when the stomach is very full; this burning sensation is due to small amounts of stomach acid seeping out past the sphincter into the esophagus.

THE STOMACH: FOOD STORAGE AND THE START OF CHEMICAL BREAKDOWN

Swallowing causes muscles in the esophagus to relax, enabling the sphincter to open and a bite of turkey sand-

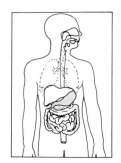
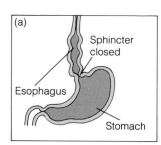
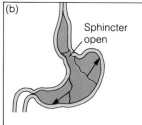
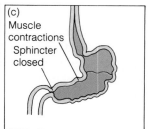
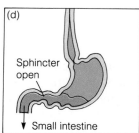

FIGURE 23.9 *Function of the Stomach: Churning, Secretion, and Initial Food Breakdown.* A food lump (a) passes through the cardiac sphincter and (b) enters the stomach, where (c) muscular churning (due to sequential contractions, or peristaltic waves) mixes the food with digestive enzymes and stomach acid. (d) The pyloric sphincter opens briefly and allows chyme to be squirted into the upper part of the small intestine for further digestion.

wich to pass into the elastic J-shaped bag called the **stomach** (Figure 23.9b). The stomach not only stores food for later processing (an adaptation that enables large animals to feed less often), but also mixes and churns its contents with digestive juices through rhythmic contractions of the stomach wall's muscle layers (Figure 23.9c). The contractions are called *peristalsis,* and the churning they produce helps expose as much of the food surface as possible to enzymes.

The average human stomach can comfortably hold about 1 liter, but the organ can stretch to accommodate much larger capacities. The notorious glutton Diamond Jim Brady reportedly ate *each day* several dozen clams and oysters, steak, roast beef, lobster, duck, several chickens, pheasant, two boxes of chocolates, several gallons of orange juice, pastries, ice cream, and more. At Brady's death, a coroner's report showed that his stomach was six times larger than that of a normal man!

When the stomach is full, waves of peristalsis in the stomach wall churn and mix the contents with an acid bath of gastric juices secreted by *gastric pits* in the stomach's innermost cell layer. This layer also secretes mucus, which protects the stomach wall from its own digestive enzymes. Gastric juices are a mixture of water, hydrochloric acid, mucus, and a precursor to the protein-cleaving enzyme called *pepsin.* The hydrochloric acid makes gastric juice acidic enough to kill off most bacteria or fungi contaminating foods. This acid also contributes to the breakdown of food pieces into constituent protein fibers, fat globules, and so on, and it allows the pepsin precursor to be converted to active pepsin. Pepsin cleaves chemical bonds in proteins. Therefore, during the time that protein-containing foods (such as the turkey in a sandwich) are in the stomach, they are partially digested to short polypeptide segments. Although starch digestion begins in the mouth, once food enters the stomach, very little additional digestion of starches and other carbohydrates (or of fats, such as those in mayonnaise) takes place. This must await passage of foods to the small intestine.

The result of the chemical activity and "mixing waves" in the stomach is a pasty, milky, and highly acidic soup called *chyme.* The mixing and churning of chyme gradually move it toward the lower stomach, where another sphincter (the *pyloric sphincter*) controls the opening to the small intestine (Figure 23.9c and d). As the chyme is pushed against that opening, the sphincter relaxes just long enough for small, carefully regulated "doses"—on average, about a teaspoonful every 3 seconds after a meal—to be squirted into the small intestine so that further digestion and absorption can take place efficiently. It usually takes between 1 and 4 hours for a meal to be processed in the stomach and delivered, spoonful by spoonful, to the small intestine—less time for a high-carbohydrate meal, more time for a fatty meal.

THE SMALL INTESTINE AND ACCESSORY ORGANS: DIGESTION ENDS, ABSORPTION BEGINS

Most of the chemical digestion of food and some of the absorption of nutrients take place in the small intestine, but this 6 m (19.5 ft) stretch of the gut tube does not accomplish these tasks alone; three accessory organs— the pancreas, liver, and gallbladder—add in substances that aid digestion and absorption.

■ **The Pancreas and Its Digestive Enzymes**   The **pancreas** is a narrow, lumpy organ situated near the junction of the stomach and small intestine (see Figure 23.8). It produces a host of digestive enzymes and secretes them into the small intestine, where food digestion continues after chyme leaves the stomach. These pancreatic enzymes help break down proteins, fats, and carbohydrates. What's more, the pancreas also secretes bicarbonate ions (the main ingredient in indigestion remedies), which help to neutralize acid entering the small intestine from the stomach. This neutralization is vital, because unlike enzymes in the stomach, pancreatic enzymes cannot function in an acidic environment and the acid could damage the small intestine.

Some pancreatic cells secrete the hormones *insulin* and *glucagon* directly into the bloodstream. These hormones help keep steady levels of glucose in the blood.

■ **The Multipurpose Liver and the Gallbladder**   The smooth, irregularly lobed and roughly hemispherical **liver** is the largest gland in the human body (see Figure 23.8). It weighs about 2 kg (4 lb) in an adult and performs biological tasks as diverse as destroying aging red blood cells, storing glycogen, and dispersing glucose to the bloodstream as circulating levels drop. One of the liver's most important functions is to produce *bile salts*—modified cholesterol molecules that act like detergents to help break up fat droplets in the small intestine. Bile salts travel through the bile duct to the **gallbladder,** a small, pear-shaped sac on the underside of the liver, where they are stored as **bile,** a yellow-green liquid containing bile salts, pigments from the breakdown of red blood cells, and other substances. The presence of fats in the small intestine stimulates the gallbladder to release drops of bile down a duct into the small intestine.

Bile salts keep the fat globules in chyme from clumping together and disperse the globules into thousands of tiny droplets, making it easier for fat-digesting enzymes to break down the lipids in the droplets into usable constituents. Bile salts also help "soak up" fully digested fat molecules and transport them to cells in the lining of the small intestine, where they pass out of the intestines and eventually enter the lymph. Doctors sometimes give people who suffer from high levels of blood cholesterol a compound called cholestyramine, which binds to bile salts and passes down the intestinal tract without being absorbed. This prevents the recycling of cholesterol from

bile salts and thus decreases the amount of cholesterol in the blood. Some researchers believe that oat bran may work in a similar way to lower blood cholesterol.

Besides manufacturing the bile stored in the gallbladder, the liver picks up, stores, and sometimes synthesizes amino acids, glucose, and glycogen and stores vitamins and other compounds that cells need to function normally. Certain liver cells also contain special enzymes that can detoxify poisons. For example, the liver transforms molecules of ammonia—a toxic, nitrogen-containing waste created by the breakdown of amino acids—into urea, which is less toxic and is excreted in urine. The liver also detoxifies alcohol; but if the organ is overloaded with the drug year after year, it can become damaged and irreversibly scarred, producing a potentially fatal condition called *cirrhosis*.

■ **The Lengthy Small Intestine** The small intestine is a remarkable coiled tube about 6 m long where carbohydrates and fats are primarily digested and where protein digestion is completed to the stage where nutrients can be absorbed into the blood. The small intestine begins just below the stomach and has three main regions along its length: the upper section, or *duodenum;* the central *jejunum;* and the remainder, or *ileum.*

The small intestine is a marvel of compact biological engineering. If the surface of its inner lining were a smooth tube like a garden hose, there would be a relatively small surface area for digestion and absorption. Instead, however, that inner lining is so convoluted that it houses a huge absorptive surface. The intestinal lining is pleated into large numbers of folds (Figure 23.10a), and each fold is covered with fingerlike extensions known as *villi* (singular, villus), which project into the lumen and come into contact with its contents (Figure 23.10b and c). Further, the outer layer of each villus is carpeted with *microvilli*, microscopic brushlike projections of the cell membrane (see Figure 23.10c). The combination of folds, villi, and microvilli creates a total surface area the size of a tennis court.

By the time a turkey sandwich reaches the small intestine, it contains partially digested carbohydrates and proteins, as well as undigested fat. Here, pancreatic enzymes and other enzymes secreted by the intestine combine with bicarbonate ions and bile salts and gradually complete the digestion of the carbohydrates in bread to simple sugars, the proteins in turkey meat to amino acids, and the fats in mayonnaise to fatty acids and glycerol. These nutrient molecules are small enough to move across the plasma membranes of the microvilli and enter the intestinal cells. The amino acids and sugars then pass into blood capillaries along with most of the available

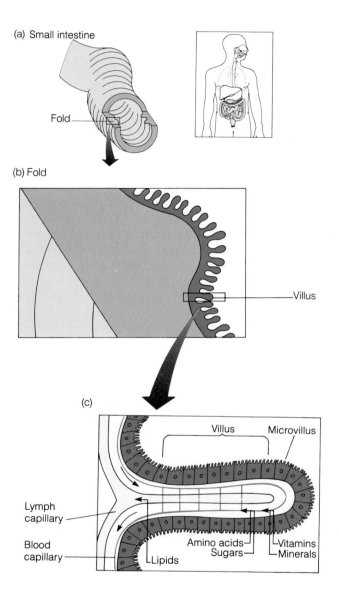

FIGURE 23.10 *Highly Absorptive Lining of the Small Intestine: Folds Bearing "Fingers" Topped with "Brushes."* Intestinal folds (a) are covered with villi (b). Each villus "finger" (c) encompasses blood and lymph capillaries that carry away nutrients absorbed through the brushlike carpet of microvilli with their enormous combined surface area.

vitamins and minerals, while lipids enter lymph capillaries (see Figure 23.10c). The remaining contents of the small intestine pass into the large intestine.

THE LARGE INTESTINE: SITE OF WATER ABSORPTION ⌐incorrect

The last 1.2 m (4 ft) of the alimentary canal is the large intestine, or **colon:** a stretch of the gut tube that ascends the right side of the body cavity, crosses the body just below the stomach, then descends the left side and ends in a short tube called the **rectum** (see Figure 23.8). The colon has two main functions: to absorb water, ions, and vitamins from the chyme and to store the semisolid undigested wastes, or **feces,** until they are excreted.

The large intestine is twice as wide as the small intestine, but its walls have a much simpler structure: They lack the many folds and villi of the small intestine. The smoother surface has a smaller surface area for absorption, but it also presents less resistance to the movement of materials through the tube.

Each day, about 0.5 liter of material (now minus the nutrients absorbed in the small intestine) reaches the colon along with 2–3 liters of water, some from food and the rest secreted by the stomach and intestines themselves. The body cannot afford to lose this much water, however, and so water is reabsorbed as the material moves slowly through the colon over a period of 12–36 hours. This absorption gradually transforms the intestine's contents from fluid to semisolid, and the wastes (including undigestible cellulose fiber from the whole wheat bread and lettuce of the turkey sandwich we started with) are stored until their pressure against the colon wall triggers a bowel movement, or muscular expulsion of wastes through the rectum and out the anus.

Nutritionists agree that a diet low in fiber produces a low volume of material in the colon, including less bound water. As a result, wastes move through the gut more slowly, toxins and carcinogens in the wastes may be in contact with the gut tissue longer, blood flow in the gut may be sluggish, and irritations and infections can develop. The answer is to get plenty of fiber in the diet, and while many people turn to oat or wheat bran, carrot fiber is five to six times more effective at binding water (and thus adding bulk to the feces), and the fiber in an apple is helpful in reducing blood cholesterol. Bran in breads and cereals is also helpful, but is no substitute for raw fruits and vegetables.

Digestion is never 100 percent efficient, and some nutrients invariably pass into the large intestine from the small. A variety of bacteria, including *Escherichia coli* and lactobacillus and streptococcus species, reside in the human intestine and live off these remaining nutrients;

in the process, they produce a number of vitamins, including thiamine (vitamin B_1), vitamin B_{12}, riboflavin, and vitamin K. The colon absorbs these vitamins along with fluids. Many animals, including termites and herbivores, which ingest large amounts of cellulose, also benefit from the metabolic activities of "live-in" microorganisms. Without its cecum and the enclosed bacteria releasing carbohydrates and vitamins from eucalyptus leaves, the koala would starve. Likewise, without their own types of fermenting chambers and enclosed bacteria, cows and horses could never extract enough nutrients and calories from grass and hay.

≋ COORDINATION OF DIGESTION: A QUESTION OF TIMING

As we have seen, different parts of the digestive system carry out different, highly specialized tasks, and the result is an efficient use of food and nutrients. These tasks must be controlled and coordinated so that the cells get their steady input of energy and nutrients, even if meals are separated by long intervals, and so that the strong stomach acid and powerful enzymes do not dissolve the gut itself.

The close control of digestion is achieved by the body's two great coordinators: the nervous system and the hormonal system with its chemical messengers. Nerves throughout the digestive tract allow communication between "downstream" and "upstream" regions so that the propulsion of food can be speeded or slowed appropriately. Nerve activity in the brain and spinal cord can also speed up or slow down digestive functions. For example, the sight, taste, smell, or even thought of food can cause signals from the brain to travel via nerves to the salivary glands, causing them to secrete saliva. The brain signals also travel to secretory glands in the stomach lining, inducing gastric juices to begin flowing into the stomach. Food arriving in the stomach and pushing against the stomach wall can of course trigger the same response.

Food in the stomach also has another effect: It initiates a feedback loop that regulates stomach acidity and helps speed the breakdown of the food. Thus, food helps to stimulate its own digestion. People who are emotionally stressed for prolonged periods are susceptible to forming *ulcers*, craterlike sores in the lining of the stomach or small intestine. Apparently, ulcer victims have less mucus protecting the stomach lining and/or excessive gastric secretions—a combination that allows small portions of the stomach or intestinal lining to be digested away.

Several hormones also help coordinate the timing and amount of enzyme secretion with the presence of food.

For example, partially digested proteins cause the small intestine to release a hormone that reaches the pancreas through the bloodstream and triggers that organ to release more digestive juices with their high concentration of protein-digesting enzymes. This same hormone can also travel from the intestine to regulatory centers in the brain and help produce the sensation of being "full." This is why nutritionists often advise dieters to eat the protein foods in a meal first so they will feel satisfied sooner.

Working together, nerves and hormones fine-tune the secretion of digestive juices so that enzymes and ions are instantly available to break food down into nutrients for the body's trillions of cells, and yet those powerful agents are present in the alimentary canal only when needed—when the animal has hunted or harvested food and consumed it.

≋ CONNECTIONS

Digestion provides nutrients for all the cells in an animal's body, enabling the organism to live from day to day, reproduce, and leave progeny. An animal's physiology and anatomy are thoroughly tied to its diet: Intense competition for nutrients selects individuals that can exploit new environments, thus driving the evolution of mouthparts, digestive organs, and behaviors that give them an advantage in obtaining their own often highly specialized "slice of the pie." The koala with its eucalyptus leaves—a diet that would starve or poison a human—is a prime example of that process at work.

This chapter has focused mainly on the sequential steps by which the digestive system efficiently extracts nutrients from food and absorbs them for use by the body's vast collection of cells. In Chapter 24, we will consider a closely related system that is directly affected by the foods and liquids an animal consumes: the kidneys or equivalent organs, which regulate the balance of water and salts in the blood and body fluids.

NEW TERMS

alimentary canal, page 394
bile, page 398
colon, page 400
digestion, page 393
esophagus, page 397
essential amino acid, page 387
extracellular digestion, page 393
feces, page 400
gallbladder, page 398

intracellular digestion, 393
liver, page 398
mineral, page 388
nutrient, page 385
nutrition, page 386
pancreas, page 398
rectum, page 400
saliva, page 397
stomach, page 398
vitamin, page 388

STUDY QUESTIONS

REVIEW WHAT YOU HAVE LEARNED

1. How do herbivores, carnivores, and omnivores differ? Give an example of each.
2. What foods provide carbohydrates?
3. What simple nutrients are produced during the digestion of carbohydrates, lipids, and proteins?
4. How do fats benefit the body?
5. How does the body use proteins?
6. What essential role do most vitamins fulfill?
7. Which six minerals does the body require? Why is each necessary?
8. What kinds of mechanical digestion occur in the mouth and stomach?
9. Give the details of chemical digestion in the stomach.
10. What is the role of the pancreas in digestion?
11. What is the liver's role in digestion?
12. How do bile salts aid in the digestion and absorption of fats?
13. What digestive events take place in the small intestine?
14. List the functions of the colon.
15. Explain how nerves and hormones coordinate digestion.

APPLY WHAT YOU HAVE LEARNED

1. How can vegetarians ensure that they take in an adequate supply of essential amino acids?
2. Are rose hips from a health food store a better source of vitamin C than orange juice from the supermarket? Explain.
3. How does the drug cholestyramine help lower the level of blood cholesterol?

FOR FURTHER READING

Campbell-Platt, G. "The Food We Eat." *New Scientist* 19 (May 1988): 1–4.

Degabriele, R. "The Physiology of the Koala." *Scientific American* 243 (July 1980): 110–118.

Frisch, R. E. "Fatness and Fertility." *Scientific American* 258 (March 1988): 88–95.

Prentice, A. "Human Energy on Tap." *New Scientist* 26 (November 1987): 40–44.

Ratto, T. "The New Science of Weight Control." *Medical Self-Care* (March–April 1987): 25–30.

① Animals must consume sufficient amounts of carbohydrates, fats, proteins, vitamins, and minerals to stay healthy.

Peter Bruegel, *The peasant wedding*

② An animal's diet must also contain enough calories; these provide sufficient energy.

③ Digestion is the mechanical and chemical breakdown of food into small nutrient molecules that can be absorbed into the bloodstream.

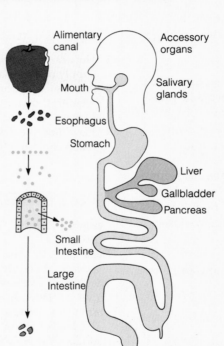

Alimentary canal

Accessory organs

Mouth

Salivary glands

Esophagus

Stomach

Liver

Gallbladder

Pancreas

Small Intestine

Large Intestine

④ Digestion occurs in steps within specialized regions of the alimentary canal and is aided by secretions of accessory organs.

⑤ Nerve impulses and hormones help coordinate the steps of digestion; as a result, body cells receive a steady supply of nutrients and energy.

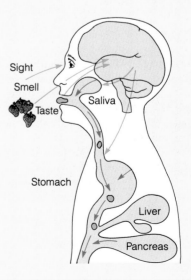

Sight

Smell

Taste

Saliva

Stomach

Liver

Pancreas

SALT AND WATER BALANCE AND WASTE REMOVAL

CHAPTER 24

KANGAROO RATS AND CAPABLE KIDNEYS

If, on some moonlit summer night, you happen to be driving along a country road in southern Arizona or New Mexico, you may chance to see a small but remarkable mammal that is a living lesson on the wonders of the kidney. To catch a glimpse, you must stop the car, sit near a low earthen mound with five or six entrance holes, and wait patiently. If you are lucky, a diminutive grayish rodent resembling a cross between a hamster and a kangaroo will pop out and begin searching for dry seeds (Figure 24.1). This animal, the kangaroo rat, moves with long leaps of its large hind feet and drags its dark-tipped tail along in the sand for balance. If it finds some seeds, it will quickly stuff the food into its cheek pouches with delicate paws, then hop off again in search of more. The truly amazing thing about the kangaroo rat is that it can live in a sweltering, parched environment, eat nothing but dry seeds, and yet *never have to drink water.* You could not survive in this way.

Living things, as we have seen, are made up primarily of water. Water helps stabilize an animal's body temperature and forms the fluid fraction of blood, transporting gases, nutrients, and gaseous wastes to and from body cells. It also contributes to the rapid movement of materials into and out of cells, which depends largely on the concentrations of water and solutes (salts, organic materials, and so on) on both sides of the cell membrane.

In this chapter, we will see that a complex and marvelous organ, the kidney, regulates the balance of salt and water throughout the vertebrate body and also presides over the excretory system, a major physiological network. While an animal's digestive system rids the body of undigested solid wastes, the excretory system, with its kidneys and associated "pipes," cleanses the blood of the nitrogen-containing organic waste molecules released by active cells and carries them out of the body in urine or its equivalent. Regardless of external environment, the kidney maintains a constant and proper internal balance of water and salt.

Two unifying themes occur in this chapter. First, natural selection has varied the shape and function of the kidney and other excretory organs to fit the animal to its environment. In the kangaroo rat, for example, fine tubules in the kidneys are thrown into extremely long loops that absorb water very thoroughly so that very little leaves the body. Second, the kidney's activities require a large expenditure of energy. Without this continuous energy flow and without a healthy functioning kidney, poisonous cellular wastes would build up, and the body fluids would become too salty or too watery for life processes to proceed normally.

This chapter describes how an animal's body excretes wastes and balances salt and water. In so doing, it answers several intriguing questions:

- What are the mechanisms for removing nitrogenous wastes?
- How do the kidney and other excretory organs function?
- How do nerve signals and chemical messengers called hormones regulate the body's water content?
- How do animals living in deserts, oceans, and ponds balance their bodies' salt and water content?

FIGURE 24.1 *Kangaroo Rat: Remarkable Desert Survivor.* Thanks to numerous behavioral and physiological adaptations, the kangaroo rat can survive in the sandy deserts of Arizona and New Mexico without drinking water.

RIDDING THE BODY OF NITROGEN-CONTAINING WASTES

Just as short ends of boards or scraps of cloth remain after you build a bookcase or sew a shirt, residues of nutrient molecules remain unused after an organism digests food chemically into carbohydrates, lipids, or proteins, then breaks these down further through cellular respiration. **Excretion** is the process of removing the by-products of metabolism from the cells, the extracellular fluid, and the blood, as well as ridding the organism of excess water and salts (Figure 24.2). Excretion differs from elimination, the expulsion of undigested food as feces (see Chapter 23), in that undigested solid materials have never been part of the body or used by its cells. Let's look more closely, now, at the molecular by-products of metabolism and the excretion process that removes them.

Cells can burn fats, carbohydrates, and proteins for energy, but the end products of their metabolism differ. After fats and carbohydrates are dismantled, water and carbon dioxide remain, and the gas is eventually expelled from the lungs or through the skin. When cells metabolize proteins, the amino acid building blocks can be used directly to synthesize new proteins or can be broken down further to release energy. Only the carbon-containing (acid) portion of the amino acid molecule, however, is shunted through the energy-producing pathway. The nitrogen-containing amino portion cannot be oxidized and becomes a waste product, like a trimmed-off piece of board or cloth. The leftovers join with hydrogen atoms to become molecules of **ammonia** (NH_3), the strong-smelling alkaline ingredient found in cleaning solutions.

Even at low concentrations, ammonia is highly toxic to cells, and all organisms must get rid of it quickly, either by excreting it directly into water or by converting it to the less toxic compound uric acid or urea (Figure 24.3). Most bony fishes and aquatic invertebrates excrete a large part of their nitrogenous wastes as ammonia, often by releasing it across their gill surfaces into the surrounding water. Animals that live on land, however, as well as some fishes, convert ammonia to less toxic compounds. In birds, the liver cells convert ammonia into insoluble crystals of the chemical **uric acid.** Uric acid is expelled from the body as a white paste called guano. The white portion of the ubiquitous bird droppings that decorate window ledges, park statues, and the occasional human head is expelled uric acid. In regions where large colonies of seabirds nest, whole islands can be whitened with a thick layer of guano.

Evolution has provided a few types of bony fishes, as well as people, kangaroo rats, and other mammals, with a third solution to the problem of ammonia: the production of **urea.** In these animals, ammonia is first detoxified by conversion to urea in the liver; next, the less toxic, water-soluble urea is removed from the blood by the kidney; and finally, it is concentrated in the **urine**—the fluid that washes nitrogen-containing wastes from the body—and excreted through the urinary plumbing system. A special adaptation in American black bears allows these hibernating animals to rid themselves of urea without awakening to urinate all winter (Figure 24.4). Our next section describes how the kidney works as a biological jack-of-all-trades in waste excretion and salt balance.

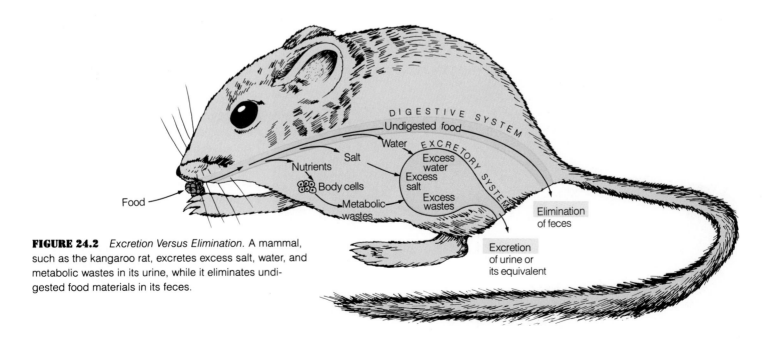

FIGURE 24.2 *Excretion Versus Elimination.* A mammal, such as the kangaroo rat, excretes excess salt, water, and metabolic wastes in its urine, while it eliminates undigested food materials in its feces.

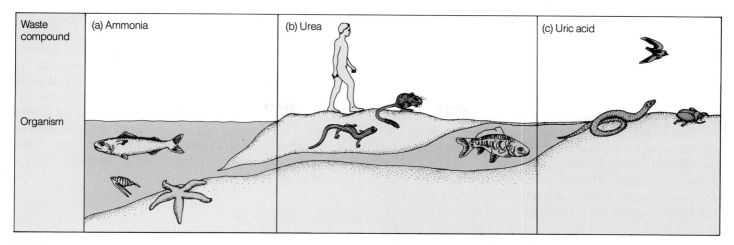

Waste compound	(a) Ammonia	(b) Urea	(c) Uric acid
Organism			

FIGURE 24.3 *Three Kinds of Nitrogenous Waste Products.* (a) Some aquatic organisms, including protozoa, sea stars, and certain fishes, excrete nitrogen-containing wastes in the form of ammonia dissolved in large quantities of water. (b) Others, including amphibians, some fishes, and most mammals, excrete nitrogen-containing wastes as urea in a smaller amount of water. (c) Still others, including birds, reptiles, and insects, excrete nitrogen-containing wastes in a paste of uric acid crystals containing almost no water.

FIGURE 24.4 *Uses for Nitrogenous Wastes.* The bladder (a urine-storing organ) of a hibernating bear absorbs urea, and the waste passes into the blood, where it is carried to the gut; there, bacteria break down the urea and use its nitrogen to make amino acids. The bear's body can then reabsorb and use the nitrogen in the amino acids to build new proteins and nucleic acids, thereby recycling the wastes into useful materials.

≋ THE KIDNEY: MASTER ORGAN OF WASTE REMOVAL, WATER RECYCLING, AND SALT BALANCE

For many diners, eating tender, pale green shoots of asparagus is a pleasurable springtime event, but the gastronomic experience has a peculiar sequel: The next time they urinate, even if just 20 minutes after eating, they notice the characteristic scent of asparagus. A chemical in the food crosses the gut, enters the bloodstream, is filtered out by the kidneys, and appears in the urine with amazing speed. Actually, the chemical is acted on no faster than any other compound. The kidneys are simply marvels at processing body fluids and filtering out the urea; the sodium, potassium, or chloride ions; and the glucose, water, and other materials that need to be excreted. The key to a kidney's rapid functioning lies in its complicated internal structure and in the efficient plumbing system of which it is a part. The human kidney serves as a good model for kidney structure and function in vertebrates.

THE KIDNEYS AND OTHER ORGANS OF THE HUMAN EXCRETORY SYSTEM

It is sometimes stated that urination is our most compelling bodily function. You will probably agree if you have ever been hungry, tired, cold, and had a full bladder simultaneously and can recall the order in which you addressed those needs. This urgency has to do with the necessary waste removal role of the excretory system and with that system's particular anatomy.

In humans, as in other vertebrates, the main organs of the excretory system are the two plump, dark red, crescent-shaped **kidneys,** each about the size of an adult's fist and located high in the abdominal cavity behind the stomach and liver (Figure 24.5). For each kidney to carry out its vital blood-filtering activities, it must receive a large, steady flow of blood. This it does through the *renal arteries* (from the Latin *renes,* meaning "kidneys"), twin branches directly off the body's main blood vessel, or aorta. Cleansed blood then leaves each kidney in a large *renal vein* that drains into the body's largest vein. As the kidneys filter and remove excess water and waste substances from the blood, these materials collect as a concentrated urine in a central cavity in each kidney, then flow down a long tube called a **ureter.** The two ureters dump urine into a single storage sac, the urinary **bladder,** which in an adult can hold about 500 ml (a pint) of fluid. The bladder can distend to hold a bit more, but when it is full, it causes considerable pressure on nearby nerves and a feeling of urgency that can temporarily eclipse all other concerns. During **urination,** the urine exits the body through the **urethra,** the single tube that drains the bladder.

A mammal's kidney has three distinct visible zones (Figure 24.6): an outer *renal cortex* zone, where initial blood filtering takes place; a central zone (*renal medulla*), divided into a number of fan-shaped pyramid regions that help conserve water and valuable solutes; and a hollow inner compartment (*renal pelvis*), where urine collects before it passes out of the kidney and down the ureter. The functional units of the kidney, twisted tubules called **nephrons,** reach from the outer cortex down into the central zone and drain into the hollow center (see Figure 24.6).

THE NEPHRON: WORKING UNIT OF THE KIDNEY

Have you ever cleaned out a desk drawer—sorting, saving certain items for future use, discarding others, and then rechecking the drawer one last time? If so, you have approximated the kidney's general sorting and cleansing action and its three basic processes—filtration, reabsorption, and secretion—all of which involve the nephron.

A human kidney contains roughly 1 million nephrons, each one a long, twisted, looping tubule. The tubule has one end that is enlarged and cup-shaped like a punched-in basketball and that lies in close contact with blood capillaries; the other end drains away urine (Figure 24.7). If all the nephrons in an adult's kidney were straightened

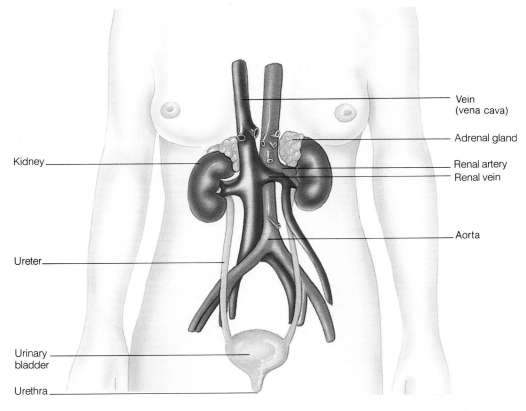

Kidney

Ureter

Urinary bladder

Urethra

Vein (vena cava)

Adrenal gland

Renal artery
Renal vein

Aorta

FIGURE 24.5 *The Human Excretory System.* The kidney, bladder, drainage tubes, and major blood vessels make up the simple but critically important excretory system, with its roles in waste removal and control of the body's internal fluids.

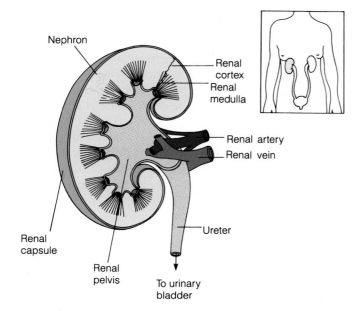

Nephron

Renal cortex

Renal medulla

Renal artery

Renal vein

Ureter

Renal capsule

Renal pelvis

To urinary bladder

FIGURE 24.6 *Anatomy of the Human Kidney: A Blood-Cleansing Organ.* A cross section of the kidney reveals millions of twisted tubules called nephrons (here, one is shown in yellow, much thicker than in reality). Each nephron stretches from the kidney's outer cortex, through its medulla, and into the renal pelvis. The renal pelvis connects to the ureter, the tube that conducts urine away from the kidney.

enclosing capsule of kidney tissue surrounding a tuft of blood capillaries forms the closest contact between the excretory and circulatory systems. A finely meshed filter lies between the blood and the cavity of the Bowman's capsule like a submicroscopic colander (see Figure 24.7b). As blood moves along through the capillaries, blood pressure forces the yellowish fluid portion of blood out of the capillaries, through the pores of this delicate filter, and into the Bowman's capsule, just as water pressure forces coffee through a coffee filter. The fineness of the

out and placed end to end, they would form a microscopically slender tube about 80 km (50 miles) long.

Nephrons are amazingly efficient at selectively removing wastes from the blood circulation while simultaneously conserving water, mineral ions, glucose, and other needed materials. In an adult human, 180 liters or more of blood—enough to fill a bathtub—pass through the 2 million nephrons in the two kidneys each day. We do not, of course, urinate 180 liters a day or even the 6 or 7 liters of fluid our bodies contain, but instead the much more reasonable amount of about 1.5 liters per day. Clearly, the nephrons must accomplish a great deal of water recycling and conservation before they produce that smaller quantity of urine. Let us now take a close look at how a nephron works.

The nephron's specialized shape enables it to carry out three basic processes—**filtration** of small molecules from the blood, **reabsorption** of useful molecules from the urine back into the blood, and **secretion** of ions and some drugs from the blood into the urine.

■ **Filtration** The portion of the nephron that filters water and solutes from the blood is the enlarged, cuplike *Bowman's capsule* in the kidney's outer cortex. Each Bowman's capsule surrounds a tight bundle of capillaries called a *glomerulus*, much the way the fingers of your right hand can wrap around your left fist (see Figure 24.7a). This

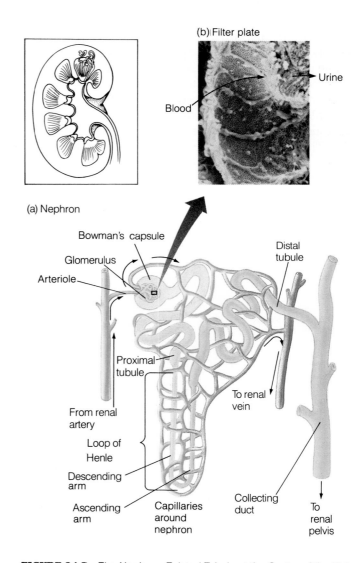

(b) Filter plate

Urine

Blood

(a) Nephron

Bowman's capsule

Glomerulus

Arteriole

Distal tubule

Proximal tubule

From renal artery

To renal vein

Loop of Henle

Descending arm

Ascending arm

Capillaries around nephron

Collecting duct

To renal pelvis

FIGURE 24.7 *The Nephron: Twisted Tubule at the Center of the Kidney's Function.* (a) Each human kidney contains 1 million nephrons, and every unit has a Bowman's capsule, a proximal tubule, an elongated loop of Henle, a distal tubule, and a collecting duct leading to the renal pelvis. (b) Small molecules filter through tiny holes in the thin membrane that separates the blood from the contents of the nephron.

pores is ultimately responsible for the nephron's blood filtration activities: The Bowman's capsule filter allows wastes, water, and other small molecules to seep through its minute openings, while the crucial components of blood—red cells, antibodies, and other large proteins—remain behind in the capillaries. The passage of water, of sodium, potassium, and chloride ions, and of sugars, amino acids, and urea out of the capillaries and into the lumen of the Bowman's capsule is passive and does not require any special output of energy other than blood pressure (Figure 24.8, step 1). The fluid, or **filtrate,** in the capsule is still very much like blood plasma, except that it contains no large proteins or cells.

■ **Tubular Reabsorption** As it leaves the Bowman's capsule, the filtrate enters the part of the nephron's twisted tubule nearest the capsule, known as the *proximal* (near) *tubule* (see Figure 24.7a). There, reabsorption begins and returns to the circulation most of the water, the sodium and chloride ions, the sugars, and the amino acids that were just filtered out of the blood in the Bowman's capsule. These materials move out through the walls of the

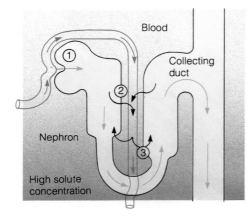

1. *Filtration:*
   Water and other
   small molecules
   filtered into nephron

2. *Tubular*
   reabsorption:
   Water, salts, and
   nutrients returned
   to blood

3. *Tubular*
   secretion:
   Some ions and
   drugs secreted
   into nephrons

Blood

Collecting
duct

Nephron

High solute
concentration

FIGURE 24.8 *Filtration, Reabsorption, Secretion: Basic Roles of the Nephron.* Water and solutes like urea are filtered from the blood and enter the nephron at the Bowman's capsule (step 1). As the filtrate flows through the nephron's looped and twisted tubule, much of the water and valuable solutes are reabsorbed into the bloodstream (step 2). The gradient of salt concentration, here denoted by deepening shades of blue, is mainly responsible for the reabsorption of water. Simultaneously, regions of the nephron secrete various wastes, drugs, and other materials into the urinary fluid (step 3). Concentrated urine then passes through the collecting duct, the renal pelvis, and the ureters to the bladder.

proximal tubule and pass back into the blood via the surrounding capillaries. Instead of being driven passively by blood pressure, reabsorption of solutes in the proximal tubule depends on energy-costly active transport across cell membranes. As the ions are being reabsorbed in this way, 80 to 85 percent of the water in the original filtrate follows them passively back into the capillaries via the process of osmosis.

Next, the filtrate passes down the *descending arm* of the nephron's U-shaped **loop of Henle,** which is bathed by an increasingly salty fluid (see Figure 24.8, step 2). Water leaves the loop of Henle by osmosis and reenters the blood of the capillaries surrounding the tubule (see Figure 24.7a). The filtrate then rounds the bend in the loop of Henle and moves up the *ascending arm* and through the *distal tubule* (far from the Bowman's capsule) toward the **collecting duct.** The walls of the distal tubule are not permeable to water, but those of the collecting duct are; and even though the filtrate passing through the duct is already quite concentrated, the surrounding tissues are even saltier. Thus, more water diffuses out of the collecting duct by osmosis and enters the blood of the surrounding capillaries.

By the time the filtrate (now called urine) has reached the part of the collecting tubule in the innermost (and saltiest) region of the kidney's medulla, 99 percent of the water originally filtered from the blood in the Bowman's capsule has been reabsorbed and returned to the body's circulation.

■ **Tubular Secretion** While the filtration and reabsorption activities of the kidneys' 2 million nephrons are removing salt and urea from the blood and forming urine, another process—tubular secretion—goes on simultaneously in both the proximal and distal tubules. Secretion removes a number of unwanted materials—including hydrogen ions, potassium ions, environmental pollutants like pesticides, and drugs such as penicillin and phenobarbital—from the blood in the capillaries surrounding the nephron and secretes them into the forming urine (see Figure 24.8, step 3).

Tubular secretion is the physiological process that makes *drug testing* possible—checking a person's urine to see if he or she has taken drugs. Drug testing employs various laboratory techniques to detect even minute traces of the metabolic breakdown products of marijuana, cocaine, heroin, sleeping pills, tranquilizers, morphine, codeine, and many kinds of prescription drugs. If a person takes a drug overdose and loses consciousness and no one is sure which drug was consumed, physicians quickly test the urine to identify the drug and determine the best treatment for saving the patient's life. Two additional uses—testing athletes and employees for drug use in the playing field or on the job—are currently quite controversial (Figure 24.9).

FIGURE 24.9 *Kidney Function Makes Drug Testing Possible.* Canadian sprinter Ben Johnson lost his gold medal in the 1988 Olympics after extremely sensitive urine tests revealed tiny amounts of muscle-building anabolic steroids.

KIDNEYS ADAPTED TO THEIR ENVIRONMENTS

So close is the association between a kidney's function and its structure that biologists have discovered nephrons of distinctly different lengths in animals with markedly different life habits. A kangaroo rat in the desert needs to conserve water to an extreme degree, and each of its nephrons possesses a very long loop of Henle. With this modification, its kidneys can reabsorb practically all the water as the filtrate moves through the long loop, and as a result, the animal can produce urine 25 times more concentrated than its own blood—the highest concentration of any mammal.

In contrast, a beaver, which swims all day in ponds and streams, takes in a great deal of water in its food and through its skin and must get rid of the fluid, not conserve it. In a beaver's kidney, the loops of Henle are short and reabsorb less water; as a result, the animal produces great quantities of urine with only twice the solute concentration of its own blood. Humans have a combination of long and short loops of Henle and produce a urine of variable concentration, depending on water availability in the person's environment.

An organ as central to homeostasis as the kidney is impossible to live without and difficult to replace. People with kidney diseases often spend many hours each week connected to a dialysis machine, a large instrument that mimics the functioning of a healthy kidney. The box on page 412 describes this machine and how it works.

≋ REGULATING THE BODY'S WATER CONTENT

While the kangaroo rat and the beaver must deal with extremely dry or wet environments, other animals must cope with surroundings that fluctuate regularly—or irregularly—from wet to dry. An impala living on the plains of East Africa, for example, experiences distinct wet and dry seasons each year, while a person working out at a health club could inhabit a number of environments—an air-conditioned room, a warm, dry tennis court, a hot and humid steam room, a swimming pool— all in a short period of time. The kidneys must be able to conserve water during the dry season or the tennis game but quickly remove it after a drink from a water hole or a glass of iced tea. Indeed, animals have two major mechanisms for regulating the activities of the kidneys and hence the water balance of the body: *thirst,* or the desire to drink, which affects the amount of water taken in, and chemical messengers called *hormones* that alter the activities of the nephrons and hence the amount of water retained or excreted at any given time.

THIRST REGULATES DRINKING AND MAINTAINS WATER BALANCE

Although the kangaroo rat never needs to drink fresh water, most terrestrial organisms do. And even the loss of a small proportion of the body's fluid content—say, 1 to 2 percent from sweating or evaporation through the lungs during a tennis game on a hot day— can cause the concentration of solutes in the blood to rise. These rising levels trigger nerve cells in the brain's thirst center, and the individual becomes aware of thirst. When a person responds by drinking a large amount of water, the water distends the stomach walls and sets up nerve impulses that inhibit the thirst center even before the water is absorbed. Once the water enters the bloodstream from the stomach and small intestine, the concentration of solutes in the blood stabilizes until more water is consumed in foods or beverages or until sweating, exhaling, and forming urine remove fluids along with wastes.

HORMONES ACT ON NEPHRONS TO CONTROL THE EXCRETION OF WATER

Just as you can turn a water faucet on or off to increase or decrease water flow, your hormones can regulate the activity of nephrons so that they remove more water from the blood to the filtrate or return more water from the filtrate to the blood. Two different hormones act to control water loss; one, called **antidiuretic hormone (ADH),** regulates the tubule's permeability to water, while the other, **aldosterone,** regulates salt reabsorption.

To see how this regulation occurs, consider what happens in your body after a salty snack of sardines and potato chips. Your brain detects the increased salt content of your blood and causes a nearby gland to release antidiuretic hormone into the blood. Upon reaching the kidney, the hormone makes parts of the tubules more permeable to water, so more water is reabsorbed into the bloodstream. This reabsorbed water makes the blood less salty and thus counteracts the original stimulus—excess salt—in a negative feedback loop.

As beer drinkers occasionally discover to their dismay, alcohol inhibits the release of this water-reabsorbing hormone, leading to the excretion of sometimes embarrassing quantities of urine. This causes dehydration and is a major reason for the "hangover" a person feels after drinking too much liquor; the body is dehydrated and must replace lost fluid.

The hormones that govern water and salt balance in the kidney can indirectly affect the heart.

THE KIDNEY'S ROLE IN HEART DISEASE

In recent years, physicians have warned people that diets high in sodium can elevate the blood pressure and strain a heart already weakened by atherosclerosis. To counteract high salt content in the blood, the kidney retains more water, thus increasing the volume of blood in the circulatory system. Like the outward-pushing force in an overfilled balloon, the larger volume of blood pushes the person's blood pressure up and forces the heart to work harder. When the blood volume and pressure increase, cells in the heart secrete a hormone (*atrial natriuretic factor*) that causes salt to appear in the urine. Water follows the salt by osmosis, thus lowering the blood volume and hence decreasing the blood pressure. Physicians hope to produce large quantities of the hormone, through genetic engineering techniques, then use it to treat people with chronic high blood pressure; some of these people do not produce enough of the hormone on their own.

The kidneys can also counteract falling blood pressure. When blood flow to kidney cells falls below a certain volume, kidney cells cause formation of a hormone called angiotensin, which makes blood vessels constrict, thereby elevating blood pressure.

The message from this section is simple: Thirst and hormones regulate how much water we drink and how much salt and water we retain, and this, in turn, affects blood content, blood pressure, and other factors vital to homeostasis and health.

≋ STRATEGIES FOR SURVIVAL: HOW ANIMALS BALANCE SALT AND WATER

"Marooned at sea" is a favorite literary theme, and as Samuel Taylor Coleridge wrote in "The Rime of the Ancient Mariner," a sailor in this situation has "water, water everywhere, nor any drop to drink." Why can't a thirsty sailor simply drink seawater? The answer lies with the salt and water balance we have been discussing: Human kidneys cannot make urine as salty as seawater. For every 1 liter of seawater humans consume, we produce 1.3 liters of urine and quickly dehydrate. If a sailor had a pet kangaroo rat, the animal could easily drink from the ocean, since its kidneys make urine more concentrated than seawater. So could a shark, flounder, or sea gull. Animals face different osmotic challenges, depending on their natural environments, and have evolved different kinds of physical adaptations for dealing with them. Let's look at some of the strategies employed in animals for regulating their salt and water balance (a process called **osmoregulation**).

SALT AND WATER BALANCE IN LAND ANIMALS

On land, animals are submerged in a dry ocean of air and thus tend to lose water through evaporation across the body and respiratory surfaces, as well as through excretion of urine or its equivalent. The result, of course, is the need to drink water; the need to dump salt in the sweat and urine and sometimes through special salt glands; and finally, the need to conserve water through physical and behavioral mechanisms. An insect's waxy cuticle, for example, helps hold in water, as do a reptile's dry, scaly skin (Figure 24.10a), a bird's feathers, and a mammal's fur. The moist skin of an earthworm, snail, or amphibian, however, is subject to tremendous evaporative water loss, and so these animals tend to remain in cool, wet places. Desert peoples tend to wear clothing that minimizes evaporation (Figure 24.10b), and they rest at midday, when the temperature is highest. All varieties of land-dwelling animals have organs for balancing and excreting salt and water: the vertebrate's kidney, the insect's

(a)

(b)

FIGURE 24.10 *Land Animals: Body Coverings and Water Regulation.* (a) This prehistoric-looking lizard of Costa Rican forests has a scaly hide that helps the animal retain much of its body moisture. (b) A Bedouin's long white robes block out heat and prevent excess evaporation as he rests and feeds his camel in Egypt's Sinai Desert.

Malpighian tubules, the earthworm's nephridia, and so on (see Chapter 17).

The kangaroo rat, with behavioral, anatomical, and physiological adaptations for conserving water and excreting salt and urea, is a prime example of salt and water balance on land. The kangaroo rat emerges from its burrow only at night when the air is cool and evaporation slow. It has a "cool nose" that helps it conserve water (Figure 24.11). Finally, the kangaroo rat also has large intestines that absorb virtually all the water from feces and eliminate only hard, dry pellets. In contrast to desert-dwelling kangaroo rats, animals that live in water face different problems.

SALT AND WATER BALANCE IN AQUATIC ANIMALS

Aquatic animals are surrounded by water, not air, and for them, evaporation is usually not a major factor. They do, however, face a continual problem of salt and water balance stemming from the difference in salt concentration between their body fluids and the surrounding water.

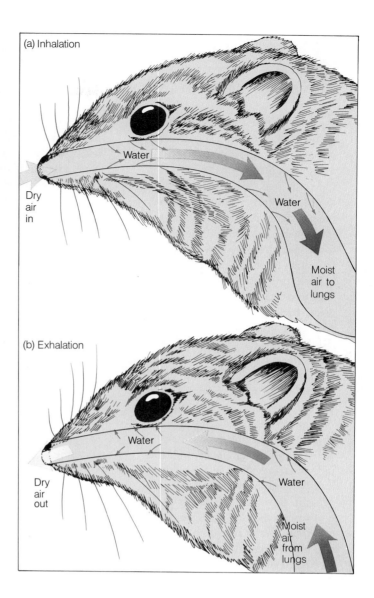

FIGURE 24.11 *A Cool Nose and Water Conservation.* The kangaroo rat's "cool nose" helps prevent water loss. (a) Water evaporates from moist nasal membranes and increases the humidity of the dry air inhaled through the nose. This cools the nose by evaporation, much as sweat cools a person's skin. (b) Upon leaving the lungs, exhaled air is very warm and carries a great deal of moisture. However, as the warm, moist air passes over the cooled nasal membranes, water precipitates out onto the nasal membranes like steam condensing on a cold water pipe. This reclaimed water is then used to help moisten the next batch of dry incoming air.

WHEN THE KIDNEY FAILS

It is easy to take the kidneys for granted: They operate silently, efficiently, and in most people continuously for a lifetime without a glitch. One only need learn about the consequences of kidney failure, however, or meet a victim of this condition, to understand how central the organs are to survival and just how difficult it is to imitate their natural functions.

The kidneys can lose their exquisite ability to cleanse the blood in several ways. Bacteria can contaminate the urinary tract, attacking the kidneys. An autoimmune attack (see Chapter 22) on the kidneys by white blood cells can block and destroy glomeruli. Finally, poisoning by mercury, lead, or certain solvents can damage the kidney tissue. A person's nephrons are so numerous and so efficient that even if two-thirds of these tubules are destroyed, the individual can still live a fairly normal life. If the number drops to 10 or 20 percent, however, the person can suffer extreme tissue swelling as a result of salt and water retention, as well as a buildup of urea and other metabolic by-products. These can cause the blood to become very acidic and can lead to coma or—if the pH drops below 6.9—to death.

Fortunately, in the 1950s, biomedical researchers invented an artificial kidney, or *kidney dialysis machine*, which works via simple diffusion to take over some of the kidney's blood-cleansing functions (Figure 1). When the machine is turned on, waste-laden blood from the patient's artery is routed through a long, porous membrane bathed by a solution much like normal blood plasma. As the blood passes through, the wastes diffuse into the solution (from the region of higher waste concentration to the region of lower waste concentration). After circulating several times through many meters of tubing and back into the body, the patient's blood is sufficiently free from wastes to permit normal activity—at least for a while.

Unfortunately, the artificial kidney has some serious drawbacks: It carries out filtration but neither the reabsorption nor the secretion activities of a living kidney. Moreover, the patient's blood must be treated with an anticoagulant so that it does not clot as it passes through the machine, then treated again with a coagulant as it reenters the person's body so that he or she won't bleed too freely. Because of the drug treatments, the artificial kidney can only be used every two or three days, and because of the slowness of waste diffusion, each session

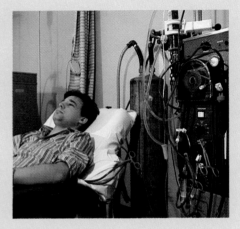

FIGURE 1 *Kidney Dialysis Machine and Patient Having Blood Cleansed.*

may last from 5 to 10 hours. People with kidney failure are literally captives of the dialyser: It extends their life but gobbles much of their time.

Kidney transplants are an alternative to this captivity. Donors are limited to blood relatives or others with closely matching tissue types. And even with this careful screening, the patient may need long-term drug therapy to suppress immune rejection. Thousands of people are saved each year by dialysis machines and transplants. Clearly, though, healthy kidneys are nothing to take for granted.

Various adaptations have evolved in different groups to excrete water, salt, or both.

■ **Aquatic Invertebrates** The easiest solution to salt and water balance is to have body fluids with the same salt concentration as the surrounding seawater, and that is exactly the situation with invertebrates such as oysters and sea stars. Other marine invertebrates, such as crabs and brine shrimp, maintain body fluids at a steady solute concentration, regardless of environment. Brine shrimp can live in fresh water, normal seawater, and even much saltier places, like the Great Salt Lake in Utah or Mono Lake in eastern California, because the animal's exoskeleton impedes water loss or salt movement. In fresh water, where the animal tends to bloat with water but lose salt, a special *excretory gland* expels water while the gills actively transport salt back in. In salty water, the animal's gills work in reverse to pump out excess salt. Because of their highly efficient gills, brine shrimp can live in a wider range of osmotic habitats than nearly any other organism.

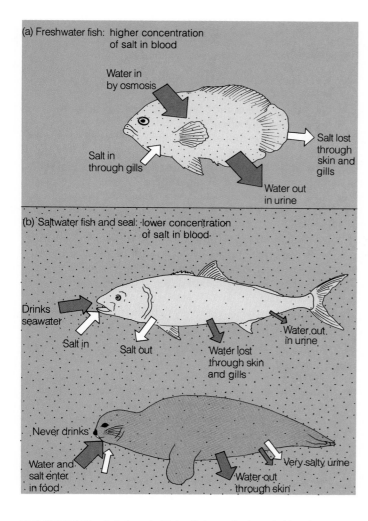

(a) Freshwater fish: higher concentration of salt in blood

Water in by osmosis

Salt in through gills

Salt lost through skin and gills

Water out in urine

(b) Saltwater fish and seal: lower concentration of salt in blood

Drinks seawater

Salt in

Salt out

Water lost through skin and gills

Water out in urine

Never drinks

Water and salt enter in food

Water out through skin

Very salty urine

FIGURE 24.12 *Solutions to Water Balance in Aquatic Vertebrates.* (a) A freshwater fish has a higher concentration of salt in its blood and thus absorbs water from, and loses salt to, the environment. To compensate, its gills actively pump in salt and its kidneys excrete large amounts of water in the urine. (b) Marine fish and marine mammals have blood containing less salt than the surrounding water, and thus, water is lost through the skin. To compensate, the saltwater fish drinks seawater, excretes little urine, and pumps out salt through the gills, whereas the marine mammal does not drink seawater and excretes very salty urine with little water content.

■ **Aquatic Vertebrates** The body fluids of aquatic vertebrates tend to maintain a constant level of salt and water independent of their surroundings. In freshwater fishes and amphibians, the blood has a higher salt concentration than the environment, and so much water flows in across the skin that these animals do not need to drink water; instead, their kidneys produce large amounts of dilute urine to discard the excess water (Figure 24.12a). Unfortunately, salt tends to leave the body as well, and so in freshwater fishes, the gills must expend energy to actively accumulate and absorb salt from the lake or river water. In amphibians, the skin itself can accomplish this active salt transport.

In most marine fishes, reptiles, and mammals, the body fluids are less salty than seawater, and thus, the animals tend to gain excess salt and lose needed water. To compensate, saltwater fishes drink seawater, retain most of the water, and produce very little urine, while their gills actively pump out excess salt (Figure 24.12b). Sea lions, whales, and other marine mammals are a bit like the kangaroo rat in that they never need to drink (taking in salt water only with food), and their very efficient kidneys conserve water strongly and excrete excess salt (see Figure 24.12b). Sea turtles, sea snakes, and other marine reptiles, as well as sea gulls and other oceangoing birds, drink seawater and pump out excess salt through special salt glands on the head (Figure 24.13).

It is clear from these descriptions that an animal's excretory system must work continuously to offset undesirable losses or gains of water or salt from the environment, and this constant work must be fueled by food energy. Homeostasis—whether of body fluid content,

FIGURE 24.13 *Salt Glands Pump Out Excess Salt.* Sea gulls, sea turtles, and many other marine vertebrates have special salt glands that secrete extremely salty droplets of water. Sea gulls can often be seen to shake their beaks rapidly from side to side; in so doing, they are usually shaking off salt droplets excreted from the salt gland.

oxygen levels in the tissues, heart rate, or body temperature—is costly but necessary to survival.

≋ CONNECTIONS

Despite a variety of environments and problems, animals have surprisingly similar means of handling water, salt, and wastes. These include kidneys, salt glands, and gills that can absorb or dump salt as needed.

Although waste removal and water and salt balance fall to the kidneys or equivalent organs, the excretory system interacts at many levels with other physiological systems. The kidney exchanges water and solutes with the circulatory system; it relies on the nervous system to initiate drinking or conserving behaviors; it depends on the digestive system to deliver water to the blood; and it is regulated by hormones that travel from the brain and other body parts. In the next chapter, we will continue our consideration of hormones and the roles they play in the chemical coordination of the body's physiological systems.

NEW TERMS

aldosterone, page 410
ammonia, page 404
antidiuretic hormone (ADH), page 410
bladder, page 406
collecting duct, page 408
excretion, page 404
filtrate, page 408
filtration, page 407
kidney, page 406
loop of Henle, page 408

nephron, page 406
osmoregulation, page 410
reabsorption, page 407
secretion, page 407
urea, page 404
ureter, page 406
urethra, page 406
uric acid, page 404
urination, page 406
urine, page 404

STUDY QUESTIONS

REVIEW WHAT YOU HAVE LEARNED

1. How do excretion and elimination differ? How are they alike?
2. List three types of nitrogen-containing wastes and organisms that excrete them.
3. What is the functional unit of the kidney, and what does it do?
4. What roles do active transport and osmosis play in the functioning of the proximal tubule?
5. How does thirst help maintain the blood's water balance?
6. Explain the connection between hormones and kidney function.
7. Do the kidneys raise or lower blood pressure? Explain.
8. How can brine shrimp live in environments as different as fresh water, dilute seawater, and salty brine?
9. How does osmoregulation differ in freshwater and saltwater fish?
10. Explain how the excretory system interacts with other organ systems.

APPLY WHAT YOU HAVE LEARNED

1. Which kidney function enables sports officials to detect drug use by athletes? Explain.
2. If you drink too much alcohol at a party, you may feel "hung over" the next day. Why?

FOR FURTHER READING

Eckert, R., and D. Randall. *Animal Physiology.* 2d ed. New York: Freeman, 1983.

Hainsworth, F. R. *Animal Physiology: Adaptations in Function.* 2d ed. Reading, MA: Addison-Wesley, 1984.

Leckie, B. "Atrial Natriuretic Peptide: How the Heart Rules the Kidneys." *Nature* 326 (1987): 644–645.

Schmidt-Nielson, K. *Desert Animals: Physiological Problems of Heat and Water.* Chap. 11, "The Kangaroo Rat: A Rat That Never Drinks." New York: Oxford University Press, 1964, pp. 149–178.

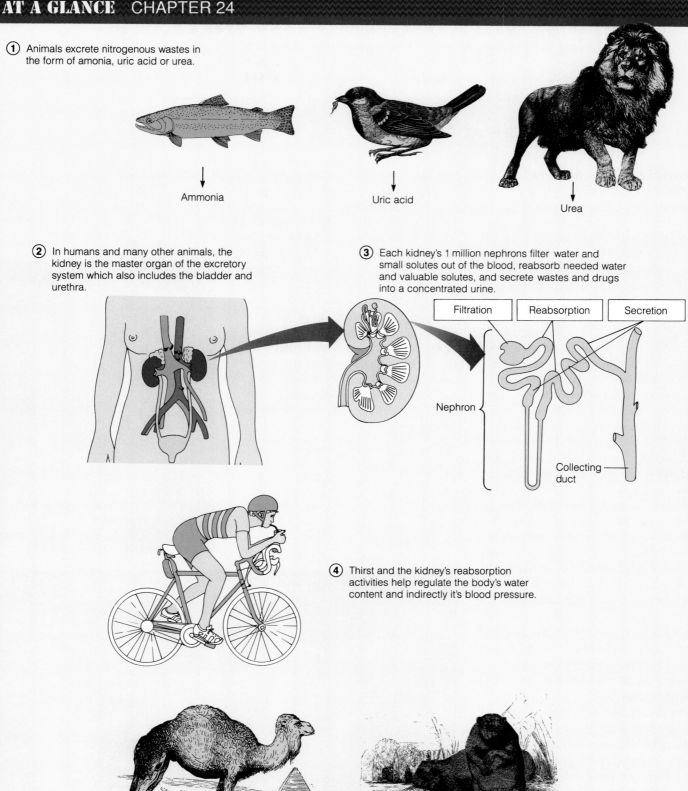

① Animals excrete nitrogenous wastes in the form of amonia, uric acid or urea.

Ammonia

Uric acid

Urea

② In humans and many other animals, the kidney is the master organ of the excretory system which also includes the bladder and urethra.

③ Each kidney's 1 million nephrons filter water and small solutes out of the blood, reabsorb needed water and valuable solutes, and secrete wastes and drugs into a concentrated urine.

Filtration   Reabsorption   Secretion

Nephron

Collecting duct

④ Thirst and the kidney's reabsorption activities help regulate the body's water content and indirectly it's blood pressure.

⑤ The way an animal's body maintains a balance of water and salts depends on it's physiology and environment.

HOW HORMONES GOVERN BODY ACTIVITIES

THE WORM TURNS

On a wild cherry tree deep in a New England forest, a voracious blue-green caterpillar the size of your little finger systematically devours one cherry leaf after another and grows fatter by the day (Figure 25.1a). After four weeks, however, the silkworm abruptly reaches a predetermined weight, stops eating, and begins to spin a slender thread of pure silk. It attaches one end to a small branch, then winds the shiny strand around its body to form a *cocoon*, a protective case within which a dramatic resculpturing soon begins. Muscle cells that moved the larva break down while other cells transform into the wings, legs, and antennae of the future adult moth. The animal, encased in a cocoon and partially transformed, is now a *pupa* and will spend the harsh New England winter in this form.

As spring approaches and the days grow longer, the pupa begins to stir. A hole appears in the cocoon, and over the next half hour, the animal—now a silk *moth* instead of a silk*worm*—extricates itself and unfurls a magnificent pair of red, black, and tan wings (see Figure 25.1b). At this stage in the life cycle, the moth is a flying reproduction machine with huge gonads but no mouth. As it pursues (or is pursued by) a mate, it simply lives off the fat stored during its ravenous youth.

The order and timing of the sequential changes that create a moth from a caterpillar must be closely regulated. A set of molecules called *hormones* do most of the regulating, and these *molecular messengers* are the subject of this chapter.

Two themes will recur as we examine how hormones regulate physical changes in insects, people, and other animals. First, molecular messengers regulate physiological systems by preventing or provoking change. For example, certain hormones in your body prevent change as they maintain a constant concentration of blood sugar, while others trigger the irreversible changes of puberty. Second, very similar hormones occur throughout the kingdoms of living things. In the animal kingdom, for example, a silkworm's transformation to a moth and a child's puberty are caused by steroid hormones. This suggests an ancient evolutionary origin for the regulatory molecules and underscores the relatedness of living organisms.

The discussion in this chapter answers several general questions:

- What are the basic principles by which hormones act in the body?
- What are the major hormones in humans and other mammals, and how do they operate?
- How do hormones prevent or provoke change?

FIGURE 25.1 *Metamorphosis: The Cecropia Silkworm Becomes a Moth.* (a) Cecropia larva devouring cherry leaves. (b) The cecropia moth emerges and its owlish wings unfurl.

(a)

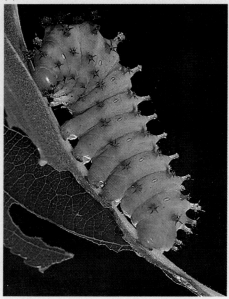

(b)

⚹ HORMONES: AN OVERVIEW

You are a composite reflection of your hormones at work. These chemical regulators control the amount you eat each day and thus your body weight; how quickly you grew during childhood and adolescence; whether you are happy, depressed, or tense; how much salt and sugar flows in your blood; and a thousand other aspects of your physiology and behavior. Hormones are clearly central to survival and normal functioning. But what are they, and how do they work?

The simplest answer is that **hormones** respond to a change or disturbance in the cell's environment by triggering some new physiological activity. More specifically, the hormone molecule is secreted by a **regulator cell,** a cell that detects perturbations in the environment and emits the hormone in response (Figure 25.2). The hormone then diffuses through a fluid medium—either air, water, blood, or the fluid surrounding cells—and acts on one or more **target cells** that receive the hormone and respond by carrying out a cellular activity that helps the organism adjust to the original perturbation.

Each target cell contains **receptors,** proteins with specific shapes, each shape complementary to that of a particular hormone. Some receptors lie in the cytoplasm or nucleus, while others are embedded in the plasma membrane at the surface of the cell. When a hormone comes along and binds to a receptor, like a key fitting into a lock, the receptor changes shape. This alteration heralds the hormone's arrival and either directly or indirectly triggers a change in cell activity.

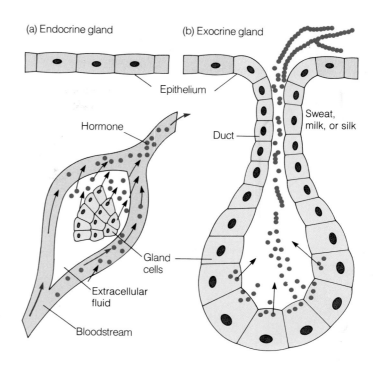

FIGURE 25.3 *Glands: With or Without Ducts.* (a) An endocrine gland secretes hormones into the extracellular fluid. From there, the chemicals pass into blood vessels and travel to distant sites in the body. (b) An exocrine gland secretes milk, sweat, or other materials into a tube or duct, and from there the substance can exit the body.

GLANDS AND HORMONE-SECRETING CELLS

Some hormones are secreted by isolated or individual cells within the animal's body, but most are produced by groups of cells organized into secretory organs called **glands.** A ductless gland that releases hormones directly into the extracellular fluid is called an **endocrine gland** (Figure 25.3a). The adrenal glands that sit atop the kidneys are endocrine glands; they secrete hormones that increase heart and breathing rate. Most hormones are produced in these and about twelve other endocrine glands. In contrast, a gland that dumps materials into a duct that generally leads out of the body is called an **exocrine gland** (Figure 25.3b). Exocrine glands make and release, among other things, milk, sweat, and the silken threads that silkworms spin into cocoons.

Hormones can be categorized by the way they are secreted. *Neurohormones* are secreted by nerve cells, but travel through the bloodstream to target cells elsewhere in the body (Figure 25.4a, page 418). Nerve cell endings in the brain's pituitary gland, for example, release the hormone oxytocin, which travels to the uterus and stim-

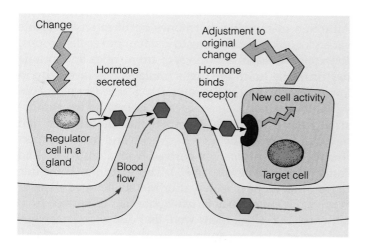

FIGURE 25.2 *A Model of Hormone Action.* The action of hormones is simple but effective. A change in the body or the environment triggers regulator cells, often in an endocrine gland, to secrete a hormone. The hormone travels from the regulator cell through fluid (usually the blood) to a target cell and induces a new activity in that cell. The new cellular activity adjusts the body to the initial change.

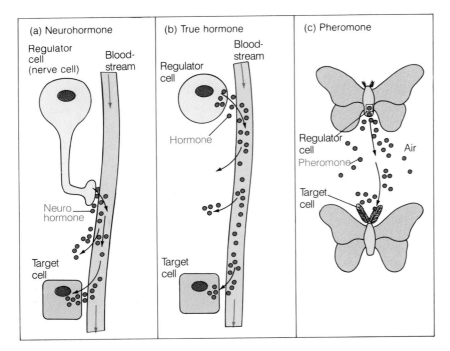

FIGURE 25.4 *Molecular Messengers: Targets and Transport.* Three kinds of molecular messengers are secreted by different types of cells and travel in different ways to their targets. (a) Nerve cells can secrete hormones called neurohormones that pass into a blood vessel, travel some distance, then diffuse out again and reach target cells. (b) True hormones are secreted by regulator cells—usually in endocrine glands—and enter the bloodstream, travel some distance, diffuse out again, and act on target cells. (c) Pheromones are secreted by regulator cells often located in exocrine glands, leave the body, travel in the air or water of the environment, and stimulate target cells in another individual's body.

ulates contractions (see Chapters 13 and 19). *True hormones* are usually secreted by endocrine glands, enter the bloodstream, and act on cells elsewhere in the body (Figure 25.4b). The silkworm hormone ecdysone, for instance, is secreted by a gland in the insect's midregion (thorax) and travels in the blood to target cells throughout the body, where it stimulates adult organs to form.

Finally, cells secrete hormonelike messengers in a third way that is an intriguing variation on the hormone theme. Exocrine glands secrete certain *pheromones*, which leave the body via ducts, travel in air or water, and stimulate target cells in other organisms located some distance away (Figure 25.4c). In a sexually mature female silk moth, for example, a scent gland in the abdomen secretes a pheromone that wafts on air currents for up to 11 km (7 miles) and randomly hits target cells in the male's antennae. This event can trigger wild sexual excitation that leads the male to fly long distances to pursue the female and attempt to mate.

≋ HORMONES AND THE MAMMALIAN ENDOCRINE SYSTEM

Mammals—especially humans—are perhaps the best-studied sources of hormones. Biologists know in great detail, for example, how hormones govern the production and release of eggs, sperm, and milk and how the messengers stimulate the heart to speed or slow. Biologists know the chemical structures of the major hormones; they have mapped out most of the glands; and they understand how these secretory organs operate collectively and in concert with the nervous system to shepherd the various stages of reproduction and development and to maintain homeostasis within each system, organ, and cell.

HORMONE STRUCTURE AND THE ENDOCRINE SYSTEM

Although mammalian hormones perform an incredible variety of tasks, most belong to just four molecular groups. *Polypeptide hormones* are strings of amino acids and include oxytocin, the hormone that stimulates milk secretion in a woman who has just given birth. *Steroid hormones* consist of four joined rings of carbon atoms and are synthesized from cholesterol. Examples include estrogen and testosterone from the human ovaries and testes, respectively. *Amine hormones* are derived from amino acids and contain an amino group (NH_2). These hormones include thyroid hormones, which alter metabolic rates. Finally, *fatty acid hormones,* such as *prostaglandins,* are derived from fatty acids. Prostaglandins were first discovered in semen and named after their supposed source, the prostate gland. They have since been discovered in most mammalian cells and tissues, however, and among other things cause muscles to contract and blood vessels to open or close (see Chapter 13). You can see the effect of prostaglandins on your own body. The next time you have a headache or menstrual cramps, consider that the pain is a result of the body's production of prostaglandins and their ability to cause blood vessels in the brain or muscle fibers in the uterus to contract. Drugs such as aspirin and ibuprofen block the production of prostaglandins and in this way relieve the pain for many people.

Mammals, as well as other vertebrates, have about a dozen major endocrine glands, similar in most cases to the ten human endocrine glands shown in Figure 25.5. Together, these ductless glands make up the *endocrine system,* one of the body's major physiological networks. The rest of this section is devoted to a gland-by-gland tour, beginning with the pituitary and hypothalamus, which together drive the activities of the ovaries, testes, thyroid, and adrenals.

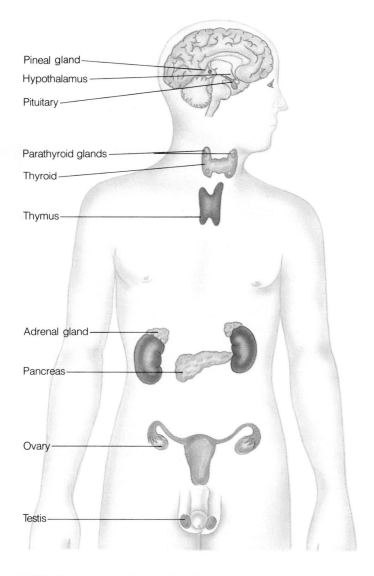

FIGURE 25.5 *Major Glands of the Human Endocrine System.* Ten different endocrine glands secrete hormones that act on target cells nearby or at a distance in the body.

PITUITARY AND HYPOTHALAMUS: CONTROLLING THE CONTROLLERS

At the base of the brain lies a collection of nerve cells called the **hypothalamus,** and hanging from this region is a fingertip-sized bulb called the **pituitary gland** (see Figure 25.5). A bit like a light bulb in structure, the pituitary develops posterior and anterior lobes as a person matures, and it makes about ten kinds of peptide hormones, many of which help control glands throughout the endocrine system. The pituitary was once considered the body's "master gland," but fairly recent evidence shows that the hypothalamus is the "master's master," controlling the posterior and anterior lobes in different ways.

■ **Posterior Pituitary** In the strictest sense, the posterior lobe of the pituitary is more of a storage depot than an actual endocrine gland. This is because the two hormones it secretes (Table 25.1, page 424) are actually made in the hypothalamus and transported to the posterior pituitary in fine extensions of the hypothalamic nerve cells called *axons.*

A nursing mother demonstrates how the posterior pituitary works (Figure 25.6a, page 420). As the baby begins to suckle, sensitive nerves in the woman's nipples send signals to the secretory cells in the hypothalamus (Figure 25.6b, step 1). These secretory cells then send nerve impulses to their axon tips in the posterior pituitary (step 2), which release the hormone oxytocin into the bloodstream (step 3). When oxytocin reaches the musclelike target cells lining the milk ducts of the woman's mammary glands (which are exocrine glands), the hormone causes the target cells to contract, forcing milk into the ducts, out the nipple, and into the baby's mouth.

■ **Anterior Pituitary** Unlike the posterior pituitary, the pituitary's anterior lobe makes and secretes its own hormones (see Table 25.1), but as in the posterior pituitary, the hypothalamus controls the timing of their secretion. The nursing mother can demonstrate how the anterior pituitary works. The anterior pituitary makes a hormone called prolactin that stimulates milk production and secretion. Most of the time, however, a woman's hypothalamus makes a hormone known as a *release-inhibiting factor,* which enters a special circulatory pathway carrying blood directly to the anterior pituitary (Figure 25.6c, step 1). This release-inhibiting factor travels in the blood to the anterior pituitary, where it prevents the cells from secreting prolactin (step 2). When a baby suckles a woman's breast, the activity stimulates nerves that send signals to the hypothalamus, and these signals stop the release of the inhibiting factor. With the chemical inhib-

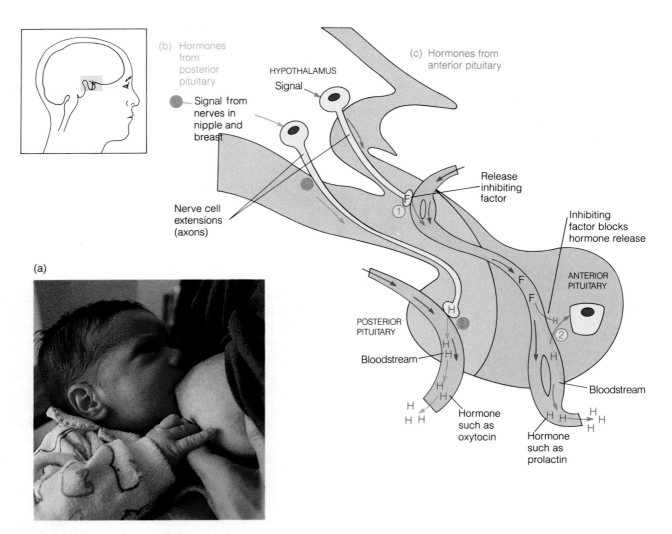

FIGURE 25.6 *How the Hypothalamus Controls the Pituitary.* (a) When a baby nurses, the sucking action signals the mother's hypothalamus and pituitary to release the milk for the current feeding and to produce more milk for a future feeding. (b) A baby's suckling stimulates nerves in the nipple and breast (1) that trigger nerves in the hypothalamus (2) to release into the bloodstream the hormone oxytocin (H) (3) which is stored in the endings of nerve cells in the posterior pituitary. This release stimulates cells in the milk ducts to contract and force milk out. (c) The baby's suckling also signals the hypothalamus to stop secreting inhibiting factors (F) (1), and the anterior pituitary can now secrete prolactin (H) (2), which stimulates milk gland cells to produce more milk. Other anterior and posterior pituitary hormones are released in similar ways, although the initial stimulus is different.

itor removed, the anterior pituitary can secrete prolactin, and the hormone can travel through the bloodstream to the mammary gland cells, which then secrete milk into the ducts. While a release-inhibiting factor from the hypothalamus blocks the anterior pituitary from secreting prolactin, other substances called *hypothalamic releasing factors* trigger the anterior pituitary to secrete other hormones.

■ **Controlling the Other Endocrine Glands** In addition to their control of nursing, the hypothalamus and pituitary participate in several regulatory loops that govern the activities of the testes, ovaries, thyroid, and adrenal glands. In each case, the hypothalamus produces a releasing or inhibiting factor that causes the anterior pituitary to secrete or not secrete a second hormone. If secreted, the second hormone travels in the blood, eventually reaching the gonad or other gland and causing it to secrete a third hormone. As the level of this third hormone builds, it eventually feeds back to the hypothalamus or pituitary and blocks further release of the first hormone. In Chapter 13, we discussed such regulatory circuits for the gonads. Here, we will focus on the thyroid, parathyroid, and adrenals (review Figure 25.5).

THE THYROID AND PARATHYROID GLANDS: REGULATORS OF METABOLISM

The hypothalamus and pituitary together control the **thyroid gland,** a small organ in the neck that acts as the body's metabolic thermostat, regulating its use of energy as well as its growth. The most abundant thyroid hormone is **thyroxine,** an iodine-containing hormone that governs both metabolic and growth rates and stimulates nervous system function. A child with an underactive thyroid gland (a condition called *cretinism*) experiences low rates of protein synthesis and carbohydrate breakdown and thus low body temperature and sluggishness. The child's growth also becomes stunted, and mental development is retarded. An adult with an underactive thyroid gland may gain weight easily and be slow mentally. Fortunately, doctors can treat these conditions by administering thyroxine.

A person can develop a *goiter,* a large lump on the neck caused by an enlarged thyroid (Figure 25.7). This enlargement may be due to feedback loops controlled by regulatory hormones from the pituitary: If the iodine concentration is low, the thyroid does not make much thyroxine, and the feedback loop causes the thyroid to respond by growing larger in a futile attempt to make more thyroxine. The iodine in iodized table salt can prevent goiters from forming.

Some people have an overactive rather than an underactive thyroid, and this can cause weight loss, nervousness, irritability, and emotional instability.

The thyroid gland and the associated **parathyroid glands,** a set of four small, dark patches of cells on top of the pale thyroid, work together to keep the level of calcium in the blood within very narrow limits. Without correct amounts of circulating calcium ions, nerves and muscles cannot function properly, and the person or other animal suffers nerve spasms, muscle contractions, and rapid death. Regulating calcium levels is so crucial that parathyroid hormone is one of only two hormones we absolutely need for survival. The other is aldosterone, a product of the adrenal glands.

THE ADRENALS: THE STRESS GLANDS

Sitting atop each of our kidneys is an **adrenal gland** that enables our bodies to react quickly to danger by fleeing or fighting (Figure 25.8a, page 422). Each adrenal gland has a middle portion, or *medulla,* with an outer covering, or *cortex.*

Cells in the adrenal medulla secrete two similar hormones: *epinephrine* (adrenaline) and *norepinephrine* (see Table 25.1, page 425). When an animal faces a sudden threat, such as a hungry lion or an angry boss, nerves trigger the secretion of the hormones (Figure 25.8b, step 1), which in turn act on target cells in the heart, lungs, intestines, and elsewhere to bring about a so-called *stress response* (step 2). The hormones cause the heart to beat faster, blood sugar levels to increase, the breathing rate to speed up, and blood to be shunted away from certain organs, including stomach and intestines. The result of all this is that the muscles receive more blood, oxygen, and sugar, and thus the animal is better able to defend itself or move away to safety (step 3).

The adrenal cortex, together with the medulla, secretes additional hormones that help an animal respond to stress. One, called **cortisol** (or hydrocortisone), speeds the metabolism of sugars as well as of proteins and fats. When an animal is stressed, perhaps by cold or by strenuous activity, its anterior pituitary secretes *ACTH* (adrenocorticotropic hormone) (Figure 25.8b, step 4), and this triggers the adrenal cortex to produce cortisol and related compounds (step 5). Acting on target cells in the liver, fat tissue, and most other organs, cortisol causes stored proteins, lipids, and carbohydrates to be broken down and glucose to be rapidly generated (step 6), providing stressed cells with a source of quick energy.

Another hormone from the adrenal cortex helps regulate salt in body fluids. The hormone *aldosterone* causes the kidney to conserve sodium and water and to excrete potassium. Without this chemical messenger, the body would excrete too much sodium and water in the urine, the blood volume would drop precipitously, and death

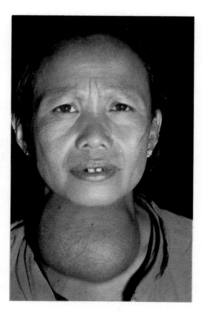

FIGURE 25.7 *Goiter and Hyperthyroidism.* Too little thyroid hormone, caused by a diet low in iodine, can lead to a goiter, a greatly enlarged thyroid gland.

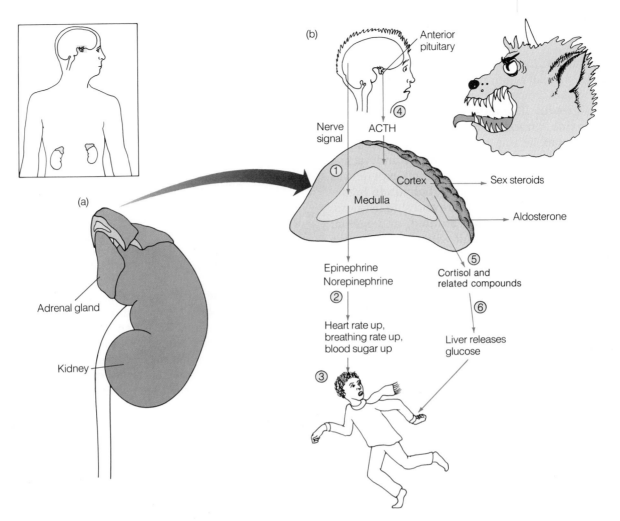

FIGURE 25.8 *Adrenal Gland Hormones and the Stress Response.* (a) The adrenal glands, masters of the fight-or-flight response, are located on top of the kidneys. (b) A sudden physical threat triggers the adrenal glands (by way of nerve signals from the brain) to release epinephrine, cortisol, and other hormones that allow the body to increase its activity and respond to the threat with self-defense (fight) or a rapid exit (flight).

would come quickly. The adrenal cortex also secretes steroid sex hormones (Figure 25.8b and Figure 25.9).

By means of hormones and feedback loops, the adrenal gland helps an animal maintain homeostasis as well as overcome stress. What's at stake here is preventing or promoting change, the major business of all glands and hormones.

≋ HORMONES, HOMEOSTASIS, AND CHANGE

Minute by minute, as the sodium, calcium, or sugar levels in an animal's blood fluctuate up and down, hor-

mones pull each one back into proper homeostatic range. Hormones also regulate daily or yearly cycles and irreversible developmental changes from child to adult or from caterpillar to moth. Because hormones, homeostasis, and change go hand in hand at all levels of physiological activity, their interactions warrant closer examination.

HORMONES AND HOMEOSTASIS

One of the most painful reminders of our need to maintain a constant internal environment is *diabetes*, the inability to maintain blood glucose levels within a normal range. Diabetics produce copious amounts of dilute, sugary urine and are always thirsty. Without treatment, they risk *coma* (unconsciousness) due to severe dehydration or the accumulation of poisons in the blood. Diabetes is the most common hormonal disorder, and it stems from defects in the *pancreas,* an elongated organ nestled near the intestines (Figure 25.10). The normal pancreas has exocrine cells that secrete digestive juices into a duct leading to the central space within the small intestine, as well as islands of endocrine cells called the *islets of Langerhans,* which secrete the hormones **glucagon** and **insulin.**

In a person with a normal pancreas, glucagon and insulin work together by means of a feedback loop to make fine adjustments in blood sugar levels. In a person with diabetes, pancreatic cells fail to generate insulin. Without sufficient insulin, body cells fail to remove glucose from the blood after a meal, so it builds up, and the kidneys remove the sugar along with lots of water from the blood. This results in copious, sweet urine, as well as thirst and danger of dehydration. What's more, without insulin, the liver, brain, and other body cells cannot remove glucose from the blood for use as fuel, and thus they "starve" in the midst of plenty. Although diabetics often crave carbohydrates, eating carbohydrate-rich foods does not help: Their glucose-starved cells burn lipids as fuel, and by-products of lipid breakdown make the blood so acidic that it can lead to coma and death. The push-pull relationship of glucagon and insulin is clearly crucial to homeostasis and day-to-day health. Fortunately, diabetes can now be partially controlled with insulin therapy.

HORMONES AND CYCLIC PHYSIOLOGICAL CHANGES

An animal's life follows daily, monthly, and yearly cycles orchestrated by hormones. Evidence is mounting that

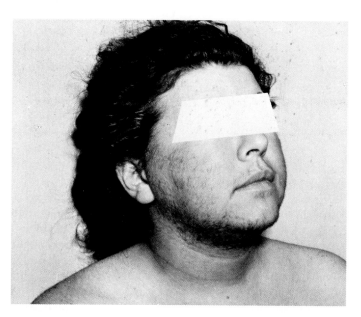

FIGURE 25.9 *Sex Hormones and Facial Hair.* A woman with a beard has an excess of androgens from an overactive adrenal cortex.

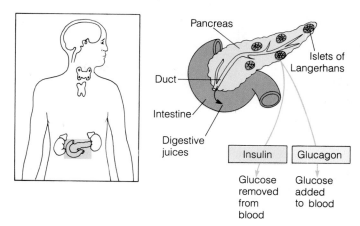

FIGURE 25.10 *Pancreatic Hormones and the Regulation of Blood Sugar.* The pancreas has islets of Langerhans, which contain some cells that secrete glucagon and other cells that secrete insulin. After a healthy person eats carbohydrates, the islets release insulin, and the liver and other cells remove and store glucose units from the food in the form of glycogen. When blood levels of glucose drop, the secretion of glucagon causes the liver to release some of its stored sugar.

such cycles are tied to a **circadian,** or daily, **rhythm** of light and dark periods and to the changing length of daylight during the four seasons. Additional evidence implicates a curious little brain structure, the **pineal gland,** in the measurement of day length and the control of reproductive cycles, onset of puberty, and moods.

The pineal gland—so named for its resemblance to a pine cone—is a bumpy, thimble-sized knob lying deep in the brain (review Figure 25.5). In lizards and certain other vertebrates, the pineal gland is called the third eye because it is structurally similar to the retina and actually perceives light directly. In humans and other mammals, the pineal gets secondhand information in the form of nerve stimulation from light receptors in the eyes. Regardless of species, however, the gland secretes a hormone called *melatonin* in response to darkness, and this hormone promotes sleep and inhibits the activity of gonads.

In people, melatonin may influence reproduction, puberty, and mood. In Finland, it is most common for couples to conceive twins and triplets (a sign of highly active ovaries) in July, when days are very long and melatonin levels are lowest, and it is least common in January, when nights are long and melatonin levels are highest. And just before puberty, melatonin levels in people suddenly drop by 75 percent, suggesting that the pineal hormone may have been inhibiting gonad development during childhood. Research with a class of morphinelike brain chemicals called *endorphins* suggests that these peptides actually regulate the activities of the pineal gland, hypothalamus, pituitary, and gonads and thus may ultimately govern the onset of puberty. Melatonin may have

TABLE 25.1 ≈ MAJOR VERTEBRATE ENDOCRINE TISSUES AND HORMONES

TISSUE	HORMONE	TARGET	MAJOR ACTIONS
Hypothalamus	Releasing and inhibiting hormones	Anterior pituitary	Stimulate or inhibit release of specific pituitary hormones
Anterior pituitary	Thyroid-stimulating hormone (TSH)	Thyroid	Stimulates synthesis and secretion of thyroxine
	Prolactin	Mammary gland	Stimulates milk synthesis
	Adrenocorticotropic hormone (ACTH)	Adrenal cortex	Stimulates synthesis of sex steroids, mineralocorticoids, glucocorticoids
	Growth hormone (GH)	Many cells	Stimulates general body growth
	Luteinizing hormone (LH)	Ovary	Stimulates ovulation and synthesis of estrogen and progesterone
		Testis	Stimulates testosterone synthesis
	Follicle-stimulating hormone (FSH)	Ovary	Stimulates growth of ovarian follicle
		Testis	Stimulates sperm production
Posterior pituitary	Oxytocin	Mammary gland	Milk ejection
		Uterus	Uterine contraction
	Antidiuretic hormone (ADH)	Kidney	Increases water absorption
Thyroid	Thyroxine	Most cells	Increases metabolic rate and growth, causes metamorphosis in amphibians
	Calcitonin	Bones	Stimulates calcium uptake
Parathyroid	Parathyroid hormone	Bones	Stimulates calcium release into blood
		Digestive tract	Stimulates calcium uptake into blood

yet another effect: Some people experience profound depression, oversleeping, weight gain, tiredness, and sadness when the days grow short in winter, then a rebound of better spirits in spring. Melatonin levels are blamed for this seasonal affective disorder syndrome (SADS), and sufferers have been successfully treated by exposure to bright lights for several hours per day in winter.

HORMONES AND PERMANENT DEVELOPMENTAL CHANGE

In addition to orchestrating cyclic change, hormones program an animal's progression through the life stages. As we just saw, a drop of melatonin precedes puberty in human teenagers, contributing to the enlargement of female breasts, the deepening of the male voice, and other accompanying physical changes. More dramatic transformations take place in insects like the silkworm and are mediated by two hormones.

TABLE 25.1 ☆ MAJOR VERTEBRATE ENDOCRINE TISSUES AND HORMONES *(continued)*

TISSUE	HORMONE	TARGET	MAJOR ACTIONS
Adrenal medulla	Epinephrine	Circulatory system	Increases heart rate, blood pressure, and blood sugar
		Respiratory system	Increases breathing rate and clears airways
	Norepinephrine	Generally same as epinephrine	
Adrenal cortex	Sex steroids	Unknown	Stimulates growth of body hair in women
	Mineralocorticoids (e.g., aldosterone)	Kidneys	Increases sodium conservation
	Glucocorticoids (e.g., cortisol)	Many cells	Stimulates carbohydrate metabolism and decreases inflammation
Pancreas	Insulin	Many cells	Stimulates glucose uptake from blood
	Glucagon	Many cells	Stimulates glucose release from cells into blood
Pineal	Melatonin	Hypothalamus	Blocks secretion of LH- and FSH-releasing factors
		Brain	Promotes sleep
Ovary	Estrogen	Many cells	Stimulates female development and behavior
	Progesterone	Uterus	Stimulates growth of uterine lining
Placenta	Chorionic gonadotropin (hCG)	Corpus luteum in ovary	Stimulates progesterone synthesis
	Placental lactogen	Mammary gland	Stimulates mammary gland development
Testis	Testosterone	Many cells	Stimulates male development and behavior
	Müllerian regression factor	Pre-oviduct cells	Kills pre-oviduct cells
Gastrointestinal tract	Gastrin	Gut cells	Stimulates hydrochloric acid secretion
	Cholecystokinin (CCK)	Pancreas	Stimulates digestive enzyme secretion
Kidney	Erythropoietin	Blood cell precursors	Stimulates red blood cell production

Silkworms pass through several larval stages, during which the insect grows, produces a new exoskeleton, then sheds its old one, or *molts* (Figure 25.11, page 426). After a final growth spurt, it spins a cocoon and secretes a dark brown pupal exoskeleton instead of another bright blue larval one. In spring, as the days lengthen, the animal, still in its cocoon, secretes one last exoskeleton—an adult exterior resplendent with glistening red, black, and tan scales (review Figure 25.1).

Two main hormones govern silkworm molting and metamorphosis from worm to moth. *Ecdysone*, a steroid, controls the *timing* of each molt. The *type* of molt, however, is the province of *juvenile hormone*, a golden oil secreted by a gland attached to the brain. If juvenile hormone levels are high, the worm molts from one larval stage to another. If juvenile hormone levels are intermediate, it molts from larva to pupa. If concentrations are very low, the pupa molts to an adult, and the juvenile disappears. Biologists do not yet know how juvenile hormone works.

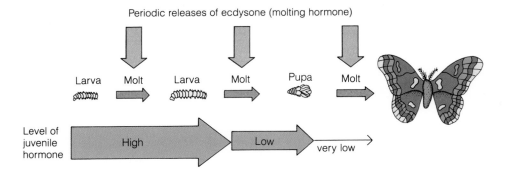

Periodic releases of ecdysone (molting hormone)

Larva — Molt — Larva — Molt — Pupa — Molt

Level of juvenile hormone: High — Low — very low

FIGURE 25.11 *From Silkworm to Silk Moth: Hormones Provoke Metamorphosis.* As a tiny caterpillar grows, pulses of ecdysone regulate the timing of each molt. Throughout early molt cycles, levels of juvenile hormone remain high and maintain the animal's larval form. Eventually, levels of juvenile hormone begin to drop, and with the next-to-last surge of ecdysone, the animal molts into a pupa. With the final surge of ecdysone, in the absence of juvenile hormone, the transformation to an adult moth is complete.

HEALTH CHOICES

ANABOLIC STEROIDS

There is an epidemic under way in North American locker rooms, gyms, and health clubs: the use of anabolic steroids to help build large muscles. The pressure to win is strong on the football field, the wrestling mat, the body building circuit, even the dating scene. That pressure has been so overpowering, in fact, that an estimated 7 to 8 percent of professional football players, 5 percent of college athletes, and 6.6 percent of high school boys are willing to take synthetic male hormones to become an overnight version of the Hulk (Figure 1). Anabolic steroids coupled with an intense weight training program can indeed help them develop extra muscle mass. But research shows that the gains are temporary, and the price is very high.

Anabolic steroids such as stanozolol (the illegal use of which cost sprinter Ben Johnson his Olympic gold medal in 1988) are artificial forms of testosterone. Just as an upsurge of the real male hormone promotes rapid muscle development in adolescent boys, taking anabolic steroids can lead to bulking that gives an athlete an advantage.

Unfortunately, the drugs also confer the other masculinizing effects of testosterone in exaggerated proportions: heavy hair growth, acne, and premature baldness. They also suppress the male's own androgens (male hormones), and this can lead to shrunken testes and enlarged breasts. And, ominously, research shows that damage to the kidneys, liver, and heart are all too common in users of anabolic steroids, as are depression, anxiety, hallucinations, paranoia, and other psychological symptoms.

For many observers, the ethical issues are as serious as the psychological ones. Pumping the body up beyond its natural maximum size and strength is really a form of cheating to one's competitors and ultimately oneself. The muscle bulk, so easily obtained through steroids and weight training, reverts quickly to that obtained by hard work alone once the drug treatments stop. But the hidden damage may already be done. Consider Steve Courson, a former football player for the Pittsburgh Steelers and the Tampa Bay Buccaneers. Anabolic steroids damaged his heart so badly that he retired early, and doctors were predicting he might live just five years longer. For too many

FIGURE 1 *Some Bodybuilders Take Anabolic Steroids to Help Generate Bulky Muscles.*

healthy athletes, this shortcut to physical fitness can become a dead end.

FOR MORE INFORMATION:

For more information, ask practitioners at your campus health center or your local physician, or read E. Marshall, "The Drug of Choice," *Science* 242 (14 October 1988): 183–184, and W. O. Johnson and K. Moore, "The Loser," *Sports Illustrated*, 3 October 1988, pp. 20–27.

In both people and insects, steroid hormones provoke changes in physical form. Good examples are ecdysone in insects and anabolic (body-building) steroids in people, such as the one that cost Ben Johnson his Olympic gold medal (see box on page 426). Such parallels reflect the retention of similar hormonal mechanisms throughout evolution.

HORMONES AND EVOLUTION

Biologists have made exciting finds in recent years that reveal startling facts unsuspected just a decade or two ago: The simplest and the most complex organisms have similar molecular messengers. Scientists have found insulin-like hormones regulating sugar metabolism in both invertebrates (like mollusks and insects) and vertebrates (such as mammals and reptiles). They have found that a mammalian hypothalamic releasing factor will turn on some sexual activities in yeasts, while the fungus mating factor will turn on some gonadal activities in rats. Clearly, hormones must have a very ancient origin and must have been conserved for hundreds of millions of years.

Some biologists suggest that the forerunners of hormones may initially have evolved in single-celled organisms to coordinate feeding or reproduction. As multicellular organisms arose, sophisticated organs evolved to govern the many individual coordination tasks, but these control centers relied on the same kinds of molecular messengers "invented" by the simpler organisms. Some of the molecules worked fairly slowly but had long-lasting action on distant cells; these became the modern hormones. Others worked more quickly, but influenced only adjacent cells for brief periods; these became involved in nerve cell communication. In the next chapter, as we see how nerve cells communicate, the familiar strategy of molecular messengers will once again emerge.

≋ CONNECTIONS

In our consideration of molecular messengers thus far, we have seen that some cells detect perturbations in the environment and respond by secreting a regulatory substance into the surrounding fluid. That substance binds to a specific receptor in a target cell and triggers some new function in the target cell that often counters the initial perturbation. This pattern holds whether the organism is a person, hamster, silkworm, or yeast and whether the activity is simple and direct or involves feedback loops and releasing factors. The end result is an animal tuned to its environment, coping effectively with change, and surviving.

Our next chapter concerns the regulators and receptors of the nervous system, which are characterized by a kind of private intercellular communication that is like a rapid-fire tête-à-tête between neighboring cells. In contrast, the hormones, with their diffuse, long-distance activity, seem more like slowly whispered rumors that echo throughout the entire body.

NEW TERMS

adrenal gland, page 421
circadian rhythm, page 423
cortisol, page 421
endocrine gland, page 417
exocrine gland, page 417
gland, page 417
glucagon, page 423
hormone, page 417
hypothalamus, page 419

insulin, page 423
parathyroid gland, page 421
pineal gland, page 423
pituitary gland, page 419
receptor, page 417
regulator cell, page 417
target cell, page 417
thyroid gland, page 421
thyroxine, page 421

STUDY QUESTIONS

REVIEW WHAT YOU HAVE LEARNED

1. How does an endocrine gland differ from an exocrine gland?
2. How do the hypothalamus and the pituitary interact?
3. Explain how an underactive thyroid affects a child. How does the same condition affect an adult?
4. What mechanism regulates the calcium level in blood? Explain.
5. What are the biological benefits of epinephrine?
6. What happens to the feedback loop that regulates the amount of glucose in your bloodstream if you are a diabetic?

APPLY WHAT YOU HAVE LEARNED

1. Will the metamorphosis of a silkworm caterpillar proceed normally if a researcher removes the glands that produce ecdysone? Explain.
2. A patient with insulin-dependent diabetes will often eat a carbohydrate snack or meal before taking a dose of insulin. How would this snacking help the insulin do its job for the patient?

FOR FURTHER READING

Gilbert, L. I., et al. "Neuropeptides, Second Messengers, and Insect Molting." *BioEssays* 8 (1988): 153–157.

Guillemin, R., and R. Burgus. "The Hormones of the Hypothalamus." *Scientific American* 227 (1972): 24–33.

Snyder, S. H. "The Molecular Basis of Communication Between Cells." *Scientific American* 253 (October 1985): 132–141.

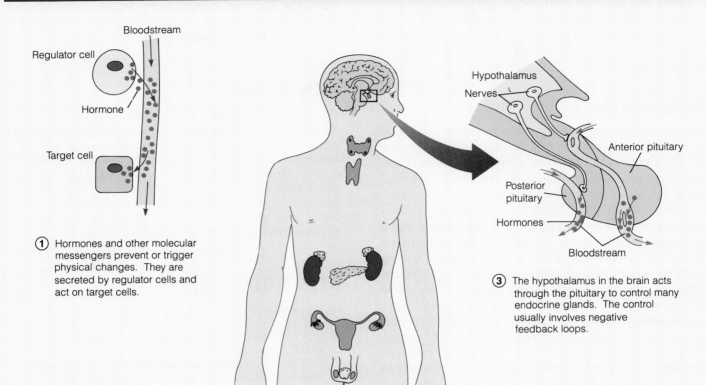

① Hormones and other molecular messengers prevent or trigger physical changes. They are secreted by regulator cells and act on target cells.

Bloodstream

Regulator cell

Hormone

Target cell

Hypothalamus

Nerves

Anterior pituitary

Posterior pituitary

Hormones

Bloodstream

③ The hypothalamus in the brain acts through the pituitary to control many endocrine glands. The control usually involves negative feedback loops.

② The glands of the human endocrine system release more than 25 different hormones, including some that control normal metabolism and some that speed it up in response to stress.

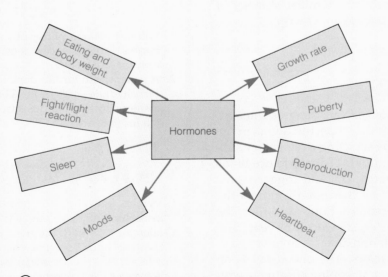

Eating and body weight

Growth rate

Fight/flight reaction

Puberty

Hormones

Sleep

Reproduction

Moods

Heartbeat

④ In general, hormones help maintain homeostasis; govern the changes that shape daily, monthly, and yearly cycles; and help control the stages of reproduction and development.

HOW NERVE CELLS WORK

Cocaine: High at What Expense?

Most of us have heard health warnings about how risky it is to take illegal mood-altering drugs. We have also read newspaper stories about the sudden death of famous young athletes or the desperate cycle of drug seeking, tolerance, craving, withdrawal, and overdosing that addicts get trapped in. Despite this information, some people still use drugs "for fun"—and often pay very dearly for the short-lived "high."

Consider, for example, a young couple from the Midwest who snorted cocaine from time to time to make themselves feel speedier, more alert, and temporarily happier. The woman even continued this "recreation" into the early months of her pregnancy, but stopped after a while as the fetus grew. Just before the baby was due, her husband gave her 5 g of cocaine as an anniversary gift, and she snorted it to "celebrate (Figure 26.1)." The drug, a nervous system stimulant, induced early labor, and she delivered a baby boy. Tragically, the newborn was paralyzed on his right side; the "hit" of cocaine the mother had taken just hours before had passed through the placenta, entered the baby's bloodstream, bound to his heart muscle, and caused his blood pressure to soar. The high pressure burst blood vessels in his brain, and this brain damage, in turn, impaired function in another part of the nervous system—the nerves that allowed movement in the baby's right arm and leg.

Our subject in this chapter is the **nervous system,** the major integrative network in an animal's body. A heartbreaking event such as a baby's irreversible nerve damage and resulting paralysis points out one of the chapter's recurring themes: The central role of the nervous system is to coordinate and control all the physiological systems in an animal's body. The cocaine the mother took altered how her own nervous system regulated behavior (her mood and alertness) as well as how the baby's nervous system controlled his circulation and blood pressure.

The way cocaine affected the baby illustrates a second general principle about the nervous system: Nerve cells communicate with each other in special ways and are organized into highly elaborate and specific networks. Deep within the baby's developing brain, the stimulant altered the way one group of nerve cells communicated with another—those that controlled blood pressure—and the consequences were dramatic.

In this chapter, we will focus on how individual nerve cells work, how they connect into an intricate system, and how addictive drugs like cocaine and nicotine work on them. In the next chapter, we will concentrate on the brain and its major sources of information, the sense organs, and how they interact to control our behavior. This chapter answers several questions:

- How does a nerve cell's structure promote communication in the body?
- What is a nerve impulse, and how does it convey information?
- How does one nerve cell communicate with another?
- How are nerve cells organized into networks that control behavior?

FIGURE 26.1 *Cocaine High: A Change in Nerve Cell Function in the Brain.* Cocaine lights up the "reward center" of the brain (bright yellow) just 15 minutes after a hit of cocaine, as shown in this PET scan, a special noninvasive view of a "slice" through a human brain.

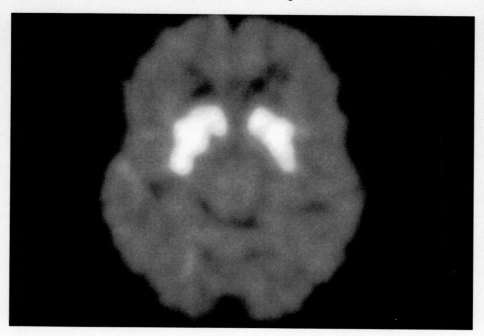

NERVE CELLS: SPECIALIZED STRUCTURE ALLOWS COMMUNICATION

A person's nervous system is made up of about 10 billion nerve cells communicating with each other in specific ways that allow a cook to pull a hand away from a hot burner, a Balinese dancer to move wrist and hand into delicate and expressive positions, and you to feel hungry, discuss a ball game with a friend, and, in fact, read this book and learn how the nervous system works. As you read this sentence, try wiggling the big toe on your right foot. Some of the nerve cells controlling that movement extend all the way from the lower part of your spinal cord to the muscles that move your toe—the longest individual cells in your body at nearly 1 m in length. A nerve cell running from a giraffe's spinal cord to its hoof can be three times as long, or nearly 3 m! Some nerve cells, in contrast, reach only a few millimeters, extending from one brain region to another. Although the human brain and nervous system contain more than a thousand different types of cells, all of them can be classified as either

neurons, the nerve cells that accomplish the actual communication tasks, or glial cells, the support cells that surround, protect, and provide nutrients to neurons and may influence their function in ways biologists do not yet fully understand. Here we focus on the neurons.

All neurons, including the ones that help wiggle your big toes, function by collecting information and relaying it to other cells in the body. Neurons are characterized by key structural features: the dendrites, soma, axon, and ends of the axon, or boutons (Figure 26.2).

Neurons gather information by a set of fine cell processes called **dendrites** (from the Greek for "little tree"). Dendrites pick up signals from the environment or from other nerve cells and pass these impulses to the soma. The **soma** is the neuron's main cell body. Housing the nucleus, mitochondria, and other intracellular organelles, the soma produces the proteins that make up the rest of the nerve cell, as well as the enzymes that determine its activity. Nerve signals pass through the soma to a long, tubular cell process called the **axon** (Greek for "axle"). The axon of some neurons extends from your spinal cord to your toe muscles, while the nerve cell's soma and dendrites are located in the spinal cord itself. Each neuron has many dendrites, but usually just one axon. A group of axons from many different neurons running in parallel bundles is called a **nerve.**

Eventually, the nerve signal reaches the often knoblike ends of an axon, the **boutons,** which terminate very close to another cell and serve as special regions of cell-to-cell communication. The place where a bouton approaches another cell is a junction known as a **synapse,** which often contains a microscopic cleft across which the nerve signal must pass at lightning speed (see Figure 26.2). The synapse is the site where cocaine acts, altering communication between nerve cells. Boutons may form synapses with other axons, with the dendrites or soma of another neuron, with muscle cells, or with the secretory cells of endocrine or exocrine glands. The neuron that sends a message down its axon and across a synapse to another cell is the **presynaptic cell** (see Figure 26.2). The cell receiving the message that crosses a synapse is the **postsynaptic cell.** A given axon may branch and rebranch, forming synaptic junctions with up to a thousand other cells. Conversely, a thousand other neurons might form synapses with the dendrites and soma of a single neuron—rather like hundreds of hands reaching out to touch a central object simultaneously. An animal's communication network is literally a net, a lacework of interconnected neurons that allows the organism to carry out such complex behaviors as wiggling the big toe on one foot. To summarize, the nervous system receives data, integrates it, and effects a change in physiology. Within a given neuron, the dendrites are the receptors, the soma is the integrator, and the axon and boutons are the effectors.

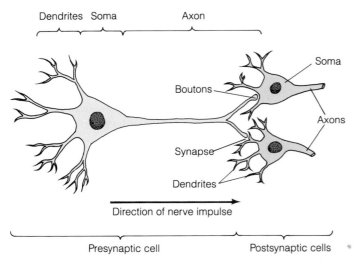

FIGURE 26.2 *Neurons Are Built to Transmit Information from One Cell to Another.* Shown here are the four parts of a neuron—dendrites, soma, axon, and axon terminals (boutons)—as well as the synapse, or junction, between one cell's axon terminal and another cell's dendrite. Most animals' nervous systems are capable of directing complex behavior because a given nerve cell can form hundreds of contacts with nearby cells, and because the branching axons of many cells can form synapses with the dendrites and soma of a large central neuron.

≋ ELECTRIFYING ACTION: THE NERVE IMPULSE

Environmental signals can differ greatly and must be received and interpreted in different ways. A chili pepper, a chocolate bar, and an aspirin tablet all produce distinguishably different taste sensations. Listening to Brahms is nothing like hearing the screech of a braking car. And none of these inputs resembles the feeling of slamming your finger in a drawer. These sensations differ because their points of origin and destinations in the brain or spinal cord vary widely. At the level of a single neuron, however, any and every sensation of sufficient magnitude produces exactly the same kind of signal: a **nerve impulse,** an electrochemical chain reaction triggered by a series of abrupt, localized changes in the neuron's membrane that allows specific ions to rush in and out of the cell.

A NERVE CELL AT REST

By focusing on a single patch of a resting neuron's plasma membrane, we can begin to see how a nervous impulse is generated. Neurons, like all cells, are bathed in an extracellular fluid that is a molecular soup containing many positively and negatively charged ions. For nerve impulses, sodium and potassium are the most important positive ions, and chloride is the most important negative ion. There is an excess of positively charged ions in a thin layer just outside the cell membrane and an excess of negatively charged ions in a thin layer just inside the cell (Figure 26.3a). This charge difference creates an *electrical potential:* a measurable amount of potential energy. In a neuron at rest, this charge difference is called the *resting potential.*

The positive ions outside the membrane are like the thousands of music fans pushing and straining to get into an amphitheater for a rock concert. And just as those rock fans represent potential energy waiting to be unleashed, the separation of positive and negative charges on either side of the cell membrane represents potential energy that, once released, can activate a nerve impulse in that small region of the cell. By controlling the traffic of ions (charged atoms) into and out of the cell, a membrane generates a nerve impulse. Now, we know that a gate keeps rock fans outside, and if a few fans sneak in early, security guards send them back over the fence. Similarly, proteins in the nerve cell membrane serve as gates and pumps controlling ion movement into and out of the cell (Figure 26.3b). A nerve cell at rest has more positive ions outside the cell membrane than inside because the protein "gates" are closed and the protein "pumps" move positive ions out. When the gates sud-

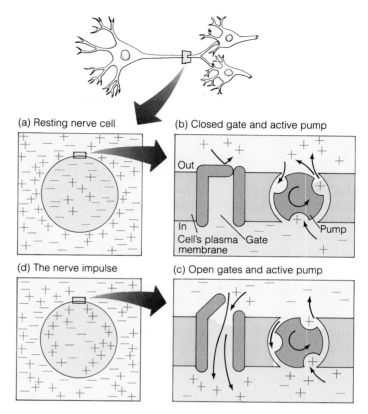

FIGURE 26.3 *The Nerve Impulse: An Electrical Switch.* (a) The thin layer of fluid outside a neuron is rich in positive ions relative to the fluid just inside the cell membrane. This is because (b) closed gates prevent an influx of certain positively charged ions. In addition, ion pump proteins in the membrane actively move these positive ions out of the cell. (c) A sudden change in the ion balance outside and inside one region of a neuron's plasma membrane leads to an action potential. During the action potential, ion channels open wide, and even though the ion pump continues to work, positive ions flood into the cell, causing (d) the inside of the cell to become positive and the outside to become negative. This charge reversal is the action potential, or nerve impulse. Finally, as the ion gates close once more and the membrane pumps continue to operate, the original ion distribution of the resting cell is restored.

denly open, the ions rush in and a nerve impulse is generated. What makes this happen?

GENERATING A NERVE IMPULSE

How, in other words, is a nerve impulse generated, allowing the animal to receive environmental information, integrate it, and react? A nerve impulse begins when a specific stimulus opens the protein gates in a small region of neuron membrane (Figure 26.3c). Ions rush through the gates, causing the outer and inner charges to reverse suddenly so that in the stimulated region, the cell is now more positive inside and more negative outside (Figure 26.3d). This charge reversal is the nerve impulse, and biologists call it the **action potential;** it is the state in which the nerve cell is firing. After the reversal of charge, gates once again close, and as membrane pumps continue working, the ions are forced back to their original distribution (more positive ions outside than inside the cell), and the cell is once again at rest, ready for another stimulus (see Figure 26.3a). The stimulus is usually a message from another nerve cell. Other specific stimuli, such as an icy rain, or a taste of horseradish, or a bird's song, can also trigger an action potential—a charge reversal—in specific neurons. Interestingly, the reverse phenomenon (the blocking of a nerve impulse) occurs when you get novocaine at the dentist's office. Novocaine enters the ion gates and blocks them. If the gates cannot open, the ions cannot pass, the positive and negative electrical charges cannot suddenly change outside and inside the cell, the action potential cannot fire, and you feel little or no pain in your teeth despite procedures that may touch exposed nerves.

Nerve impulses have two properties that help explain how nerve cells can coordinate body activities. First, for a split second after firing, a neuron cannot respond to a new stimulus. This state of temporary shutdown is called the *refractory period;* it limits the number of action potentials a neuron can experience (i.e., the number of times it can fire) each second. Action potentials have another peculiar property: They are *all-or-none* responses; that is, either they do not occur at all, or when they do, they are always the same size for a given neuron, regardless of how powerful the stimulus may be. A jab to the ribs is more intense than a tender caress, not because the jab causes larger or faster action potentials, but because it causes more cells to fire more impulses.

PROPAGATING THE NERVE IMPULSE

Once a neuron receives a stimulus at a particular patch of its membrane, the resulting action potential generated in that local region is communicated along the length of the cell. To wiggle your toe, for example, nerve cells send (or "propagate") an action potential all the way down the axon from the spinal cord to the nerve cell boutons in your toe. That communication depends on the action potential passing from one patch of the membrane to an adjacent patch, then to the next, in a process called **propagation** of the nerve impulse.

Interestingly, the impulses in a given neuron travel in one direction only (from spine to toe, for example, but not from toe to spine), and the explanation for this is that the refractory period prevents reverse propagation. For a split second after an action potential, a membrane patch cannot experience another action potential; thus, in nature, the signal can be propagated forward to the next patch, but never moves backward toward the previous one. And unlike the electric current in a copper wire, which dies out over distance, a nerve impulse in a living neuron remains just as strong when it reaches the axon terminal in the tip of the toe as when propagation began back in the spinal cord, because of the all-or-none response.

In many invertebrates, a nerve impulse traveling down a neuron moves only about 2 m per second. A tennis match would be a sluggish affair indeed if players had to wait 1 second for an impulse to travel from their brains to their toes. One way the propagation of nerve signals can be speeded up is for the diameter of the axon to increase, since a larger axon will have less resistance to the flow of an action potential, just as a wider pipe has less resistance to water flow than a narrower one. A few invertebrates, such as the fast-moving squid, have in fact evolved this nervous system strategy; the axons that innervate the squid's muscular body are well over 1 mm in diameter and can carry impulses rapidly. A person, however, could never make do with giant axons: To contain the huge number of interlocking nerve cells we use to produce complex actions and thoughts, our heads would have to be ten times bigger. Instead, humans and other vertebrates have developed a means of rapid impulse propagation without giant neurons.

In vertebrates, fatty sheaths insulate neurons and speed impulse travel. Schwann cells, special supportive glial cells, extend along an axon like fatty sausages on a string and wrap their plasma membranes around the axon surface like sticky straps (Figure 26.4a and b). Together, the Schwann cells form the **myelin sheath,** a lipid-rich layer that insulates the axon from the extracellular fluid (see Figure 26.4b). In the tiny gaps between these end-to-end Schwann cells lie bare regions known as *nodes of Ranvier,* and only in these uninsulated nodes can ions flow across the axon membrane. As a result of this cellular arrangement, the impulse fairly leaps from one node to the next, bouncing along the axon 20 times faster than if the axon lacked the myelin sheath. This permits the rapid reactions and movements we need to return a smashing tennis serve or perform a Rachmaninoff concerto on the piano. To see the value of myelin insulation, notice the mouse in Figure 26.4c that is shivering uncontrollably. This mutant mouse lacks myelin sheaths on its axons, and as a result, signals move erroneously between the

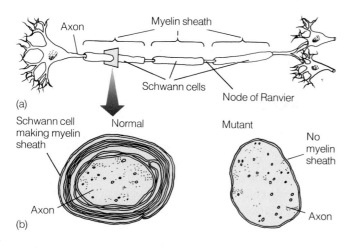

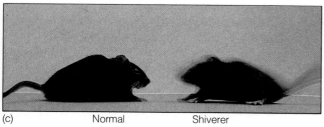

FIGURE 26.4 *Myelin Sheath: An Insulated Axon Propagates Impulses More Efficiently.* (a) Vertebrate neurons may be wrapped in Schwann cells, which make up the insulating layers called the myelin sheath. Schwann cells are separated by bare regions, or nodes of Ranvier. The action potential can jump from one node to the next, allowing rapid propagation of the nerve impulse. (b) Drawing of a cross section of neurons from a normal mouse (left) shows Schwann cells wrapped around axons. Cross section of neurons from a mutant *shiverer* mouse (right) reveals no encircling Schwann cells. (c) A normal mouse (left) and a mutant shiverer mouse (right). The shiverer is unable to control its body movements because without myelin sheaths, adjacent axons are not electrically insulated.

axons from side to side as well as from end to end. Likewise, the human disease *multiple sclerosis* (MS), whose symptoms usually first appear in early adulthood, causes some nerve cells to lose their myelin sheath. Without the sheath, nerve impulses can no longer leap along the cell, and transmission slows, leading to double vision, weakness, and wobbly limbs.

In summary, then, a stimulus causes positive and negative electrical charges to reverse across a nerve cell's membrane, thereby generating an all-or-none action potential that propagates down the axon. The impulse travels rapidly if the axon is wrapped in insulating myelin, but this carries the message only to the end of the first neuron. For the message to be integrated in the brain or spinal cord and stimulate an appropriate reaction in the organism, it must cross from that initial cell to another cell and then another and another in a molecular relay. The next section explains how such relaying works.

HOW A NERVE CELL COMMUNICATES WITH OTHER CELLS

As we saw in the chapter's introduction, cocaine crossing the placenta caused an increase in a fetus's blood pressure. For such a reaction to occur, cells in the brain passed information to other neurons that in turn signaled the cells controlling blood pressure. This type of cell-to-cell signal transmission required the signal to cross the specialized junction, or *synapse*, between a neuron and a neighboring cell (review Figure 26.2). A typical human neuron has so many branches that it forms between 1000 and 10,000 synapses with other cells. These junctions usually connect the axon of one cell to the dendrites or soma of another, but synapses sometimes join axon to axon or dendrite to dendrite. In any case, synapses generally conduct signals in only one direction, from the sending (presynaptic) cell to the receiving (postsynaptic) cell.

Neurons have developed different ways of conveying messages to other cells. Some pairs of neurons are electrically coupled and communicate directly via this electrical connection, while other neuron pairs are separated by a cleft and rely on chemicals that drift across the cleft through a fluid.

ELECTRICAL JUNCTIONS: RAPID RELAYERS

Animals, from jellyfish to slugs to humans, have some neurons that connect to other cells through special pores that serve as *electrical synapses* (Figure 26.5, page 434). These pores (also called gap junctions) allow propagation of the impulse in a direct electrical connection, somewhat like a phone line.

Message relay is especially rapid at these electrical junctions, and that speed is a valuable trait for certain kinds of neurons. In the giant motor neurons of the crayfish, for example, lightning-fast relay allows the animal to flip its tail instantly and escape a predator. Unfortunately, tight interconnections that allow such rapid transmission also leave little room for the neuron to modify the ways it relays a signal. It is no surprise, therefore, that electrical junctions are vastly outnumbered in higher animals by chemical synapses, which function more slowly and require more steps, but also have more potential places to modify messages for specific conditions.

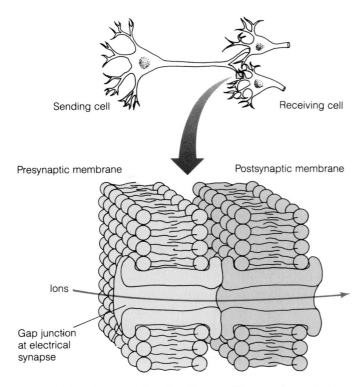

Sending cell

Receiving cell

Presynaptic membrane

Postsynaptic membrane

Ions

Gap junction
at electrical
synapse

FIGURE 26.5 *Electrical Junctions Between Neurons.* At an electrical synapse, the plasma membranes of two adjacent neurons are tightly joined by gap junctions that allow ions to flow from one cell to another and thus to propagate an action potential smoothly and immediately.

CHEMICAL SYNAPSES: COMMUNICATION ACROSS A CLEFT

In vertebrates, most neuron-to-neuron and neuron-to-muscle communication takes place across junctions called *chemical synapses*. Between the knoblike bouton of a sending cell's axon and the flat surface of the receiving cell lies a minute but distinct cleft, or space (Figure 26.6a). A molecular messenger, or **neurotransmitter,** must cross this cleft to relay the nerve signal from sender to receiver.

■ **How Neurotransmitters Work** A neuron makes neurotransmitters in its cytoplasm, and these neurotransmitters accumulate in the knoblike bouton (see Figure 26.6a). When an action potential reaches a bouton, the sending cell liberates thousands of neurotransmitter molecules into the cleft. These molecules rapidly diffuse across the small cleft and bind like keys to locklike receptor proteins embedded in the membrane of the receiving cell. This binding causes the receiving cell to respond with another action potential (Figure 26.6b). Biologists

know that the receptor proteins are important because people with a condition called myasthenia gravis have few receptors for neurotransmitters on their muscle cells; as a result, the muscle cells cannot respond to nerve signals, and the patients experience muscular weakness and fatigue.

Once a neurotransmitter enters the cleft and triggers the receiving cell, it rapidly disappears; it is either degraded by enzymes or reabsorbed into the sending cell's bouton. Without this rapid "cleanup," the messenger molecules would remain, continuously stimulating the receiving cell. This is what happens in a cocaine high. The drug blocks the rapid cleanup of a specific neurotransmitter (dopamine) in certain brain cells, and as a result, the transmitter activates these receiving cells for a longer period of time. Evidence indicates that cocaine exerts its effects in the brain's so-called "central reward system," producing euphoria and a feeling of excitement (Table 26.1, page 436). In response to the sluggish cleanup of transmitter, which cocaine causes, homeostatic mechanisms generate more efficient removal of dopamine. But then, because of the more effective removal of transmitter, the person craves more cocaine; without it, he or she loses interest in activities that normally stimulate the central reward system. Thus begins the destructive cycle of craving and drug-seeking called **addiction.** In fact, cocaine addiction takes hold faster than for any other drug, including heroin: Numerous animal species, from mice to monkeys, quickly learn to push a lever to obtain cocaine (Figure 26.7). Laboratory mice will experience their first "high," crave the drug, become addicted, and die from a self-delivered overdose, all within one week's time.

■ **Types of Neurotransmitters** Dopamine, the neurotransmitter involved in cocaine addiction, is just one of more than 50 different chemicals that can serve as neurotransmitters in the nervous systems of various animals. The best understood are *norepinephrine* and *acetylcholine.* Norepinephrine, found in synapses throughout the brain, seems central to keeping mood and behavior on an even keel. Acetylcholine plays a role in transmitting nerve impulses within the brain and also relays messages from neurons to the skeletal muscles involved in maintaining posture, breathing, and the movement of limbs. Unfortunately, many countries in the world, including Iraq and the United States, have developed chemical weapons—nerve gases that act via the neurotransmitter acetylcholine. These weapons prevent the breakdown of acetylcholine at synapses, and so muscles are continuously stimulated, which causes paralysis, convulsions, and death within minutes. Detailed knowledge of this aspect of nerve cell functioning is helping scientists to develop protection against or antidotes to these terrible weapons.

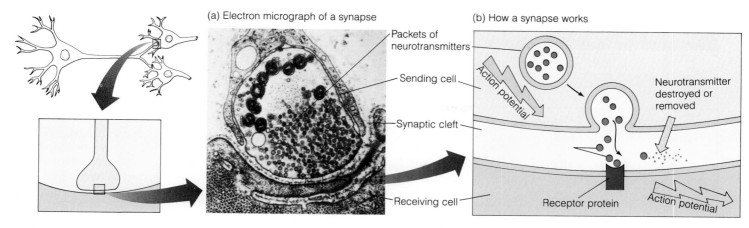

(a) Electron micrograph of a synapse

(b) How a synapse works

Packets of neurotransmitters

Sending cell

Synaptic cleft

Receiving cell

Action potential

Neurotransmitter destroyed or removed

Receptor protein

Action potential

FIGURE 26.6 *At a Chemical Synapse, Neurons Transmit Action Potentials Across a Cleft.*
(a) An electron micrograph of a synapse reveals that a cleft separates the sending (presynaptic) cell, with its bouton-containing packets of neurotransmitter, from the receiving (postsynaptic) cell.
(b) An action potential traveling down a sending cell and reaching the bouton triggers the release of neurotransmitter; this messenger then diffuses across the cleft, binds to a receptor protein, and triggers an action potential in the receiving cell. Cleanup activities then remove excess neurotransmitter, and as a result, the receiving cell fires a single time. This normal cleanup activity fails to occur in the "reward centers" of the brain in the presence of cocaine.

The growing list of neurotransmitters includes glycine and other amino acids, many peptides, and the specific substances serotonin and gamma-aminobutyric acid (GABA), as well as dopamine. Each of these plays specific roles in behavior, and their absence can lead to disease. Decreased quantities of dopamine, for example, are associated with the rigidity, weakness, and shaking of Parkinson's disease. These symptoms sometimes diminish when the patient takes oral doses of L-dopa, a chemical that brain cells can modify into dopamine. The take-home message here is this: Neurons generally communicate with other cells by releasing a neurotransmitter at a synapse. The neurotransmitter diffuses toward a receptor on another cell and causes an alteration in the receiving cell's behavior.

Many commonly prescribed drugs—and many over-the-counter drugs as well—affect the release and uptake of neurotransmitters. For example, at 3 in the morning, a college student cramming for final exams may reach for another amphetamine tablet to keep her awake. Her drowsiness dissolves, but her roommate finds her jumpy and irritable. This is because amphetamines increase the release of norepinephrine and dopamine at brain synapses and slow their removal from the cleft. One result of these "overactive synapses" is a general increase in arousal manifest as an enhancement of mood and increased power of concentration and alertness. But blurred vision, insomnia, anxiety, and increased blood pressure are some negative manifestations of this increase in arousal.

Understanding how neurons work at the cellular level can help researchers design drugs that relieve clinically

FIGURE 26.7 *Animal Addict.* If given unlimited access to cocaine and left alone, a mouse will learn to take doses of cocaine, become addicted, and eventually consume a lethal overdose of the drug—all in less than a week's time. The mouse shown here has learned to press the lever repeatedly with its paw, and this activates a spout to the left. When the mouse touches its tongue to the spout, a small amount of the drug is delivered directly into its mouth. This continues until the animal overdoses from constant self-administration.

TABLE 26.1 ⚭ SOME COMMON DRUGS AND HOW THEY AFFECT THE NERVOUS SYSTEM

DRUG CLASS/DRUG	COMMON NAME OR SOURCE	ACTIONS ON BODY	EFFECTS ON MOOD	DANGERS OF ABUSE
Opiates				
Morphine, heroin, codeine	Horse (heroin)	Causes drowsiness	Causes euphoria	Physical dependence, nausea, vomiting, constipation, death from respiratory failure in fatal doses
Stimulants				
Cocaine	Coke, crack	Raises heart rate and body temperature, dilates pupils	Causes euphoria, excitation	Convulsions, hallucinations, cardiovascular damage
Amphetamine, methamphetamine	Dexedrine, crystal meth, crank, cross tops	Increases heart rate, respiration, and blood pressure	Causes wakefulness, depresses appetite, increases alertness	Irregular heartbeat, chest pain, dizziness, anxiety, paranoia, hallucinations, convulsions, coma, cerebral hemorrhages in fatal doses
Nicotine	Tobacco	Causes vasoconstriction, racing heart, increased blood pressure	Acts as stimulant, causes euphoria	Dizziness, nausea, vomiting, withdrawal in addicted users
Caffeine, theophylline	Coffee, tea	Stimulates central nervous system and visceral muscles, increases heart rate and urine production	Causes some mood elevation, increases alertness	Anxiety, nervousness, insomnia, convulsions

important maladies. Table 26.1 summarizes the known effects of many nonprescription, prescription, and illegal drugs on the human body.

■ **Summation of Excitatory and Inhibitory Synapses**
Synapses can either "encourage" or "discourage" the firing of the receiving cell. At an *excitatory synapse*, the molecular messenger brings the postsynaptic cell closer to firing a nerve impulse. In contrast, at an *inhibitory* *synapse*, the neurotransmitter decreases the likelihood that the postsynaptic cell will fire a nerve impulse.

Since most neurons receive synaptic input from the axons of possibly thousands of other cells, the ultimate activity in any given region of the cell results from the cumulative impact of all the molecular messages—some excitatory, some inhibitory—ferried across the synaptic cleft. Whereas the firing of many impulses at an excitatory synapse might be enough to cause a receiving neuron to fire, an inhibitory synapse nearby that receives an equal number of impulses could counteract the excitatory effect. In a process called *summation*, the receiving cell sums up all the information impinging on it in a nearby region and over a short time interval. If the "encourage-

TABLE 26.1 ≋ SOME COMMON DRUGS AND HOW THEY AFFECT THE NERVOUS SYSTEM

DRUG CLASS/DRUG	COMMON NAME OR SOURCE	ACTIONS ON BODY	EFFECTS ON MOOD	DANGERS OF ABUSE
Sedatives				
Alcohol	Beer, wine, liquor	Decreases respiration and body temperature, causes vasodilation, impairs vision, depresses central nervous system, numbs pain	Causes euphoria	Damage to liver, heart, brain, pancreas; respiratory failure in fatal doses
Methaqualone	Quaalude	Sedates, causes sleep, depresses central nervous system	Quieting	Nausea, vomiting, delirium, convulsions, coma, slow heart rate
Benzodiazepine	Valium, Librium	Relaxes skeletal muscles; decreases circulation, respiration, and blood pressure	Reduces anxiety, elevates mood	Depressed heart and lungs, decrease in muscular coordination, drowsiness; withdrawal can trigger seizures
Hallucinogens				
Cannabinoids	Marijuana, pot, grass, hash	Increases heart rate, causes vasodilation	Elevates mood, causes euphoria, sensory distortions	Lung damage from smoking; reduced sperm count, low testosterone in males
Lysergic acid diethylamide, psilocybin, mescaline	LSD, acid, mushrooms, peyote	Acts as a stimulant, raises heart rate and blood pressure, dilates pupils	Causes euphoria, hallucinations, sensory distortion	Irrational behavior
Dissociative Anesthetic				
Phencyclidine	PCP, angel dust	Kills pain, increases heart rate and blood pressure, causes swelling and fever	Elevates mood, causes perceptual disturbances	Coma, convulsions, psychosis, respiratory depression

ments" outnumber the "discouragements," an action potential is triggered, and the neuron fires an impulse in an all-or-none fashion.

Within every nervous system, there are thousands, millions, or billions of connections between neurons. The variety of neurotransmitters at work, as well as the countless ways neurons link together, gives neuronal networks a rich repertoire of potential responses despite the fact that the nerve impulse itself is exactly the same in each neuron and its neighbors. Our final section focuses on how networks of neurons—which are still relatively simple systems compared to the unbelievable intricacy of the brain and nervous system—interact to control behavior.

≋ HOW NETWORKS OF NEURONS CONTROL BEHAVIOR

Although we have been discussing neurons as if they sit end to end like matchsticks on a table, they are actually arranged in delicate circuits. This spatial organization allows you and other animals to receive, integrate, and act on information from your surroundings and thus to survive and reproduce; it also helps explain your thoughts, language, and sensations.

In many different animals, a basic type of nerve circuit called the **reflex arc** underlies simple behaviors. This simple neural loop links a stimulus to a response in a very direct way. If you step on a tack with your bare toe, for example, a reflex arc drives muscles in your leg to pull your toe away immediately, even before you consciously realize what has happened (Figure 26.8). Reflex arcs mediate behaviors that are generally rapid, involuntary, and nearly identical each time the stimulus is repeated, requiring no input from the brain.

Most nerve circuits, even simple ones like reflex arcs, contain three kinds of neurons: sensory neurons, interneurons, and motor neurons (see Figure 26.8). A **sensory neuron** receives information from the external or internal environment, such as the stab of a tack, and transmits it toward the spinal cord or brain, the nervous system's integrating centers. In those centers, **interneurons** relay messages between nerve cells and integrate and coordinate incoming and outgoing messages. While some reflex arcs do not have interneurons, an animal must have a network of many interneurons to carry out complex behaviors. Interneurons often connect with **motor neurons,** cells that send messages from the brain or spinal cord out to muscles (such as the ones that raise your foot from the floor) or secretory glands. Motor neurons allow an organism to act—to respond to the information the sensory neurons bring in.

LEARNING: CHANGES AT THE SYNAPSE

Organisms must constantly adjust to their environments. When conditions change, an animal's behavior must also change so that the animal can meet new challenges or seize new opportunities. One phenomenon that allows an organism to adapt to a changing environment is **habituation:** a progressive decrease in the strength of a behavioral response to a constant weak stimulus when that stimulus is repeated over and over again with no undesirable consequences. Habituation takes place when you get dressed in the morning. When you first put on your clothes, you can feel them against your skin, but within a few minutes, you become unaware of the constant light pressure. Your body has "learned" to ignore a stimulus with trivial consequences. Researchers studying the simple learning response of habituation in a sea slug have discovered that habituation is due to a physical change at the synapse. After habituation, the sensory

neurons release less neurotransmitter, and as a result, the motor neurons are less likely to fire.

The opposite of habituation is **sensitization,** an increase in a behavioral response to a mild stimulus following a strong or harmful stimulus. Experiments, again with the sea slug, show that sensitization involves the release of more neurotransmitter at synapses participating in the response. The phenomena of habituation and sensitization help explain how animals learn to sense and interpret harmless or harmful stimuli and respond appropriately for their own survival. Such simple learning is due

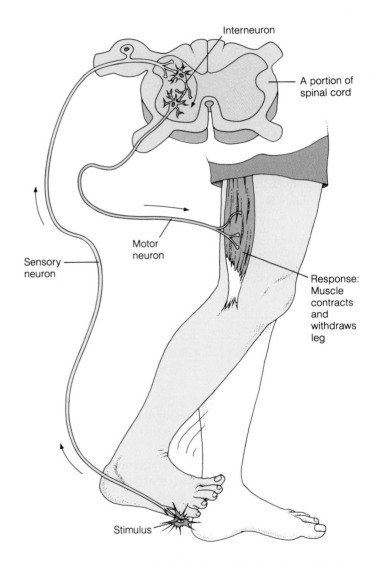

FIGURE 26.8 *Anatomy of a Reflex Arc.* Jerking back your foot after stepping on a tack is based on a reflex arc that links the spinal cord and the muscles in the back of the thigh. A sensory neuron detects the tack and sends an action potential along its axon toward the spinal cord. The sensory neuron forms a synapse with an interneuron within the spinal cord. The interneuron links to a motor neuron that runs to certain leg muscles, causing the muscles to contract and lift the foot away from the tack.

to changes in how much neurotransmitter crosses various synapses. Can more sophisticated human learning be explained in the same way? Even simple behaviors, like scratching the end of your nose, must require the proper activation of a great many cells to ensure that the finger lands on the end of your nose and not in your eye and that the movement is a gentle scratch and not a stab. The more sophisticated the action—a ballet pirouette, say, or a reverse slam dunk of a basketball—the more cells and organization are needed to carry it out. Future researchers must determine whether straightforward rules that explain a simple behavior hold for all other learning as well. Perhaps the cellular events that control how you become accustomed to the feel of your clothes also control the way you come to understand principles for a biology quiz.

≋ CONNECTIONS

The nervous system, along with the endocrine system, allows the animal body to work as an integrated whole. This chapter moved from the events within a single nerve cell to the communication between two neurons to neural networks like reflex arcs and how they bring about and control meaningful behavior in animals. Underlying that integrated operation is the basic evolutionary principle that the nervous system, like the other physiological systems we have discussed, enhances an animal's chances of surviving and reproducing.

Simple behaviors controlled by reflex arcs depend on just two or three interlinked neurons. More complex behaviors depend on millions of neurons organized into an intricate nervous system. The most intricate neural control center of all, the human brain, is also the most fantastically organized tissue that has yet evolved. Complex behaviors and their complex neural control organs are the subject of our next chapter.

STUDY QUESTIONS

REVIEW WHAT YOU HAVE LEARNED

1. List and describe the parts of a typical neuron.
2. Trace the events that lead to a nerve impulse.
3. True or false: A neuron can "fire" virtually without stopping. Explain your answer.
4. What is a myelin sheath? How does it affect transmission of an impulse along an axon?
5. How does an action potential cross a synapse?
6. Name two neurotransmitters, and state the function of each.
7. How do excitatory and inhibitory synapses differ?
8. Make a chart listing the three categories of neurons and the main function of each.
9. What is habituation?
10. What is sensitization?

APPLY WHAT YOU HAVE LEARNED

1. A middle-aged woman with Parkinson's disease takes L-dopa. How does this substance relieve her symptoms?
2. Late one night, while driving down a lonely highway in the Nevada desert, a truck driver takes an amphetamine-laced "pep pill" to keep him alert. How does the pill function? What biological hazards does the driver risk in taking this pill?

FOR FURTHER READING

Goelet, P., V. F. Castellucci, S. Schacher, and E. R. Kandel. "The Long and the Short of Long-term Memory—A Molecular Framework." *Nature* 322 (1986): 419–422.

Gottlieb, D. I. "GABAergic Neurons." *Scientific American* 258 (February 1988): 82–89.

Holden, C. "Street-Wise Crack Research." *Science* 246 (15 December 1989): 1376–1381.

Kuffler, S. W., J. Nicholls, and A. Martin. *From Neuron to Brain*. 2d ed. Sunderland, MA: Sinauer Assoc., 1984.

NEW TERMS

action potential, page 432
addiction, page 434
axon, page 430
bouton, page 430
dendrite, page 430
glial cell, page 430
habituation, page 438
interneuron, page 438
motor neuron, page 438
myelin sheath, page 432
nerve, page 430
nerve impulse, page 431

nervous system, page 429
neuron, page 430
neurotransmitter, page 434
postsynaptic cell, page 430
presynaptic cell, page 430
propagation, page 432
reflex arc, page 438
sensitization, age 438
sensory neuron, page 438
soma, page 430
synapse, page 430

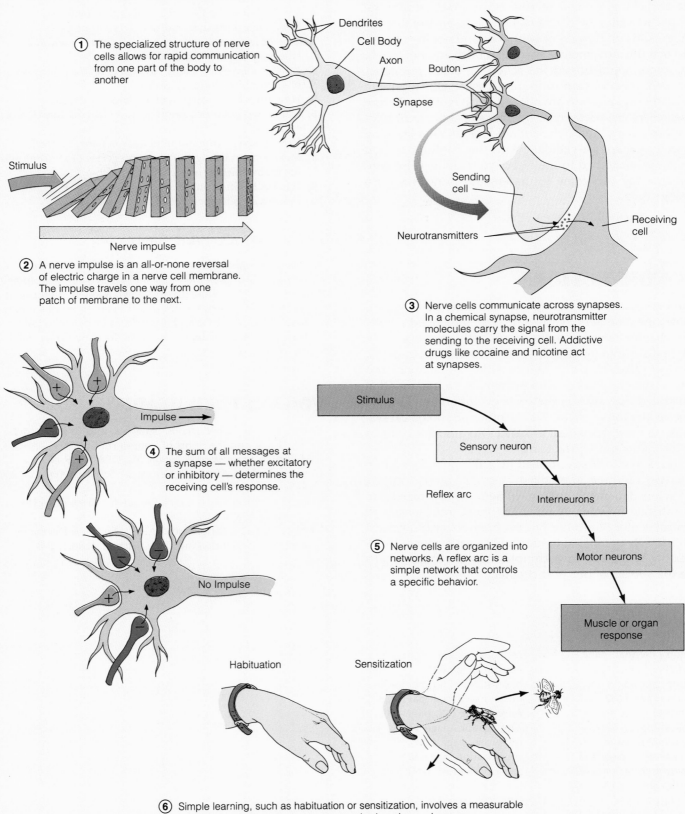

① The specialized structure of nerve cells allows for rapid communication from one part of the body to another

Dendrites

Cell Body

Axon

Bouton

Synapse

Sending cell

Receiving cell

Neurotransmitters

Stimulus

Nerve impulse

② A nerve impulse is an all-or-none reversal of electric charge in a nerve cell membrane. The impulse travels one way from one patch of membrane to the next.

③ Nerve cells communicate across synapses. In a chemical synapse, neurotransmitter molecules carry the signal from the sending to the receiving cell. Addictive drugs like cocaine and nicotine act at synapses.

Impulse

④ The sum of all messages at a synapse — whether excitatory or inhibitory — determines the receiving cell's response.

No Impulse

Stimulus

Sensory neuron

Reflex arc

Interneurons

⑤ Nerve cells are organized into networks. A reflex arc is a simple network that controls a specific behavior.

Motor neurons

Muscle or organ response

Habituation

Sensitization

⑥ Simple learning, such as habituation or sensitization, involves a measurable change at one or more synapses; complex learning and memory involve similar changes.

SENSES AND THE BRAIN

CHAPTER 27

KEEN SENSES OF A NIGHT HUNTER

Perched on an oak limb in the moonlight, a reddish brown owl rotates its head slowly. Like a ship with sonar, it scans the darkened English woods, listening for the slightest indication of a mouse scuffling across the forest floor at night. When it senses the rustle of leaves, the predatory bird swoops

FIGURE 27.1 *A Night Hunter Swoops to Catch Prey.* The tawny owl (*Strix aluco*) can navigate in the dark by sounds and "mental maps" to catch a rodent and fly its prey back to the nest.

silently from its perch, deftly avoids a large branch looming out of the dim shadows, and pounces on the source of the sound with sharp talons outstretched. Grabbing a gray mouse with its feet and then snapping the rodent's skull with a powerful stroke of its hooked beak, the owl returns to its perch and shares the meal with its owlets.

This night hunter, the tawny owl, is the most common owl in British forests and inhabits the woods throughout much of Europe, northern Africa, and western Asia (Figure 27.1). Like other owls, it has large, front-facing eyes, but a simple experiment shows that vision is not the primary sense in the owl: If an experimenter ties a dry leaf to a mouse's tail and places the rodent in a dimly lit room with an owl, the rodent will scurry about in fright and the bird will pounce—not on the prey but on the rattling leaf! Clearly, a tawny owl's hearing is very acute; in fact, it is sufficiently sharp to allow the bird to capture prey in absolute darkness. Curiously, though, its ability to perceive sound is not superior to yours. Both you and the tawny owl can detect the smallest sound in a quiet environment.

In addition to keen senses, an owl also has a precise mental map of where tree limbs project near its perch, the location of rocks that could shelter small rodents, and which trails the mice usually follow. Similar mental maps of your own room allow you to walk to the bathroom in the dark without bumping into your desk. Such maps are based on the function of the brain, an organ that integrates information from the senses and controls the animal's reaction to it.

This chapter describes how sense organs work and how nervous systems integrate sensory information to coordinate behaviors that help animals survive and reproduce. Three main themes will recur in our discussion. First, the nervous system's major role is to monitor internal needs and external conditions, process the information, and initiate appropriate behavioral responses. Because of its nervous system, the tawny owl can sense hunger, detect a mouse scampering through fallen leaves, and respond by pouncing on the prey.

Second, sense organs contain nerve cells modified in ways that increase their sensitivity to one physical aspect of the environment. Thus, your eye is most sensitive to light, your ear to sound.

Finally, nervous systems become more complex in more complicated animals. The nerve cells of a jellyfish resemble those of a tawny owl or person, but in the higher animals, sense organs and nervous integrators become more centralized and contain many more cells. In fact, the human brain is the most complex and most intricately organized entity in the biological world.

Our survey of sense organs and nervous system function will answer several questions:

- How do animals hear and see?
- How did the brain and senses evolve?
- What are the main parts of the vertebrate's nervous system, and where in a person's brain are the seats of motor coordination, sensation, emotion, facial recognition, and language?

WINDOWS ON THE WORLD: THE SENSE ORGANS

A cyclist hears a car approaching from behind and steers her bike toward the highway's shoulder. A batter sees the pitch, leans toward a ball barreling in at 80 mph, swings, and connects. A chef smells warm cinnamon spice and removes a perfectly browned apple crisp from the oven. A child bites into an unripe cherry and rejects the sour taste. People and other animals detect all such events in the body and its surroundings by **sense organs:** groups of specialized cells that receive a stimulus (such as light or sound waves, odor or flavor molecules, or pressure) and convert it to a kind of energy that can trigger a nerve impulse. That impulse then travels to the brain and delivers two types of information: the *kind* of stimulus the sense organ received (the pitch of the sound, the color of the light, the nature of the chemical) and the *strength* of the stimulus (loud or soft, bright or dim, concentrated or dilute). Our day-to-day survival depends on accurate interpretation of this information.

A sense organ's receiving cells have bare nerve endings not covered by a myelin sheath (Figure 27.2). There are several different kinds of receptor cells, each one a specific shape and able to make special proteins. Other affiliated cells screen out extraneous forces and amplify the effects of the specific stimulus before it reaches the receptor. For example, the eye's light receptor cells make light-sensitive proteins, while nearby support cells focus light on the receptors. In contrast, pressure-sensing cells in the skin detect physical bending in surrounding tissue, and the support cells magnify the bending (see Figure 27.2b). The two senses that people rely on most heavily in their day-to-day lives are hearing and vision—the same ones the tawny owl employs in its night hunting. For these reasons, we focus primarily on the mechanisms of hearing and vision in the following sections. Because smell and taste are so crucial to the survival of so many other vertebrates, however, we also describe those senses in some detail.

THE EAR: THE BODY'S MOST COMPLEX MECHANICAL DEVICE

With over a million moving parts, the ear's **cochlea** is the most complex mechanical apparatus in the human body and in the bodies of most other vertebrates. Cells in this coiled apparatus detect sound by first sensing subtle movements in the fluid that surrounds them. The stimulated cells then send a nerve impulse to the brain. Finally, the brain interprets the sound of hands clapping, the shrill scolding of a bluejay, or the deep musical "hoo-hooo-hoooo" of the tawny owl.

■ **How the Ear Detects Sound** A sound is really a set of compressed air molecules that pushes adjacent molecules so that a wave of compressed air travels along and eventually strikes the ear (Figure 27.3a). This compression could begin with two hands clapping together or with a mouse's foot scraping a leaf. The sculptured outer ear funnels sound waves to the eardrum, a taut membrane that is stretched like a drum skin across the ear canal and that vibrates in time with the sound wave frequencies. A chain of three bones in the middle ear in turn transmits eardrum vibrations to the small, coiled cochlea, the sensory receptor in the inner ear that converts vibrations into nerve impulses. Overall, the cochlea has a snaillike shape, compact and coiled. Inside, elastic membranes partition it into three fluid-filled chambers, and attached to one membrane partition are rows of hair cells, each cell crowned by a bundle of little hairlike projections called *stereocilia* (Figure 27.3b). Each hair cell bears about 100 little hairs; in total, each human ear contains 1 million.

How do the structures of the inner ear enable us to detect sound? When a sound wave vibrates the eardrum and bones of the middle ear, movement of those bones compresses the fluid in the cochlea. This, in turn, causes the little hairs on the hair cells to bend; and that bending

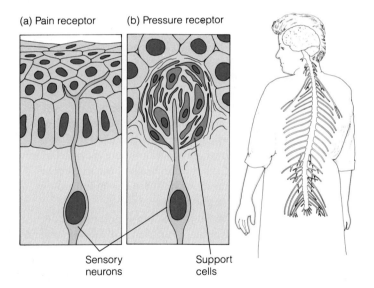

(a) Pain receptor (b) Pressure receptor

Sensory neurons

Support cells

FIGURE 27.2 *Sense Receptors.* (a) The simplest sensory neurons, touch or pain receptors, have nerve endings surrounded by a thin layer of supporting cells. Pressure on those supporting cells stimulates the receptor nerves directly. (b) Pressure receptors project from a capsule of supporting cells and amplifying cells. The spatial arrangement of amplifying cells increases the effects of pressure. They trigger an action potential in the pressure receptor neuron itself.

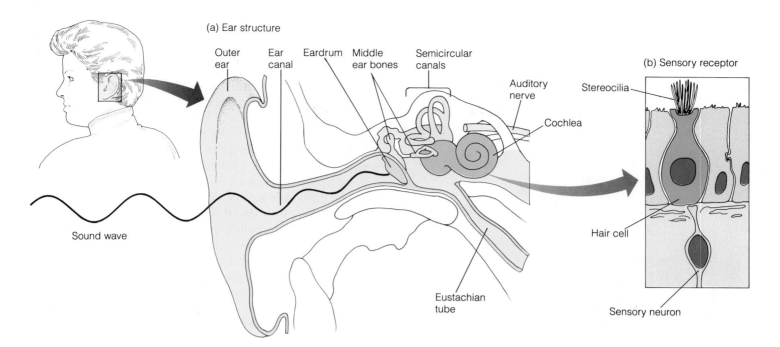

FIGURE 27.3 *How the Ear Works.* (a) As the text explains, sound waves vibrate the eardrum, then three small middle ear bones, then fluid in the coiled cochlea. (b) Stereocilia on the hair cells of the cochlea move in response to sound, and the hair cells convert this mechanical movement into an electrical signal that crosses a synapse and triggers a sensory neuron. This neuron, in turn, sends a message to the brain that a sound has been received.

generates a nerve impulse that crosses synapses to a sensory neuron leading to the brain.

Receiving the signal, how does the brain know whether the incoming sound is the high-pitched whistle of a bull elk or the low growl of a grizzly bear? Once again, the answer involves the mechanical properties of the membrane lined with hair cells. The membrane vibrates in different places when stimulated by different pitches. Since each region of the membrane connects to a different part of the brain, different pitches stimulate different brain cells. In essence, a musical staff is inscribed in space across a certain part of the brain. A baby can distinguish the characteristic tones of its mother's voice because that voice stimulates hair cells at specific places in the cochlea, which in turn correspond to specific brain areas. While a soothing voice causes the stereocilia on the hair cells to sway gently, loud sounds, such as those from a jackhammer or a heavy metal rock band, can break off the stereocilia and thus permanently damage the hair cells and along with them the sense of hearing (Figure 27.4, page 444). We can distinguish loud sounds from soft ones because loud sounds cause more neurons to fire and cause them to fire more frequently.

To fully sense its environment, an animal must know the direction of a sound as well as its tone qualities. Owls, for example, can localize sounds in the horizontal plane about as well as people, but the birds are much more accurate than people at locating sounds above or below them. This is because an owl's left ear is aimed higher than its right and also because the owl has an internalized map in which each brain cell responds to sound from a different point in space.

■ **How the Ear Detects Balance and Motion** While hair cells are crucial to hearing, they also figure prominently in the detection of the body's position in space and the proximity of other animals and objects. In a tawny owl, the **semicircular canals**—three curving, fluid-filled tubes that function as organs of balance and motion detection (see Figure 27.3a)—register the bird's acceleration as it swoops down toward a scurrying mouse. As the owl banks and dives, the detection of this motion is based on the swaying of stereocilia on the hair cells in the semicircular canals, much as passengers standing on a bus sway when the vehicle rounds a corner. Hair cells thus provide vital information and enhance the survival of the tawny owl—and indeed the survival of various other organisms in their own special environments.

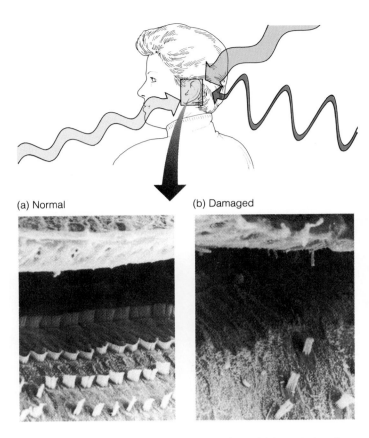

(a) Normal   (b) Damaged

FIGURE 27.4 *Loud Sounds Damage Hair Cells.* (a) Intact hair cells from the human cochlea, highly magnified. (b) Violent vibrations from extremely loud sounds break the little hairs from the hair cells, causing irreparable damage and permanent hearing loss.

THE EYE: AN OUTPOST OF THE BRAIN

The colors of the rainbow. A loved one's smile. Breathtaking mountain scenery. These delightful visual sensations are all possible because the eye changes electromagnetic radiation in the form of light into neural energy, just as the ear changes mechanical energy into nerve impulses. Vision is our most important sense, and not surprisingly, the human eye is large—about the size of a golf ball. For a tawny owl to see slightly better than the average person, its eyes must be a bit larger than golf balls, even though its head is far smaller than a person's. If our eyes were proportionately as large, they would be the size of softballs!

■ **How the Eye Detects Light** Light bouncing from the leaves of a seedling and striking the human eye first passes

through the protective transparent outer layer, or *cornea* (Figure 27.5a). If you gently place your finger on a closed eyelid and turn your eye from left to right, you can feel the bulge your cornea makes. After penetrating the clear cornea, light enters the *pupil*, which is a black, circular, shutterlike opening in the *iris* (the eye's colored portion), then traverses the *lens*, a circular, crystalline structure that focuses light onto the **retina.** The retina is a multilayered sheet that lines the back of the eyeball and contains light-sensitive **photoreceptor cells** (Figure 27.5b); these specialized neurons begin changing light energy to electrical energy that can eventually form brain patterns and create a visual image of an object. Because the lens is curved, it bends the light rays. Thus, the image of the object it projects on the retina is upside down and backward (with left and right reversed; see Figure 27.5a). Once the visual impulses reach the brain, however, they are integrated and sorted out so that the image conforms to reality.

Within the multilayered retina, the rear layer (closest to the back of the eye and just inside the eye's tough outer covering) consists of jet black pigment cells that protect the photoreceptors from extraneous light (see Figure 27.5b). Nestled just in front of the pigment cells are two types of photoreceptor cells called **rods** and **cones** because of their distinctive shapes (Figure 27.5c). Rods are very sensitive to low levels of light, but cannot distinguish color, while cones need more light but can detect color. This is the reason you have trouble seeing colors in dim light. As one might expect, the retina of a tawny owl is jam-packed with rods, allowing it to pick up the dimmest light during night hunting; but since the retina contains few cones, the animal's color vision is probably very poor. In the retina's front layer, rods and cones synapse with sensory neurons that send axons toward the brain. Because these sensory neurons lie in front of the rods and cones, light must pass through the neurons before reaching the photoreceptors. The axons leading from individual retinal neurons collect into a bundle called the **optic nerve,** which exits the eye and leads to the brain (see Figure 27.5a and b).

Recall that each kind of sensory neuron cell can tell the brain the strength of the signal and its type. In the case of neurons in the eye, strength means the brightness of the light. Brighter light stimulates more sensory neurons to generate more nerve impulses in a shorter period of time. Photoreceptors can also convey to the brain the position of the light source in the visual field. This is because photoreceptors in different parts of the retina receive light from different parts of the visual field and link up with different parts of the brain. For example, a horizontal line in the visual field stimulates a line of photoreceptors in the retina. This line on the retina links to a line of neurons in the visual center at the rear of the brain, allowing you to see a horizontal line.

THE TONGUE AND NOSE: OUR CHEMICAL SENSE ORGANS

The eyes and ears, as well as touch receptors in the skin, receive and detect energy—light energy, acoustical vibrations, and physical force—and are thus our *physical senses*. The tongue and nose, on the other hand, receive and detect flavor and odor molecules—actual chemical tidbits of the environment. Thus, biologists consider taste and smell to be our *chemical senses*.

■ **How the Tongue Detects Taste** People often think of the tongue as a perceptual genius and the nose as a sensory dullard. However, precisely the opposite is true. The tongue is studded with small conical bumps, which house the taste buds, and each bud consists of a pore

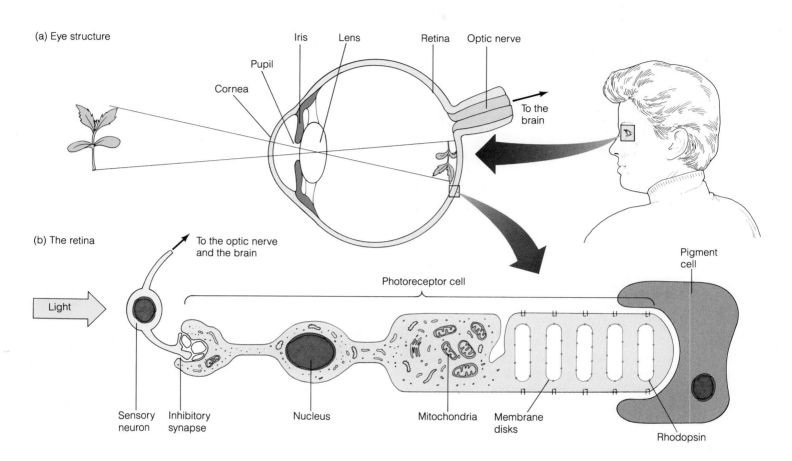

(a) Eye structure

(b) The retina

(c) Rods and cones

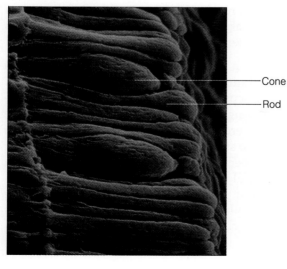

FIGURE 27.5 *How the Eye Works.* The text explains in detail how the eye works. (a) A cross section of the eye reveals the pupil, cornea, lens, retina, optic nerve, and other structures. The fovea has the highest concentration of rods and cones and is our region of greatest visual acuity. You can demonstrate this by choosing a word on this page and focusing on it. Then, without moving your eyes, try to read the surrounding words. You cannot read many, can you? To read effectively or view a scene, we must constantly shift our eyes over the words or details, and in so doing, continually focus light on the fovea. (b) A view of a sensory neuron, photoreceptor cell (here a rod cell), and pigment cell of the retina. (c) This scanning electron micrograph shows that the very sensitive rod cells are tall and numerous, and this explains why our eyes can detect even extremely dim light. Cones are shorter and club-shaped, and each one can detect either blue, green, or red light.

FIGURE 27.6 *The Nose Knows and So Does the Tongue.* A taste bud on the tongue (a, b) and odor-detecting cells in the nose (c, d) consist of receptor cells that generate a nerve impulse when stimulated by a specific chemical.

leading to a nerve cell and surrounded by accessory cells arranged in an overlapping pattern that resembles an artichoke (Figure 27.6a and b). The nerve cells in taste buds have receptors capable of receiving flavor molecules, but they can only distinguish four general classes of flavor: sweet, salty, bitter, and sour. We can tell similar foods apart—beef from pork, beets from turnips, pickles from sauerkraut, honey from sugar—and sense the subtlety of their flavors for two reasons: (1) These foods stimulate the four receptor types to different degrees, and (2) the volatile aroma molecules from the food travel into the nose or up the back of the throat and bind to receptors in the olfactory epithelium. This epithelium consists of button-sized patches of yellowish mucous membrane high in the nasal passages.

■ **Smell, Taste, and Memory** Neurons called olfactory cells lie embedded in these epithelial patches. One end of the nerve cell is hairlike and binds odor molecules (Figure 27.6c and d). The other is a long, spindly axon that feeds into the olfactory nerve, which carries signals directly to the smell region of the brain. Some scientists believe that olfactory receptors can distinguish a minimum of 32 primary odors, including sweaty, spermous, fishy, urinous, musky, minty, malty, and camphoraceous. If you take a bite of a sandwich and simultaneously pinch your nose shut, you can still perceive the four primary tastes, but not primary odors—and thus very little of the food's complex flavor.

People have long noted peculiarly intimate connections between smells, memories, and emotions. Who hasn't been temporarily overcome by a flood of memories, complete with appropriate emotions, when catching a whiff of a Christmas tree, or a puppy's fur, or a pipe like Grandfather smoked? The explanation for such odor déjà vu lies in the anatomy of the nose and brain. The olfactory lobes are closely connected to a series of small structures in the brain that are largely responsible for generating fear, rage, aggression, and pleasure and for regulating sex drives and reproductive cycles. A smell, therefore, stimulates the brain's centers of memory, emotion, and sexuality as well as the seat of conscious thoughts and learning that surrounds the olfactory lobes.

Olfactory research is a far smaller and younger field than vision or auditory research, but experimenters are uncovering many intriguing facts about our primal sense.

- A person can smell as little as a few molecules of ethyl mercaptan (the essence of rotten meat)—literally a few molecules of the material.
- An experimental subject tends to recall the visual details of a given painting with almost 100 percent accuracy, but will forget the details within three months. The same subject will recall a series of odors with only 80 percent accuracy, but the accuracy remains at that level for a year or more. An odor, once remembered, is rarely forgotten!
- Roommates in a women's dormitory often have synchronized menstrual cycles, and women with irregular periods often grow more regular when in a man's company on a routine basis. In both cases, the effects seem to be largely olfactory and due, perhaps, to pheromone-like molecules in the sweat.

Future research is sure to explain more about the role of olfaction in human emotion, sexuality, and evolution.

So far, our discussion has centered on the sensing portion of the sensing-integrating-reacting process, and the main message here is straightforward: An animal's brain can decipher the stimulus it receives as light or sound or taste or smell because each sense organ is tuned to a different physical or chemical stimulus and sends a signal to a different part of the brain. The strength of the signal simply reflects the number of sensory neurons activated. To discuss the next step, neural integration, we need a more detailed knowledge of the nervous system and how it is organized.

TRENDS IN THE EVOLUTION OF NERVOUS SYSTEMS

Although an animal's eyes, ears, and other sense organs are crucial for detecting stimuli from within the body as well as from the environment, the animal's survival often depends on the ability to act—perhaps escape from a predator, pounce on prey, scout the area in search of flowers, or select a mate. Action requires both an integration of the sensory input and a coordination of different body parts. When an owl, for example, hears a rustling on the forest floor, it must be able to determine whether the sound comes from a mouse the owl could eat or a prowling fox that would eat the owl. Those kinds of discriminations are made by neurons arranged in complex ways in the animal's central nervous system.

Whereas the survival of a complicated animal depends on the integrated function of a complex nervous system, a simple animal like a sponge can get along with no specialized nerve cells at all. Each sponge cell communicates directly only with its immediate neighbors. Lacking overall neural integration, however, a sponge is little more than a reproducing, filter-feeding vase.

The nervous systems of animals more complex than sponges show three general evolutionary trends: concentration of the nervous system in the head; a body organization with left-right symmetry; and an increase in the number of *interneurons,* the nerve cells that connect one neuron to another. Radial animals such as hydras (see Chapter 17) have the simplest form of nervous organization in the animal world, the **nerve net:** a lacework of nerve cells and fibers that permits the conduction of impulses from one area to another. A stimulus anywhere on the body prompts a nerve impulse to spread across the network to other regions. More complex animals have a concentration of nerves in the animal's front end, the body region that first encounters new environments. A flatworm's nerve net, for example, contains **ganglia** (singular, ganglion): distinct clumps of nerve cell bodies in the head region that act like a primitive brain. Distinct nerves arranged in mirror-image symmetry on either side of the body carry sensory information from the periphery to the head ganglia and carry motor commands from the head ganglia back to the muscles; this facilitates coordinated walking, flying, crawling, or climbing.

In addition to a centralized "brain," more complex animals have smaller ganglia that help coordinate outlying regions of the animal's body. These ganglia contain many interneurons and thus reflect the third evolutionary trend: an increase in the number of nerve cells that connect one kind of neuron with another. Since interneurons do much of the integrating that takes place in nervous systems, the more interneurons, the more complex behavior patterns an animal can display. Vertebrate brains, which we turn to next, are the culmination of this trend.

ORGANIZATION OF THE VERTEBRATE NERVOUS SYSTEM

Complex animal behavior—from the grouse's elaborate mating dance to the cobra's hypnotic weaving and the aerobic dancer's vigorous kicks and turns—depends on a highly organized nervous system. A vertebrate's nervous system is organized into two main units based on location and function: a peripheral nervous system and a central nervous system (CNS) (Figure 27.7, page 448). The peripheral nervous system is the "actor" or "doer"; it includes the sensory and motor neurons, and it connects the CNS with the sense organs, muscles, and glands of the body. The CNS is the nervous system's "thinker," or "information processor," and consists of the brain, which performs complex neural integration, and the spinal cord, which carries nerve impulses to and from the brain and participates in reflexes. The interplay between the central and peripheral systems allows an animal to sense environmental stimuli, integrate the information, respond appropriately, and in so doing, carry out the fascinating range of behaviors one sees in the animal kingdom.

THE PERIPHERAL NERVOUS SYSTEM: THE NEURAL ACTORS

Peripheral neurons sense changes and cause actions, but do little information processing. Peripheral neurons include *sensory neurons* like those in the eyes, ears, and skin, which carry information about the environment to the CNS, and *motor neurons,* which convey information from the CNS to muscles and glands and trigger some activity.

■ **Sensory Neurons** The peripheral nervous system is intimately connected with the spinal cord, a bone-encased trunk of nerves carrying messages to and from the brain (see Figure 27.7b). Attached to the spinal cord are 31 pairs of spinal nerves, each a cable of thousands of axons reaching out to different regions of the body. To see how sensory and motor neurons are organized within a spinal nerve, think about the reflex that causes you to pull your foot from a tack (review Figure 26.8). The dendrites of pain receptors in your skin send action potentials to the spinal cord, where they connect with interneurons. These interneurons are part of the CNS and relay a message to the brain, where it is interpreted as pain. Other stimulated interneurons relay signals within the spinal cord to motor neurons whose axons leave the CNS via the spinal nerve and signal your muscles to quickly pull your foot away from the tack.

(a) Organization of the human nervous system

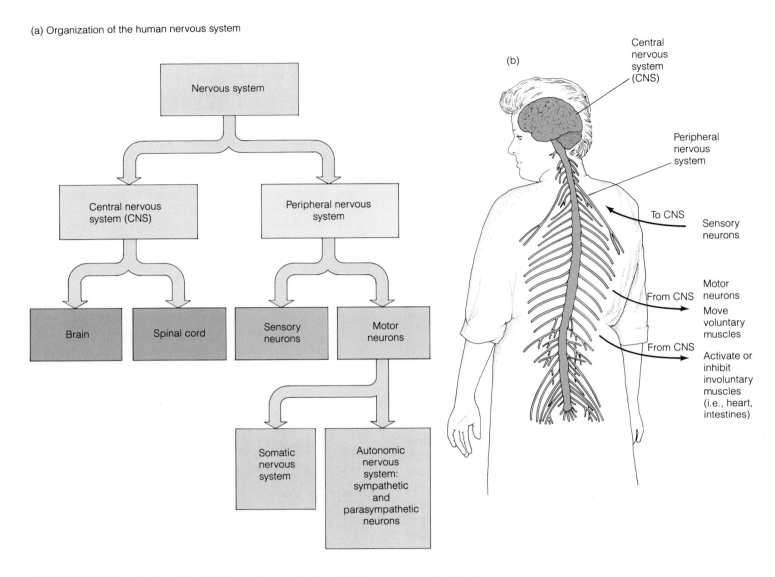

FIGURE 27.7 *Organization of the Vertebrate Nervous System.* (a) Your nervous system has two main branches, the *central nervous system* (*CNS*), which controls the processing and integration of information, and the *peripheral nervous system*, which connects the CNS with sensory receptors, muscles, and glands. The CNS consists of the *spinal cord*, which transports information to and from the brain and makes reflexes possible, and the *brain*, which performs complex neural integration. The peripheral nervous system includes both *sensory neurons*, which carry information about the body and the environment to the CNS, and *motor neurons*, which convey instructions from the CNS to muscles and glands. Motor neurons are of two types: Those of the *somatic nervous system* control voluntary muscles, while those of the *autonomic nervous system* control internal organs, muscles, and glands. Finally, the autonomic nervous system itself has two components: the *sympathetic nervous system*, which controls energy-expending activities (such as a rapid heart rate or an actively secreting adrenal gland), and the *parasympathetic nervous system*, which supplies or conserves energy (e.g., in a churning intestinal tract). In (b) we see the information-sending and -receiving pathways for different parts of the body.

FIGURE 27.8 *Mind Over Matter: This Yogi Can Consciously Lower His Metabolic Rate.* This yogi, of Madras, India, is meditating and has learned to consciously control his autonomic nervous system.

■ **Motor Neurons** The motor neurons in spinal nerves form two important systems: the *somatic nervous system*, which in general activates muscles under voluntary control (like those that allow you to ride a bicycle), and the *autonomic nervous system*, which regulates the body's internal environment by controlling glands, the heart muscle, and smooth muscles in the digestive and circulatory systems (see Figure 27.7a).

A person's autonomic nervous system operates primarily at the subconscious level, performing many of its duties through the spinal cord without input from the brain. There are two sets of autonomic neurons that function in opposition, a bit like the accelerator and brakes of a car. One set of neurons (called *sympathetic neurons*) speeds up and strengthens the heartbeat, for example. Another set (called *parasympathetic neurons*) slows the heartbeat.

In general, the parasympathetic set of autonomic neurons serves as a "housekeeping system" for the body, stimulating the stomach to churn, the bladder to empty, and the heart to beat at a slow and even pace during most daily and nightly activity. When an emergency arises, however, generating intense anger, fear, or excitement, the sympathetic set of autonomic nerves dominates, and its neural signals increase heart rate, dilate the pupils of the eyes (to let in more light), expand tubules in the lungs (to improve gas exchange), and slow down nonessential digestive activities until the emergency is past. Some medications for high blood pressure (so-called *beta-blockers*) specifically inhibit the sympathetic synapses that stimulate the heart and blood vessels so that the heart rate is slower and steadier.

Some cultures have explored conscious control over the autonomic nervous system with intriguing results.

Some yogis in India, for example, permit themselves to be buried in sealed underground chambers, and they claim they can survive such entombment for many days by lowering their metabolic rate and thereby using far less oxygen (Figure 27.8). Many such claims are surely exaggerated. But by measuring the yogis' breathing, pulse, and oxygen utilization with reliable instruments, Western scientists have concluded that some yogis can indeed consciously lower their metabolic rate below normal resting levels. The Western practice of biofeedback training, which uses devices to indicate changes in blood pressure, blood flow to extremities, or other external states, is another example of a crossover between the conscious and autonomic nervous systems.

THE CENTRAL NERVOUS SYSTEM: THE INFORMATION PROCESSOR

Although the peripheral nervous system collects data about the environment and directly controls the actions of the body's organs, all would be chaos if the central nervous system could not make sense of the data and coordinate the actions. The CNS can integrate sensing and reacting because of the intricate construction of the spinal cord and brain and because of the millions of communication pathways within and between them.

■ **The Spinal Cord: A Neural Highway** A cross section taken from a human spinal cord reveals a butterfly-shaped core of gray material surrounded by a bean-shaped field of white (Figure 27.9, page 450). The gray matter contains neuron cell bodies and nerve fibers not surrounded by a myelin sheath; it is a zone of many synapses, where local neural traffic occurs (such as the reflexes discussed earlier). The white matter, on the other hand, is an interstate freeway; it consists mainly of thin axons that transport information long distances up and down the spinal cord to and from the brain. Since these long axons are insulated with a white, fatty myelin sheath, the entire region has a whitish color. The spinal cord is responsible for preliminary integration of signals from sensory neurons and interneurons. For example, when a flame touches your finger and you withdraw your finger reflexively, synapses between the sensory neuron, interneuron, and motor neuron are taking place in the gray matter. Despite the preliminary role of the spinal cord in neural integration, however, most major data processing occurs in the brain.

■ **The Brain: The Ultimate Processor** The human brain—a reddish brown organ that weighs about 1.4 kg (3 lb), contains 100 billion neurons, and has the consis-

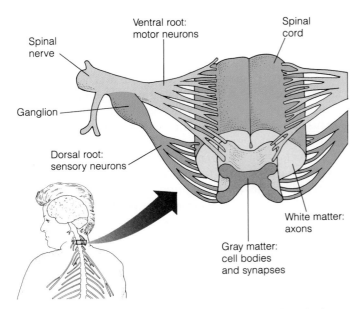

FIGURE 27.9 *Organization of the Spinal Cord.* Each major spinal nerve has both a dorsal and a ventral root as it emerges from the central spinal cord. The dorsal root carries sensory information to the brain, while the ventral root conveys signals for action out to muscles and glands.

tency of vanilla custard—embodies our feelings and strivings, our knowledge and memories, our musical and verbal abilities, and our sense of past and future. These complex behaviors and emotions are localizable to specific regions or groups of regions in the brain, as even minor damage to a given area can reveal. Thus, the anatomy of the brain provides a map for behavior, and certain behavioral disorders can be clues to the activities of specific brain regions.

The human brain has three interconnected parts (Figure 27.10): the **brain stem,** an extension of the spinal cord; the highly rippled **cerebellum,** attached to the brain stem; and the large, folded **cerebrum,** sitting atop the brain stem and spanning the inside of the head from just behind the eyes to the bony bump at the back of the skull.

■ **The Brain Stem**   The brain stem plays three crucial roles in sustaining life: It helps integrate sensory and motor systems, it regulates body homeostasis, and it controls arousal. The lowest part of the brain stem, the *medulla oblongata,* lies in the upper part of the neck at about the level of the mouth. The medulla helps keep body conditions constant by receiving data about activities in various physiological systems and by regulating numerous subconscious body activities, such as respi-

ratory rate, heart rate, and blood pressure. In the brain stem above the medulla lie the *pons* and *midbrain,* regions that relay sensory information from our eyes and ears to other parts of the brain (see Figure 27.10).

The *hypothalamus,* still farther up the brain stem, regulates the pituitary gland and provides a link between the nervous and endocrine systems (see Chapter 25). The hypothalamus also has important roles in maintaining homeostasis, helping to regulate body temperature, water balance, hunger, and the digestive system. By electrically stimulating different regions of the hypothalamus, researchers can make an animal behave as if it feels alternately hot and cold, hungry and satisfied, angry and content. Some regions seem to be pleasure centers: A rat with electrodes permanently implanted in such areas will continue to press a foot pedal that turns on the current and stimulates the pleasure centers until the animal drops from exhaustion. Just above the hypothalamus is the **thalamus,** a relay area to the highly convoluted cerebrum that surrounds it (see Figure 27.10).

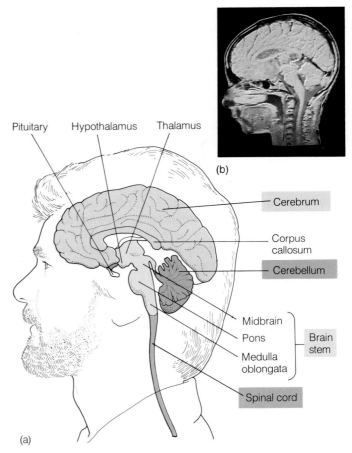

FIGURE 27.10 *Major Regions of the Brain.* (a) The drawing reveals the brain stem, cerebellum, and cerebrum, as well as major parts of each section. (b) A photo of the brain taken with magnetic resonance imaging reveals the living tissues in remarkable detail.

Extending throughout the brain stem from the thalamus down to the spinal cord is the *reticular formation,* a network of tracts that reaches into the cerebellum and cerebrum. When specific neurons of the reticular formation are actively firing, a person is awake, and when those neurons fall silent, the person sleeps or loses awareness. An intermediate amount of stimulation leads to a dulling or sharpening of our senses. This selective shaping of neural input by the reticular formation and the thalamus helps a person to concentrate on a dinner partner's conversation in a noisy restaurant and plays a part in the groggy incoherence of an early-morning phone conversation.

■ **The Cerebellum: Muscle Coordinator** Attached to the midbrain, or middle of the brain stem, is the *cerebellum,* a convoluted bulb that serves as a complex computer, comparing outgoing commands with incoming information about the status of muscles, tendons, joints, and the position of the body in space (see Figure 27.10). The result is a sculpting and refining of motor commands. A person whose cerebellum is damaged has jerky and exaggerated muscle movements. The cerebellum modifies motor commands initiated elsewhere in the brain until they are smooth and coordinated. Learning the finer points of a basketball pattern or a graceful ballet turn involves the cerebellum. Birds, including owls, which rely on precise muscle coordination to avoid tree limbs and other obstacles, have relatively large cerebellums. Although muscle movement is refined in the cerebellum, it is initiated in the largest part of the brain: the cerebrum.

■ **The Cerebrum: Seat of Perception, Thought, and Humanness** Two side-by-side hemispheres of the brain's cerebrum fit like a cap over the brain stem. This mass of tissue embodies not only all the attributes we consider human, such as speech, emotions, musical and artistic ability, and self-awareness, but also traits we share with other vertebrates, such as sensory perception, motor output, and memory. In humans, the **cerebral cortex,** the cerebrum's highly convoluted surface layers, contains about 90 percent of the brain's cell bodies. During vertebrate evolution, this cap increased in size relative to other brain regions. One of the great unsolved questions of modern biology is how these cells can control complicated traits. For now, biologists have many ideas but no single answer. One principle has been well established, however: All our basic cerebral functions, including sensations (like seeing or hearing), motor ability (movement), cognitive functions (like language and perception), affective traits (emotions), and character traits (like friendliness or shyness), can be traced to specific regions of the cerebrum (Figure 27.11, page 452). Let's first look at the sensory and motor functions of the cerebral cortex and then turn to more complex behaviors.

■ **Motor and Sensory Cortices** Brain surgery on patients with conditions such as epilepsy (a disorder that leads to seizures, or convulsions) has helped researchers map the cerebral cortex. Despite its billions of neurons, brain tissue has no pain receptors, so surgery can be carried out with only a local anesthetic for the scalp incision. By inserting extremely fine low-current electric wires into the exposed brain, neurosurgeons can stimulate specific regions of the brain and observe which muscles the patient moves or ask the alert patient to report the sensations he or she feels. With studies like these, scientists have mapped a projection of the body onto the surface of the cerebral cortex—that is, a map showing which brain regions control the sensations and motor functions of which parts of the rest of the body. If you run your right index finger from your left ear straight up to the top of your head, the arc traces the surface area devoted to the left **motor cortex,** the part of your brain that controls muscles on the right side of your body, including the ones that move your right hand. For example, the cortex region just above your left ear moves the muscles of your right jaw, and a region a bit higher moves the muscles of your right arm, while the part nearest the top of your head moves your right hip muscles. Likewise, the corresponding arc on the surface of the right side of your brain moves the left side of your body (see Figure 27.11a and b).

Just behind the motor cortex lies the **sensory cortex,** which registers and integrates sensations from body parts (see Figure 27.11a and c). (Note that the surface regions of the motor and sensory cortices are responsible for similar but not identical parts of the body, since motor nerves go to muscles, while the sensory nerves go to sense organs such as touch receptors in the skin.) Again, the left side of the brain receives sensations primarily from the right side of the body, and vice versa.

For the maps of both motor and sensory cortices, the representation of body parts does not match their true proportions in the body: Lips, face, and fingers appear too large, and the trunk too small. This disproportionate mapping reflects the exceptionally large number of delicate touch receptors in face and fingers and the large number of muscles necessary for speech and manual dexterity relative to the small numbers needed for sensation and movement of the trunk. Other specific brain regions are dedicated to vision, hearing, and smelling. Finally, the map can change with use. This could explain the acquisition of skills such as learning Braille, the raised-dot writing system of the blind.

■ **Higher Cerebral Function** Many brain capacities in addition to sensory and motor functions can be mapped to specific regions, and once again, brain injuries have

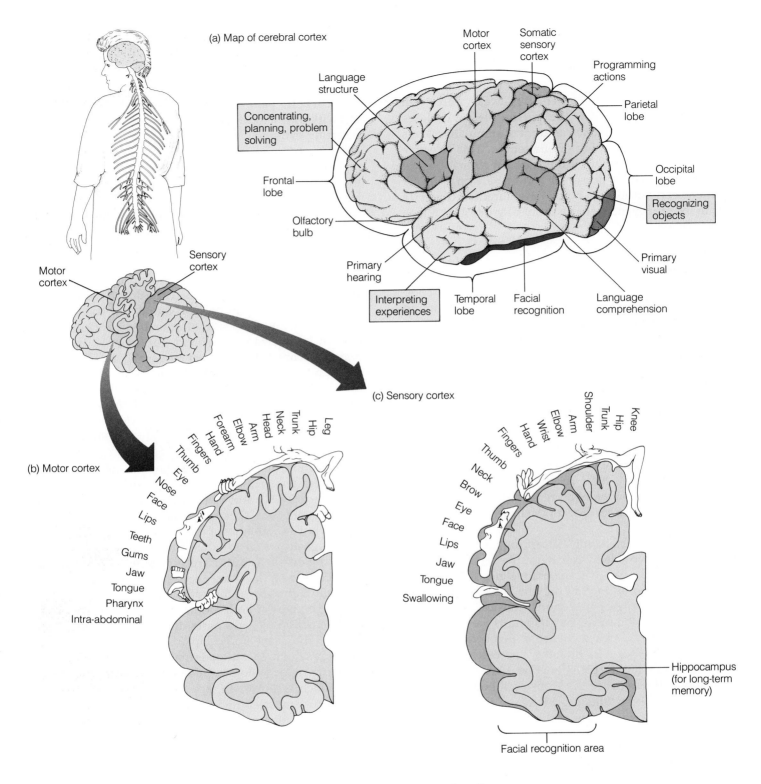

FIGURE 27.11 *Mapping the Human Brain: Specific Functions and Behaviors Reside in Specific Surface Regions of the Cerebral Cortex.* (a) A map of the cerebral cortex reveals the mental functions associated with various regions of the brain's convoluted surface. The motor cortex can be traced to one "slice" and the sensory cortex to another. (Association cortex is shown tan.) (b) If one were to lift the motor "slice" and map the regions that controlled the movements of corresponding body parts, the map would look like the strange distorted person shown here. (c) If one did the same with the sensory "slice," the map of regions and corresponding body sensations would look somewhat different.

led researchers to connect function and location in the cortex. One area of the cerebral cortex, for example, is dedicated to recognizing faces (see Figure 27.11a and c), and a person who suffers an injury to this part of the brain can no longer recognize the faces of loved ones, although the injured person can still recognize their voices. A person with damage to the right side of the parietal lobe just behind the sensory cortex may completely ignore the left half of the body, failing to wash or dress the affected side. If the patients' left arm is passively placed before them in full view, they will deny that the limb belongs to them. Regions of the frontal lobe near the forehead seem to be involved in short-term memory, anxiety, and the ability to weigh consequences and act accordingly.

Deep in the cerebrum lies a small structure called the **hippocampus,** which plays a crucial role in the formation of long-term memory (see Figure 27.11c). Have you ever tried memorizing a new phone number, only to have someone interrupt you even for an instant? You probably forgot the number because your hippocampus failed to establish it in long-term memory before the interruption occurred. A patient with a damaged hippocampus can recall events that happened before the injury—even decades before—but fails to remember new facts and experiences (including the shapes of objects, phone numbers, and people's names) for more than a few moments. The patient may see the same doctor a dozen days in a row, but will greet the physician each time as a stranger.

Patients with Alzheimer's disease lose neurons in the hippocampus. Currently, Alzheimer's affects 2 million Americans and accounts for over half of the recorded cases of dementia (mental deterioration). It attacks otherwise healthy middle-aged and older people and causes a progressive loss of mental function. While the behavioral patterns (forgetfulness, inattention to hygiene, and so on) are characteristic, a physician can make a positive diagnosis only by examining brain tissue after the patient's death. The brains of Alzheimer's victims reveal a number of specific features, including the already mentioned loss of neurons from the hippocampus. Researchers are learning to transplant brain tissue experimentally in the hope of helping such patients someday (Figure 27.12).

Researchers have also made other important findings on memory. For example, brain cells must express certain genes that were not being expressed and synthesize new proteins to form a new memory. These events alter the strength or the number of connections (synapses) between cells and hence communication within the brain. These long-lasting changes seem to occur in many different brain regions and suggest that memory involves far more than just the hippocampus.

Speech—a quality that separates us from all other animals—is localized in two areas of the cerebral cortex, one for language structure and the other for language comprehension (see Figure 27.11a). If a stroke (damage to a brain region caused by a blood clot) disrupts the area of language structure, the patient may understand spoken and written language but be unable to speak well themselves. One such patient, talking about a dental appointment, said haltingly, "Monday—Dad and Dick— Wednesday nine o'clock—ten o'clock—doctors—and— teeth."

Brain areas responsible for the comprehension and structure of language are usually located in the cerebrum's left hemisphere. Corresponding regions of the right hemisphere regulate emotional gesturing and the musical intonation of speech. Thus, mapping shows that the brain is asymmetrical. In most people, language abil-

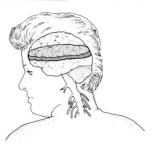

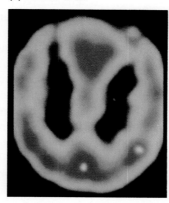 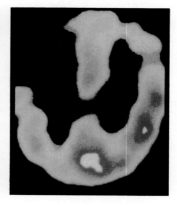

(a) Normal brain   (b) Brain of Alzheimer's victim

FIGURE 27.12 *Alzheimer's Disease: Replacement Parts for Damaged Brains?* Comparing a normal brain (a) and the brain of an Alzheimer's victim (b) reveals diminished blood flow to critical brain regions. How can scientists and physicians treat people with Alzheimer's disease before it destroys their mental abilities? Although it may sound like science fiction, researchers are now exploring the possibility of transplanting brain tissue. Brain implant experiments with rats provide hope that brain transplants in people may someday help to alleviate the progressive mental deterioration of the Alzheimer's patient, and in so doing ease the pain and disruption they and their families experience. Since the source of cells for the implants might be the brains of still-born human embryos or aborted fetuses, society must carefully consider the issue, just as we did before passing laws to allow corneas, kidneys, hearts, and livers to be transplanted from dead people to the living.

IMPROVING YOUR—WHAT WAS IT?

You lock your keys inside the car. You call your mother-in-law "What's her name." You know the right answer but can't produce it for the midterm. What's wrong with your memory? Probably nothing. We all have lapses of memory. The brain, after all, is a living organ, subject to the ups and downs of body condition, age, and environment. Luckily, researchers have discovered many of the factors that affect brain function as well as numerous ways to boost your information retrieval.

Illness. Alzheimer's disease (dementia) is probably the best-known cause of major memory loss, but Parkinson's disease, alcoholism, head injuries, and Lyme disease can also affect normal recall. Conversely, general good health is usually accompanied by a good memory.

Drugs. Several drugs can interfere with normal memory, including sleeping pills, tranquilizers, and alcohol. Happily, the flip side is also true: Certain compounds improve memory, including lecithin, a major component of the extensive cell membranes so important for nerve cells, and choline, which forms part of the neurotransmitter acetylcholine (see Table 23.1 and

Chapter 26). Several pharmaceutical companies are working on new lecithin/choline drugs to help Alzheimer's patients and other sufferers of major memory impairment.

Nutrition. Vitamin deficiencies can impede your memory. The best studied is vitamin B_{12}, but severe shortages of thiamine (vitamin B_1), folic acid, and vitamin C can all influence normal recall. There is some evidence that mild, unnoticed vitamin deficiencies can erode the memory just enough to make life a little harder. The usual nutrition advice applies here: Eat a varied diet from the four basic food groups.

Exercise. Aerobic exercise brings more blood and more oxygen to brain cells, and there is solid evidence that when out-of-shape people begin to exercise, their memory improves. Thrilling sports—skiing, windsurfing white-water rafting, skydiving—trigger the release of stress hormones, which also enhance memory. Ironically, mental stress and anxiety do the opposite. A comfortably stressed body and a relaxed mind seem to be the best combination.

Memory Aids. The old trick of tying a string around your finger has been replaced by the portable appointment

calendar. But they are both external aids to jog the memory and take their rightful place beside stick-on notes and a special corner to put your keys, gloves, and glasses. External aids aren't always possible, though; you can't take notes at a party or meeting to remember people's names, and you can't use a "cheat sheet" during a test. And that's why people have invented internal memory aids.

Memory experts suggest that you pay special attention to anything you want to remember—say, a new name—by repeating it deliberately. "Sam Smith, how nice to meet you." Later, "I'm curious, Mr. Smith, how you chose this field of study." And finally, "I hope to see you again, Sam." Mental images can help you remember long names: a barking cow on skis for Bob Bowcowski. Abbreviations are useful for lists of information. Recall the CHNOPS acronym in Chapter 2? No? Oh, well. Rhyming can help: Gail the Whale. Art the Cart. And when it comes to memorizing details, several short periods on consecutive days are better than one long cram session the night before.

Happy memories!

ities reside mainly on the brain's left side (remember, *language*, *left*), as do the underpinnings of analytical thought and fine motor control. The right hemisphere, on the other hand, is mainly responsible for intuitive thought, musical aptitude, the recognition of complex visual patterns, and the expression and recognition of emotion. Almost all right-handers and 75 percent of left-handers have left-brain language areas. Many left-handers, however, have language areas on *both* sides of the cerebral cortex.

If separate functions are encoded in each hemisphere, how do the two sides of the brain communicate? Once

again, neurosurgeons attempting to help epileptics contributed to our understanding of this problem. In epileptics, nerves fire back and forth across the brain in a crescendo of positive electrical feedback. This firing can continue to escalate, like feedback from a loudspeaker, resulting in a seizure. Surgeons thought that they could interrupt the loop by "splitting the brain," or severing the bridge between the left and right cerebral hemispheres. These "split-brain" patients improved, and they seemed completely normal to friends. But when neurobiologists tested such patients in controlled experiments, they found that the patients' right visual cortex could not communicate what it saw to the region of language ability in the left brain. Thus patients *knew* what they saw but could not *say* it. The split-brain procedure is used much less frequently today to control epilepsy, but it dramatically underscores the principle that specific abilities are localized in particular regions of the brain. Even so, no function is the exclusive property of one brain region; rather, each function requires integrated actions of neurons located in several discrete regions, each region simultaneously carrying out parallel computations to coordinate behavior. While neurobiologists have made considerable progress in mapping functions to different parts of the brain, they have yet to learn how the parts operate. For example, each of the main cerebral lobes contains regions called *association cortex* (see light tan areas of Figure 27.11a). These areas appear to integrate information from other areas and reconstruct it at varying levels of consciousness. But exactly how these areas reassemble information and coordinate it remains unclear. To a large extent, these areas of association cortex contain the physiological underpinnings of emotions, memory, personality, reasoning, judgment, and the conglomeration of traits we call intelligence and remain for biologists of your generation to untangle.

≋ CONNECTIONS

This chapter discussed how neurons and nervous systems operate to sense the environment—to hear, see, smell, taste, and feel the world around us and to sense balance, position, and internal body states. The nervous systems that make such sensation possible became more centralized and more complex during evolution, culminating in the most complicated organizational unit in the known universe, the human brain. Our discussions have covered the sensing and integrating functions of an animal's physiology, but not the reacting: how information finally flows out again to the motor neurons of the peripheral nervous system to activate and coordinate the work of the body. Chapter 28 takes up at this point and then goes on to investigate how muscles, glands, and other organs work together to produce the leap of the long jumper and the pounce of the night hunter—and in general to enhance an organism's survival.

NEW TERMS

brain stem, page 450	nerve net, page 447
cerebellum, page 450	optic nerve, page 444
cerebral cortex, page 451	photoreceptor cell, page 444
cerebrum, page 450	retina, page 444
cochlea, page 442	rod, page 444
cone, page 444	semicircular canal, page 443
ganglion, page 447	sense organ, page 442
hippocampus, page 453	sensory cortex, page 451
motor cortex, page 451	thalamus, page 450

STUDY QUESTIONS

REVIEW WHAT YOU HAVE LEARNED

1. Give an example of how amplifying cells augment the activities of a sensory receptor.
2. What sequence of events in the ear results in sound detection?
3. How do the ear's semicircular canals function?
4. Explain how the eye detects light.
5. Describe the reflex pathway at work when you jerk your toe away from a tack.
6. Which body regions are stimulated by parasympathetic nerves? Which by sympathetic nerves? Give an example of the action of each.
7. Name the parts of the brain stem and the function(s) of each part.
8. Summarize the major roles of the cerebellum and the cerebrum.
9. How are the motor cortex and sensory cortex alike? How are they different?
10. Describe the functions of the right and left brain hemispheres.

APPLY WHAT YOU HAVE LEARNED

1. Explain how a college student who regularly plays her stereo at top volume may be damaging her sense of hearing.
2. Why is it often harder to taste food when you have a bad head cold?
3. Recent studies suggest that fetal tissue implants may help people suffering from Alzheimer's disease. What problems might be associated with this approach?

FOR FURTHER READING

Calvin, W. H. *The Cerebral Symphony: Seashore Reflections on the Structure of Consciousness.* New York: Bantam Books, 1989.

Fine, A. "Transplantation in the Central Nervous System." *Scientific American* 255 (August 1986): 52–67.

Kennedy, M. B. "How Does the Brain Learn? A Molecular View." *Engineering and Science* (September 1986): 4–9.

Knudsen, E. I. "The Hearing of the Barn Owl." *Scientific American* 245 (December 1981): 113–125.

Livingstone, M. S. "Art, Illusion and the Visual System." *Scientific American* 258 (January 1988): 78–85.

Mishkin, M., and T. Appenzeller. "The Anatomy of Memory." *Scientific American* 256 (June 1987): 80–89.

Sacks, O. *The Man Who Mistook His Wife for a Hat.* New York: Perennial Library, Harper & Row, 1987.

Sacks, O. *Seeing Voices: A Journey into the World of the Deaf.* Berkeley: University of California Press, 1989.

Schnapf, J. L., and D. A. Baylor. "How Photoreceptor Cells Respond to Light." *Scientific American* 256 (April 1987): 40–47.

Wurtman, R. J. "Alzheimer's Disease." *Scientific American* 252 (January 1985): 62–74.

① The nervous system receives and processes information from the internal and external environment and initiates appropriate behavioral responses.

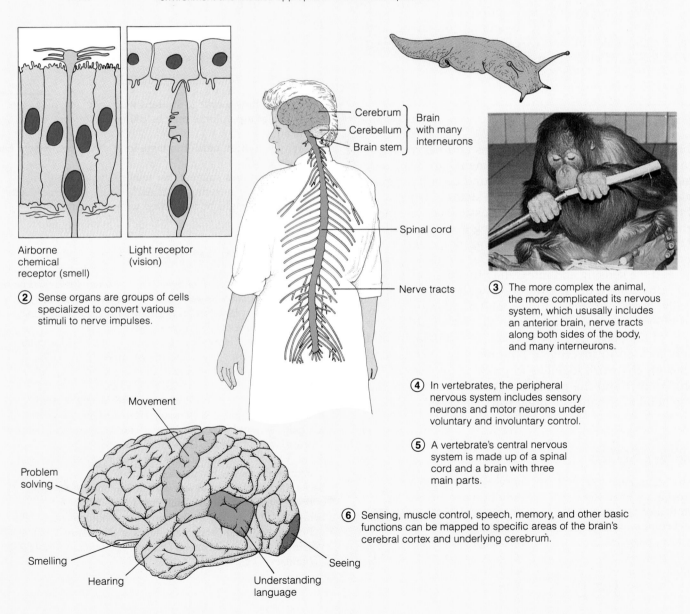

Airborne chemical receptor (smell)

Light receptor (vision)

② Sense organs are groups of cells specialized to convert various stimuli to nerve impulses.

Cerebrum
Cerebellum } Brain with many interneurons
Brain stem

Spinal cord

Nerve tracts

③ The more complex the animal, the more complicated its nervous system, which ususally includes an anterior brain, nerve tracts along both sides of the body, and many interneurons.

④ In vertebrates, the peripheral nervous system includes sensory neurons and motor neurons under voluntary and involuntary control.

⑤ A vertebrate's central nervous system is made up of a spinal cord and a brain with three main parts.

⑥ Sensing, muscle control, speech, memory, and other basic functions can be mapped to specific areas of the brain's cerebral cortex and underlying cerebrum.

Movement

Problem solving

Smelling

Hearing

Understanding language

Seeing

THE BODY IN MOTION

CHAPTER 28

BATTLE FOR AN EQUINE HAREM

Wild horses roam the Granite Range of northern Nevada, wandering many miles each day in search of green grass and fresh water. The horses usually live in stable bands, with a single stallion protecting and directing a harem of females and young. The leftover bachelors are excluded from the band, but try, from time to time, to wrest control of a harem from the existing leader male. Such challenges can lead to furious contests, and nearly all stallions have wounds and vicious battle scars on their muscular bodies (Figure 28.1).

In May of 1981, researchers watched a 6-year-old bachelor they called Harry wrangle for three straight days with Moscha, a 21-year-old stallion whose small harem included two mares. At times, the two combatants engaged in direct hoof-to-hoof combat, each kicking at his competitor's head, nipping at his legs, and trying to sever the tendons that join leg muscles to bone and thus cripple his adversary. To defend against a leg bite, a horse would immediately collapse to the ground, and this fall would misdirect the attack. As the three-day struggle ended, the weary Moscha stood with head lowered, too exhausted to protest, while the heavily muscled Harry escorted his new harem off into the desert.

The clash between Harry and Moscha helps illustrate the topics of this chapter: how the muscles work to drive the skeleton and how the body's physiological systems are integrated to maintain homeostasis during the stress of heavy exercise.

Two themes will emerge as we discuss muscles and skeletons. First, muscles and skeletons adapt to stresses both over the lifetime of the individual and over the evolution of the entire species. A horse's bones have become highly specialized during evolution for forward running, while a person's skeleton is less specialized, capable of running, lifting, carrying, and a range of other activities. Nevertheless, the muscles and skeletons of both horse and person are quite adaptable, and exercise can cause an individual's muscles and bones to enlarge and grow stronger.

Second, rigorous physical activity involves not just the muscles and skeleton, but all of the body's physiological systems, so that homeostasis can be maintained even during radical changes. When horses engage in protracted contests like the titanic battle between Harry and Moscha, or when a person swims the English Channel or runs a marathon, the stress of the event activates heart, lungs, gut, kidneys, nerves, and hormones, in addition to the muscles and skeleton. The activation of these systems requires close integration so that the individual functions as one coordinated unit that fights, swims, or runs with power and grace—a dynamic animal.

This chapter will answer several questions:

- How does the shape and internal structure of a bone allow it to perform many functions?
- What are muscles, and how do they work?
- How does exercise affect the body's homeostatic mechanisms?

FIGURE 28.1 *Muscular, Battle-Scarred Stallions Fight Over Leadership of the Harem.*

FIGURE 28.2 *A Sea Anemone's Hydroskeleton: Muscle Contraction Against an Incompressible Internal Liquid.* A Beadlet anemone shortens when longitudinal muscles contract (left) and lengthens when circumferential muscles contract (right).

❈ THE SKELETON: A SCAFFOLD FOR SUPPORT AND MOVEMENT

A primary tactic of fighting horses is to bite at tendons in the legs. This maneuver works because it can sever a muscle's attachment to the skeleton. Just as a stretched rubber band does no work unless it is attached to a stationary object at one end and to a movable object at the other, a muscle is useless without bones or other hard structures to pull against. The rigid body support to which muscles attach and apply force is called a **skeleton** in both vertebrates and invertebrates. Skeletons are of two basic types. The most familiar is a *braced framework* that provides solid support; the hard shell of a shrimp or the stiff rods of bone in your arms and legs are examples of braced frameworks. A less familiar but no less effective type of skeleton depends on liquid to transmit force, like the hydraulic brake system in a car. Let's look at this simpler type first.

WATER AS A SKELETAL SUPPORT

Animals as diverse as sea anemones, earthworms, and snails have a form of internal support called a *hydroskeleton,* composed of a core of liquid (water or a body fluid such as blood) wrapped in a tension-resisting sheath. A hydroskeleton is like a balloon filled with water: If you squeeze on one end, force is transmitted through the fluid to the other end. Likewise, contracting muscles can

push against a hydroskeleton, and the transmitted force generates body movement.

A sea anemone illustrates how a hydroskeleton works (Figure 28.2). This denizen of tide pools has a central digestive cavity filled with seawater and surrounded by the body wall. The wall contains two layers of muscles: one layer (*circumferential muscles*) encircles the wall, and the other layer (*longitudinal muscles*) extends from the animal's base to its tentacles. These two muscle layers act as **antagonistic muscle pairs**—groups of muscles that move the same object in opposite directions. When the longitudinal muscles contract, the animal becomes shorter and wider (as when you push a water-filled balloon from both ends simultaneously). When the circumferential muscles contract, the animal becomes longer and narrower (as when you wrap your hand around a balloon and squeeze).

Hydroskeletons are common among invertebrates, with their lack of bone, but even a system as complex as the human body relies on the principles of hydroskeletons under some situations. People lifting heavy weights off the floor tend to hold their breath and tighten their abdominal muscles. This helps support the load, not just with the backbone, but also with the compressed fluid of the abdominal cavity adding to the rigidity. In many mammals, including humans, a hydroskeleton also stiffens the penis; valves allow blood to flow forcefully into the organ but not to flow out as freely, and the trapped fluid causes the penis to expand and grow rigid.

BRACED FRAMEWORK SKELETONS

The simple hydroskeleton is perfectly adequate for an aquatic animal like a sea anemone, which has the extra buoyant support of a watery environment, or for certain land animals like the earthworm, whose locomotion requires only moderate versatility, speed, or coordina-

FIGURE 28.3 *A Cicada's Exoskeleton: External Support for the Body and Attachment Sites for Muscles.* A cicada nymph leaves behind its old, tight exoskeleton (brown) as it molts. The soft, new iridescent blue one will quickly become more rigid and protective.

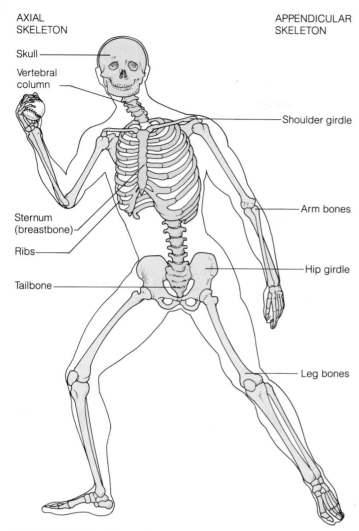

AXIAL
SKELETON

APPENDICULAR
SKELETON

Skull

Vertebral
column

Shoulder girdle

Arm bones

Sternum
(breastbone)

Ribs

Tailbone

Hip girdle

Leg bones

FIGURE 28.4 *The Human Skeleton: Internal Support and Muscle Attachment.* The adult body has 206 bones grouped into two portions: the axial skeleton (80 bones, shown in yellow), including the skull, vertebral column, ribs, sternum, and tailbone; and the appendicular skeleton (126 bones, shown in tan), including the 64 bones of the arms and hands, the 60 bones of the legs and feet, and the hip and shoulder girdles.

tion. A rapidly moving land animal, however, requires a braced framework of solid material made of interlocked proteins and minerals. It can be either an *exoskeleton*—a stiff sheet or other external covering that surrounds the body (Figure 28.3)—or an *endoskeleton*—a set of rigid rods and plates of bone inside the body (Figure 28.4 and Chapters 17 and 18). Here we will concentrate on the endoskeletons of vertebrates.

Mammalian endoskeletons are made of **bone,** a living tissue that contains the long, stringy protein collagen hardened by a calcium phosphate salt. Mammals have an **axial skeleton,** which supports the main body axis, and an **appendicular skeleton,** which supports the appendages—the arms and legs (see Figure 28.4). Together, these two skeletons have 206 individual bones.

■ **The Axial Skeleton** The skull, vertebral column, breastbone, ribs, and tailbone make up the axial skeleton. The *skull* is made up of the *cranium,* which encloses and protects the brain, as well as several bones of the face and inner ear. The skull also contains the tongue bone, the only human bone not jointed with other bones. (Instead, it is suspended at the back of the mouth by ligaments and muscles.) The thin curved bones of the cranium are laid down by membranes during embryonic development. Areas of fibrous membranes (the "soft spots," or *fontanels*) remain at birth and allow the baby's head to change shape slightly and pass through the birth canal more easily. The fontanels will later become hardened into bone, and the cranium will then be a continuous bony cap around the brain.

The head sits on top of the backbone, or **vertebral column,** made up of 33 *vertebrae.* Each vertebra is a bony box that fits together with other vertebrae in a stack, encasing and protecting the spinal cord and giving support to the trunk. This movable stack of boxes gives flexibility to the body axis, allowing a shortstop to bend forward and scoop up a line drive, or a gymnast to do a back walkover. The vertebral column, however, is a bit fragile and requires special precautions as you lift, stand, or sit to avoid developing lower back problems. Curving forward from the vertebrae are 12 pairs of ribs, which meet and attach at the breastbone (*sternum*) and form a protective compartment around the heart and lungs.

■ **The Appendicular Skeleton** Other bones are attached to the axial skeleton; these include the **shoulder girdle** and **hip** (pelvic) **girdle,** the parts of the appendicular skeleton that support the arms and legs and allow them to rotate (see Figure 28.4). The human hip girdle forms a rigid bowl that supports the internal organs. Attached to the shoulder and hip girdles are the bones of the limbs. If we compare our forelimb (arm) bone to a horse's forelimb (foreleg) bones, it becomes clear that both species show variations on a common anatomical theme (Figure 28.5, page 460). Our arms are specialized for grasping but are poorly adapted for running (think of a circus performer walking on her hands), while a horse's forelegs are specialized for rapid forward motion but are poorly adapted for movement in other directions (picture the awkwardness of a fallen horse trying to stand up again). The horse's bones that correspond to our palm and fingers are elongated; this adds an extra functional segment to the limb and means that the toe will touch only at its tip, with the middle "fingernail" being the hoof. The stresses of natural selection acting over thousands of generations have molded the skeleton of the horse for superb running ability, just as entirely different stresses have led to our adaptations for grasping, carrying, writing, and holding books in our arms.

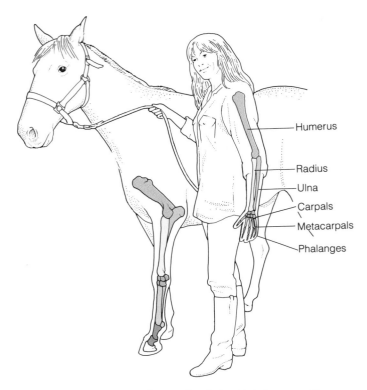

FIGURE 28.5 *Horse and Human Forelimbs: Variations on a Theme.* Although the human forelimb is specialized for grasping and the horse forelimb is specialized for running, the upper bone in each is fairly similar, and the human wrist bones are equivalent to the horse's "knee." The human hand, however, consists of 5 metacarpals and 14 phalanges, while the horse has the equivalent of one metacarpal and one main phalange with an enlarged nail—its hoof. The horse also has two tiny vestigial phalanges that have helped biologists trace the evolution of the modern horse from three-toed progenitors a bit like the modern tapir.

■ **Bones, a Living Scaffold** The bones of the skeleton not only support the body and anchor the muscles, but also encase vital organs, form blood cells, and store calcium and phosphate ions. A closer look at a large bone such as the femur in our leg shows how these tasks are accomplished (Figure 28.6). The long shaft of the femur (Figure 28.6a) allows it to support the body and transmit the weight of the animal to the ground. To help accomplish the support function, the femur has at each end an expanded portion covered with cartilage that forms a joint with other bones. Projections from the bone called *processes* serve as attachment sites for ligaments and tendons. **Ligaments** are connective bands linking bone to bone, while **tendons** are tissue straps that connect bone to muscle. It is easy to find and feel examples of these skeletal structures. You can feel a large process (a bump) where your head joins the back of your neck and attaches to muscles that keep your head erect; you can feel the ligament that attaches your kneecap to your shinbone (tibia) just below your knee when you tighten your thigh muscles (quadriceps); and you can easily feel tendons between the muscles in your upper arms and the bone in your forearm by putting your hand beneath the edge of a desk, lifting, and, with the other hand, touching the "cables" that now project on the inside of your elbow.

These sculptural details of the bones reveal how the stresses on an individual animal can mold the shape of its skeleton. Processes, for example, become larger when the muscles to which they are attached become stronger. Anthropologists excavating Italian cities buried centuries ago under ash and cinders from the eruption of the volcano called Mt. Vesuvius are able to distinguish the skeleton of a slave girl from that of a nobleman's daughter

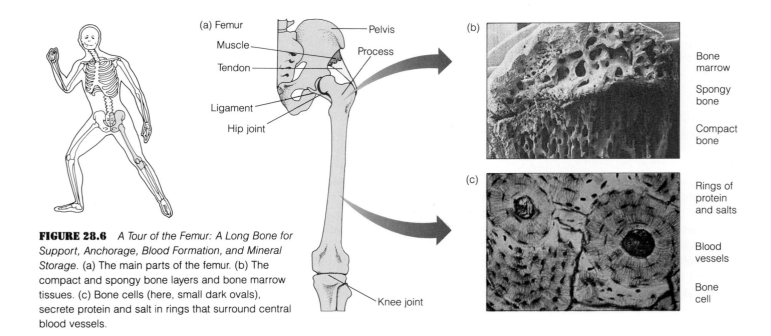

FIGURE 28.6 *A Tour of the Femur: A Long Bone for Support, Anchorage, Blood Formation, and Mineral Storage.* (a) The main parts of the femur. (b) The compact and spongy bone layers and bone marrow tissues. (c) Bone cells (here, small dark ovals), secrete protein and salt in rings that surround central blood vessels.

simply by comparing the size of the processes on arm and leg bones.

A longitudinal section of the femur reveals how a bone can anchor, strengthen, store minerals, and perform its other functions (Figure 28.6b). The outer layer of *compact bone* is thick, solid, strong, and resistant to bending; this enables it to provide support. At each end of the femur is a region of *spongy bone,* a looser area crisscrossed by girders to provide strength near the joints. The space between girders down the center of the bone's shaft is called the *marrow,* a site of blood cell production. Embedded within the bone are cells that secrete the proteins and minerals of the bone in rings surrounding a central channel filled with blood vessels (Figure 28.6c); bone cells also sculpt our bones as we develop and knit bones back together after a break.

Bone cells are especially active during periods of rapid growth. Bones elongate at a region near their ends called the growth plate. As an animal grows, bone is added to this plate, and when the organism matures, the plate disappears. If the bones are jarred by too much physical activity during childhood (a result of running marathons, for example, or of carrying heavy loads), the growth plates can be damaged and overall bone growth may be retarded.

If bone cells remove too much calcium, the bones become weak and fragile. This condition, known as *osteoporosis,* is relatively common in women after menopause and in inactive men. Recent research shows that the hormone estrogen helps many women absorb calcium from their diets. Some doctors now prescribe small doses of estrogen in conjunction with another hormone to help menopausal women avoid osteoporosis. Research also shows that regular exercise stimulates stronger bones, and this, too, can help prevent osteoporosis.

■ **Joints: Where Bones Come Together** Bones can move with respect to each other because of **joints,** points of contact between bones. Nearly immobile joints, such as those in the skull, have just a thin layer of connective tissue to separate adjacent bones. Slightly movable joints, such as those between adjacent vertebrae in the spine, have pads of cartilage that absorb shock but allow limited mobility. Freely movable joints, such as the shoulder, hip, and knee (Figure 28.7), have pads of cartilage at the ends of the two adjoining bones, but also have a flattened sac (a bursa) filled with synovial fluid between the bones that acts to cushion the joint and ease the gliding of bones across each other. Temporary inflammation of this joint sac is called *bursitis.* Long-term inflammation of the joint can result in *arthritis,* a painful swelling in which white blood cells infiltrate the sac and degrade cartilage and other connective tissues. Figure 28.7 describes how the knee joint works and why it is a source of trouble to joggers and professional athletes alike.

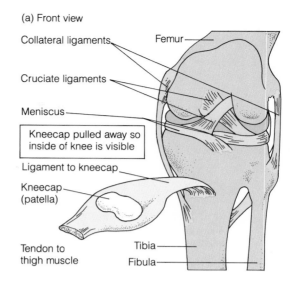

(a) Front view

Collateral ligaments
Femur
Cruciate ligaments
Meniscus
Kneecap pulled away so inside of knee is visible
Ligament to kneecap
Kneecap (patella)
Tendon to thigh muscle
Tibia
Fibula

(b) Side view

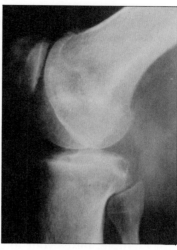

FIGURE 28.7 *The Knee.* (a) Anatomy of the knee joint. Your knee joint lies where the thighbone (femur) meets the shinbone (tibia). Lying between these two bones is a space filled with fluid that acts like a sponge to cushion the pounding of running and jumping. Two pillows of cartilage (each called a meniscus) lie inside the fluid-filled space and muffle shocks to the area. The overuse syndrome called runner's knee can injure these pads, causing severe pain and limiting joint mobility. Your knee resists twisting and extending forward because tough bands of tissue form a pair of crossed braces (the cruciate ligaments) that hold the femur and tibia tightly together. Violent forces (like kicking a soccer ball while off balance) can converge to shred these crossed braces. Finally, straps of tissue (the collateral ligaments) on each side of your knee prevent the joint from bending to the side. A clip on the football field, however, can rip one of these side straps in two. (b) X-ray of the knee joint as it bends. To see how the knee works, sit on a low chair and fully extend your right leg, then relax your thigh muscles. Now feel the position of your kneecap with your hand; the cap will rest in the position shown, but you can move it around with your fingers. While keeping your leg extended and your hand on your kneecap, tighten your thigh muscles and notice how the ligament lengthens and your kneecap moves up your leg.

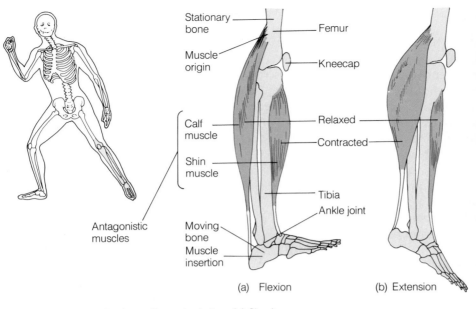

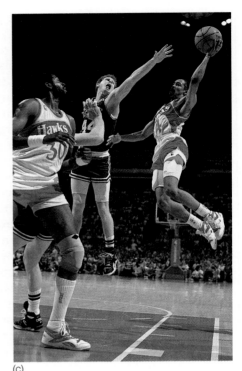

(a) Flexion   (b) Extension

(c)

FIGURE 28.8 *Muscles Lever Bones to Action.* (a) Simultaneous contraction of the shin muscle and relaxation of the calf muscle flex the foot, while (b) contraction of the calf muscle and relaxation of the shin muscle extend the foot. Because the ankle joint acts as a fulcrum, people with longer heels (levers) can jump much higher. (c) Here, 5 ft 6 in. Spud Webb soars up to the basket above players well over a foot taller than he is.

■ **How Bones Act as Levers to Transmit Force** Bones can pivot about joints because of the particular way muscles attach to bones. As in sea anemones, with their opposing circular and longitudinal muscles, a vertebrate's muscles are arranged in antagonistic pairs. The shin muscle, for example, flexes your foot toward your body (Figure 28.8a), while its antagonist, the calf muscles, extends your foot to point away from the body (Figure 28.8b). These movements occur because one end of the muscle, the *origin,* attaches to a bone that remains stationary during a contraction, while the other end, the *insertion,* attaches to a bone that moves. Your calf muscle attaches to your femur just above the back of your knee and at the opposite end to your heel bone. When the calf muscle contracts, the shortening forces the foot to rotate around a pivot, the ankle joint, much as a lever rotates around a fulcrum. Because of this arrangement, a small contraction of the muscle transmits a large movement to the bone. Spud Webb, standing a mere 5 ft 6 in., won the National Basketball Association's slam dunk contest. Webb, who soared above men more than a foot and a half taller (Figure 28.8c), probably has very efficient muscle and skeleton levers to complement his tough training schedule.

≋ MUSCLES: MOTORS OF THE BODY

A horse can gallop across the prairie and a person can perform aerobic dance exercises because a type of muscle called **skeletal muscle** propels the skeleton (Figure 28.9). Shortening (contraction) of a skeletal muscle starts after the brain sends an action potential down a nerve cell to its junction with the muscle (see Chapter 26). When a nerve impulse reaches the junction, it provokes an electrical activity in the muscle cell that causes the muscle to contract. But how does the muscle cell shorten? The answer lies in the unique structure of muscle cells.

Skeletal muscles are made up of *muscle fibers* (Figure 28.10a and b), giant cells with many nuclei that may extend the full length of the muscle—often several centimeters. Each muscle fiber cell has a system of protein filaments, a system of membranes, and an energy system that act, respectively, as the muscle's engine, its electrical ignition switch, and its fuel.

PROTEIN FILAMENTS: THE MUSCULAR ENGINE

Within each muscle cell lie threads called **myofibrils** that do the actual job of contracting the muscle (see Figure 28.10b). A myofibril has repeating units of dark and light bands. Each repeating unit is called a *sarcomere* (from the Greek words for "flesh" and "part of") (Figure 28.10c).

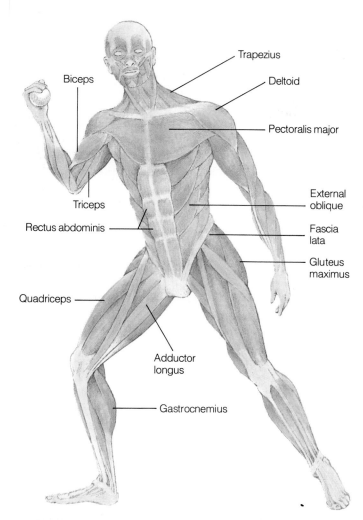

FIGURE 28.9 *Muscles of the Human Body.* The body's 600 or so muscles attach to and move the bones of the human endoskeleton. Although it's not actually possible because muscles work in antagonistic pairs, if all 600 muscles could contract simultaneously and pull in the same direction, they could lift the weight of about 25 compact cars. This anatomical chart identifies only the largest muscles of the torso and limbs.

Within each unit, thin actin filaments interlock with thick myosin filaments, and their overlapping arrangement creates the visible bands, a bit like the pattern you create when you interlace the outstretched fingers of your two hands. Actin and myosin are not exclusive to muscle, but are found in most eukaryotic cells, where their contraction changes the shape of the cell during embryonic development, localizes materials, and causes dividing cells to pinch in two (review Figure 3.14).

When a muscle fiber contracts, each of its sarcomere units becomes shorter, thus causing the whole fiber to shorten. The band of myosin within each sarcomere, however, does not shorten during contraction (Figure 28.11, page 464). To picture this, try thinking of a woman putting her hands into a furry muff; her hands slide past the insides of the muff, her elbows come closer together,

but the muff itself does not decrease in length. Likewise, in a muscle, the actin filaments (the woman's hands) and the myosin filaments (the muff) must slide past each other as the sarcomere (the distance between elbows) shortens. This remarkable process, called the *sliding filament mechanism* of muscle contraction, is powered by ATP. In fact, *rigor mortis*, the stiffening of the body shortly after an animal dies, occurs because the supply of ATP has run out, the muscle fibers cannot slide past each other, and the muscles can no longer contract or relax. The mechanism of sliding protein filaments powered by ATP explains how muscles shorten. It does not, however, explain how muscles know when to contract. That requires signals from the cell membrane.

MEMBRANE SYSTEM: THE IGNITION

Muscle contraction enhances an animal's survival only if the organism can control when contraction takes place. This control is exerted by muscle cell membranes, which

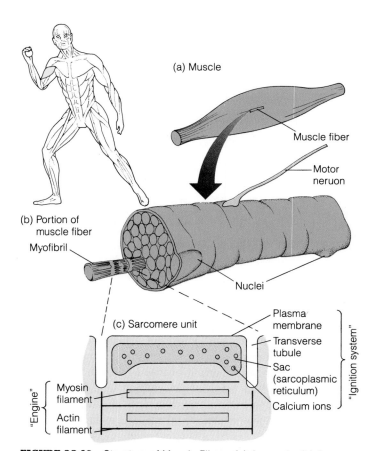

FIGURE 28.10 *Structure of Muscle Fibers.* (a) A muscle. (b) A muscle fiber and myofibrils. (c) A sarcomere with myosin and actin filaments.

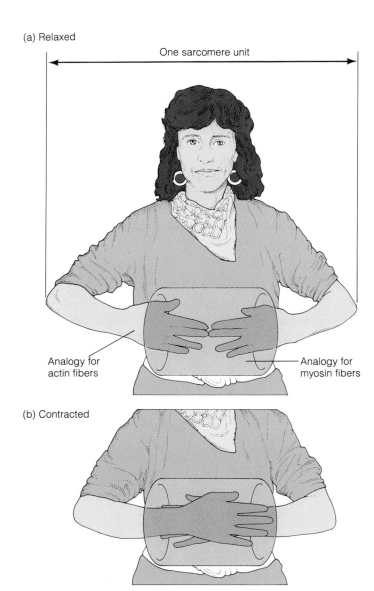

(a) Relaxed

One sarcomere unit

Analogy for
actin fibers

Analogy for
myosin fibers

(b) Contracted

FIGURE 28.11 *Muscle Contraction: Actin Slides in, Myosin Remains in Place.* The sliding filament mechanism of muscle contraction is a bit like a woman's hands sliding into a muff; her arms (actin) slide in, but the muff (myosin) remains unchanged.

act to ignite the muscular engine. Three sets of membranes help stimulate muscle fibers to contract (see Figure 28.10c): the plasma membrane surrounding the huge fiber cell; the *transverse tubules*, which lead like tunnels from the plasma membrane deep into the cell's interior; and a sac (*sarcoplasmic reticulum*) that stores large amounts of calcium ions.

When a nerve impulse reaches the junction between a nerve cell and a muscle cell, it causes the neuron to release the neurotransmitter acetylcholine onto the cell's plasma membrane. The binding of neurotransmitter to

cell membrane receptors generates an action potential in the muscle cell (similar to the action potential in a neuron), and electrochemical activity on the cell surface is channeled toward the cell's interior by the transverse tubules, like a rabbit scurrying over the ground and disappearing down its tunnel. Next, a signal passes from the transverse tubules to the sac, causing the sac to release some of its stored calcium into the cell's cytoplasm.

It is the jolt of released calcium ions interacting with muscle proteins that causes the muscle cell to contract. After the cell shortens, an ATP-driven pump quickly restores calcium ions to the sac, preparing the cell for the next contraction. ATP is thus used in two ways during muscle contraction: to help myosin slide past actin and to pump calcium ions back into the sarcoplasmic reticulum. The message here is clear: Muscles can generate force against a skeleton because protein filaments expend energy to slide past each other and shorten the muscle cell.

The entire process is smooth and extremely rapid: An action potential fires, calcium is released, and protein filaments slide past each other. The repetition of this process hundreds of times in millions of molecules inside muscle cells help explain how a stallion can rear up, how a dancer can leap and spin, and how a harpsichordist can play a Bach concerto with exquisite finger control.

HOW CARDIAC MUSCLE AND SMOOTH MUSCLE MOVE BODY FLUIDS

The mechanisms just discussed apply to skeletal muscles, those muscles attached to bones and which function to move the skeleton (see Figure 28.9 and Figure 28.12a). Vertebrates, however, have two additional types of muscles that move fluids rather than bones. **Cardiac muscle,** found only in the heart, drives blood through the circulatory system, while **smooth muscle** propels food through the digestive tract and provides tension in the urinary bladder, uterus, and arteries.

Cardiac muscle fibers consist of networks of muscle cells that are electrically connected to each other directly by tiny holes in their membranes (Figure 28.12b). Thus, an impulse initiated any place on the heart muscle quickly propagates through the entire organ, causing the whole fist-sized heart muscle to contract. Like skeletal muscle, cardiac muscle is striped (*striated*) because its protein filaments are organized into repeating units.

The smooth muscle cells surrounding the digestive tract, bladder, blood vessels, and other hollow body organs lack stripes because their actin and myosin filaments are not as well ordered as those in skeletal and cardiac muscle (Figure 28.12c). Smooth muscle cells usually have a single nucleus and communicate electrically with other smooth muscle cells via gap junctions. This communication allows the rhythmic pushing of food down the digestive tract or the pushing of a baby through the birth canal.

Despite the organizational differences between skeletal, cardiac, and smooth muscle, all three types contribute to an animal's success. Skeletal muscles allow a wild horse to lope across the desert in search of green grass or fight to win mares and mate (thus aiding its reproduction); cardiac muscle propels blood through vessels to every body tissue; and smooth muscle propels food through the digestive tract and helps in various other ways to keep this complex system operating smoothly. All this muscle contraction—so vital to survival—depends on a substantial amount of energy in the form of ATP.

SOURCES OF FUEL FOR MUSCLE CELLS

While actin and myosin filaments enable muscle cells to contract, and membrane systems ignite muscle activity, the fuel for muscle cell contraction is ATP. The sources of ATP are the mitochondria and certain enzymes in the cytoplasm. Three energy systems—immediate, inter-

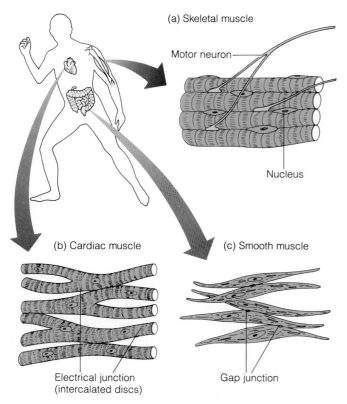

FIGURE 28.12 *Three Types of Muscle: Skeletal, Cardiac, and Smooth.* The three types of muscle differ at the cellular level. (a) Several skeletal muscle fibers can contract together because branches of the same motor neuron contact them all. (b) Cardiac muscle cells have electrical junctions that coordinate their contraction. (c) Smooth muscle cells communicate through gap junctions (see Figure 3.27) and tend to contract in sequential waves.

mediate, and long-range—supply ATP. The duration of physical activity dictates which system the body uses. The immediate energy system is instantly available for a brief explosive action, such as one heave of a shot put; it depends on a muscle cell's stores of ATP plus a high-energy molecule called creatine phosphate, and it can fuel muscle contraction for several seconds. The intermediate energy system, based on glycolysis (see Chapter 5), depends on glucose in the muscles and can sustain heavy exercise for a few minutes. After that, the long-range, or oxidative (aerobic), system takes over. In the presence of sufficient oxygen, this system can produce energy by breaking down carbohydrates, fatty acids, and amino acids mobilized from other parts of the body. Clearly, anyone interested in melting away body fat should engage in aerobic (oxygen-utilizing) activities like rapid walking, swimming, bicycling, or jogging, which rely primarily on the oxidative energy system and its ability to use fats as fuel (see the box on page 467).

The immediate, glycolytic, and oxidative energy systems do not contribute equally to the energy budgets of all muscle fibers. Instead, some types of skeletal muscle fibers rely more heavily on one energy system than another. Slow-twitch muscle fibers, also called *slow oxidative muscle fibers*, obtain most of their ATP from the oxidative system. Slow-twitch fibers require about one-tenth of a second to contract fully; they are packed with mitochondria; they receive a rich supply of blood; and they have large quantities of a red protein (*myoglobin*) that stores oxygen in muscle cells (Figure 28.13, page 466). These characteristics make slow-twitch fibers deep red, like the dark meat of a chicken. Slow-twitch fibers are resistant to fatigue and thus able to contract for long periods of time. Athletes trained for endurance sports have a large proportion of slow-twitch muscles. Slow-twitch muscles also provide functions critical for survival, such as maintaining posture. Without the contraction of slow-twitch muscles in your jaw muscles, your mouth would be wide open as you read this sentence, and a marine clam would be unable to keep its shell closed for hours to protect against predatory sea stars.

Whereas slow-twitch fibers bestow endurance, fast-twitch fibers, also called *fast glycolytic fibers*, provide power (see Figure 28.13). Fast-twitch fibers derive most of their ATP from glycolysis, and they reach maximum contraction twice as quickly as slow-twitch fibers. They soon grow fatigued, however, since they run through their limited stores of glycolytically generated ATP in short order. Fast-twitch fibers are white because they are jammed with white actin and myosin proteins (which maximize contractile force), and they contain very little of the red protein myoglobin for oxygen storage. The white meat in the breast of a chicken is the fast-twitch muscle that

FIGURE 28.13 *Slow-Twitch and Fast-Twitch Muscle Fibers Use Different Energy Sources.* With a particular biological stain, the myosin-rich fast-twitch fibers in a cross section of muscle tissue stain dark, while the slow-twitch fibers appear as light-colored patches. A person highly trained for and naturally good at shot putting is more likely to have a high proportion of fast-twitch muscle fibers, while a person trained for aerobic endurance events like cross-country skiing is more likely to have a high proportion of slow-twitch fibers.

 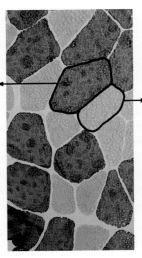 

powers the wings and enables the chicken to suddenly burst away from a fox and fly to safety in a tree. A third kind of muscle fiber has characteristics midway between fast and slow fibers.

Most animals have a combination of fast, slow, and intermediate fibers distributed according to the animal's requirements for survival. American quarter horses are rapid sprinters and can outrace thoroughbred horses over short distances owing to differences in ratios of fast- to slow-twitch fibers in the two horse breeds. Likewise, weight lifters, shot-putters, and sprinters have a higher proportion of fast-twitch fibers, while cross-country skiers and long-distance runners have more slow fibers.

Although fiber types are genetically determined to a large degree, intense training can change their characteristics. For example, endurance training can cause both fast and slow fibers to develop increased oxidative capacity but reduced explosive strength. In contrast, strength training stresses immediate energy supply systems and suppresses oxidative capacity.

Training affects much more than the fast and slow muscle types; as the next section explains, it also has dramatic consequences for the body's homeostasis.

≫ EXERCISE PHYSIOLOGY AND SURVIVAL

Heavy exercise or strenuous work simulates the same fight-or-flight response that characterizes the escape of a field mouse from a diving hawk. When an animal needs to move—and quickly—homeostatic control mechanisms vault into action so that the circulatory system can pump blood faster; the respiratory system can deliver oxygen more rapidly; the immune system can suppress inflammatory responses; and the digestive system can provide immediate energy. All this activity is orchestrated ultimately by the endocrine and nervous systems to ignite the muscles that move the skeleton—and of course the entire animal—away from danger or toward a needed commodity. An investigation of exercise physiology allows us to integrate many of the principles we have discussed in the past few chapters.

ESCAPE OR SKIRMISH: A SURVIVAL RESPONSE

A sparrow that has just spotted a cat crouching nearby, a wild horse galloping across the desert, and a student waiting to give an oral report in front of the class are all experiencing stresses that kindle the fight-or-flight response, an automatic set of events managed largely by the hypothalamus at the base of the brain. In times of stress, as during heavy exercise, the hypothalamus dispatches signals to various body tissues. These signals raise blood glucose levels and increase the amount of energy-rich fatty acids in the blood; elevate heart rate and boost blood pressure, circulating fuel to muscles; escalate breathing rate and dilate air passages to provide ample oxygen; and divert blood from the skin and digestive organs into the skeletal muscles, thus supplying food and oxygen where they are needed the most. The hypothalamus also triggers the secretion of epinephrine (adrenaline) from the adrenal glands, and this nervous system stimulant intensifies and prolongs the various effects. In exercising horses like Moscha and Harry, the activation can be unusually pronounced. The pulse rate of a healthy horse can skyrocket from 25 beats per minutes to 250 in the heat of a race. A well-trained human athlete, in contrast, might experience a climb from about 45 beats per minute to about 190 in a full-out footrace.

RUNNER'S FUEL

Runners. You see them at all hours of the day; on tracks, trails, streets, and sidewalks; and in all states of physical fitness, from the sinewy to the soft and flabby (Figure 1a). Perhaps you are one of them. If so, in all likelihood, you have thought about how to improve your performance—that is, how to run farther, faster, and with less discomfort. Whether you sprint, jog a few miles, or run marathons, recent findings about muscles and metabolism can provide you with sound training tips.

The immediate fuel for muscle power is ATP, but your muscles can store only enough ATP to sustain heavy work for about 1 second. This is long enough for a power lifter to perform one clean and jerk (Figure 1b), but not enough time to complete any running event.

A fast sprinter can cover 100 m in under 10 seconds, with an explosion of muscle activity fueled mostly by stored energy. But since muscles can store enough ATP to last the sprinter only about 1 second, the athlete's muscle cells must generate ATP from other fuel supplies. These include creatine phosphate, a short-term energy storage molecule smaller than ATP and specific to muscle cells. Your muscle cells can use the phosphate of creatine phosphate to quickly generate ATP, and when your muscles are working their hardest, creatine phosphate is the favorite fuel. Even so, the muscles store only about 6 or 7 seconds worth of creatine phosphate, so between it and ATP, there is still not quite enough stored fuel to prevent an average sprinter from getting muscle fatigue during a 100 m dash. Clearly, it would help if the sprinter could increase his or her stores of creatine phosphate, and some researchers now think you can do this through weight training.

In people who run a few miles at a

FIGURE 1 *Exercise: Fueling for the Long and Short Haul.* (a) Thousands of people run the 7-mile Bay-to-Breakers race each year in San Francisco. (b) A weight lifter stores enough ATP in his muscles to perform this strenuous clean and jerk.

time, the muscles tend to burn glycogen, a huge, highly branched molecule made up of glucose subunits (review Figure 2.20). Exercising muscles break down glycogen to glucose, which then enters the cells' energy-producing pathway (glycolysis) and produces ATP. If muscles are contracting so rapidly that they use oxygen faster than it is delivered by the lungs, heart, and blood, then muscle cells cannot carry out aerobic respiration to produce extra ATP and must rely on anaerobic metabolism (glycolysis plus fermentation). But anaerobic metabolism releases lactic acid, a waste product that alters the cellular environment and causes fatigue. The best way to fight this fatigue is to rinse lactic acid out of the muscles and into the blood as fast as possible. For this reason, milers often train in a way that will increase the blood supply to their leg muscles. What they do is push themselves hard enough during training so that their muscles are working anaerobically; this encourages the growth of blood vessels into new parts of the muscles. Such training also increases the capacity of the heart and lungs so that more oxygen will be delivered to the hard-working muscles during a medium-length run.

A marathon runner has the largest, longest-lasting energy needs of all.

Since a race can take several hours, the body will use a combination of stored glycogen and fat. For muscles to use fat as fuel, the fat stored in special fat cells must be broken down to fatty acids and then transported in the blood to the muscles. Muscles can trap this large supply of fat energy, however, only if they are working slowly enough so that they have an adequate supply of oxygen. In a sense, a marathoner must "run a tightrope." If he or she runs too fast, the muscles will work anaerobically, and the body's supplies of glycogen will be depleted long before the race ends. If the runner slows the pace by about half so that the muscles can work aerobically and get some of their energy from fat, he or she will probably lose the race. To run at their fastest times, marathoners must pace themselves so that they use a precise mix of glycogen (at a fast pace) and fat (at a slower pace). People can learn this balance only by repeated long-distance running. This is why marathoners usually include one 20-mile run each week during peak training.

Research on muscles and energy harvesting will no doubt continue to provide new hints on how to train to improve athletic performance. In the meantime, keep on exercising.

During a fight-or-flight response, the hypothalamus orchestrates the release of the steroid *cortisol*, a so-called stress hormone. Cortisol causes the breakdown of storage proteins, the liberation of amino acids, and the formation of sugar from amino acids. The result of all this activity is that as the fight-or-flight response primes the body for physical activity (perhaps necessary for survival), cortisol fortifies the system with amino acids and energy sources that help heal damaged tissue. Growth hormone, which also facilitates repair of injured tissues, is also secreted by the pituitary under control of the hypothalamus during stress.

Heavy exercise places special demands on body temperature regulation—a heating and cooling system that relies on the blood vessels to route hot blood to the body surface, and on sweat glands to dump moisture on the skin surface and carry away body heat through evaporation. Stress triggers still another physiological mechanism—the secretion of antidiuretic hormone (ADH; see Chapter 24). ADH causes the kidney to retain water, which may be essential in cases of heavy sweating or blood loss. We can summarize this way: Athletic training mimics the fight-or-flight response. It calls into action all the major physiological systems of the body, whose combined effects are integrated by the nervous and endocrine systems.

HOW ATHLETIC TRAINING ALTERS PHYSIOLOGY

Since heavy exercise evokes the fight-or-flight response, people who exercise several times a week enter this state of stress repeatedly. What effects do such repeated challenges have on the body's homeostatic mechanisms? Training works by causing "breakdown" followed by "overshoot": a breakdown of stored fuel, for example, followed by the increased deposition of fuel molecules; or a slight breakdown of muscle tissues, with an overall strengthening after the tissues are repaired. To increase fitness without injury, therefore, one must gradually and progressively augment the intensity, frequency, and duration of workouts.

One goal of human athletic training is to boost the amount of oxygen a person can deliver to working muscles (the so-called *maximal oxygen uptake*). To show how this uptake occurs, exercise physiologists in Dallas measured oxygen utilization in sedentary men immediately before and after three weeks of solid bed rest. The researchers measured oxygen uptake again over an eight-week training period in which each subject ran from 2.5 to 7 miles on 11 different occasions each week. As the graph in Figure 28.14 shows, lying in bed all day long caused a substantial drop in the body's ability to take up and utilize oxygen. After a few days or weeks of training, however, oxygen utilization increased dramatically: The

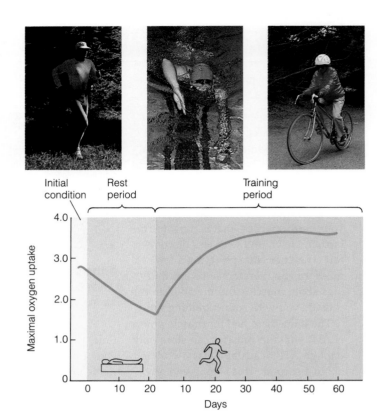

FIGURE 28.14 *Training Increases the Body's Ability to Use Oxygen.* When experimental subjects stayed in bed for three weeks, the ability of their tissues to take up and use oxygen dipped dramatically. But they greatly exceeded their original levels by exercising regularly after confinement. This experiment suggests that activity dramatically improves the body's oxygen utilization in the formerly bedridden or sedentary.

men doubled their ability to use oxygen and perform physical work.

What changes in the men's bodies can account for their increased ability to take up oxygen? Two main factors appeared responsible. First, the men's hearts pumped more blood per minute after training than before. This was due not to a more rapid heart rate but to an increase in the amount of blood pumped per beat (the stroke volume). The more one exercises, in fact, the slower the heart beats at rest: Cross-country skiers and bicycle racers often have resting heart rates of about 45 beats per minute as compared to an average of about 70 beats per minute in the general population. They have enlarged hearts with stronger walls, which pump more blood per beat. Second, physically fit people can use more oxygen because their muscles extract more oxygen from the blood. Energy-supplying mitochondria grow larger, the muscles generate more of the red oxygen-storing protein myoglobin, there is an increase in the number of blood capillaries in the muscles, and all these phenomena work together to provide the muscles a larger blood supply from which to extract more oxygen.

Exercise does more than stimulate fight-or-flight mechanisms and increase oxygen intake: It stimulates the release of *endorphins* from the pituitary and other brain regions. These peptide hormones act a bit like the addictive drug morphine in that they reduce pain and enhance a feeling of well-being. The release of endorphins during strenuous exercise may explain, in part, why many people experience feelings of relaxation and contentment after a workout. In recent years, researchers have found that regular exercise does even more than improve the body's fitness and make you feel good: It also has dramatic positive effects on the heart and vascular system and on a person's chances of living a long and disease-free life.

EXERCISE AND HEART DISEASE

A physically inactive person is about twice as likely as an active one to die of heart disease. Considerable evidence suggests that heredity, diet, obesity, cigarette smoking, high blood pressure, and high blood levels of cholesterol may all increase one's risk of heart disease. Nevertheless, regular exercise can lower body weight, decrease one's desire to smoke, lower blood pressure slightly, and increase blood flow to the heart. While there is no guarantee against heart disease, medical researchers believe that a person can lower his or her risk of developing this killer by remaining smoke free, eating a balanced diet low in saturated fats, and exercising regularly. It is ironic that many people recognize the importance of diet and exercise to their pets' health, but fail to apply the same principles to their own human bodies. This has led Swedish exercise physiologist Per-Olof Åstrand to make the following suggestion: Walk your dog whether you have one or not.

≋ CONNECTIONS

Self-generated movement is a hallmark of animal life. This chapter investigated how animals move and how that movement relies on (1) organized arrays of protein filaments that slide past each other and cause muscles to shorten and (2) a rigid support, a skeleton, to which the shortening muscles transmit force. Skeletons can be water-filled tubes (as in sea anemones and worms), hard exoskeletons (as in pill bugs and beetles), or bony internal frameworks (as in people and horses). But the muscles that move them contract in the same way in all animals.

The ultimate source of energy that moves muscles in a horse, in a rider, or in a cougar hunting a foal is the sunlight captured by green plants. The structures that allow plants to capture the sun's energy, the principles that direct the plant's growth, and the ways that plants cope with environmental stress are all subjects of the next part of this book.

NEW TERMS

antagonistic muscle pair, page 458
appendicular skeleton, page 459
axial skeleton, page 459
bone, page 459
cardiac muscle, page 464
hip girdle, page 459
joint, page 461

ligament, page 460
myofibril, page 461
shoulder girdle, page 459
skeletal muscle, page 461
skeleton, page 458
smooth muscle, page 464
tendon, page 460
vertebral column, page 459

STUDY QUESTIONS

REVIEW WHAT YOU HAVE LEARNED

1. Describe the structure of bone.
2. What do bone cells do?
3. What are joints, and how do they work?
4. Explain how bones act as levers to transmit force.
5. Describe the structure of a myofibril.
6. Explain the sliding filament mechanism of muscle contraction.
7. Compare cardiac muscle, smooth muscle, and skeletal muscle.
8. Which energy system most effectively powers sustained muscle contractions? Explain.
9. Name and describe the three types of skeletal muscle fibers.
10. What are the health benefits of regular exercise?

APPLY WHAT YOU HAVE LEARNED

1. A racehorse falls, breaks its neck, and dies. A few hours later, the once-limp corpse has stiffened. What has happened?
2. A tired student feels revitalized after playing tennis in the early evening. Explain.

FOR FURTHER READING

Birke, L. "Equine Athletes: Blood, Sweat and Biochemistry." *New Scientist* 22 (May 1986): 48–52.

Bonn, D. "Hormones for Healthy Bones." *New Scientist* 19 (February 1987): 32–35.

Cunningham, C., and J. Berger. "Wild Horses of the Granite Range." *Natural History* 95 (April 1986): 132–138.

Hildebrand, M. "The Mechanics of Horse Legs." *American Scientist* 75 (1987): 594–601.

Newsholme, E., and T. Leech. "Fatigue Stops Play." *New Scientist* (22 September 1988): 39–43.

Timmerman, M. "Nerve and Muscle: Bridging the Gap." *New Scientist* (10 September 1987): 63–66.

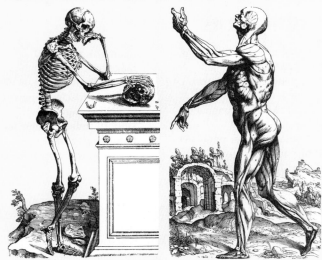

① Mammals have a braced framework skeleton and muscles that help support and move the body.

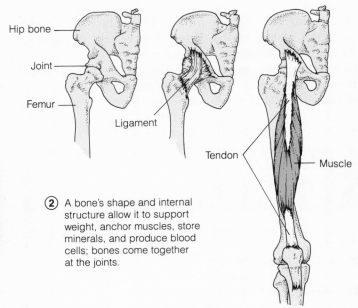

Hip bone

Joint

Femur

Ligament

Tendon

Muscle

② A bone's shape and internal structure allow it to support weight, anchor muscles, store minerals, and produce blood cells; bones come together at the joints.

One sarcomere

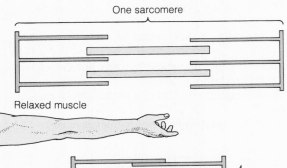

Relaxed muscle

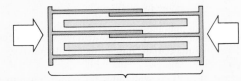

Contracted muscle

Shortened sarcomere

③ The three kinds of muscle cells contain thick and thin protein filaments that slide past each other as the muscle shortens (contracts).

Carbohydrates

Stored ATP

Lipids

Nerve impulse

Muscle contraction

④ Muscle contractions are triggered by nerve impulses and powered by ATP from three different energy sources.

⑤ Virgorous physical activity not only involves the bones and muscles; it also calls into action all the major physiological systems.

Edgar Degas, *Ballet Rehearsal on Stage*

PART FIVE

HOW PLANTS SURVIVE

PLANT FORM AND FUNCTION

CHAPTER 29

A BLOOMING PEAR TREE

One of the loveliest sights of spring is a mature fruit tree in full bloom. The tall, dark trunk of a pear tree, for example, supports an enormous crown of large and small branches, as well as a cloud of white flowers. Each branch is a study in contrasts, with its delicate, fragrant blossoms growing directly from sturdy twigs (Figure 29.1) and with an entourage of bees flying actively from flower to flower, bristled bodies and legs loaded down with pollen grains. By summer, the flowers will have dropped their petals and the branches will bend under the weight of the ripening fruit that encloses the seeds of the next generation.

A pear tree can begin producing fruit just a few years after sprouting from a seed, when its trunk is still slender and its branches small. The tasty fruit is the reason birds, deer, and other animals are drawn to pear trees in late summer and early autumn. Each pear seed an animal inadvertently swallows and deposits in a new location is capable of initiating the growth of a new organism—a pear tree, complete with roots, a tall woody trunk, branches, twigs, broad green leaves, fragrant blossoms, and juicy pears.

Using the pear tree as a central example, this chapter explores the form and function of plants, focusing specifically on flowering plants. As Chapter 16 revealed, not only do flowering plants make up the largest and most diverse of all plant groups, but they have had an immense impact on people. Civilization depends on grains, fruits, vegetables, animal fodder, wood, fiber, spices, beverages, chemicals, drugs, and myriad other plant by-products.

Most plants, of course, are stationary organisms that cannot move about in pursuit of water, energy, a carbon source, and mineral nutrients. But a plant's anatomy, way of life, and mode of reproduction are all beautifully adapted to meeting its needs while it remains rooted to one spot.

Three unifying themes will emerge in this chapter. First, a plant's form and function strike a compromise between conflicting needs. Plants often have large, flat leaves for collecting maximum sunlight, and the leaves usually have tiny openings that allow gas exchange. Both large surface area and openings, however, conflict with the need to prevent excessive water loss. Second, the anatomical parts that we examine separately make up tissues, organs, and systems that function collectively as whole plants. Transport pipelines, for example, are continuous from roots to stems to leaves. Third, plants have open growth, allowing them to continually produce new organs and larger body size throughout life. This mode of development helps overcome the limitations imposed by rigid cell walls and hardened mature tissues.

Our discussion of how a plant develops the structures it needs for survival will answer several questions:

■ What are a plant's main tissue types and overall growth patterns?
■ What are the main structures in flowers, fruits, and seeds, and how do they contribute to reproductive success?
■ How does the root anchor the plant, store starch, and channel water and minerals?
■ How does the stem support the plant and transport materials?
■ How does the leaf collect solar energy and carry out other vital tasks?

FIGURE 29.1 *Pear Branch in Bloom.* A flowering tree in spring is a reminder of the angiosperm life cycle.

THE PLANT BODY: PLANT TISSUES AND GROWTH PATTERNS

Pear trees, with their clouds of white flowers in spring-time and their loads of golden fruit in autumn, are just one species (*Pyrus communis*) among nearly half a million vascular plants. As a group, vascular plants are distinguished by their vascular tissue, or internal transport tubes. This mammoth group dominates the land and includes ferns, conifers, and flowering plants. The conifers and flowering plants produce seeds, and these seed plants have two additional adaptations for life on land: sexual reproduction with internal fertilization, and a tough, waterproof seed coat that protects the developing embryo. These last two characteristics are similar to the innovations of internal fertilization and eggshells we saw in the reptiles, in Chapter 18. The vast majority of vascular plants also have a main axis made up of root and shoot.

THE PLANT'S MAIN AXIS: ROOT AND SHOOT

Pear trees, like most typical vascular plants, have roots below the ground and a shoot above the ground. The **shoot** includes the stem, branches, leaves, flowers, and fruit (Figure 29.2). Lifting the rest of the shoot is the stem, a stiff bundle of internal transport tubes plus surrounding tissues, all enclosed within an external waterproof coating that minimizes water loss. In pear trees as well as other trees and shrubs, the main stem matures into a large, bark-covered trunk with many side branches supporting leaves and flowers.

Roots are branching organs that grow downward into the soil and help support the plant both physically, by spreading out through the soil to provide a solid base of attachment, and nutritionally, by absorbing and transporting water and mineral nutrients that the plant cannot take in from the air. Pear trees, dandelions, and carrots each have a thick, trunklike *taproot* that grows straight down into the soil (Figure 29.3a, page 474). Taproots store water, as well as food in the form of starch. In contrast, grasses and a few other kinds of plants have a mass of narrow *fibrous roots* (Figure 29.3b) that branch downward and outward from the plant's stem and efficiently anchor the plant and absorb water and nutrients. In general, roots that arise from aboveground structures (such as the stem) are called *adventitious roots*. Examples are the thick, aerial prop roots of a corn plant or banyan tree (Figure 29.3c) and the fine roots that allow ivy plants to cling to walls and fences.

Both roots and shoots are made of the same three kinds of tissue systems.

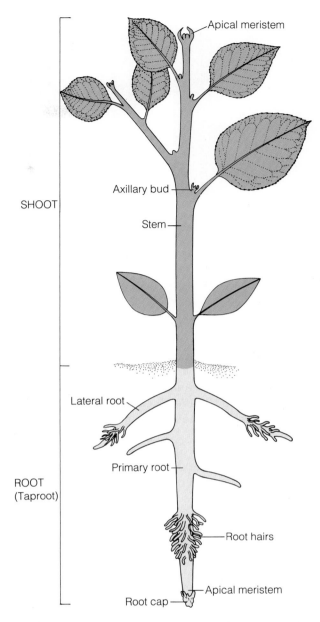

FIGURE 29.2 *The Plant Axis: Root and Shoot.* The above- and below-ground portions of the plant have vastly different functions and hence very different anatomies. Shoot structure is appropriate for its functions in photosynthesis, support, and sexual reproduction, while root structures anchor the plant and absorb water and minerals.

TISSUE SYSTEMS AND TISSUE TYPES

Complex flowering plants, such as pear trees and pumpkin vines, have tissue systems with important physiological roles, made up of a few important tissue types. Each tissue, in turn, is composed of different cell types with distinctive characteristics appropriate to their function. The three main tissue systems in a mature vascular

(a)

(b)

(c)

FIGURE 29.3 *Patterns of Root Growth.* (a) A dandelion has a central taproot with fine lateral roots. (b) Grass has numerous fibrous roots that anchor the plant very firmly. (c) A single banyan tree has many aerial prop roots to help support the plant and absorb water and minerals.

plant are the **dermal tissue** system, which, like skin, protects the plant from water loss and injury to internal tissues; the **ground tissue** system, which provides support and stores starch; and the **vascular tissue** system, which conducts fluids and helps strengthen the roots, stems, and leaves (Figure 29.4a). These tissue systems are continuous throughout the plant.

■ **The Dermal Tissue System: Protection and Waterproofing** Pear trees and other vascular plants have a dermal system that covers every part of the plant and is analogous to an animal's skin (Figure 29.4b). In a young seedling, the dermal tissue consists of just an *epidermis*, or protective outer covering. When you peel a pear, you remove its epidermis. Epidermal cells in stems and leaves secrete a waxy waterproof coating, the cuticle. In trees and woody shrubs, the outer areas of the bark (called the periderm) replace the epidermis.

■ **The Ground Tissue System: Storage and Support** A plant's root and shoot contain a ground tissue system—a kind of background tissue that packs around the vascular pipelines. Ground tissue makes up the bulk of most plant organs; it stores the starchy products of photosynthesis and helps keep the plant from collapsing into a formless heap (Figure 29.4c). The starchy cells of a baked potato, the stringy fibers near the surface of a celery stalk, and the gritty particles in a pear fruit are all cells of the ground tissue system.

■ **The Vascular Tissue System: Material Transport and Vertical Strength** Within a vascular plant, water and materials travel in two kinds of tubular tissues, xylem and phloem, each of which forms a continuous vascular

system extending from root tips to leaves. In general, **xylem** transports water and minerals absorbed from the soil up through the roots, stems, and leaves and forms what we call wood. **Phloem** transports dissolved sugars and proteins from "source to sink"—that is, from cells that produce sugars, like those in the leaves, or store sugars, like those in the root, to cells that use sugars rapidly, such as those in a rapidly growing shoot.

The xylem of most flowering plants is composed principally of *tracheids* and *vessel members* (Figure 29.4d). Both tracheids and vessel members transport water only after they have died: Each cell type becomes hollow at maturity, when the cell contents disintegrate and leave behind empty cell walls. These hollow cylindrical cells, stacked end to end, form efficient transport pipelines.

Phloem transport tissue is composed of *sieve tube members* (see Figure 29.4d), cells that are arranged end to end to form pipelike tubes similar to those found in xylem; but phloem and xylem differ in three important respects. First, phloem cells contain living cytoplasm and are thin-walled, while xylem cells are dead and thick-walled. Second, phloem transports sugars and amino acids dissolved in small quantities of water, while xylem transports minerals and large quantities of water. Third, while xylem transports materials from roots to leaves, phloem moves sugars from a region of high concentration—say, photosynthesizing leaves in summer—to areas using sugars rapidly—say, enlarging fruits in summer. Chapter 31 explains how phloem and xylem transport substances.

OPEN GROWTH: A PLANT'S PATTERN OF PERPETUAL GROWTH AND DEVELOPMENT

Imagine the chaos that would ensue if an animal—a dog, let's say—were to continue growing throughout life, producing new eyes, ears, legs, feet, livers, and other organs.

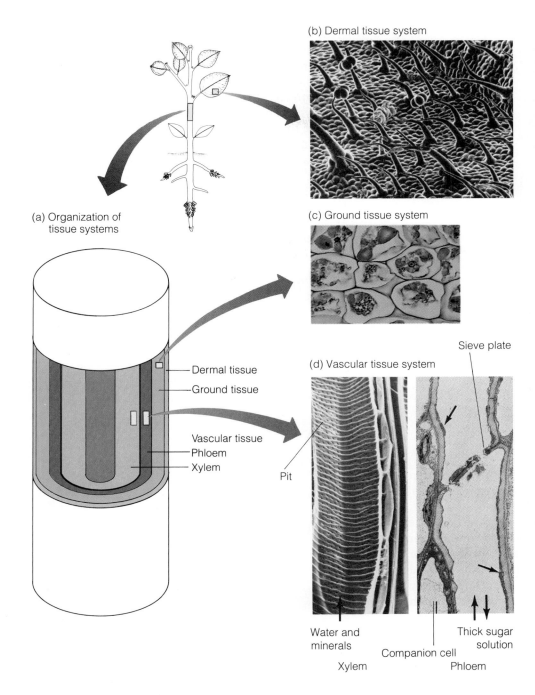

(a) Organization of tissue systems

(b) Dermal tissue system

(c) Ground tissue system

(d) Vascular tissue system

Dermal tissue
Ground tissue
Vascular tissue
Phloem
Xylem

Pit

Sieve plate

Water and minerals

Xylem

Companion cell

Thick sugar solution

Phloem

FIGURE 29.4 *The Three Tissue Systems in Plants.* (a) Plant tissue systems are generally organized in concentric circles. (b) The dermal system is a visible protective shield covering all parts of root and shoot. This scanning electron micrograph shows the epidermis covering a tomato leaf, with hairlike, protective trichomes. (c) Cells of the ground tissue system store, strengthen, and lend hardness to plant parts. Most ground tissue cells are loosely packed cells (*parenchyma*), such as these potato cells with their lavender starch grains. The stringy fibers of a celery stalk are made up of tough cells (*collenchyma*) with their thickened corners and irregular shapes; and the hard, gritty structures in a pear's flesh are thick-walled ground tissue cells (*sclerenchyma*) that lack cytoplasm when mature. The fibers that make up the strong, slender threads of hemp and flax (and, in turn, ropes and linen) are also composed of sclerenchyma ground tissue cells. (d) The vascular tissue system provides transport and support. Hollow xylem cells transport water and inorganic nutrients in all types of vascular plants. Xylem cells contain pits—tiny holes that allow lateral movement of water from one cell to the neighboring cell. Some xylem cells, like the one shown here, are called *vessel members;* they stack end to end, have plates with holes at each end, and channel water up the central cavity. Other xylem cells (*tracheids*) have pointed ends and transfer water from one to another through their pits. The functional cells of phloem are *sieve tube members;* they have a thin layer of cytoplasm but lack a functioning nucleus and lie next to a small *companion cell* whose nucleus directs the activities of both cell types. The perforated end walls of sieve tube members are called *sieve plates.*

The result might be a bizarre creature with three tails, seven legs, and four eyes. Conversely, think what would happen if a pear plant grew like an animal, keeping its original root, stem, and two initial leaves, each simply getting bigger and bigger as the plant grew. Clearly, neither growth pattern would suit the other type of organism.

A plant has a growth pattern called **open growth.** Throughout life, the plant adds new organs, such as branches, leaves, and roots, enlarging from the tips of the root and shoot. Since they cannot move from place to place seeking favorable conditions, plants grow toward light, water, and mineral nutrients and away from harm-

ful situations. This continual growth is based on **meristems,** tissue that remains perpetually embryonic and that gives rise to new cells and cell types throughout the plant's life. Meristems allow adult plants to generate eggs and sperm as well as new tissues and organs.

Plants have two types of meristems. The first type, **apical meristems,** are perpetual growth zones at the tips (*apices;* singular, apex) of roots and stems (Figure 29.5, page 476). Apical meristems allow shoots to grow upward

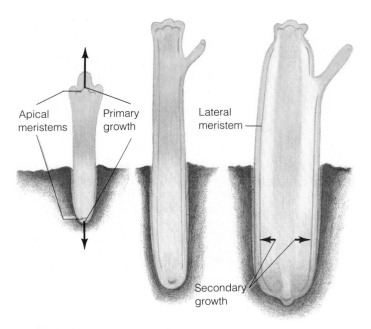

Apical meristems | Primary growth | Lateral meristem

Secondary growth

FIGURE 29.5 *Primary and Secondary Growth from Apical and Lateral Meristems.* Primary growth from apical meristems at the tips of root and shoot causes the plant to lengthen. Secondary growth from lateral meristems causes the shoot and root to increase in diameter. This drawing shows a longitudinal section.

toward the light and allow roots to push ever deeper into the soil to a water source. Growth arising from apical meristems is called **primary growth.** The second type, **lateral meristems,** are cylinders of dividing cells in the stems and roots that cause these parts to become thicker (see Figure 29.5). The enlarged diameter of a stem or root due to cell divisions in the lateral meristems is called **secondary growth.** The distinction between primary growth from the apical meristems and secondary growth from the lateral meristems explains why a nail pounded into the bark of a young pear tree will never be lifted above its original distance from the ground: Primary growth, which lengthens the tree, occurs only at the apical meristems at the ends of branches, far above the nail. The nail will eventually be buried in bark and wood as secondary growth from lateral meristems adds new tissue. *Woody* plants, such as pears, palms, and pines, have secondary growth. *Herbaceous* plants, such as daisies and dandelions, have only primary growth and slender, generally flexible, green stems.

Although flowering plants have an open growth pattern based on meristems, some live only one season,

whereas others live from 2 years to 5000. In a single season, an **annual plant** sprouts from a seed, matures, produces fruit and new seeds, and dies (Figure 29.6a). Marigolds, zinnias, petunias, poppies, and soybeans are familiar annuals. **Biennials** are plants that have a two-year life cycle: They grow from seeds to adults in the first year; then, in the second year, the adults produce flowers, fruit, and seeds and then die (Figure 29.6b). Celery, cabbage, and carrots are all biennials. Many biennials and most annuals are herbaceous, lacking secondary growth. Finally, **perennials** live for many years and typically bloom and set seeds several times before the adult plant dies. Some perennials, like tulips and dahlias, are herbaceous; others, like rosebushes and pear trees, have extensive secondary growth and are woody.

FIGURE 29.6 *Annuals and Biennials.* (a) Annuals, such as the California poppy shown here, have one-year life cycles and often pass the winter in seed form. (b) Biennials, like this Queen Anne's lace, grow the first year and reproduce the second.

FLOWERS, FRUITS, SEEDS, AND PLANT EMBRYOS: THE ARCHITECTURE OF CONTINUITY

How many times has each of us chosen a plump, golden pear, eaten the sweet, soft fruit, and tossed out the core without further thought? That pear, however, represents much more than just a luscious snack. The rounded fleshy part, remarkably like a womb in shape, is actually the swollen stalk of the flower (Figure 29.7). The tough inner ring surrounding the core is the wall of the flower's ovary, and it encircles the small, dark seeds. Each seed contains a tiny living embryo; nourishment to sustain the embryo until it can emerge and take in water, minerals, and solar energy; and all the genetic information needed to generate a tall, productive tree.

The organs of a tree or other plant—the stems, leaves, fruit, and flowers—unfold over time, beginning in the seed and continuing throughout life. Since the adult plant's tissue systems get their start through fertilization and development, our discussion must begin at the site of fertilization—the flower.

FLOWERS: SEX ORGANS FROM MODIFIED LEAVES

Flowers are highly modified shoots that contain the reproductive organs of trees and other flowering plants.

While most of the plant's cells have two sets of chromosomes (are diploid), certain cells in the flowers undergo meiosis and produce pollen grains and embryo sacs that consist of a few haploid cells, each with just a single set of chromosomes (review Figure 16.23). The various parts of the flower help ensure the transfer of pollen, with its sperm nucleus, to the egg inside the embryo sac. Flowers consist of four rings of structures called sepals, petals, stamens, and carpels (see Figure 29.7b). The sepals in the outer ring and the petals in the next ring often have showy colors and shapes and fragrances that attract animal pollinators, animals that transfer pollen grains containing sperm from male flower parts to the female flower parts. A pear blossom's five sepals are green, but its five petals are bright white and attract bees. Nutritious pollen grains and sweet nectar are often the pollinators' reward. The third ring of these structures consists of the pollen-producing *stamens* (20 in pears), within each of which the *anther* (containing two pollen sacs) sits high on a *filament* (see Figure 29.7b). This long stalk aids pollen dispersal and is especially helpful in wind-pollinated plants. Making up the innermost ring are one or more *carpels*, which may fuse together, as do the five carpels in a pear blossom. The base of the carpel, or *ovary*, houses the *ovules*, and the "neck," or *style*, supports the *stigma*, a sticky surface that catches pollen grains. Pollen grains form a tube that grows downward through the carpel and allows sperm to fertilize the egg (review Figure 16.23).

The numbers and sizes of flower parts vary from species to species; lilies, for example, have flower parts in

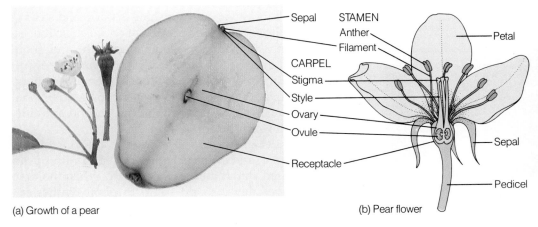

(a) Growth of a pear

Sepal
STAMEN
Anther
Filament
CARPEL
Stigma
Style
Ovary
Ovule
Receptacle
Petal
Sepal
Pedicel

(b) Pear flower

FIGURE 29.7 *Pear Fruit and Flower.* (a) The remnants of a pear flower, the styles, filaments, and sepals, decorate an inverted pear fruit. The flower stalk (receptacle) expands enormously into the fleshy, succulent fruit. (b) Each part of the flower has a specific role in pollination, fertilization, or seed protection. A complete perfect flower like a pear blossom has four concentric rings of flower structures: for pears, 5 sepals, 5 petals, 20 stamens, and 5 carpels. Many kinds of apples also have five fused carpels. The five bumps on the bottom of a red Delicious apple reflect that organization.

multiples of three, pears in multiples of five. In both lilies and pears, each flower contains both stamens and carpels (and is called a *perfect* flower). In other species, such as corn or willows, the flowers are missing either male or female parts (and are *imperfect* flowers). Corn has separate male and female flowers (the tassel and the ear) on each individual plant, while willows have two types of individual plants, some with only male flowers and others with only female flowers.

POLLINATION, FERTILIZATION, AND SEED FORMATION

Pollination and fertilization are important steps in the life cycle of flowering plants, and the pear is a good case study for these events. Pear pollination takes place in spring, when bees carry pollen from one blossom to another. Each pollen grain contains two sperm cells. In a fertilization event unique to flowering plants, one sperm nucleus fuses with two nuclei in the embryo sac, forming a triploid cell—that is, a cell with three sets of chromosomes. This cell develops into a tissue (the *endosperm*) that will nourish the embryo within the seed. The other sperm nucleus fuses with the egg nucleus and forms the diploid fertilized egg, as in all sexually reproducing organisms. The fertilized egg divides into a group of cells that become organized into an embryonic root (*radicle*) at one end and a tiny shoot (*plumule*) at the other (Figure 29.8). These early divisions thus establish the root-shoot axis that will dominate the pear plant's anatomy for the rest of its life. The tiny embryonic root will eventually develop into the plant's entire root system, while the embryonic shoot will become the plant's trunk, branches, leaves, blossoms, and fruit.

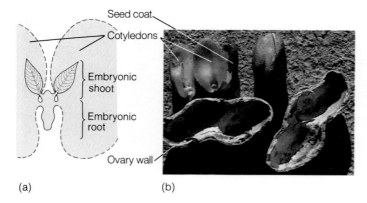

(a)　　　　　　　(b)

FIGURE 29.8 *Seed of a Peanut Plant.* (a) A peanut embryo's tiny shoot is visible above the embryonic root and is surrounded by the two large first leaves (cotyledons), which are tan in this drawing. The first leaves would turn green only after germination, when photosynthesis begins.

An embryo plant is mostly *cotyledons*, the first leaves growing from the embryonic shoot. Many plants, including pears and peanuts, have embryos with two first leaves—two cotyledons—and are therefore called dicotyledonous plants (*dicots*). Others, including grasses, lilies, and palms, have only one first leaf—one cotyledon—and are called monocotyledonous plants (*monocots*). The first leaves often absorb most of the nutritive tissue (endosperm) as the embryo develops inside the seed coat.

To understand seed structure, we can take a roasted peanut and carefully separate its parts (see Figure 29.8). The woody peanut shell is actually the fruit—the enlarged ovary wall—while the reddish papery coating around each peanut is the seed coat, the wall of the ovule. The two oval halves of the peanut are the dried, salted remains of the two first leaves (peanuts are dicots). The tiny fleck in the middle is the dried embryo, and close inspection will reveal the embryonic root and the leafy embryonic shoot.

It is easy to see how the good taste of a peanut or the sweet fragrance and bright color of a fleshy fruit like a pear might attract deer, birds, or other animals and how these hungry animals, in turn, will swallow seeds along with fruit. Later, they inadvertently disperse seeds to new areas when they excrete feces—a rich supply of natural fertilizer. As a result of seed dispersal, new little plants can emerge and become rooted a good distance away from the parent plant and perhaps away from direct competition for sunlight, water, and minerals.

GERMINATION: THE NEW PLANT EMERGES

To survive in many regions of the world, seeds must withstand the cold temperatures of winter or the lack of water in a dry season. When water is plentiful from rain or snowmelt, however, and sunshine is warm and steady, the new embryo begins the process of *germination:* It uses nutrients at a higher rate, resumes growth, and soon presses out against the seed coat's inner surface, eventually cracking the coat. The tiny root of the new seedling can then slip out and push downward into the soil. Soon the shoot pushes upward through the ground and lifts the leaves toward the sun (Figure 29.9).

The plantlet grows as cells divide at the tips of the shoot and root, and eventually the cotyledons disappear as their stored starch fuels early growth. When the first normal leaves develop and begin to photosynthesize, the young plant becomes truly independent.

DEVELOPMENT FROM SEEDLING TO PEAR TREE

As the spring days lengthen, the shoot of the pear seedling advances skyward, and the root burrows deeper into the ground. Growth in both directions is driven by

(a)   (b)

FIGURE 29.9 *Germination: The Seedling Emerges.* In a monocot such as corn (a), the plumule bearing the first true leaf bursts upward and emerges from the soil. In many dicots, such as the bean (b), the lower shoot arches upward and breaks the soil first, drawing the cotyledons behind it.

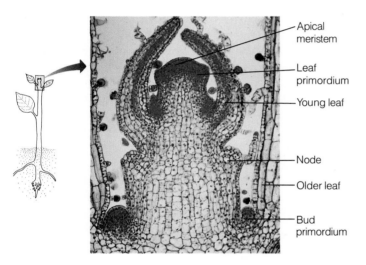

Apical meristem

Leaf primordium

Young leaf

Node

Older leaf

Bud primordium

FIGURE 29.10 *Primordia: New Organs Unfold.* The longitudinal section of the growing tip of a coleus reveals the primordia that will give rise to new leaves, flowers, and branches.

the elongation of cells produced in the rapidly dividing apical meristems. Tiny swellings on the pear shoot called *primordia* flank the apical meristem (Figure 29.10). These primordia contain undifferentiated meristematic tissue and will become **buds** that differentiate into leaves, new shoots (branches), or flowers. The very first primordia to form, however, are always leaf primordia, since the seedling must quickly develop and unfurl energy-collecting surfaces.

About the fourth or fifth year in a pear tree, flower buds appear in many places where buds for new branches would have formed. With the opening of blossoms and the pollination and fertilization that may follow, new life cycles begin and new individuals appear with their roots and shoots. (The box on page 480 introduces some exotic foods derived from seeds, flowers, and other plant organs.)

≋ ROOTS: ABSORBING ANCHORS

Roots are familiar plant organs with important functions: They anchor plants firmly to one spot, probe the soil to absorb water and minerals, and often store starch. This section tours root anatomy, beginning deep in the soil at the very tip of the root.

THE ROOT: FROM TIP TO BASE

■ **Root Cap** At the lowest tip of a typical root is a dome-shaped *root cap* (Figure 29.11). As the root grows downward, rough soil particles damage and scrape off cells at the surface of the root cap. These damaged cells slough off and cover the rest of the root with a slime that eases penetration through the soil.

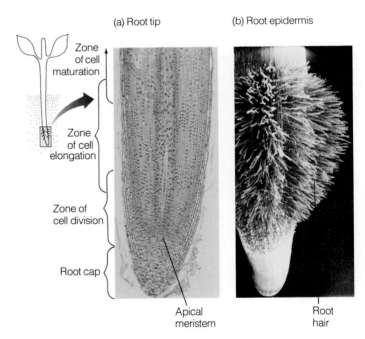

(a) Root tip   (b) Root epidermis

Zone of cell maturation

Zone of cell elongation

Zone of cell division

Root cap

Apical meristem

Root hair

FIGURE 29.11 *The Growing Root Tip.* (a) A longitudinal section of an onion root reveals, from the tip backward, the protective root cap just below the apical meristem; the small, new cells of the zone of cell division; the lengthening cells of the zone of cell elongation; and the zone of maturation. (b) The epidermal cells and root hairs in the zone of maturation.

EXOTIC PLANTS FOR A HUNGRY PLANET

Our lives depend, quite literally, on the products of the ancient enterprise of agriculture. Despite this, however, our dependence rests on a fairly limited range of plant species. Of the 450,000 plants currently recognized, people have learned to use only about 4000 for food (about a tenth of 1 percent), and of those, we rely heavily on just 30 or so grains, fruits, and vegetables.

Every few decades, American farmers become enthusiastic about a "new" crop. In the 1930s, for example, soybeans were practically unknown outside of Asia, but today, the soybean is the third largest U.S. crop. Likewise, in the past 20 years, kiwifruit has risen from obscurity to prominence in American fruit markets.

Many plant researchers are investigating little-known and "forgotten" plant species to recommend new resources for feeding, clothing, and housing the world's burgeoning human population. The suggestions encompass each of the major plant organs discussed in this chapter, and the list that follows, organized by plant part, is just a small sample. Many of these will sound bizarre today. But who knows? Some of them may find their way to your table or your garden in the not-so-distant future.

FRUITS

- The sweetsop, or sugar apple (*Annona squamosa*), of South America and the carambola (*Averrhoa carambola*) of Malaysia have luscious flavors and highly unusual shapes that many customers would like.
- The pummelo (*Citrus grandis*) is the ancestor of the grapefruit but is hardier and lacks any trace of bitterness.

SEEDS

- The seeds of jojoba (*Simmondsia chinensis*), a desert shrub native to

FIGURE 1 *A Florida Winged Bean Harvest.*

Mexico and the southwestern United States, contain a high-quality wax that is already being used in place of sperm whale oil in lotions and shampoos as well as in machine oils and transmission fluids.
- Grain amaranth, or quinoa (species of *Amaranthus*), once a staple for the Aztecs and Incas, produces tiny seeds that, when heated, burst and taste like popcorn. The seeds contain high levels of the essential amino acid lysine as well as high levels of protein—nutrients often lacking in common grains.

ROOTS AND TUBERS

- Buffalo gourd (*Cucurbita foetidissima*) produces heavy, starchy roots (up to 40 kg, or 88 lb) in just a few growing seasons, with a flavor and food value much like cassava.
- The groundnut (*Apios americana*) is a common native of eastern North American forests, and the tasty small, round tubers were long a protein staple of American Indians.

LEAVES

- An Australian tree called leucaena (*Leucaena leucocephala*) is a legume a bit like the American black locust. Its roots associate with another organism that can take nitrogen from air

and convert it to amino acids. It grows very rapidly, and the protein-rich leaves make good animal fodder.
- Dry-adapted shrubs of Central America and Mexico, chayas (*Cnidoscolus* species) also grow quickly and serve as an attractive hedge, and people can eat the leaves as a tasty vegetable.

STEMS

- An amazingly fast-growing annual from East Africa, kenaf (*Hibiscus cannabinus*) grows 6 m (nearly 20 ft) in a year, and pulp from its thick stem can be used to make good-quality paper and cardboard.
- The sap from the stems of the guayule plant (*Parthenium argenteum*) can be used to make rubber. This is being considered as a substitute for products of the rubber tree and petroleum-based synthetic rubber.

WOOD

- A hardwood tree from Australia's tropical rain forest, mangium (*Acacia mangium*) fixes nitrogen, grows as fast as pine, and has the wood quality of walnut.
- Bracatinga (*Mimosa scabrella*) is also a nitrogen-fixing tree, but is native to Brazil, grows up to 15 m (nearly 50 ft) tall in just three years, and is used to reforest logged areas of Costa Rica.

MULTIPLE PLANT PARTS

- The winged bean (*Psophocarpus tetragonolobus*), an annual plant from the Philippines, has edible flowers, seeds, seed pods, leaves, tendrils, and tubers (Figure 1). The seeds and tubers are rich in protein, and the ruffled pods can be steamed, stir-fried, or eaten raw. Over the next few years, watch for the winged bean in your grocer's produce section . . . just to the left of the kiwifruit.

■ **Apical Meristem** Damaged root cap cells are replaced by the apical meristem, a little disc just above the root cap that produces new cells from both its surfaces. The cells from the lower surface of the disc become root cap cells, the cells from the opposite edge become part of the lengthening root. Moving up a root, we can see a *zone of cell division*, which includes the apical meristem protected by the root cap cells; a short *zone of elongation*, where individual cells lengthen and force the tip to move through the soil; and a *zone of maturation*, where cells develop their specialized roles as members of the dermal, ground, or vascular tissue systems (see Figure 29.11).

THE ROOT: FROM OUTSIDE TO INSIDE

■ **The Dermal System: Epidermis and Root Hairs**
Looking at a cross section of a mature root cut just above the zone of maturation, we can see an outer protective layer just one cell thick—the epidermis, derived from the outer cells in the zone of maturation (Figure 29.12a). The root epidermis absorbs water and minerals from the soil, and tiny extensions of the root epidermis called *root hairs* extend the root's absorptive capacity (see Figure 29.11b). Each root hair is an extension of a single epidermal cell, but their numbers can be very great, and their collective surface area can be amazingly large: A single ryegrass plant has about *14 billion root hairs*, with a combined surface area the size of a tennis court!

Root hairs are delicate and short-lived, breaking off as the root tip pushes deeper into the soil. New root hairs continually arise, however, in the region of maturation. Since most water absorption occurs in the root hairs near root tips, gardeners must be careful not to tear off young root tips when transplanting flowers or shrubs, and they often get the best results by fertilizing fruit trees several feet out from the trunk, nearest the root tips.

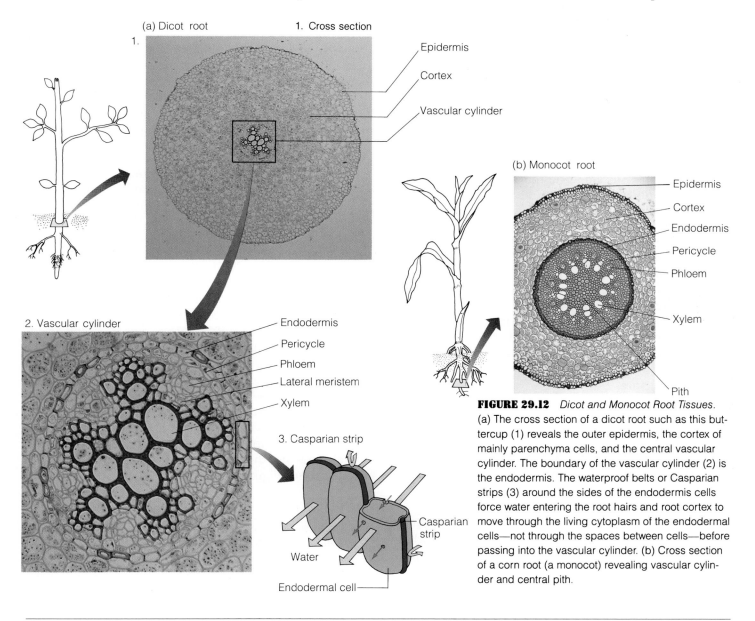

(a) Dicot root   1. Cross section

Epidermis
Cortex
Vascular cylinder

(b) Monocot root

Epidermis
Cortex
Endodermis
Pericycle
Phloem
Xylem
Pith

2. Vascular cylinder

Endodermis
Pericycle
Phloem
Lateral meristem
Xylem

3. Casparian strip

Casparian strip
Water
Endodermal cell

FIGURE 29.12 *Dicot and Monocot Root Tissues.* (a) The cross section of a dicot root such as this buttercup (1) reveals the outer epidermis, the cortex of mainly parenchyma cells, and the central vascular cylinder. The boundary of the vascular cylinder (2) is the endodermis. The waterproof belts or Casparian strips (3) around the sides of the endodermis cells force water entering the root hairs and root cortex to move through the living cytoplasm of the endodermal cells—not through the spaces between cells—before passing into the vascular cylinder. (b) Cross section of a corn root (a monocot) revealing vascular cylinder and central pith.

■ **The Root's Ground Tissue System** A cross section of a bright orange carrot or a dandelion root reveals a thick layer of cells called the *cortex* lying just inside the epidermis and surrounding the central core of the vascular system (see Figure 29.12a, diagram 1). The cortex makes up most of the root's bulk and often stores excess starch. Water moving inward from the soil passes freely between most cortex cells.

The innermost layer of the cortex is the *endodermis* ("inner skin"), a cylinder of tightly packed cells just one cell thick through which water and minerals must pass to reach the rest of the plant. Each cell in the endodermis is a bit like a brick in a circular brick wall (see Figure 29.12a, diagram 2). The outside surfaces of the "bricks" face the rest of the cortex, the inside surfaces face the vascular tissue in the root's center, and, like the mortar surrounding a brick in a wall, a waxy water-resistant substance encircles each cell in a narrow water-resistant belt—the *Casparian strip* (see Figure 29.12a, diagram 3). Because of the waterproof belts, water cannot diffuse in between endodermal cells, but must pass through the cytoplasm of the cells to reach the vascular tissue, and from there move throughout the rest of the plant. Thus, the cytoplasm of the endodermal cells is the portal through which all minerals and water must pass to reach the rest of the plant. Just inside the endodermis is a cell layer (the *pericycle*) that gives rise to new lateral roots that grow parallel to the surface of the ground.

■ **The Vascular System** The root's ground tissue surrounds yet another cylinder—a central zone of vascular tissue consisting of xylem, which carries water and minerals up to the stem and leaves, and phloem, which transports sugars and amino acids between the leaves, stems, and roots (see Figure 29.12a, diagram 2). The xylem and phloem tubes in plant roots are arrayed in various ways, with distinct differences between dicots and monocots (compare Figure 29.12a and b).

The root's simple anatomy reflects its relatively simple tasks of anchoring, starch storage, and absorption of water and minerals (described further in Chapter 31). The shoot, however, with its stems, leaves, and flowers, has a more complex structure.

≋ STRUCTURE AND DEVELOPMENT OF THE SHOOT

The tallest California redwood trees have trunks that may reach a tenth of a kilometer (100 yd) into the sky, while the delicate stems of some wildflowers on the ground below may be mere fractions of an inch tall. Regardless of size, stems have two fundamental tasks: supporting the plant and acting as a central corridor for the transport of water, minerals, sugars, and other substances.

Modified stems can also store starch and allow plants to adhere to vertical surfaces, like ivy growing on a wall. But with or without such specializations, stems share the same basic structure.

THE YOUNG STEM

■ **The Dermal System: Epidermis** As in the root, the dermal system of the shoot consists of a layer of epidermis one cell thick (Figure 29.13). Unlike root epidermis, which absorbs water, shoot epidermis resists water loss. Just as a plastic bag keeps a sandwich from drying out, stem epidermal cells have a water-resistant waxy coating, the *cuticle*, which keeps the plant from losing water.

■ **The Ground Tissue System: Cortex and Pith** As in the root, the ground tissue system of the stem is primarily cortex made up of thin-walled cells (see Figure 29.13). Stems, however, need more structural reinforcement than roots, and many stems also have strands of thick-walled cells around the outer edge of the cortex, just inside the epidermis. Also, a stem's ground system generally has a central core of *pith*, which provides some storage and support, while a root's may lack it.

■ **The Vascular System of Young Stems** In the stem of a young pear seedling or in a nonwoody green-stemmed plant, such as a zinnia, xylem and phloem are organized in groupings called *vascular bundles*. Within each bundle, the xylem lies inside the phloem. In the stems of most dicots, vascular bundles form a ring around the stem's central core of pith, and a layer of lateral meristem (*cambium*) lies between the xylem and phloem (see Figure 29.13a). The lateral meristem in each vascular bundle extends to the adjacent bundles, forming a complete ring around the pith. In conifers, such as pine trees, and woody dicots, such as pear trees, the arrangement of lateral meristem sandwiched between xylem and phloem is profoundly important to future development, because it can lead to the secondary growth of wood and bark and then to the plant's considerable enlargement. In monocots and a few dicots, by contrast, the vascular bundles are scattered throughout the cortex, there is no lateral meristem, and the stems are incapable of true secondary growth (see Figure 29.13b).

SECONDARY GROWTH IN STEMS

Late in the first summer of a pear seedling's life, lateral meristem cells divide and produce additional xylem and phloem cells (Figure 29.14a and b, page 484). The new conducting vessels are called *secondary xylem* and *secondary phloem*, since they arise not from the apical meristem,

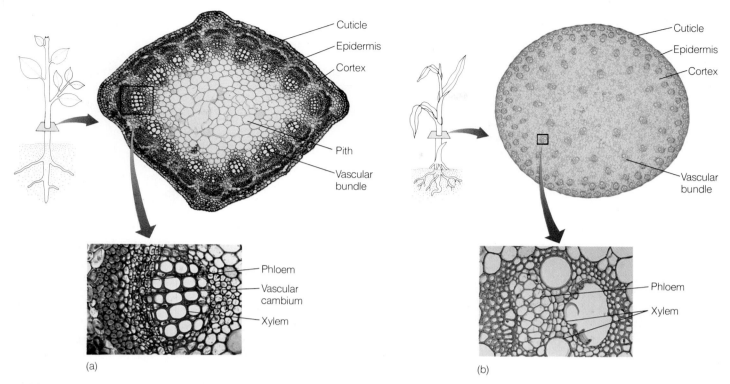

FIGURE 29.13 *Stem Structure in Dicots and Monocots.* (a) The development of stem tissue in annual dicots such as alfalfa is characterized by a circle of vascular bundles; a lateral meristem, or vascular cambium, within each bundle gives rise to xylem toward the stem's interior and phloem toward its exterior. (b) Annual monocots, such as corn, have scattered vascular bundles.

as in primary growth, but instead from a lateral meristem (vascular cambium), itself derived from primary growth. These secondary growth tissues also transport water and other materials. Because the cambium makes far more secondary xylem than secondary phloem, most of the stem tissue in a sapling, bush, or older tree is made up of dead xylem cells (**wood**) in a thick interior rod.

■ **Wood** A tree cut down in a temperate region—say, in a North American hardwood forest—will have concentric rings of lighter and darker wood (Figure 29.14c). These *growth rings* occur because newly produced xylem cells enlarge more in the spring and summer, when water and sunlight are plentiful, than in the fall and winter, when growth conditions are harsh. In moist, tropical regions, on the other hand, where a tree can grow year round, the wood may have no obvious growth rings. Scientists use the width of growth rings in ancient trees such as the 5000-year-old bristlecone pines of eastern California to reconstruct historic weather patterns. They have dated a massive volcanic eruption on the Greek island of Thera, which probably destroyed the Minoan civilization and must have darkened skies around the world with ash, to approximately 1625 B.C. They did this by identifying very narrow growth rings in bristlecone pines growing at that time and still alive today.

As a tree grows older, the xylem at the center gradually ceases to conduct water and minerals, and newer xylem nearer the periphery of the trunk takes over. The nonconducting central wood, or *heartwood*, becomes infiltrated with oils, gums, resins, and tannins, all of which make it dark, aromatic, and resistant to rot. The *sapwood* around the outside continues to transport water in the plant (see Figure 29.14d).

■ **The Bark** When the stem expands during secondary growth, it ruptures the seedling's original cortex and epidermal layers. A pear or other sapling still needs a protective waterproof covering, however, and another lateral meristem—producing waterproof **cork** cells—develops here, as it does in the roots. Together, this layer, the underlying phloem, and many layers of cork cells that die but remain in place make up the outer protective **bark.** Sheets and plugs of cork for flooring, bulletin boards, and bottle stoppers come from the cork of cork oak trees native to Spain, Portugal, and Algeria. Cork cutters are careful to leave the sugar-conducting phloem intact. Without a functioning phloem all the way around the trunk, sugars cannot move downward from the leaves to the roots, and the tree will eventually die.

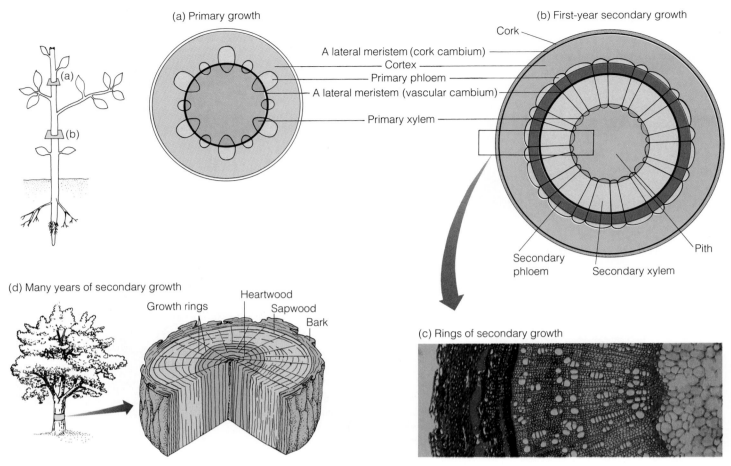

(a) Primary growth

A lateral meristem (cork cambium)
Cortex
Primary phloem
A lateral meristem (vascular cambium)
Primary xylem

(b) First-year secondary growth

Cork

Pith

Secondary phloem

Secondary xylem

(d) Many years of secondary growth

Growth rings
Heartwood
Sapwood
Bark

(c) Rings of secondary growth

FIGURE 29.14 *Secondary Growth in Stems.* (a) During primary growth, the apical meristem lays down the cortex, pith, and primary xylem and phloem. As a plant matures, the root primary xylem and primary phloem differentiate with a lateral meristem between them. (b) During the first year of secondary growth, the lateral meristem (vascular cambium) adds secondary xylem and secondary phloem, the trunk expands, and another lateral meristem (the cork cambium) adds the cork. (c) A cross section of a maple stem reveals two years of secondary growth. (d) A closeup view of wood from the trunk of a sugar maple.

Bark does more than keep water in; it helps keep plant-eating insects, fungi, viruses, and other parasites out. Healthy trees also secrete resins that ooze out of holes bored by insects and often force the intruders back out. Frankincense and myrrh are fragrant tree resins, and the treasured jewel amber is nothing but fossilized resin.

≋ THE STRUCTURE AND DEVELOPMENT OF LEAVES

The leaves that grow on plant stems can be smaller than a fingernail or larger than a table top and shaped like a needle, a knife, a plate, a fan, or a hand. Regardless of form, however, nearly all leaves share the same basic functions: exposing a photosynthetic surface area that is usually large, flat, and catches maximum sunlight; obtaining carbon dioxide from the atmosphere as a carbon source for photosynthesis, yet retaining as much water vapor as possible; and helping to draw water and nutrients up through the vascular system.

LEAF BLADE AND PETIOLE

Most leaves are composed of a broad, flat portion, called the *blade,* and a *petiole* or a *sheath* connecting blade and plant stem (Figure 29.15). Leaf blades are often flat and thin, which maximizes surface area for absorbing light and carbon dioxide. Structural modifications, however, can help balance the conflicting needs of obtaining carbon dioxide but avoiding water loss, as well as serving a variety of other functions. Cactus leaves, for example, are modified into small, dry, needlelike spines that defend rather than photosynthesize (Figure 29.16a). Other plants have leaves modified for clinging and climbing, for capturing little pools of water, and for attracting animals toward inconspicuous blossoms (Figure 29.16b). In a few species, including pitcher plants and sundews, the leaves assist predation, trapping a visiting insect, then enzymatically digesting its proteins (Figure 29.16c).

Leaves may be *simple,* like those of the pear, and consist of a single blade, or they may be *compound,* like those of the ash tree or pea, and consist of many small leaflets.

INTERNAL ANATOMY OF A LEAF

Like its external form, a leaf's internal anatomy is intimately associated with its many functions. Most leaves have an outer layer of epidermis protecting internal ground cells that photosynthesize and a vascular system to bring in water and minerals and carry away the products of photosynthesis.

■ **The Dermal Tissue** A leaf's epidermis has a waxy coating, the cuticle. In fact, the leaf is so well sealed that it needs tiny openings called **stomata** (singular, stoma), surrounded by guard cells, to admit carbon dioxide for photosynthesis and to release water vapor and oxygen (Figure 29.17, page 486; also see Figure 31.6). Stomata are most numerous on the undersides of leaves, and their opening and closing are based on the plant's conflicting needs to conserve water and take in carbon dioxide.

■ **The Ground Tissue** Sandwiched between the upper and lower epidermal layers is the leaf's *mesophyll* layer, made up of ground tissue cells surrounding transport vessels. The *palisade* layer, the main photosynthetic tis-

(a)

(b)

(c)

FIGURE 29.16 *Specialized Leaves Can Protect, Attract, or Kill.* (a) The leaves of this prickly pear cactus are modified into barbed spines that protect the succulent stem. (b) Brilliantly colored leaves like those of the coleus can help attract insect pollinators to the plant's less conspicuous purple flowers. (c) Sticky stalks projecting from the leaf of a sundew plant entrap insects, such as the fruit fly shown here; the stalks then secrete digestive enzymes and absorb the products of digestion, supplementing their supply of nutrients from other sources.

(a)

(b)

FIGURE 29.15 *The Architecture of Leaves: Dicot Versus Monocot.* (a) Dicot leaves are often broad and net-veined and have a stalk, or petiole, and a blade. (b) Monocot leaves have parallel veins, and often the slender blade is attached directly to a stem surrounded by a sheath, rather than being attached by a petiole.

sue, lies just beneath the upper epidermis. It usually consists of one or more rows of vertically oriented, column-shaped cells, each enclosing dozens of chloroplasts (see Figure 29.17). A layer of rounder cells (the *spongy* layer) lies between the upper photosynthesizing cells and the lower epidermis. The loosely packed spongy layer provides a huge surface area (analogous to an animal's lung) for absorbing carbon dioxide from the air that enters the stomata. Carbon dioxide entering the stomata can move rapidly through the spongy layer to the palisade layer above, where most of the photosynthesis takes place.

■ **The Vascular Tissue** Leaves need an elaborate "plumbing" system to distribute the products of photo-

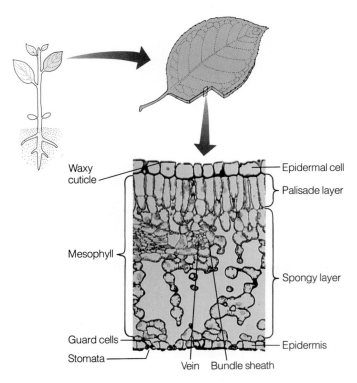

Waxy cuticle — Epidermal cell
— Palisade layer
Mesophyll
— Spongy layer
Guard cells
— Epidermis
Stomata
Vein   Bundle sheath

FIGURE 29.17 *Leaves: An Inside View.* A small section taken from a leaf of a privet bush reveals the cuticle-covered upper epidermis, layers of palisade and spongy mesophyll cells, a vein, or bundle of vascular tissue surrounded by a sheath, and a lower epidermis ventilated by openings—stomata—that are flanked by guard cells.

synthesis and bring in water and minerals, and so the vascular system is continuous between leaves, stem, and roots. Monocot leaves have bundles of xylem and phloem called **veins** running parallel to each other along the long axis of the leaf; dicot leaves have veins that form a branching pattern (see Figure 29.15). All summer long, mesophyll cells in the leaf store the sun's energy in sweet sugar molecules, and the phloem transports sugars from the leaf to the developing fruit. The end result is a delicious pear—the plant's evolutionary solution to dispersing the species.

≋ CONNECTIONS

Since plants cannot move to solve problems, they grow to improve their situation. Dermal, ground, and vascular tissue systems help the immobile plant to function efficiently by allowing sugar, water, and minerals to move even though individual cells remain in place. The downward growth of the root and upward growth of the shoot at apical meristems exemplify plant growth strategy.

There is no germ line in plants. The gametes of flowering plants arise from the same line of cells that makes stems and leaves. And a plant grows only from meristems, continually adding new organs as it grows upward, downward, and outward. Our next chapter examines how hormones control a plant's open growth.

NEW TERMS

annual plant, page 476
apical meristem, page 475
bark, page 483
biennial, page 476
bud, page 479
cork, page 483
dermal tissue, page 474
ground tissue, page 474
lateral meristem, page 476
meristem, page 475
open growth, page 475

perennial, page 476
phloem, page 474
primary growth, page 476
root, page 473
secondary growth, page 476
shoot, page 473
stomata, page 485
vascular tissue, page 474
vein, page 486
wood, page 483
xylem, page 474

STUDY QUESTIONS

REVIEW WHAT YOU HAVE LEARNED

1. List the types of roots, and give an example of each type.
2. Describe where the three tissue systems in a plant are found, and discuss their functions.
3. Describe the structure of xylem and phloem, and state the ways in which phloem cells differ from xylem cells.
4. Why is a plant's meristematic tissue so important?
5. Name the three general regions of a growing root, and briefly describe each region.
6. Where is a plant's endodermis? What is its function?
7. What is the main function of lateral meristems?
8. Where does most of a plant's photosynthesis take place?

APPLY WHAT YOU HAVE LEARNED

1. When a gardener transplants a shrub, why must he or she exercise care in digging up the roots?
2. Botanists were able to correlate unusually narrow growth rings in a bristlecone pine in California with the eruption of a Greek volcano in 1625 B.C. What is the connection between the two?
3. Counselors at a summer camp warn playful campers not to peel bark off the camp's birch trees. What might be the result of this vandalism?

FOR FURTHER READING

Raven, P. H., R. F. Evert, and H. Curtis. *Biology of Plants.* 4th ed. New York: Worth, 1986.
Walbot, V. "On the Life Strategies of Plants and Animals." *Trends in Genetics* (June 1985): 165–169.
Wayward, S. Y. "Genes Play the Field." *New Scientist* (9 September 1989): 49–53.

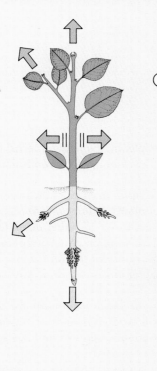

① Plants have the potential for perpetual growth and development—upward and outward in the shoot, and downward and outward in the root.

② Plants have three tissue systems: a protective dermal tissue, a supportive ground tissue, and a cargo-carrying vascular tissue.

Dermal tissue

Ground tissue

Vascular tissue

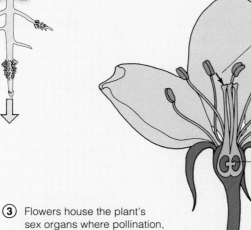

Pollination

Fertilization and seed formation

③ Flowers house the plant's sex organs where pollination, fertilization, and seed formation take place.

④ At germination, a new plant emerges from the seed and begins the transformation from seedling to adult; the earliest primordia give rise to leaves, and later primordia give rise to buds that become flowers or branches.

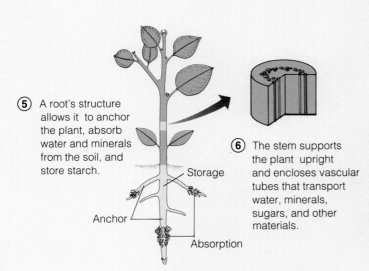

⑤ A root's structure allows it to anchor the plant, absorb water and minerals from the soil, and store starch.

Storage

Anchor

Absorption

⑥ The stem supports the plant upright and encloses vascular tubes that transport water, minerals, sugars, and other materials.

Henri Rousseau, *Snake-charmer*, 1907

⑦ Leaves expose a broad surface area to sunlight, exchange gases with the atmosphere, and help draw water and nutrients up through the plant.

HOW PLANTS GROW

FOOLISH SEEDLINGS

So many people on this planet depend on rice that 20 percent of all human activity is rice powered! Asians began cultivating rice in Thailand nearly 6000 years ago (Figure 30.1); and Japanese farmers, in particular, began, very early on, to breed improved varieties of the rice plant (*Oryza sativa*) and to record and study its diseases.

One disease that has plagued rice growers for hundreds of years causes afflicted plants to grow far taller and faster than normal and in the process to become spindly and weak and blow over in the wind and die. The name for this disease is fitting: foolish seedling disease (see Figure 30.2a, page 490). The symptoms of disease are caused by chemicals that a fungus (*Gibberella fujikoroi*) releases when it infects rice plants. These disease-causing chemicals are called **gibberellins,** after the fungus.

In the mid-1950s, researchers were surprised when they isolated these chemicals from a healthy bean plant. Soon it became clear that the chemicals are plant growth regulators, or hormones, that occur naturally in many plants. The fungus had simply evolved a way to make and release the hormones in quantities that can cause the rice plants to literally grow themselves to death.

In this chapter, we will see that a plant's growth, development, and survival depend on *internal regulators*—primarily plant hormones such as gibberellins—and *external regulators*, including temperature, light, gravity, and the number of daylight hours. Agricultural researchers have used their knowledge of these regulators to control weeds and increase food production for the rapidly expanding world population.

As we consider internal and external plant growth regulators, we will encounter three unifying principles. First, plants generally adjust to changing environmental conditions via growth. A plant cannot move away from an environmental problem, but it can grow toward high light levels, grow a shorter or longer stem as needed, or send roots deeper into soil.

Second, changes in plant shape or function are often regulated by plant hormones produced in response to environmental factors such as temperature, light, or gravity. For example, plant hormones can promote or inhibit growth, depending on season. Third, plant hormones act at the level of cells to induce cell division, enlargement, or cell maturation. One hormone promotes growth, for example, by triggering internal cellular changes that make cell walls more flexible.

This chapter introduces the major types of plant hormones, then follows the life cycle of a flowering plant, focusing at each stage on how hormones and external regulators control the events. The chapter answers several questions:

- What does each plant hormone do?
- How is seed germination regulated?
- How are plant growth and development controlled?
- What governs flowering and fruit development?
- What controls dormancy, dying, and the falling of leaves and fruit?
- How do plants defend themselves from infection and wounds?

FIGURE 30.1 *Cultivating Rice: An Ancient Practice.* For about 6000 years, people have been cultivating rice (*Oryza sativa*) by traditional methods, such as those depicted in this painting by Chiao Ping-chên.

PLANT HORMONES: FIVE MAJOR KINDS

The foolish seedling phenomenon brought on by high levels of gibberellins is characteristic of how hormones function in plants. Three of the five main classes of plant hormones promote and regulate growth, while the other two classes inhibit growth or promote maturation. The names and activities of all five classes are given in Table 30.1. In general, plant hormones are small molecules that probably act on target cells much the way animal hormones do. Let us survey the growth promoters and growth inhibitors separately, then see how they interact to regulate a plant's life cycle.

HORMONES THAT PROMOTE GROWTH

■ **Gibberellins** Gibberellins are a good example of the widespread effects of particular plant hormones. They are made in a variety of organs, such as young leaves, embryos, and roots, and move passively throughout the plant. Gibberellins are primarily involved in regulating plant height. Too little gibberellin results in dwarf plants, but too much results in long, pale, "foolish" stems (see Figure 30.2a, page 490). Gibberellins also play important roles in *bolting* (sudden stem lengthening) in plants such as cabbage (Figure 30.2b) and in inducing the seeds of rice, barley, and other grasses to germinate.

■ **Auxins** Like gibberellins, **auxins** generally promote growth. The name, in fact, comes from the Greek *auxein*, meaning "to increase, or augment." Auxins have an even broader set of activities than gibberellins; and instead of being produced in various tissues and traveling up and down through the vascular system, most auxins are made in shoot apical meristems and developing leaves and diffuse from the site of production downward toward the roots.

Auxins promote growth by triggering enzymes that loosen the tightly woven fibers in cell walls, allowing the cell to expand from within. This action leads to the elongation of cells in stems, leaves, and the wall of the ovary. Auxins also act in other ways to prevent leaves, fruits, or flowers from falling off prematurely and to inhibit growth in lateral buds and roots in favor of growth at the apical meristem. In high concentrations, auxins can cause uncontrolled growth and plant death. Garden stores sell the synthetic auxin 2,4-D, which causes the leaves and stems of broad-leaved plants like dandelions to droop and curl within hours (Figure 30.3, page 490). Within days, the chlorophyll degrades and the plant dries out and dies. During the Vietnam War, the U.S. government sprayed millions of acres with Agent Orange, a mixture containing a chemical related to auxins, to kill forests and expose enemy soldiers and equipment. These plant-killing substances (*herbicides*) appear to induce such rapid cell division and tissue proliferation that the phloem becomes plugged. Unfortunately, Agent Orange is con-

TABLE 30.1 ≋ INTERNAL REGULATORS OF PLANT GROWTH AND DEVELOPMENT		
REGULATOR	TRANSPORT	ACTION
Hormones		
Gibberellins	Upward and downward in vascular system	*Promote growth;* promote stem lengthening in dwarf and rosette plants and seed germination in grasses; involved in flowering and fertilization; involved in growth of new leaves, young branches, and fruits
Auxins	From tip of shoot downward in vascular system	*Promote growth;* augment growth by cell elongation; inhibit growth of lateral buds; foster growth of ovary wall; prevent leaf and fruit drop; orient root and shoot growth
Cytokinins	From root upward in vascular system	*Promote growth;* stimulate cell division; kindle growth in lateral buds; block leaf senescence
Abscisic acid	Diffuses short distances in leaf and fruit	*Inhibits growth* by opposing the three growth-promoting hormones; induces and maintains dormancy
Ethylene	Through air, as a gas	*Promotes maturation;* enhances fruit ripening; promotes dropping of leaves, flowers, and fruits
Pigment		
Phytochrome	Not transported; remains in cell that produces it	*Detects light;* changes form in response to light; mediates flowering, germination, growth, and plant form

(a)   (b)

FIGURE 30.2 *Gibberellins Cause Growth.* (a) Foolish seedling disease is a curious effect of gibberellins on plant growth. When a young rice plant is infected by the fungus *Gibberella fujikoroi* (left), the parasite releases gibberellins, which cause the stems to elongate rapidly, grow weak and spindly, and blow over and die in even a gentle breeze. A healthy dark green plant is on the right. (b) Bolting. Gibberellins cause the short stem in a plant with rosette form to greatly elongate. On the right are cabbage plants *(Brassica oleracea)* in normal rosette form. On the left are plants of the same species after treatment with gibberellins. Bolting takes place naturally in a biennial before flowering in the second year.

taminated with *dioxin,* a compound that many suspect is highly toxic to people. Newer herbicides, such as glyphosate, kill all green plants but are much safer and rapidly degrade completely to harmless compounds.

■ **Cytokinins** Members of this third class of growth-promoting plant hormones generally stimulate cell division, including cytokinesis (see Chapter 7). **Cytokinins** move through the plant less readily than auxins or gibberellins and appear to move in opposition to auxins—from root upward to shoot, not shoot downward to root. Also unlike auxins, cytokinins promote the growth of lateral buds, not shoot tips. Like most plant hormones, cytokinins also have other effects, including the prevention of leaf aging, or senescence.

HORMONES THAT INHIBIT GROWTH

If plants had only gibberellins, auxins, and cytokinins, they would grow constantly. Sometimes, however, it is more advantageous for the plant to stop growing—to close its stomata, decrease the level of photosynthesis, drop aging leaves, or become dormant.

■ **Abscisic Acid** **Abscisic acid** (ABA) counteracts the growth hormones, apparently by indirectly blocking protein synthesis and new growth. ABA moves only short distances from its site of production. For decades, scientists believed that ABA's main role was to accelerate the dropping of leaves and fruit *(abscission).* Now, however, they think that the main role of ABA is to induce and maintain metabolic slowdown, or **dormancy,** especially in buds, and the closing of a leaf's gas exchange pores (stomata) to prevent excess water loss.

■ **Ethylene** The fifth major plant hormone is a small, simple molecule that exists as a gas at normal temperatures and is dispersed from one plant or plant part to another by air. **Ethylene** is behind the old adage about one rotten apple spoiling the whole barrel: The hormone is produced by ripening fruits, and it stimulates ripening in nearby fruits. Ethylene also stimulates the aging and dropping of leaves and fruits and may have an important role in plant self-protection (see page 499).

FIGURE 30.3 *Synthetic Auxin Causes Some Plants to Grow Themselves to Death.* In this photo, the broad-leafed plants in the foreground were killed by herbicides, but the narrow-leafed corn in the field behind is unaffected.

Our survey has revealed that the five major types of plant hormones are small, mobile molecules that promote or inhibit growth or maturation. Now let us see how the plant hormones interact with one another and with environmental cues to regulate a plant's life cycle.

※ INTERNAL AND EXTERNAL REGULATORS OF GERMINATION

After animals or air currents carry seeds to new locations, the seeds germinate, and tiny new plants emerge. But what is the best time for germination? Autumn? Winter? Spring? And what triggers germination at a time when the seedling's chances of survival are greatest? Although specific answers depend on the plant species, environmental cues generally act through plant hormones that in turn regulate the timing and sequential events of germination.

Some seeds germinate shortly after they reach a new location and imbibe enough water for the seed coat to soften and the tiny plantlet to burst out. Rice seeds are in this category, as are willows, poplars, and silver maples. This pattern is especially common in tropical plants, with their relatively mild, moist, and stable environments.

SEASONAL INCREASE IN TEMPERATURE AND DAY LENGTH

In temperate zones, delayed germination is often more advantageous. The seedlings of an annual that emerged in the late summer or early autumn might not complete an entire life cycle before the short days, cold temperatures, and dry conditions of winter set in, and the seedlings of biennials and perennials might not survive that first harsh winter season. Many temperate plants, therefore, germinate in spring or early summer after surviving winter as an embryo encased and protected within a seed coat. In spring, increasing temperatures, lengthening days, and melting snow or rain trigger hormonal activities within the embryo and, in turn, germination.

MOISTURE, COLD, AND LIGHT

For some seeds, especially in dry areas, moisture is the external trigger, counteracting a growth-inhibiting hormone such as ABA. When enough moisture leaches away the inhibitor, growth resumes. This chain of events helps guarantee that the ground will be damp enough for the seedling to survive.

For some kinds of seeds, light is a determining factor. Lettuce seeds, for example, are so tiny that seedlings buried deeper than a few millimeters will run out of food reserves before reaching the sunlit surface. Certain lettuce seeds, therefore, germinate only when they detect light. In contrast, the seeds of some desert plants germinate only in deep, dark, moist soil.

In a few remarkable plant species, seeds can lie dormant for immense periods until conditions are just right. Some seeds from the Japanese lotus (Nelumbo nucifera), for example, can germinate successfully after lying dormant for 2000 years. And some lupine seeds frozen in the arctic permafrost have reportedly remained viable for 10,000 years. Nevertheless, most seeds remain dormant and viable for only a few years at best. Even with greatly slowed metabolism, their food reserves are eventually used up and the embryo dies.

※ REGULATION OF PLANT GROWTH AND DEVELOPMENT

Although it might surprise you, farmers in California and along the Gulf Coast grow one-twelfth of the world's rice, and like other rice farmers, they start new crops in springtime. Rice farmers soak rice seeds, wait for the rice grains to germinate, then set the tiny seedlings in rows in the fields. As they grow, rice plants pass through a series of developmental stages, each controlled by external and internal regulators, so that the plants enlarge, develop new leaves and roots, then flower and set seed. This section considers the external and internal triggers of such life cycle events and shows how plants adapt to their surroundings through changing physiology and altered shape and orientation in space.

INFLUENCES ON A PLANT'S ORIENTATION

A rice seedling in a field must orient its main axis so that the roots grow down, not up or sideways, and the shoot grows up, not in some other direction. This seems logical enough, but how does a tiny plant manage to grow in the correct orientation? A plant's orientation is based on **tropisms:** bending, turning, or directional growth in response to external and internal stimuli. The term *tropism* comes from the Greek word for "turning toward." A tropism is considered *positive* if the organism's orientation is *toward* the stimulus, and *negative* if the organism's orientation is *away* from the stimulus. Botanists have named a number of tropisms by the external stimulus that causes them. For example, the orientation of roots toward water is called *hydrotropism* (turning toward water);

the orientation of a plant part toward the ground is *grav-itropism* (turning toward gravity or earth, sometimes called *geotropism*); and the orientation toward light is *phototropism*. These last two tropisms help explain how a rice seedling's roots grow down and its shoot grows up.

■ **Gravitropism** When a seed germinates underground, both root and shoot are surrounded by soil, yet the seedling, once it has emerged, achieves the proper root-down and shoot-up orientation. This feat is accomplished through the plant's sensitivity to gravity. The root is *positively gravitropic,* since it grows down toward the pull of gravity, and the shoot is *negatively gravitropic,* since it grows up against the pull of gravity.

Investigations of the root tip of plants laid on their sides have revealed that dense starch granules *(amyloplasts)* rapidly sink to the bottom of the cells (Figure 30.4). Plant physiologists suggest that these starch granules may act like little rocks to detect the pull of gravity and

(a) Shoot grows up; root grows down

(b) Root tip bends

Cells elongate

Cells do not elongate

(c) Root tip cells

Amyloplasts (dense starch granules)

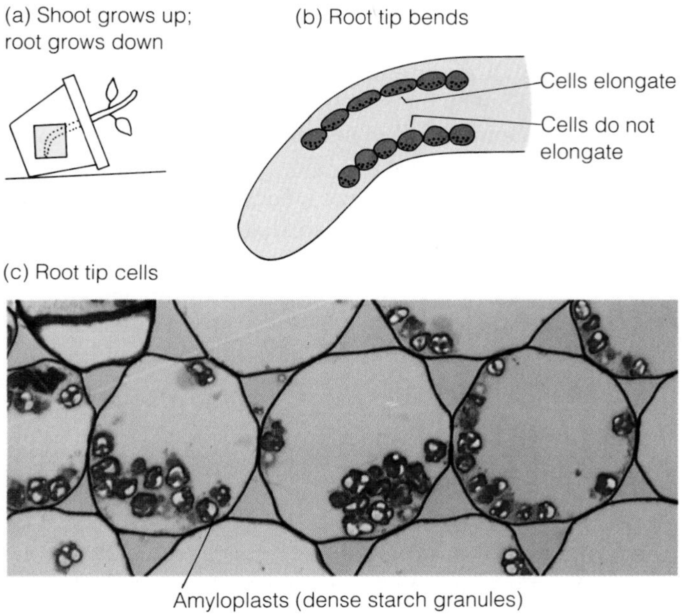

FIGURE 30.4 *Gravitropism: Growth Toward or Away from Gravity.* (a) A plant's shoot will grow upward and its roots downward no matter how the organism is oriented. (b) Cells on the upper root surface elongate and cause the root to grow down. (c) The downward growth of roots occurs because gravity pulls heavy amyloplasts to the bottom of certain stem and root cells, and the settling of these organelles triggers gravitropism. The photo shows cells from the root of a buttercup plant. After growth reorients the root, the amyloplasts again sink to the bottom of the cell. Amyloplasts appear to work the same way in stem and root cells, but research suggests that in the stem, hormones stimulate cell growth on the organ's downward side, making the stem grow upward.

may induce auxin to move to the zone of elongation along the lower side of a root and block the elongation of lower cells. This would allow roots to grow downward, thrusting toward a more likely source of moisture and minerals.

■ **Phototropism** Plants need light for photosynthesis, but they may lack sufficient access to it. For example, a wild rice seedling might by chance germinate beneath the trees and bushes in a dense forest. To maximize its exposure to the sun, the plant will grow toward light, and its leaves will adjust and turn constantly to face it.

A series of experiments on oat plants showed that some chemical substance (rather than an electrical signal, for example) moves from a plant's tip to its stem, causing the stem to bend toward light. In one study, the researcher cut off the tip of an oat seedling and placed the tip on a block of jello-like agar (Figure 30.5a). After a while, he discarded the plant tip and placed the agar on one edge of the cut surface of a second decapitated oat seedling (Figure 30.5b). Although the second oat seedling was kept in the dark throughout the experiment, the stem began to bend away from the side with the agar block (Figure 30.5c). A material the researcher named "auxin" had obviously moved from the first tip to the agar and from the agar to the second seedling, causing the second

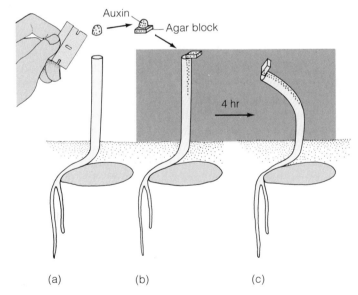

Auxin

Agar block

4 hr

(a)   (b)   (c)

FIGURE 30.5 *An Experiment Showing That Auxin Controls Phototropism in Oat Seedlings.* (a) A researcher removed the tip of an oat seedling and placed it on an agar block. Substances from the tip diffused into the block. (b) The researcher placed the agar block on one side of a decapitated seedling, and (c) the seedling's stem curved away from that side. The conclusion: A substance the researcher called auxin (red dots) diffused from the tip to the block, and from the block to the second plant, causing cells on the receiving side to elongate and force the stem to bend in the opposite direction.

plant to bend. Later researchers isolated the growth hormone auxin and have continued to study precisely how light affects auxin and how auxin, in turn, triggers differential growth and phototropism. So far, evidence suggests that auxin causes the internal cell environment to become more acidic; this change allows an enzyme to become active, the enzyme weakens cellulose in the cell walls, and these walls then stretch and enable the entire cell to elongate.

PLANT MOVEMENTS OTHER THAN TROPISMS

Although tropisms are immensely important to a plant's orientation in space, not all plant movements are tropisms oriented toward or away from the stimulus that causes them. A leaf of a sensitive plant (*Mimosa pudica*; see Figure 4.20), for example, will droop if touched anywhere along its length, and the leaves of the Venus flytrap will close if a fly triggers any of the little hairs inside the trap (Figure 30.6). Such plant movements that are in response to a stimulus, but not oriented with respect to the stimulus, are called *nastic responses*. These responses are not always slow, irreversible, and based on the growth of cells, as tropisms tend to be. Nastic responses may be rapid and irreversible, rapid and reversible (as in the sensitive plant), or even slow and reversible (as shown by the Venus flytrap). Moreover, the so-called sleep movements some plants display when kept in total darkness (Figure 30.7) provide strong evidence that plants, like animals, have internal biological clocks that meter the passage of time and trigger regular responses.

INFLUENCES ON A PLANT'S SHAPE

Although rooted in place, plants have a tremendous capacity to turn, bend, and grow slower or faster, which allows them to move part of their body toward or away from light, water, nutrients, gravity, or substrates. Just as external factors influence plant growth, they also affect shape in ways that adapt the plant to its environment.

FIGURE 30.6 *The Venus Flytrap: A Slow, Reversible Nastic Response.* When an insect or frog lands on a Venus flytrap, an electrical stimulus (rather than a hormonal one like auxin) triggers an irreversible change in acidity in the walls of cells where the leaves are hinged. The cell walls weaken, the cells expand permanently, and this eventually forces the leaf to snap shut. Later, the cells on the inside of the hinge enlarge and force the trap open again in preparation for another fly.

■ **Shape Changes in Response to Light** Light can have a profound effect on a plant's shape. A crabgrass plant growing in a dark spot—under a board, let's say—is a good example. The plant looks spindly and pale yellow rather than healthy and green, because, deprived of direct light, it wastes little energy making chlorophyll and instead grows longer and longer in a search for light (Figure 30.8a, page 494). Even in less extreme shade, plant survival depends on the ability to detect light levels and adapt by growth to this changing environmental stimulus.

■ **How Plants Detect Light** Like some of your eye cells, plants have a light-sensitive pigment that allows them to detect the quality or quantity of light in the environment. This pigment in plants, called **phytochrome** (meaning "plant color"), regulates plant growth, but it is not a hormone because it does not diffuse from cell to cell (see Table 30.1). Phytochrome allows a light-sensitive

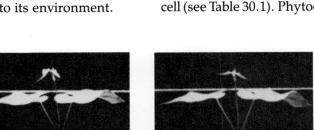

2:00 A.M. 12:00 P.M. 9:00 P.M. 11:00 P.M.

FIGURE 30.7 *Sleep Movements in Plants.* Leaves of the bean (*Phaseolus vulgaris*) droop at night and fully extend by noon, even when kept in the dark. These "sleep movements" are evidence of a biological clock in plants, probably based on hormones.

(a) Plants grown in darkness

(b) Phytochrome conversions

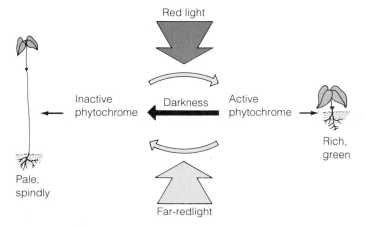

(c) The hour glass

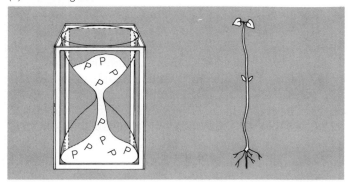

FIGURE 30.8 *The Phytochrome Light Timer.* (a) Plants grow spindly and pale when deprived of light. In grass under a board, chlorophyll is degraded, and the plants grow toward any light source (here, the edges). (b) If a plant is exposed to far-red light or remains in the dark long enough, the phytochrome molecules in its cells convert spontaneously from the active to the inactive form, and the plant becomes pale and spindly. When exposed to red light, in contrast, phytochrome switches to active form, and the plant stays healthy and green. The ratio of active phytochrome (red P's) to inactive (orange P's) acts like a biochemical hourglass that times periods of darkness.

seed, for example, to detect whether light is available and thus whether to germinate. Phytochrome also spurs flowers to emerge at appropriate times for pollination.

Phytochrome works through a change in physical form. The molecule is present in a plant in either an inactive form or an active form. Red light of a very dark shade (called far-red light) causes the active form to change to the inactive form, but light of a brighter red shade turns the inactive form into the active. Normal daylight contains high levels of bright red light, but little far-red light; it causes most phytochrome molecules to convert to the active form. When a plant is exposed to darkness, however, the phytochrome molecules slowly convert to the inactive form, like sand pouring through an hourglass. The molecules are thus a natural timing device, with the ratio of the two forms (active to inactive) signifying the length of time a plant has been deprived of light (Figure 30.8b and c).

The link between phytochrome and plant function is this: When the phytochrome inside plant cells is in the active form, the plant grows normally, but when the molecules are converted to the inactive form, the plant loses chlorophyll and becomes pale and spindly like the crabgrass under the board. The control of plant growth by phytochrome is just one aspect of how growth regulators affect plant shape. Another control comes from substances at the tips of the plants.

■ Apical Dominance, Hormones, and Overall Plant Shape
Plants can take on a wide variety of shapes, but there are, in general, two major forms, represented by the spruce and maple pictured in Figure 30.9a. Spruce trees usually have a single main trunk with numerous side branches—long at the base and shorter at the top—so that the whole tree has a conical shape. In contrast, maples often branch low, with several main branches and no central trunk, and the overall shape is roughly spherical, not conical.

A major factor in these plant shapes is **apical dominance** (literally, domination by the tip of the plant), an inhibition of lateral buds by auxin transported down the shoot from the apical bud at the plant's tip. Auxin promotes growth in the stem but inhibits growth of lateral buds. Because auxin is continuously broken down as it moves down the stem, its concentration drops off (Figure 30.9b). Thus, buds closest to the tip of the main stem are most inhibited (and smallest), while branches farther from the tip of the plant are the least inhibited (and oldest and largest). The result is a conical shape. At the same time, cytokinin also comes upward from the roots, counteracts the effects of auxin, and causes the lateral buds and the lowest branches to grow more strongly.

In a plant with a more spherical shape, the apical bud may produce less auxin, and lateral buds and branches may be less sensitive to inhibition by the hormone; thus, the branches both high and low on the stem can grow

(a) Spruce and maple

(b) Apical dominance

Inhibited
lateral bud

Auxin

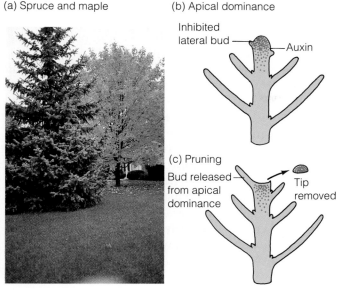

(c) Pruning

Bud released
from apical
dominance

Tip
removed

FIGURE 30.9 *Apical Dominance: Hormones from Growing Tips Can Affect Overall Plant Shape.* (a) A conical spruce tree and a roughly spherical maple tree demonstrate the dramatic effects of apical dominance. (b) Auxin produced in the plant's apex travels downward and inhibits the growth of lateral branches. (c) If that apex is removed, lower branches are released from apical dominance and begin to grow.

strongly. If a gardener or hungry animal removes the apical bud of a plant, the buds farther down the stem are released from apical dominance and begin to grow, producing a bushy appearance (Figure 30.9c). Commercial Christmas tree growers take advantage of apical dominance by pruning back the tips of branches. This allows buds toward the inside of the tree to grow, yielding a fuller, more desirable tree.

It is clear from the foregoing information that plants generally respond to environmental hardships or opportunities by growth and change in form and that hormones mediate these responses. Life cycle events are part of this responsiveness, and as in germination and subsequent growth, flowering and fruit formation depend on a mixture of environmental and hormonal triggers.

≋ CONTROLLING INFLUENCES ON FLOWERING AND FRUIT FORMATION

Like germination and growth, flowering and setting fruit must be timed to the appropriate seasons so that various aspects of reproduction occur in seasons when the species has the greatest potential for success. Environmental factors that change with the seasons—temperature,

moisture, and the number of hours of sunlight in a day—act as external regulators of the plant life cycle. Seasonal timing is equally important for annual plants such as rice, which flower and produce seeds once in a single season, and for perennials. An exceptional example of the latter is the rare *Puya raimondii*—a large plant of the windswept Andean high plain that can live for nearly two centuries and that produces the world's largest floral structure (Figure 30.10)—but only once every 30 to 150 years. Seasonal timing appears to depend on both environmental cues (such as temperature and light) and internal hormonal triggers.

FIGURE 30.10 *Timing Is Everything: Hormones and Environmental Cues Regulate Flowering in* Puya raimondii. Light, temperature, and internal hormones interact to control flowering in this giant member of the pineapple family. The plant produces a 4.5 m (15 ft) flower spike—the largest floral structure in the plant kingdom—just once every 30 to 150 years.

TEMPERATURE AND FLOWERING

Just as some seeds will germinate only after a long period of cold, certain plants, including biennials like beets, carrots, and turnips, flower only after a period of exposure to cold, a process called *vernalization*. Such exposure to cold signifies that winter has come and gone. Because flowering comes early in the spring, the plant produces seeds early in its second summer, so the seedlings have plenty of time to grow into hardy plants that can withstand the following winter.

LIGHT AND FLOWERING

For many plants, temperature alone is too risky a trigger for flowering. A cold September followed by a warm October and a harsh November could find a plant covered with flowers but buried in snow. For this reason, many plants track the seasons by **photoperiod,** the length of light and dark periods each day, rather than by daily temperature. Early researchers noticed that some plants, like rice, seem to require short days to bloom, and they called these plants *short-day plants.* Other plants seem to require long days and were therefore called *long-day plants.* (They called plants that flower without regard to photoperiod *day-neutral plants.*) For example, ragweed blooms in the fall and requires *less than* 14½ hours of daylight to blossom; thus, it is a short-day plant. In contrast, spinach needs *more than* 14 hours of light per day to flower, so it is a long-day plant. At 16 hours of light, the spinach blooms but the ragweed does not, and at 8 hours of light, the ragweed will blossom but the spinach won't (Figure 30.11a and b).

Generally, long-day plants blossom in spring or early summer, when the days are becoming longer, and short-day plants flower in the late summer or fall, when the days are becoming shorter. For example, poinsettias are short-day plants because they will flower only when the days are less than 10 hours long (and the nights are more than 14 hours long) as in autumn. People with indoor potted poinsettias must therefore put them in a dark closet for 14 or more hours each day during the fall in order to enjoy their showy red and yellow displays by Christmas.

To probe how plants measure critical photoperiod, botanists exposed ragweed plants to the short-day length that should cause flowering, but interrupted their period of darkness with a short blast of light (Figure 30.11c). The result: No flowers formed. Since interrupting the night blocks flowering, plants must measure night length rather than day length—in other words, short-day plants are actually "long-night plants." This explains why even a quick peek at a poinsettia plant in the middle of the

night would disrupt the dark period this short-day (long-night) plant needs to set flowers. For a long-day (short-night) plant like spinach, however, a flash of light in the middle of the night promotes flowering.

Such light flash experiments suggested that plants might measure night length and hence photoperiod via phytochromes. Recall that when a plant is surrounded by darkness, phytochrome converts spontaneously from the active to the inactive form (review Figure 30.8b). A light flash in the middle of the night, however, can regenerate the active form. In short-day plants, the active form *blocks* flowering; these plants will flower only after a long night allows active phytochrome molecules to disappear. When a flash blocks the disappearance of the active form, the active phytochrome molecules, in turn, block flowering (see Figure 30.11c). In long-day plants, on the other hand,

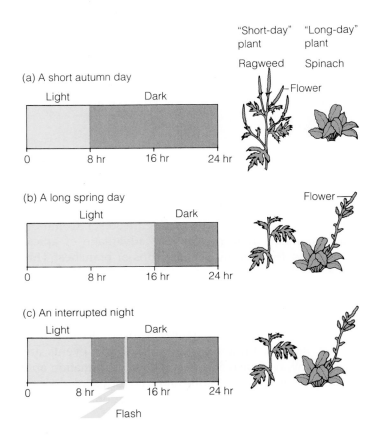

FIGURE 30.11 *The Light/Dark Cycle Regulates the Onset of Flowering.* (a) On an 8-hour autumn day, ragweed—a short-day plant that flowers if there are fewer than 14½ hours of light—will flower. On the same day, however, spinach—a long-day plant that needs more than 14 hours of light a day—will fail to flower. (b) In contrast, on a spring day, ragweed (a short-day plant) will not flower, but spinach (long-day) will. (c) If an experimenter interrupts a long autumn night with a brief flash of light, the plants respond as if the day were long and the night short. This suggests that light resets the "clock" that times the length of the night. Further studies have shown that phytochrome molecules mediate these timing actions.

1. OBSERVATION: The top bud of a chrysanthemum becomes a flower in the fall when the days get shorter.

2. QUESTION: There must be some "clock" in the plant that measures the light/dark cycle. I wonder where it is?

3. HYPOTHESIS: I suspect that the clock is in the bud itself.

4. EXPERIMENT AND PREDICTION: To find out, I'll expose the bud on one group of chrysanthemums to short days, but their leaves to long days. As a control, I'll do the reverse for a second group of plants. If my hypothesis is right, only the plants with buds exposed to short days will flower.

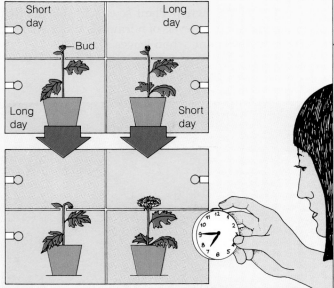

5. RESULTS: Surprisingly, the plants with buds exposed to short days did not flower, but the plants with *leaves* exposed to short days did flower.

6. CONCLUSION: Apparently, a plant's "clock" is in its leaves, and the leaves send a signal to the bud triggering the formation of a flower. I'll have to do new experiments to find out more.

Adapted from Chailakhyan, M. K. (1937)
Hormon Theory of Plant Development.
Moscow, Leningrad: Akad, Navk. USSR

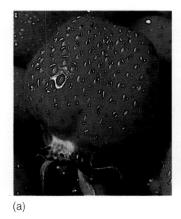
(a)

(b)

FIGURE 30.12 *Seeds Liberate Auxin, Which Triggers Fruit Development.* (a) Strawberry seeds secrete auxin, and this promotes fruit development around the seeds. (b) If all the seeds are removed, the fruit is not stimulated to enlarge or ripen, but if a few seeds are left, fruit will develop just beneath them.

active phytochrome *promotes* flowering, instead of *inhibiting* it.

The phytochrome "hourglass," reset by light and gradually flowing from active to passive molecules in the dark, allows such accurate timing of photoperiod and thus of flowering initiation that all the individuals of a species can flower within a day or two of each other. This provides a ready source of pollen that wind or animals can convey to other blossoms of the same species.

Once researchers had determined that phytochrome controls flowering, they went on to show that it exerts its effect through hormones (see the box on the left), probably through the combined effects of several already familiar plant hormones. In summary, botanists now agree that plants detect changes of seasons by means of a light-sensitive pigment that acts via hormones, causing flowers to appear at favorable times of the year.

TRIGGERS TO FRUIT DEVELOPMENT

As Chapter 29 explained, the fruit—with its cargo of seeds—begins to develop once the flower has been pollinated and seeds start to form. Fruits, as we have seen, help protect the seeds and disperse them. But fruits are also energy-expensive organs to produce—especially the large, juicy ones—and thus, some plants ensure that fruits develop only where there are viable seeds. The dozens of tiny seeds on the outside of a strawberry, for example, help control the development of a normal fruit (Figure 30.12). If all the seeds are removed when the fruit is still small and green, it never enlarges and ripens. The strawberry plant puts its resources only into fruit still covered with viable seeds. However, deseeded strawberries treated with auxin will grow normally, suggesting that the seeds promote fruit growth by secreting auxin.

The ethylene hormone also plays an important role in fruit development: It accelerates ripening. When a peach begins to ripen, it gets softer through the partial breakdown of cell walls, sweeter through the conversion of organic acids and starches to sugars, and more fragrant through the production of aromatic molecules; and it turns from green to rosy yellow through the formation of carotenes and other pigments. All these changes enhance the fruit's appeal to animals that eat and distribute its seeds, and ethylene is behind most of these changes. Most fruits release ethylene gas as they begin to ripen, and the gas further accelerates ripening. This explains why green fruits ripen faster if placed in a plastic bag with ripe fruit.

≋ SENESCENCE, LEAF DROPPING, AND PLANT DORMANCY

Plant life cycles are strongly synchronized with the seasons. Annuals, for example, set seed and die before winter, while perennials often drop their leaves and become dormant during the cold months. In plants, the process of growing old and dying is called **senescence,** while the process that allows mature fruit or dying leaves to fall is called **abscission.**

SENESCENCE AND ABSCISSION

Senescence and abscission are technical terms for the events of autumn we all know and enjoy: the turning of green leaves to glorious shades of red, gold, orange, and purple, followed by the fading and finally the falling of the leaves into large fragrant drifts beneath the trees and bushes (Figure 30.13). Several external and internal events control this sequence. In annuals, the life cycle ends after seed production and dispersal, and the adult plants are programmed to die. Fields of soybeans, for example, will grow yellow and die within a few days of each other— evidence that external cues (cold, dryness, diminishing light) trigger mass senescence. Tomatoes, zinnias, and marigold plants in the garden will also die back after they have produced flowers and seeds, each species on its own schedule and dependent on environmental conditions. Flowering itself is a key to this senescence, since in many annuals, the apex of each shoot converts into a

flower bud and ends the open growth potential of that branch. When the last branch tip has converted to a flower, the plant stops growing and soon begins to senesce.

In perennials, the plant must conserve resources over the dry, cold months of winter, and the **deciduous** growth habit (the dropping of leaves) is very common. In the cold and dark of winter, leaves are much less efficient photosynthetic organs and provide a dangerous avenue for water loss. The leaves may also serve as organs of excretion, storing wastes which then fall from the plant when the leaves drop.

In fall, a deciduous tree such as a sugar maple withdraws valuable nutrients—including sugars and amino acids—from the leaves and transports the nutrients to the trunk and roots for storage. One of the last molecules to be broken down and withdrawn is chlorophyll, the main photosynthetic pigment. However, with an environmental trigger such as a night or two of subfreezing temperatures (or in warmer areas an extended period of dryness), the chlorophyll, too, breaks down and the plant stockpiles the subunits. With the green pigment disappearing over a period of days, the other brilliantly colored red, yellow, orange, or purple leaf pigments that remain can show through. Eventually, these pigments also degrade, the leaves fall, and the plant survives winter in a state of dormancy. Plant scientists suspect that fall conditions influence levels of hormones and that these, in turn, may determine when leaves senesce.

FIGURE 30.13 *Senescence Unmasks the Flaming Colors of Fall.* The breakdown of chlorophyll allows yellow, red, orange, and purple light-gathering pigments to show. Eventually, these, too, senesce, and the colors also fade. A scene in New York State's Catskill Mountains.

Leaf abscission is also governed by environmental cues and the ratios of plant hormones. During the long, sunny days of summer, each leaf produces large quantities of auxin, and this hormone inhibits the formation of an abscission zone, an area of relatively weak cells at the base of the stem, where the leaf will eventually separate from the branch. In the fall, auxin levels drop, enzymes begin to break down the cell walls in the abscission zone one by one, and eventually, a gust of wind or a few pelting drops of rain will send the leaf fluttering to the ground.

DORMANCY: THE PLANT AT REST

Once the leaves have fallen, the plant lives through winter in a state of greatly reduced metabolic activity known as *dormancy*. Dormancy can take several forms. Some perennial plants, like irises and daffodils, die back to roots or bulbs that are protected from desiccation and extreme heat or cold by the earth around them, while others, like chestnuts and linden trees, stand leafless from late autumn until early spring (Figure 30.14). And many plants rest over winter in embryonic form inside dormant seeds. Regardless of exterior form, dormancy is characterized by greatly diminished use of energy and slowed protein synthesis, cell division, and maintenance. Light cycles, perhaps timed by phytochrome, may be involved in the onset of metabolic slowdown, and hormones no doubt relay the message to individual cells and help maintain the state of dormancy.

☷ PLANT PROTECTION: DEFENSIVE RESPONSES TO EXTERNAL THREATS

Plants, as we have seen, usually adjust to changing environmental conditions by growth. One of the biggest environmental threats of all, however, is the onslaught of viruses, bacteria, animal predators, and fungi like the cause of foolish seedling disease in rice. Plants can often protect themselves quite effectively from microbes and other predators by producing physical barriers, such as thorns, leaf hairs, and sticky sap (see Chapter 29); by generating various toxic compounds; and by walling off injured areas.

CHEMICAL PROTECTION AND COMMUNICATION

Most plants produce an arsenal of molecules that help ensure survival by repelling, killing, or interfering with the normal activities of plant-eating organisms. Plants can't run from their enemies, but they can resort to

FIGURE 30.14 *Dormancy: A Key to Winter Survival.* Many perennial plants, such as this American linden, survive the bitter cold and drought of winter by entering a state of dormancy triggered by environmental cues and maintained by plant hormones, including perhaps abscisic acid (ABA).

chemical warfare! In some tree species, up to 50 percent of the plant matter is secondary defensive compounds. And these are so effective that in a temperate forest in a typical year, predators consume only 7 percent of the total leaf surface area.

■ **Four Defense Strategies** Protective compounds with familiar names include cyanide, camphor, cocaine, caffeine, and nicotine—all of which are toxic to animals. There are over 10,000 more such compounds, and they act in four general ways.

Repellents, including volatile compounds (such as camphor) and bad-tasting compounds (such as tannins), can discourage insect pests from eating a plant's leaves or laying eggs on them. Poisons, such as the nicotine in tobacco that can kill various insects or the antibiotics some plants produce when attacked by fungi and bacteria, are another defense against predators. Digestion inhibitors are a third weapon in the protective arsenal. When a hungry caterpillar attacks a tomato plant, for example, it somehow triggers the tomato to produce compounds that block the action of the caterpillar's digestive enzymes. This leads to the predator's malnutrition and death, or at least to its exodus to another plant species. Likewise, citrus fruits make bitter terpenes that stunt the growth of various insects. Finally, some plants generate compounds that disrupt normal insect metamorphosis. The bugleweed, for example, produces a compound that can cause developmental abnormalities and subsequent death in larvae that attack the plant. The flossflower makes a substance that causes insect larvae to metamorphose too early and die.

Almost every plant in nature is at least 2 percent toxic chemicals, and in some insect-damaged fruits and vegetables, that figure may reach 10 percent. Since the protective chemicals produced by plants under stress can harm other animals as well as insects (some may even cause cancer), it is best to stay away from diseased or damaged celery, apples, peanuts, and so on. Interestingly, we take in more toxins with damaged "organic" fruit than with fruit containing a very slight residue of synthetic pesticide.

WALLING OFF INJURED AREAS

Besides producing protective chemicals, plants have a second general response for protection from viruses, bacteria, fungi, animals, or simple physical injury, such as loss of a limb in a storm. Plants cannot heal themselves the way animals can, but they can wall off the damaged area to prevent invaders from gaining access to healthy tissues. This walling off process is called *compartmentalization,* and it involves the production of toxic chemicals in the invaded area and the plugging of nearby xylem and phloem tubes with thick saps or resins to prevent invaders from spreading to other parts of the plant. With the injury walled off, the tree can then continue growing, surviving, and defending itself against new attacks.

≋ CONNECTIONS

To survive, a plant must adapt to, make use of, and defend itself from its environment. It usually flowers when other members of its species do, which leads to successful reproduction. Its seeds normally germinate when weather will favor the tender young sprouts. It usually becomes dormant before conditions turn harsh. And it generally grows toward adequate sources of water and light and fends off attackers. The control of all these events involves a combination of external cues (light, temperature, moisture, gravity, seasonal changes) and internal growth regulators.

Gibberellins, auxins, cytokinins, abscisic acid, and ethylene have dozens of regulatory roles in plants, and alone or in combination, they control the timing of germination, tropisms, light-regulated growth, apical dominance and overall shape, flowering, fruit development and abscission, senescence, leaf drop, and dormancy. Since most of these events involve a change in the plant's physical form, plant hormones are both a practical and a conceptual bridge between anatomy, environment, and survival. In Chapter 31, we explore that day-to-day survival in more detail by focusing on how plants collect nutrients and energy.

NEW TERMS

abscisic acid, page 490
abscission, page 498
apical dominance, page 494
auxin, page 489
cytokinin, page 490
deciduous, page 498
dormancy, page 490

ethylene, page 490
gibberellin, page 488
photoperiod, page 496
phytochrome, page 493
senescence, page 498
tropism, page 491

STUDY QUESTIONS

REVIEW WHAT YOU HAVE LEARNED

1. Name three kinds of plant hormones, and describe the activities of each.
2. How does abscisic acid differ from hormones that promote growth in plants?
3. Explain how moisture, cold, and ground cover affect seed germination.
4. Define tropism, and give two examples.
5. The "sensitive plant" *(Mimosa pudica)* and the Venus flytrap can move. How?
6. If most of a plant's phytochrome has converted to the active form, is the plant likely to be green and healthy? Explain.
7. How does apical dominance help determine a plant's shape?
8. Which hormones are involved in fruit growth, fruit ripening, and fruit abscission?
9. Name four types of compounds that help protect plants.

APPLY WHAT YOU HAVE LEARNED

1. A gardener sprays the herbicide 2,4-D on a patch of poison ivy. What mechanism causes the weed to die?
2. The Christmas cactus is a short-day plant. To induce blooming in late December, you place the plant in early November in a lightproof box and water it only during the day. Does this strategy work? If so, how?

FOR FURTHER READING

Evans, M. L., R. Moore, and K. Hasenstein. "How Roots Respond to Gravity." *Scientific American* 255 (December 1986): 112–119.

Moses, P. B., and N. Chua. "Light Switches for Plant Genes." *Scientific American* 258 (April 1988): 88.

Raven, P. H., R. F. Evert, and H. Curtis. *Biology of Plants.* 4th ed. New York: Worth, 1986.

Rosenthal, G. A. "Chemical Defenses of Higher Plants." *Scientific American* 254 (January 1986): 96.

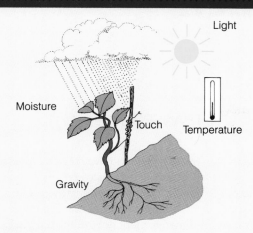

(1) Plants detect environmental stimuli. In response, they grow, change form, and exhibit slow movements.

Hormones	Main activity	
Gibberellins Auxins Cytokinins	Promote or inhibit growth	
Abscisic acid Ethylene	Inhibit growth or promote or inhibit maturation.	

(2) Five kinds of plant hormones promote growth, inhibit growth, or promote or inhibit maturation.

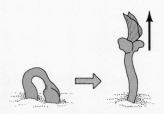

(3) Hormones regulate germination, and this helps ensure a seedling's survival.

Vincent Van Gogh, *Wheatfield with Cypresses*

(4) Hormones regulate growth, maturation, and flowering and seed production. This helps ensure successful reproduction.

(5) Hormones control dormancy, dying, and the falling of leaves and fruit.

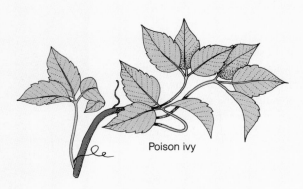

Poison ivy

(6) Plants are protected by means of chemical weapons and the walling off of injured areas.

CHAPTER 31

THE DYNAMIC PLANT

TOMATOES: A CASE STUDY FOR PLANT TRANSPORT

On a golden autumn afternoon in 1820, Robert Gibbon Johnson, a young colonel in the U.S. Army, stunned the people of Salem, New Jersey. He stood on the steps of the town's courthouse, and with a crowd gathered before him, he committed what seemed at the time a suicidal act: He ate a tomato. Tomatoes are members of the plant family that includes poisonous mandrake and deadly nightshade, and this connection must have created real skepticism in nineteenth-century New Jersey. Colonel Johnson, of course, survived, and his act—perhaps his bravest!—convinced the townspeople that the scarlet fruits were not only safe but downright delicious. In fact, the once reviled tomato is now a billion-dollar industry, and because North Americans eat more than 8 million tons of tomatoes each year, this single species contributes more vitamins and minerals to our diet than any other (Figure 31.1).

The tomato is a good case study for an important question: How does a plant procure the energy, water, and materials required for growth, development, self-protection, and reproduction? The answer involves photosynthesis (see Chapter 6) and two topics that are central to the functioning of a dynamic plant: the transport of sugars and water and plant nutrition.

As we consider the uptake and transport of water and nutrients in plants, four unifying themes will emerge: First, plants depend on the special physical properties of water that enable water and nutrients to be pulled from the soil and continuously carried to individual cells throughout the plant body. Second, the avenue for this flow of materials is a network of transport tubes—the vascular system—which integrates all the parts of the plant into a smoothly functioning unit. Third, as in animals, plants have mechanisms for maintaining homeostasis, especially a constant fluid environment inside plant cells. Finally, soil quality is crucial to the growth and survival of plants because each individual must procure all its inorganic nutrients from the soil in which it germinates and grows.

The detailed understanding of how plants function and survive has supported an exciting trend in modern agriculture: the development of improved crops through plant engineering. With the tools of biotechnology, for example, plant scientists have created a pomato—a plant that produces potatoes belowground and tomatoes above.

As this chapter examines how a dynamic plant gets all the materials needed for daily survival and reproduction, it answers the following questions:

- How does a plant take up and transport water to all body cells?
- Which nutrients does a plant need, and how does it get them?
- How does a plant move the products of photosynthesis from areas of production and storage to areas of use?
- How are people improving crops through traditional breeding methods and biotechnology?

FIGURE 31.1 *Tomatoes: America's Favorite Garden Vegetable.* With its deep green leaves, red fruit, and toxic compounds to ward off predators, the tomato is a model of plant survival.

☰ HOW PLANTS TAKE IN AND RETAIN WATER

Over the course of a summer, each tomato plant in a typical backyard garden will consume about 32 gal of water, which is over half the water a gardener weighing 30 times more will require! The explanation for this apparently insatiable "thirst" lies in a fundamental difference between your body and a plant's body: In most animals, fluids tend to recycle through a circulatory system, while in plants, water travels in a one-way path from roots through stems to leaves, then out into the environment again (Figure 31.2). A steady supply of water enables a plant to carry out photosynthesis, to remain crisp and erect with water-filled cells, to stay cool despite a baking sun on outstretched leaves and stems, and to transport substances through the plant.

Land plants would quickly lose all the water they take in if it weren't for waterproof coatings on the aboveground portions of the plant, such as the waxy cuticle or bark layers. But if such layers made an absolutely watertight and airtight seal, the plant could not take in the carbon dioxide required for photosynthesis. The evolutionary solution to this dilemma takes the form of tiny access ports in the leaves and stems, called *stomata*. The opening and closing of stomata are regulated by adjacent *guard cells*, and this system allows sufficient air (with its carbon dioxide) to enter the plant while preventing excess water loss through evaporation.

Let us trace the one-way flow of water from soil to roots to leaves to air and see how the physical properties of water and plant tissues—including stomata—make this life-sustaining movement possible.

HOW ROOTS DRAW WATER FROM THE SOIL

A garden plant such as a tomato absorbs water from the soil into the roots. Soil is a combination of organic matter and weathered particles of the earth's crust. A film of water and dissolved minerals coats these soil particles, and air fills many of the larger spaces between them. Hundreds of thousands of root hairs (see Figure 29.11b) as well as the root-fungal extensions called mycorrhizae (see Figure 16.5) project from the cells of lateral roots into the spaces in soil (Figure 31.3, page 504). These projections have a huge combined surface area and can take up water from the minute reservoirs all around them.

Despite the tremendous importance of water to a plant's survival, water enters roots by passive diffusion. Water diffuses across root hairs and then passes through the root cortex. Some moves through the cytoplasm of root cells, but most passes between the walls of adjacent cells (see Figure 31.3). When it reaches the endodermis (the cell layer that separates the root cortex from the central

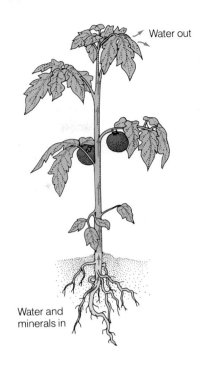

FIGURE 31.2 *One-Way Flow of Water.* Xylem tissue provides a one-way route for the flow of water and inorganic nutrients from the roots through the stem and to the leaves. The transport system and its one-way flow represent an energetically inexpensive way to move materials through a complex organism.

Water out

Water and minerals in

vascular cylinder), water can no longer pass between cell walls because of the waxy belts (Casparian strips) that surround each endodermal cell; the belts act like gaskets that prevent water from flowing from cell to cell, forcing it to pass through the cytoplasm of endodermal cells in order to enter the xylem.

Water diffuses into the root's xylem tissue by osmosis because the fluid inside xylem has a higher concentration of dissolved particles than the surrounding cells. The reason xylem fluid contains a high concentration of dissolved particles is that surrounding cells use energy to actively pump ions into the fluid. The original source of the pumped ions was the soil. Root hair cells absorb the ions, and the ions pass through the cortex and endodermis into the vascular cylinder, primarily via the cytoplasmic route shown in Figure 31.3.

What happens to the water drawn into the xylem by the osmotic mechanism just discussed? The process tends to increase the volume of fluid in the xylem, which must move somewhere. It cannot flow back between the endodermal cells because the waxy Casparian strip prevents both the ions and the water from leaking back out of the vascular cylinder. The only place the fluid can move, therefore, is up the xylem, where it creates a force called *root pressure*. This force can push water out of the leaves of grass, tomatoes, strawberries, and numerous

FIGURE 31.3 *How Water Enters Root Hairs and Crosses into the Vascular Cylinder.* Water and mineral nutrients (arrows) enter root hairs and move either through the cytoplasm of root cells or within the cell walls of root cells. Most of the water flows in the cell walls, while most of the ions flow through the cell cytoplasm. Materials reaching the endodermis via the cell walls are blocked from entry into the root's center by the Casparian strip. Instead, materials must pass through the cytoplasm of endodermal cells. Once in the xylem, water and nutrients can move up to the rest of the plant.

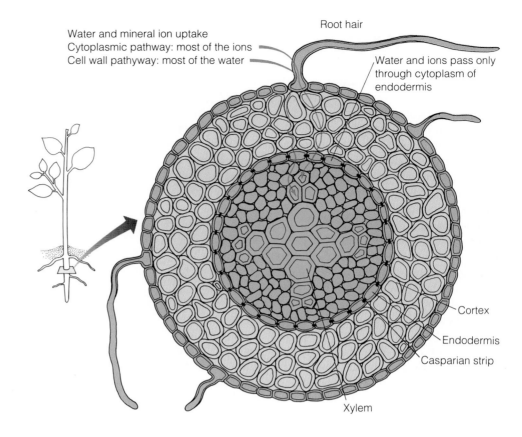

Water and mineral ion uptake
Cytoplasmic pathway: most of the ions
Cell wall pathyway: most of the water

Root hair

Water and ions pass only through cytoplasm of endodermis

Cortex

Endodermis

Casparian strip

Xylem

FIGURE 31.4 *Guttation: Water Forced Upward from Below.* A strawberry leaf in early morning is rimmed with water droplets because of a process called guttation. This oozing of water based on root pressure occurs only in relatively short plants.

other small plants—a process called *guttation* (Figure 31.4)—where the water forms droplets if it does not evaporate quickly enough. Interestingly, measurements of root pressure show that it is much too weak to push water up to the tops of tall trees, and many plants, including pines, do not develop root pressure at all. How, then, do water and nutrients reach the tops of tall plants?

WATER TRANSPORT WITHIN THE PLANT

The tallest living thing, a California redwood, must transport water more than 100 m from roots to highest leaves. Biologists have long known that atmospheric pressure alone can push water up a tube only about 10 m high, and root pressure is even weaker. Conceivably, such phenomena could explain how water reaches the tops of tomato plants, rosebushes, and even small trees, but what allows water to reach the redwood's highest leaves? Experiments show that the main force is not water pushed up from below, but rather water pulled up from the top of the plant.

■ **Transpiration: The Life-Giving Chain of Water** Even though water droplets on tomato leaves prove that root pressure exists in some plants, the major explanation for water transport in plants lies in a fascinating bit of natural engineering: evaporative pull. Tests in plants reveal that when xylem is cut, air moves into the stem and fills the space vacated as water moves up the stem away from the wound, while water moving up from the roots stops at the cut. This observation and numerous other tests led plant scientists to pose the *transpiration pull theory* (also called the *cohesion-adhesion-tension theory*) for water transport in plants. **Transpiration** is the loss of water by evaporation mainly through the stomata on stem and leaves. When stomata are open, water molecules move from a region of high concentration (inside the leaf cells) to a region of lower concentration (the air surrounding the plant) (Figure 31.5a).

The physical properties of water help explain how transpiration moves fluid from roots to leaves. Within a given xylem pipeline, water molecules cling to one another in a long, unbroken liquid chain (cohesion) (review Figure 2.13). When a water molecule evaporates from an open stoma, it is replaced by the next water molecule in line, which pulls up the next water molecule, and so on, until the entire liquid column moves up in the xylem tube one link, and a new water molecule then moves into the roots below (see Figure 31.5a). Because the water molecules pull one another up from above, the entire chain is under constant tension. So long as this tension remains unbroken and evaporation occurs, the water column will keep rising, and moisture will continue to move up through the plant's vascular tissues.

■ **Water Stress: A Break in the Chain** On a hot, dry day, water can be lost from the leaves faster than it can be gained by the roots, and tension on the chain of water molecules becomes greater and greater. Under such circumstances, tension on the chain of water molecules can become so strained that the column simply snaps. In fact, plant physiologists with sensitive microphones can actually hear the snapping and popping inside plants on a hot, dry day!

Once the water column breaks, an air bubble forms inside the xylem tube, and transpiration can no longer pull water from above. If within a short period of time the temperature drops, evaporation slows, or the soil becomes damp again, root pressure from below as well as capillary action within the narrow xylem tubes can help rejoin the ends of the broken water column. If conditions are not reversed quickly enough, however, wilting can destroy part or all of the plant. Likewise, cutting off the bottom 2 inches of the stems in a bouquet of flowers removes ends of xylem tubes blocked by air bubbles. This reestablishes an intact water column and enables the flowers to stay crisp and beautiful (Figure 31.5b).

STOMATA AND THE REGULATION OF WATER LOSS

Since the degree of transpiration is so crucial to a plant's daily survival, water loss is carefully regulated, mainly through the opening and closing of stomata. In the leaf epidermis, each small opening, or stoma, is enclosed by two kidney-shaped guard cells (Figure 31.6a, page 506). Like curved water balloons, guard cells arch away from each other when they swell with water, increasing the size of the opening between them (Figure 31.6b). When guard cells lose water and become flaccid, they slump

FIGURE 31.5 *Transpiration Pulls Water to the Top of the Plant.* (a) Water moves upward through a plant as a result of transpiration, the evaporation of water from leaves, and an upward pull on unbroken chains of water molecules. Water molecules remain strongly linked in chains as a result of cohesion. The evaporation of each water molecule from the leaf creates tension that helps pull the entire chain upward through the stem and bring a new water molecule into the root below. (b) Transpiration helps cut flowers stay crisp. Trimming the stems of cut flowers will remove any xylem containing air bubbles drawn in after the stems were first cut. When the trimmed stems are then placed in a vase of water, a complete water column will be reestablished and the bouquet will last longer.

(a) Tension / Transpiration / Cohesion

(b)

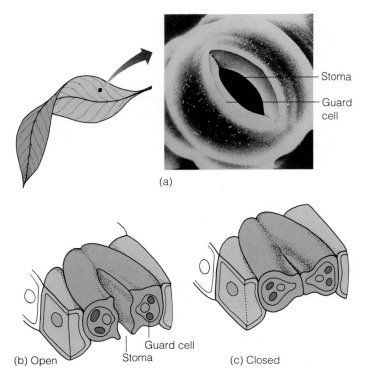

(a)

(b) Open · Guard cell · Stoma · (c) Closed

FIGURE 31.6 *Stomata: Regulators of Water Loss.* (a) The epidermis on the lower side of a leaf is ventilated by stomata. Each tiny opening is bounded by two guard cells. (b) When the guard cells swell with water, the stoma opens. (c) When the guard cells lose water, they slump together, closing the stoma.

together and close off the stoma, blocking transpiration (Figure 31.6c). Although the overriding factor controlling stomatal opening and closing is water gain and loss, several other factors can play a role, including carbon dioxide, light, temperature, and daily rhythms in the plant. These factors affect the level of calcium inside plant cells, and the calcium acts as a molecular messenger for changes inside target plant cells.

Together, the mechanisms of transpiration and guard cell swelling help maintain a constant supply of water to plant cells—a major aspect of homeostasis. For these mechanisms to work, plant cells must absorb appropriate quantities of ions and other mineral nutrients—our next topic.

≈ HOW PLANTS ABSORB MINERAL NUTRIENTS

Around 1600, a Dutch physician performed what may have been the first quantitative experiment in biology.

Jan Baptista van Helmont questioned an assertion made by Aristotle centuries earlier that a plant's body derives most of its substance from soil. To test this idea, he filled a container with 91 kg (200 lb) of dry soil and planted in it a 2.2 kg (5 lb) willow tree. For five years, van Helmont watered the tree with rainwater. At the end of that time, he dug the tree out of its pot, removed the soil and dried it, weighed both soil and tree, and discovered that while the tree had gained 75 kg (164 lb), the soil had lost a mere 27.5 g (2 oz). Thus, van Helmont succeeded in showing quite clearly that the main mass of the willow tree came not from the soil, as Aristotle had suggested, but from some other source.

Today, plant scientists know that plants synthesize organic compounds from the carbon and oxygen in carbon dioxide, from the hydrogen in water, and from small amounts of minerals in soil. Fully 96 percent of a plant's dry weight is made up of carbon, oxygen, and hydrogen. The dozen or so chemical elements that make up the remaining 4 percent, however, are equally essential to plant survival.

WHAT ARE THE PLANT NUTRIENTS?

Plants require at least 16 different chemical elements, the so-called *essential elements*, which different species require in different amounts. All plants require the **macronutrients** in relatively large amounts and the **micronutrients** in much smaller amounts. By drying many types of plants, weighing the dried plant matter, and analyzing it for chemical content, plant physiologists were able to tabulate information on the nine macronutrients and seven micronutrients needed by all plants (Table 31.1). Plants use both macro- and micronutrients in amounts roughly comparable to the levels animals require. And just as animals suffer vitamin and mineral deficiency diseases if deprived of essential nutrients, plants also show specific symptoms relating to specific deficiencies.

■ **Macronutrients** The macronutrients are a plant's fundamental constituents, making up most of the atoms in carbohydrates, proteins, lipids, and nucleic acids (see Chapter 2). Because we have already discussed the roles of carbon, oxygen, and hydrogen in biological molecules and reactions in Chapters 2 through 6, we will concentrate here on the other six macronutrients.

After carbon, oxygen, and hydrogen, nitrogen is the most important macronutrient, being an essential component of amino acids (and hence proteins), chlorophyll, coenzymes, and nucleic acids. Nitrogen is frequently the most important growth-limiting nutrient: The less nitrogen available, the slower the plant grows. And although gaseous nitrogen (N_2) makes up 78 percent of the earth's atmosphere, plants cannot use it directly. Instead, **nitrogen fixation** must occur: The nitrogen must be *fixed*, or

TABLE 31.1 ≋ PLANT NUTRIENTS AND THEIR FUNCTIONS

NUTRIENTS (PERCENT OF DRY WEIGHT)*	LOCATION/FUNCTION
Macronutrients	
Carbon (45.0)	In all organic compounds
Oxygen (45.0)	In most organic compounds, including all sugars and carbohydrates
Hydrogen (6.0)	In all organic compounds
Nitrogen (1.0–4.0)	In proteins, nucleic acids, chlorophyll, and coenzymes
Potassium (1.0)	Involved in activating enzymes, protein synthesis, and regulation of osmotic balance
Calcium (0.5)	In cell walls and starch-digesting enzymes; regulates cell membrane permeability
Magnesium (0.2)	In chlorophyll and many cofactors
Phosphorus (0.2)	In nucleic acids, some coenzymes, ATP, and some lipids
Sulfur (0.1)	In proteins, some lipids, and coenzyme A
Micronutrients	
Iron (0.01)	In electron transport molecules; involved in synthesis of chlorophyll
Chlorine (0.01)	Essential to photosynthesis
Manganese (0.005)	Essential to photosynthesis; activates many enzymes
Boron (0.002)	Unknown
Zinc (0.002)	In some enzymes; involved in protein synthesis
Copper (0.0006)	In some enzymes
Molybdenum (<0.0001)	Essential to nitrogen fixation and assimilation
Elements Essential to Only Some Plants	
Silicon (0.25–2.0)	In the cell walls of grasses and horsetails
Sodium (Trace)	Essential to a few desert and salt-marsh species
Cobalt (Trace)	Essential for nitrogen fixation

*Concentration of nutrients expressed as percentage of dry weight in a typical plant. Actual proportions vary greatly from species to species.

converted from the simple gas N_2 to some other form, such as *ammonia* (NH_3) or *nitrate ions* (NO_3^-), which can then be modified further and incorporated into plant compounds.

Farmers and home gardeners add fixed nitrogen to the soil in the form of nitrate-containing commercial fertilizers. Wild-growing plants, however, do not have such a benefit and rely instead on nitrogen-fixing bacteria, including cyanobacteria, which can convert molecular nitrogen to usable forms. Many of these microorganisms live independently in the soil. Some of the most interesting ones, though, live in the root cells of certain vascular plants (Figure 31.7, page 508). Peas, beans, alfalfa, clover, lupine, and other members of the pea family (known as **legumes**) have swellings, or **nodules,** on their roots that house nitrogen-fixing bacteria. If root nodules produce ammonium ions in excess of the plant's needs,

they are released into the soil, where they join the ammonium ions generated by the free-living nitrogen-fixing bacteria. Plants that do not house nitrogen-fixing bacteria can take up ammonium ions in the soil from these sources. Once in the plant, ammonium is the starting point for the biosynthesis of amino acids and many other nitrogen-containing biological molecules.

The formation of root nodules is a classic example of symbiosis: The plant supplies the bacteria with carbohydrates, while the bacteria provide fixed nitrogen. The excess ammonium ions released from nodules is a classic example of community interrelationships among plant species. For centuries, farmers have rotated crops to take advantage of such relationships. Although they may have been unaware of the microbiological basis, they observed

(a)   (b)

FIGURE 31.7 *Nitrogen-Fixing Bacteria in Root Nodules Help Nourish Peas, Beans, and Alder Trees.* Nodules on the roots of this alfalfa plant (a) look granular when magnified about 400 times by a scanning electron microscope (b). Each nodule contains thousands of cells of the *Rhizobium* bacterium, capable of fixing atmospheric nitrogen to ammonia (NH₃).

that if they grew clover or alfalfa one year, the following year's crop of corn or wheat would grow more luxuriantly. Likewise, rice farmers have encouraged the growth of water ferns in their flooded rice paddies because cyanobacteria living symbiotically in the ferns fix atmospheric nitrogen and enrich the growth of the rice plants. In waterlogged bogs, where soils tend to be too acidic for bacteria to survive, *insectivorous plants,* such as the Venus flytrap and the pitcher plant, have evolved the ability to gain needed nitrogen by trapping and digesting insects.

Plants often require 4 to 40 times more nitrogen than they do the remaining five macronutrients, but these elements are still essential to normal plant growth and development. Potassium helps regulate osmosis in plant cells such as guard cells and also helps activate enzymes, including those involved in protein synthesis. Potassium deficiency causes curled, mottled, or burnt-edged leaves (Figure 31.8a). Calcium acts as an intracellular messenger to control cell membrane permeability and thus plays a role in opening and closing the stomata, directional growth in plant cells, responses triggered by light-absorbing pigments (see Chapter 30), and gravitropism. Calcium also has a structural role; it is an important component of *pectin*, a substance that glues adjacent cells together, prevents young plants from making brittle cell walls, and puts the "jell" in jams and jellies. Experiments document that tomato plants deficient in calcium produce fruit susceptible to blossom-end rot disease.

Magnesium is considered a macronutrient because atoms of the element occur in chlorophyll molecules and

in the cofactors of many kinds of enzymes, while phosphorus occurs in the backbone of DNA and RNA molecules, in ATP and other high-energy compounds, and in membrane phospholipids. Tomato seedlings deficient in phosphorus have purple leaves (Figure 31.8b). Because sulfur is an important component of two amino acids, plants require it for building most proteins, as well as for manufacturing some fats and coenzymes (see Chapter 5).

■ **Micronutrients**   Micronutrients are required in very small amounts to support healthy growth (see Table 31.1). Because plants require such tiny quantities, deficiencies of the micronutrients are rare, but they do occur. Iron, for example, is involved in the synthesis of chlorophyll. A tomato plant with an iron deficiency does not make enough chlorophyll to mask the yellow pigments in leaves (Figure 31.8c). A deficiency of copper (present in chloroplasts and certain enzymes) can result in severe deformation of stems, leaves, and fruits in many plant species. A deficiency of chlorine in a tomato plant will stunt the roots and fruit and wilt the entire plant. Because zinc plays a role in protein synthesis, zinc-deficient apple and peach trees become stunted and grow miniature leaves. Soil is the natural source of the minerals that prevent these conditions.

SOIL: THE PRIMARY SOURCE OF MINERALS

Soil is the source of all macro- and micronutrients beyond the carbon, oxygen, and hydrogen that plants take in from air and water. Soil is so fundamental and ubiquitous that aside from farmers and geologists, few of us give it any real thought. But soil composition has a major influence on the kinds of plants (and indirectly, the kinds of animals) that can grow in a particular region of our planet.

Soil is a mixture of organic and inorganic materials that starts with bedrock, relatively unbroken and

(a)   (b)   (c)

FIGURE 31.8 *Nutrient Deficiencies in Tomato Plants: Mineral Deficiencies Produce Distinctive Symptoms.* (a) A potassium deficiency causes curled, mottled leaves. (b) Phosphorus deficiency leads to purple leaves in a tomato seedling. (c) Iron deficiency renders the fine leaf veins green, but leads to yellowing in the rest of the leaf.

unweathered rock. Over time, the action of water, wind, heat, and cold disintegrates and decomposes bedrock to produce the inorganic parts of soil—particles varying in size from coarse sand and silt to small clay particles (Figure 31.9). Bacteria, fungi, algae, lichens, and plants extract minerals from rocks, sand, silt, and clay and convert them to organic material, or *humus*, which ultimately becomes part of the soil when the organisms excrete waste products and when they die.

Air spaces between soil particles are crucial to plant life (see Figure 31.9). In a soil with a mixture of particle sizes, the spaces contain about half water and half air, with the water forming a continuous film over the soil particles. The size of the soil particles, however, determines the soil's water-holding capacity, so that soil made up primarily of coarse sand tends to hold water poorly, while soil made up mostly of clay particles holds water tenaciously. Plants such as cacti that grow in sandy soil must absorb water quickly and store it, while plants that grow well in clay soil must have roots resistant to dense, soggy, unaerated surroundings. The best agricultural soils are deep layers of *loam* possessing a high mineral and humus content and a mixture of particle sizes.

PATHWAYS OF NUTRIENTS INTO THE PLANT: ROOT HAIRS AND MYCORRHIZAE

Several things must happen before a plant can utilize mineral nutrients from soil. Mineral molecules must first dissolve in the layer of water surrounding soil particles. Then the minerals must move into the root and be distributed to all parts of the plant. Specifically, minerals must pass through the plasma membranes of the root hairs, move through the cells of the root cortex and the endodermis cytoplasm, and be secreted into the xylem (review Figure 31.3). There, the powerful pull of transpiration can lift them (along with water) through the xylem pipelines, and they can be transported to all plant organs. The active transport of ions against a gradient of concentration requires a good deal of energy. Expending this energy to procure necessary inorganic nutrients is simply a cost of staying alive.

The development of millions of fine roots hairs provides an enlarged surface area for absorbing water and nutrients. In addition, the symbiotic association of roots and highly specialized mycorrhizal fungi can expand the absorptive surface area still further. Most land plants have such associations, as Chapter 16 explained. When mycorrhizal fungi infect plant roots, they often cover the roots with a spongy mantle of threadlike hyphae that resembles a string sculpture. The hyphae can extend outward up to 8 m (25 ft) from the root and even penetrate the roots of nearby plants. In nutrient-poor soils, mycorrhizal fungi provide their hosts with greater concentrations of inorganic nutrients as well as growth-promoting hormones. Some plants, including cultivated citrus and pine, grow far more efficiently with mycorrhizae.

Mycorrhizae occasionally interconnect unrelated plants and even allow a few odd plant species to parasitize others. Indian pipe, a small ghostly white plant of the forest floor that lacks chlorophyll, has mycorrhizae on its roots that connect it to the roots of nearby plants; the fungi absorb nutrients from the parasitized plant (Figure 31.10, page 510).

Since absorbing macro- and micronutrients requires a considerable expenditure of energy, plant roots must have a reliable energy supply. Unable to photosynthesize, root cells must receive energy compounds from the leaves. Thus, the downward transport of sugars is crucial to survival and a subject of great importance to understanding plant physiology.

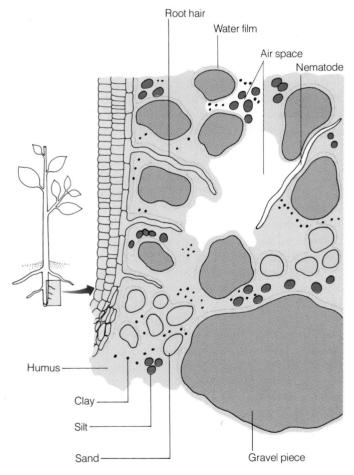

FIGURE 31.9 *Soil: A Source of Air, Water, and Nutrients for Plants.* Soil has inorganic and organic particles, as well as pore spaces filled with air and lined with water. Root hairs probe those spaces and take in water and minerals that dissolve from soil particles. Burrowing animals such as nematode worms and earthworms make tunnels that help aerate the soil.

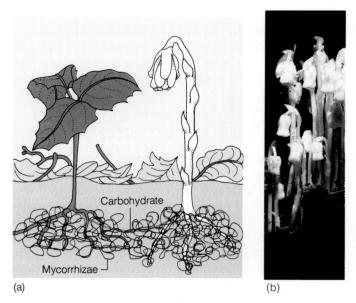

(a)   (b)

FIGURE 31.10 *The Fungal Connection: A Mycorrhizal Bridge Transfers Sugars from Green Plant to Indian Pipe.* (a) The Indian pipe plant (*Monotropa uniflora*) has an obligate relationship with mycorrhizal fungi that are associated with a second plant, in this case a green, actively photosynthesizing angiosperm. The fungus forms a bridge that transfers carbohydrates from the photosynthetic plant to the Indian pipe. (b) Indian pipes pop up on the forest floor, and lacking chlorophyll, look ghostly white.

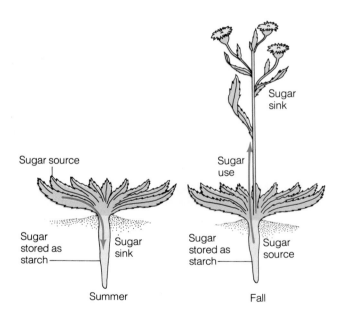

FIGURE 31.11 *Seasonal Sugar Transport in the Tarweed.* In summer, sugars produced by photosynthesis in the tarweed's leaves are transported downward to the roots and stored as starch. In fall, the transport is reversed: Sugars are chemically cleaved from the stored starch and carried upward to fuel the biosynthesis of materials for an elongating stem, flowers, and seeds.

☰ MOVING SUGARS AND OTHER NUTRIENTS THROUGHOUT THE PLANT

Plants have an inherent distribution problem: They generate sugars in their leaves (the *source* of sugar), but tend to store organic nutrients in their roots and use those nutrients in their growing tips, flowers, and fruits (the *sinks* for sugar). What's more, sources and sinks can change as the season progresses. In the tarweed, the root is a sink early in the growing season, receiving and storing sugar produced in the leaves; but in the fall, the root turns from sink to source, as stored carbohydrates are transported up from the root and used to grow a tall stem, flowers, and seeds (Figure 31.11). Since plants lack a circulatory system, what mechanism do they have for disbursing photosynthetic products? The answer is translocation.

TRANSLOCATION: HOW SUGAR MOVES FROM SOURCE TO SINK

Plant scientists devised a clever method for studying **translocation,** the movement of nutritional materials in the phloem of plants. Phloem tubes are so delicate that puncturing them even with a fine glass probe or needle stops the flow of solutes. The small greenish insects called aphids, however, can suck plant saps from phloem tubes without disrupting them; and plant physiologists found that if they allow an aphid to insert its feeding tube into a phloem pipeline and cut away its body, the phloem's contents will ooze out of the feeding tube and can be studied (Figure 31.12). With this technique, the scientists discovered that plant sap contains up to 30 percent sugars (mostly sucrose) and about 70 percent water.

Plant scientists also discovered that they could expose a plant's leaves to radioactive carbon dioxide, then trace the path of the carbon through the phloem once it was incorporated into sugars via photosynthesis. In this way, they found out that sugar flow involves a mass movement of phloem fluid based on **bulk flow**—the overall movement of a fluid from an area of high pressure to an area of low pressure, with all the molecules moving together in the same direction. Plant scientists think that sugars produced in source regions such as photosynthesizing leaves are loaded into the phloem's sieve tube elements by the companion cells (Figure 31.13). This active transport increases the solute concentration in the phloem, and water follows by osmosis from the nearby xylem cells. The influx of water plumps up the phloem cells, increasing their turgor pressure and forcing the sugary solution out of the sieve plates at the ends of each sieve tube member and away from the leaf.

Meanwhile, root cells remove organic solutes from the phloem, and the solute content becomes so low that water

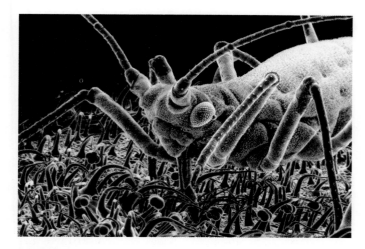

FIGURE 31.12 *Aphid Taps Contents of Phloem.* In this scanning electron micrograph, an aphid pierces the veins of a tomato leaf with its hypodermic syringe-like feeding tube.

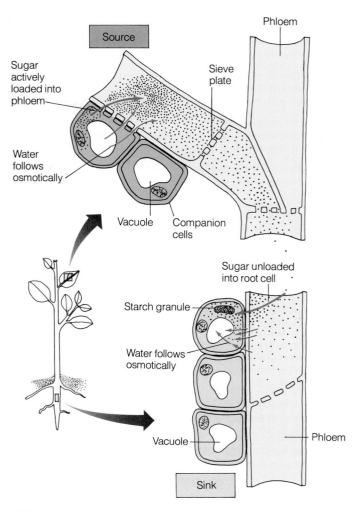

FIGURE 31.13 *Translocation: How Plants Transport Sugars.* In leaves, companion cells load sugars into phloem tubes, water follows osmotically, and sugar solution is forced through the sieve plates and down a pressure gradient. This gradient is formed because root cells with lower concentrations of sugars and amino acids take in the nutrients, making the solution in the phloem tubes more dilute. As water leaves the phloem in a sink area, the turgor pressure in nearby phloem cells also drops.

now flows out of the phloem tubes and back into the xylem tubes, where it is carried upward again by transpirational pull. Ordinary water pressure and the loading activities of companion cells are thus behind the movement of sugars, as well as amino acids and a few mineral ions, from sources to sinks.

In the early spring, a similar mechanism probably causes plant sap to rise in the phloem of biennials and perennials. At that time, the roots begin breaking down stored starch into sugar and loading it into the phloem's sieve tube members. Water from the xylem follows, and as the roots bring in more water from the soil, enough pressure is created to push the sap up the phloem into the trunk of the tree. New Englanders carefully tap into the phloem of maple trees in early spring, drain off some of the sweet maple sap, and boil it down to make maple syrup.

In a sense, sugar translocates itself, since its production and use create osmotic pressure and bulk flow. Other nutrients, however, are moved passively by the flow of fluid in either the xylem or the phloem.

TRANSPORT OF INORGANIC SUBSTANCES

Nutrients other than carbon, oxygen, and hydrogen—including all the other nutrients listed in Table 31.1—can be transported in two different ways. Some, such as calcium, move only in the xylem; they are simply carried along in the column of water that moves upward from the roots to the leaves via transpiration. These nutrients are deposited permanently wherever they leave the xylem and enter a living cell, and they cannot be redistributed to the other parts of the plant. Others, such as sulfur or phosphorus, can be redistributed, via the phloem, to new tissues as needed, carried along passively with the sugars and other phloem contents.

※ ENGINEERING USEFUL PLANTS

With their knowledge of plant anatomy, plant development, plant hormones, and the role of water and nutrients in plant survival, modern plant scientists are approaching—and entering—a new frontier: biological engineering of plant species with dramatically different traits.

Because so many of the world's people are hungry, and because remaining unplanted lands are only marginally useful, plant breeders are constantly working to increase the yield and nutritional content of crops grown under increasingly adverse conditions. They are attempting to combine the hardiness of wild plants that grow in dry, salty, cold, or barren areas with the traits of the best agricultural crops to create, for example, wheat, rice, corn, or tomato plants with the drought resistance of cactus, the salt tolerance of marsh grass, the nitrogen-fixing ability of legumes, and the ability to resist insect pests, diseases, and herbicides. Let us explore the techniques and early achievements of this new field.

NEW TECHNIQUES OF ARTIFICIAL SELECTION

The choice foods we have today—the sweet pears, high-protein grains, juicy strawberries, and others—are the result of farmers carefully selecting seeds from plants with the best individual traits and planting the seeds for the next generation. New techniques such as tissue culture, however, are speeding up the process of artificial selection.

Plant breeders are often eager to select traits displayed by a particular individual, but these traits can disappear owing to the genetic recombination that occurs during normal sexual reproduction. **Tissue culture,** or growing new identical plantlets from body (somatic) cells, avoids the problem of genetic recombination. Researchers begin the tissue culture process by placing a bit of plant tissue cut from a leaf or stem in a nutritious culture medium in a warm, well-lighted room (Figure 31.14a). Soon the cells of the tissue form an unorganized mass of cells called a *callus* (Figure 31.14b). The plant breeder then breaks up the callus and places bits of it in a new culture medium with hormones that promote differentiation. There, some of the callus cells grow and differentiate into embryos and then tiny plants (Figure 31.14c). The plants can be raised and studied directly, or the embryos can be packaged as synthetic seeds for later use. Either way, since all these plants are derived from the body cells of the parent plant, theoretically they all have identical genes and provide numerous copies of the desired original. Some plant nurseries already propagate certain commercial crops using tissue culture—for example, strawberry plants that are free of viral infections.

For unknown reasons, the DNA of many plants grown from the initial tissue-cultured callus contains mutations. Wheat, carrots, celery, and tomatoes produce all sorts of genetic variants in tissue culture—variants that would require years to find by traditional selection techniques. This tendency of tissue-cultured plants to display new and heritable traits is called *somaclonal variation* ("soma" referring to the body cells of the plant, and "clonal"

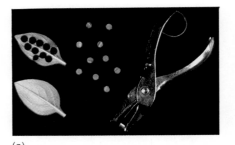

(a)

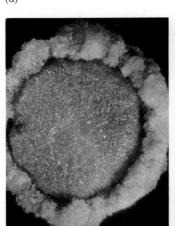

(b)

(c)

FIGURE 31.14 *Plant Tissue Culture: Growing Whole Plants from Individual Cells or Tissues.* (a) Researchers remove a bit of tissue from a plant (in this case a petunia) with desirable traits by punching out small disks of leaf tissue. (b) Next, they culture clumps of callus. (c) Then they separate individual cells from the clump and induce some of the cells to grow into new petunia plants.

because each variant population arises from a single initial mutant cell). From such an array of varying cells, breeders select those that express a desired trait. If they are looking for plants with tolerance to saline soils, for example, they can treat the culture with salt. Those cells with genes for salt tolerance live; the rest die. In some species of plants, the surviving salt-tolerant cells can then be treated with hormones so that they grow and differentiate into whole plants, from which the experimenters can collect seeds.

In one of the earliest somaclonal variation experiments, researchers grew 230 plants from the cells of a single tomato plant. Among the variants, they found 13 separate mutations, several of which turned out to be commercially useful. Some mutations produced novel bright orange and yellow tomatoes (Figure 31.15), while others produced tomatoes with unusually high solid content, making them valuable to canners of tomato paste.

Selecting new plant lines from cell cultures has certain limitations. With present technology, biologists cannot grow some kinds of plants from single cells. Nor can

FIGURE 31.15 *Building New Tomato Varieties.* In a somaclonal variation experiment, researchers used tissue culture techniques to generate 230 tiny tomato plants. From these they searched out variants and found 13 mutants with interesting traits, including bright orange and yellow fruits and larger, brighter tomatoes with a high content of solids and less juice than a normal tomato.

geneticists select for traits that are not expressed in single cells, such as yield or fruit color. Finally, both traditional selection techniques and modern tissue culture techniques require that some cells in the population already have genes for the desired traits. To select a salt-tolerant tomato from somaclonal variants, for example, requires that some tomato cells already be somewhat salt-tolerant. What happens, though, if tomato cells invariably lack genes for salt tolerance? The answer is that plant researchers can use the same kind of gene transfer techniques that allowed geneticists to develop giant mice (see Chapter 9).

GENETIC ENGINEERING IN PLANTS

Genetic engineering offers a pathway to varieties with more profound alterations than traditional techniques are ever likely to produce. In a single generation, a plant geneticist can introduce into a species the extreme expression of a trait or a completely novel trait that might have taken hundreds of generations to develop through standard breeding approaches. Work is now under way on several fronts, and researchers are already reporting initial successes.

Plant geneticists have outfitted petunia plants with a gene for resistance to the widely used herbicide glyphosate. Workers can now spray a field of the resistant plants with the herbicide, and weeds will die off but the petunia crop will remain unharmed. Geneticists have also produced and field-tested an engineered variety of the plant-infecting bacterium *Pseudomonas syringae*. Normal *P. syringae* bacteria trigger the formation of ice crystals on crop plants such as potatoes and lead to millions of dollars of damage from early frosts in autumn before the tubers are harvested. Now, genetic engineers have produced a variety of *P. syringae* that, through the deletion of a single gene, does not induce frost formation and can help protect plants in the field. Some environmentalists expressed concern over the spraying of the engineered organisms onto potato plants in open fields (see the box on page 514), but tests seem to show that the altered bacteria remain on the intended host plants.

Ultimately, genetic engineers hope to develop varieties of wheat, corn, and rice that, like meat and dairy products, contain high quantities of all essential amino acids. They also hope to engineer nonleguminous crop species that can form root nodules filled with nitrogen-fixing bacteria, plus dozens of other plant species with increased resistance to insects, diseases, and herbicides, increased nutritional value, and increased tolerance to harsh environments.

Each of the three ways to produce new plant varieties—standard plant breeding, artificial selection from tissue cultures, and genetic engineering—has advantages and disadvantages. Traditional plant breeding is slow, but is well understood, predictable, and controllable. Tissue culturing is much faster, but traits selected for in tissue culture often bring along undesirable "riders" such as sourness or susceptibility to disease. Genetic engineering has unlimited potential for combining traits that never occur together in nature, but only traits controlled by single genes or very small groups of genes can be added or subtracted. Clearly, the modern tools of plant breeders supplement—but do not replace—traditional breeding techniques.

≋ CONNECTIONS

The tomato plant may look static from the outside, but it is an undeniably dynamic organism: Water and nutrients move the length of the plant from roots to stems and leaves and sometimes back again. Leaves track the sun's movement through the sky. Flowers emerge from buds. Fruits grow from flower bases, becoming plump and red. Toxic compounds form in the leaves and stems and fend off potential predators. And plant hormones coordinate all these activities in response to environmental cues, including light, temperature, day length, and water availability. The next part of this book, dealing with evolution and ecology, fixes our attention more directly on the dynamic interactions of plants, animals, and other organisms with each other and with the physical world.

NEW CROPS AND NEW RISKS

Genetic engineers are promising a new era for agriculture: genetically improved crops that will grow faster and bigger, require fewer fertilizers and chemical sprays, and feed people and livestock more nutritiously. The specific claims are impressive:

- Corn, wheat, and beans with high-quality proteins that could reduce our reliance on meat
- Potatoes with new genes that protect the plant against viral diseases
- Tomato and tobacco plants that produce insect-fighting compounds
- Soybeans, cotton, corn, and sugar beets that can resist plant-killing sprays and thus allow farmers to weed their fields by spraying herbicides on crops and weeds alike.

Each year, the list of promises grows longer and more impressive. Many biologists feel that a major hope for feeding the world's hungry rests with such new-age crops. At the same time, certain respected plant scientists and ecologists have raised environmental concerns about recombinant plants.

Will engineered plants remain safe to eat? Traditional plant breeders in the 1960s generated a strain of potatoes that was resistant to insects because of a toxin made by the plant. Unfortunately the toxin harmed people as well. Some observers worry that such mistakes could occur more easily as we shift genes from plant to plant with our sophisticated recombinant DNA techniques, others point out that genetic engineers may be less likely to make such errors because they work with well-defined genes.

Will engineered plants be as nutritious as their nonengineered counterparts? Some tomato strains have lost vitamins and minerals as plant breeders have produced hybrids with tougher skins for shipping, more pulp

for cooking, and similar improvements. Again, such nutritional losses have occurred through conventional breeding techniques, but could they happen more easily when crop plants are artificially fitted with new, radically different genes?

Giving plants a new genetic ability to resist weed killers could indeed allow farmers to weed their crop fields by spraying with herbicides. But some ecologists worry that this will encourage farmers to use more herbicide sprays, with their sometimes damaging environmental side effects, and will subtly discourage the development of nonchemical alternatives.

Likewise, it seems a good idea to give crops new genetic resistance to insect pests. But some observers point out that such resistance is never permanent: Natural selection will favor those few individual insects that can attack the resistant plants, and those pests will eventually proliferate and become a problem anyway. Still, some argue that we could stay one step ahead of evolution by continually engineering plants with new kinds of resistance to replace those to which insects develop their own tolerance.

Perhaps most significantly, some ecologists are skeptical about the wisdom of introducing recombinant organisms (in this case, crop plants with genes spliced in from other species) into new ecological areas. Field tests of engineered potatoes, strawberries, and a few other plants have not shown significant dangers so far. However, ecologists have reams of gloomy data on the intentional introduction of so-called exotics—species that are brand new to an area. In the United States, for example, people have introduced about 5800 species of nonnative agricultural and ornamental plants, and of those, 128, or 2 percent, have become serious pest

weeds. Moreover, of the 40 introduced birds and mammals, 19, or nearly 50 percent, have displaced or destroyed native species and become environmental problems. Many ecologists are worried that engineered crop plants—genetic exotics—could also escape and become dangerous. One fear is that an engineered plant could cross-pollinate with a weed growing near the crop field, and then the crop/weed hybrid—perhaps now bearing a new gene that gave it resistance to herbicides or natural insect predators—could begin to spread and take over large areas of cultivated land. Others argue, however, that the risks are really no greater than with normal exotics, and the potential benefits are much larger.

Because of these and other ecological concerns, some biologists are urging caution in the development, regulation, and release of engineered crops. They do not want to keep farmers—and hungry people—from realizing all the potential benefits of genetically improved grains, fruits, and vegetables. At the same time, however, they want to avoid additional ecological risks in a world already beset with environmental insults.

You can learn more about the benefits and risks of engineered crop plants from the following articles:

FOR FURTHER INFORMATION

Ellstrand, N. C., and C. A. Hoffman. "Hybridization as an Avenue of Escape for Engineered Genes." *Bioscience* 40 (June 1990): 438–442.

Pimentel, D., M. S. Hunter, et al. "Benefits and Risks of Genetic Engineering in Agriculture." *Bioscience* 39 (October 1989): 609–614.

Wickelgren, I. "Please Pass the Genes." *Science News* 136 (19 August 1989): 120–124.

NEW TERMS

bulk flow, page 510
legume, page 507
macronutrient, page 506
micronutrient, page 506
nitrogen fixation, page 506

nodule, page 507
soil, page 508
tissue culture, page 512
translocation, page 510
transpiration, page 504

STUDY QUESTIONS

REVIEW WHAT YOU HAVE LEARNED

1. What is the function of root hairs? What is the role of the plant's endodermis?
2. How does root pressure cause water to rise in a plant?
3. Explain how transpiration works to move water up through a plant.
4. What factors affect the opening and closing of stomata?
5. How does a legume get needed nitrogen? How does a nonlegume get nitrogen?
6. List five plant macronutrients and five micronutrients, and give the major roles of each.
7. How does soil form?
8. Trace the pathway by which minerals enter a plant.
9. What is a source and what is a sink of sugars in a plant?
10. Explain how a biologist can use tissue culture to grow genetically identical plants; to select variations.
11. Name three accomplishments and three risks of plant genetic engineering.

APPLY WHAT YOU HAVE LEARNED

1. A farmer decides to plant alfalfa in a field where he previously grew corn. Why?
2. A Vermont family makes maple syrup early in the spring. What raw material do they gather, and from where?
3. A plant breeder tries using tissue culture to develop salt-tolerant tomato plants. What might cause the experiments to fail?

FOR FURTHER READING

Crosson, P. R., and N. J. Rosenberg. "Strategies for Agriculture." *Scientific American* (September 1989): 128–135.

Fischhoff, D. A., K. S. Bowdish, F. J. Perlak, P. G. Marrone, S. M. McCormick, J. G. Niedermeyer, D. A. Dean, K. Kusano-Kretzmer, E. J. Mayer, D. E. Rochester, S. G. Rogers, and R. T. Fraley. "Insect Tolerant Transgenic Tomato Plants." *Bio/technology* 5 (1987): 807–813.

Gasser, C. S., and R. T. Fraley. "Genetically Engineering Plants for Crop Improvement." *Science* 244 (16 June 1989): 1293–99.

Hitz, W. D., and R. T. Giaquinta. "Sucrose Transport in Plants." *Bioessays* 6 (1987): 217–221.

Morrison, R. A., and D. A. Evans. "Haploid Plants from Tissue Culture: New Plant Varieties in a Shortened Time Frame." *Bio/technology* 6 (1988): 684–689.

Strange, C. "Cereal Progress via Biotechnology." *Bioscience* 40 (January 1990): 5–9.

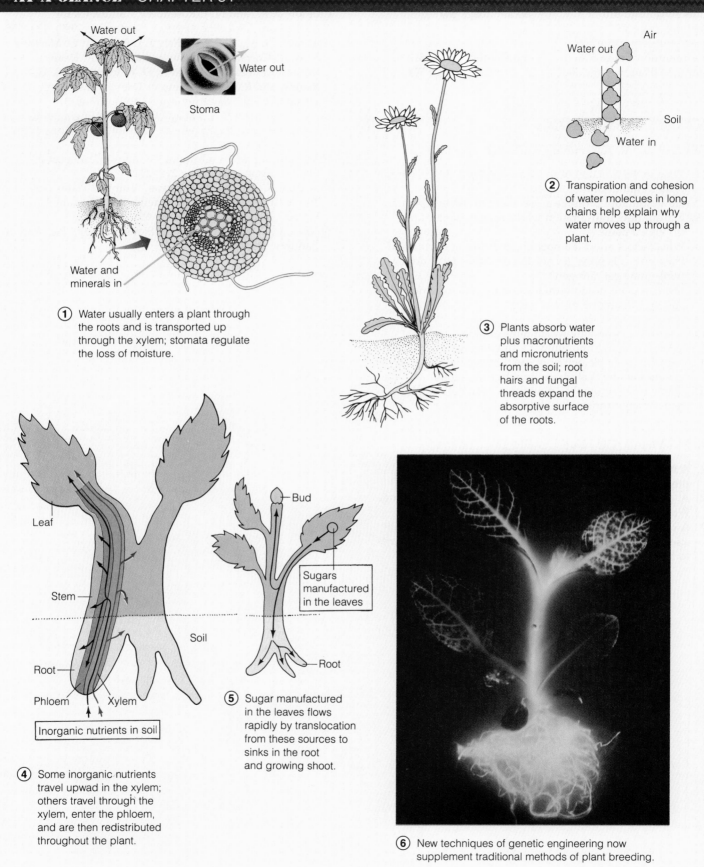

Water out

Water out

Stoma

Water and minerals in

① Water usually enters a plant through the roots and is transported up through the xylem; stomata regulate the loss of moisture.

Air

Water out

Soil

Water in

② Transpiration and cohesion of water molecues in long chains help explain why water moves up through a plant.

③ Plants absorb water plus macronutrients and micronutrients from the soil; root hairs and fungal threads expand the absorptive surface of the roots.

Leaf

Stem

Soil

Root

Phloem   Xylem

Inorganic nutrients in soil

④ Some inorganic nutrients travel upwad in the xylem; others travel through the xylem, enter the phloem, and are then redistributed throughout the plant.

Bud

Sugars manufactured in the leaves

Root

⑤ Sugar manufactured in the leaves flows rapidly by translocation from these sources to sinks in the root and growing shoot.

⑥ New techniques of genetic engineering now supplement traditional methods of plant breeding.

PART SIX

INTERACTIONS: ORGANISMS AND ENVIRONMENT

CHAPTER 32

THE GENETIC BASIS FOR EVOLUTION

CHEETAHS: SPRINTING TOWARD EXTINCTION

Cheetahs of the African savanna search for prey from an elevated lookout, such as a rock or fallen tree limb. When a young zebra or Thompson's gazelle strays too close to the predator's perch, the cheetah noiselessly descends and approaches the prey, well camouflaged in the long, uneven grassland shadows by its buff and brown-spotted coat. The cheetah then crouches and springs toward the startled animal. In just one second and with two bounding strides, the world's fastest sprinter is running 72 km/hr (45 mph), closely pursuing the terrified antelope as it cleaves a zigzag path through the grass (Figure 32.1). Chances are good that the feline hunter will soon bring down the prey with a powerful swipe of a forepaw, then strangle the antelope with a killing bite to the throat.

Compared to the cheetah—the world's fastest-running animal—even a Corvette seems sluggish. While the lithe cat can be running at its top speed of 116 km/hr (72 mph) within seconds, the Corvette takes three times as long to reach that same speed. (The car would eventually overtake the cat because the cheetah doesn't have enough strength to sustain the chase.)

Surprisingly, despite the cheetah's near perfection as a sprinter and its enormous success as a hunter, the species is dangerously near to extinction. While the animals once enjoyed a worldwide distribution, they are now limited to a few sites in southern and eastern Africa and to a total population of less than 20,000. What's more, among the remaining cheetahs, every individual is very much like every other. The lack of genetic diversity is similar to what one would expect if humans married their cousins generation after generation, and

it has serious negative consequences: Cheetahs are highly susceptible to disease; they have a low birthrate; and although they can outrun a fast sports car for the first mile or two, they become so exhausted by the chase that nearly any competing predator (lion, leopard, hyena, or even a person with a stick) can drive them off and steal their catch.

The story of how cheetahs became such beautifully adapted hunters and yet have reached the current point of near extinction serves as an excellent case study for the genetic basis of evolution. As we study the role of genetics in the evolution of cheetahs and other living organisms, we will see three unifying themes emerge. First, a group of interbreeding organisms evolves as its constellation of genes changes over the generations. This chapter explores the mechanisms that cause these genetic changes to occur. Second, genetic diversity is a precondition for evolution, and

most populations of organisms possess a remarkable amount of genetic diversity. Modern cheetahs, though, have little opportunity for evolution because of their paucity of genetic variability. Finally, evolution occurs because certain individuals survive (are selected by) environmental pressures or chance events to be parents for the next generation. The fastest cheetahs, for example, were better able to feed their young and leave more offspring. Unfortunately, some past catastrophes eliminated most cheetahs and left the few remaining ones with harmful alleles.

Our consideration of the genetic basis of evolution will answer these questions:

- What produces and maintains genetic variation?
- What are the agents of evolution?
- How do new species arise?
- What accounts for the observed patterns of evolution and extinction?

FIGURE 32.1 *The World's Fastest Runner: Handsome but Endangered.* The cheetah (*Acinonyx jubatus*) evolved with a slender body, camouflage markings, and high-speed gait. Its extremely flexible vertebral column allows the cheetah to bound in 7 m strides and to remain airborne half the time during a high-speed chase.

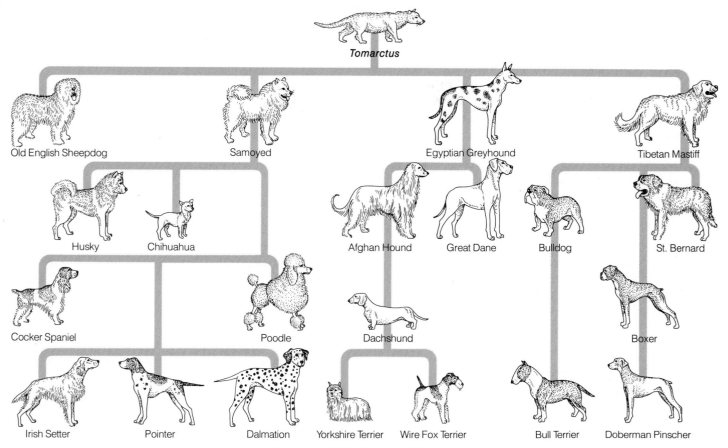

FIGURE 32.2 *The Diversity of Domestic Dogs.* An ancient progenitor in the genus *Tomarctus* gave rise to dozens of dog breeds—from the mastiff and the St. Bernard to the Yorkshire terrier and Chihuahua—all members of the same modern species, *Canis familiaris*.

≈ GENETIC VARIATION: THE RAW MATERIAL OF EVOLUTION

Over several centuries, dog breeders have produced canines as different as Chihuahuas and St. Bernards simply by selecting for desired traits from the tremendous amount of heritable variation that existed in the original dogs (Figure 32.2). Charles Darwin, in fact, argued that the numerous varieties of domestic plants and animals prove that most species have extensive heritable variation. But Darwin had to base his theories on simple observations of organisms and on inferences that their variations were somehow heritable. Although Gregor Mendel was working out the principles of genetics and the mechanisms of inheritance about the same time Darwin was studying evolution (see Chapter 8), Darwin knew nothing of Mendel's genetic theory.

Modern evolutionary biologists, however, have combined Darwin's ideas with modern genetics and have devised a set of principles they call **population genetics** to clarify what happens at the genetic level in populations. A *population* is a group of interacting individuals of the same species that inhabit a defined area. While the sum total of all alleles (two alleles per gene) carried on an individual's chromosomes is its genotype (see Chapter 8), the sum total of all alleles carried in all members of the population is the **gene pool**. (Recall that alleles are alternative forms of the same gene.) Population geneticists study how the frequencies of certain alleles in gene pools change over time. An individual's genotype, of course, does not change, from conception to death. But the genetic makeup of a population does change over time—for example, as new alleles arise by mutation and as old but rare alleles disappear if all individuals that have those rare alleles die.

WHAT IS THE EXTENT OF GENETIC VARIATION?

The actual degree of genetic variation between individuals of the same species is even greater than Darwin could have guessed. As we saw in Chapter 8, an organism may have any one of several alleles at any given gene locus on a chromosome. If the two alleles at a chromosome location are different, the person is heterozygous for that gene. In heterozygous individuals, one allele is often dominant, the other recessive, and only the dominant allele is expressed in the organism's physical makeup (phenotype). This means that a large degree of variation is hidden, transmitted from generation to generation through recessive genes. Most of the variation in genetic makeup is not visible in the phenotype.

In the early 1960s, molecular biologists developed simple and sensitive techniques that enabled them to estimate the extent of genetic variation in a population's gene pool. With these techniques, they could look at the structure of proteins encoded by the same gene but occurring in different individuals. At first, biologists were shocked to find a huge amount of genetic variation in populations. Each person, for example, has different alleles (is heterozygous) at about 10 percent of the gene sites on their chromosomes (Figure 32.3). Many plant species are

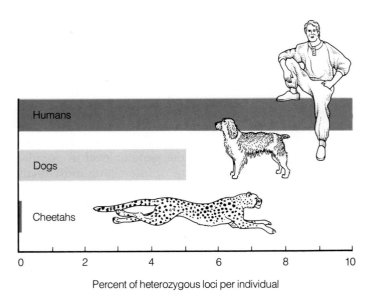

FIGURE 32.3 *People and Dogs Have Far More Genetic Variation Than Do Cheetahs.* People have about 100,000 genes, and each person is heterozygous (has two different alleles) at 10 percent, or 10,000, of those genes. Dogs also have about 100,000 genes and are heterozygous at about 5 percent, or 5000, genes. By contrast, cheetahs are heterozygous at only 0.067 percent of their genes, the equivalent of only 67 genes in 100,000.

even more variable, having a heterozygosity of 18 percent (they are heterozygous at 18 percent of their genes).

Looking at proteins directly gives evolutionary biologists a valuable way to probe beyond traits visible to the eye such as tail length, eye color, or flower shape. These visible traits can vary because of environmental rather than genetic factors and obscure an organism's actual underlying genetic makeup. Biologists have, for example, probed human racial groups with interesting results. It turns out that the correlation is very small between racial and genetic differences. People have about 100,000 genes, and since each person is heterozygous at about 10 percent, each person has two different alleles for 10,000 genes on average. When biologists compare the distribution of these heterozygous genes *within* a given racial group with the distribution *between* racial groups, they find that nearly all genetic variation (90 percent) exists between individuals of the same race for traits like height, weight, and the amino acid sequences of individual proteins, while only a small amount (10 percent) of the variation accounts for all the racial differences between European, African, Indian, East Asian, New World, and Oceanic peoples. In other words, it would be easy to find individuals from different races who—except for superficial traits like skin color or facial features—are more genetically similar to you than are many individuals of your own race.

The extent of genetic variability in populations is critical for a species' survival because populations that fall below a necessary minimum variability have difficulty adjusting to a constantly changing environment and so become endangered. The cheetah is a prime example.

Cheetahs today have one of the lowest rates of heterozygosity among mammals: 0.067 percent (see Figure 32.3). Thus, they have 100 times less genetic variability than people and less even than certain highly inbred strains of livestock or laboratory mice. Partly because of this genetic uniformity, they face extinction. A single new virus, which the cheetah's immune system lacks sufficiently varied alleles to combat, could wipe out their entire population.

Before we can understand how the cheetah came to have such a small amount of genetic variation, we need to consider the sources of genetic variation and what maintains it in populations.

WHAT ARE THE SOURCES OF GENETIC VARIATION?

Genetic variation can increase in a population in several ways. The first source is *point mutation*, a random change in the DNA sequence of a single gene (Figure 32.4a). As Chapter 10 discussed, some mutations are neutral; their random and spontaneous appearance produces neither harmful nor advantageous effects. Mutations in genes

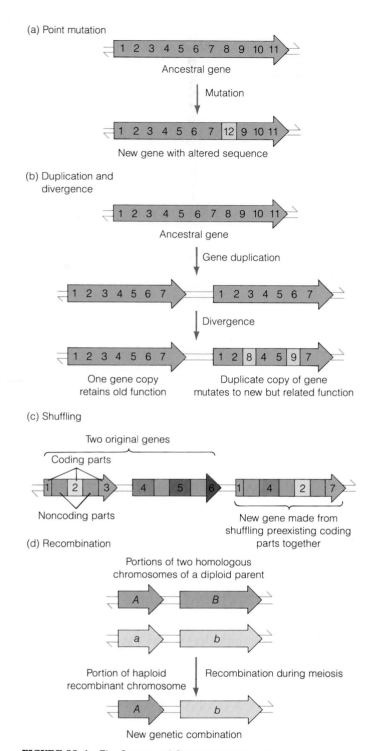

(a) Point mutation

Ancestral gene

Mutation

New gene with altered sequence

(b) Duplication and divergence

Ancestral gene

Gene duplication

Divergence

One gene copy retains old function

Duplicate copy of gene mutates to new but related function

(c) Shuffling

Two original genes

Coding parts

Noncoding parts

New gene made from shuffling preexisting coding parts together

(d) Recombination

Portions of two homologous chromosomes of a diploid parent

A   B

a   b

Portion of haploid recombinant chromosome

Recombination during meiosis

A   b

New genetic combination

FIGURE 32.4 *The Sources of Genetic Variation.* (a) In a simple point mutation, a nucleotide (here, labeled 8) in a gene can be replaced with an entirely different one (here, labeled 12). (b) In duplication and divergence, an ancestral gene duplicates, and one copy remains identical (left side) while the other diverges in sequence by simple mutation (bottom right). (c) In shuffling, copies of separate pieces of two genes (here, 1–6) get shuffled into a different set and sequence (1, 4, 2, 7) with a new function. (d) In recombination, random assortment and crossing over lead to the rearrangement of chromosome parts and hence new genetic combinations.

controlling quantitative traits, like bushels of corn per acre or gallons of milk per cow per day, can provide favorable genetic variation. Many mutations, however, are either harmful or lethal; one specific mutation, for example, disrupts a muscle protein gene and causes the human disease muscular dystrophy. What makes truly new genes with new, useful functions is gene duplication and exon shuffling.

Gene duplication and *divergence* can produce new genes without destroying the original function (Figure 32.4b). A chance error in DNA replication or recombination can duplicate an original gene, leaving two identical copies arranged in tandem along the chromosome. These two genes are then free to undergo random point mutations independently, so that their sequences will come to differ, or diverge. Thus, one copy can remain unmutated and maintain the original function, while the other can mutate to a new form that may, by chance, encode a slightly new function. This process has resulted, for example, in multiple copies of human hemoglobin genes; embryos, fetuses, and adults use different hemoglobin genes, since different developmental stages have different oxygen requirements.

Shuffling is a third source of new genes (Figure 32.4c). Recall from Chapter 10 that eukaryotic genes are often split into parts that encode proteins (*exons*) and intervening parts that do not encode proteins (*introns*). As a result of chromosome rearrangements, different protein-encoding parts can be positioned by chance next to each other and can produce new genetic combinations. The gene for the protein that removes cholesterol from a person's blood and hence lowers the risk of heart disease, for example, contains one part from an immune system gene, another part from a skin cell growth factor gene, and still other portions from genes for blood-clotting proteins. It is as though evolution picked gene parts off the shelf, pasted them together, and provided a new gene with a new function.

Mutation, duplication and divergence, and shuffling can create new alleles, such as *a* from *A* and *b* from *B*, as well as new genes, such as *AE* from *A* and *E*. This increases genetic variability. Once these processes produce a large number of alleles, *random assortment* and *recombination* during sexual reproduction can rearrange the new alleles and genes even further, providing the fourth source of genetic variation (Figure 32.4d). For example, the chromosome containing alleles *A* and *B* can recombine with the chromosome containing *a* and *b* and form recombinant chromosomes carrying *Ab* and *aB*. Novel phenotypes may arise from these new allele combinations and enable the organism to cope better with some environmental stress.

Together, recombination, point mutations, gene duplications, and exon shuffling generate the huge amount of genetic variation found within species. But other principles are required to explain how genetic variation can be maintained once it arises. As it turns out, these principles fooled many early geneticists.

HOW IS GENETIC VARIATION MAINTAINED?

The gene for short stubby fingers (a defect known as *brachydactyly*; Figure 32.5) is dominant, and the gene for normal fingers is recessive. Since the mutation is dominant, why doesn't everyone have brachydactyly within a few generations after the mutation arises? The answer is the **Hardy-Weinberg principle** named after the British mathematician and German physician who discovered it independently. According to this principle, in the absence of any outside forces, the frequency of each allele in a population will not change as generations pass. The brachydactyly example is instructive because it shows the importance of not confusing the frequency of the phenotype (short stubby fingers) with that of the genotype (the actual allele that causes the condition). For example, if a person homozygous for the dominant trait

of brachydactyly (*BB*) marries a person homozygous for the recessive trait of normal fingers (*bb*), then all their offspring will be heterozygous (*Bb*), and all will show the phenotype for brachydactyly. Considering only phenotype, it appears that the dominant trait has increased and the recessive trait has disappeared. But by considering the genotype, it is clear that the parents (*BB* × *bb*) have four alleles, half of which are *B* and half *b*. Since the children's genotypes are all *Bb* (again, half *B*, half *b*), the allele frequency has not changed. In fact, an analysis of matings through several generations shows that the allele frequencies remain the same. Geneticists say that the population is in *Hardy-Weinberg equilibrium*.

The Hardy-Weinberg principle is a useful model for predicting allele frequencies in populations, but only so long as the population is free of outside influences that can change allele frequencies. Five conditions must hold for allele frequencies to remain constant over generations and for the Hardy-Weinberg principle to correctly predict the number of individuals with each different genetic makeup in a population:

1. *No natural selection.* Recall from Chapter 1 that natural selection allows individuals with certain alleles to reproduce more successfully; thus, natural selection increases the frequency of individuals bearing successful alleles in the next generation. This results in a population in which the allele frequencies are changing.

2. *No mutation.* Since mutation changes one allele into another, mutation would increase the frequency of the new allele and decrease the frequency of the old allele and once again result in changed allele frequencies.

3. *No gene flow.* When an individual leaves one population and migrates to another, alleles are removed from the original population and transferred to another. This movement of alleles from one population to another is called *gene flow,* and it explains why migrations disturb allele frequencies.

4. *Random mating.* If individuals do not mate randomly, that is, if they mate preferentially with others bearing certain alleles, then genotype frequencies will change.

5. *Large population size.* If a population is not large, chance fluctuations can cause significant deviation from the ratio of alleles that the Hardy-Weinberg principle would predict.

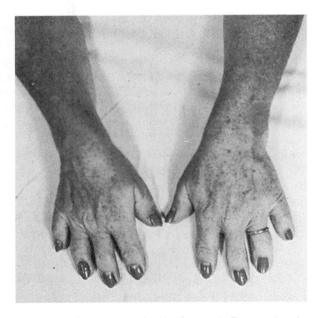

FIGURE 32.5 *Brachydactyly and Its Genotypic Frequencies.* A person with one or two alleles of the dominant gene for brachydactyly has hands with very stubby fingers; in each finger, the bone nearest the hand is fairly normal in length, but the second and third bones are greatly foreshortened. The dominant allele does not "take over" in the population, however, because according to the Hardy-Weinberg principle, the frequency of alleles remains unchanged in a population over the generations.

As long as all five conditions are met, allele frequencies in a population will remain unchanged, and the Hardy-Weinberg principle can accurately predict these frequencies from one generation to the next. A population in equilibrium, however, would be static and unchanging. But as we have seen again and again, populations evolve. **Evolution** can be defined as changes in allele frequencies within a population over time. Thus, the Hardy-

(a)

(b)

FIGURE 32.6 *Selection of Traits That Promote Survival.* (a) The stone plant (*Lithops divergens*) is so well camouflaged that an animal is unlikely to find and consume it. (b) These fleet antelopes may escape attackers on the African savanna by hiding in tall grass and outrunning predators if they have a slight head start.

Weinberg predictions provide a theoretical standard against which to compare real populations influenced by natural selection, mutation, gene flow, nonrandom mating, and small size.

In the next section, we will see how each violation of the Hardy-Weinberg principle affects real populations, and in so doing, we will see the actual agents of evolution at work.

≋ AGENTS OF EVOLUTION: WHAT CAUSES ALLELE FREQUENCIES TO CHANGE?

Biologists think that cheetahs and other living cats evolved several million years ago from an ancient population of cats with considerable genetic variation. But while such variation is the raw material of evolution, it could no more change an ancient cat into a cheetah, or an ancestral dog into a St. Bernard, than flour could make itself into a bran muffin. Something has to work on the raw materials first. Of the many agents capable of this action, natural selection is probably the most important.

HOW NATURAL SELECTION CHANGES ALLELE FREQUENCIES

Question: What do cactuses, brown stone plants, poison ivy, and Thompson's gazelles have in common? Answer: All have adaptations to avoid being eaten. The cactus has thorns; a stonelike appearance camouflages the stone plant (Figure 32.6a); poison ivy leaves produce a toxic alcohol that poisons and repels hungry animals; and the gazelle has the speed and agility to stay a few paces ahead of a sprinting cheetah—at least for a while (Figure 32.6b). In each case, nature has selected among the plants' and animals' ancestors for these traits; the thornier, more camouflaged, more poisonous, and faster of the organisms survived better and left more offspring. This exemplifies *natural selection,* the changing of allele frequencies by differential reproduction and survival.

Biologists once used phrases like "the survival of the fittest" to describe natural selection. In Darwinian terms, **fitness** means the ability to survive to reproductive age and pass genes to the next generation. If an allele causes an individual to be healthier, for example, it increases the individual's chances of surviving to reproductive maturity. And alleles that increase fertility increase the number of offspring the individual leaves. Since individuals with more offspring pass beneficial alleles on to an enlarged number of descendants, the frequency of the beneficial allele will increase in the population. For evolution to occur as described, alleles with different effects on health, developmental rate, and fertility must preexist in the population, originating from mutation and recombination. Natural selection merely chooses parents with the most suitable phenotypes (encoded by specific genotypes) for the next generation.

The result of natural selection is *adaptation:* the accumulation of structural, physiological, or behavioral traits that increase an organism's fitness, that is, its ability to survive and reproduce in its environment. Now let's look at some examples of how natural selection has affected populations of different organisms, sometimes favoring one extreme form, sometimes favoring an intermediate form, and sometimes favoring two extreme forms.

(a)   (b)

FIGURE 32.7 *The Peppered Moth: Selection in Action.* A dark peppered moth on a clean tree trunk (a) is conspicuous to birds, while a light gray one on a sooty tree trunk (b) is equally conspicuous. The light moth, however, is barely visible on the light trunk (a), and the dark moth is well camouflaged on the dark trunk (b). For the peppered moth, selection favored wing color the color of tree trunks in a given locale.

In some cases, natural selection favors one extreme form of a trait over all other forms. Consider the weight of cheetahs over the centuries. When the first cheetahs appeared about 4 million years ago, they weighed more than twice as much as modern cheetahs. Fossils show that within a large population of these giant cheetahs, however, the weights were variable. Some of the cats weighed only 91 kg (200 lb), some 136 kg (300 lb) or more, and the rest between those two extremes. Over time, light, fast-running animals reproduced more successfully, and thus, natural selection favored cheetahs with alleles causing lighter weight.

Another classic case of natural selection favoring an extreme form is the English peppered moth. An ancient mutation produced two alleles of this insect's color gene: a dominant allele for black and a recessive allele for gray. Before the advent of the Industrial Revolution in the 1840s, genetically light-colored moths could easily rest on gray tree trunks, and so camouflaged, escape being eaten by birds (Figure 32.7). After industrial soot darkened urban trees, however, the tables were turned, and light-colored moths were the easy targets. Moths bearing the "black" allele, on the other hand, blended in perfectly and survived, contributing their alleles for dark color to succeeding generations. In urban areas, therefore, the frequency of alleles for the extreme trait of dark color increased, and the moth population evolved.

To the dismay of farmers and health authorities, natural selection is also responsible for the resistance insects and microorganisms acquire to pesticides and antibiotics. For example, in 1946, half a kilogram (1 lb) of a new insecticide killed all but a few of the insects in a farm's population, allowing the farmer to produce 30,000 bushels of corn. But a few insects survived because by chance they already contained mutated genes that rendered them "immune" to the toxic effects of the poison. Over the years, these genetically tolerant insects passed the resistance allele on to their offspring, and eventually, a large population of resistant individuals arose. Today the farmer uses 140 times as much pesticide, and yet his yield of corn is less than it was before he started using pesticide. Likewise, the microorganisms that cause syphilis and gonorrhea have evolved so much resistance to penicillin that it now takes ten times as much of the drug to fight these infections as it did just 30 years ago.

In some situations, natural selection favors intermediate individuals rather than the fastest or darkest or most resistant. For example, far more human babies are born weighing about 3.2 kg (7 lb) than any other weight. Very heavy or very light babies have lower chances of surviving, since heavy babies complicate delivery, while lightweight babies tend to be premature and less ready for life outside the womb. Height in adult humans is a similarly stabilized trait, with most people falling toward the center of the bell curve for height (Figure 32.8). Natural selection can help explain why certain organisms—so-called living fossils—have persisted for millions of years with little or no outward change in form. Examples include scorpions, which have looked very similar for 350 million years; and ginkgo trees, which hark back 200 million years (Figure 32.9). Each organism probably arose in an environment that remained unchanged for immense time spans, such as the rain forests. Alleles causing a particular phenotype especially suited to the stable surroundings would be selected for, while other alleles would eventually disappear.

FIGURE 32.8 *Selection for an Intermediate Form.* Natural selection can favor the middle ground between extreme expressions of a trait. People are more likely to be of average height, and less likely to be very tall or very short, as these turn-of-the-century army recruits illustrate.

(a)

(b)

FIGURE 32.9 *Similarity of Ginkgo Fossil and Fresh Leaf Reflects Selection in a Stable Environment.* For 200 million years, selective pressures have favored retention of graceful, fan-shaped leaves in the ginkgo. Compare the fossil imprint (a) and the fresh leaf from a ginkgo tree (b).

While natural selection sometimes favors intermediate forms, as in human height and birth weight, in other situations, the intermediate forms are selected against and the two extremes are selected for. In parts of Africa, for example, females of the butterfly *Pseudacraea eurytus* occur in two forms (Figure 32.10), each looking exactly like one or another butterfly species (*Bematistes epaea* or *B. macaria*) that tastes offensive to birds. *P. eurytus* individuals that are good mimics of either foul-tasting species will evade predation and be selected for, while *P. eurytus* individuals with alleles for an appearance intermediate between the two extremes will mimic neither bad-tasting species and will be eaten by predators.

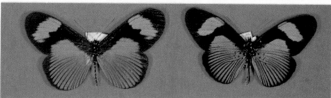

FIGURE 32.10 *Selection for Two Extreme Forms.* Natural selection can favor two extreme expressions of a trait. The two butterflies on the left belong to two different species (*Bematistes epaea* and *B. macaria*), and both species are foul tasting to birds. The blue and the orange butterflies on the right belong to the same good-tasting species, *Pseudacraea eurytus*. This mimicry protects the tasty species from predation. But an intermediate form of the good-tasting species—say, part blue and part orange—would not mimic a foul-tasting species, would not be protected, and would thus be selected against.

The different patterns of natural selection are clearly means of changing allele frequencies in a population. But while natural selection can favor extremes or intermediate forms, it can operate only if a population already contains alleles for different phenotypes, which will have arisen via mutation.

MUTATION AS AN AGENT OF EVOLUTION: REPLACING ONE ALLELE WITH ANOTHER

Mutation, a prime source of genetic variation, can alter allele frequencies within a population by changing one allele into a different allele. For example, the most common eye tumor in children, retinoblastoma, acts in about 30 percent of all cases as if it were due to a newly arisen dominant mutation. Each time a new mutation occurs in the gene leading to this tumor, one normal allele is removed from the gene pool and replaced with the tumor allele. When this happens, the allele frequency changes a tiny amount and the Hardy-Weinberg equilibrium is disturbed slightly. The primary importance of mutation to evolution, however, is not this small change in allele frequency, but rather the opportunity the new mutation may provide for natural selection.

GENE FLOW: MIGRATION ALTERS ALLELE FREQUENCIES

If individuals migrate from one population to an adjacent population, they may remove alleles from one group and introduce them into a second group. This change in allele frequencies due to immigration or emigration is called **gene flow,** and few populations are so isolated that they escape it entirely. Prairie dogs in Kansas, Nebraska, and other states, for example, live in tight-knit populations separated from one another by both geographical distance and strong social ties to group members (Figure 32.11, page 526). Prairie dogs do not tolerate immigration and drive out any strange prairie dogs that attempt to enter their populations. During the late summer, however, when the pups reach maturity, social restrictions on immigration are relaxed, and dispersing prairie dogs are, for a short time, permitted to establish themselves in new populations without reprisals. Many are picked off by predators during their emigration, but survivors become breeding members of a new colony. Their alleles are thus added to the gene pool of an existing population and are likewise removed from their parents' population.

Another instance of gene migration is detectable in human populations. Researchers compared four groups of people—whites from Georgia, blacks from Africa, and blacks from Oakland and Detroit—for one allele of a cer-

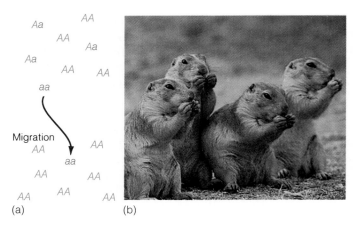

FIGURE 32.11 *Prairie Dogs and Gene Flow.* (a) Migration can lead to gene flow. (b) During most of the year, prairie dogs (*Cynomys ludovicianus*) fend off immigrants from other areas. In the late summer, however, they allow new members to join established populations. Immigrants may introduce new alleles into a population's gene pool.

tain blood group called *Duffy*. The allele's frequency was 42 percent in the whites, 0 percent in the African blacks, 22 percent in the Oakland blacks, and a bit higher in those from Detroit. The researchers interpreted this to mean that because of intermarriage, the allele has migrated from white populations to black populations in America.

GENETIC DRIFT: CHANCE CHANGES IN POPULATIONS

Biologists use the term *genetic drift* to refer to unpredictable changes in allele frequency due to small population size. The Hardy-Weinberg principle can accurately predict allele frequencies only for large populations, because as with all probabilities, it is much easier for nonrandom events to occur in small populations. To see why, consider a single die (from a pair of dice) with its six sides numbered 1 through 6. By rolling the die a million times, you would come very close to rolling each number in one-sixth of the throws. But if you threw the die only six times, the results might not be so evenly distributed. For example, you might throw a 4 on half of the six tosses, but no 2s on any toss.

Now, relating this example to allele frequencies and populations, suppose that a specific allele exists in one-sixth of the individuals in a population of cherry trees. If a chance occurrence like a severe spring ice storm were to strike an area with 2 million cherry trees and half died, a million would still survive, and the probability is high that one-sixth of the survivors would still bear the specific allele. If that same storm hit an area with just 6 trees, however, and only 3 survived, it is much more likely that the single tree bearing the specific allele would be among

the dead. If this happened, the population would lose one allele completely, and the allele frequency will have changed in an unpredictable way.

A natural disaster that drastically reduces a population's size, such as severe weather, widespread disease, or excessive predation, can bring about the phenomenon biologists call a **population bottleneck.** Just as only a small amount of liquid can move through a narrow-necked bottle in a short time, only a small number of organisms survive a population bottleneck, and for the reason we just discussed, the survivors might have a nonrandom sample of the alleles present in the original population. When the bottlenecked population enlarges once again, it lacks the genetic variability of the original population, and its potential for adapting to further environmental changes may be greatly reduced. Cheetahs experienced a population bottleneck due to disease or drought or overhunting by people about 10,000 years ago, and biologists suspect that the small surviving population had, by chance, alleles for high disease susceptibility and low reproductive rates. Inbreeding among remaining cheetahs must have created the gene pool we see today, with its high numbers of detrimental alleles (Figure 32.12).

Genetic drift can also be important in a situation called the **founder effect,** which occurs when a few individuals separate from a large population and establish a new one. Since the small group of founders bear only a fraction of the alleles from the original large population, they may represent a nonrandom genetic sample. The 12,000 or more Amish people who live in thriving communities in eastern Pennsylvania are all descendants of about 30 individuals who emigrated from Switzerland beginning about 1720. Some of these "founders" had a recessive allele for a gene that causes short forearms and lower legs. As a result, in the Amish living around Lancaster, Pennsylvania, the frequency of this allele is 1 per 14, instead of the 1 per 1000 found in other populations (Figure 32.13). Later in this chapter, we will see how the founder effect can help give rise to entirely new species and thus how important genetic drift is to the process of evolution.

NONRANDOM MATING AND EVOLUTION

Besides the processes of natural selection, mutation, gene flow, and genetic drift, a population's mating tendencies can also alter the frequencies of heterozygotes and homozygotes from the frequencies predicted by the Hardy-Weinberg principle. Theoretically, every individual in a population has an equal chance of mating with any other individual but often it does not work this way. Consider a human population in which all short adults chose to mate only with other short adults, and all tall adults with tall adults. If height were controlled by two separate alleles, one for tall stature and one for short,

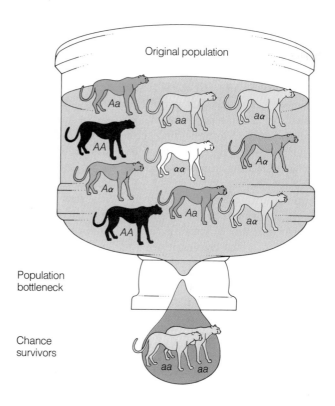

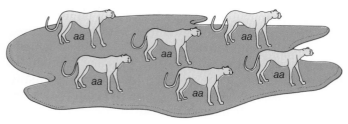

New population of descendants

FIGURE 32.12 *Bottleneck Effect: A Dramatic Reduction in Population Size Leads to Reduced Genetic Variability.* Ancestral cheetahs probably had normal levels of genetic variability, represented in the bottle by the three hypothetical alleles *A*, *a*, and *α* and the six genotypes *AA*, *Aa*, *aa*, *Aα*, *aα*, and *αα*. Ever since some genetic catastrophe perhaps 10,000 years ago, however, each surviving cheetah carries the homozygous recessive genotype *aa*, which may well allow the expression of a maladaptive trait.

then such nonrandom mating would lead to few heterozygotes—far fewer than predicted by random mating. The proportion of "short" and "tall" alleles would remain the same; only the frequencies of homozygotes and heterozygotes in the population would change. Random mating does not hold in people because tall women tend to marry tall men, and short women tend to marry short men.

One type of nonrandom mating, called **sexual selection,** is a type of natural selection in which individuals select a mate on the basis of physical or behavioral characteristics, regardless of the effect of that characteristic

on general fitness. Take the bowerbird of Australia, for example. The male bowerbird constructs a nestlike "bower" of grass and plant material and then proceeds to decorate it with flowers, shells, teeth, bottle caps, and just about any shiny object it can find (Figure 32.14, page 528). Female bowerbirds mate only with males that can lure them into a highly decorated bower, and the male's ability to build his nest depends on the specific alleles he inherited for the nest-building instinct. Thus, inheritance affects his reproductive success, since the nest-building genes are selected for through the female's willingness to mate, even though these alleles may decrease fitness by exposing the bird to predators and decreasing the time he has for feeding.

We have now seen the five agents of genetic change in populations—natural selection, mutation, gene flow, genetic drift, and nonrandom mating—which can act singly or in concert to violate the Hardy-Weinberg principle. Biologists tend to agree that each of these factors plays a role in evolution, but they hotly debate the relative importance of each evolutionary agent—particularly that of genetic drift and natural selection—in altering allele frequencies.

FIGURE 32.13 *The Founder Effect.* One of the original members of the Amish community in Pennsylvania by chance carried a recessive allele for a rare kind of dwarfism. Inbreeding among members of the colony produced homozygous individuals who show the trait.

FIGURE 32.14 *Sexual Selection: The Bowerbird. This male, resplendent in his metallic plumage, has built and decorated a nest with bottles, blossoms, and shiny shards that may help attract a female.*

NATURAL SELECTION VERSUS GENETIC DRIFT

Many evolutionary biologists are **selectionists:** They suggest that natural selection is the primary agent of evolution. They contend that different alleles of a gene affect an individual's health, survival, or fertility differently and that natural selection acts on these differences to choose the parents for the next generation. Some of their colleagues, however, question this idea. These critics have pointed out that most alleles found in natural populations seem to be equivalent and that most mutations are "neutral," conferring neither advantages nor disadvantages on an organism's fitness. Thus, these critics of the selectionist viewpoint are called **neutralists.** The neutralists argue that many different alleles may ensure an organism's survival and reproduction equally well. If a mutation leads to a slightly different protein, that protein need not be identical to the original as long as it works about as well. Neutralists maintain that most genetic variation is the result of neutral mutations, unpredictable genetic drift, and gene flow rather than the various forms of natural selection. To support their argument, neutralists point out that mutations seem to accumulate at a constant rate during the evolution of new species, whether they are evolving rapidly or slowly. For example, in each of the lineages that gave rise to people and goldfish, one amino acid replaced another in a certain protein every 7 million years on average. This steady tempo of change has been called a *molecular clock* because it seems to allow a new mutant allele to replace other alleles in the population every so many years, as long as the mutation does not destroy protein function.

Many biologists take a view that combines the best arguments of the selectionists with the sound observations of the neutralists. The intermediate view is that even if most mutations are neutral, there is still enough nonneutral variation to affect evolution.

By now, we have seen that genetic variation is primarily the result of chance physical processes like mutation and recombination. We have also seen that a number of evolutionary agents can change allele frequencies in breeding populations, again, with some changes being influenced by chance and others by the demands of the environment and natural selection. With this background, we are now ready to consider how the agents of evolution operate to transform a population of one species into groups of two or more species.

≫ HOW NEW SPECIES ARISE

WHAT IS A SPECIES?

Every zoo goer knows that cheetahs, tigers, and lions are three different species of cats, one slender and spotted, one powerful and striped, and one large and tan with a dark mane. Biologists who have tested these three cats have also found that each has its own unique genetic makeup, and in the wild, each breeds and reproduces only with others of its species. The criteria of unique genetic makeup and absence of interbreeding place the three kinds of cats unambiguously in one species or the other: There are no half tigers/half cheetahs or half cheetahs/half lions in nature, and tiger/lion hybrids are produced only in zoos.

According to modern evolutionary theory, species are groups of populations that interbreed with each other in nature to produce healthy and fertile offspring. Each species, in short, is *reproductively isolated* from (cannot make fertile progeny with) every other species in nature. In some tropical frogs, identical-looking species can live side by side and not interbreed. Conversely, if very different-looking organisms (like the butterflies shown in Figure 32.10) can interbreed in nature, they are considered races or varieties of a single species. We may be fooled by physical looks but not by interbreeding.

The definition of a species as a shared, reproductively isolated gene pool has been the most satisfactory to date, but even it has limitations. The biggest drawback is that it applies only to sexually reproducing organisms. Asexual organisms, such as some bacteria, reproducing only by cell division, must be excluded because each individual is, in effect, reproductively isolated from all others.

(a)

(b)

(c)

FIGURE 32.15 *Sterile as a Mule: Crossing Two Species Can Lead to Hybrid Vigor and Hybrid Sterility.* The cross-breeding of a male donkey (a) and a female horse (b) yields a mule (c), which is stronger and more vigorous than either of its parents. The failure of horse and donkey chromosomes to pair during meiosis in the mule's gonads, however, prevents the hybrid from reproducing.

With organisms like these, a species must be identified by phenotypic and biochemical traits, rather than by the potential for interbreeding.

To see how a species arises, we must first consider the factors that keep separate species like lions and cheetahs or sets of identical-looking tropical frogs from interbreeding and producing fertile offspring.

REPRODUCTIVE ISOLATING MECHANISMS

Any structural, behavioral, or biochemical feature that prevents the members of a species from successfully breeding with those of another is considered a **reproductive isolating mechanism.** Some such mechanisms prevent members of different species from mating in the first place, while others render matings that do take place between members of different species unsuccessful at producing fertile offspring.

Some isolating mechanisms prevent crossbreeding. For example, dozens of frog species may inhabit the same area, but one species may mate only in ponds, another only in flowing streams, and a third in shallow pools and puddles. Biologists call this mechanism *habitat isolation.* In a similar fashion, some species mate only in the spring and others only in the fall, and this is considered *seasonal isolation.*

Different courtship and mating behaviors can block crossbreeding between species, as in identical-looking tropical frog species, each with distinctive mating calls. This is *behavioral isolation.* Finally, specific physical structures can hinder mating. In many spider mites, for example, the male genitalia are shaped like "keys" that open the genital plates of their species' females; one species' "keys" cannot open another's "locks." This is called *mechanical isolation.*

Sometimes, the males of one species do mate with the females of another species, but a second set of mechanisms prevents them from producing fertile offspring.

In some cases, the gametes of the two species are so different that fertilization is impossible. In other cases, fertilization does take place, but the **hybrids,** the offspring of the two crossbreeding species, are not viable individuals. The two sets of genes fail to cooperate harmoniously during development, leaving hybrids weak and malformed. With many American frogs, genetic differences between species lead to abnormal hybrid embryos and freakish individuals.

Occasionally, the cross-mating of two species produces extremely healthy offspring (an example of *hybrid vigor*), but these offspring are themselves nonreproductive (they experience *hybrid sterility*). Breeders produce mules by crossing a male donkey (31 chromosomes in each sperm) with a female horse (32 chromosomes in each egg), and although mules (with their 63 chromosomes per somatic cell) are noted for strength and endurance, they cannot produce further offspring with other mules, with horses, or with donkeys because chromosomes fail to pair normally in meiosis (Figure 32.15).

Since reproductive isolating mechanisms separate the gene pools of two related species, we can begin to understand **speciation,** the emergence of new species, by learning how reproductive isolating mechanisms develop.

THE ORIGIN OF NEW SPECIES

Most biologists suppose that the majority of species arise after populations become geographically isolated and evolve in separate ways. A physical barrier, such as a river, a desert, different vegetation belts, or even a new highway or pipeline, can separate populations of a single species and prevent gene flow between the groups. The split populations slowly diverge as mutation, genetic drift,

(a) Geographical speciation   (b) Speciation in overlapping territories   (c) Speciation within the same area

FIGURE 32.16 *How New Species Emerge.* (a) Geographical barriers can isolate populations and lead to speciation. (b) Speciation can occur among populations with overlapping territories. (c) Reproductive isolating mechanisms arising from genetic, behavioral, or ecological barriers can lead to speciation within a population living in the same area.

and adaptation cause different sets of characteristics to accumulate. Eventually, barriers to reproduction emerge and prevent matings even if, in the future, the two populations once again come into contact. This mechanism is called *geographical speciation* (Figure 32.16a).

For example, 7 million years ago, the Colorado River began carving out the Grand Canyon in an area inhabited by one squirrel species. Today, the Kaibab squirrel occupies the north rim of the canyon, while the closely related Abert's squirrel inhabits the south rim. The two species most likely arose through geographical speciation, gradually diverging as the canyon became an uncrossable barrier. Cheetahs may be an example of a population beginning to be split by geographical speciation. The remaining populations of cheetahs in East and South Africa can no longer experience gene flow because unpassable terrain separates them. Perhaps with time, therefore, genetic differences will begin to arise between the two populations, and eventually they will no longer be able to produce a fertile hybrid.

Populations in adjacent areas sometimes slightly overlap territories and experience some gene flow, but still diverge anyway. This speciation occurs if, despite the contact, strong reproductive isolating mechanisms develop as the population slowly splits into two separate ecological roles. Such speciation is common among plants, snails, flightless insects, and other organisms that can travel only short distances (Figure 32.16b). Some of the best examples involve plants that can grow on contaminated mine dump soils; within this group, certain species have rapidly evolved tolerance to toxic metals.

Occasionally, a single population will diverge into two species after a genetic, behavioral, or ecological barrier to gene flow arises between subgroups of the population. Most likely, reproductive isolating mechanisms arise

that prevent crossbreeding, and part of the population starts a different way of life amid the old (Figure 32.16c). More than 500,000 plant and animal species, for example, are parasites that live and reproduce on host species. Many of the parasites are able to select an appropriate host because a few genes allow them to recognize the host chemically. Single mutations in the genes that encode the "recognizer proteins" could cause a parasite to select a different host and create sudden isolation. This happened when American growers introduced English walnut trees to the United States; the coddling moth, which parasitizes apple and pear trees, rapidly produced a species that exploits walnuts.

We have seen how variation, heritability, differential survival, and chance all set the stage for organisms to evolve. These principles can clearly explain how populations of peppered moths change colors and why cheetahs are endangered. The mechanisms of genetic drift, natural selection, and other evolutionary agents also explain how new species form—how a population of ancient cats, for example, can speciate into lions, cheetahs, and other feline species. But how well can the modern synthesis of genetics and evolution explain the immense diversity of life on earth—from worms and beetles to toadstools and tulips—with all its patterns and trends?

≋ THE MODERN SYNTHESIS AND MAJOR TRENDS IN EVOLUTION

Biologists have a term for small changes in allelic frequencies within species, such as those that led to an average human birth weight of 7 lb or that brought about a color change in peppered moths: **microevolution.** For the large-scale evolutionary changes that differentiate taxonomic categories above the species level—pines from spruce, birds from mammals, fungi from plants—biologists use the term **macroevolution**. Biologists have grappled with two important questions in recent decades: Could the agents of small change (microevolution) have produced the big changes (macroevolution) that we see among the genera, families, orders, classes, phyla, and kingdoms of living things? And are the rules of microevolution sufficient to explain the observed pattern of species appearance and extinction in the fossil record?

The diversity we see on our planet is a product of billions of years of evolution. Paleontologists (biologists who study fossils to learn about extinct forms of life) have shown that the earth formed 4.6 billion years ago and life appeared about 4 billion years ago (see Chapter 14). This fossil record serves as our only long-term history of evolution, and within its patterns lies part of the answer to questions surrounding evolution.

HISTORY IN STONE: LIFE'S FOSSIL RECORD

Fossils are the only tangible evidence of what past organisms looked like and when they appeared, but unfortunately, these impressions and mineralized bones occur only in sedimentary rocks, which in turn form only under certain conditions. In any local area, episodes of sedimentary rock formation may last only a few days—as when a river overflows its bank and covers a bone with mud—or tens of thousands of years—as when a river deposits a huge delta. Between such episodes there are gaps, times when no sediments were laid down and no fossils formed in that area.

Despite such gaps, paleontologists can examine fossils from many locations and piece together a picture of evolution's overall course. Radiodating also allows scientists to determine the speed with which extinct species changed from one form to another. Evolutionists make branching diagrams called *phylogenetic trees* to show evolutionary relationships within and between lineages of organisms. Such trees can show which major group gave rise to which other groups above the species level (review Figure 14.13).

PATTERNS OF DESCENT

There are several patterns of evolutionary descent. One is the gradual change in allele frequencies in a population (Figure 32.17a, page 532), such as the alterations that little by little led to the diminishing size of the cheetah over geological time. Another is *divergent evolution*, whereby reproductive isolating mechanisms split two populations off from a common ancestral population, and genetic differences then accumulate in the two

descendant groups as in the two squirrel species separated by the Grand Canyon (Figure 32.17b).

Sometimes, divergence occurs simultaneously among a number of populations in ways that produce a variety of phenotypes. This fan-shaped branching pattern of evolution is called **adaptive radiation** (Figure 32.17c). Radiations may also be triggered when an ancestral species invades new territories that allow new and different ways of life. The colonization of the Hawaiian Islands by colorful birds called honeycreepers led to a spectacular example of this type of radiation. Descendant species had beaks of various shapes that allowed them to eat diets as different as tiny insects, hard seeds, soft fruits, and the nectar from deep, tubular flowers.

Another evolutionary pattern occurs when two or more dissimilar and distantly related lineages evolve and become more similar in certain superficial ways. Since the phenotypes "converge," biologists call this phenomenon *convergent evolution* (Figure 32.17d). A striking example at the class level is the similarity between sharks and dolphins. Both evolved a streamlined body shape because they adapted independently to a similar way of life in the open ocean, not because they inherited the same set of streamlining traits from a common ancestor. Each shows distinctive features that identify it as a fish or a mammal. Designers of submarines picked up a few tips from the sleek marine architecture, as well.

Parallel evolution involves two or more physically similar and genetically related lineages independently changing in the same direction and away from the ancestral phenotype (both toward larger size and away from smaller, for example) (Figure 32.17e). Sometime during the last 35 million years, for example, four genera of catlike predators separately evolved saberlike teeth and other adaptations for killing large prey. Since the cats shared a common ancestor, their parallel resemblances were far more than superficial. Each lineage of sabertoothed cat also shared the shortened face, retractile claws, and other features common to most living cats.

In the evolutionary pattern called *coevolution*, two species interact so closely that each one's evolutionary fitness depends on the other. An evolutionary change in one, therefore, creates a selective pressure on the other, so that both evolve together. As Chapter 16 explained, flowering plants and their animal pollinators show many classic coevolutionary relationships. For example, yucca plants and the moths that pollinate them have coevolved to such an extreme that yucca plants can be pollinated solely by the yucca moth, and those animals, in turn, can reproduce only in yucca flowers and seeds. Coevolution is so widespread and important that Chapter 34 discusses it in greater detail.

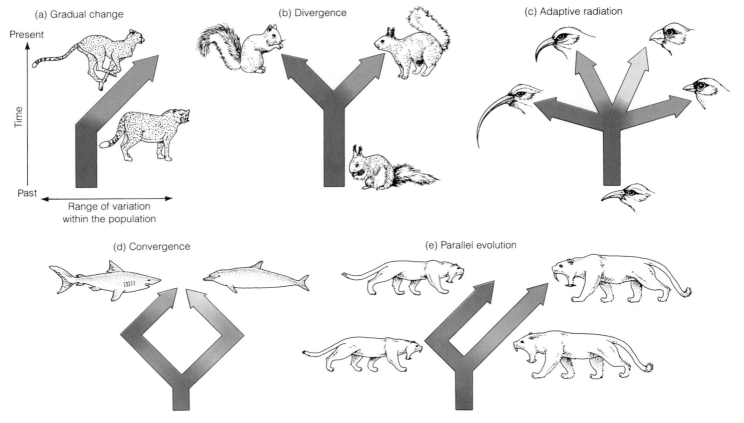

FIGURE 32.17 *Patterns of Descent in Evolution.* Over time, (a) a population's gene pool may gradually change in a certain direction, as when cheetahs became smaller, or (b) the population may split into two populations, each changing in a different way, as when two species of squirrels evolved from one original species when the Grand Canyon formed. (c) Adaptive radiation results in several new species arising from a single common ancestor, as with the small honeycreepers of Hawaiian forests. (d) In convergent evolution, separate lines converge toward similar forms because they are exploiting similar environments, as with sharks and dolphins. (e) In parallel evolution, two similar and related ancestors lead independently toward new species with similar traits, as in the evolution of different species of saber-toothed cats.

TRENDS IN MICROEVOLUTION: GRADUALISM AND PUNCTUATED EQUILIBRIUM MODELS

The mechanisms just discussed (adaptive radiation; convergent, divergent, and parallel evolution; and coevolution) imply two basic evolutionary processes. Genetic modification can occur within a single line of descent, making a current population look and act differently from its ancestor population, and genetic changes can split one population into two or more different groups. Traditionally, evolutionary biologists have agreed that after a group splits in two—whether because of geological or biological barriers—both subgroups diverge from each other at about equal rates as natural selection, genetic drift, or gene flow cause modifications within each line. This view is called **phyletic gradualism,** since it assumes that each group gradually becomes different from the original group and from each other (Figure 32.18a).

Adherents of phyletic gradualism suggest that variation in an organism's form occurs gradually during evolution and that in some instances, steady changes in body form can accrue within an entire population over time without the population splitting into two distinct species.

If new species do indeed form by means of phyletic gradualism, then the fossil record should show numerous intermediate species, as one species gradually transforms into the next species, then the next. While Darwin propounded this view, there was little fossil evidence to back it up in his time, and despite more than a century of additional fossil collection, paleontologists still have few records of gradually transformed lineages with numerous intermediate species. Scientists traditionally attribute such gaps in fossil lineages to gaps in the fossil record, explaining that intermediate fossils are missing because the rocks that contained them have eroded away, are still undiscovered, or were never formed.

In 1972, paleontologists Niles Eldridge and Stephen Jay Gould suggested that we should not make the data fit our ideas, but rather make our ideas fit the data. They proposed that the gaps might themselves be telling us how speciation takes place. According to the fossil record, the typical species is not in a constant process of change; instead, it exists for millions of years without significant alteration. These long periods of phenotypic equilibrium are interrupted, or "punctuated," only rarely by great phenotypic changes that result in new species.

Eldridge and Gould called this alternative model of how evolution occurs **punctuated equilibrium** (Figure 32.18b); it is based on three assumptions: (1) Alterations in body form evolve very rapidly in evolutionary time; (2) during speciation events, changes in form occur almost exclusively in small populations, and the result is that new species are quite different from their ancestral species; and (3) after the burst of change that results in speciation, species retain much the same form until they become extinct, perhaps millions of years later.

To get a feel for the different patterns of descent predicted by punctuated equilibrium and phyletic gradualism, consider two contrasting family trees for the giraffe and its smaller forest-dwelling relative, the okapi. Phyletic gradualism predicts that the two evolutionary lines began to diverge slowly before speciation occurred; then after reproductive isolating mechanisms developed, modifications in form gradually continued until the

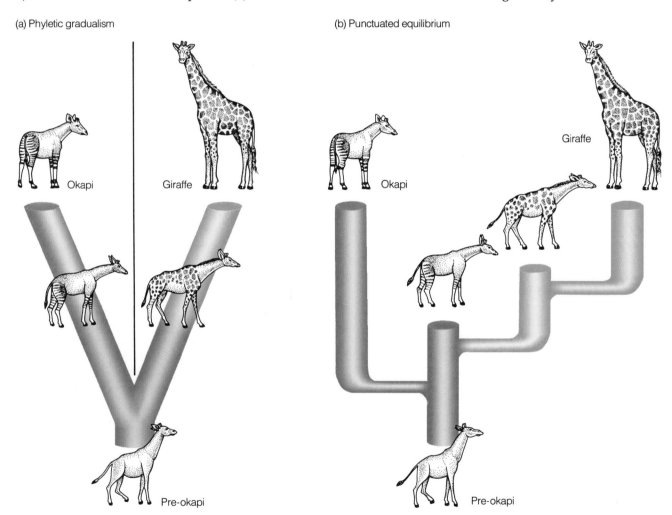

(a) Phyletic gradualism

(b) Punctuated equilibrium

FIGURE 32.18 *Phyletic Gradualism and Punctuated Equilibrium: Contrasting Models of Evolution.* How do new species arise? (a) Proponents of gradualism see new species arising as a population slowly and gradually diverges and differentiates into two daughter species (here, okapi and giraffe), each different from the original common ancestor (here, pre-okapi). (b) Proponents of punctuated equilibrium see the rapid budding off of a small divergent population while the original population remains unchanged. In this view, a small population budded off from the ancestral pre-okapi group and led to modern okapis, while a number of more giraffelike species arose and became extinct as time passed.

present-day okapi and giraffe emerged (review Figure 32.18a). In contrast, punctuated equilibrium predicts that offshoots repeatedly emerged as small, isolated populations split from the main group, underwent bursts of evolutionary change, and then maintained their new features for a long or short period of time before the line died out (see Figure 32.18b).

In fact, both processes probably contributed to this evolutionary tree. Proponents of the phyletic gradualism and punctuated equilibrium models clash partly because they define some terms differently. To begin with, the debated hypotheses emphasize different ways of seeing the rates of evolutionary change. One thousand generations is a short time to paleontologists, who often support the punctuated model, but a very long time to geneticists, who often favor the gradual model. What's more, the two groups define species in different ways: Geneticists use reproductive isolation as the sole criterion for a species, whereas paleontologists distinguish species by physical forms in the fossil record that differ substantially from other similar related forms. Obviously, a geneticist cannot test whether one extinct mollusk could have mated successfully with another fossil of different shape. Nor can paleontologists assign separate species status to two extinct mollusks that look nearly identical but were, in fact, reproductively isolated.

As a compromise position, we can view evolution as often involving a small population becoming isolated geographically or ecologically from other members of its species and accumulating significant genetic changes over many generations (but a short stretch of geological time). Once sufficient change has occurred to fit the new species to its environment, very little additional change may take place. The principles of population genetics can thus explain geologically rapid shape change followed by long periods of stasis. In addition, those principles do not require a strict correspondence between morphological evolution and speciation; changes in shape can come before or after the evolution of a new species.

TRENDS IN MACROEVOLUTION: PHYSICAL CHANGES, SELECTION, AND EXTINCTION

Theories of macroevolution seek to explain evolution above the level of species and hence the origin of major distinguishing features (reptiles' scales, birds' feathers, mammals' hair). These theories seek to answer the question posed earlier: Can microevolutionary mechanisms explain the origin of taxa above the species level, or are different evolutionary mechanisms required for such major differences?

■ **Functional Changes** Biologists have observed that evolutionary innovations usually follow from a change in size or function in a preexisting body part, rather than from the sudden appearance of something totally new. For example, horses' high crowned molars, which they require for chewing silica-laden grasses, arose by gradual increases in a small tooth bump already present in their ancestors. And during the evolution of snakes, limbs decreased in size, presumably because snakes with smaller legs could slither faster than snakes with protruding legs, and hence legs were selected against.

Other evolutionary innovations arose when a body part performing one function changed in shape and was able to perform a new function. Biologists call the original structure or function a *preadaptation*. The original structure did not, of course, evolve in anticipation of a future need; but once it had arisen, the structure could be modified for a new use. For example, *archaeopteryx*, the earliest known bird (see page 318), is distinguished from small, flightless reptiles that lived at the same time solely by its feathers. Mutation and selection changed reptilian scales (a preadaptation) into lightweight feathers, an evolutionary novelty that allowed better gliding and heat retention. While this flying reptile had feathers like a bird, its skeleton still resembled that of a reptile, lacking the specialized breastbone and powerful flight muscles of modern birds. A case like this, where some characteristics evolve without simultaneous changes in other body parts, is called *mosaic evolution.*

■ **Regulatory Genes and Developmental Changes**
Sometimes, the evolution of a dramatic change in form involves an entire suite of genes and can have its roots in a simple developmental event. For example, a single mutation event in a fruit fly can cause four wings to form instead of the usual two. Since the mutation is in a regulatory gene that acts early in development to control the function of many other genes, the result is a startlingly novel effect—four wings, not two.

Clearly, the principles of microevolution, including mutation and adaptation, can account for changes that occur during long reaches of time. But can they also account for the patterns of extinction that mark macroevolutionary trends?

■ **Species Selection and Extinction** Modern horses run about on a single toe per foot, but their early ancestors had the primitive condition of three toes per foot. Some macroevolutionists account for such evolutionary trends by invoking the principle of *species selection:* the idea that certain species continue to break up into new species, while other species become extinct. To see how species selection works, consider the diagrammatic view of horse evolution shown in Figure 32.19. The change from three toes to one toe did not occur smoothly in a single lineage.

Instead, various species arose, gave rise to new species, and then became extinct in a great evolutionary bush with dozens of branches. Sometimes, the trend toward fewer toes reversed, and sometimes it moved forward. Our view of this trend is colored because only the genus *Equus* happened to escape extinction. We do not know whether *Equus* survived because of the structural features that distinguish it from other genera or because it was simply lucky and its special environment persisted.

Microevolutionary theory—including genetic variation, drift, natural selection, and speciation—can explain the changes in form that took place during horse evolution. But since extinction is not part of microevolution, those models cannot fully explain why living horses, zebras, and donkeys walk about on a single toe per foot, while those with three toes per foot died out. It is left to macroevolutionists to attempt to explain why some lineages branch and form many species and why some lineages become extinct.

Ultimately, extinction is the fate of all species. Most mammal species, for example, survive no longer than 2 to 5 million years. And the number of all living species today—plants, animals, fungi, and microbes included—is probably less than 0.01 percent of the estimated 500 million species that have existed since life began on earth. In the most recent major extinction, nearly 80 percent of the large mammal species of the Western Hemisphere disappeared. Some paleontologists attribute this extinction to *prehistoric overkill*—overhunting by prehistoric people—and note that the extinction occurred in stages: A fair number of species died out in Europe and Africa about 80,000 years ago; massive extinctions took place in North and South America between 10,000 and 12,000 years ago; and still more extinctions occurred on Madagascar and other large islands between 1000 and 6000 years ago. In each period, extinctions followed human colonization of the region. The overkill theory is far from unanimously accepted. Many biologists, in fact, think

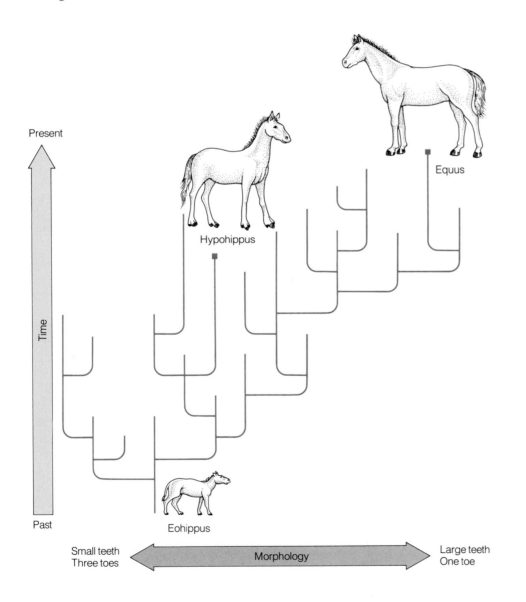

Present

Time

Past

Equus

Hypohippus

Eohippus

Small teeth
Three toes

Morphology

Large teeth
One toe

FIGURE 32.19 *Species Selection, Extinction, and Evolutionary Trends.* The tiny, dog-sized progenitor of modern horses, *Eohippus*, had three toes, and so did the larger, ponylike *Hypohippus*. Modern *Equus* has one toe. The lineage from *Eohippus* to *Equus* was not a gradual change in a single lineage, but rather a bush with many branches, all of which have died out except modern horses.

that rapid climatic and environmental change at the end of a long period of glaciation caused these mass extinctions.

Cheetahs have experienced a number of extinctions. At one time, there were five species ranging over Africa, Europe, Asia, and North America. Such wide distribution is rare among land mammals and shows that cheetahs at one time had the extensive genetic variation necessary to exploit a wide range of environments. One large North American species disappeared 2.5 million years ago, followed by the European cheetah about 1 million years ago; then, about 20,000 years ago, a Chinese species and a second North American species became extinct. As we have seen, the last cheetah species, *Acinonyx jubatus* of Africa, suffered the ill effects of a population bottleck about 10,000 years ago and is now hanging on by a thread with just two remaining populations in East and South Africa. Animal conservationists have various plans for breeding cheetahs in zoos and protecting them in the wild to prevent further loss of population. While extinction is the ultimate evolutionary fate of all species, one hopes that the disappearance of the world's fastest-running animal comes later rather than sooner.

≋ CONNECTIONS

Ten thousand years ago, cheetahs leapt after prey in many parts of the globe. Today, they are limited to just two parts of Africa, with one population of interbreeding cats in each area. Populations like these constitute the basic units of evolution—the units in which allele frequencies change and allow the gene pool to evolve. Events that bring about these changes in allele frequencies select the parents for the next generation. Some of these events influence alleles that lead to an organism's reproductive success; other genetic events occur by chance and have less predictable—but still powerful—consequences.

Each individual's genetic constellation is just part of the flow of genes in populations. Populations, however, are not islands; they interact daily with other populations of plants, animals, fungi, and microbes, as well as with their physical environments. The next several chapters describe these interactions and address a basic question: What makes organisms live where they do, the way they do, and in the numbers in which they are found?

NEW TERMS

adaptive radiation, page 531
evolution, page 522
fitness, page 523
founder effect, page 526

gene flow, page 525
gene pool, page 519
Hardy-Weinberg principle, page 522
hybrid, page 529
macroevolution, page 531
microevolution, page 530
neutralist, page 528
phyletic gradualism, page 532
population bottleneck, page 526
population genetics, page 519
punctuated equilibrium, page 533
reproductive isolating mechanism, page 529
selectionist, page 528
sexual selection, page 527
speciation, page 529

STUDY QUESTIONS

REVIEW WHAT YOU HAVE LEARNED

1. Define gene pool.
2. Name four causes of genetic variation.
3. State five conditions that limit the Hardy-Weinberg principle.
4. Define evolution in genetic terms.
5. List three ways in which natural selection can affect a population, and give an example of each effect.
6. Explain how each of the following serves as an agent of evolution: mutation, gene flow, genetic drift, nonrandom mating.
7. Define species.
8. How do microevolution and macroevolution differ?
9. Compare phyletic gradualism and punctuated equilibrium.

APPLY WHAT YOU HAVE LEARNED

1. What aspects of the cheetah's population genetics cause it to be listed as an endangered species?
2. Opponents of evolution point to the theory of punctuated equilibrium as proof that even scientists doubt Darwin's ideas. Do you agree? Explain your answer.
3. The Hardy-Weinberg principle applies only under very special conditions. What is its value, given that real populations rarely fit these conditions?

FOR FURTHER READING

Cook, L. M., G. S. Mani, and M. E. Varley. "Postindustrial Melanism in the Peppered Moth." *Science* 231 (1986): 611–613.

Koehn, R. K., and T. J. Hilbish. "The Adaptive Importance of Genetic Variation." *American Scientist* 75 (1987): 134–141.

Mayr, E. *Toward a New Philosophy of Biology: Observations of an Evolutionist.* Cambridge, MA: Harvard University Press, 1988.

O'Brien, S. J., D. E. Wildt, and M. Bush. "The Cheetah in Genetic Peril." *Scientific American* (May 1986): 84–95.

Stebbins, G. L., and F. J. Ayala. "The Evolution of Darwinism." *Scientific American* (July 1985): 72–82.

Wilson, A. C. "The Molecular Basis of Evolution." *Scientific American* (1985): 164–173.

Recombination

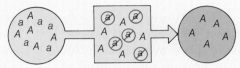

① Genetic variation is the raw material of evolution; its sources include mutation and recombination.

Mutation

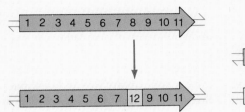

② Evolution can be defined as changes in allele frequencies in a population over time. The agents of evolution include natural selection, gene flow and genetic drift.

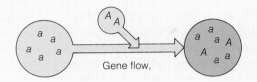

Natural selection

③ New species arise when one population of a species becomes reproductively isolated from other populations.

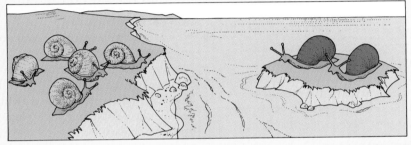

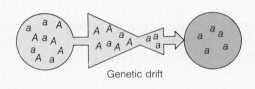

Gene flow,

Genetic drift

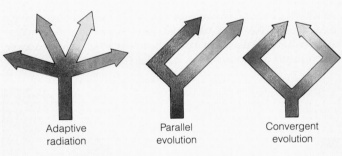

| Adaptive radiation | Parallel evolution | Convergent evolution |

④ There are several patterns of evolutionary descent.

⑤ Biologists have proposed two models of evolution: the gradual accumulation of small differences and bursts of rapid change punctuating long periods with little change.

Gradualism

Punctuated equilibrium

⑥ Extinction is the ultimate fate of all species.

CHAPTER 33

POPULATION PATTERNS IN SPACE AND TIME

THE RISE AND FALL OF A DESERT POPULATION

More than 2000 years ago, a band of native Americans migrated from Mexico to one of the world's most forbidding deserts, the Salt River valley of central Arizona, where Phoenix sprawls today. Using sharp sticks and manual labor, small teams of the immigrants (and later their descendants) dug an immense network of irrigation canals over 1600 km (1000 miles) long (Figure 33.1). These ditches captured winter runoff from the surrounding mountains and in summer carried water from intense local thunderstorms. With this precious redirected resource, the Hohokam Indians each year were able to grow two full crops of corn, squash, beans, barley, cotton, and tobacco over an area of more than 25,600 km² (10,000 square miles).

Archaeological evidence shows that the Hohokam people were probably the first to irrigate in North America, and with their amazing agricultural success, they flourished in the desert valleys. Their population doubled and redoubled, and their settlements grew from clusters of simple mud huts to at least 22 cities of large, multistoried earthen buildings. Some experts believe that the adobe dwellings housed 1 million people—nearly the population of present-day Phoenix.

Then, about 600 years ago, the Hohokam's huge desert culture suddenly vanished, leaving nothing but ruins, grave markers, and bones. Archaeologists think that alkali salts may have built up in overirrigated croplands, poisoning the plants and causing agricultural production to plummet. The Hohokam may also have exhausted the surrounding desert of its limited natural resources—firewood in particular.

Was the rise and fall of the Hohokam special? Or was it governed by the same biological principles that regulate population growth in other species? What are the implications of the Hohokam's boom and crash cycle for today's human population growth? And how does such rapid growth affect the quality of our lives? As this chapter explores questions like these, two themes will emerge. First, closely interacting biological and physical factors govern the abundance and distribution of any species in any area. For example, the amount of rainfall influenced the number of trees available to the Hohokam for fuel, which in turn affected the tribe's population size in the Arizona desert. Second, the distribution and abundance of populations in space and time are closely tied to the evolution of species. Genetic adaptations determine a population's size and distribution, and changes in population size or location can affect the prevalence of specific genes.

This chapter will answer the following questions:

- At what levels do organisms interact with their surroundings?
- Why are particular organisms found only in specific places?
- Why are organisms plentiful at one time but not at another?
- How has human population growth become the most important ecological factor on earth?

FIGURE 33.1 *Canals of the Vanished Hohokam Tribe Helped Support a Large Population in the Arizona Desert.* An archaeological expedition in 1964 and 1965 uncovered miles of these 1000-year-old hand-hewn "lifelines" over 2 m (7 ft) deep and 9 m (30 ft) across, which conveyed water from the surrounding hills to corn, bean, and squash fields in the Salt River valley.

THE SCIENCE OF ECOLOGY: LEVELS OF INTERACTION

Early peoples amassed a simple form of natural history. Their survival depended on knowing the whereabouts of food sources, and they could describe, often through legends or rituals, the seasons when deer were abundant and the valleys where huckleberries grew along with other edible fruits, nuts, and plants. The modern science of ecology has its roots in natural history, but it applies research tools and the scientific method to probing the explanations behind observed phenomena. What factors cause deer populations to increase in the summer, for example, and what allows a particular valley to bristle with huckleberries?

Ecology is the scientific study of how organisms interact with their environment. Ecologists use their science to determine the distribution and abundance of organisms, and the tenets of ecology have become as crucial to human survival as natural history once was to our ancestors. The more people know about the factors that influence plant growth, the better they can manipulate those factors to produce additional food. That is what the Hohokam did when they channeled water to the desert, and it is what modern plant scientists seek to do with plant breeding and genetic engineering.

A key word in the definition of ecology is *interact*. Organisms interact with the other living things that collectively constitute their *biotic* environment, as well as with the nonliving physical surroundings that make up their *abiotic* environment. For a squirrel, the biotic environment includes the acorns it eats, the blue jays that compete for the nuts, the ticks that parasitize and weaken the squirrel, and the siblings that share the squirrel's territory. The abiotic environment includes the rainfall, sunlight, and soil that regulate acorn production and the hot and cold temperatures that the squirrel must endure in summer and winter.

So numerous are each organism's interactions with other living things and the physical environment that biologists organize ecology, the science of interactions, into a hierarchy of four levels: populations, communities, ecosystems, and the biosphere (Figure 33.2).

A **population** is a group of interacting individuals of the same species that inhabit a defined geographical area. The Hohokam Indians of central Arizona, the saguaro cactus of that same region, the saber-toothed cats of Rancho La Brea in prehistoric Los Angeles, and the alligators of a Louisiana swamp are all examples of populations.

A **community** consists of two or more populations of different species occupying the same geographical area. The Hohokam and the giant saguaro cactus whose fruits they gathered and ate made up a community, and so do

alligators and the fish they consume and orchids and the bees that pollinate them.

Populations and communities include only biotic factors. Such groupings always exist within a physical setting, however, and so ecologists have a third hierarchical level: the **ecosystem,** made up of interacting living things together with the physical factors of the environment.

FIGURE 33.2 *Ecological Hierarchy.* The Hohokam Indians were a population of organisms in a desert environment (lowest level). The Hohokam interacting with their living environment, including cactuses, corn, and jackrabbits, made up a community. That community, in its nonliving environment—the dry physical setting—constituted an ecosystem. And all of earth's ecosystems combined make up the biosphere.

The Hohokam's ecosystem included saguaro cactus, corn, and black-tailed jackrabbits, as well as sparse rainfall and the searing summer temperatures of the Arizona desert.

Ecologists recognize that ecosystems are further influenced by global phenomena, such as climate patterns, wind currents, and nutrient cycles. Thus, in the four-tiered hierarchy, groups of ecosystems make up the highest level, the **biosphere**—our entire planet with all its living species, its atmosphere, its oceans, the soil in which living things are found, and the physical and biological cycles that affect them.

The next few chapters examine ecological interactions at increasingly higher levels of organization. Here, we begin with the level of population and discuss the influences on a population's location in time and space, as well as the question of success as measured by a population's size.

(b)

(a)

FIGURE 33.3 *Temperature: A Physical Factor That Limits Plant Distribution.* Pine trees of the central Alps (a) grow best at a maximum of 15°C (59°F), while dry-adapted *Hammada* bushes of Israel's Negev Desert (b) grow best at 44°C (111°F).

≋ DISTRIBUTION PATTERNS: WHERE DO POPULATIONS LIVE?

Organisms tend not to be scattered evenly, but to exist in only certain spots and be absent entirely from others. What limits where an organism lives?

LIMITS TO GLOBAL DISTRIBUTION

Why is it that saguaro cactus and mule deer appear naturally in Arizona, where the Hohokam lived, but oak trees and squirrels occur more commonly in the areas where most North Americans live? Questions and answers involving why an organism's range extends to one part of the world but not another are especially important to us when they involve organisms we use for food or materials or when they involve species that cause disease in people and domestic plants and animals. In general, three conditions limit the places where a specific organism might be found: physical factors, other species, and geographical barriers.

■ **Physical Factors** Organisms may be absent from an area because the region lacks the proper sunlight, water, temperature, mineral nutrients, or any one of a host of physical or chemical requirements. For example, pine trees from the Austrian Alps photosynthesize best at a cool 15°C (59°F), but the *Hammada* bush of the Israeli desert photosynthesizes best at a scorching temperature of 44°C (111°F) (Figure 33.3). Neither plant could survive in the other's environment because genetic adaptations

fostered by natural selection have left each plant specialized for a particular set of physical conditions.

■ **Other Species** Other species may block survival and limit a population's distribution. If certain species are already firmly established in an area, they may prevent the incursion of new species by monopolizing food supplies or acting as predators or parasites. Large regions of Africa, for instance, are nearly uninhabitable to people or cattle because the resident tsetse fly transmits the protist that causes sleeping sickness.

■ **Geographical Barriers** A species may be absent from even highly favorable areas because a geographical barrier blocks access. Seas, deserts, and mountain ranges can be so wide or high that an organism cannot crawl, swim, fly, or float across the barrier. Europeans artificially bridged such a gap when in 1890 they introduced about 80 starlings into New York City's Central Park. Now, millions of the dark, speckled birds chatter throughout America's cities and countrysides. North America turned out to be a prime habitat for these organisms; the obstacle to their earlier dispersal was simply the ocean.

The factors that limit an organism's global range are only part of what we must know to explain its distribution. The other part is how the organism is dispersed within its range.

LOCAL PATTERNS OF DISTRIBUTION

Within their ranges, organisms can have a uniform, random, or clumped distribution (Figure 33.4). Organisms have a *uniform* distribution if they are spaced at regular intervals. The apple trees in an orchard have uniform

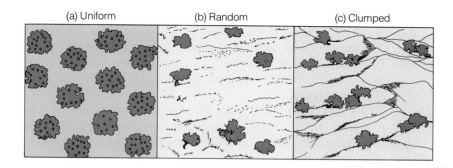

(a) Uniform   (b) Random   (c) Clumped

FIGURE 33.4 *Distribution Patterns in Local Populations.* (a) Apple trees planted in an orchard have a uniform distribution, but the pattern is uncommon in nature. (b) A random distribution, as displayed by manzanita bushes, is often found when resources are uniform, but (c) a clumped pattern is the most common distribution, as seen in live oaks growing on the rolling yellow hills due east of San Francisco.

distribution. Such a pattern is rare in natural populations and usually occurs only where physical factors like water and soil quality are also uniform and where organisms compete strongly for some limiting resource.

By contrast, organisms will have a *random* distribution if individuals do not influence each other's spacing and if environmental conditions are uniform. Randomly distributed populations include plant species whose fruits are eaten by animals that later drop the seeds haphazardly in their feces. Like uniform distributions, random distributions occur infrequently in nature because resources are seldom uniformly available.

The most common pattern of population distribution in space is a *clumped* distribution, with several members of the population occurring close to each other but a long distance from other groups. Clumping occurs because resources are almost always limited to certain **habitats,** or special areas within a range where an organism can actually live. The Hohokam did not spread out across all of what is now Arizona; instead, they occupied their habitat—river valleys with adequate water and fertile flatlands.

Herds of caribou, prides of lions, and flocks of great blue herons also show clumped distribution, as do plants whose seeds drop and sprout near the parent plant, such as daisies and poppies.

By describing a species' global range and its local distribution, ecologists define the organism's habitat. Then, by identifying the various living and nonliving factors that allow a population to exploit that habitat, they take a big step toward understanding where organisms live and why. The next step is to determine the factors that limit a population's size.

≋ CONSTRAINTS ON POPULATION SIZE

Like modern-day farmers and foresters, the ancient Hohokam were concerned with a population's density—the number of individual organisms in a certain amount of space (Figure 33.6). They could use this knowledge to

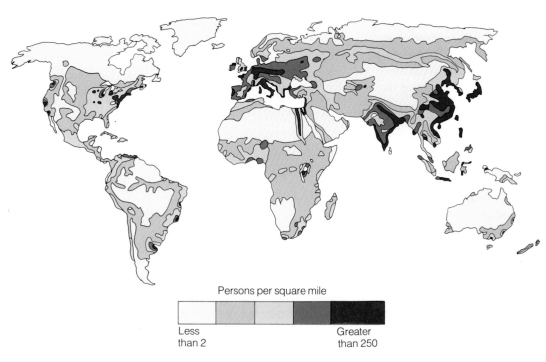

Persons per square mile

Less than 2   Greater than 250

FIGURE 33.5 *Worldwide Distribution of Human Population.* The human population is not distributed equally around the globe. There is a greater population density—more people per square mile—in much of Asia, India, Europe, and eastern North America and around coastlines, rivers, and inland bodies of water than in other parts of the world.

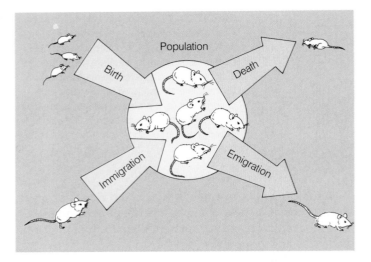

FIGURE 33.6 *Birth, Death, Immigration, and Emigration: Factors Affecting Population Size.* Birth and immigration tend to increase the size of a population, while death and emigration decrease population size.

maintain the crops they harvested at optimal levels so that there would be as much grain or lumber available as possible on a continuing basis. The curious biologist, however, is usually not content just to describe a population's density at one point in time, but wants to understand how that population density changes over time.

BIRTHS, DEATHS, AND POPULATION SIZE

Population size changes over time when individuals enter or leave the population. Clearly, if more members enter than leave it, the population will grow. Individuals can enter the population by either birth or immigration, and members can leave the population by either death or emigration (Figure 33.6). If the number of individuals gained from births and immigration is exactly equal to the number lost to deaths and emigration, then there is *zero population growth*.

■ **How Survival Varies with Age** Death and birth are major parameters affecting population size. And for many species, the likelihood that an individual will die depends on its age. The chances that a 90-year-old person will live for one more year are much slimmer than the chances that a 20-year-old will live for one additional year. A *survivorship curve* is a plot of the data representing the proportion of a population that survives to a certain age (Figure 33.7). A table of the numbers from a survivorship curve is called a *life table*, and it shows the *life expectancy* (average time left to live) and probability of death for given ages.

Insurance companies use life tables to determine the policy costs for customers of different ages. From the survivorship curve in Figure 33.7, you can see why a 70-year-old man must pay more for insurance than a 70-year-old woman: He is more likely to die in the next year than she is.

As one might expect, different species have survivorship curves of different shapes. People and most large mammals, such as rhinoceroses, have a low mortality in early and middle life and an increasing death rate in old age. In contrast, many bird species have a fairly constant death rate at all ages from birth to the end of the life span, while in many species of fish and invertebrates, very young individuals have a high probability of dying, but those that survive this dangerous initial period have a good chance of reaching old age (Figure 33.8).

■ **Innate Capacity for Population Growth** Given plenty of nutrients, space (or shelter), water, and benign weather, and assuming the absence of predators or agents of disease, every population will expand infinitely, because all organisms have an innate reproductive capability when conditions are ideal. The capacity for reproduction under idealized conditions is amazing. Darwin calculated that a hypothetical pair of impossibly long-lived elephants could give rise to 19 million descendants in 750 years! And most organisms reproduce even faster. If a female cockroach had about 50 surviving daughters each month, the number of female roaches would rise to 800 billion after just seven months. Women seldom approach the maximum human fertility of about 30 children per female. The highest fertility rate recorded for any human population was among the Hutterite communities of Canada's prairie provinces in the early twentieth century, where the average family had 12 children.

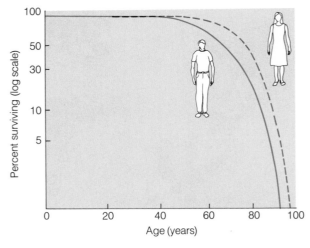

FIGURE 33.7 *Human Survivorship Curve: Most Men Live to Age 70 and Most Women to Age 80, but Succumb Before 90.* Each year after 80, the chances of living another year go down substantially. Beyond 90, the drop is precipitous.

Rapid population growth is easier to visualize when plotted on a graph such as the one in Figure 33.9. Let's say that the individuals of a long-lived mouse species grow to reproductive age in one year and that, ignoring males, a population initially consists of ten female mice that each produce one female offspring per year on average. (Since only females bear young, ecologists often view populations as females giving rise to more females.) At the end of the first year, the population will include 20 female mice, and if each of those has one more female per year, there will be 40 female mice at the end of two years, 80 after three years, and so on. The explosive growth results in a **J-shaped curve** representing **exponential (logarithmic) growth.**

We did not consider the death rate in this mouse example of a J-shaped curve, but even with losses due to death, the growth curve would still be J-shaped; it would simply have taken longer for the population to reach a given number of individuals on its rise to infinity. Under these artificially ideal conditions, any population would follow the J-shaped curve of exponential growth if the birthrate exceeded the death rate by even a small amount.

Fortunately, real organisms in natural situations do not rigidly follow the J-shaped exponential growth pattern. Organisms in nature cannot sustain continued, limitless growth at the full force of their reproductive potential because food supplies and living space are finite. Hence, our planet is not covered with elephants neck-deep in roaches. The realities of supply and demand explain this

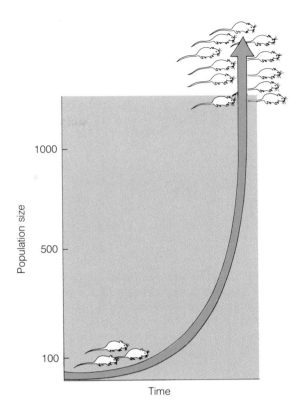

FIGURE 33.9 *Exponential Growth Results in a Population Explosion.* Every species has the innate reproductive potential to found its very own population explosion—a J-shaped curve, heading for an infinity of mice or snails or pear trees or people.

curb on population growth despite the organism's reproductive capabilities.

■ **Actual Capacity for Population Growth** With the J-shaped curve, ecologists have an idealized standard against which to measure actual growth in real populations. A classic case was the growth of the sheep population on the island of Tasmania, south of Australia, in the early nineteenth century (Figure 33.10, page 544). When English immigrants first introduced the sheep to the new environment, resources were abundant, and the sheep population expanded nearly exponentially for several decades. As the density of sheep on the island rose, competition for limited resources increased, and by 1840, each sheep had a smaller share of food and living space. As a result, each individual was less likely to survive and more likely to die, and each had a smaller chance of reproducing. After 1850, the total growth rate decreased, and the population size leveled off at about 1.6 million sheep.

As Figure 33.10 shows, the graph representing population growth began like a J-shaped curve, but flattened into an **S-shaped curve**, representing a situation in which a large population grows more slowly than a small pop-

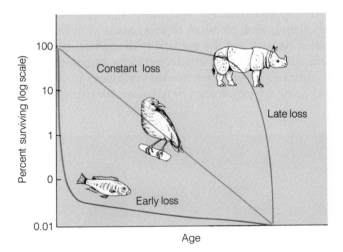

FIGURE 33.8 *Survivorship Curves for Various Species.* With species like the rhinoceros, mortality is low in early life but increases for older individuals; these are called "late-loss" species. For most birds, the perils of living are permanent; mortality occurs steadily, regardless of age; these are "constant-loss" species. For species like most fishes, surviving early life is quite unlikely, but once past the first months as an easy target for predators, the organism's chances of surviving remain stable until old age; these are considered "early-loss" species.

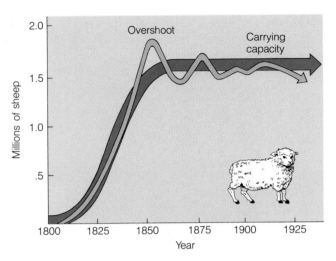

FIGURE 33.10 *Sheep on Tasmania: The Growth of Real Populations Can Follow an S-Shaped Curve.* Starting in 1800, sheep populations on the island of Tasmania rose exponentially, but by 1855, they overshot the carrying capacity of the land. Eventually, their numbers oscillated and stabilized, and this plateau is reflected by the S-shaped curve.

ulation would in the same area. The density at which a growing population levels off—1.6 million sheep on Tasmania in the preceding example—is called the **carrying capacity** (represented by the symbol *K*). Carrying capacity is the population density at which all the available resources are being used by the organisms under study. The addition of more organisms would cause others to die for want of sufficient resources. Since the S-shaped curve takes into account the birthrate, death rate, and proportion of resources that remain unused (or available), it is a more accurate reflection of what happens in natural populations than the J-shaped curve.

Archaeological data show that the Hohokam population probably fit an S-shaped curve, growing from a small initial size 2000 years ago to a relatively stable maximum number and remaining there for two or three centuries before disappearing about 1450 A.D. Examples like sheep on Tasmania and Hohokam in Arizona cause ecologists to ask, what factors prevent unlimited growth?

HOW THE ENVIRONMENT LIMITS GROWTH

The limited growth that real populations display comes from a phenomenon called *environmental resistance to growth:* factors such as limited food supplies and limited living space that prevent exponential growth. Ecologists call limiting factors like these *population-regulating mechanisms* and classify them as either *density-dependent mechanisms* or *density-independent mechanisms.* Density-dependent mechanisms become more influential as the population's density increases, and they can have absolutely catastrophic effects. They work by increasing death and limiting reproduction, and disease is a classic example. In dense populations, disease spreads more rapidly; in sparse populations, it spreads more slowly. For instance, in prairie dog towns of low density, the incidence of flea-transmitted bubonic plague is very low. But when the dogs are densely packed, outbreaks of plague often wipe out entire populations (Figure 33.11). On the other hand, density-dependent factors can stabilize population size, as when a group of squirrels is just large enough to use all the nuts available in a particular forest.

Density-independent mechanisms, as the name implies, exert their effects regardless of population density. These are best illustrated by adverse weather conditions—floods, droughts, or freezing temperatures. These also increase death and lower reproduction, either outright or by affecting food supplies, and take their toll in sparse and dense populations alike. In actual practice, it can be difficult to separate the effects of density-dependent and density-independent mechanisms, because they often work together. For example, a severe drought would kill some individuals in a desert population of any size, but the effects would be far worse if the density were greater. This interaction leads many ecologists to classify population-regulating mechanisms by an additional criterion: whether the mechanism originates outside or inside the population.

Mechanisms that originate outside the population (*extrinsic mechanisms*) include biological factors, such as food supplies, natural enemies, and disease-causing organisms, and physical factors, such as weather and shelter. Rainfall and the availability of saguaro cactus fruits were outside factors that affected Hohokam populations. In contrast, *intrinsic mechanisms* originate in an organism's anatomy, physiology, or behavior. For example, under

FIGURE 33.11 *Dense Populations Are More Likely to Be Plagued by Disease.* Among these prairie dogs (*Cynomys ludovicianus*), overcrowding allows flea-borne bubonic plague to spread more easily.

FIGURE 33.12 *Contest Competition: A Clash Over Resources.* Male caribou (*Rangifer arcticus*) in arctic Alaska clash over access to mates, a limited resource.

crowded conditions and depletion of resources, many marsupials, such as kangaroos and koalas, resorb already developing embryos; this physiological response lowers the rate of population growth. In another example, mouselike creatures called lemmings will migrate away from food-depleted regions, a behavior that lowers the local population density.

■ **Competition** The most important intrinsic regulating mechanism is competition among members of the same species; such competition depends, in part, on population density. As the population grows and resources diminish, competition for food and space becomes intense among population members. *Scramble competition* occurs when resources are equally available to all members of a population but individuals must rush for their share or risk losing it. A flock of pigeons descending on a grain lot illustrates scramble competition beautifully: Birds that don't push, peck, and aggressively pursue their morsels risk going hungry.

Contest competition involves literal clashes, usually among males, for social position, mates, or territory. The clashing of antlers as male caribou engage in tournaments for leadership of a herd (Figure 33.12) and the jousting of rams in spring are awesome displays of contest competition.

■ **Population Crashes** While competition and other intrinsic and extrinsic mechanisms can cause population growth to slow and level off, as it did with sheep in Tasmania, in some species, these forces cause a rapid decline to follow a period of intense population growth. For example, the number of reindeer introduced onto an island off the southwest coast of Alaska (Figure 33.13) grew from an initial population of 25 in 1891 to about 2000 and then crashed to 8 by 1950. When the reindeer were first introduced, lichens and other food sources were plentiful, having accumulated for centuries without predation. After the deer ate the accumulated food, however, new food appeared too slowly during each short

growing season to support the booming reindeer population, and so the population crashed. This is called a "boom-and-crash" sequence of population growth. Some populations routinely boom and crash as they repeatedly overshoot the carrying capacity of their environment, deplete critical resources, and then crash (see Figure 33.13).

To summarize, populations in nature usually follow one of two patterns: the S-shaped growth curve, in which density slowly reaches the carrying capacity of the environment and then remains stable for long periods (as with the Tasmanian sheep), or the boom-and-crash pattern of nearly exponential (J-shaped) growth followed by an overshoot of the carrying capacity and a precipitous decline in density (as with the island reindeer). Indi-

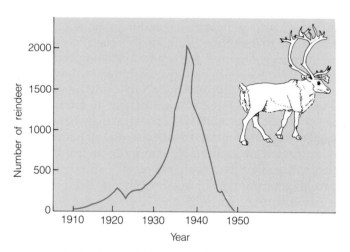

FIGURE 33.13 *Overexploiting Limited Resources Can Lead to a Population Crash: Boom and Crash in the Island Deer.* Reindeer introduced to a small island in the Aleutian chain southwest of Alaska experienced a boom—exponential J-shaped growth—by eating centuries' worth of slow-growing lichens in a few decades. Once the food ran out, the population crash was rapid and spectacular.

(a)

(b)

FIGURE 33.14 *Rapid Reproduction Versus Careful Investment.* Dandelions and rhinoceroses symbolize the life history traits associated with *r*- and *K*-selection. (a) Dandelions produce many small seeds, grow rapidly to sexual maturity, and reproduce rapidly, filling up a newly opened habitat quickly before suffering a population crash. (b) The rhinoceros, on the other hand, has a single huge baby that grows slowly to maturity, maintaining the population size near the maximum its environment can hold.

vidual species clearly respond differently to their environments, but what characteristics of individual species cause them to follow one pattern or the other?

POPULATION GROWTH AND STRATEGIES FOR SURVIVAL

Evolution acts on organisms to maximize their individual genetic contributions to future generations. To successfully grow, survive, reproduce, and thus make a genetic contribution to the future, individuals must allocate their limited energy supplies. A very fast growing organism that expends most of its energy enlarging may have little energy left over for reproducing. Conversely, an individual that expends a huge amount of energy attracting a mate or producing thousands of eggs may have little energy remaining for day-to-day survival activities. The way an organism allocates its energy is its **life history strategy.** There are two basic life history strategies, one exemplified by dandelions and the other by rhinoceroses (Figure 33.14).

A dandelion's life history strategy is to reproduce as rapidly as possible. Ecologists call this *r-selection* because in these organisms, natural selection has favored genes for *r*apid *r*eproduction. Dandelions quickly fill an environment—a newly plowed field, for example—before the environment changes—that is, before winter comes, before the corn grows too large, or before the farmer plows again. Natural selection has caused dandelions to

maintain genes that allow rapid embryonic development, early reproduction, and large numbers of small seeds containing few stored nutrients. Dandelions experience an *early-loss* type of survivorship because most of the light, wind-borne seeds die shortly after germination. Our expression "to grow like a weed" reflects the life history strategy of dandelions. Small invertebrates, such as fruit flies and water fleas, also show *r*-selection, and like dandelions and other such organisms, often have a boom-and-crash growth curve.

Rhinoceroses, on the other hand, inhabit more stable environments and tend to maintain stable populations near the habitat's carrying capacity. Ecologists call this life history strategy *K-selection* because natural selection has favored genes that keep the population near the carrying capacity (abbreviated *K*). Rhinoceroses, most other mammals, and most large, long-lived plants, such as the saguaro cactus, reflect this life history strategy. In rhinoceroses, embryonic development is slow (gestation takes about 15 months), and reproduction is delayed until the age of about five years. Although rhinoceroses have only one calf at a time, the newborns are huge (the weight of an average male college student). Once born, they survive for about 40 years, experiencing a *late-loss survivorship* schedule.

Like other species with stable population sizes, rhinoceroses are highly specialized to compete for resources in their environments. Specializations include immense size, thick, armor-plated skin, and a fingerlike extension of the upper lip that pushes grasses and twigs into the mouth. Rhinoceroses are fast approaching extinction because people slaughter them for their nasal horns. While these protuberances are made of the very same protein found in our hair and fingernails, some people have a superstition that the powdered horn bestows aphrodisiacal properties, and they are willing to pay more for the material than for gold, ounce for ounce. Knowing that rhinoceroses are *K*-selected organisms with slow rates of reproduction, you can appreciate how hard it is for populations of the fascinating beasts to become reestablished once decimated.

In both dandelions and rhinos, genes favored by natural selection determine the life history traits. These genes can, for example, decrease the weight of the seeds or delay reproductive maturity. A population of any organism could theoretically follow a J-shaped, S-shaped, or boom-and-crash growth pattern; but in reality, while dandelions flourish under boom-and-crash conditions, rhinos would quickly become extinct. The Hohokam population also seemed to rise, reach a plateau, and then crash. Does this suggest that humans are *r*-selected or *K*-selected? And what does an analysis of our own species' population patterns predict for the future? The final section of the chapter addresses these questions.

﹏ THE PAST, PRESENT, AND FUTURE OF THE HUMAN POPULATION

People have achieved unparalleled mastery over their environment through agriculture, medicine, sanitation, transportation, and industrialization. Are we—with our gleaming cities, our gigantic corporate farms, our burgeoning global population—governed by ecological rules? Or have we somehow moved beyond booms, crashes, and growth curves? Stated more formally, to what extent do general ecological principles on the distribution and abundance of organisms apply to populations of people? The answer involves a backward glance at human history, long-term population trends, and some predictions for our future population growth.

TRENDS IN HUMAN POPULATION GROWTH

You can see the history of human population growth and its staggering current proportions at a glance (Figure 33.15). In the first phase of human history, from our species' origin to about 10,000 years ago, the population grew slowly as people existed by hunting animals and gathering naturally occurring roots and fruits. Population density worldwide was probably about 10 million by 8000 B.C. As a species, we seem to fall into the *K*-selection category because of our slow development, long

FIGURE 33.15 *Human Population Bomb: Our Species' Exponential Growth Rate Is Based on Cultural Advances in the Use of Environmental Resources.* (a) Human population growth since prehistoric times displays a classic J-shaped exponential curve. Our doubling rate is currently about 40 years. (b) Throughout most of human history, we lived as hunter-gatherers like this South African boy today. (c) In our agricultural phase, from about 8000 B.C. to about A.D. 1750, populations grew steadily and lived much as these Bolivians do now. (d) The industrial phase, beginning in eighteenth-century England with urban sweatshops like this hat factory, ushered in the exponential growth that continues today.

lives, large bodies, relatively few offspring, extended and intensive parental care, and highly specialized brains that helped us compete for resources with cunning efficiency.

Growth accelerated during a second phase beginning about 10,000 years ago, when people started planting and tending crops and domesticating animals in the so-called *agricultural revolution*. The shift to agriculture was rapid and worldwide, perhaps because people are so adaptable and can transmit their culture, or ways of living, to others. As agricultural techniques spread and improved between about 8000 B.C. and A.D. 1750, the world population increased slowly from 10 million to about 800 million. Since agriculture allows more efficient use of resources, its practice increases the environment's carrying capacity for humans. The Hohokam, with their intensive irrigation of desert river valleys, exemplify this stage in human cultural development. In its natural state, the Arizona desert has a very low carrying capacity, but irrigation increased the amount of corn the Hohokam could grow and the number of people the desert could support.

A third phase of growth began in eighteenth-century England with the *industrial revolution*. Inventions and scientific advances triggered vast social changes that changed a populace living mainly as farmers, craftspeople, and merchants to a population working mainly in factories and living in crowded cities (see Figure 33.15). In the next 250 years, much of the world would follow this industrialization and social upheaval. The steam engine was a key invention, and its impact was enormous. A farmer with a steam engine attached to a tractor could accomplish the work of dozens of people in a single day and thus increase food production. A steam-driven train or ship could rapidly distribute food and other necessities of life and thus blunt the impact of local famine.

The towering ascension in the graph of human population growth should unnerve you. It is the familiar J-shaped pattern of exponential growth, much like that of the Alaskan island reindeer just before they overexploited their environment and suffered a population crash. By analyzing the causes of our own population boom, ecologists hope to learn how humans can avert a crash in the future.

UNDERLYING CAUSES OF CHANGE IN HUMAN POPULATION SIZE

How did the agricultural and industrial revolutions quicken the pace of the human population explosion? Ecologists and historians alike have wondered whether the invention of agriculture *allowed* human populations to increase, or whether people were *forced* to invent agricultural practices to help support population densities that were already exceeding the carrying capacity of the land where they lived. Many observers believe the latter and suggest that population growth has been a constant feature of the human experience, continually forcing people to adopt new strategies for increasing the amount of food their land could produce. If true, this necessity has led to some marvelous inventions, including "green revolution" supercrops and genetically engineered hybrid food plants. Carrying capacity, however, cannot be increased forever; the productivity of the land must, at some point, be reached and exceeded.

BIRTHRATES AND DEATH RATES IN DEVELOPING NATIONS

To understand the causes of human population increase, particularly the tremendous surge after the industrial revolution, we must recall that the growth of a population is determined by the number of people born each year minus the number of people who die. If more people are born than die, the population grows. Before 1775, in relatively well-off countries like Sweden, about 39 people were born each year and 35 died for every 1000 people in the population, and so the population enlarged at a constant low rate of about 4 people per 1000 per year. After 1775, as industry expanded, people enjoyed improved nutrition, better personal and public hygiene, protection of water supplies, and the reduction of communicable diseases such as smallpox. With these factors came a gradual decline in the death rate. While the death rate began to decline in 1775, the birthrate did not start to drop in developed countries until about 100 years later. Consequently, each year many more people were born than died, and this translated into an increase in the rate of population growth. But, by the last decade of the twentieth century, both the birth and death rates in industrialized nations had dropped to all-time lows: For every 1000 Swedes, about 10 died and 10 were born per year. A changing pattern from high birth and high death rates (39 and 35 per 1000) to low birth and low death rates (about 10 per 1000) is called a **demographic transition.**

In industrial Europe and North America, the demographic transition occurred in the first half of the twentieth century and reduced overall growth rates to low levels. The populations of Africa, Asia, and Latin America, however, have continued to grow at immense rates for much of this century (Figure 33.16). The concept of a demographic transition helps explain why. The death rate in those countries remained high until the mid-twentieth century, when the importation of Western medicine and public health technology helped spare lives in record number and lowered the death rate to about 12 per 1000 in the mid-1970s. Simultaneously, however, the

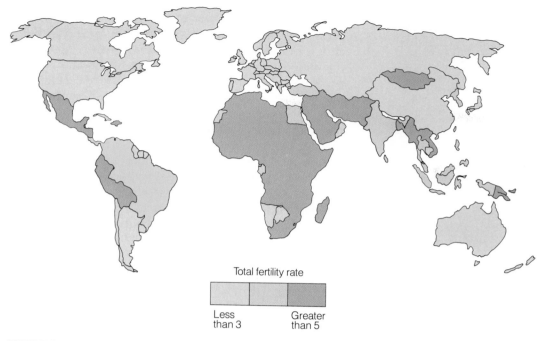

Total fertility rate

Less than 3 | | Greater than 5

FIGURE 33.16 *Global Patterns of Human Birthrate.* In virtually all developed (highly industrialized) nations, the total fertility rate is less than three children per woman, while in less-developed countries, the rate is three to five or more. A notable exception is China, where institutionalized policies for late marriage and one-child families have slashed the birthrate in just a few decades.

birthrate remained high, about 35 per 1000, based largely on traditional cultural practices and beliefs. Thus, many more people were born than died (35 minus 12 equals 23 per 1000 each year in this example). The huge disparity between death rate and birthrate caused an enormous net growth rate and hence an immense population boom in the twentieth century. One of the biggest and most potentially disruptive consequences of this alteration in Third World population size is an increase in the proportion of young people.

GROWTH RATES AND AGE STRUCTURE

A sure sign of a population's growth rate is its **age structure:** the number of people in each age-group. Age structure is significant because it can be used to judge a population's growth status. Rapidly growing populations generally have many young individuals and hence a pyramid-shaped age profile, like the one depicting Mexico in 1977 (Figure 33.17a). In contrast, profiles of stable or declining populations tend to be more bullet-shaped, like the one depicting Sweden in 1977 (Figure 33.17b), because for each new person born (thus entering the chart from the bottom), another person dies (generally leaving the chart from the top). With one look at data graphed this way, you can infer the kinds of social services that will be needed by slowly or rapidly growing populations: schools for pyramid-shaped populations, health-care facilities for the elderly in bullet-shaped ones.

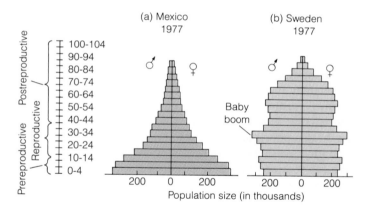

FIGURE 33.17 *Age Structure Diagrams Help Reveal Human Population Growth and the Potential for Future Explosions.* In these graphs, males are represented on the left, females on the right, and the length of each bar signifies a given age-group's proportion of the population. (a) Note the pyramid-shaped age structure of modern Mexico, with its high birthrate and low death rate and the huge population segment that will soon be reaching reproductive age. (b) The Swedish population in the same year had a very different age structure because the birthrate had dropped very close to the death rate, and as a result, each age class was only slightly different in size from the one below it.

FIGURE 33.18 *A Successful Species?* Will human populations level out from exponential (J-shaped) growth to S-shaped growth, or will a crash follow our species' 250-year boom?

WHAT CAUSES BIRTHRATES TO DECLINE?

By analyzing the demographic transition, it is clear that the key to population control is a low birthrate. At the heart of controversies surrounding population control is a dispute over the factors that lead to low birthrates. Some sociologists feel that birthrates decline because people who have moved to the cities view large families as an economic burden, while those in rural areas continue to see children as potential farmhands and a source of security in old age. Others contend that people have always wanted fewer children, but only widespread knowledge and availability of contraceptives make low birthrates possible.

If improved socioeconomic conditions are a prerequisite for reduced birthrates, then government education programs on contraceptives and the benefits of reduced family size can only be effective after a population experiences an improved standard of living. Most populations with high growth rates, however, cannot afford to wait for improved economic conditions before curtailing their growth, because the added population actually impedes further development. Hypotheses aside, everyone agrees that even if developing nations immediately instituted stringent regulations over birthrates, the global population would continue to expand for some time.

POPULATION OF THE FUTURE

The number of people alive today is greater than the total number who have ever lived and died before us. The press of humanity totals over 5 billion people, and if present growth rates continue, another 5 billion will be added in the next 40 years—perhaps before you retire (Figure 33.18). Even if, from this year forward, family size were reduced to the replacement rate of 2.2 children per woman, the global population would continue to grow for decades, because the youthful citizens of countries like Mexico and Nigeria will soon attain reproductive age. Estimates for a stable size for the human population by the year 2040 range from a low of 8 billion, assuming rapidly implemented birth control programs, to a high of 14 billion, with less successful birth control campaigns. Regardless, our planet's human population will either double or triple in the not-so-distant future.

These astounding figures lead us, quite logically, to ask whether the planet can support 8 to 14 billion people at a reasonable standard of living. At present, no one can accurately predict what the earth's ultimate carrying capacity will be. Scientists do suspect that coal, oil, and some of the other resources on which we now depend may well become exhausted early in the twenty-first century. One thing is certain: Without dramatic steps taken immediately and decisively, the crush of humanity will diminish our standard of living and reduce or forever destroy complex and delicate biological systems such as tropical forests, as well as millions of individual species that have taken entire geological eras to evolve on our planet.

≈ CONNECTIONS

The study of ecology, with its organized approach to understanding population sizes, densities, and growth rates, allows us to reconstruct a plausible explanation for the rise and fall of the Hohokam population. These native Americans flourished in the harsh Southwestern desert, in part because their enormous agricultural achievements, including an immense network of irrigation canals, raised the carrying capacity of their desert environment for both corn and people. Then, however, as the density of their settlements grew in a slow but relentless upswinging J-shaped curve, environmental resistance must have caused hardships, increased mortality, a declining birthrate, and a flattening of the growth curve to an S-shaped model of population growth. There simply could not have been enough good cropland to keep up with a continued exponential rate of growth. If the tribe experienced a few years of drought, as may have happened around 1450, the desert's carrying capacity would have dropped drastically, leaving the population density well above the environmental saturation point. Resources would have diminished in relation to popu-

(1) Organisms interact with the environment and each other as members of populations, communities, ecosystems, and the biosphere.

(2) Climate and competition for resources help determine how populations are distributed in space.

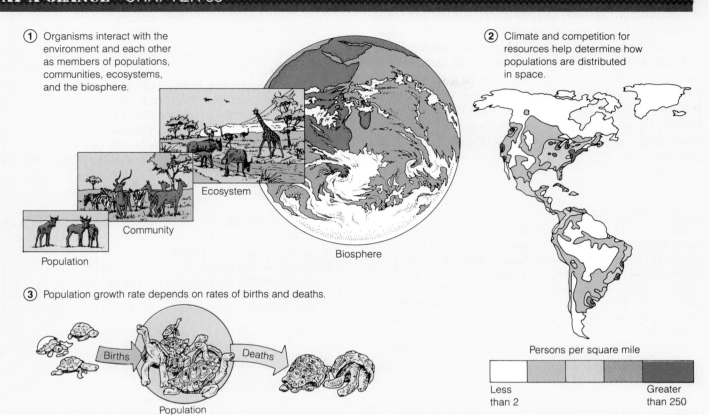

Population

Community

Ecosystem

Biosphere

Persons per square mile

Less than 2

Greater than 250

(3) Population growth rate depends on rates of births and deaths.

Births

Deaths

Population

(4) A low-density population can grow exponentially in a J-shaped curve. Then population growth either overshoots carrying capacity and crashes or stabilizes at the carrying capacity.

(5) Species that bear many small young but lose many tend to experience boom-and-crash cycles; species that produce fewer, larger, better-surviving young tend to have stable populations.

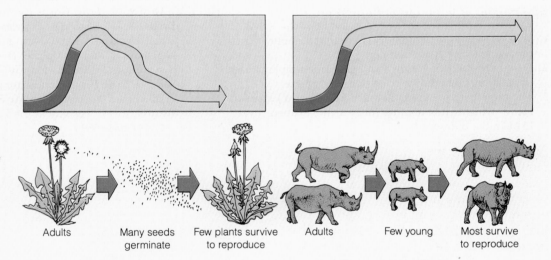

Adults

Many seeds germinate

Few plants survive to reproduce

Adults

Few young

Most survive to reproduce

(6) The human population is growing exponentially, and without dramatically lowered birthrates, there will be a massive population crash.

lation, and disease and mortality would have struck, causing the death rate to exceed the birthrate. Eventually, the population crashed. Despite great cultural advances, people are clearly constrained by ecological principles.

In this chapter, we focused on the dynamics of populations of individual species—sheep, reindeer, dandelions, rhinoceroses, and people. Species, however, do not exist as ecological islands. In the next chapter, we will study communities and see how the distribution and abundance of a particular species depends on interactions with many other species sharing the same environment.

NEW TERMS

age structure, page 549
biosphere, page 540
carrying capacity (*K*), page 544
community, page 539
demographic transition, page 548
ecology, page 539

ecosystem, page 539
exponential (logarithmic) growth, page 543
habitat, page 541
J-shaped curve, page 543
life history strategy, page 546
population, page 539
S-shaped curve, page 543

STUDY QUESTIONS

REVIEW WHAT YOU HAVE LEARNED

1. What is in your living (biotic) environment? Your nonliving (abiotic) environment?
2. Define the following terms, and give an example of each: population, community, ecosystem, biosphere.
3. The growth of one insect population follows a J-shaped curve, while that of another follows an S-shaped curve. Explain the difference.

4. Do "surplus" resources influence a population's size? Explain.
5. How does a density-independent mechanism of population regulation differ from a density-dependent one?
6. Give an example of how a population can "boom" and then "crash."
7. Name a species that shows *r*-selection and one that shows *K*-selection, and explain the difference.
8. What main factors have contributed to the J-shaped growth pattern of the human population?
9. Compare the overall demographic changes in industrialized countries with those in Third World nations.

APPLY WHAT YOU HAVE LEARNED

1. Some people believe that powdered rhinoceros horn is an aphrodisiac. How does this superstition affect the growth of the world's rhino population?
2. As the human population explosion continues into the twenty-first century, do you believe its J-shaped curve will shift to an S-shaped curve or a boom-and-crash curve like that of the reindeer in Figure 33.13? Support your answer.

FOR FURTHER READING

Ehrlich, P. R., and A. H. Ehrlich. *The Population Explosion.* New York: Simon & Schuster, 1990.

Haury, E. W. "The Hohokam: First Masters of the American Desert." *National Geographic* (May 1967): 670–701.

Krebs, C. J. *Ecology: The Experimental Analysis of Distribution and Abundance.* 3d ed. New York: Harper & Row, 1985.

Newman, J., and G. Matzke. *Population: Patterns, Dynamics and Prospects.* Englewood Cliffs, NJ: Prentice-Hall, 1984.

Sai, F. T. "The Population Factor in Africa's Development Dilemma." *Science* 226 (1984): 801–805.

Westoff, C. W. "Fertility in the United States." *Science* 234 (1986): 554–559.

THE ECOLOGY OF LIVING COMMUNITIES

CHAPTER 34

THE INTERTWINED LIVES OF FLOWER AND MOTH

The yucca, a tall, stately member of the lily family, grows in hot, dry regions of the western United States. In California's Mojave Desert, the pointed leaves of the plant jut upward from the parched ground like a sheaf of swords. Above the leaves and stem rises a single flower stalk 4 m (13 ft) tall and loaded with more than 1000 white blossoms. Every spring, dark-spotted female moths arrive at the younger flowers near the top of the stalk (Figure 34.1). Each moth visits several blooms and collects a ball of pollen, which she carries beneath her chin with specially modified mouthparts. Bearing this pollen ball, she then flies to another plant and enters an older flower near the base of the stalk. Here she extends a long, sharp drill from the tip of her abdomen, pierces the older flow-

er's ovary, and pumps her eggs into the flower. Egg laying completed, the moth moves up to the flower's green stigma, separates a bit of pollen from the pollen ball with her specialized mouthparts, then rubs the pollen directly across the stigma. This ensures that the flower will be fertilized and that fruit and seeds will develop, with some serving as a food source for the moth's offspring. Fortunately for the plant, the growing caterpillars do not eat all of the seeds from fertilized flowers. The plump mature yucca caterpillars eventually become adults and repeat the cycle.

This chapter describes how organisms like the yucca plant and moth interact in *communities*—assemblies of populations of different species in a particular area at a particular time (Figure 34.2, page 554).

It also explains how the associations may have developed. In our discussion, three themes will recur. First, interaction with other species is a major limiting factor in the abundance and distribution of organisms. Beneficial interactions, such as the reproduction-food link between yucca plant and yucca moth, may encourage the growth and spread of both species, while harmful interactions—competition between the moth and the beetles or deer that also eat the yucca plant, for example—may limit a species' success. Second, in extremely tight interactions, one species can influence another's evolutionary fitness, so that both kinds of organisms evolve together in the process called coevolution. Genes governing the shape of the yucca flower and genes affecting the moth's behavior pattern must have evolved together, guaranteeing the survival of both species. Finally, interaction with the human species is the most powerful biological factor in the world today. As people encroach upon the desert, for example, they destroy the yucca's habitat. In fact, our human species has become a major force in shaping the earth's ecological communities.

Our exploration of how organisms interact in communities will address these questions:

- Where and how do populations live in communities?
- How does competition between species affect community structure?
- How do predators affect a prey's population density, and vice versa?
- What relationships between two species benefit one or both of them?
- How does a community's structure change over time and space?

FIGURE 34.1 *Coevolutionary Partners: The Yucca Moth and Yucca Flower.* One kind of yucca plant (*Yucca whipplei*, also called Our Lord's Candle) growing in the Mojave Desert. The yucca moth (see inset) lays its eggs in the ovary at the base of a yucca flower and then pollinates the same flower. Moths depend on the plant for food and reproduction; the plant depends on the moth for pollination. This mutually beneficial relationship is called mutualism.

WHERE ORGANISMS RESIDE AND HOW THEY LIVE

To understand the intricate web of relationships between populations of organisms in a community, the observer must discover where the organisms live and how they get needed energy and materials. The general physical place in the environment where a certain kind of organism resides is its **habitat.** A habitat is analogous to an organism's "address," or "home." In describing the general places where aquatic organisms live, for example, an ecologist might speak of an open-water habitat, a shore habitat, a muddy bottom, or a surface-film habitat. Likewise, certain birds may be found in grassland, pinyon-juniper, or tropical habitats—all places where birds live.

Whereas a species' physical home is its habitat, its functional role in the community is its **niche.** The niche is analogous to the organism's "job"—how it gets its supply of energy and materials and what it does in and for a living community. The yucca's niche, for example, is that of *primary producer,* deriving energy directly from the physical environment. The niche of the yucca moth is that of *herbivore,* obtaining nourishment from plants like the yucca that grow in its habitat.

A field ecologist might investigate a warbler's niche in the forests of New England and observe that the bird eats insects wherever they occur in trees, at any height and at any distance from the trunk, and that the bird nests any time in June or July. The potential range of all conditions under which an organism can make a living is called its *fundamental niche.* If a warbler could eat insects any place in the tree, it would be operating in its fundamental niche for prey location.

In nature, however, a warbler cannot find insects just anywhere because several species of insect-eating warblers compete for food in eastern forests, each species performing a slightly different but specialized role in the community (Figure 34.3). The different species eat insects at different heights in the trees and at different distances from the trunk, and their heavy eating comes at slightly different times during the year, depending on when they nest. The myrtle warbler, for instance, eats insects at the base of trees and in lower branches, while the bay-breasted warbler specializes in insects in middle branches, and the Cape May warbler seeks insects at the outer edges of the top branches. The part of the fundamental niche that a species actually occupies in nature is its *realized niche,* and interactions with other organisms often force a species into a realized niche that is more restricted than its fundamental niche. In the rest of the chapter, we look at four kinds of interactions that can limit an organism's realized niche, including competition and predation, wherein one or both of the species suffer, and mutualism and commensalism, in which neither species is harmed by the interaction.

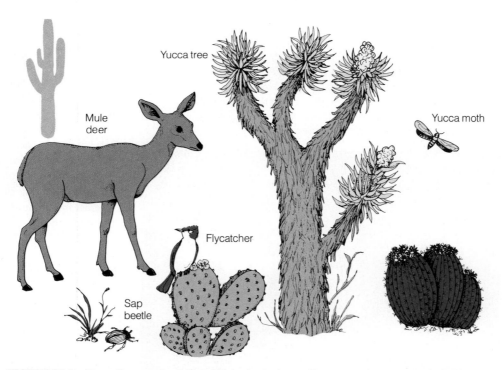

FIGURE 34.2 *Yucca Community: A Web of Interdependence.* The mule deer and sap beetle eat yucca flowers, while the ash-throated flycatcher picks off yucca moths. Belowground, the yucca roots release a soaplike substance that makes scarce water more available for itself and its nitrogen-fixing neighbors.

COMPETITION BETWEEN SPECIES FOR LIMITED RESOURCES

There are never enough good things to go around—good sunny spots in which to germinate and put down roots, good places to build nests, good berry patches, or good places to hide. Because of this, different species often compete for the same limited resources, and this interaction restricts the abundance of the competing species. The key feature of **interspecific competition,** the use of the same resources by two different species, is that one or both competitors have a negative effect on the other's survival or reproduction.

MODELS FOR COMPETITION IN COMMUNITIES

Let us imagine a community in which red birds and blue birds compete for the same food, mosquitoes. How would competition between the two kinds of birds limit each other's population growth rates? Recall that in a population that has saturated its resources, growth is limited by the environment's carrying capacity. If such a population of red birds has to share some of its limited resources with a second species, the blue birds, the carrying capacity of the environment must drop for the first population. How far the carrying capacity drops depends on how well each species competes. Mathematical models reveal that if a high density of red birds affects the growth of the blue birds more than it affects itself, the red birds will eventually take over the niche. But if the reverse occurs, then the blue birds will eradicate the red birds. A situation where one species eliminates another through competition is called **competitive exclusion.**

A third possibility is that competition hinders each species' own population more than it affects the other species. In one scenario, the density of the red bird population inhibits its own further growth, and the blue bird population slows its own growth; as a result, the two species end up coexisting.

These possibilities are the result of mathematical models and look good on paper or on a computer screen. But do they happen with real organisms? Laboratory experiments show that they do (Figure 34.4, page 556). When competing species of the slipper-shaped protist *Paramecium* were cultured together, the smaller-celled species drove the larger-celled species to extinction. Evidently competitive exclusion can happen with real species. Experiments with organisms as different as protists, beetles, water fleas, fruit flies, and aquatic plants give similar results and show not only that competition takes place, but that its most likely outcome is competitive exclusion, at least in the laboratory.

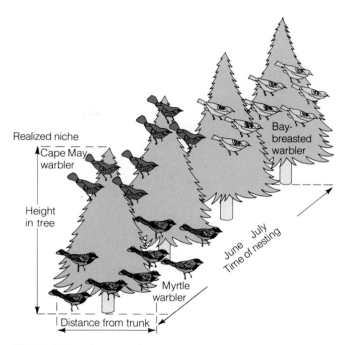

FIGURE 34.3 *Niche: An Organism's Role in the Community.* The realized niche is an organism's role in its normal environment. The Cape May warbler catches insects only in the outer branches toward the top of the tree, and it nests only in June. The bay-breasted and myrtle warblers feed in different places and nest at different times, thus occupying separate realized niches. The realized niche contrasts with the fundamental niche, which is the full range of conditions under which a given organism could operate if certain restrictions imposed by other organisms were absent. If a forest warbler could catch insects in a tree at any height and on branches any distance from the trunk, and if it could nest any time between early June and late July, it would have a larger fundamental niche than the realized niches shown here.

COMPETITION IN NATURAL COMMUNITIES

It is easy to see competitive exclusion in laboratory culture dishes and to trace the factors behind the victory of one species and the demise of another. In natural communities, however, it is much harder to pin down the factors critical to each competitor's success. In the western United States, for example, wild burro populations have increased dramatically since laws were passed in the 1970s that prohibit their killing. As the burros have multiplied in the arid open lands of Arizona and southern California, however, the number of desert bighorn sheep has rapidly declined. Many biologists suspect that the introduced burros are somehow outcompeting the native bighorn sheep, but they are not sure exactly how. Sheep and burros eat some of the same desert plant spe-

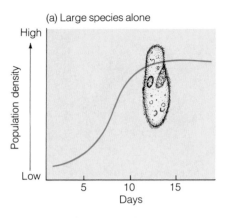
(a) Large species alone

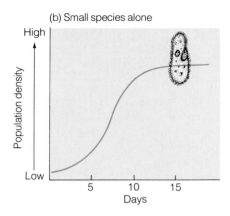
(b) Small species alone

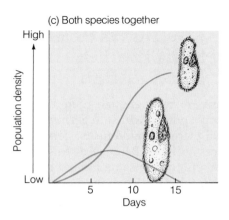
(c) Both species together

FIGURE 34.4 *Competition Between* Paramecium *Species*. (a) A large-celled species (*P. cauda-tum*) experiences S-shaped growth in a culture. (b) A small-celled species (*P. aurelia*) experiences similar population growth. (c) When the protists compete for the same resources in the same culture, the success of the smaller-celled species spells disaster for the larger-celled species. This is competitive exclusion.

cies, but burros may be able to consume or use these plants more efficiently than bighorn sheep (Figure 34.5).

Two kinds of competition may come into play. When two species exploit and have equal access to identical resources, it is called *exploitation competition,* and this may be happening with burros and bighorn sheep. Burros also tend to congregate around desert water holes, however, and often chase other mammals away. The use of aggressive behavior to keep competitors from a resource is called *interference competition,* and perhaps *this* explains the sheep's demise. Ecologists do not yet know whether competitive exclusion is unavoidable among burros and bighorn sheep, resulting in the inevitable extinction of the sheep, or whether a stable coexistence will be the outcome. One thing, however, seems certain: Wild burros do indeed restrict bighorn sheep to a smaller realized niche.

COMPETITION CAN ALTER A SPECIES' REALIZED NICHE

Many species respond to the threat of competitive exclusion by a slight alteration of their niche. Figure 34.6 depicts an experiment with two species of *Paramecium,* one of which was the winner in the experiment of Figure 34.4; the other was a new challenger. This time, the two kinds of paramecia coexisted in the same test-tube culture because they split up the territory, with the smaller species moving to occupy a different niche than the larger challenger species. While the entire test tube was the fundamental niche for both *Paramecium* species, in a mixed

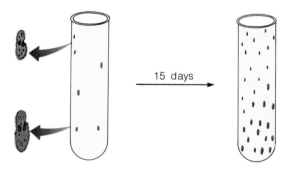

FIGURE 34.5 *Burros and Bighorn Sheep: Competitive Exclusion in Nature.* Non-native burros may be eating rings around bighorn sheep and excluding them from their native territories.

15 days

FIGURE 34.6 *Dividing the Spoils: Resource Partitioning in Two Paramecium Species.* After introduction into a culture tube (left), two species of large- and small-celled paramecia can coexist by partitioning resources (right). The smaller cells get the top territory with greater oxygen penetration, and the larger cells get the leftovers.

culture, each came to fill a more restricted realized niche (only part of the tube), and this minimized direct competition. Ecologists use the term **resource partitioning** to indicate the process of dividing up resources so that species with similar requirements use the same resources in different areas, at different times, or in different ways. The warblers of New England forests minimize the harmful effects of competition by resource partitioning—in this case, by feeding in different parts of the trees.

In the portions of a species' range that overlap the range of a strong potential competitor, hereditary changes often evolve in the species' physical or behavioral characteristics. Ecologists call such changes **character displacement** and think that it brings about a partitioning of resources by otherwise competing species. Character displacement is an example of **coevolution,** hereditary changes in two or more species as a consequence of their interactions within a community. For example, imagine an area with huge numbers of elephants, horses, and antelope, as in Africa today or North America some 25,000 years ago. Those animals would serve as an immense food resource for predators like lions, cheetahs, and mountain lions. Cats kill prey with their pointed canine teeth, and animals with larger canines can kill larger prey. Now, if several cat species with approximately equal-sized canines attack and consume prey animals in a given area, they will be limited to prey of the same size, and they will end up in vigorous interspecific competition. Natural selection (see Chapter 32) would therefore favor different species of cats with different-sized canines. The expected result is indeed the reconstructed history of big cats in North America 25,000 years ago. Huge, saber-toothed cats became specialized for elephants and giant ground sloths, and lions for grazers like buffalo; and the weak-toothed American cheetahs evolved the speed and smaller teeth necessary to kill small gazelles (Figure 34.7).

Evolutionary ecologists contend that competition between species and subsequent character displacements are probably one source of adaptive radiation (review Figure 32.17c). Whether competition leads to competitive exclusion or character displacement and coevolution, it generally slows population growth in both species. In contrast, a second major type of community interaction—predation—harms only prey populations and is thus negative for only one of two interacting species.

⚹ THE HUNTER AND THE HUNTED

In several kinds of interactions, one species consumes the other—a situation clearly beneficial to the consumer but harmful to the consumed. Animals like lions or Cape May warblers that kill and eat other animals are **predators,** their food is **prey,** and the act of procurement and consumption is **predation.** Mule deer and yucca moths are *herbivores.* They eat plant parts and often harm the plant but do not kill it. *Parasites* like tapeworms feed on a host organism, usually without killing it. Disease-causing organisms, or *pathogens,* are usually fungi, bacteria, or protists that obtain nourishment from a plant or animal host and weaken or kill it. Here we focus on how predation affects the population size of both prey and predator on a short time scale, how hunter and hunted evolve strategies to outwit each other on a longer evolutionary time scale, and what makes parasites special kinds of predators.

(a)

(b)

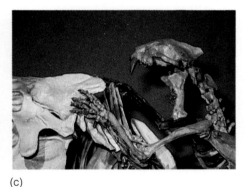
(c)

FIGURE 34.7 *Character Displacement in Cats Leads to Resource Partitioning.* (a) Cheetahs, with their short canines, take down small, fast-running prey—here, a gazelle. (b) Lions, with their medium-sized canines, specialize in larger prey like this Cape buffalo. (c) Extinct saber-toothed cats had huge, daggerlike canine teeth and brought down prey as large as woolly mammoths.

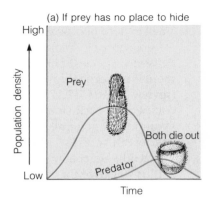

(a) If prey has no place to hide

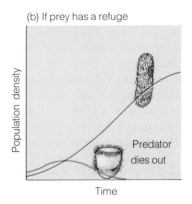

(b) If prey has a refuge

FIGURE 34.8 *How Populations of Predators and Prey Interact.* (a) If the predatory protist *Didinium* is grown in culture with its favorite prey species, *Paramecium*, both populations increase until the predator population is so large that it eats the prey faster than the prey population can grow. Then both populations crash. (b) If the *Paramecium* cells have a refuge inaccessible to the predator, the *Didinium* cells may starve and die.

POPULATIONS OF PREDATOR AND PREY

In Chapter 15, we encountered a minute but voracious predator called *Didinium* that hunts down *Paramecium* cells and swallows them whole (see Figure 15.1). Laboratory experiments on *Didinium* and *Paramecium* have revealed several principles about predator and prey populations. First, as the density of the prey population increases, the predators find more food to eat, which in turn increases the size of the predator population. Eventually, however, the large number of predators will eat so many prey that the prey population begins to fall. If *Didinium* cells devour all the *Paramecium* cells, then the predator population will

crash, too, because its food supply will have been exhausted, and both organisms will finally become extinct in their laboratory setting (Figure 34.8a). If, however, the prey have a *refuge*, a safe haven out of the predators' reach, then when the predators have eaten all the accessible prey, they will die out and the prey can take over the environment (Figure 34.8b). Finally, if the environment is complex enough to offer many partial refuges where prey can survive for a while before successive pulses of predators migrate in and discover them, then the populations of predator and prey will tend to oscillate up and down. In nature, populations of the snowshoe hare and Canadian lynx periodically rise and fall in this way (Figure 34.9); we do not yet know if these wide swings in lynx and hare populations occur because hares have partial refuges or because changes are taking place in the hare's food supply.

Whatever drives the hare-lynx cycle, it appears that predators can regulate their prey's population density. The only way for an ecologist to test a predator's true regulatory effects on prey populations is to remove the predator and watch the results. Wildlife officials in the state of Arizona performed such an operation—not as an experiment, but to control predation on ranchers' cattle. Beginning in 1907 on the Kaibab Plateau, north of the Grand Canyon, sharpshooters removed practically all of the area's mountain lions, wolves, and coyotes (Figure 34.10). Within 11 years, Kaibab deer herds, no longer controlled by their predators, had grown to ten times their original size.

Clearly, predators can help regulate populations of their prey, and people have used this fact to control populations of pest species. The prickly pear cactus, a native of American deserts, became a pest in Australia after escaping gardens and producing impenetrable thickets over vast expanses of land (Figure 34.11). Australian officials

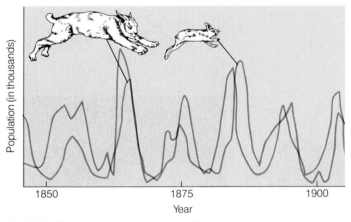

FIGURE 34.9 *The Hare-Lynx Cycle: Responsiveness of Predator and Prey Populations.* Populations of snowshoe hare and its predator, the lynx, oscillate in northern Canada. Hare populations boom when vegetation is abundant and "crash" when lynx populations rise in response to the increased availability of their prey; hares may also have partial refuges that affect the oscillations.

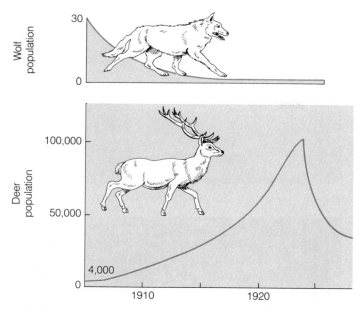

FIGURE 34.10 *Prey and Predators Near the Grand Canyon: An Inadvertent Experiment.* Sharpshooters removed the lions, wolves, and coyotes from the Kaibab Plateau in the early 1900s, and the size of the area's deer herds increased first tenfold and then twentyfold. But the large herds of hungry deer soon overbrowsed and irreparably damaged the area's vegetation, and during the harsh winters of 1925 and 1926, more than 60,000 deer starved to death. The surviving deer population never bounced back to peak levels, but remained higher than if predators were present, suggesting that predators can control prey populations.

imported a moth from Argentina whose larvae burrow into the cactus pads for food, and soon the cactus population fell to an acceptable level.

As with prickly pear and moth, hare and lynx, populations of predator and prey may grow or shrink over the short term, but over the long term, genetic changes can influence the evolutionary balance between hunter and hunted.

THE COEVOLUTIONARY RACE OF PREDATOR AND PREY

Genetic changes in response to natural selection result in a grand coevolutionary race, with predators evolving more efficient ways to catch prey and the prey evolving better ways to escape. It is hard to tell which set of adaptations—the predator's or the prey's—are more fascinating.

■ **Predators' Strategies** Many predators can simply amble, swim, or fly up to a stationary prey target and start to eat. If the prey is mobile, however, then the predator must catch its food through one of two main options: *pursuit* or *ambush*. Predators who pursue their prey are selected for speed and often for intelligence, as well.

Carnivores store information about the prey's escape strategies and must make quick decisions while in pursuit. In keeping with their strategic demands, vertebrate predators generally have larger brains (in proportion to their body size) than the prey they catch. Fossils reveal that 60 million years ago, carnivorous predators like the cats in Figure 34.7 had larger brains than their hoofed prey and that while selective pressures forced the prey to become more wary, the predators (and their more cunning descendants) always stayed one step ahead.

For some predators, ambushing is an effective strategy for capturing prey. A familiar example is the frog that ambushes flying insects by snapping out a sticky tongue, hinged at the front of the mouth. An ambush can be even more effective if the predator can lure the prey: The alligator snapping turtle not only blends in with its river-bottom habitat, but has a worm-shaped tongue that entices prey right into its mouth (Figure 34.12, page 560).

■ **Prey Species' Countermeasures** To avoid being eaten, prey have evolved not only obvious strategies like rapid running or flying, but also some remarkably devious tricks. One defense strategy is to avoid confrontation altogether through *camouflage*: the occurrence of shapes, colors, patterns, or even behaviors that enable organisms

FIGURE 34.11 *Prickly Pear Pests Down Under.* (a) Species of *Opuntia*, or prickly pear, escaped gardens and spread across Queensland and other parts of Australia in the 1800s. (b) Three years after releasing the natural predator *Cactoblastis* (an Argentinean moth that lives off cactus pads), the scourge had receded.

(a)

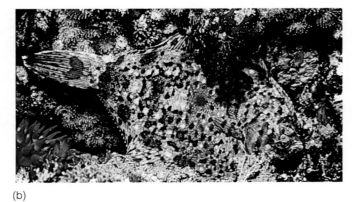

(b)

FIGURE 34.12 *Predatory Strategies.* The alligator snapping turtle (*Macrochelys temminckii*) lures prey into its powerful hooked jaws by lying camouflaged as detritus on the pond bottom and dangling a wormlike tongue.

FIGURE 34.13 *The Countermeasures of Prey.* (a) A dead-leaf-mimicking moth, complete with phony leaf veins, and (b) a turbot fish blend so well with their backgrounds that if they freeze in place, they are nearly invisible.

to blend in with their backgrounds and escape predation (Figure 34.13).

Chemical warfare is another common defense strategy (see Chapter 30). Eucalyptus and creosote plants, for example, produce distasteful oils or toxic substances that kill or harm herbivores. People sometimes plant olean- ders (Figure 34.14a) as decorative shrubs because they resist insect pests, but the leaves are so poisonous that chewing one can kill a small child. Animals are not with- out their own arsenals: Stinkbugs, lacewings, and bom-

bardier beetles (Figure 34.14b) produce highly offensive chemicals that repel attackers. The Colorado River toad produces alkaloids so toxic that a small nibble on one of them can seriously injure a predator as large as a coyote or dog.

Poisonous prey species usually evolve brightly colored patterns, enabling the experienced predator to recognize

FIGURE 34.14 *Chemical Warfare in Plants and Animals.* (a) The color- ful, fast-growing, dry-adapted olean- der (*Nerium oleander*) has poison- ous leaves and stems. (b) The bombardier beetle (*Brachinus* spe- cies) defends itself by spewing a volatile irritant from special glands.

(a)

(b)

and avoid them. This is called **warning coloration.** The brilliantly colored but poisonous strawberry frogs of South America (see Figure 1.1) have evolved this strategy, as have the boldly patterned harlequin beetles on this book's cover and the monarch butterfly pictured in Figure 34.15a. Many nonpoisonous prey species masquerade as poisonous species that fool predators in a process called **mimicry.** For example, the adult monarch butterfly is foul-tasting and dangerous to predators, who soon learn to leave it alone. Meanwhile, the smaller viceroy butterfly looks strikingly like a monarch and is protected by its warning coloration, even though it tastes good to blue jays and other birds (Figure 34.15b).

Plants and animals also evolve with thorns, spines, sharp spikes, and horns—other weapons that discourage predators (Figure 34.16).

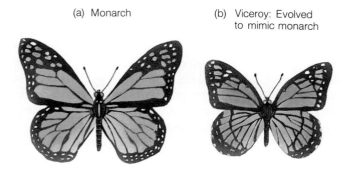

(a) Monarch (b) Viceroy: Evolved to mimic monarch

FIGURE 34.15 *Warning Coloration and Mimicry: More Defense Strategies.* (a) A monarch butterfly (*Danaus plexippus*). If a blue jay eats even part of a monarch, it becomes sick. After one sickening taste, the bird will never again eat a monarch or any butterfly that resembles a monarch. (b) The flamboyant viceroy (*Limenitis archippus*), mimic of the unrelated monarch butterfly.

PARASITES: THE INTIMATE PREDATORS

Parasites are insidious kinds of predators; they are usually smaller than their hosts, often live in close physical association with individual victims, and generally just sap their victims' strength rather than killing them outright (review Figure 17.8). Some parasites, like fleas, ticks, and leeches, live on the host's exterior, while others, like tapeworms, liver flukes, and some protozoa, inhabit internal organs or the bloodstream. Parasites and their hosts probably coevolve in such a way that the parasite becomes relatively harmless to the host; if the parasite were more dangerous, it would destroy its own habitat.

≫ SHARING AND TEAMWORK

In our discussion so far, most of the relationships between species in a community have involved harm to at least one of the parties. Sometimes, however, one or both species benefit from their interactions. In **commensalism,** one species benefits from the alliance, while the other is neither harmed nor helped; in **mutualism,** both species are helped.

COMMENSALISM

Commensalism is common in tropical rain forests, and the most easily observed examples are the "air plants" (*epiphytes*) that grow on the surfaces of other plants. The air plants include mosses, large and small bromeliads, and many beautiful orchids that festoon the branches or decorate the forks of trees (Figure 34.17a, page 562). Using the tree merely as a base of attachment, the plants take

FIGURE 34.16 *A Spiny Arsenal for Self-Protection.* Porcupines (a) and some kinds of tropical plants (b) have evolved sharp spines that ward off predators.

(a) (b)

FIGURE 34.17 *Commensalism: A Little Harmless Help from a Host.* In the tropics, orchids (a) often rely on tree trunks for support. On coral reefs, clown fish (b) find safe haven within the folds of a sea anemone.

(a)   (b)

no nourishment from the "host" and do no harm unless their numbers become excessive. Other commensal relationships include birds that nest in trees, algae that harmlessly grow on a turtle's shell, and small fish that live among the stinging tentacles of sea anemones—unharmed and safe from predators (Figure 34.17b).

MUTUALISM

In a mutualistic interaction, both species benefit, and the yucca and yucca moth are a classic example. Plants such as the yucca provide pollinators such as the yucca moth with a high-energy nectar "reward," and as the pollinator visits other individuals of the same plant species in search of more nectar, pollination is accomplished. Mutualism between plants and pollinators is so common that one can almost predict that a bizarre flower will have a bizarre pollinator. Charles Darwin once amused his skeptical contemporaries by predicting that somewhere in the rain forest of Madagascar there must fly a moth with a 30 cm (foot-long) tongue, because a flower lives in those same jungles with a 30 cm floral tube. Forty years later, naturalists caught the moth with the longest known tongue and have recently shown how it could have evolved.

Sometimes, both species in a mutualistic relationship will benefit, but neither will be wholly dependent on the other for survival. Many species of aphids, for example, excrete large quantities of sweet, saplike fluid, and ants harvest this sap. The ants receive nutrients, and in return, their presence keeps predators away from the aphids. The ants do not depend on the sap, however, and the aphids produce it even when the ants are absent.

In contrast, some interacting organisms need each other to survive. A relationship of this type, called *obligatory mutualism*, binds the yucca plant to the yucca moth. The moth is the plant's sole agent of fertilization, and the pollen and seeds of the yucca plant are the sole food source for yucca moth larvae and adults. Obligatory mutualism is involved in only a small number of existing interspecies interactions, perhaps because it is so risky: No doubt many past examples of this mutual dependence have disappeared because both species became so specialized and so interdependent that when one member could no longer survive in a changing environment, the other became extinct, too.

≈ HOW COMMUNITIES ARE ORGANIZED IN TIME AND SPACE

The principles we have studied—competition, predation, commensalism, and mutualism—affect not just pairs of species, but entire communities consisting of up to hundreds of species. One of the biggest challenges faced by ecologists has been to understand the web of interdependencies in a community and to see how the community itself changes over time and space. A preliminary task was to discover whether communities are fixed groups of species unfailingly bound together, or whether they are more fluid entities. Research projects undertaken nearly 50 years ago confirmed that a community is not a "superorganism," a package of highly specific groupings of plants and animals (like an animal with one heart, two lungs, and a precise assortment of other organs). Rather, communities consist partially of species that happened to immigrate into the area and can survive under the available physical conditions and partially of some species that will grow only if other species are also present. Given this premise, a goal of ecologists is to learn how the addition or subtraction of a species will affect the whole community in the short and long term.

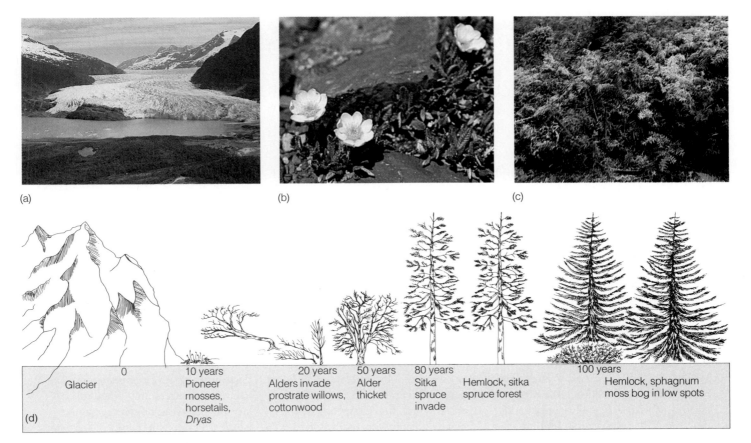

FIGURE 34.18 *Succession in the Path of an Arctic Glacier.* (a) As Alaska's massive Mendenhall Glacier receded, a succession of plants followed, from (b) the rockrose (*Dryas* species) to cottonwood and alder and eventually (c) sphagnum moss and hemlock. (d) A time line of the glacier's recession and the plant's succession. With the passing decades, the soil gains nitrogen and becomes more acidic.

COMMUNITIES: CHANGES OVER TIME

When volcanic explosions, floods, fires, hurricanes, glaciers, hungry elephant herds, or human farmers, miners, builders, or loggers strip an area of its original vegetation, a regular progression of communities regrows at the site in a process called **succession.** Soon after the denuding, a variety of species begin to colonize the bare ground. These species make up a *pioneer community* and modify environmental conditions like soil quality at the site. These modifications can either inhibit additional species or allow additional species to establish themselves and form a *transition community.* The change is rapid at first as more species invade. Finally, an assemblage of plants and animals becomes stable and perpetuates itself as a *climax community.*

A well-studied example of ecological succession can be seen at Glacier Bay, Alaska, where a glacier that devastated thousands of square kilometers of land has melted back about 100 km in the last 200 years, exposing the ground below (Figure 34.18). The most recently exposed areas are inhospitable piles of rock, sand, and gravel lacking usable nitrogen and essentially devoid of plant or animal life. The first plants to colonize this barren scene form a pioneer community of wind-dispersed species: mosses, horsetails, fireweed, willows, cottonwood, and a matlike rose called *Dryas.* Most are severely stunted and grow close to the ground as a result of nitrogen deficiency, but *Dryas* has nitrogen-fixing nodules on its roots that provide the growth-limiting nutrient. Within a few years, it crowds out other plants and forms a dense mat over the soil.

In the areas exposed for 10 years or so, seeds of alder trees begin to arrive from other sites. Alder roots have nitrogen-fixing nodules, so these trees can also grow rapidly. Dead alder leaves add nitrogen to the soil and stimulate the growth of willows and cottonwoods. In areas exposed for about 20 years, dense thickets of these plants shade the pioneer species and kill them. Eventually Sitka spruce invades, and in those areas exposed for 80 years, nitrogen released by the alders enables the spruce to form dense forests that shade out the alders and willows.

INTEGRATED PEST MANAGEMENT: COMMUNITY ECOLOGY APPLIED TO AGRICULTURE

Competition is a powerful force in modern agriculture. Species such as corn smut, the cotton bollworm, and the cattle screwworm compete with us for the bounties of farmland, and while people have had some real victories in controlling "the competition," we have also had some real disasters.

In 1949, Peruvian cotton farmers began using large quantities of DDT and other synthetic chemicals to control pests. Cotton production immediately soared as populations of the cotton bollworm, a serious insect pest, fell, but after the third year of spraying, the insects evolved resistance: Individuals that happened to already have pesticide-resistance mutations survived the insecticide treatment and multiplied, and soon larger and larger doses of pesticide were required. By 1956, despite immense doses of toxic insecticides, resistant bollworms were again chewing away at the cotton crops, accompanied by six new cotton predators with resistance to the poisons. That year, the cotton farmers suffered their worst losses ever: Insects ate half their crops. Faced with economic disaster, the Peruvian farmers were forced to reevaluate their pest

control strategy, and they decided to ban synthetic pesticides and return to mineral insecticides. In addition, they imported 130 million wasps that attack the eggs of the bollworm. Small numbers of bollworms continued to survive in their fields, but cotton yields reached new highs.

The experience of the Peruvian cotton farmers led to the first success of *integrated pest management*, a system based on the principles of community ecology and aimed at keeping pest populations below economically harmful levels with a minimum of chemical pesticides. Integrated pest management requires a broad view of all interactions between the crop and other species: weeds that compete for sun and nutrients; insects that eat the crop; predators that eat the insect pests; animals that pollinate the crop; and other wildlife living in the area. Pest managers have found that by carefully controlling the timing, spacing, and intermixing of crops, they can enhance the activities of natural predators of crop pests. They can also introduce biological control agents from other regions that feed exclusively on the pest, such as wasps. In addition, they can release into the environment millions of steri-

FIGURE 1   *A Serious Pest of Cotton.*

lized adult male insects of the pest species, which mate with natural females but produce no offspring (Figure 1). Finally, integrated pest managers use chemical controls only when insects appear poised to do damage above some preestablished level of economic injury. And when spraying is necessary, they choose and apply compounds to hit target species, leaving "innocent bystanders" alone.

Integrated pest management is more difficult than simply spraying the fields with chemicals, but because it avoids a whole host of problems, such as pests evolving resistance, it offers an ecologically and economically acceptable alternative to losing much of a crop to pests.

Finally, after 100 years or so, shade-tolerant hemlock trees invade, grow below the canopy of spruce branches and needles, and become the most frequent tree in the climax community. In low places, however, sphagnum moss invades the forest floor, soaks up large amounts of water, and kills trees by choking off their roots' oxygen supply. This leaves a *mosaic climax* consisting of patches of spruce-hemlock forest intermixed with sphagnum bog. But even after the climax community becomes established, it will probably not remain stable for very long. Local climates change, and a cycle of growth and decay pervades all communities.

TRENDS IN SPECIES DIVERSITY IN COMMUNITIES

Coral reefs and tropical rain forests are dense with interacting species, while arctic tundra or deserts have far fewer. The total number of species found in a community is its **species richness,** or diversity. In most communities, there are few common species and many rare types of organisms. In a local area, for example, English ecologists captured a group of almost 7000 moths and identified individuals of 197 different species. Fully one-quarter of the moths belonged to a single species, and another quarter belonged to just five other species. The

(a)

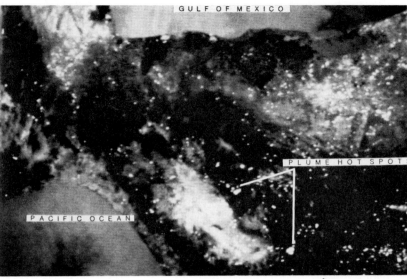

(b)

FIGURE 34.19 *Diversity and Destruction in the Tropics.* (a) Farmers in the tropics often carry out a traditional method of agriculture: "slash and burn" forest clearing, followed by planting, then moving on to new areas when the soil is depleted. (b) A satellite photo of agricultural areas in Mexico taken in 1984 shows the glowing lights of deliberately set fires, as well as white plumes of smoke. Some represent the annual clearing of established fields, but others are the ongoing destruction of virgin forest.

remaining half of the moths fell into the other 191 species—some represented by just one or two individuals.

Both *latitude* (north/south position) and *isolation* (occurrence on peninsulas, island chains, or other out-of-the-way locales) influence the species richness of an area. Some communities in the tropical latitudes, for example, have about 600 types of land bird species, while an area of similar size in the arctic tundra may have only 20 to 30 species of land birds at many sites. Species diversity on island chains is limited in such specific ways that ecologists are now applying the principles of island ecology to the design of nature preserves, which are often islands within a sea of human development (see the box on page 566).

Recent observations support the conclusion that the more resources available in an area—water and solar energy, for instance—the greater the species richness the area can support. This helps explain the numerous varieties of plants found in tropical forests. Other factors influence species richness, however, including competition. Competition can increase diversity because resource partitioning divides up niches into more and more specialized compartments. For example, a lizard species that has an island all to itself will eat insect prey of any size, but a lizard species that shares an island with four other lizard species will specialize on prey of a particular size. With competition forcing smaller niches, a community can accommodate more species.

SPECIES DIVERSITY, COMMUNITY STABILITY, AND DISTURBANCES

Some evidence suggests that highly diverse communities, such as those in tropical forests and coral reefs, would be the most *stable* or *resilient*—the best able to return to normal after disturbances like fires, storms, frosts, plagues of insects, or human land-clearing activities. Other analyses, however, suggest that when more species are interwoven into complex networks of competition, predation, and mutualism, the community becomes more fragile. A complex community might be less stable because the loss of each species may affect many other organisms. For instance, if the yucca moth were exterminated, the yucca plant would die out, too, and so might the sap beetle and many other species whose lives depend on the yucca in unknown ways. Ecologists do not yet know whether diverse or simple communities are more stable, but are convinced that community richness and stability are not related in a very simple way.

Perhaps the greatest ecological challenge facing modern civilization is the fact that the immense increase in human population is leading to larger and more frequent disturbances in the species-rich communities of tropical latitudes and is threatening their destruction. As Figure 34.19a shows, tropical peoples have traditionally cleared

DESIGNING NATURE PRESERVES TO SAVE SPECIES

Islands are microcosms of evolution and ecology, isolated from invasion and relatively unperturbed by the outside world. Many of our nature preserves are islands, too, enveloped not by water but by farms, roads, and towns—hostile environments, zones of human development. These surrounding regions fail to provide proper habitat and present fatal dangers to organisms that stray beyond the safety of the refuge. National parks and other nature preserves represent our best hope for saving many species from extinction. Since a nature preserve is truly an island, community ecologists are now applying the principles of island ecology to their design in the hopes of conserving life's irreplaceable diversity.

Researchers studying island ecology have tried to understand how an island's size and its distance from sources of colonization affect its species richness. By cataloging species richness on islands—whether reptiles in the West Indies, ants in Melanesia, or vertebrates on Lake Michigan islands—naturalists have found that larger islands have more species than smaller islands. This is probably because population sizes are larger on large islands, and so chance extinctions are less likely. The lesson ecologists draw from this is that large nature preserves can harbor more species than small ones. Unfortunately, the land area set aside for most preserves today is quite limited.

To determine how the species richness of an island is affected by the distance from the mainland, researchers devised some clever experiments on mangrove islands located at various

FIGURE 1 *Mangrove Tent Experiment.*

distances from the mainland off the Florida Keys. The experimenters sealed the small islands with plastic sheets (Figure 1), fumigated to kill the existing insects, spiders, scorpions, and other arthropods, then removed the tents and monitored the immigrations of arthropod species. They found that the closer an island was to the mainland, the higher its final species richness. Evidently, species have difficulty immigrating to a distant island, and thus such an island reaches a lower steady-state density of species. Studies like these had an additional message for ecologists who work with nature preserves: If preserves must be separated from each other, they should be as close together as possible to maximize immigration and species diversity.

Based on the principles of island ecology, it is clear that preserves should be large enough to hold sizable populations; but if there must be several small regions, they should be clumped close together so that migration between the preserves can remain high. If human society is to help pre-

serve nature's exquisite diversity, we must act immediately to set aside as many large preserves as possible. Extinctions will no doubt proceed anyway. But if we are too slow to preserve our islands of refuge, if we are more concerned with human developments than evolutionary ones, then a large measure of ecological complexity may be lost forever.

FOR MORE INFORMATION

The Nature Conservancy, 1815 North Lynn Street, Arlington, VA 22209.

Clark, W. C. "Managing Planet Earth." *Scientific American* (September 1989): 47–54.

Fearnside, P. M. "Extractive Reserves in Brazilian Amazonia." *BioScience* 39 (June 1989): 387–395.

Ruckelshaus, W. D. "Toward a Sustainable World." *Scientific American* (September 1989): 166–174.

Westman, W. E. "Managing for Biodiversity." *BioScience* 40 (January 1990): 26–33.

land for agriculture by burning the forest, planting crops for a few years, and then moving on when the soil becomes depleted of nutrients. So large are the human populations in these areas now, and so immense are their survival needs, that vast areas of the rain forest are going up in smoke to provide farmland for crops and pastureland on which to graze cattle for export to developed nations. These fires are so huge and so common that they are clearly visible on satellite photos from space (Figure 34.19b). Many ecologists fear that the plant and animal communities of the tropics may not be able to recover from such extreme disturbances and that by the year 2000, no tropical rain forests will remain intact. Still worse, many fear that this disruption may cause up to 20 percent of all living species to become extinct in our lifetime.

Ecologists consider it an urgent research priority to learn what makes communities resilient and how they may (or may not) be able to persist in the face of human encroachment (see the box on page 564 for an important agricultural application of this knowledge). We must solve this problem (discussed in more detail in Chapter 36) if our most diverse and interesting communities are to survive through the twenty-first century.

≋ CONNECTIONS

Our exploration of ecology has shown that populations of individual species are inevitably linked to other species in the community, even though each species has its own habitat and niche. Despite chains and webs of interconnections, however, communities are not prepackaged groups of species. Instead, species act as individuals, living where their special environmental requirements are met. The world is in danger of losing the species richness of many communities because as a species, we humans are massively disturbing the globe, yet we do not understand how community stability is maintained or whether it might bounce back. Biologists do know more about ecological stability than we have so far discussed, however, and in our next chapter, we will examine the interactions of communities with sunlight, temperature, rainfall, nutrients, and other physical aspects of the universe.

NEW TERMS

character displacement, page 557

coevolution, page 557

commensalism, page 561

competitive exclusion, page 555

habitat, page 554

interspecific competition, page 555

mimicry, page 561

mutualism, page 561

niche, page 554

predation, page 557

predator, page 557

prey, page 557

resource partitioning, page 557

species richness, page 564

succession, page 563

warning coloration, page 561

STUDY QUESTIONS

REVIEW WHAT YOU HAVE LEARNED

1. Define habitat and niche.
2. What is competitive exclusion? Does it occur in nature? Explain.
3. Use an example to show how natural selection can lead to resource partitioning.
4. What do biologists mean by coevolution? Give an example.
5. How do predator populations regulate prey populations?
6. Give two examples of predator strategies.
7. Name five adaptations for defense.
8. Is it true that parasites inevitably kill their hosts? Explain.
9. Compare commensalism with mutualism.
10. How does a pioneer community differ from a climax community?
11. How can competition increase species richness?
12. Why do biologists consider a tropical rain forest a fragile community rather than a stable one?

APPLY WHAT YOU HAVE LEARNED

1. When predators on the Grand Canyon's Kaibab Plateau were exterminated, the area's deer population eventually began to starve. Why?
2. Charles Darwin confidently predicted that investigators would one day find a moth with a foot-long proboscis in a South American rain forest. What made him so sure?
3. Some Alaskan glaciers have been receding for 200 years. Recently uncovered land is barren, but trees and shrubs now flourish in the first areas to be freed from the ice two centuries ago. Describe the ecological sequence that has led to this restored plant cover.

FOR FURTHER READING

Ehrlich, P. R. *The Machinery of Nature*. New York: Simon & Schuster, 1986.

Lewin, R. "In Ecology, Change Brings Stability." *Science* 234 (1986): 1071–1073.

Moore, P. D. "What Makes a Forest Rich?" *Nature* 329 (1987): 292.

Rickleff, R. E. *The Economy of Nature*. 2d ed. New York: Chiron Press, 1983.

Sattaur, O. "A New Crop of Pest Controls." *New Scientist* 14 (1988): 48–54.

Sutherland, W. J. "Predation May Link the Cycles of Lemmings and Birds." *Trends in Ecology & Evolution* 3 (1988): 29–30.

① When two species compete for limited resources, one can eliminate the other, or both can coevolve and coexist.

Competition

Elimination: competitive exclusion

Coexistence: character displacement and resource partitioning

② Evolutionary changes can affect the balance between predator and prey species.

③ Some interactions between two species benefit one or both and harm neither.

④ The species in a community arrive in an area by chance and survive the local conditions. Communities undergo a natural succession of species.

⑤ People are destroying many communities, especially in the tropics, and diminishing species richness worldwide.

ECOSYSTEMS: WEBS OF LIFE AND THE PHYSICAL WORLD

A VOYAGE OF DISCOVERY

In the spring of 1977, two scientists and a pilot climbed into the *Alvin*, a small, spherical submarine, plunged into the Pacific, and descended to a seldom-seen world of frigid, inky black waters and strange fluorescent fish and other sea life (Figure 35.1). After an hour of slow free-fall, 2.5 km (1.5 miles) of dark water lay above them. Only *Alvin*'s bright floodlights pierced the abyssal gloom as they started to creep forward along the deep-sea floor.

For several minutes, they moved across a monotonous gray desert, collecting brown, pillow-shaped rocks with the mechanical arm of the submersible. As their circle of light crept forward,

FIGURE 35.1 *The Alvin's Descent.* Researchers in the Alvin descend into the deep ocean.

they were surprised to see a pair of large, purple sea anemones loom out of the blackness and were even more perplexed to notice that warm water streamed from every bottom crack and fissure over an area larger than four football fields. They were soon transfixed by a fabulous scene that no humans before them had ever witnessed or even suspected (Figure 35.2, page 570). There on the lifeless floor of the ocean, groups of yellow mussels and fields of giant clams grew in clusters all around the deep, warm vents. Pure white crabs scuttled about through the shimmering water. Bizarre-looking worms nearly 1 m long lay encased in white tubes with frilly red plumes waving in the slow current. The two scientists worked quickly to collect samples of the water, sediments, and animals and to photograph the amazing spectacle before them. Before long, they once again entered the bright surface world, carrying an unprecedented collection of new and unusual organisms.

The team aboard the *Alvin* had inadvertently discovered a new *ecosystem:* a community of organisms interacting with a particular physical environment. As with more familiar ecosystems, deep-sea vents must possess an energy source as well as a supply of carbon, nitrogen, and other materials that make up the bodies of living organisms. In the absence of solar energy (and the green plants that capture it), what substitute energy source could be driving the deep-sea vent ecosystem? And how could energy and materials be flowing through the organisms in these bizarre and remote outposts of life? This chapter will answer these questions while also

addressing the broader issue of how all biological communities receive and use energy and materials. The answers will provide a basis for understanding the ramifications of the "greenhouse effect," acid rain, nuclear winter, and other serious ecological concerns.

Two unifying themes will emerge in our discussion: Because the organisms in an ecosystem have an absolute need for energy and materials, they are dependent on each other and on the physical environment. Plants and certain bacteria capture energy and inorganic compounds from their surroundings; animals depend on those autotrophs for their nutrients; and plants depend on bacteria and fungi that decompose dead organisms and return certain of their elements to the soil.

We will also see how human activities drastically affect the health of ecosystems by altering the flow of energy and the recycling of materials. For example, people are destroying entire tropical rain forest ecosystems in many parts of the world. The consequences of our activities are difficult to predict, but their long-term ecological effects may ultimately endanger us and the other living organisms that share our planet.

This chapter answers several questions:

- What are the pathways of energy flow and material cycling in ecosystems?
- How do ecosystems change energy from one form to another?
- How do materials like carbon and nitrogen cycle in ecosystems?
- How have human activities altered the earth's ecosystems?

FIGURE 35.2 *Strange Community of Aquatic Organisms Discovered in the Deep Ocean Abyss.* In 1977, scientists discovered an entirely new ecosystem based on the heat energy and chemicals from hydrothermal vents. The living community in this physical setting included yellow vent mussels, crabs (here, looking like ghostly white spiders), large vent clams, tube worms with red plumes extended, and chemosynthesizing bacteria too small to be seen.

⚞ PATHWAYS FOR ENERGY AND MATERIALS: WHO EATS WHOM IN NATURE?

Energy flows through an ecosystem in a one-way path, entering living things from the physical world, passing from one organism to another, and finally escaping back to the physical environment (Figure 35.3). In most ecosystems, energy originates in the sun in the form of light, but in the deep-sea vent ecosystem, the source of energy is heat generated by the decay of radioactive elements deep in the earth's core. In contrast to energy, materials cycle through an ecosystem, passing from one organism to another, to the physical environment, and then back through the organisms once more (see Figure 35.3). The carbon in carbon dioxide, for example, passes from seawater to bacteria, to the tissues of deep-sea vent clams and tube worms, and then back to the water as the organisms burn organic molecules and give off carbon dioxide.

In all ecosystems—whether remote ones like the deep-sea world that the *Alvin* first explored or familiar ones like forests and grasslands—organisms have two basic strategies for obtaining energy and materials: generating organic molecules from nonliving sources (autotrophy) or taking in preformed organic molecules made by other living things (heterotrophy). The strategy that each species uses defines its place in feeding levels operating within the community.

FEEDING LEVELS: STRATEGIES FOR OBTAINING ENERGY

In any ecosystem, organisms like plants and certain bacteria are the primary **producers;** they produce all the biological molecules they need for their growth from nonliving substances. On the other hand, organisms such as animals, fungi, and many kinds of microbes obtain their biological molecules by consuming other living things; they are an ecosystem's **consumers.** Ecologists assign every organism in a community to a *feeding level* (*trophic level*), depending on whether it is a producer or a consumer and depending on what it eats (Figure 35.4).

At the lowest feeding level lie the producers—the plants and certain bacteria that support all other organisms directly or indirectly. In most terrestrial ecosystems, green plants are the producers. By collecting solar energy and carbon dioxide, they build energy-rich sugar mole-

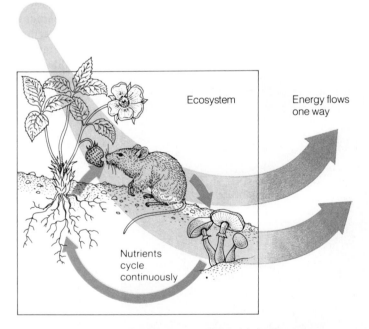

FIGURE 35.3 *Pathways of Energy and Materials in Ecosystems.* Energy flows in a one-way path through an ecosystem, usually from the sun through living things and into the environment, while nutrients and other materials continually cycle from the physical world to the living world and back again.

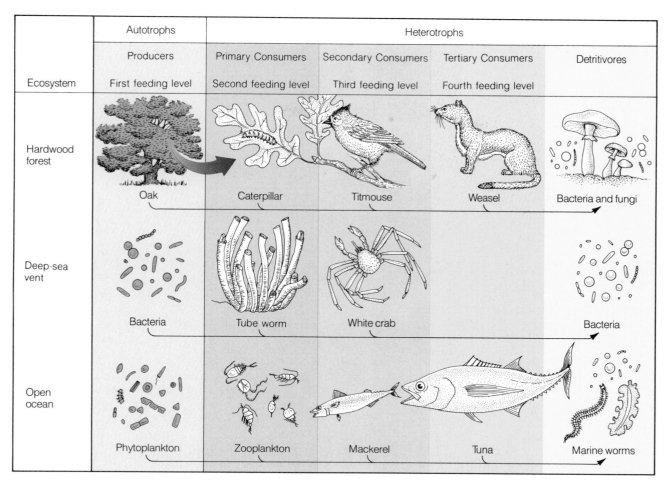

Ecosystem	Autotrophs	Heterotrophs			
	Producers	Primary Consumers	Secondary Consumers	Tertiary Consumers	Detritivores
	First feeding level	Second feeding level	Third feeding level	Fourth feeding level	
Hardwood forest	Oak	Caterpillar	Titmouse	Weasel	Bacteria and fungi
Deep-sea vent	Bacteria	Tube worm	White crab		Bacteria
Open ocean	Phytoplankton	Zooplankton	Mackerel	Tuna	Marine worms

FIGURE 35.4 *Feeding Levels and Food Chains: Producers and Consumers in Nature.* The organisms may vary from one ecosystem to another (shown here are those for forest, deep-sea vent, and open ocean), but in each ecosystem, autotrophic producers form the first feeding level, while consumers form the next three feeding levels, and detritivores utilize the wastes from all levels. Within each ecosystem, we can identify chains of producers and consumers. In the hardwood forest, for example, oak trees are producers, caterpillars consume oak leaves, titmouse birds consume caterpillars, weasels consume the birds, and fungi and soil bacteria live on plant and animal wastes.

cules—the main source of carbon atoms in all types of biological molecules. Producers also absorb nitrogen, phosphorus, sulfur, and other needed atoms from the environment and fix them into biological molecules. In the deep-sea vent ecosystem, certain bacteria are the sole producers, since no light penetrates to these depths and plants cannot survive. These bacteria harvest energy stored in the bonds of hydrogen sulfide (H_2S) molecules dissolved in the water, and they use this energy to fix carbon dioxide from seawater into sugar molecules. The bacteria then incorporate the carbon from the energy-rich sugars into the other biological molecules they require for growth and reproduction. The process is called *chemosynthesis* (rather than photosynthesis).

At the second feeding level are the *primary consumers*, the organisms that eat the producers. Herbivores, such as caterpillars or cows, efficiently digest plant matter for energy and serve as ecological links between the producer level and all other levels. Near deep-sea vents, huge tube worms act as a special kind of primary consumer. Although they lack a mouth or other way of eating, the tube worms have a special internal organ packed with chemosynthesizing bacteria. Red blood coursing through the worm's scarlet plumes delivers hydrogen sulfide to the bacteria, which in turn produce sugars that the tube worm uses for energy and carbon.

At the next highest level, *secondary consumers* are carnivores (meat eaters) that consume the herbivores. A titmouse eating a caterpillar is a carnivore, and so is a white crab feeding on a tube worm. In the next feeding level,

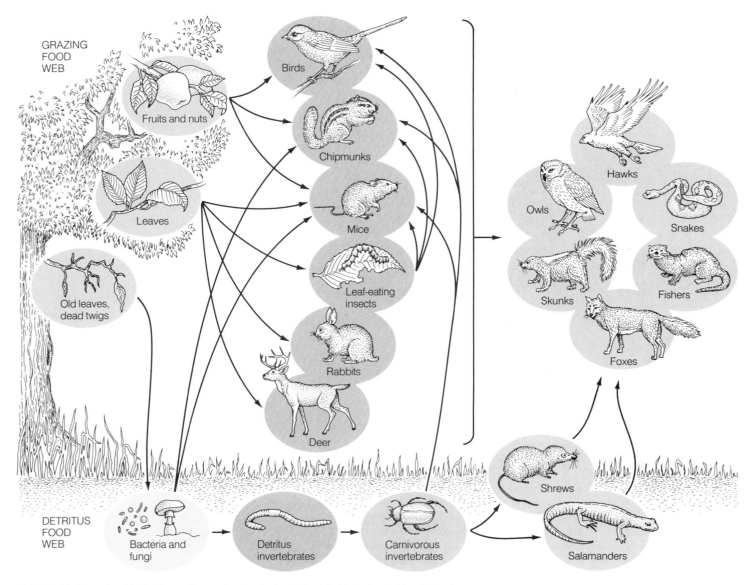

FIGURE 35.5 *A Food Web: Cross-Dependencies in the Living World.* In the Hubbard Brook Forest ecosystem, dozens of species derive energy from each other in a complex grazing food web, while others are actors in the detritus food web. Both webs are linked at numerous points by the activities of specific organisms. Birds, for example, can feed on berries, leaf-eating insects, and insects from the detritus layer. This simplified diagram omits hundreds of additional species and their tangled network of interrelationships.

tertiary consumers are carnivores that eat other carnivores; a weasel may eat swifts, for example, and a shark may eat mackerel. Finally, a few ecosystems have one more level containing carnivores that eat tertiary consumers, such as cougars eating weasels and sharks eating other sharks.

A special class of consumers, the **detritivores,** or decomposers, obtain energy and materials from *detritus,* organic wastes and dead organisms that accumulate from all trophic levels. Fungi and bacteria are the most ubiquitous detritivores, but worms, nematodes, many kinds of insects, and carrion feeders like vultures are also important detritus consumers.

The simplified diagram in Figure 35.4 suggests that each feeding level leads directly to the next in a simple chain, and indeed, **food chains** do exist in nature, with groups of organisms involved in linear transfers of energy from producer to primary, secondary, and tertiary consumers. More commonly, however, feeding relationships resemble not chains but complex interwoven webs.

FOOD WEBS IN NATURE

Organisms usually consume more than one other species, and some animals feed at several levels. A high-level carnivore like a cougar may eat herbivores, such as rabbits, as well as other carnivores, such as foxes or hawks. And as an omnivore, you yourself eat vegetables (producers), poultry (herbivores), tuna fish (carnivores), and mushrooms (detritivores). Ecologists call complicated interconnected feeding relationships **food webs.** In a northern hardwood forest ecosystem of New Hampshire, for example, the major producers are sugar maple, beech, and yellow birch trees. These producers support two main food webs, a *grazing food web* that stems directly from the living plants, and a *detritus food web* that begins with dead plant parts and animal wastes (Figure 35.5). In the grazing food web, herbivores like jays and chipmunks consume fruits and seeds, while herbivores such as mice, deer, and caterpillars graze on leaves. These primary consumers then fall prey to hawks, skunks, snakes, and other carnivores. In the detritus food web, fungi and bacteria break down dead animals and fallen leaves, twigs, and fruits and are eaten in turn by grubs and earthworms. Grazing and detritus food webs are themselves linked, as detritivores become food for other consumers, like salamanders and shrews.

Feeding levels and food webs describe the general routes of energy flow and material cycling in ecosystems. They do not, however, portray the *amounts* of energy or materials that pass through each level. To fully understand how ecosystems function, ecologists measure the energy transfers.

※ HOW ENERGY FLOWS THROUGH ECOSYSTEMS

Whether an organism is a producer or a consumer, it needs energy for movement, for active transport of nutrients and ions, and for synthesis of proteins, nucleic acids, and other large molecules necessary for growth and repair. Producers obtain their energy directly from the environment in the form of light (in most ecosystems) or inorganic molecules (in deep-sea vents and a few other ecosystems). Consumers, however, can get their energy only from producers. Hence, the activities of producers in a community set a limit for the amount of energy that can be captured and channeled throughout the entire ecosystem. Ecologists have closely studied the hardwood forest ecosystem to precisely measure available energy and how it is spent.

ENERGY BUDGET FOR AN ECOSYSTEM

Within the White Mountains of New Hampshire lies the Hubbard Brook Experimental Forest, a fragrant zone of tree-studded rolling hills that has fascinated researchers for more than 25 years (Figure 35.6). Researchers have focused on a few small *watersheds* (regions drained by a single stream or river), because the ridges that separate watersheds provide convenient dividing lines between adjacent ecosystems. Since impermeable bedrock lies below each watershed, researchers can measure all the groundwater leaving the ecosystem (as well as its load of dissolved nutrients) by monitoring the single stream that drains it. By studying the input of energy, water, mineral nutrients, and organic matter into these watersheds and tracing their incorporation into both living things and the physical environment, workers have learned how energy flows through an entire forest.

During the growing season (June–September), about 500,000 kcal of solar energy, including both heat and light, strike each square meter in the Hubbard Brook Forest. (To put this in perspective, you would need this much energy to jog 5000 miles!) Over 83 percent of the solar energy that reaches the hardwood forest is released as heat and another 15 percent as reflected light—for a total of 98 percent that returns rapidly to the physical environment (Figure 35.7, page 574). The 2 percent or so remaining is the amount of energy that producers convert by photosynthesis to chemical energy in the form

FIGURE 35.6 *Hubbard Brook Experimental Forest: A Living Laboratory for Studying Energy Flow in Ecosystems.* Ecologists have clear-cut bands and zones of the Hubbard Brook Forest to determine the effects of logging on soil erosion and nutrient loss and have monitored energy flow and material cycling through hundreds of specific experiments.

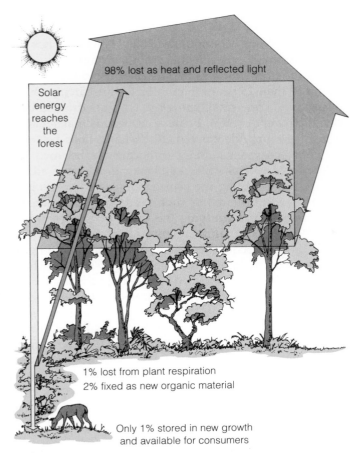

98% lost as heat and reflected light

Solar energy reaches the forest

1% lost from plant respiration
2% fixed as new organic material

Only 1% stored in new growth and available for consumers

FIGURE 35.7 *Energy Budget of a Hardwood Forest.* This diagram of the Hubbard Brook Experimental Forest traces the flow of energy through an ecosystem. Most of the sunlight that strikes the forest is returned to the environment as light and heat, and only 2 percent of the original energy is fixed in the plants as new organic material—the gross primary productivity. Because of energy loss from respiration in the plants, only 1 percent of the original energy from sunlight is stored (the net primary productivity) and becomes available for consumers, such as deer, or eventually falls to the forest floor as dead leaves and twigs.

cells, leaves, roots, stems, flowers, and fruits (see Figure 35.7). Of all the energy impinging on the ecosystem, only the net primary productivity is available to consumers.

During a growing season, plants retain some of the net primary productivity in permanent organs (new stems and roots, for example), but most becomes litter on the forest floor. In fact, nearly twice as much energy in a forest is stored in litter and decomposing humus as in the majestic banks of leaves overhead! Most of the energy contained in the litter fuels the detritus food web (see Figure 35.5). Only a small fraction of the energy stored aboveground in the forest enters the grazing food web, and some of that eventually enters the detritus food web owing to the excretion and mortality of consumers from the grazing food web. Energy dissipation continues as the consumers that graze plants or eat detritus radiate heat via respiration. Only a tiny amount of energy then remains in the soil or exits the ecosystem in the stream that drains the watershed. Clearly, despite the essentially limitless power flowing from the sun to the earth, plants can store only a small fraction of that energy, and that fraction sets up an upper limit to the energy available to all other organisms in the ecosystem.

Such information about energy use allows ecologists to formulate general principles about the energy budget of a hardwood forest like Hubbard Brook. First, even in a lush, leafy green forest, plants and other producers convert only a small fraction (2 percent or less) of the solar energy that enters the ecosystem into stored chemical energy, and they use half of that in their own respiration. Second, animals ingest an even smaller amount (in this case, 0.01 percent of the energy) in the grazing food web. Finally, as energy flows through the feeding levels of the ecosystem, metabolic activities (mostly respiration) release it back to the air, where it ultimately returns to space as heat.

of sugars and other organic compounds. Ecologists call this small fraction an ecosystem's *gross primary productivity,* and ultimately, it limits an ecosystem's structure, including how many birch trees will grow and how many chipmunks will thrive.

Not all the chemical energy that a plant initially traps will be stored in newly formed leaves, roots, and fruits. Plant cells themselves use a little more than half of this energy to fuel their own cellular respiration, eventually losing it as heat. The small amount of energy remaining after respiration is called the *net primary productivity,* the amount of chemical energy that is actually stored in new

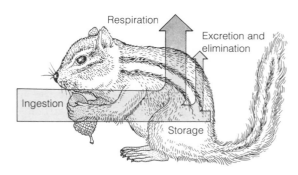

Respiration

Excretion and elimination

Ingestion

Storage

FIGURE 35.8 *Energy Budget of a Forest Herbivore.* Energy flows through individual consumers like this chipmunk, as well as through the forest ecosystem it inhabits. Of the energy the chipmunk ingests from seeds and leaves, most is used to fuel its continued respiration; some is lost in the waste products it excretes; and a very small amount (less than 2 percent) is stored in new tissue.

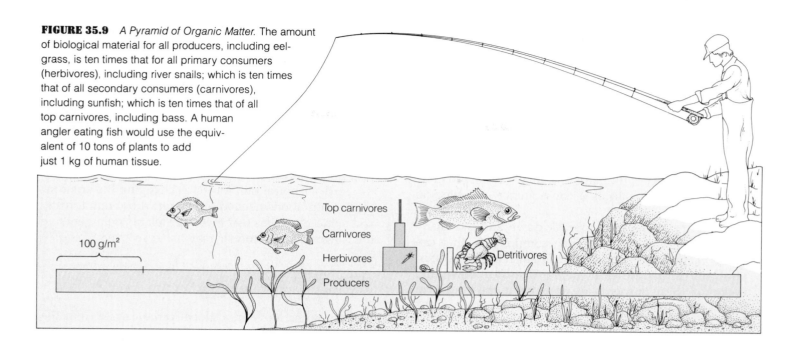

FIGURE 35.9 *A Pyramid of Organic Matter.* The amount of biological material for all producers, including eelgrass, is ten times that for all primary consumers (herbivores), including river snails; which is ten times that of all secondary consumers (carnivores), including sunfish; which is ten times that of all top carnivores, including bass. A human angler eating fish would use the equivalent of 10 tons of plants to add just 1 kg of human tissue.

Labels in figure: Top carnivores, Carnivores, Herbivores, Detritivores, Producers, 100 g/m²

ENERGY AND ECOLOGICAL PYRAMIDS

Ecologists have discovered another important fact in their experiments on energy flow from one feeding level to the next: Food chains rarely have more than four links. To see why, consider the fate of the energy that enters the grazing food web. A chipmunk, for example, one of the most important herbivores of the Hubbard Brook Forest, ingests energy in the form of seeds or leaves, assimilates some of the energy, excretes some in liquid wastes, and eliminates some in solid wastes (Figure 35.8). Most of the assimilated energy escapes as heat through respiration. Chipmunks store less than 2 percent of ingested food energy in new tissues or offspring. The consequence of a huge loss through respiration and a small net increase in growth is that chipmunks store very little energy in a form that can be used at the next feeding level, say, by a red fox.

Ecologists portray the energy relationships between different feeding levels in a pyramid-shaped diagram. The building blocks of this kind of graphic are proportional in size to the amount of either energy or organic matter (**biomass**) available from the level below. Figure 35.9 shows such a pyramid for a river ecosystem in Silver Springs, Florida, which includes eelgrass and algae (producers); turtles, snails, and caddis flies (herbivores); and beetles, sunfish, and bass (carnivores). At each feeding level, the amount of organic matter present in the organisms is about a tenth that of the level below it, which accounts for the pyramid shape (a **pyramid of biomass**). By moving up four levels in a given linear food chain, one finds very little biomass in the top carnivores such as large, meat-eating bass fish. Taking it one step further (and another 10 percent reduction in biomass), an angler would need to catch and eat 10 kg of bass fish to put on 1 kg of human tissue. The 10 kg bass tissue comes from 100 kg of smaller carnivorous fish, 1000 kg of insect herbivores, and 10,000 kg (about 10 tons) of plants. All that for 1 kg (2.2 lb) of human being!

The lesson from the pyramid of organic matter was summarized by a popular 1970s phrase: "Eat low on the food chain." This refers to the fact that it takes 10 kg of grain to build 1 kg of human tissue if the person eats the grain directly, but it takes 100 kg of grain to build 1 kg of human tissue if a cow eats the grain first, and the person eats the beef. Eating lower on the food chain—eating producers, not consumers—saves precious resources on a small planet.

The pyramid pattern of energy flow in ecosystems has an important ramification: the tendency for toxic substances to increase in concentration in progressively higher levels of a food chain, a process called **biological magnification.** Many chemical insecticides such as DDT and chlordane resist degradation in the environment and tend to be stored in body fats. If farmers spray DDT on their cabbage plants to control cabbage looper caterpillars, some of the chemical inevitably runs off into streams and lakes. There, instead of breaking down, some of it may enter water plants later eaten by herbivorous fishes. Fishes and other animals cannot break down or excrete the toxin, and it is instead stored in their body fats.

The magnification continues when carnivorous fishes, such as pike, eat herbivorous fishes and when top carnivores, such as ospreys, devour pike. The concentration of DDT in the birds' bodies can reach levels 10,000 times greater than in the water plants that originally took it up.

This amount of DDT interferes with calcium metabolism and results in thin-shelled, easily broken eggs. During the two decades or so that DDT was commonly used in the United States (early 1940s to late 1960s), the numbers of ospreys, falcons, hawks, eagles, and condors dropped significantly as a result of this lethal chain of events. Although the use of DDT was banned in 1968, the chemicals are still widely used in overpopulated and developing nations to control mosquitoes that spread malaria, the greatest killer of humans, and they continue to magnify in the food chain as a consequence of ecological pyramids.

While organisms cannot break down and use many toxic chemicals and recycle their components, they can utilize many compounds that contain nitrogen, sulfur, phosphorus, and other elements. The breakdown and cycling of these nutrients are essential to the long-term health of ecosystems.

≋ HOW MATERIALS CYCLE THROUGH ECOSYSTEMS

In contrast to the essentially inexhaustible stream of solar energy striking our planet, the physical materials on which life depends are limited to what is presently in the soil, water, and air around our globe. Therefore, they must be recycled for ecosystems to survive. Our planet has several *biogeochemical cycles* (literally, "life-earth-chemical" cycles), global loops of material utilization. In these cycles, substances enter organisms from the atmosphere or soil, reside temporarily in those organisms, then return to the nonliving world when the organisms respire or decompose (review Figure 35.3). Repositories of materials, including the atmosphere and living organisms, are called *pools* or reservoirs, and in general, the organismal pool is much smaller than the nonliving pool. Pools remain constant in size as long as a substance's rate of entry equals its rate of departure.

The most important biogeochemical cycles are those for water, carbon, nitrogen, and phosphorus. These underlie the health of all ecosystems, but human activities can alter them in profound and often undesirable ways.

THE WATER CYCLE: DRIVEN BY SOLAR POWER

In the global **water cycle,** water circles from the atmosphere to the earth's surface as rain or snow, and back again to the atmosphere by evaporation from puddles, ponds, rivers, and oceans and by transpiration from plants (Figure 35.10). (Recall that transpiration is the loss of water, mainly through stomata on leaves and stems.)

The role of vegetation in promoting transpiration, evaporation, and hence rain is especially evident in the tropical rain forests (Figure 35.10). When people cut down a rain forest in one area, rainwater drains off and eventually reaches the sea instead of rising from the plants to the clouds and falling once again on this or another part of the forest. The ongoing massive destruction of the earth's tropical rain forests is changing the environmental conditions needed to support those rain forests, and some ecologists fear that since plants help generate their own rain, the forests may never recover.

THE CARBON CYCLE: COUPLED TO THE FLOW OF ENERGY

Carbon atoms move in a global **carbon cycle** from the physical environment through organisms and back to the nonliving world, just as moisture moves through the

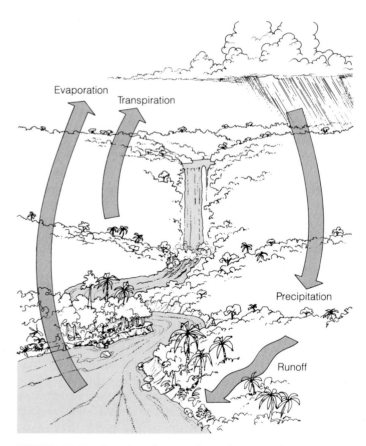

FIGURE 35.10 *The Water Cycle: A Global Exchange from Atmosphere to Earth's Surface to Plants and Back Again.* Water leaves the air and falls to earth as rain or snow. This precipitation then either evaporates into the air or enters living things. Water leaves organisms by evaporation or transpiration (from plants), enters the air, and the cycle is complete.

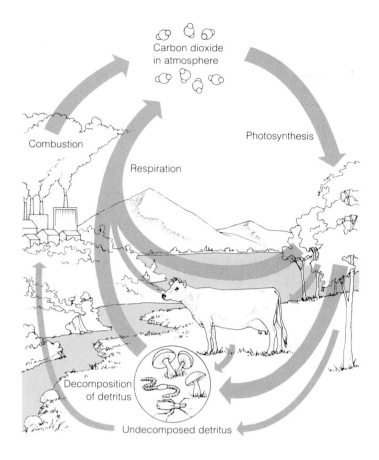

FIGURE 35.11 *The Carbon Cycle: From Atmosphere to Plants, Animals, Decomposers, Human Activities, and Back.* Carbon cycles through a number of forms, from atmospheric carbon dioxide to biological molecules to organic molecules in the soil to geological deposits of fossil fuels, and back to carbon dioxide after combustion of the fuels or respiration of the biological molecules.

water cycle. And like the water cycle, the carbon cycle is linked to energy flow because producers—including the photosynthetic plants of the forests and oceans and the chemosynthetic bacteria of deep-sea vents—require environmental energy (either sunlight or inorganic hydrogen compounds) to trap carbon into sugars. The trapped carbon comes from carbon dioxide in the surrounding air or water (Figure 35.11). As the cycle proceeds, consumers devour the organic carbon compounds that producers manufacture. Then, via respiration, both consumers and producers return carbon to the non-living environment in the form of carbon dioxide. Some carbon accumulates for many years in wood and is eventually returned to the atmosphere in fires or by consumption and respiration of fungi, bacteria, and other detritivores. Organic carbon can leave the cycle for even longer periods of time after sediments bury organic litter, which decomposes and gradually transforms into coal or oil. Carbon also leaves the cycle when cast-off calcium

carbonate shells of marine organisms sink to the ocean floor and become covered with sediments that compress them into limestone. Eventually, however, even these carbon deposits are recycled into atmospheric carbon dioxide as the limestone erodes and the fossil fuels are burned. The more we humans burn gasoline to run automobiles, and coal to produce electricity for homes and factories, the more we disturb this carbon cycle. Our activities add more carbon dioxide to the atmosphere than plants remove through photosynthesis. Many scientists think that the resultant accumulation of carbon dioxide in the atmosphere contributes to the warming of the earth through the greenhouse effect (see the box on page 596 of Chapter 36).

THE NITROGEN CYCLE DEPENDS ON NITROGEN-FIXING BACTERIA

Nitrogen gas makes up 79 percent of our atmosphere, and it is an essential component of every protein and nucleic acid in every living thing. Ironically, though, most organisms cannot use nitrogen in its gaseous form. Instead, they depend on a few species of nitrogen-fixing bacteria (review Figure 31.7) to trap nitrogen in a biologically accessible form that organisms can use in building proteins and nucleic acids. When these nitrogen-containing biological molecules are destroyed, other bacterial species return the nitrogen to the atmosphere as nitrogen gas and complete the **nitrogen cycle.** Figure 35.12, page 578, depicts the major steps in this cycle of transformations.

In agricultural ecosystems, where the net primary production helps feed the human race, nitrogen is often the factor that limits productivity. Because of this, farmers have long fertilized their fields to increase the soil's content of ammonium and nitrate, the ionic forms of nitrogen that plants convert to biological molecules. In the seventeenth century, the Pilgrims observed native Americans burying fish with their corn seeds. Soil bacteria produced ammonium from the fish proteins and nucleic acids, the corn assimilated the available nitrogen, and the plants grew taller and faster. Many contemporary farmers continue to practice *crop rotation*. They plant leguminous crops such as beans, clover, or alfalfa one year and corn, wheat, or sugar beets the next year to take advantage of the nitrogen that the nitrogen-fixing bacteria in the legumes' root nodules release into the soil.

Today, most large farms around the world depend on nitrogen fertilizers produced through industrial nitrogen fixation rather than a bacterial process. In fact, nitrogen fixed in chemical factories may now represent 30 percent of the input to the global nitrogen cycle, a truly great shift in the natural cycle. Unfortunately, industrial nitro-

1. OBSERVATION: The lake has a thick growth of slimy green algae, and the fish are dying.

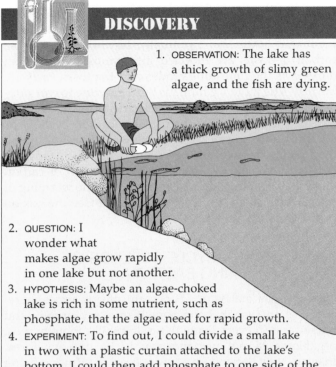

2. QUESTION: I wonder what makes algae grow rapidly in one lake but not another.

3. HYPOTHESIS: Maybe an algae-choked lake is rich in some nutrient, such as phosphate, that the algae need for rapid growth.

4. EXPERIMENT: To find out, I could divide a small lake in two with a plastic curtain attached to the lake's bottom. I could then add phosphate to one side of the lake but not the other; this second side would be a control.

Phosphate

5. RESULTS: A few weeks later, aereal photos clearly reveal heavy growth of algae on the phosphate-seeded side.

6. CONCLUSION: This suggests that phosphates in detergents and fertilizers that pollute some lakes probably encourage heavy algal growth.

Adapted from: P. W. Schindler, *Science*, 184 (1974):857–899

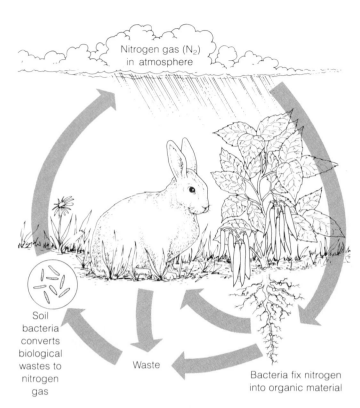

FIGURE 35.12 *The Nitrogen Cycle: From Air to Organisms and Back.* Bacteria in root nodules and in the soil fix atmospheric nitrogen into forms that plants can then use to make proteins. Animals, by eating the plants, can also gain needed nitrogen. Soil bacteria eventually release the nitrogen, allowing it to return to the atmosphere.

gen fixation requires tremendous heat and pressure, and this is usually produced by burning great quantities of fossil fuels. In some areas, people dump more energy into the soil in the form of nitrogen fertilizers than they extract from the soil in food calories. Since reserves of fossil fuels are limited, this enormous reliance on industrially fixed nitrogen fertilizers cannot go on indefinitely. Many biologists think a far better solution would be to genetically engineer plants and soil bacteria to increase natural nitrogen fixation (see Chapter 31).

The nitrogen cycle is sensitive to human activities such as deforestation. To measure this sensitivity, ecologists removed all trees from one watershed in the Hubbard Brook Forest and sprayed the area with herbicides to block regrowth (see Figure 35.6). In their test area, 60 times more nitrogen drained away in the stream than in the control watersheds where the forests remained undisturbed. Such severe nitrogen loss drastically limits regrowth of the forest.

The nitrogen cycle, like the water and carbon cycles, has an atmospheric phase. While nitrogen gas, water vapor, and carbon dioxide are airborne, they can be blown anywhere on earth by the wind. This mobility makes the

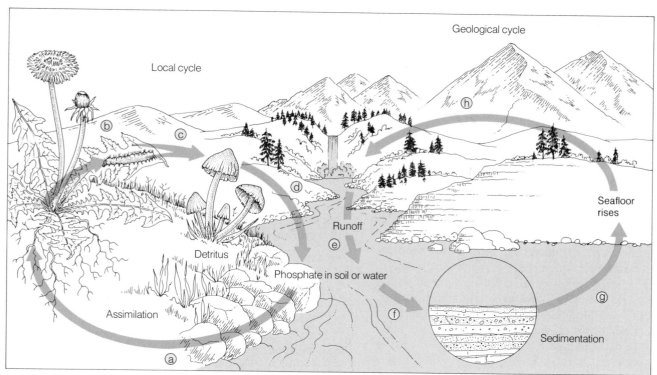

FIGURE 35.13 *The Phosphorus Cycle: Short-Term and Local, Long-Term and Global.* Plants assimilate phosphates from soil or water (a), animals can gain phosphates by eating the plants (b), and when the autotrophs and heterotrophs die or give off wastes (c), these return to the soil or water in a local cycle (d). Some phosphates become tied up in sedimentary rocks at the bottom of the seas (e, f), but when the climate changes and the seafloor rises after thousands or millions of years (g), the phosphates from this long-term cycle can once again enter the local, short-term cycle (h).

nitrogen, water, and carbon cycles truly planetary. In contrast, some substances, like phosphorus, lack a gaseous phase and thus, in the short run, cycle locally.

PHOSPHORUS CYCLES LOCALLY AND IN GEOLOGICAL TIME

Phosphorus is essential for life. It is a component of cell membranes, nucleic acids, and ATP, with its capacity to transport energy within the cell. The **phosphorus cycle** consists of two interlocking circuits, one that acts locally during short stretches of time and another that operates more globally over vastly longer time periods. In the local cycle, phosphorus moves from the rocks or soil into organisms and back to the soil (Figure 35.13a–d). Such cycling of phosphorus can cause great fluctuations in populations of local organisms. In the Arctic, for example, cycles of phosphorus and other nutrients in the tundra ecosystem trigger periodic population booms and crashes of the furry tundra rodent called the lemming.

Although phosphorus does not leave local ecosystems by way of the air, it *can* leave terrestrial ecosystems dissolved in streams and rivers (Figure 35.13e). Thus, the second phosphorus cycle, if viewed over geological time, has a global aspect (Figure 35.13e–h). After phosphate

eventually washes from terrestrial waterways down to the sea, it can form insoluble compounds that fall as sediments and become incorporated into rock. Eons later, when climates change and the seafloor rises, exposing new land, these phosphorus-containing rocks may form the base of a terrestrial ecosystem, and the phosphorus can once again enter the local ecological cycle.

As with other biogeochemical cycles, human activities have altered the dynamics of the phosphorus cycle—especially in aquatic ecosystems, for which phosphorus is often the factor limiting primary productivity. Phosphates are major ingredients of agricultural fertilizers and until recently were also main components of detergents. When phosphates from these sources run off into lakes, algae and other aquatic plants grow faster, and the ecosystem becomes oversupplied with primary producers (see the Discovery box on page 578). As individual algal plants die in an oversupplied lake, decomposing bacteria feed on the dead algal cells and use up so much dissolved oxygen that fishes and other animals may suffocate. If the source of excess phosphate dries up, however, the algal bloom will fade and the lake will recover its previous condition. The possibility of spontaneous recovery suggests that people can decrease the destruction of lake and stream ecosystems by restricting the use of phosphate-containing detergents and fertilizers.

Unfortunately, ecological studies reveal that human activities are altering not only the dynamics of biogeochemical cycles in local regions, but in some cases, their global operations—and our own future—as well.

≋ HOW HUMAN INTERVENTION ALTERS ECOSYSTEM FUNCTION

Human activities have the potential to drastically modify the nutrient cycles that support life on earth and to alter physical features of the natural environment as well, including air and water temperatures and acidities and the amount of available solar energy. Let us consider two of the biggest current ecological concerns: acid rain and the threat of nuclear winter.

WHEN IT RAINS, IT POURS SULFURIC ACID

About 35 years ago, anglers began to note declining fish populations in formerly productive lakes in Sweden, Ontario (Canada), and upper New York State. Swedish ecologists traced the cause to increased acidity in the lake water and in turn to abnormally acidic precipitation: In short, it was raining sulfuric acid! Observers soon noted that acid rain was also killing trees from Colorado to the Carolinas to Vermont, stunting the growth of lake trout in northern Ontario, and eroding stone monuments that are centuries old (Figure 35.14).

Clues to the origin of acid rain lay in the distribution of the affected areas: They were all downwind from the industrial centers of Europe and North America. The burning of coal and oil to generate electric power and run factories and the operation of automobile engines all create nitrogen- and sulfur-containing compounds that acidify the rain. In many instances, these airborne pollutants escape from extremely tall industrial smokestacks and can travel hundreds of kilometers, suspended on air currents. Up in the clouds, they combine with water molecules to form sulfuric and nitric acids. When rain precipitates from these clouds, the acids are carried with it.

Since the major source of acid rain is pollution from coal-fired plants that generate electricity, the immediate solution would seem to be the removal of sulfur and nitrogen oxides from their exhausts. This can be done with smokestack "scrubbers," but it raises people's power

FIGURE 35.14 *The Effects of Acid Rain.* After gracing a stone cathedral for hundreds of years, this northern European sculpture has begun to dissolve from the highly corrosive effects of acid rain.

bills by about 5 percent, and so far, electric power utilities have stubbornly resisted this solution. An informed public must decide whether it is willing to pay slightly higher bills for electricity to preserve its forested recreational lands and wood products industry.

Clearly, human activities can unfavorably alter the nutrient cycles that support life, as well as the flow of the sun's energy that drives life functions; and only a change in human activities can reverse the situation.

NUCLEAR WINTER: A SERIOUS ECOLOGICAL THREAT

Whenever a natural calamity throws a substantial amount of dust into the air, the sunlight reaching the earth's surface is cut precipitously, and this decline has a major impact on ecosystems. In 1815, for example, when the volcano Tambora erupted in Indonesia, dust particles drifted around the globe and affected plant growth as far away as New England and Switzerland. Millions of years earlier, a meteor hit the earth and threw up an even more gigantic cloud of dust and smoke. Some scientists hypothesize that this cloud darkened skies all around the world, blocked photosynthesis, and may have contributed to the demise of the dinosaurs (see Figure 18.11). As frightening as the prospects may seem, our species now possesses the technology to cause similar global disasters through nuclear war.

A large-scale nuclear war employing even a fraction of the warheads now stockpiled by the United States and the Soviet Union would directly kill nearly everyone in North America, Japan, Europe, and the Soviet Union. Most of the survivors would need medical attention that would no longer be available, and thus they, too, would succumb in days or weeks. Horrendous as such a scenario may seem, the longer-term consequences to the earth's ecosystems would be even more devastating. Nuclear warheads detonated at ground level would inject enormous quantities of fine, radioactive dust into the atmosphere, and bombs exploding over urban and forested areas of the Northern Hemisphere would ignite widespread fires that would heave huge clouds of smoke into the air.

What would the climatic effects be from such massive clouds of dust and smoke? Physicists, atmospheric scientists, and ecologists predict that nuclear smoke clouds would drift from place to place in the Northern Hemisphere, and beneath those clouds, temperatures would plummet to well below freezing, a condition called **nuclear winter**. Plant species that can normally survive cold winters would probably be killed if the nuclear war occurred during spring or summer in their active growing periods. Monsoon rains would be severely disrupted in the Southern Hemisphere. Rainfall over burned-out forests would trigger massive flooding and erosion. Radiation levels would remain lethal to most animals and many plants for long periods. Finally, poisonous substances released by urban fires would impede the regrowth of vegetation. The lesson here is straightforward: Even if a nuclear war did not extinguish the human species immediately, any nuclear strike would be suicidal.

Only those organisms that do not depend on sunlight for energy—the bacteria, tube worms, clams, crabs, fishes, and other residents of the deep-sea vent communities—would remain largely unaffected. Independent of light and warmed from within the earth, these assemblages of creatures could survive even if all other living things were destroyed. Perhaps over an enormous span of geological time, these organisms of the deep-sea vents would leave their refuges and evolve, recolonizing the globe. But perhaps they wouldn't.

≋ CONNECTIONS

Energy flow and nutrient cycles in ecosystems are fundamental biological principles governing life. As history plainly shows, no organisms are exempt from these laws or their implications. We humans are unique among living organisms only in the degree and speed with which we can modify the earth's ecosystems, in our ability to evaluate the consequences of our actions, and in our potential to take measures as individuals and as nations to deal with the ecological imbalances and threats we now face.

In the next chapter, we will examine the global forces that link ecosystems into one all-encompassing biosphere.

NEW TERMS

biological magnification, page 575
biomass, page 575
carbon cycle, page 576
consumer, page 570
detritivore, page 572
food chain, page 572
food web, page 573

nitrogen cycle, page 577
nuclear winter, page 581
phosphorus cycle, page 579
producer, page 570
pyramid of biomass, page 575
water cycle, page 576

STUDY QUESTIONS

REVIEW WHAT YOU HAVE LEARNED

1. Define producer and consumer.
2. How do chemosynthesis and photosynthesis differ? Name one type of organism that employs each method.
3. Name several types of detritivores, and give the energy source of each type.
4. Create a diagram showing as many organisms as you can in the food web of a forest ecosystem from your geographical region.
5. True or false: Only 1 percent of the solar energy that reaches the earth is stored via photosynthesis, and yet this provides energy for virtually all other life in the biosphere. Explain your answer.
6. What do the various levels in a pyramid of biomass represent?
7. Describe the water cycle.
8. Describe the steps of nitrogen transformation during the nitrogen cycle.
9. Trace the steps in the local phosphorus cycle.
10. How does acid rain form?
11. Discuss the ecological hazards of a nuclear war.

APPLY WHAT YOU HAVE LEARNED

1. Farmers, gardeners, and forest managers are no longer allowed to use DDT as an insecticide. Why not?
2. Native Americans taught the Pilgrims to plant fish with corn seed. Why was this planting strategy successful?
3. The labels of some laundry detergents state that the products contain no phosphates. Why is the absence of phosphates important?

FOR FURTHER READING

Childress, J. J., H. Felbeck, and G. N. Somero. "Symbiosis in the Deep Sea." *Scientific American* 256 (May 1987): 114–120.
Gosz, J. R., R. T. Holmes, G. E. Likens, and F. H. Bormann. "The Flow of Energy in a Forest Ecosystem." *Scientific American* 238 (March 1978): 93–102.
Mohnen, V. A. "The Challenge of Acid Rain." *Scientific American* 259 (August 1988): 30–39.
Western, D., and M. C. Pearl, eds. *Conservation for the Twenty-First Century.* New York: Oxford University Press, 1989.

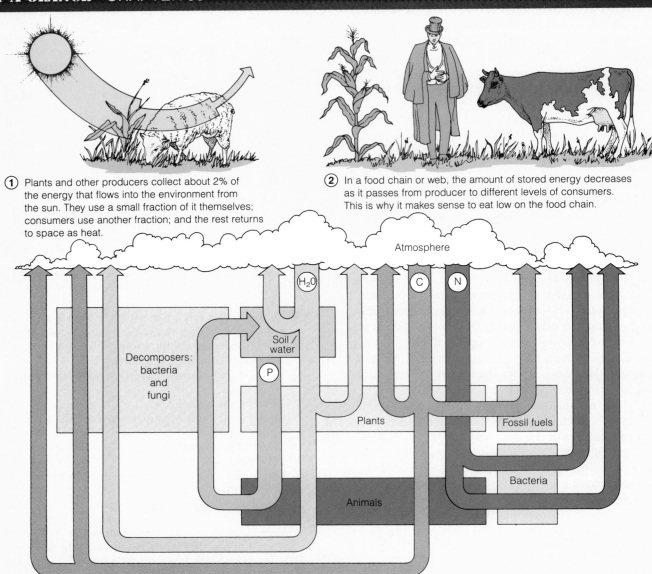

① Plants and other producers collect about 2% of the energy that flows into the environment from the sun. They use a small fraction of it themselves; consumers use another fraction; and the rest returns to space as heat.

② In a food chain or web, the amount of stored energy decreases as it passes from producer to different levels of consumers. This is why it makes sense to eat low on the food chain.

③ Water, carbon, nitrogen, and phosphorus help sustain life as they cycle locally through ecosystems and globally through the atmosphere and oceans.

④ Individuals as well as governments must make ecologically sound choices to reverse the current planetary trend toward environmental pollution and destruction.

THE BIOSPHERE: EARTH'S FRAGILE FILM OF LIFE

CHAPTER 36

AN ILL WIND

A traveler to the Galápagos Islands today will witness a phenomenon that Charles Darwin himself probably viewed: large colonies of Galápagos fur seals—270 kg (600 lb) bulls, much smaller cows, and tiny, furry pups—dozing together on the sandy beaches of the black volcanic islets lying due west off the coast of Ecuador. While their mothers plunge into the chilly surf in search of food, the seal pups stay on shore in the safety of the colony. The female usually spends about a day and a half in the open ocean swimming, diving, and consuming many kilograms of fish and squid. When she returns, she nurses her hungry pups with milk that is seven times richer than cow's milk, and the young grow rapidly.

In early 1983, the air and cold ocean currents streaming past the Galápagos Islands suddenly reversed their normal westward flow and began to roll east toward South America. This caused the ocean around the Galápagos to warm by 5°C (9°F), and because of this warming, the ocean could no longer support the normal growth of plankton. Since this decreased the populations of fish and squid, most seal cows could not eat well enough to provide their pups with sufficient milk, and eventually the baby seals starved (Figure 36.1). The pups, along with the other residents of the area, were victims of El Niño southern oscillation: an unexpected warming of the ocean waters off the coast of Peru and Ecuador that recurs every five years or so.

Interestingly, an El Niño causes effects far beyond the zone of ocean warming. In the El Niño of 1983, unusual torrential rains pelted the normally dry Galápagos Islands as well as the west coasts of North and South America, while parts of Africa, India, and Australia suffered the worst droughts in decades. The El Niño reminds us that the earth is a large, integrated system.

In this chapter, we describe how worldwide currents of air and water produce regional climates and how these climates, in turn, help determine the abundance and distribution of living organisms. Two main themes will emerge from our discussion. First, as the broad-leaved trees of the rain forest and the spiny plants of the desert demonstrate so clearly, different climates promote different communities of organisms. Second, communities with similar physical environments contain organisms with similar evolutionary adaptations, even when the individual species are unrelated. Plants inhabiting American deserts tend to have thick body parts, small leaves, and spines, and so do many plants from African deserts, even though the two plant groups evolved from totally different ancestors.

As this chapter explores the global physical forces that determine how the world's major communities of plants and animals are distributed, it addresses several questions:

- How do sunlight and the earth's rotation generate global climatic zones?
- What features characterize the major terrestrial communities of plants and animals?
- What common attributes do aquatic organisms share?
- How has the biosphere changed in the past, and how might human activities affect it in the future?

FIGURE 36.1 *Galápagos Fur Seal Pups: Victims of the 1983 El Niño.* This young fur seal was born a different year and survived El Niño's ill wind, but pups born that season were not so lucky.

WHAT GENERATES THE EARTH'S CLIMATIC REGIONS?

The starvation of fur seal pups during the 1983 El Niño reveals just how powerfully weather can affect biological systems. **Weather** is the condition of the atmosphere at any particular place and time, including its humidity, wind speed, temperature, and precipitation, whereas **climate** is the accumulation of seasonal weather events over a long period of time. What forces bring about normal patterns of weather and climate, so that some parts of the planet are desertlike, while others are covered with jungles, tundra, grasslands, or forests—each with plant and animal species typical to its terrain? The answer to this question lies in the global factors affecting the *biosphere*, that portion of our planet, including water, air, and soil, that supports life. The biosphere is the highest level of organization in the natural world (review Figure 1.5) and includes every body of water (rivers, ponds, lakes, and oceans), the atmosphere to a height of about 10 km, the earth's crust to a depth of several meters, and all the living things within this collective zone. As huge as the biosphere may seem, it is actually a delicate veneer at our planet's surface. If you inflated a round balloon to a diameter of 20 cm (8 in.) to represent the earth, then the thickness of its taut rubber skin would be proportional to the biosphere, the thin film of life encircling our world. Within the biosphere, weather has temporary, local effects, but climate is the major physical factor determining the abundance and distribution of living things. What, then, causes climate? The answer involves the uneven way that sunlight heats our planet; the behavior of water and air at different temperatures; and the rotation of the earth.

A ROUND, TILTED EARTH HEATS UNEVENLY

The tropics are warm, the poles are cold, and other regions of the globe have their own characteristic climates, partly because the earth rotates on a tilted axis and partly because the sun heats the earth's surface unevenly. Like a flashlight beam shining directly down onto a table from above versus one coming in obliquely from the side, sunlight hitting the earth directly is more intense than sunlight striking at an angle, where the same amount of light fans out over a larger area (Figure 36.2a). The angle of sunlight is therefore the issue, and anyone who has traveled extensively and has also experienced changing seasons knows the importance of *latitude* (the distance between any point and the equator) and *time of year.*

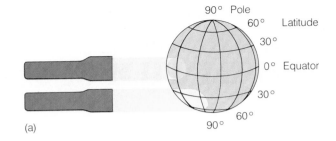

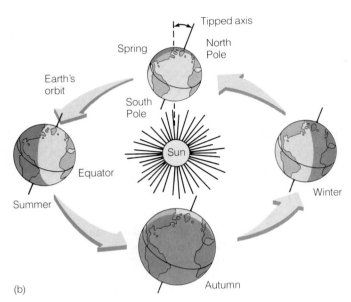

FIGURE 36.2 *Seasons: Light Shining on a Tilted Globe.* (a) The sun's angle determines the intensity of solar radiation at a given latitude. Consequently, sunlight is five times stronger at the equator than at the poles. (b) Since the earth spins on a tilted axis, the Northern Hemisphere is tipped closest to the sun in summer and away in winter, and the differences in light intensity generate the seasons.

About five times more light and heat energy fall at the earth's equator than at higher latitudes (farther from the equator), and the amounts of energy decrease the closer one gets to the poles. These differences in incoming solar power help explain why equatorial regions are warm, why polar regions are cold, and why tropical forests are usually so productive.

Temperatures also grow warmer and colder with the seasons, because our spherical planet spins on a tilted axis. The tilt is maintained all year long, even as the earth revolves around the sun (Figure 36.2b); consequently, during the summer, the Northern Hemisphere is tipped toward the sun, while in winter, it is tipped away and receives less solar energy. The seasons are reversed in the Southern Hemisphere. In both hemispheres, this tilting of the earth and uneven illumination help explain the warm temperatures and abundant growth of organisms in summer and the cold temperatures and dormancy of living things in winter.

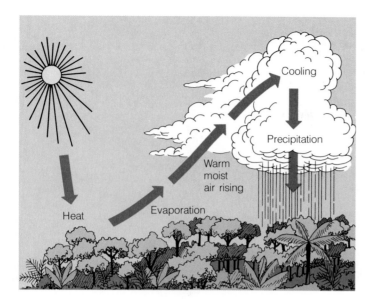

FIGURE 36.3 *Let It Rain: The Generation of Precipitation.* When heat strikes the land or ocean surface, the warm, moist air rises; the air then cools and releases its moisture content as rain or snow.

THE FORMATION OF RAIN

Sunlight striking the earth's tilted sphere at different places and different times heats the air and oceans unequally. Air has different properties when warm and cool, and these account for the distribution of air masses and the formation of rain and snow. Cold air is dense: It weighs more per cubic meter and so tends to sink through lighter, warmer air. This is why New England farmers plant cold-sensitive fruit trees on the sides of valleys instead of on the valley floor, where cold, heavy surface air settles at night. Warm air, in contrast, is less dense than cool air and tends to rise. This is why smoke, steam, and hot-air balloons drift upward.

Dense, cool air and light, warm air hold different amounts of moisture—cold air less, warm air more—and this principle ultimately brings about precipitation. You can distinguish a hot water pipe from a cold water pipe in a steamy shower room because water condenses into droplets on the cold pipe. Likewise, when warm, moist air rises high into the atmosphere, it cools, and its capacity to hold water decreases. The water condenses into droplets, forms clouds, and when the droplets become large and heavy, they fall to the earth as rain (Figure 36.3). This explains why a region like the tropics is so wet. Powerful sunlight at the equator heats the air, which picks up moisture from evaporation and plant transpiration and rises. Once aloft, the air cools, and the moisture precipitates out, falling onto the lush tropical rain forest.

Because sunlight strikes the earth unequally, large masses of air heat up, rise, and cool, then flow to new areas where the cool air falls and flows over the earth in a motion we call *wind*. The earth's natural rotation deflects the wind, influencing climate and weather.

AIR AND WATER CURRENTS: THE GENESIS OF WIND AND WEATHER

The earth's rotation sets up major currents in the atmosphere and oceans, and these lead to our climate and weather patterns.

■ **Air Currents: Global Treadmills** Hot air rises at the equator and moves north and south at high altitudes. As it moves aloft, the air cools, becomes heavy, descends, and then flows back to the equator near the surface (Figure 36.4a, page 586). Because the earth rotates, this moving, thermally generated air mass breaks up into six coils (Figure 36.4b). In each of these coils, the air swirls through a corkscrew pathway. It is as if six "slinkies" of moving air were stretched out around the earth at different latitudes. The movement of air through these coils establishes the direction of prevailing winds at different latitudes and in turn the direction that weather fronts normally travel.

The direction of air flow and the ascent and descent of air masses in their giant coils determine the earth's climatic zones and their vegetation types. Let's consider coil 1 north and coil 1 south, for example. Warm, moist air rises at the equator and drops its moisture in heavy tropical rains as it cools, creating the conditions for rain forests around the equator (Figures 36.3 and 36.4a). The cooler, drier air now travels at high altitude and descends at latitude 30° (both north and south). This descending dry air creates the great deserts of Australia, North and South Africa, and North America. At ground level, the dry air moves toward the equator, causing surface winds. These predictable breezes, called *trade winds*, propelled traders' sailing ships in past centuries. The other two pairs of coils work in a similar fashion, creating the general flow of air from west to east over the United States and the snow and rain that falls over the great forests of Canada and northern Europe. Together, the six coils of circulating air help us understand why deserts and forests occur where they do; they also drive the circulation of ocean currents.

■ **Ocean Currents: Patterns of Circulation** Wind blowing over the ocean surface sets the sea's upper layers moving, and the continents redirect the flow in slow, circular patterns (Figure 36.5, page 587). Ocean currents, like air currents, redistribute heat and hence influence climate and the distribution of plants and animals. The Gulf Stream, for example, carries warm water from the tropics up the eastern coast of North America, then east-

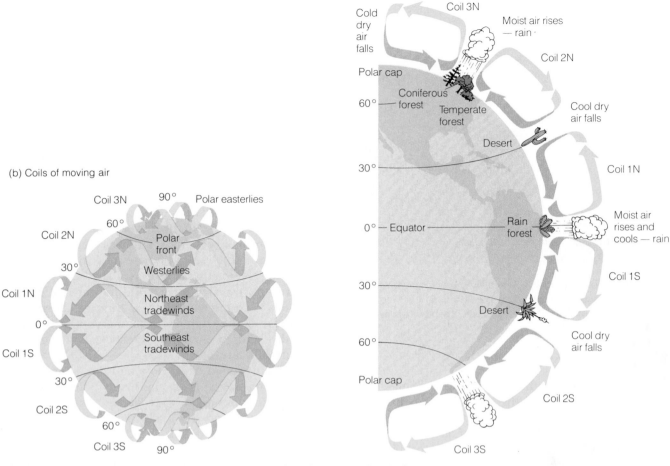

(b) Coils of moving air

(a) Air movement creates climatic zones

FIGURE 36.4 *Air Ascending and Descending Through Massive Coils Creates the Earth's Climatic Zones.* (a) Hot air rises at the equator, moves toward the poles, sinks, and returns to the equator in six loops. The air movements create global climatic zones with their characteristic vegetation, such as the tropical rain forests near the equator, where hot, moist air rises, and the deserts of 30° latitude, where dry air descends. (b) Because the earth rotates, the loops are drawn out into six coils. The air swirls up into the atmosphere and down toward the planet's surface, as indicated by the arrows. The direction of movement sets up prevailing wind patterns.

ward across the Atlantic, warming northern Europe (see Figure 36.5a).

■ **El Niño Explained** The principles of air and ocean currents just described help explain the dramatic weather changes that accompanied the strong El Niño event in 1983. Under normal conditions, cold currents flow north along the west coast of South America and move water out to sea near Peru (see Figure 36.5c and d). To replace the water moving away from the coast, deep waters laden with nutrients well upward and support the prolific growth of plankton and huge populations of anchovies and other fishes. During an El Niño event, for reasons scientists do not fully understand, warm water from around Tahiti suddenly flows back toward South America and warms the ocean around the Galápagos, Ecuador, and Peru. The thick surface layer of warm water

hinders the normal upwelling of nutrient-rich water from the ocean depths, the ocean's productivity falls, mother fur seals have unsuccessful hunting trips, and their pups perish.

Despite short-term irregularities such as El Niño, the effect of the sun on the tilting, rotating earth creates patterns of rainfall and temperatures that determine the general character of the earth's major communities of plants and animals.

≋ LIFE ON THE LAND

Wherever similar climatic conditions exist in the world—the deserts of Australia and Africa, the rain forests of Indonesia and Brazil—similar forms of plant life have evolved to cope with the climate's benefits and draw-

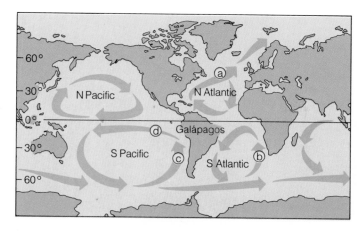

FIGURE 36.5 *Ocean Currents Flow in Four Major Circles, Redistributing Heat.* The major warm and cold ocean currents and their directions of flow result from the planet's rotation, the push of prevailing winds across the ocean surfaces, and the position of the continents. These currents, in turn, influence climate and hence the form of plants and animals on the landmasses. Four major currents are (a) the Gulf Stream, (b) the Benguela current, (c) the Humboldt current, and (d) the south equatorial current.

backs. Ecologists use the term **biome** to designate any of the major communities of organisms on land. Biomes are defined mainly by their vegetation and are characterized by specific adaptations to each climate. Major plant types characterize biomes because plants best reflect adaptations to rain, temperature, light, and other specific climatic conditions, and as primary producers, plants influence the consumers and decomposers that coexist in the biome.

It becomes clear that climate determines biomes when we study a biome distribution map (Figure 36.6). Such a map reveals orderly patterns of forests, deserts, and tundra lying roughly in bands stacked south to north that correspond with patterns of atmospheric circulation and climate.

Ecologists recognize as few as 7 or as many as 15 or more biomes, depending on how they classify the areas of species mixing between biomes. Here we consider eight biomes arranged roughly in order from equator to poles—tropical rain forests, savannas, deserts, temperate grasslands, chaparral, temperate forests, coniferous forests,

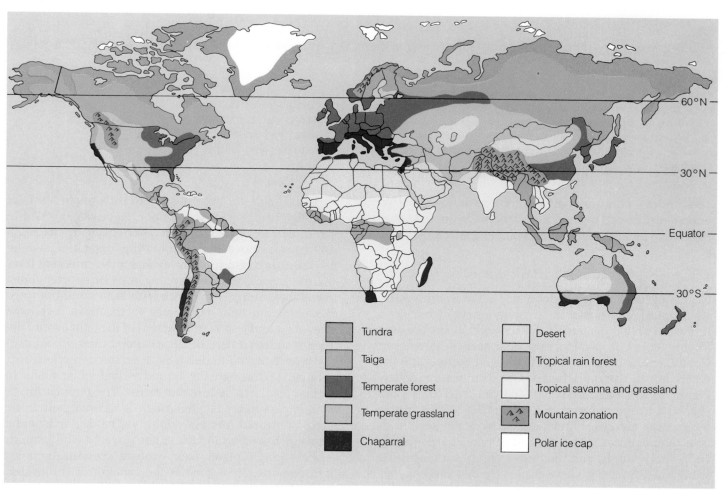

FIGURE 36.6 *The World's Major Biomes.* The eight major biomes, plus mountains and polar ice caps, are depicted here.

(a)

(b)

FIGURE 36.7 *Tropical Rain Forest: Lush Equatorial Biome.* (a) A tangle of large leaves, vines, and tree trunks shades out the rain forest floor in La Selva Biological Station, Costa Rica, in Central America. (b) Levels of life in the rain forest. Only a few tall trees and vines reach the emergent layer of the rain forest (top), while shorter forest trees form a canopy (upper middle) in which epiphytes and animals thrive. Low trees, shrubs, and underbrush grow in the shady understory (lower middle), and fallen leaves and a few seedlings carpet the scarcely illuminated forest floor (bottom).

and tundra—plus mountains and the polar ice caps. This arrangement helps reveal the links between air and water currents, climatic patterns, and biome characteristics.

TROPICAL RAIN FORESTS

Nearly half of all living species reside in the world's warm, wet **tropical rain forests,** even though these lush regions represent only 3 percent of the continental surface area (Figure 36.7a). Tropical forests occur near the equator in Central and South America, Africa, and Southeast Asia, where hot air rises and then dumps its moisture (review Figure 36.4a). Rainfall is 200–400 cm (80–160 in.) per year, and temperatures average about 25°C (77°F). Jungle inhabitants are clearly not limited by water supply or cold temperatures, but light and mineral nutrients can be in short supply, and the dense vegetation blocks the wind. These limitations have shaped the rain forest biome.

The struggle for light has led to three main levels of plant height above the forest floor (Figure 36.7b). Within the upper story, or *emergent layer*, trees up to 50 m (160 ft) rise above the surrounding vegetation, where they capture direct sunlight. Beneath the emergent layer lies the *canopy*, the overlapping tops of shorter forest trees. The canopy is so dense with leaves and branches that only dim light penetrates to the third level, the *understory*, and less than 2 percent of the light eventually reaches the *forest floor*. Because canopy trees do not usually branch in the understory, a person can often walk unimpeded through the humid twilight that filters to ground level in a mature rain forest. The popular image of an impenetrably tangled jungle is accurate only at the edges of cleared forests, along river banks, or where a large tree has fallen and the sunlight reaches the ground.

Many jungle plants have evolved specializations for light gathering that combat the twilight of the understory. Vines, also called *lianas*, have roots in the soil and long, spindly stems that thrust their leaves into the canopy. Orchids cling to the trunks and limbs of canopy trees and catch water and nutrient debris in specially shaped leaves. Many plants that dwell in the understory's pale light, like philodendrons, produce huge, dark green leaves capable of capturing the maximum amount of available light, a trait that makes them good houseplants.

Because tropical rain forests have poor soil, roots tend to be shallow, and the trunks of large trees are often expanded into *buttresses*, thick flanges that provide support (review Figure 36.7b). Roots occupy the thin upper layer of soil, where mineral nutrients occur in highest concentrations. Heavy rains tend to leach nutrients from the deeper soils, making the biomass of the living forest itself the biggest source of nutrients. Fallen trees, dropped leaves, and dead animals quickly decay in the warm, damp environment, and their nutrients rapidly recycle.

Because the overhanging canopy is so dense, air in much of the tropical rain forest is deathly still, and a plant that relied on wind pollination might have difficulty reproducing. Therefore, many of the plants have evolved beautiful and elaborate flowers that attract insects, birds, or bats as animal pollinators. Not surprisingly, a Sumatran rain forest is home to the largest individual flower on earth: a 1 m (39 in.) blossom that weighs 7 kg (15 lb) and smells like rotten meat.

Wind is also undependable for dispersing the seeds of tropical plants, and those plants tend to produce large, succulent, showy fruits that attract animals. Indeed, much of the forest activity takes place in the canopy, where leaves, flowers, and fruits abound.

Despite the timeless beauty and incredible species richness of tropical rain forests, people are rapidly destroying them to help feed their own expanding populations or to harvest hardwood trees or graze cattle for export (see Figure 34.19). This is a threat to all biomes. Because tropical forests are the earth's biggest repositories for carbon, the forest fires people are deliberately setting to clear land for agriculture and ranching are significantly increasing atmospheric carbon dioxide. This, in turn, is contributing to global warming through the "greenhouse effect" (see the box on page 596). In fact, burning the great jungles into carbon dioxide and ashes could alter climatic patterns all over the globe.

TROPICAL SAVANNAS

Whereas tropical rain forests grow only where it is constantly wet and warm, year-long warmth with an extended dry season results in a tropical **savanna.** Situated between tropical forests and deserts, savannas contain stunted, widely spaced trees with tall grasses growing in between (Figure 36.8). The world's major savannas lie in Africa, South America, Australia, and parts of Southeast Asia. While the center of life in the rain forest remains high in

FIGURE 36.8 *Savanna: Tropical Drylands.* Rare Grevy's zebras graze on golden savanna grass beneath flat-topped thorn trees in southern Kenya in central Africa.

the canopy, the savanna's productivity is concentrated near the ground, where the leaves and seeds of grasses and short trees are well within reach of grazing animals like cape buffalo and kangaroos. The herbivores, in turn, support lions, dingos, and other carnivores. Many savanna species, including the rhinoceros and the elephant, are threatened with extinction because people kill them for their coats, ivory tusks, or spectacular horns.

DESERTS

Desert regions receive less than a tenth the annual rainfall of tropical rain forests; hence, desert plants are widely spaced and cover less than one-third of the ground surface (Figure 36.9a, page 590). Some desert areas in northern Africa have not seen rain in over a decade and are essentially lifeless tracts of shifting sand. Many deserts, like those of Mexico, Australia, and North and South Africa, are dry because they lie directly below the zones where dehydrated air—rained out in the tropics—descends back toward earth (see Figure 36.4a).

During the day, deserts are generally hot as well as dry, because on cloudless afternoons, the relentless, blistering sunlight can send temperatures soaring to 48°C (120°F). On cloudless nights, however, heat radiates from the desert back to space and leaves a biting chill to the air. To cope with searing heat and the potential loss of precious moisture, many desert plants, including Afri-

(a)

(b)

FIGURE 36.9 *Deserts: Parched Zones Where Life Is Sparse.* (a) The Namibian Desert of southwestern Africa receives almost no rainfall and supports few plants save scrubby drought-tolerant trees and a handful of odd, dry-adapted species. (b) Desertification claims precious grazing land south of the Sahara. This satellite photograph of Africa in springtime shows a strip of green grassland traversing central Africa. This green zone now lies 200 miles south of its former position, and the intervening grazing land has become desert. Tan areas represent extremely dry regions, while green to red to white indicate regions of increasing moisture.

can succulents like ice plant and American cactuses like prickly pear, have numerous adaptations, including thick, waxy cuticles; fleshy or absent leaves; and sunken stomata. The roots of desert plants may penetrate the sandy soils to a depth of several meters and obtain water, or they may remain shallow and widely spread, quickly soaking up any rain that falls. Many desert plants also utilize a pathway of photosynthesis that conserves water (the C_4 pathway; see Chapter 6). And others simply avoid harsh desert conditions by germinating, growing, flowering, and setting seed all within the short periods of rainfall. This sets some deserts aglow with dazzling red, gold, and purple flowers for a few brief weeks.

Desert animals have also evolved water conservation strategies. Most are active only in the evening or early morning hours, keeping to their burrows, crevices, or shaded areas during the heat of the day. As Chapter 24 described, the North American kangaroo rat has adaptations that allow it to survive without drinking (see Figure 24.1). By convergent evolution, the jerboa of Saudi Arabia has many similar adaptations. And small crustaceans such as fairy shrimp have evolved the ability to survive for years as dry embryos that hatch quickly when temporary pools form.

Ironically, while human activity is decreasing the areas of tropical rain forests, our actions are also causing the world's desert biomes to expand in a process called **desertification.** Deserts enlarge because overgrazing and poor irrigation practices remove grass from the grasslands that rim the deserts. The loss of ground cover allows fine, nutrient-bearing soil particles to blow or wash away, and this leaves only sand, gravel, and other coarse materials behind that can no longer hold water. Satellite photography provides a sobering view of desertification in the Sahel region south of the Sahara (Figure 36.9b).

TEMPERATE GRASSLANDS

Bordering many of the world's deserts are **grasslands,** treeless regions dominated by dozens of grass species (Figure 36.10). Known as *prairies* in North America, the *pampas* in South America, the *steppes* in Asia, and the *veldt* in Africa, these regions are wetter than deserts but drier than forests. The flat expanses of Kansas and the gently rolling hills of Nebraska are typical grassland terrains. Unlike savannas, grasslands lack trees and have seasonal extremes of hot and cold rather than wet and dry. Nevertheless, the windswept plains tend to dry out in the summer and fall, so fires are a recurrent feature.

As a result, grassland species have evolved the ability to regrow rapidly after a fire. Often touched off by lightning, these fires prevent grasslands from turning into forests. Open grasslands support animals like bison, antelope, prairie dogs, anteaters, armadillos, coyotes, snakes, and hawks.

Grassland soils are richer in organic matter than the soils of other biomes. Grasses have pervasive roots and underground stems, often penetrating to 2 m and weighing several times more than the aerial leaves and stems. The accumulated debris from generations of these plants thus makes grassland soils thick and fertile. While grasslands make rich agricultural fields and pastures, plowing breaks up the complex soil structure and leaves it vulnerable to erosion. With proper farming techniques, such as contour farming (plowing at right angles to the slope) and strip cropping (leaving broad strips of grass between fields), former grasslands can probably support agriculture indefinitely. Unfortunately, farmers in many parts of the world—including the United States—are often not sufficiently careful to conserve the precious natural resource of topsoil.

CHAPARRAL

A biome called the **chaparral,** or temperate scrublands, borders grasslands and deserts along the shores of the Mediterranean and along the southwest coasts of North and South America, Africa, and Australia. Chaparral has

FIGURE 36.10 *Temperate Grasslands: Treeless Seas of Grass.* Vast herds of bison once grazed the extensive grasslands of central and western North America. Today, just a few protected remnant herds remain.

FIGURE 36.11 *Temperate Forest: Deciduous Hardwoods Bordering Grasslands.* The new-green leaves of spring brighten Tennessee's Great Smoky Mountain National Park.

hot, dry summers and cool, wet winters. Chaparral plants are generally less than 2 m (6.5 ft) tall and have hairy, leathery leaves that stay green year round. Many chaparral plants, like sage and manzanita, have spicy, aromatic odors; the natural alkaloids probably help deter insect herbivores. Because this biome, like the grasslands, experiences frequent fires, the belowground portions of chaparral perennials have evolved fire resistance, and the seeds of some chaparral annuals must be seared by fire before they germinate.

TEMPERATE FORESTS

Continuing to move farther north and south from the equator, we leave the grasslands and enter the **temperate forests,** such as those of eastern North America (Figure 36.11), Europe, China, Japan, New Zealand, Australia, and the tip of South America. These forests are dominated by broad-leaved trees, and as the name suggests, they have intermediate amounts of rainfall and fairly moderate temperatures that fluctuate between summer highs and winter lows. They are about twice as productive as grasslands.

Seasonal temperature changes have strongly shaped the evolution of temperate-forest organisms. The dominant trees, including oak, maple, birch, and hickory, drop their leaves in late fall and remain dormant until spring.

FIGURE 36.12 *Coniferous Forest: Fragrant Evergreens in the Northern Latitudes.* A stand of deep-green conifers grows on the shores of Philip's Lake in Teton National Park.

The fallen leaves that accumulate on the forest floor allow for the recycling of nutrients and produce an excellent topsoil. In early spring, the leafless trees permit sunlight to fall unobstructed on the forest floor, and spring wildflowers like wood sorrel, bluebells, and violets bloom. Later, when the canopy leaves enlarge and shade the forest floor, only shade-tolerant plants such as ivy and honeysuckle thrive.

The animals of temperate forests must also cope with the changing seasons. Many, like bears and snakes, hibernate in winter, while others, like robins, migrate to warmer regions.

CONIFEROUS FORESTS

At latitudes with cold, snowy winters and short summers, vast forests of conifer trees grow in the **coniferous forest** (also called *taiga* or *boreal forest*) (Figure 36.12). This broad band of mixed pine, fir, spruce, and hemlock trees stretches across much of Canada, Northern Europe, and Asia. A few broad-leaved trees such as aspens do survive in these areas around streams or lakes. The southern border of the coniferous forest tends to fall at the southern limit of the polar front, where cold air sweeping down from the pole meets warm air pushing up from the south (review Figure 36.4). The Southern Hemisphere has little coniferous forest because ocean rather than land occupies appropriate latitudes. Coniferous forests are less productive than temperate forests, but are more productive than grasslands.

During winter in the coniferous forest, ice and snow lock up water, making it unavailable to trees. Conifer leaves, which are needle-shaped and have thick, waxy cuticles, combat water loss. Since these needles are not shed each winter, they can begin collecting sunlight as soon as the short growing season begins. Many conifers have sloping limbs that shed snow easily and tend not to break under the weight. Coniferous forests support bark beetles, elk, porcupines, wolves, and lynx.

A related minor biome is the *moist coniferous forest* (or *temperate rain forest*), which grows along the west coast of the United States from Canada to northern California. Very large hemlock trees, Douglas firs, and giant redwoods—the tallest living things—grow in these areas, festooned with ferns, mosses, and lichens. The trees depend on frequent rains and the summer fog that rolls in off the Pacific, collects on the feathery leaves and needles, and drops to the ground. Both kinds of coniferous forests contain the world's greatest lumber reserves.

TUNDRA

At the northern boundary of the coniferous forest, the polar high-pressure center blows cold, dry air down from the poles, even in summer (review Figure 36.4). Here, the fragrant, giant conifers give way to the low vegetation of the **tundra,** a cold, treeless plain (Figure 36.13). The annual temperature in this frosty biome averages 15°C (23°F), and even in the summer, the tundra soil thaws to only about 1 m (39 in.). The deeper, permanently frozen soil is called *permafrost*, and this subterranean zone prevents the establishment of trees. The soil that does thaw in summer stays soggy and dotted with pools because the land is largely flat and runoff is slow.

Frigid weather and a short growing season have heavily influenced the evolution of tundra organisms. Plants, including grasses, sedges, mosses, lichens, and heathers, grow low, where they can absorb warmth reradiated from the solar-heated ground. These plants support herbivores like caribou, arctic hare, and lemmings, and they, in turn, are prey for foxes, weasels, and snowy owls. During summer, clouds of mosquitoes and flies fill the air, reproducing in the temporary pools. Many birds migrate to the tundra in the summer to feast on these insects and breed during the long summer days.

The tundra is only one-sixteenth as productive as the tropical rain forest; the track of a single vehicle can mar the terrain for decades.

POLAR CAP TERRAIN

The Arctic ice cap, a vast frozen ocean covering the planet's North Pole (Figure 36.14), and the continent of Ant-

FIGURE 36.13 *Tundra: Cold, Treeless Plain in the Far North.* The low vegetation in Alaska's Denali National Park erupts with the russets and golds of autumn.

arctica, a landmass lying beneath a great raft of ice and surrounded by frigid seas, take up millions of square kilometers of land and ocean surface. These icy regions are not considered true biomes because they lack major plants; only animals and microbes survive in these regions. In the Arctic, polar bears hunt seals and other aquatic animals, while penguins and seals inhabit the ice shelf at the edge of the Antarctic Sea. The only true terrestrial

FIGURE 36.14 *Polar Caps: Icy Regions at the Highest Latitudes Lack Major Plants.* A huge polar bear stalks a walrus on a windswept ice flow near the Arctic Circle.

animals of Antarctica are a few insects, the largest of which is a wingless mosquito.

LIFE ZONES IN THE MOUNTAINS

Just as ice covers both poles, glaciers and snow fields cap high mountains—even near the equator. In fact, as one goes down in elevation from a high peak, one can often experience a series of biomes that reverses the order we have just reviewed from the equator to the poles (Figure 36.15). Leaving the snow fields at the top of a mountain in Mexico—let's say the ancient volcanic cone called Citlaltepetl, which is 5754 m (18,700 ft) tall—the traveler descends through an alpine tundra, then a coniferous forest similar to Canada's boreal zones, followed by a temperate forest like that of the eastern United States, and finally enters a lush tropical rain forest at the mountain's foot. Regions of varying elevation, climate, and organisms in the mountains are called *life zones.* Life zones occur because precipitation and temperature vary with altitude, just as they do with latitude, and each combination of temperature and water influences the kinds of organisms that can survive in those conditions.

FIGURE 36.15 *Life Zones in the Mountains: Altitude Mimics Latitude.* Descending a high mountain in the tropics, one passes through altitude zones that correspond to latitudes. Shown here is the correspondence between mountaintop and ice cap, alpine region and tundra, and so on, down to the tropical rain forest shrouding the mountain's base.

Just as climate helps shape terrestrial biomes, it has had a similar shaping influence on life in the water—a realm that is less familiar to most of us, but is central to our understanding of the biosphere.

≋ LIFE IN THE WATER

Three-fourths of our planet's surface is covered with water, and most of that area is *marine*—salty oceans and seas. Only 0.8 percent is covered by freshwater lakes, rivers, ponds, streams, swamps, and marshes, but this still represents tens of thousands of square kilometers. Ecologists do not regard marine or freshwater communities as biomes, but their studies show that aquatic ecosystems differ from each other in significant ways.

PROPERTIES OF WATER

As we saw in Chapter 2, water has unique physical characteristics, and these properties strongly influence aquatic organisms. For example, it takes much more energy to heat or chill water than it does to change air temperature. Aquatic organisms therefore suffer fewer rapid temperature changes than their terrestrial counterparts. In addition, air grows steadily denser as it cools, but water has its greatest density at a temperature slightly above freezing. Because of this, ice floats. Were this not the case, lakes would freeze from the bottom up, only the upper layers would thaw in summer, and bodies of fresh water outside of the tropics would support little life.

FRESHWATER COMMUNITIES

Ecologists divide freshwater ecosystems into bodies of running water and bodies of standing water.

■ **Running Water** A stream that originates from a ground spring or from melting snow is cold and clear but contains few nutrients. As the water tumbles downhill across rocks, it picks up oxygen. Species that live in these turbulent zones include algae, mosses, and trout, which thrive only where oxygen is plentiful and water temperatures are low. Farther downstream, the water becomes cloudy and enriched with nutrients as leaves from nearby plants fall into the stream, and the water moves more sluggishly. In these areas, plants grow near the banks, and microbes and invertebrates live on detritus in the bottom sediments. Catfish and bass replace trout as the amount of dissolved oxygen decreases in the slower-moving stream.

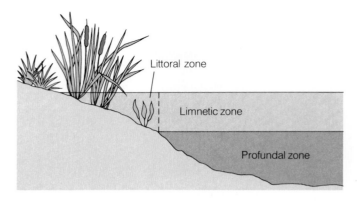

FIGURE 36.16 *Freshwater Lakes: Zones of Light Support Life.* Cross-sectional view of a lake, showing the littoral, limnetic, and profundal zones with their varying water depths and degrees of light penetration.

■ **Standing Water** As a stream enters a lake, the current slows still further, and most suspended particles settle to the bottom. Those particles that remain suspended block light penetration, determining the maximum depth at which photosynthesis can occur. In shallow areas, light can penetrate to the lake's bottom. This well-lit zone (*littoral zone*) has abundant producers like water lilies, cattails, and algae and consumers such as insects, snails, amphibians, fishes, and birds (Figure 36.16).

Farther away from the shoreline, the standing water has two main regions: the top layer through which light can penetrate (the *limnetic zone*) and the dark bottom region (the *profundal zone*). The top layer supports huge populations of phytoplankton, such as algae, and zooplankton, such as small crustaceans, while the deep zone supports mainly insect larvae, scavenger fishes, and decomposers.

Dissolved nutrients are as important as light in lakes. Under normal circumstances, bottom sediments represent a natural source of nutrients such as phosphates. In areas where the winters are cold, seasonal changes in water temperature churn up nutrients from the bottom in a *water turnover* each spring and fall, which nourishes a quick bloom of algae.

SALTWATER COMMUNITIES

■ **Estuaries** Fresh and salt water mingle in **estuaries,** where rivers meet oceans (Figure 36.17). In these areas, temperature and salt concentrations vary widely because of the daily rhythm of the tides and seasonal variations in stream flow. Estuary organisms like brine shrimp can often tolerate wide ranges of salinity (see Chapter 24). Constant water movements stir up nutrients and make estuaries some of the earth's most productive ecosystems and nursery grounds for many species of fish. In many populated areas, people dump urban and industrial

FIGURE 36.17 *Estuaries: Confluences of Life.* In Australia, a silted, richly productive estuary marks the confluence of a freshwater stream and the ocean.

effluents into estuaries and fill them with rocks and soil to claim new waterfront property. Such activities are significantly decreasing the great biological productivity of our estuaries.

■ **The Oceans** Water eventually passes through estuaries and out to sea. Although the ocean may look undifferentiated from the surface, various parts receive different amounts of sunlight and nutrients. The bottom can be muddy, sandy, or rocky; the shorelines can vary from smooth, sandy beaches to jagged, rocky cliffs; and the water can be centimeters or kilometers deep. These factors combine to dictate the types of organisms that occupy each region. Ecologists classify ocean habitats and their organisms on the basis of depth, type of bottom, and light levels (Figure 36.18).

The **intertidal zone** extends along the coastline between high and low tide levels and is periodically exposed to the air. In many intertidal environments, waves and wind create violent pounding. Most producers (algae and kelp) have structures that anchor them to rock surfaces (see Chapter 16), while most animals have tough bodies or shells as well as underwater glues. Intertidal organisms must also tolerate dry spells as the tide rolls out each day. A group of burrowing organisms, including clams, snails, and worms, live beneath the extensive tidal mudflats. Many visitors to the seaside enjoy investigating **tide pools,** depressions formed in rock by wave action, and observing the sea anemones, sea urchins, crabs, sea stars, and fishes that inhabit them.

The region near shore (*neritic zone*) is the ocean's most productive region, extending from the intertidal zone to the edge of the continental shelf, the submerged part of the continents. Producers near shore include many species of large algae living in extensive kelp beds, as well

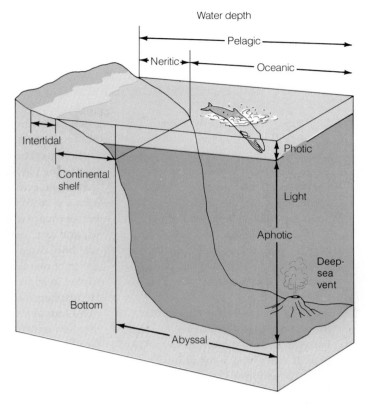

FIGURE 36.18 *Ocean Zones: Depth, Bottom, and Light Determine Habitats of the Sea.* The ocean is a complex aquatic realm with several zones based on several parameters: water depth; bottom characteristics; and light penetration.

as huge populations of phytoplankton and the zooplankton that serve as a nutrient base for the rest of the ocean's consumers. The largest animals that have ever lived, the blue whale (see Chapter 19) and the whale shark, feed entirely on microscopic plankton.

Shallow tropical oceans are home to reef-building corals and the photosynthetic bacteria within them, which construct stony mounds with exotic shapes (Figure 36.19, page 597). Eventually, these reef builders alter their own environment; the crannies and caves in their stony colonies create sheltered nesting and hiding places for the most diverse communities in the seas. Per unit area, coral reefs are the most productive ecosystems on earth; they are 10 times more productive than freshwater regions and 20 times more so than the open ocean.

The open ocean (*oceanic zone*), beyond the continental shelf and covering the deep ocean's floor (abyssal plain), is nearly deserted, even near the surface (see Figure 36.18). While the surface of the open ocean may have plenty of light, oxygen, and carbon dioxide, it often lacks nutrients such as phosphate and usable nitrogen and overall is less

DOUBLE TROUBLE FOR THE ATMOSPHERE: DECREASING OZONE AND INCREASING CO₂

Our atmosphere is under siege: Decades of human activity have created measurable changes in the envelope of gases that surrounds our planet—changes that are causing great scientific concern. Luckily, we understand the source of the problems and can choose a different course, as individuals and societies, to prevent global devastation. The question is, Will we make the necessary sacrifices, and will we do it in time?

One change involves a molecular form of oxygen called ozone (O_3) that hangs in a thick band from 15 to 50 km (9 to 31 miles) above the ground surface. Ozone forms when ultraviolet light streaming in from the sun strikes regular oxygen molecules (O_2) in the atmosphere. This layer of O_3 molecules absorbs about 99 percent of the ultraviolet light that would otherwise strike the earth's surface and destroy many biological molecules. The shielding effect of the ozone layer probably allowed living things to leave the seas and invade land hundreds of millions of years ago, and it continues to protect us today from most of the DNA-damaging effects of ultraviolet light (see Chapter 14).

Ominously, about 1974, atmospheric chemists began to detect slight decreases in the ozone layer, and with it, increases in the amount of ultraviolet light reaching the earth. More recently, scientists have detected a huge, seasonal hole in the ozone layer over much of Antarctica. Research has shown that the organic compounds we use to propel droplets from spray cans, to create styrofoam, and to cool refrigerators and air conditioners (so called chlorofluorocarbons, or CFCs) can float into the atmosphere and break down

ozone, allowing more UV light to sear the earth's surface. Globally, the ozone shield has already thinned by 3 percent during the past 20 years, and by more than 50 percent over Antarctica. For each 1 percent drop, approximately 2 percent more ultraviolet light reaches us, and as a result, scientists expect to see more cases of skin cancer (the incidence of one deadly form has already tripled in some parts of the Southern Hemisphere), more eye cataracts, and more immune system problems, not to mention mutations, stunting, and surface damage to plants, animals, and other organisms.

In the summer of 1990, 93 nations signed an international agreement to phase out the use of CFCs by the year 2000, but experts are already predicting that no replacement compounds will be as effective as chlorofluorocarbons.

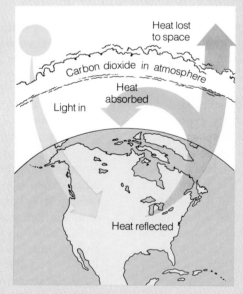

FIGURE 1 *The Greenhouse Effect.* Like sun shining through the glass windows of a greenhouse, light penetrates the atmosphere and warms the earth, but much of the reflected heat becomes trapped. The more carbon dioxide in the air, the more heat is trapped.

Will the people of the world, ourselves included, be willing to forgo conveniences like smoothly operating cooling systems and aerosol products in order to solve the problem of the thinning ozone layer? Only time and the dedication of an informed public will tell.

In the meantime, we have an even bigger problem to face. The atmosphere is about 0.03 percent carbon dioxide, and this gas has increased by roughly 25 percent in the past century. The increase is of real concern, because atmospheric carbon dioxide is analogous to the glass in a greenhouse (Figure 1). Light energy passes through the carbon dioxide and warms the earth, but some of the energy becomes trapped inside the layer as infrared rays, or heat. The twentieth-century increases have meant more trapping of infrared rays, and with that, the average atmospheric temperature has risen 0.5°C. Much of the extra carbon dioxide originates with the burning of coal, oil, and gasoline in our factories and automobiles and with the burning of tropical rain forests (review Figure 34.19). Other pollutants, such as methane (CH_4), nitrous oxide (N_2O), CFCs, and certain sulfur compounds, also contribute to the greenhouse effect.

Scientists predict that sometime between the years 2025 and 2075—well within the expected lifetime of those reading (and writing!) this book—levels of carbon dioxide in the atmosphere will be twice what they were before the industrial revolution. This accumulating blanket will trap more and more heat, and global temperatures will climb an average of 1.5–5.5°C (2.7–9.9°F). Even relatively small increases like this would cause ice at the poles to melt and sea levels to rise from 1 to 3 m, inundating low-lying cities such as New York, London, and Tokyo, as well as Bangladesh, the Netherlands, and other flat countries. Catastrophically, the increased temperatures would also change the great grain-producing regions of the American Midwest, Canada, and the Soviet Union into deserts. Hints of this are already with us: The late 1980s saw some of the hottest years in the hottest decade on record, with widespread droughts and resultant forest fires and crop losses. These, some say, were a mild preview of what could ultimately happen.

To reverse this frightening trend, some scientists think that human societies will have to quickly reduce the global consumption of fossil fuels by 50 percent; to stop deforesting the tropics; and to start a vigorous campaign of tree planting there and in other denuded areas (recall that trees and plants in general take up CO_2 during photosynthesis).

While nations and municipalities are notoriously slow to initiate and carry out such measures, we can immediately begin to act as individuals.

■ We can lower the temperatures of our homes in winter, use less air-conditioning in summer, and rely on public transportation, bicycles, and our own two feet more often to lower our fuel consumption by half.
■ We can steadfastly refuse to buy teak and other tropical wood products.
■ We can eat less meat, since beef cattle are often raised on newly stripped tropical lands and since, in general, the production of animal protein requires far more energy than the production of plant protein.
■ We can avoid CFC propellants and styrofoam containers and use smaller, more efficient air conditioners and refrigerators.

■ We can plant trees in and around our towns, campuses, and homes.
■ We can inform ourselves about these issues and stay active at the local and national levels.

Here are some sources of reading material:

FOR FURTHER INFORMATION

El-Sayed, S. Z. "Fragile Life Under the Ozone Hole." *Natural History* (October 1988): 73–80.
Houghton, R. A., and G. M. Woodwell. "Global Climatic Change." *Scientific American* (April 1989): 36–44.
Jones, P. D., and T. M. L. Wigley. "Global Warming Trends." *Scientific American* (August 1990): 84–91.
Manzer, L. E. "The CFC-Ozone Issue: Progress on the Development of Alternatives to CFCs." *Science* 249 (6 July 1990): 31–35.
Schneider, S. H. "The Greenhouse Effect: Science and Policy." *Science* 243 (10 February 1989): 771–780.
White, R. M. "The Great Climate Debate." *Scientific American* (July 1990): 36–43.

FIGURE 36.19 *Coral Reef: Fantastic Productivity Beneath the Waves.* A school of bright orange fish dart above a dozen kinds of coral on this reef in the Red Sea.

productive than even the arctic tundra. The open ocean is fruitful only in regions where currents move offshore, as they do west of Ecuador and Peru. The associated upwelling of nutrient-rich deep waters sweeps nutrients far offshore toward the Galápagos and its colony of fur seals.

☰ CHANGE IN THE BIOSPHERE

Change is part of nature. Climates shift, continents drift, and organisms evolve, influencing their immediate surroundings and the larger biosphere around them. Recall from Chapter 14 how the evolution of photosynthesis decreased atmospheric carbon dioxide and increased oxygen gas. Unfortunately, as we saw in Chapter 35, certain human activities are causing equally pervasive but much faster modifications in the biosphere—modifications that outstrip the speed with which organisms usually evolve adaptations. The greenhouse effect and the degradation of the earth's ozone layer are prime examples of our species' collective impact on the biosphere (see the box on pages 596 and 597).

Recent global phenomena, such as rising carbon dioxide levels and the dwindling ozone layer, contradict our traditional assumption that nature will somehow neutralize all the wastes we release. So great is our impact that some ecologists estimate that we humans are currently using about 40 percent of the potential primary productivity on all earth's land area! Our species is the only one capable of changing the global environment in a short time period, capable of understanding that we are in fact mismanaging the earth's physical and biological resources, and capable of foreseeing the consequences. The survival of our species and millions of others depends on the responsibility we take for these capacities.

≋ CONNECTIONS

The biomes that make up the earth's thin film of life are fashioned by the atmosphere, soil, and water; by the planet's daily rotation on its tilted axis; by its yearly trip around the sun; by spontaneous events like the warming of the eastern equatorial Pacific during El Niño; and by continents drifting and climates changing from hot to cold, dry to moist, and back again. Plants and animals in regions with similar climates have slowly evolved very similar adaptations, even though they may not be closely related genetically.

Human activities all over the globe are rapidly changing the earth's physical environment, causing desertification in dry areas and damaging the temperate forests through acid rain. What's more, some of our activities threaten to affect the entire biosphere in a short time through the greenhouse effect, the loss of ozone, and nuclear winter. We must devise and institute solutions with equal speed, so that our all-pervasive influence does not unbalance the complex interdependencies of nature.

All of our foregoing subjects—cell biology, genetics, development, physiology, evolution, and ecology—help us understand how organisms survive in their environments. With this background, we are ready to consider our final topic: a fascinating set of adaptations called animal behavior, which allows animals to respond to each other and to their surroundings in a range of ways from the innate and automatic to the learned and deliberate.

NEW TERMS

biome, page 587
chaparral, page 591

climate, page 584
coniferous forest, page 592

desert, page 589
desertification, page 590
estuary, page 594
grassland, page 590
intertidal zone, page 595
savanna, page 589

temperate forest, page 591
tide pool, page 595
tropical rain forest, page 588
tundra, page 592
weather, page 584

STUDY QUESTIONS

REVIEW WHAT YOU HAVE LEARNED

1. Define biosphere and biome.
2. Describe the six coils of air currents that circulate around the earth.
3. What factors influence ocean currents?
4. What are the basic features of a tropical rain forest and a tropical savanna?
5. What conditions create a desert?
6. What are the basic features of grassland and chaparral?
7. How do a temperate forest and a coniferous forest differ?
8. What prevents trees from growing in the tundra?
9. List the living organisms that inhabit the polar ice caps.
10. How do nutrients circulate in a lake?
11. What is an estuary? What is its ecological role?
12. Explain and support the following contention: The open ocean is a biological desert.

APPLY WHAT YOU HAVE LEARNED

1. Philodendrons grow wild in tropical rain forests, but they are also easy to cultivate as houseplants. What makes them so successful indoors?
2. Snow-capped Mt. Kilimanjaro is located at the equator in Africa. Explain why you would or would not expect to find tundra-type plants growing on its slopes.
3. A trout fisherman standing in the rapids of a mountain stream finds the cold, tumbling waters uncomfortable, so he moves downstream to a calmer spot. Have his chances of catching trout improved? Explain.

FOR FURTHER READING

Cloud, P. "The Biosphere." *Scientific American* 249 (September 1983): 176–189.

de Blij, H. J., ed. *Earth '88: Changing Geographic Perspectives.* Washington, DC: National Geographic Society, 1988.

Diamond, J. M. "Human Use of World Resources." *Nature* 328 (1987): 479–480.

Frederick, J. E., and H. E. Snell. "Ultraviolet Radiation Levels During the Antarctic Spring." *Science* 241 (1988): 438–440.

Graham, N. E., and W. B. White. "The El Niño Cycle: A Natural Oscillator of the Pacific Ocean–Atmosphere System." *Science* 240 (1988): 1293–1302.

Kerr, R. A. "No Longer Willful, Gaia Becomes Respectable." *Science* 240 (1988): 393–395.

Nebel, B. J. *Environmental Science: The Way the World Works.* 2d ed. Englewood Cliffs, NJ: Prentice-Hall, 1987.

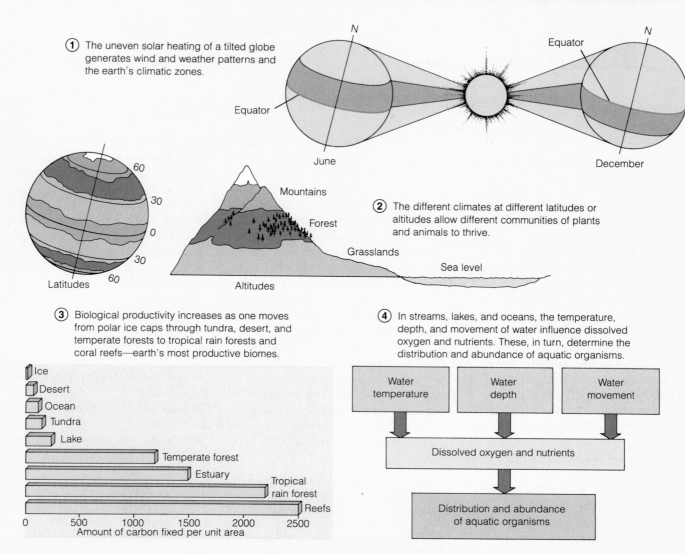

① The uneven solar heating of a tilted globe generates wind and weather patterns and the earth's climatic zones.

N

Equator

June

Equator

N

December

60

30

0

30

60

Latitudes

Mountains

Forest

Grasslands

Sea level

Altitudes

② The different climates at different latitudes or altitudes allow different communities of plants and animals to thrive.

③ Biological productivity increases as one moves from polar ice caps through tundra, desert, and temperate forests to tropical rain forests and coral reefs—earth's most productive biomes.

④ In streams, lakes, and oceans, the temperature, depth, and movement of water influence dissolved oxygen and nutrients. These, in turn, determine the distribution and abundance of aquatic organisms.

Ice

Desert

Ocean

Tundra

Lake

Temperate forest

Estuary

Tropical rain forest

Reefs

0   500   1000   1500   2000   2500

Amount of carbon fixed per unit area

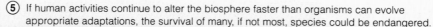

Water temperature

Water depth

Water movement

Dissolved oxygen and nutrients

Distribution and abundance of aquatic organisms

⑤ If human activities continue to alter the biosphere faster than organisms can evolve appropriate adaptations, the survival of many, if not most, species could be endangered.

CHAPTER 37

ANIMAL BEHAVIOR: PATTERNS FOR SURVIVAL

MENACE OF THE BROKEN SHELL

Each spring, white sea gulls with coal black faces and white-rimmed eyes migrate to the coastlines of northern Europe. Males of this species (the black-headed gull) arrive earliest, select their positions on the dunes, and begin stretching forward and calling out raucously in a stereotyped ritual (Figure 37.1). If a newly arrived male tries to lay claim to another male's territory, the occupant rushes at the invader, strikes him with a wing, and drives him off.

Males often react quite differently to approaching females, initiating a stylized mating dance in which the pair stretch forward, strut side by side, then, as if by prearranged choreography, turn their faces away from each other. Next the female begs food from her suitor, and he regurgitates fish into her mouth. The dancing and dining over, the pair may mate, select a nesting place, then noisily defend it from all comers. The female deposits khaki-colored eggs with dark speckles in the loose mound of grass and straw; then the parents take turns warming and turning the eggs in the nest. After each chick hatches, the parent on duty dumps the empty eggshell far from the nest.

Over time, the chicks begin to wander about their parents' territory. Black-headed gulls recognize their own chicks, but harshly peck little ones that wander in from other territories. The chicks, in turn, recognize their parents' attraction calls at mealtimes and ignore those of neighboring adults.

As summer wears on, the chicks mature, and as winter approaches, the entire colony of black-headed gulls migrates to a warmer winter home in Africa. The following spring they will return once again and enact a new drama of territorial marking, mating, and brooding.

The complex behaviors of the black-headed gull help the organism obtain energy, reproduce more effectively, and survive more successfully in nature. In this chapter, as we explore the fascinating range of simple and complex behaviors in the animal kingdom—from mating to migrating, from fighting to befriending—two main themes will recur. First, the ability to perform behaviors originates in genes, arises during development, and may be modified by the environment. Genes, for example, specify neural connections that allow a gull to recognize and remove broken eggshells; the neural patterns take shape as the embryo develops into a chick and are expressed by the mature adult. Second, natural selection shapes behavior. Gulls that remove broken eggshells lose fewer chicks to predators than gulls that leave broken eggshells in their nests; experiments have shown that this is because flashy, white, jagged eggshells attract predators. The genes that program eggshell removal therefore increase in frequency in the population of black-headed gulls.

Our discussion of behavior answers several questions:

- How do genes determine behavior?
- How does an animal's interaction with the environment modify its behavior?
- How does natural selection shape behaviors?
- Why do some animals live in groups?

FIGURE 37.1 *A Raucous Male Gull Defends His Territory.* This black-headed gull (*Larus ridibundus*) has carved out a small patch of grass on the beach and defends it from other males while awaiting a female's approach.

GENES DIRECT DEVELOP-MENT AND BEHAVIORAL CAPABILITIES

The scientific study of how animals behave in their natural environments is called **ethology,** and Niko Tinbergen, a pioneer in this field, conducted some of the first experiments into the heritable basis of behavior. He wondered what properties prompt black-headed gulls to remove broken eggshells, so he placed various objects into gull nests alongside the eggs. The birds accepted and incubated round rocks, flashlight batteries, corks, and light bulbs as if they were eggs. But gulls ejected from the nest bottle caps, toy soldiers, seashells, and paper squares. Tinbergen concluded that when the birds saw or felt an object with a jagged edge, they removed it from the nest. Apparently, sensory receptors in the gull's eyes or skin detect sharp edges, and the nervous system directs the bird to remove the object. Since genes direct the development of the bird's sensory and nervous system and set up specific nerve connections, or *neural programs*, it seemed reasonable to suspect that genes direct the behavior.

Other experimenters following Tinbergen's lead conducted breeding experiments with garter snakes to test the hypothesis that genes direct behaviors that adapt individuals to specific environments. Biologists have observed a population of garter snakes that inhabits the foggy forests of coastal California and eats mainly slugs, and another population of the same snake species that lives in higher, drier areas several miles inland and feeds on fish or frogs from lakes and streams. Researchers wondered whether the distinctly different food choices are due to genetic differences between the two snake populations or whether inland snakes will eat slugs if given the opportunity.

To find out, experimenters captured pregnant snakes from each population, raised the young in the laboratory, and offered them small chunks of fresh banana slug. Baby snakes from inland regions flicked their forked tongues only a few times at a cotton swab soaked in slug juice and then avoided the slimy slug meat (Figure 37.2), while coastal snakes repeatedly flicked their tongues and ate slugs heartily. This result suggests that different genes in the two populations control slug refusal and slug feasting, and crossbreeding showed that alleles for refusal are dominant to alleles for feasting.

It is easy to see how natural selection would favor preexisting mutations that led snakes of the damp coastal forests to eat slugs: Slug eaters would have an abundant and nutritious food supply that could help them produce more offspring, and through these young, alleles for slug eating would increase in frequency. It is also easy to see how alleles for slug eating would be selected against in

FIGURE 37.2 *Tongue Flicking in Garter Snakes.* Both coastal and inland garter snakes (*Thamnophis elegans*) will flick their tongues to "smell" slug juice offered to them by an experimenter. Only coastal garter snakes, however, will eat bits of slug meat. Instinctive slug refusal may help protect the inland snakes from swallowing the leeches that occur in their territory.

snakes of the dry inland mountains: The same genes enhance leech eating, and leeches, although absent from coastal areas, inhabit inland lakes. However, it would be disadvantageous for garter snakes to eat leeches on a regular basis, since when swallowed alive, these parasites can sometimes eat a hole right through a snake or other predator. Hence, alleles that allow snakes to accept both slugs and leeches would be selected against and decrease in frequency in inland populations.

The garter snake study points out two distinct causes of behavior. The immediate (or *proximate*) cause for refusing to eat a slug is in certain dominant alleles that direct the construction of the individual snake's nervous system. In contrast, the evolutionary (or *ultimate*) cause for refusing to eat a slug is the selective advantage gained by avoiding harm, in this case, the eating of leeches. Ethologist Niko Tinbergen, in his studies of broken shell removal by black-headed gulls, was able to build on his knowledge of the proximate causes (perception and removal of jagged-edged objects) to understand ultimate causes (protection from predators). When Tinbergen placed eggshells at various distances from gull nests, he found that nests with eggshells nearby were three times more likely to lose eggs to predators than nests without nearby eggshells. Evidently, natural selection has favored alleles that cause black-headed gulls to remove predator-attracting eggshells from their easily accessible nests. Similar alleles are lacking, however, and appear to be unnecessary in cliff-dwelling kittiwakes (close relatives of black-headed gulls), since foxes, crows, and hawks cannot gain access to the narrow ledge where kittiwakes build their nests.

In summary, experiments with gulls and garter snakes helped identify the proximate causes of animal behavior, such as perceiving environmental signals and responding through neural programs that develop under the direction of genes. They also helped reveal the ultimate causes: the force of selection exerted on individuals in their natural environments.

≋ EXPERIENCES IN THE ENVIRONMENT CAN ALTER BEHAVIOR

Animals display a staggering range of behaviors, from the instinctive to the highly complex and learned: Birds sing specific songs; gulls do precise mating dances; salmon migrate hundreds of kilometers to spawn where they were born; and college students study and remember the intricacies of biology. Some behaviors, as we have seen, have strong genetic determinants and change only over many generations as natural selection winnows out specific alleles but leaves others. For many other behaviors, however, interactions with the environment can lead to modifications during the lifetime of a single individual. There is, in fact, a continuum of behavioral types, from **instincts,** which are actions that are innate and unchangeable, to actions with a limited capacity to be changed, to behaviors that can be *learned,* or modified by experience during an individual's lifetime.

INNATE BEHAVIORS: AUTOMATIC RESPONSES TO THE ENVIRONMENT

The very first season a black-headed gull tends a nest, it responds to the jagged edges of broken eggshells. Its behavior is instinctive: It is inherited; it does not change with experience; it includes a set sequence of events; and each member of the species performs the sequence in the same manner. An innate series of precise physical actions of this sort is called a **fixed-action pattern.**

Fixed-action patterns are characterized by five features. First, they are triggered by specific environmental signals called **releasers.** Sights, sounds, smells, tastes, or any information picked up by the sense organs may function as releasers. For the black-headed gull, the releaser for eggshell removal is an object with a jagged edge.

Second, once a releaser prompts a fixed-action pattern, the behavior must proceed from start to finish in an all-or-none fashion and in a particular sequence. The swallowing reflex is a classic fixed-action pattern. Have

you ever tried to stop swallowing a mouthful of food midway down? It can't be done; more than a dozen muscles fire in coordinated sequence, each step unalterably triggered by the preceding step in a classic all-or-none pattern.

Third, since a fixed-action pattern is innate, it is fully formed and functional the first time it is used. Fourth, all species members of the same age and sex perform the behavior the same stereotyped way under similar environmental conditions. And fifth, an animal's instinctive response to a releaser depends on its stage of anatomical and physiological maturation and on physiological states of motivation, or *drive,* which are influenced by hormones or other internal stimuli. In the chameleon, for example, a drop in glucose level in the blood may motivate the animal (physiologically, not consciously) to catch insects by tongue-flicking behavior, another fixed-action pattern (Figure 37.3).

While fixed-action patterns are triggered by environmental cues and depend on the animal's stage of development plus internal states such as blood chemistry, once begun, they are stereotyped and constant. Other behaviors on the continuum from instinct to learning, however, can be modified by information from the environment.

BEHAVIORS WITH LIMITED FLEXIBILITY

Black-headed gulls frequently go on feeding forays, return to their nests, and regurgitate the food for their chicks. Gull parents of many species perform the same sequence,

FIGURE 37.3 *Internal Drive Causes a Chameleon to Flick Its Tongue.* This lizard (*Chamaeleo gracilis*) from Kenya's Tsavo National Park can swivel its eyes independently, spot insects or other morsels, and, if hungry (an internal drive), instantly lash out its long, sticky tongue. The programming for such behavior is innate, and chameleons no more than 1 hour old can perform it.

and upon their return, the chicks instinctively peck at the parents' wagging bills to beg for food. In the herring gull, adults have a distinctive red spot on their bill that serves as the releaser for bill-pecking behavior, a fixed-action pattern (Figure 37.4). This behavioral sequence, however, has some flexibility, or potential for modification. At first, the chicks will peck at anything resembling the parent's red bill spot and will actually peck more excitedly at a swinging stick with red bars painted on it than at a live gull's bill. As the chicks mature, however, begging behavior becomes more directed. Eventually, they will beg only from their own parents. In these gull chicks, information acquired from the environment (about their parent's physical characteristics) can modify what is initially a fixed-action pattern. The most common categories of behavior with flexibility are imprinting and learning.

IMPRINTING: LEARNING WITH A TIME LIMIT

In the 1930s, Austrian ethologist Konrad Lorenz discovered the phenomenon called **imprinting:** the recognition, response, and attachment of a young animal to a particular adult or object. At the time, Lorenz was studying ducks and geese, and he knew that goslings normally follow their mother from the nest to a nearby pond or lake within a day after hatching. The researcher tried walking away from a group of newly hatched goslings in the nest while he made gooselike honking sounds. To Lorenz's delight, the gaggle of goslings followed him around as readily as they would have their real mother. He speculated that to survive, newly hatched geese must

FIGURE 37.4 *Red Spot Pecking in Young Gull Chicks: A Behavior with Limited Flexibility.* A hungry herring gull chick (*Larus argentatus*) pecks at the red spot on its parent's bill and triggers a regurgitated meal.

FIGURE 37.5 *Imprinted for Life.* Imprinting has an inherent time limit; geese form parental attachments in the first two days after hatching. The attachments, however, are lifelong. Here, two imprinted geese preen Konrad Lorenz's "feathers."

quickly learn to recognize and follow their parents, and they are apparently programmed to trail after the first large moving object they see, even if it is a gray-bearded, pipe-puffing ethologist!

Imprinting has an inherent time limit—a *critical period,* or specific developmental stage, during which the imprinting can occur and after which it cannot. In geese, this period lasts from the time of hatching to the second day; imprinting is impossible from the third day on. Imprinting requires specific releasers, in this case large objects moving away from the nest making goose sounds. Finally, imprinting is irreversible: The imprinted goslings recognized Lorenz as "mother" for the rest of their lives and could not be coaxed to imprint on their real parents later (Figure 37.5).

Biologists suspect that imprinting can take place in baby birds because development "primes" their nervous systems to change in a defined way based on environmental information shortly after hatching. Imprinting is thus a form of learned behavior, but one limited to a narrow time period. In most forms of learning, behavioral patterns can be repeatedly modified throughout an individual's lifetime.

LEARNING: BEHAVIOR THAT CHANGES WITH EXPERIENCE

A hungry toad that has never eaten a millipede will, upon seeing the many-legged arthropod, flip out its sticky tongue and snap up the prey. The captured millipede quickly retaliates, however, and exudes a foul-tasting toxin. The poison works quickly, and the nauseated toad spits out the would-be prey unharmed. A toad is far from clever, but it will remember this single disagreeable experience forever and will henceforth reject all millipedes, even when hungry. The toad's nervous system has cataloged the shape and markings of a millipede and associated them with a repulsive taste that should be avoided—a one-trial learning event.

The toad's aversion to millipedes encapsulates the essential elements of **learning:** an adaptive and enduring change in an individual's behavior based on personal experiences in the environment. In our example, the toad's learning is adaptive because it allows the animal to avoid poisonous food; the learning is based on a personally experienced event rather than an inherited trait; and the behavioral change endures, in this case the toad's entire life. One-trial learning is just one type of trial-and-error learning. Other kinds of learning include habituation, classical conditioning, and insight learning. As we shall see, each form of learning can increase an animal's fitness—its ability to survive and reproduce in nature.

■ **Habituation** *Habituation* is a simple kind of learning in which the animal comes to ignore a repeated stimulus that is never followed by reward or punishment (see Chapter 26 for the physical basis of habituation). Without habituation, an animal might become overloaded with sensory information. Young black-headed gull chicks, for example, instinctively crouch down when objects pass overhead, but this response wastes time and energy if the flying object is a harmless leaf or robin. It is adaptive, therefore, for chicks to habituate and learn to ignore silhouettes of common harmless species, but to continue to cower at the distinctive shapes of predatory birds, such as hawks. Since these predators pass overhead infrequently, habituation for them and other rarely seen objects never occurs, and this, too, is clearly adaptive.

■ **Trial-and-Error Learning** More complex than habituation is **trial-and-error learning.** Also called *operant conditioning,* it is a form of learning in which an animal associates a response, or *operant,* with a *reinforcer,* a reward or punishment. The millipede-munching toad, for example, experienced trial-and-error learning: It learned to associate a response, capturing the millipede, with a reinforcer, a foul taste. Likewise, a pet dog will learn to relate the response of begging to a reinforcer, a handout of a dog biscuit.

Behaviorist B. F. Skinner studied trial-and-error learning under rigorously controlled laboratory settings. He suggested that virtually any response can be conditioned, or trained, by a proper schedule of reinforce-

ments. More recent experiments show that operant conditioning works readily only for stimuli and responses that have meaning for animals in nature. Rats rapidly learn to press a lever with their front paws to obtain a food reward, but they cannot learn to push a lever to avoid an electric shock. On the other hand, rats quickly learn to jump off the ground to escape a shock, but they cannot learn to jump to obtain a food reward. These results make sense when viewed in an ecological context; in nature, rats obtain food by manipulating items with the paws, but this is not how they avoid harmful stimuli. These experiments show that animals are innately prepared to learn some things via trial and error more easily than others, and in ways that appear to increase survival and hence reproductive success.

■ **Classical Conditioning** While operant conditioning involves an association between a response and a reinforcer, **classical conditioning,** or *associative learning,* involves the association of two separate stimuli. Russian behaviorist Ivan Pavlov developed the concept from what has become the classic case of classical conditioning. He repeatedly presented a dog with meat, a natural stimulus for salivation, and at the same time rang a bell, an arbitrary, unnatural stimulus. Through classical conditioning, the dog learned to associate the natural stimulus and the substitute and began to salivate at the sound of the bell. In nature, a predatory hawk can learn to associate the presence of a broken eggshell with the presence of a defenseless, tender gull chick and act to collect a reward: dinner. Advertising agencies have learned to manipulate our own species' susceptibility to classical conditioning by associating commercial products with positive images: fancy cars with successful business executives; quarter-pound hamburgers with happy family occasions; toothpaste brand with social popularity (Figure 37.6).

We can conclude from the study of classical and operant conditioning that an animal's genes program it to learn different things in different ways that fit the animal to its environment Many animals are adept at both operant and classical conditioning. One final type of learning, however, is common only in primates and reaches its peak in humans.

■ **Insight Learning** At this moment, as you read, you are acquiring knowledge through **insight learning,** or reasoning. By this means, people and other primates formulate a course of action by understanding the relationships between the parts of a problem. Insight learning allows an animal to encounter a new situation (including a new printed sentence) and, based on similar past experiences, to figure out how to respond without actually trying out all possible solutions. A chimpanzee that enters a room where boxes litter the floor and bananas hang from the ceiling out of reach can figure out how to stack

FIGURE 37.6 *Classical Conditioning: Madison Avenue Exploits Our Evolutionary Tendency to Associate Separate Stimuli.* This advertisement links security, love, and owning a brand-new 1940 Buick Roadmaster.

the boxes and climb up to grab the food. And experimenters have taught chimpanzees how to use American Sign Language and how to operate computer keyboards—remarkable examples of the chimpanzee's innate capacity to learn (Figure 37.7).

In contrast, a dog may wrap its leash around a tree so that it can no longer reach its food and water, but still it cannot conceptualize and address the cause of the problem. The dog could be trained to keep its leash untangled through operant conditioning, but a primate would conceive of the problem beforehand and simply avoid it.

Behavioral ecologists suspect that animals have evolved the particular forms of behavior—from fixed-action patterns to insight learning—that provide the best chance of surviving and propagating. And natural selection seems to have resulted in an immense variety of developmental systems and behavioral strategies, as the next section shows.

≋ HOW NATURAL SELECTION SHAPES BEHAVIORS

A black-headed gull selects a particular nesting site at its summer home on a European beach. The bird feeds itself and its young, detects and drives away predators, selects a mate, and cares for its chicks. As part of all these activities, it communicates to its family and colony mates

through a variety of sounds and movements. And it has an innate sense of time and season that allows it to fish at the best time of day and migrate at the right time of year. In gulls, as in thousands of other animal species, such behaviors have been shaped by natural selection and markedly influence the organism's chances of obtaining energy (overcoming disorder) and producing progeny (overcoming death).

LOCATING AND DEFENDING A HOME TERRITORY

Just as people choose apartments, homes, and building sites, gulls find protected nesting spots on the beach, beavers locate the best site along a stream to build a lodge, and other animals locate territories that suit their needs. What mechanisms allow them to do this?

■ **Taxes** Every animal has sensory receptors suitable for detecting environmental stimuli. A female tick, for example, must locate a mammal and drink its blood before laying her eggs. Light-sensitive receptors in the tick's skin lead her to a well-lit leaf or branch, and then olfactory receptors help her detect butyric acid with its odor of rancid butter coming from a nearby mammal. Once triggered by smell, the tick drops from the leaf, and if she

FIGURE 37.7 *Conversations with a Chimp.* Researcher Herbert Terrace taught the chimpanzee "Nim Chimpsky" to write a few American Sign Language gestures. Here, Terrace signs his name "Herb," and Nim copies him.

FIGURE 37.8 *A Migration of Monarchs to Mexico.* Each autumn, 100 million monarch butterflies migrate from the United States to the mountains of western Mexico and roost on Oyamel fir trees. Scientists are still studying how they navigate the nearly 2000 km to this particular region.

chances to land on a mammal, her heat detectors allow her to crawl to a warm spot. There she bites a hole, drinks the warm blood, and her eggs begin to mature.

Just as a plant's orientation toward light is called a tropism (see Chapter 29), an animal's movement toward light, specific chemicals, or heat is called a *taxis*. The tick's movements involve positive *phototaxis* (movement toward light), positive *chemotaxis* (movement toward a chemical), and positive *thermotaxis* (movement toward heat). Natural selection has favored alleles that direct the development of special sense organs attuned to different physical stimuli; these stimuli, in turn, trigger behaviors that help the animal locate an appropriate home.

■ **Homing and Migration** Many animals routinely travel great distances in search of an appropriate habitat. Whales, sea turtles, eels, and butterflies, for instance, travel great distances to find suitable winter habitat or breeding areas (Figure 37.8). But the champion voyagers are birds. Black-headed gulls trek from Africa to northern Europe and back annually, and the small arctic tern flies 16,000 km (10,000 miles) from Greenland and Alaska to Antarctica and back in a search of endless summer.

While migrations are seasonal, homing behavior can take place anytime certain kinds of animals are displaced from their home territories. When experimenters trans-

ferred an individual shearwater, a long-winged seabird, from its home in Great Britain to a new location in Massachusetts, the remarkable bird was back on its nest in 12 days, having crossed 4800 km (3000 miles) of trackless ocean! Likewise, wildlife managers in Yosemite National Park routinely trap troublesome black bears and move them dozens of kilometers away from heavily used campgrounds, only to find the animals have returned in a few days. How do animals find their way, whether homing or migrating? Ethologists are not entirely sure, but tests confirm that some animals do have an inborn sense of **orientation,** or direction finding, and a partially learned sense of **navigation,** the more exacting ability to move from one map point to another.

Which sensory systems help an animal to navigate? In daytime, birds, ants, and bees orient via the sun's position as it sweeps across the sky relative to their path. For example, a pigeon released west of its birdhouse in the morning will fly toward the sun to get home, while a pigeon released west of its birdhouse in the evening will fly away from the sun to get home (Figure 37.9). Pigeons use an internal "timepiece," or *circadian clock*, to help compute orientation relative to the sun's path.

This is only part of the story, though, because homing pigeons can find their way home even in the dark or when researchers have fitted them with tiny, thickly frosted contact lenses that reduce their vision to a murky haze. Remarkably, birds still orient and navigate under such conditions by detecting the earth's magnetic field. Biologists have discovered a few tiny crystals of magnetic iron oxide in tissues near the brains of birds and porpoises. They suspect that these crystals align with the earth's magnetic field, like the iron needle in a compass, and cause neurons to fire. The researchers have concluded that the sun is a pigeon's primary cue for orientation, but the earth's magnetic field acts as a substitute cue in the dark.

These studies on homing pigeons and other animals help explain how animals migrate. Animal species can

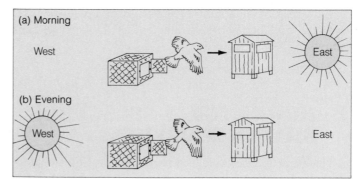

FIGURE 37.9 *A Pigeon's Internal Clock Allows It to Navigate by the Sun.* As the text explains, a bird will home to its birdhouse in the east by flying (a) toward the sun in the morning and (b) away from the sun in the evening.

travel short distances or long, in isolation or in large groups, but once they reach their destination, they often claim a territory and defend it, even against their own former traveling companions.

■ **Territoriality** Once a male gull claims a spot on the beach, he rebuffs other male interlopers by assuming stylized body positions and issuing specific calls that proclaim his turf (see Figure 37.1). Behavior that defends living space from intruders is called **territoriality.** Sleek male impalas of the African savanna are good examples of animals competing for territories rich in food resources. The richer the territory, the more females it entices, and more females means greater reproductive opportunity for males. The males stake out the richest territories they can find, and when a competing male arrives, they clash, interlocking their graceful horns and thrusting furiously, trying to force each other out. Bands of roving females stay in each territory only so long as the food holds out, and during their stay, the resident male mates with all receptive females. The stakes are clear: Males guarding more food have access to females longer, mate more frequently, and hence tend to have more offspring. A male without fertile territory to defend will have fewer contacts with females and thus far fewer offspring.

In most territories, the most valuable commodity is a ready source of food, and a second group of behaviors has arisen around obtaining nourishment.

FEEDING BEHAVIOR: ANIMAL "ENTREPRENEURS"

Is it more advantageous for a black-headed gull to expend the energy to fly several kilometers out to sea, where herring and other fish are bountiful, or to travel a shorter distance but forage where food is less abundant or less nutritious? What fraction of its feeding time should a small bird spend scanning the skies for a hungry hawk instead of searching the ground for seeds? To answer such questions, behavioral ecologists have devised and tested the *optimality hypothesis,* the notion that animals behave in ways that allow them to obtain the most food energy with the least effort and the least risk of falling prey to a predator.

Ethologists studied crows, for example, as the birds preyed on whelks—large, snail-like mollusks (Figure 37.10). A crow will typically search the intertidal zone and select a big whelk over several smaller ones, grasp it, fly it to a height of 5 m (16 ft), then drop it onto the rocks below. If the shell shatters, the crow consumes the tender flesh; if it doesn't break, the crow lifts and drops it again. Researchers predicted that if the crow were following optimal foraging strategy, large whelks should be more likely to smash than small ones; a drop of 5 m should be enough to break most whelk shells; and dropping a whelk a second time should be just as likely to lead to a

FIGURE 37.10 *Crow Dropping a Whelk: Minimizing Effort, Maximizing Gain.* A crow hunting whelks will choose large prey and smash them on the rocks by dropping the shells from 5 m (16 ft). If the shell doesn't break, the crow tries again. This precise strategy supports the optimality hypothesis; animals feed in such a way that feeding effort is minimized while energy gain and safety from predators are maximized.

meal as searching for a new whelk. In fact, each of these predictions proved true. Considered in evolutionary terms, a crow with optimal feeding behavior spends less time and energy foraging, but has more time and energy left over for producing offspring, and thus sends more genes into the next generation.

The optimal foraging solution is neither conscious nor deliberate nor universal. There is no evolutionary advantage to gorging one minute only to be devoured by a predator the next, so animals must often feed in ways that are nutritionally far from optimal in order to avoid the dangers of predation. The dark and light birds called juncos interrupt their search for seeds on the ground about every 4 seconds to scan the sky for hawks. This behavior sacrifices immediate food gain for long-term survival. If food is abundant, juncos form flocks; this cuts the time each bird must spend being watchful by spreading the vigilance over a larger group. Juncos in flocks, however, must share the available seeds with other birds. The optimality hypothesis predicts that individual juncos will alter their behavior to adjust for food availability and imminence of attack and will thus maximize their personal biological fitness.

While territoriality and feeding strategies help animals take in energy, every animal faces a second and equally important challenge: overcoming death by bearing and perhaps caring for young.

REPRODUCTIVE BEHAVIOR: THE CENTRAL FOCUS OF NATURAL SELECTION

Successful organisms reproduce effectively, and different sexes have different strategies for achieving reproductive success. The males of most species make many highly mobile, lightweight gametes and behave in ways that increase the number of eggs they fertilize. In contrast, most female animals make fewer but much larger gametes. A woman, for example, produces a few hundred eggs in her lifetime, while a man, in a similar period, produces enough sperm to inseminate all the women in the world. Not surprisingly, such differing reproductive strategies shape the reproductive behaviors of the sexes.

■ **Parental Investment** The fact that males and females produce gametes of different sizes and numbers has major consequences, including parental investment and mate selection. *Parental investment* is the allocation of a parent's resources, such as food, time, and effort, to each individual gamete and the young that arises from it. Since each large egg represents more resources than each small sperm, the female sacrifices opportunities to make more eggs in favor of increasing the chance that each egg will become a viable descendant. An extreme example of this is the kiwi bird, which lays an egg weighing one-quarter her body weight—the equivalent of a woman having a 13.5 kg (30 lb) baby! Most female mammals shelter and nourish their offspring in the uterus and feed them with milk long after birth. Such behaviors extract a heavy cost in physiological stress and increased exposure to predators. Yet, the extra care pays dividends in terms of the chances that each offspring will survive to contribute genes to the next generation.

Males usually exhibit much less parental investment. Behaviorists have deduced that male parenting is likely to evolve only in those ecological situations where females cannot provide all the care necessary for raising a brood successfully or where a male who abandons his mate is unlikely to find another partner. In such cases, evolution will favor behavioral mechanisms that ensure paternal proximity and investment, such as establishing and defending a nesting territory, as do black-headed gulls.

■ **Mate Selection** The second consequence of gender differences in gamete size is a sex difference in *mate selection*. Typically, males—the sex with the lesser investment—compete among themselves to fertilize many females, while females, with their heavier investment, select the male of choice to father their offspring. As a result of competition within a sex and mate selection between sexes, members of the same species exert selective forces on each other—a pressure called **sexual selection.**

In species where males make little parental investment, the males compete to inseminate as many females as possible. Biologists call competitive pressure between members of the same sex *intrasexual selection*. In some species, this leads to a **dominance hierarchy,** a ranking of group members based on past success in aggressive encounters over access to food or mates. Bull elephant seals, for example, are massive beasts that threaten each other with tremendous roars and bite and batter competitors. The loser of a skirmish lowers his head and retreats, with the victor in pursuit. A few years ago, an ethologist ranked ten bulls living on a small island in the Atlantic by their track records in winning such encounters, and then counted the successful copulations each achieved during the breeding season. The top-ranking male accomplished nearly 40 percent of all the copulations in the group, the second male less than 20 percent, and the other males only about 5 percent each. Thus, natural selection in elephant seals (and in many other animals, as well) favors those behaviors that win dominance contests and hence mates.

In many species, females do the mate choosing. A female generally picks males that provide the "best" genes or material benefits to enhance her offspring's survival. Females thus engage in *intersexual selection*, wherein one sex (the female) chooses desirable mates, and in so doing, acts as an agent of natural selection on the behavioral and physical traits of the other sex.

By choosing dominant or propertied males, females increase their own biological fitness. Elephant seal cows, for example, create intersexual selective pressure by simply screaming loudly whenever a bull attempts to couple

FIGURE 37.11 *Attempted Mating by Bull Elephant Seal Sets Off the Female's Automatic Alarm Call.* A casual advance by an amorous male—a flipper caress—sets the female elephant seal (*Mirounga leonina*) to screaming at high decibel. She calls this way regardless of the suitor's stature in the colony, but if the male is subordinate, the king bull will probably chase him off.

(Figure 37.11). Her cries alert the harem master, and if the courting male is low on the dominance hierarchy, the dominant bull sprints to the site of the tryst and chases away the intruder. The female's automatic cries greatly increase her chances of being inseminated by the dominant male rather than by a subordinate and hence of having sons with aggressive behaviors that will favor high social rank.

Elaborate courtship rituals probably also evolved under the pressure of females exercising mate choice. A male that is too decrepit to perform a ritual dance like that of the black-headed gull is unlikely to provide the favorable genes or male parental care the female needs for successful reproduction.

COMMUNICATION: MESSAGES THAT ENHANCE SURVIVAL

Easterners and midwesterners enjoying the fragrant air and trilling insect sounds of a warm summer night can often catch a wonderful aerial display as "lightning bugs," or fireflies, flash their luminous yellow abdomens on and off while they dart about the green lawns, fields, or trees. The flickering streaks they create are based on flashing patterns with a rhythm characteristic for each species (Figure 37.12). Some species give long pulses of continuous light, others blink three times in a row, and still others give a fluttering flicker like a light bulb about to burn out. The airborne flashers are males communicating their availability for mating, and when a female sees the right pattern of flashes, she responds with her own burst of light, revealing her position to the male.

FIGURE 37.12 *Luminous Communication: Fireflies on a Summer Night.* Fireflies such as the common eastern species *Photurus pennsylvanicus* are actually soft-bodied beetles that light up the countryside with glowing abdomens. This drawing depicts the male courtship displays of four species: One emits three long bright flashes in a row; one emits eight long bright flashes; one a swooping continuous burst of light; and one a zigzagging burst. Each communicates mating interest to waiting females.

The firefly's blinking light is a form of **communication,** a signal produced by one individual that alters the behavior of a recipient individual and improves both participants' biological fitness.

■ **Communication by Sight, Sound, and Scent** Each animal uses one or more channels of communication that fit the species' ecological circumstances. In open areas where individuals live near each other, visual displays such as the courtship dances of black-headed gulls are common. Animals that hide in thickets or pond vegetation are invisible not only to predators but to potential mates, as well, and for them, auditory cues, like the chirp of a cricket or the croak of a bullfrog, are typical. Sound communication, such as the alarm call of a gull or prairie dog, is also advantageous as a warning to colony members who may not see the approach of a predator.

Some animals live such solitary lives that they are unlikely to see or hear signals from other members of their species. For them, odor molecules can convey information. The sex pheromones released into the air by female moths (see Chapter 25) can travel for miles on the wind, announcing the females' location and sexual readiness. Chemical signals not only disperse farther than auditory or visual messages, but also can persist longer in the environment. Ants establish durable trails this way.

■ **Communication by Touch** Groups of animals that live in dark hives, caves, or underground chambers often communicate by touch, as well as by sound and smell. The classic example of touch communication is seen in the "dances" bees perform in the dark interiors of their hives, communicating the location of rich food sources. If a foraging bee finds flowers that are laden with pollen and nectar and are situated fairly close to the hive (within 80 m, or 260 ft), the bee flies home, enters the hive, and immediately performs a *round dance* by walking in tight circles (Figure 37.13a, page 610). Observer bees follow the dancer in the dark, sense its movements by touch, and learn that a nearby food source exists. Since worker bees can also smell bits of food clinging to the forager's body, they usually have no trouble locating the specific flowers.

If the forager discovers a food source far from the nest, she returns with a rich load of pollen, climbs into a vertical wall of the hive, and begins a *waggle dance*, moving in a figure eight while wagging her abdomen from side to side and making an occasional burst of sound (Figure 37.13b). The dancer's animation indicates the ease or difficulty of getting to the food: A lethargic dance with few waggles and little sound indicates a long or uphill trip to the food source; an excited dance with many waggles and much sound suggests a more readily accessible site. Simultaneously, the dancer's orientation on the hive wall

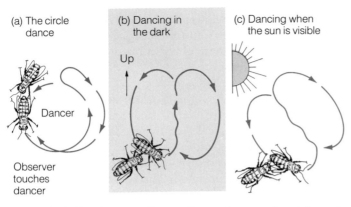

(a) The circle dance

(b) Dancing in the dark

Up

(c) Dancing when the sun is visible

Dancer

Observer touches dancer

FIGURE 37.13 *Honeybee Dances.* Forager bees perform (a) round dances when a discovered food source is close to the hive and (b) waggle dances when the source is more distant. Orientation of the dancing bee on the wall of a dark hive signifies the direction to fly relative to the sun's position, with the understanding that a symbol for the sun is straight up. If sunlight shines into the hive (c), the bee orients to the light.

reflects the angle of the necessary flight path with respect to the sun. If the hive is dark and the food lies in the sun's direction, the bee waggles while moving straight up (see Figure 37.13b). If the hive is dark and food lies 90° to the right of the sun, the bee waggles while moving horizontally to the right. If sunlight enters the hive, bees orient their dance directly with respect to its rays (Figure 37.13c).

Whether an animal communicates via sight, sound, scent, touch, or body positions, the exchange of information has adaptive value: It allows individuals to find homes, food, and mates more successfully or to evade predators and thus to send their alleles on to future generations. While solitary animals like fireflies, moths, and frogs exchange information from time to time, and those living in small family groups, like black-headed gulls, do so more regularly, communication is most complex in animals that live together in large social groupings. By their very complexity, animal societies pose an interesting challenge to evolutionary ecologists: How could such social groups with their roles and rules have evolved?

≋ THE EVOLUTION OF SOCIAL BEHAVIOR

A fascinating minority of animal species live in groups—either as giant colonies, like those of ants, bees, and termites; as large groups, like flocks of gulls; as small, stable

packs, like those of wolves (Figure 37.14); or as individuals with only seasonal associations, such as the graceful manatees of Florida. What biological mechanisms cause gulls to live in cooperative societies while turtles live as lone individuals? And how does an individual's evolutionary fitness increase from, say, cooperative behavior, such as many gulls mobbing a predator, if this behavior puts it in danger? Questions like these are the focus of **sociobiology,** the study of social behavior.

SOCIAL LIVING: ADVANTAGES AND DISADVANTAGES

What does an individual gain and what does it sacrifice by living in a group? To answer this question, ethologists have compared the societies of white-tailed and black-tailed prairie dogs. Regardless of species, a prairie dog has a characteristic response to the approach of a badger, weasel, or other predator: It freezes for a moment, then gives a loud alarm call. This dangerously betrays its location to the predator, but it also warns family members and neighbors to dash off to the safety of their burrows. Animal behaviorists noted that black-tailed prairie dogs live in larger groups than white-tailed prairie dogs. That differential gave them a means of probing the advantages of group living for this species.

Experimenters pulled a stuffed badger at constant speed through different-sized prairie dog colonies and recorded the time it took before a prairie dog would sound the alarm. They found that among black-tails, with their larger colonies, predators were detected sooner than among the smaller white-tail colonies. With the support of so many watchful colony mates, black-tails could spend less time scanning for predators and more time foraging.

FIGURE 37.14 *Safety in Numbers.* Many animal species live in social groups. Some groups are large and permanent, such as colonies of birds. Some are small and intimate, such as this pack of gray wolves. Some are only seasonal, like the manatee, which typically nurses one offspring each year (see Figure 18.15).

Larger colony size is clearly advantageous for providing a means of early predator detection plus more foraging time. So why don't white-tailed prairie dogs live in large colonies too?

The answer to that question lies with the disadvantages of living in larger societies. Observations reveal that prairie dogs in larger groups spend more time aggressively bickering over territory than those in smaller groups. In addition, plague-carrying flea populations are four times as dense in the larger black-tailed colonies as in the smaller white-tailed groups. Apparently, in avoiding predators, black-tails pay the costs of infighting and disease. One could predict from this that the black-tails' environment exposes them to more predators. Indeed, large colonies of black-tails live exposed on the open plains of the American West, while the smaller colonies of white-tails inhabit shrubland and can hide in the bushes.

From experiments like these, sociobiologists conclude that animals will socialize only when the benefits of group living outweigh the costs. In each case, the benefits require cooperation between individuals, and the cooperation can occasionally cost an individual its feeding time or even its life. How can natural selection explain such seemingly altruistic behaviors?

THE EVOLUTION OF ALTRUISTIC BEHAVIOR

Cooperative behavior often favors both helper and helped, as when a lioness chases a wildebeest toward her hunting partners and all share the kill. Sometimes, however, the helper suffers, as when a worker bee stings a raccoon to protect the hive but rips out her stinger and subsequently dies. Ethologists have long wondered how such **altruism,** or unselfish acts for the welfare of others, could evolve, since natural selection would not favor the individual altruist. Animals displaying the selfless behavior would tend to reproduce less frequently. So why don't the alleles that foster altruism gradually disappear?

Behaviorists solved the puzzle of altruism by recognizing the goal of reproduction: to multiply one's own alleles. The obvious way to accomplish that goal is by reproducing oneself, but there is another way, too: encouraging the reproduction of family members that share the same alleles. Evolutionary biologists recognize a specific type of natural selection called **kin selection,** in which an individual increases its genetic contribution to the next generation by helping relatives—who share its alleles—to reproduce.

You can see how kin selection works by considering how many alleles you share with your family members. Because of meiosis and fertilization, you inherited half your alleles from your mother and half from your father. Since you and your siblings share the same parents, half of your alleles will be carried by each brother or sister. Your aunts, uncles, and grandparents, in turn, carry one-quarter of your alleles, and your cousins one-eighth.

The proposer of kin selection, W. D. Hamilton, once quipped: "I'd lay down my life for two brothers or eight cousins." By this he meant that an act of altruism toward close relatives promotes survival and transmission of one's own alleles. His theory helps explain how something as drastic as the suicidal sting of a honeybee could have evolved.

■ Kin Selection and the Evolution of Social Insects

Before Hamilton proposed the concept of kin selection, the rigid caste system of social bees, wasps, and ants was quite puzzling. Such societies often consist of a single reproducing female, the *queen*, plus a few *drones*, or males, and up to 80,000 sterile female helpers, or *workers* (Figure 37.15). But why would so many sterile workers relinquish their own reproduction to help the queen reproduce?

Geneticists and animal behaviorists have discovered that in bees and wasps, females are diploid, but males are haploid. As a result, daughters share half their mother's alleles but *all* their father's alleles instead of just half of their father's, as in most organisms. As a result, each female bee shares more alleles with her sisters than you do with your sister(s). By helping the queen reproduce more sisters (including future queens), she is really favoring the preservation of her own alleles.

Complex caste societies evolved among ants, bees, and wasps favored by their special genetic system. In contrast, many mammalian societies evolved because group hunting provides advantages, as in wolves, or because relatives could assist with the prolonged developmental needs of young ones, as in primates.

FIGURE 37.15 *The Queen and Her Court: How Kin Selection Evolves.* A circle of worker bees tend their regal sister, the queen bee, with her dark, shiny thorax and abdomen enlarged by eggs.

SOCIOBIOLOGY AND HUMAN BEHAVIOR

Biologists can point to thousands of examples from the animal kingdom in which social behavior seems to promote the survival of an animal's alleles through a combination of individual and kin selection. But what about human behaviors? Do similar genetic principles apply to us? In the mid-1970s, Harvard biologist E. O. Wilson stirred a heated controversy by cautiously suggesting that human behavior has evolutionary underpinnings and can be analyzed in terms of individual and kin selection. Some scientists and lay people alike issued strenuous objections to Wilson's ideas, mistakenly thinking that if human behavior is genetically determined and evolutionarily adaptive, our actions cannot and should not be changed. Some critics were justifiably concerned that theories of genetic predeterminism could be twisted to promote fascism, sexism, and racism. Wilson had intended merely to suggest a way of analyzing the causes of human behaviors, not justifying them from an ethical standpoint.

Debates continue on the evolution and adaptiveness of specific human behaviors, but it is clear that some human behavioral patterns are innate. The tiny hand of a human baby grasps tightly to a parent's finger, its little lips suck rhythmically at a nipple, and it communicates forcefully with a cry that a parent cannot ignore. A baby born blind still flashes an alluring smile at the sound of its parents' voice—an innate behavior pattern triggered by a releaser. A sociobiologist interpreting such inborn human behavior would suggest that infants inheriting alleles that foster smiling or crying are more likely to solicit parental care and hence more apt to survive and pass on such alleles to the next generation.

While there are many innate behaviors, and while genes underlie at least some aspects of human behavior, many more behaviors are passed along through cultural traditions. Genes determine the native ability to learn language, for example, and this unfolds as neural pathways develop in the growing embryo and make it capable of learning. But whether a child learns to speak English or Polish or Yanomamö after it is born depends on the language its parents teach it.

Sociobiologists propose that this flexibility directs people to accept cultural practices that enhance their reproductive fitness and reject those customs that reduce genetic success. For example, Brazilian tribes living in areas with poor soils and overhunted forests adopt Western agricultural techniques, while clans inhabiting less disturbed areas cling to traditional habits. In both cases, the human nervous system seems to respond to the environment with an innate flexibility that is naturally selected to elevate an individual's economic condition and hence provide for offspring. A few other animals also seem to accept new adaptive cultural practices and transmit them to their

FIGURE 37.16 *Cultural Transmission in Primates: The Case of the Clean Sweet Potatoes.* A young female macaque accidentally learned to wash the sand off sweet potatoes she picked up from the beach, and soon others in the troop copied her. They began teaching their young to wash off the sand , and the young later grew and taught their own offspring. This learned behavior has become a permanent part of the troop's culture, much the way reading, writing, metallurgy, and agriculture have become part of human culture.

offspring. Macaques who learned to wash sweet potatoes is probably the most famous example (Figure 37.16).

Psychologists have conducted many recent studies on identical and fraternal twins raised together in the same family or apart by adoptive families to explore the contention that genes help shape human personality and behavior (Figure 37.17). These twin studies show that many human personality traits are influenced about equally by genes and the environment. Such traits include social effectiveness, achievement, social closeness, alienation, aggression, respect for authority, and avoidance of harm. A single gene, of course, encodes a protein, not a trait like aggression or achievement. Behavioral geneticists suggest that hundreds of genes combine to determine a person's range of emotions; the thresholds for anger, hostility, and sexual arousal; and the inclination to learn one type of behavior versus another.

Sociobiologists conclude that as the human brain evolved, social behaviors that enhance the bearer's reproductive fitness emerged. Those behaviors evolved when people lived in small, isolated family groups, but in the meantime, we have been enormously—perhaps disastrously—successful as a species. As the biosphere threatens to collapse under the pressure of human population, our gene-encoded behavioral plasticity remains. Thus, the hope remains, as well, that we will not destroy the quality of life for the animal called *Homo sapiens*, nor for any of the living species that share our planet, and that the natural world will remain, for us, a challenge and a source of perpetual wonder.

FIGURE 37.17 *Reunited Twins: Something Was Missing.* Identical twins Jerry Levey and Jim Tedesco were separated as infants and long felt something was missing in their lives. Reunited in middle age, they discovered not only a close physical resemblance, but identical vocations—firefighting—and avocations—flirting, telling jokes, and drinking beer.

⚹ CONNECTIONS

Natural selection has tuned an animal's genes to direct the development of its nervous system in ways that promote specific behaviors. Some behaviors are rigid and stereotyped, like a baby gull pecking at its mother's beak, and others are plastic and variable, like a macaque learning to wash potatoes in the sea. But regardless of flexibility level, the behaviors that have evolved help increase an individual's survival and genetic perpetuation. All of the fixed and plastic behaviors that suit an organism to its environment have immediate and ultimate causes, and these help us understand how and why animals act as they do—even in puzzling instances of altruism.

Most sociobiologists believe that the principles of kin selection apply to many aspects of human behavior, as well as to the behavior of prairie dogs and honeybees. By understanding the evolution of human nature, we stand a better chance of solving our current global crises of overpopulation, ecological degradation, disease, starvation, and warlike aggression. Well-informed citizens are in the best position to apply biological knowledge to these issues and perhaps to help make our world a safer place for life.

NEW TERMS

altruism, page 611
classical conditioning, page 604
communication, page 609
dominance hierarchy, page 608
ethology, page 601
fixed-action pattern, page 602
imprinting, page 603
insight learning, page 604
instinct, page 602

kin selection, page 611
learning, page 604
navigation, page 606
orientation, page 606
releaser, page 602
sexual selection, page 608
sociobiology, page 610
territoriality, page 607
trial-and-error learning, page 604

STUDY QUESTIONS

REVIEW WHAT YOU HAVE LEARNED

1. Ethologists distinguish between the proximate cause and the ultimate cause of a behavior. Explain.
2. How does an instinct differ from a learned behavior?
3. List five features of a fixed-action pattern.
4. Name four types of learning.
5. What factors influence animal navigation?
6. Give an example of territoriality.
7. Which sex generally invests more resources in their offspring, males or females? Explain.
8. List four types of animal communication, and give an example of each type.
9. What is kin selection?
10. What proportion of alleles do you share with your father, sister, grandmother, uncle, and cousin?

APPLY WHAT YOU HAVE LEARNED

1. Will a gull be likely to incubate an infertile chicken egg or a cola bottle cap placed in her nest? Explain.
2. You discover a nest of three-day-old goslings and decide you'd like to have them imprint on you and follow you home. Are you likely to succeed? Why or why not?
3. An advertisement for the Nature Nuggets Company features a hiker eating a bowl of cereal on a mountaintop. What type of learning does the advertisement rely on?

FOR FURTHER READING

Breed, M. D. "Genetics and Labour in Bees." *Nature* 333 (1988): 299.

Huntingford, F., and A. Turner. "Aggression: A Biological Imperative?" *New Scientist* (4 August 1988): 1988.

Peck, J. R., and M. W. Feldman. "Kin Selection and the Evolution of Monogamy." *Science* 240 (1988): 1672–1674.

Rosen, C. M. "The Eerie World of Reunited Twins." *Discover* (September 1987): 36–42.

Tinbergen, M. "The Shell Menace." *Natural History* 72 (1963): 28–35.

① Genes direct the neural programs that give rise to behavior.

② Some behaviours are innate and performed in a fixed-action pattern; still others can be learned from experience through habituation, imprinting, trial and error, classical conditioning, or insight.

Fixed action

Learned action

③ Behaviours influence an organism's chances of surviving to reproduce and are shaped by selective forces, including kin selection.

④ For animals that live in groups, the evolutionary benefits of social behavior outweigh the costs.

⑤ People behave the way they do because of their genetic programs, environmental conditions, and cultural traditions.

Vincent van Gogh, *The Potato Eaters*, 1885

Auguste Renoir, *Le Moulin de la Galette*, 1876

GLOSSARY

Abscisic acid (ABA) A plant hormone that inhibits growth, apparently by blocking protein synthesis and new growth. (30)

Abscission (L. *ab*, away, off + *scissio*, dividing) In vascular plants, the dropping of leaves or fruit at a particular time of year, usually the end of the growing season. (30)

Acid Any substance that gives off hydrogen ions when dissolved in water, causing an increase in the concentration of hydrogen ions. Acidity is measured on a pH scale, with acids having a pH less than 7; the opposite of a base. (2)

Action potential A temporary all-or-nothing reversal of the electrical charge across a cell membrane; occurs when a stimulus of sufficient intensity strikes a neuron. (26)

Activation energy The minimum amount of energy that a molecule must have in order to undergo a chemical reaction. (4)

Active site A groove or a pocket on an enzyme's surface to which a specific reactant or set of reactants binds. (4)

Active transport Movement of substances from areas of low concentration to areas of high concentration, requiring the expenditure of energy by the cell. (4)

Adaptation (L. *adaptere*, to fit) A particular form of behavior, structure, or physiological process that makes an organism better able to survive and reproduce in a particular environment. (1)

Adaptive radiation Evolutionary divergence of a single ancestral group into a variety of adaptive forms, usually with reference to resources or habitat. (32)

Addiction The process by which an individual becomes physically or psychologically dependent on a substance, such as a drug. (26)

Adrenal gland (L. *ad*, to + *renes*, kidney) A hormone-producing endocrine gland located on top of the vertebrate kidney; secretes hormones that are mainly involved in the body's response to stress, such as epinephrine (adrenaline) and norepinephrine. (25)

Aerobic respiration Pathways of carbohydrate metabolism requiring oxygen that take place within cell mitochondria. (5)

Age structure In populations, the number of people in each age-group. (33)

Aldosterone A hormone that controls water loss by regulating salt reabsorption in the kidney's nephrons. (24)

Alga (pl. alagae) (L. seaweed) In the kingdom Plantae, multicellular aquatic organisms with simple reproductive structures; also includes aquatic photosynthesizing organisms in the kingdoms Monera and Protista. (16)

Alimentary canal A system of tubes and chambers associated with the digestive system. Includes the mouth, esophagus, stomach, and intestines. (23)

Allele One of the alternative forms of a gene. (8)

Altruism Self-sacrifice for the benefit of others; a form of behavior that results in increased fitness for the individuals for whom the altruistic act is performed while the individual performing the act reduces its own fitness; for example, a bird giving a warning cry to tell other birds of an approaching predator increases their chances of escape while drawing attention to itself and increasing its own chances of being eaten. (37)

Alveolus (L. small cavity) A thin-walled, saclike structure in the vertebrate lung in which gas exchange takes place. Each lung contains thousands of alveoli. (21)

Amino acid (Gr. Ammon, the Egyptian god near whose temple ammonium salts were first prepared from camel dung) A molecule consisting of a central carbon atom bound to four key groups of atoms, including an amino group (NH_2), an acid group (COOH), a single hydrogen atom (H), and a side group of atoms that differs in each type of amino acid. The 20 types of amino acids are the basic building blocks of proteins. (2)

Ammonia The molecule NH_3, formed from the breakdown of amino acids; it is excreted into water or converted to uric acid or urea. (25)

Amniocentesis A procedure for obtaining fetal cells from amniotic fluid for the diagnosis of genetic abnormalities. (11)

Amphibian A cold-blooded vertebrate that starts life as an aquatic larva, breathing through gills, and metamorphoses into an air-breathing adult. Examples include frogs and newts. (18)

Anaerobic metabolism Biological reactions that do not require oxygen, such as glycolysis. (4)

Anatomy The science of biological structure. (19)

Angiosperm A flowering plant. (16)

Animal A multicellular heterotroph that moves about under its own power at some point during its life cycle; includes vertebrates and invertebrates. (17)

Animalia The kingdom comprising muticellular life forms that feed on other organisms and exhibit movement, such as insects, clams, birds, reptiles, and mammals. (1, 14)

Annual plant (L. *annos*, year) A plant that completes its life cycle within one year and then dies. (29)

Antagonistic muscle pairs Groups of muscles that move the same object in different directions. (28)

Anther In flowers, the terminal portion of a stamen that contains pollen in pollen sacs. (16)

Anthropoid A suborder of the primate order; includes Old and New World monkeys and humans. (18)

Antibody A specific protein produced by B cells of the immune system in response to the entry of a specific antigen (foreign substance) into the body. (22)

Antidiuretic hormone (ADH) A hormone that acts to control water loss; regulates the permeability of water in the kidney's nephrons. (24)

Antigen (Gr. *anti*, against + *genos*, origin) Any substance, including toxins, foreign proteins, and bacteria, that when introduced into the body cause antibodies to form. (22)

Aorta The main artery of the body; leads directly from the heart's left ventricle. (20)

Apical dominance The inhibition of lateral buds or meristems by the apical meristem. (30)

Apical meristem A zone of continual growth located at the tip of roots and stems. (29)

Artery A large blood vessel with a thick, multilayered muscular wall, which carries the blood from the heart to the body; divides into smaller vessels called arterioles. (20)

Arthropod An invertebrate with an exoskeleton, modified special segments, rapid movement and metabolism due to specialized respiratory structures, and acute sensory system, in the phylum Arthropoda; members include fossile trilobites, spiders, crabs, and insects. (17)

Asexual reproduction A type of reproduction wherein new individuals arise directly from only one parent. (7)

Atom (Gr. *atomos*, indivisible) The smallest particle into which an element can be broken and still retain the properties of that element. (2)

ATP Adenosine triphosphate, a molecule consisting of adenine, ribose sugar, and three phosphate groups. ATP can transfer energy from one molecule to another. ATP loses a phosphate to form ADP, releasing energy in the process. (5)

Autonomic nervous system The portion of the nervous system that regulates the heart, glands, and smooth muscles in the digestive and circulatory systems. (28)

Autosome Any chromosome other than the sex chromosomes. (8)

Autotrophs (Gr. *auto*, self + *trophos*, feeder) Organisms such as plants that can manufacture their own food. (5)

Auxin (L. *auxein*, to increase or augment) A plant hormone involved in elongating cells in stems, leaves, and the ovary wall; prevents premature falling of leaves, fruits, and flowers. (30)

Axon The portion of a nerve cell that carries the impulse away from the cell and also stores hormones. (26)

Bark Protective, corky tissue of dead cells present on the outside of older stems and roots of woody plants. (29)

Base Any substance that accepts hydrogen ions when dissolved in water and has a pH greater than 7; the opposite of an acid. (2)

B cell (B lymphocyte) A type of lymphocyte in the blood that makes and secretes antibodies in response to foreign substances (antigens). Each B cell is capable of secreting only one type of antibody, which is different from another B cell's antibody. (22)

Biennial A plant that completes its life cycle in two years and then dies. (29)

Bilateral symmetry Describes a body plan in which the right and left sides of the body are mirror images of each other. (17)

Bile An enzyme-containing fluid produced by the liver, stored in the gallbladder, and released into the small intestine. (23)

Biological molecules The large organic molecules found in living organisms, including carbohydrates, lipids, proteins, and nucleic acids. (2)

Biology (Gr. *bios*, life) The scientific study of life. (1)

Biological magnification The tendency for toxic substances to increase in concentration as they move up the food chain. (35)

Biomass The total dry weight of organic matter present at a particular feeding level. (35)

Biome A major ecological community type such as a desert, rain forest, or grassland. (36)

Biosphere (Gr. *bios*, life + *sphaira*, sphere) That part of the planet that supports life; includes the atmosphere, water, and the earth's crust. (33)

Bird A warm-blooded vertebrate with feathers and lightweight, hollow bones that lays hard-shelled eggs; most species are capable of flight; includes flamingoes, condors, and cassowaries. (18)

Bladder A storage sac, for example, for urine. (24)

Blastocyst A stage in early embryonic development of mammals resulting from cleavage (divisions of the fertilized egg) and similar to the blastula. Consists of a thin-walled hollow sphere of cells called the trophoblast and a mass of cells at one side that are destined to become the embryo. (12)

Blastula A stage of embryonic development, at or near the end of cleavage and immediately preceding gastrulation, that consists of a hollow ball of cells (blastomeres). (12)

Body cavity A fluid-filled space contained in the mesoderm; found in animals more complex than the roundworms, such as mollusks and segmented worms. (17)

Bone The main supporting tissue of vertebrates, composed of a matrix of collagen hardened by calcium phosphate. (28)

Bouton The end of an axon that terminates close to another cell, often shaped like a knob; serves as a special region of cell-to-cell communication. (26)

Bowman's capsule In the vertebrate kidney, the bulbous unit of the nephron that surrounds the glomerulus, the region that filters water and solutes from the blood. (25)

Brain stem The most posterior portion of the vertebrate brain, which forms an extension of the spinal cord and consists of the medulla, pons, and midbrain. (27)

Bronchiole One of the thousands of small branches from the bronchi that lead to the alveoli in the lungs. (21)

Bronchus One of the two hollow passageways branching from the trachea, each of which enters a lung. (21)

Bryophyte The division of the plant kingdom comprising mosses, liverworts, and hornworts. (16)

Bud Meristematic tissue at the tips of shoots that differentiates into leaves, new shoots, or flowers. (29)

Bulk flow The overall movement of a fluid from an area of high pressure to an area of lower pressure. (20, 31)

C$_3$ plant A plant in which the light-independent reactions of photosynthesis start with a three-carbon compound. Most plants are C$_3$ plants. (6)

C$_4$ plant A plant such as corn in which the light-independent reactions of photosynthesis start with a four-carbon compound. (6)

Cancer A disease characterized by uncontrolled cell division. (7)

Capillarity The tendency of a liquid substance to move upward through a narrow space, as when water moves up through the narrow "pipeline" in a plant's stem. (2)

Capillary Any of the tiny blood vessels with walls one cell thick that permeate the tissues and organs of the body. Substances such as oxygen diffuse out of capillaries to the tissues, and waste products diffuse into capillaries from the tissues. (20)

Carbohydrate (L. *carbo*, charcoal + *hydros*, water) An organic compound composed of a chain or ring of carbon atoms to which hydrogen and oxygen atoms are attached in the ratio of approximately two hydrogens to one oxygen. Carbohydrates include sugars and starch. (2)

Carbon cycle The global flow of organic molecules from plants through animals to the atmosphere, soil, and water and back to plants (6, 35)

Carbon fixation The sequence of reactions that converts atmospheric gaseous carbon dioxide into carbohydrates. (6)

Cardiac muscle The specialized muscle of the heart. (28)

Carrier A heterozygous individual not expressing a recessive trait but capable of passing it on to her or his offspring. (11)

Carrying capacity The density at which a population ceases to grow due to the limitation imposed by resources. (33)

Cartilage A dense, fibrous connective tissue in the skeletons of higher vertebrates; in sharks, rays, and skates, most of the skeleton is made of cartilage. (18)

Catalyst A substance that speeds up the rate of a chemical reaction by lowering the activation energy and without being changed or used up during the reaction. (4)

Cell cycle The process by which a cell grows and divides in two; the progeny grow and divide in two, thereby continuing the cycle. (7)

Cell division The splitting of one cell into two cells. (7)

Cell growth An increase in cell size. (7)

Cell membrane A biological envelope surrounding a cell; also called the plasma membrane. (3)

Cell theory A major doctrine of biology proposed over a century ago stating that all living things are composed of cells; cells are the basic living units within organisms; the chemical reactions of life take place within cells; and cells arise only from other cells. (3)

Cell wall A fairly rigid structure that encloses prokaryotic, fungal, and plant cells. (3)

Central vacuole In plants, cellular organelle with a single membrane that appears empty through a microscope but contains water and various storage products and can take up 5 to 95 percent of a cell's volume. (3)

Cerebellum The hindbrain region of the vertebrate brain between the cerebrum and pons; it integrates information about body position and motion, coordinates muscular activities, and maintains equilibrium. (27)

Cerebral cortex The highly convoluted surface layers of the cerebrum, consisting of neurons and glial cells; contains about 90 percent of the brain's cell bodies. It is well developed only in mammals and is particularly prominent in humans, and it is the region involved with conscious sensation and voluntary muscular activity. (27)

Cerebrum (L. *brain*) The forebrain of the vertebrate brain; the largest part of the brain, consisting of right and left cerebral hemispheres. It coordinates and processes sensory input and controls motor responses. (27)

Cervix The narrow necklike portion of an organ, in particular of the uterus. (13)

Character displacement The process by which one species evolves different physical or behavioral characteristics to avoid competing for resources with similar species in the same range. (34)

Chemical reaction The making or breaking of chemical bonds between atoms or molecules. (4)

Chemosynthesis The metabolic process in which organisms trap chemical energy from inorganic molecules (like hydrogen sulfide) dissolved in water and use that energy to fix carbon dioxide to make molecules needed for growth and reproduction. (35)

Chlorophyll (Gr. *chloros*, green + *phyllon*, leaf) Light-trapping pigment molecules that act as electron donors during photosynthesis. (6)

Chloroplast A cell-like organelle present in algae and plant cells that contains chlorophyll and is involved in photosynthesis. (3, 6)

Chordate An animal with a notochord, a spinal cord, gill slits, a tail, and muscle blocks at some time during its life cycle; in the phylum Chordata; includes both invertebrates (tunicates and lancelets) and vertebrates (fish, amphibians, reptiles, and mammals). (18)

Chorion A fluid-filled sac that surrounds the embryos of reptiles, birds, and mammals. In placental mammals it gives rise to part of the placenta and produces HCG, a hormone that maintains pregnancy by preventing menstruation. (13)

Chorionic villus sampling A procedure for removing fetal cells from the developing placenta for the diagnosis of genetic defects. (11)

Chromosomal mutation A mutation affecting large regions of a chromosome or even an entire chromosome, thereby affecting many genes; includes changes in the number and shape of chromosomes. (10, 11)

Chromosome A self-duplicating body in the nucleus of a cell that is made up of DNA and proteins; contains genes and a centromere region. Chromosomes are transmitted from one generation to the next via gametes. A human cell contains 23 pairs of chromosomes, one of each pair received from the mother (egg) and the other inherited from the father (sperm). (3, 7, 9)

Chromosome translocation Occurs when part of one chromosome moves spontaneously to a new location on a different chromosome. (11)

Cilium (pl. cilia) (L. eyelash) A short, centriole-based hairlike organelle used for movement by many unicellular organisms. Cilia also aid the movement of substances across epithelia surfaces of animal cells, such as cells that line the airways leading to our lungs. (3)

Circadian rhythm (L. *circa*, about + *dies*, day) A regular rhythm of an activity that occurs on about a 24-hour cycle. (25)

Circulatory system An organ system consisting of a heart, blood vessels, and blood, which transports substances around the body; also a system that helps maintain homeostasis in mammals. (20)

Classical conditioning A form of trial-and-error learning that involves the association of two separate stimuli with the same response. (37)

Cleavage The early, rapid series of mitotic divisions of a fertilized egg, which results in a hollow sphere of cells known as the blastula. (12)

Climate The accumulation of seasonal weather patterns over a long period of time. (36)

Cnidarian An invertebrate that possesses radial symmetry, a three-layered body wall, a gastrovascular cavity, and nerve cells and grows as a polyp or medusa in the phylum Cnidaria; includes jellyfish, hydras, corals, and sea anemones. (17)

Cochlea A spiral tube in the inner ear that contains sensory cells involved in detecting sound and analyzing pitch. (27)

Codon Three adjacent nucleotide bases in DNA or mRNA that code for a single amino acid. (10)

Coevolution Natural selection on two organisms for increased interdependence based on selective advantages for both organisms. (16, 34)

Cold-blooded Refers to animals that grow active in warm environments and sluggish in cold ones, such as fishes, amphibians, and reptiles. (18)

Collecting duct Any duct that drains an organ; in the kidney, ducts that drain the renal pelvis of the kidney and carry urine to the ureters. (24)

Colon The last portion of the large intestine, the wide part of the alimentary canal that leads to the rectum. (23)

Commensalism Relationship between two species in which one species benefits and the other suffers no apparent harm. (34)

Communication The process by which a signal produced by one individual alters the behavior of a recipient individual, thus improving both participants' biological fitness; in cells, the process by which electrical currents and/or molecules pass from cell to cell and cause a change, or response. (3, 37)

Community Two or more populations of different interacting species occupying the same area. (33)

Competitive exclusion A situation in which one species eliminates another through competition. (34)

Compound A mixture of two or more elements. NaCl (sodium chloride), or table salt, is an example of a compound. (2)

Cone In plants, the reproductive structure of a conifer; in animals, a photoreceptor cell of the eye that can detect color but is not sensitive to low light levels. (16, 27)

Consumer In ecology, an organism that eats other organisms. (35)

Contraception The prevention of conception. (13)

Cork An external secondary plant tissue impermeable to water and gases. (29)

Cortisol Hormone secreted by the adrenal cortex and medulla that helps animals respond to stress by speeding metabolism, thus providing quick energy. (25)

Covalent bond (L. *co*, together + *valere*, sharing) A form of molecular bonding characterized by sharing a pair of electrons between atoms. (2)

Cultural transmission The process of teaching future generations about cultural practices, such as hunting, cooking, and sharing, through actions and language. (18)

Cyanobacteria Blue-green algae in the kingdom Monera. (15)

Cytokinesis The process of cytoplasmic division accompanying nuclear division. (7)

Cytokinin A plant growth hormone that stimulates cell division; prevents leaf aging or senescence. (30)

Cytoplasm The part of a cell between the outer membrane and the nuclear envelope. (3)

Cytoskeleton Found in the cells of eukaryotes, an internal framework of microtubules, microfilaments, and intermediate filaments that gives the cell and its components structural supports. (3)

Deciduous (L. *decidere*, to fall off) Plants that shed leaves at a certain time of year. (30)

Demographic transition A changing pattern from high birth and death rates to low birth and death rates within a population. (33)

Dendrite (Gr. *dendron*, a tree) The branching processes of a neuron, or nerve cell, that transmit the nerve impulse to the body of the cell. (26)

Dermal tissue system The protective and waterproofing tissues of a plant; consists of epidermis and bark. (29)

Desertification A process by which desert biomes expand; occurs due to the removal of grass from grasslands surrounding deserts by overgrazing and poor irrigation practices. (36)

Determination In embryonic development, the process by which cells are triggered to express certain traits when mature. (12)

Detritivore A consumer organism that obtains energy from dead or waste organic matter. (35)

Development The process by which an offspring increases in size and complexity from a zygote to an adult. (1, 12)

Developmental determinant A substance within the egg cytoplasm that activates the expression of certain genes in each cell, causing cells to develop differently. (12)

Diaphragm The muscle that separates the thoracic cavity from the abdominal cavity; used during breathing. (21)

Differentiation The process by which a cell becomes specialized for a particular function. (12).

Diffusion (L. *diffundere*, to pour out) The tendency of substances to move from areas of high concentration to areas of low concentration. (4)

Digestion The mechanical and chemical breakdown of food into small molecules that the organism can absorb and use. *Extracellular digestion* occurs in a special body organ or central body cavity in complex animals; *intracellular digestion* occurs within cells, in simple animals like sponges. (23)

Diploid A cell that contains two sets of chromosomes in its nucleus. (7)

Diversity Refers to the wide variation present within communities of living species. (1)

Dominance hierarchy A behavioral phenomenon in which members of a group are ranked from highest to lowest. Ranking is based on past success in aggressive encounters with the other members of the group; the individual winning all fights is the dominant member. Dominant individuals have greater access to food and mates. (37)

Dominant Said of an allele or the corresponding phenotypic trait that is expressed in the heterozygote. (8)

Dormancy (L. *dormire*, to sleep) A state of reduced physiological activity that occurs in seeds and buds, particularly over the winter. Normal physiological activities such as growth only resume if certain conditions—for example, increased temperature—are met. (30)

Double helix A structural form resembling a ladder twisted along its long axis; the form taken by DNA molecules. (9)

Down syndrome A syndrome resulting from abnormal meiosis and failure of a paired chromosome to separate, causing a child to inherit an extra chromosome 21. Symptoms include mental retardation, malformations of the heart, a special type of eye fold, short stature, stubby hands and feet, and abnormal palm prints. (7)

Echinoderm An invertebrate with a gut tube, a body cavity, a head end in some larvae and adults, no segmentation, an endoskeleton, and a unique water vascular system for locomotion; in the phylum Echinodermata; includes sea stars, sea urchins, and sea cucumbers. (17)

Ecology The scientific study of how organisms interact with their environment and with each other. (33)

Ecosystem A community of organisms interacting with a particular environment. (33)

Electron A subatomic particle with a negative charge. (2)

Electron transport chain The second stage in aerobic respiration in which electrons are passed down a chain of electron-removing (oxidizing) agents to produce a rich harvest of energy in the form of ATP. (5)

Element A pure substance that cannot be broken down into simpler substances by chemical means. (2)

Embryo An organism in the early stages of development. In humans this phase begins at conception and lasts for about two months, ending when all the structures of a human have been formed; then the embryo becomes a fetus. (12, 13)

Endocrine gland A ductless gland that secretes its hormones into extracellular spaces from where they diffuse into the bloodstream. In vertebrates these glands include the pituitary, adrenal, and thyroid glands. (25)

Endoderm The inner layer of the embryo, which gives rise to the inner linings of the gut and the organs that branch from it, including the lungs. (12)

Endoplasmic reticulum A system of membranous tubes, channels, and sacs that form compartments within the cytoplasm of eukaryotic cells; functions in protein and lipid synthesis and in the manufacture of proteins destined for secretion from the cell. (3)

Endoskeleton (Gr. *endos*, within + *skeltos*, hard) An internal supporting structure, such as the vertebrate's bony skeleton. (17)

Energy The ability to do work. (4)

Entropy A measure of disorder of randomness in a system. (4)

Enzyme A protein that facilitates chemical reactions by lowering the required activation energy but is unaltered itself in the process; enzymes are also termed *biological catalysts*. (2, 4)

Enzyme-substrate complex A complex that forms between the active site of an enzyme and an appropriate reactant, or substrate. (4)

Epidermis The outer layer of cells of an organism. (17)

Epithelial tissue One of the primary tissues that covers the body surface and lines the body cavities, ducts, and vessels. (19)

Esophagus The muscular tube leading from the pharynx to the stomach. (23)

Essential amino acid Any one of eight amino acids that the human body cannot manufacture and that must be taken in the diet. (23)

Estrogen A female steroid hormone produced in the ovary involved in the regulation of the menstrual cycle; causes the uterine lining to grow thick and more heavily supplied with blood. (13)

Estuary The area where a river meets an ocean. Estuaries are some of the earth's most productive ecosystems. (36)

Ethology (Gr. *ethos*, habit or custom + *logos*, discourse) The scientific study of animal behavior. (37)

Ethylene A plant hormone that stimulates ripening; exists as a gas at room temperature and is dispersed through the air between plants. (30)

Eukaryotic A cell whose DNA is enclosed in a nucleus and associated with proteins and that contains other membrane-bounded organelles. (3)

Evolution The changes in gene frequencies in a population over time. (1, 32)

Excretion The elimination of metabolic waste products from the body. In vertebrates the main excretory organs are the kidneys. (24)

Exocrine gland A gland with its own transporting duct that carries its secretions to a particular region of the body, such as the digestive and sweat glands; contrasts with endocrine glands. (25)

Exon The sequence of bases in a eukaryotic gene that end up in the mature messenger RNA. (10)

Exoskeleton (1) The thick cuticle of arthropods, made of chitin impregnated with calcium. (2) Any stiff sheet forming an external covering around the body. (17)

Exponential or logarithmic growth Growth of a population without any constraints. Hence the population will grow at an ever-increasing rate. (33)

Extracellular fluid Fluid inside the body but outside the cells; it surrounds the cells and is not contained within vessels. (4)

Fat An energy storage molecule that contains a glycerol bonded to three fatty acids. (2)

Fatty acid A molecule with a carboxyl group at one end and a long hydrocarbon at the other. Fatty acids are components of many lipids. (2)

Feces Semisolid, undigested waste materials that are stored in the colon until they are excreted. (23)

Feedback loop A system in which the result of a process influences the functioning of the process. (13, 19)

Fermentation Extraction of energy from carbohydrates without oxygen, generally producing lactic acid or ethyl alcohol as a byproduct. (5)

Fertilization The fusion of two haploid gamete nuclei (egg and sperm) to form a diploid zygote. (5)

Filtrate A fluid resembling plasma containing water, mineral ions, sugars, amino acids, and urea that diffuses into the kidney's Bowman's capsule from the capillaries. (24)

Filtration The process by which small molecules are removed from the blood in the kidney's nephron. (24)

First filial (F_1) generation The first generation in the line of descent arising from a cross between two individuals in the parental (P) generation. (8)

First law of thermodynamics The first energy law, which states that energy cannot be created or destroyed but can be converted from one form into another, such as potential to kinetic energy, or chemical bond energy to heat. (4)

Fitness Biological fitness; the ability to survive and produce offspring that can in turn reproduce. (32)

Fixed-action pattern In behavioral science, an action that continues to completion even if the stimulus causing the behavior is removed; for example, a bird that rolls an empty eggshell out of its nest will continue the rolling process to the edge of the nest even if the eggshell is removed during the procedure. (37)

Flagellum Long whiplike organelle protruding from the surface of the cell that either propels the cell and so acts as a locomotory device, or moves water and so becomes a feeding apparatus. (3)

Flatworm An invertebrate with bilateral symmetry and a head, three body layers, true organs and organ systems; in the phylum Platyhelminthes. (17)

Flower The reproductive structure of angiosperms; a complete flower includes sepals, petals, stamens, and carpels.

Fluid mosaic model Current model of cell membrane structure in which proteins are embedded in the lipid bilayer and can move freely through the fluid lipid structure. (3)

Follicle (L. *folliculus*, a small ball) In the mammalian ovary, an oocyte surrounded by follicular cells. When the oocyte has developed into a mature ovum (egg), it is released from the follicle by rupture of some of the follicular cells. (13)

Food chain The levels of feeding relationships among organisms in a community; who eats whom. (35)

Food web Complex, interconnected feeding relationships between all the species in a community or ecosystem. (35)

Founder effect In evolutionary biology, the principle that individuals founding a new colony carry only a fraction of the total gene pool present in the source population. (32)

Fruit A ripe, mature ovary containing seeds. (16)

Fungi The kingdom comprising multicellular heterotrophs, such as mushrooms or mold, that decompose other biological tissues. (14, 16)

Gallbladder The sac beneath the right lobe of the liver; stores bile. (23)

Gamete (Gr. wife) A specialized, haploid sex cell such as an ovum (egg) or a sperm. A male gamete (sperm) and a female gamete (ovum) fuse to give rise to a diploid zygote, which develops into a new individual. (7)

Gametogenesis The formation of haploid gametes, eggs, and sperm by the process of meiosis and maturation of the cells. (12)

Gametophyte The haploid gamete-producing phase in the life cycle of plants. (16)

Ganglion (pl. ganglia) A distinct clump of nerve cells that acts like a primitive brain, found in the head region of many invertebrates. In vertebrates the term is restricted to aggregations of nerve cell bodies located outside the central nervous system. (27)

Gastrulation The movement of cells in the embryo to form three concentric circular layers—the ectoderm, mesoderm, and endoderm—each layer in turn giving rise to specific body organs and tissues. (12)

Gene (Gr. *genos*, birth or race) The biological unit of inheritance that transmits hereditary information from parent to offspring and controls the appearance of a physical, behavioral, or biochemical trait. A gene is a specific discrete portion of the DNA molecule in a chromosome that encodes a protein, tRNA, or rRNA molecule. (1, 8)

Gene flow The incorporation into a population's gene pool of genes from one or more other populations through migration of individuals. (32)

Gene mutation A mutation that affects individual genes; can result when one base pair substitutes for another. (10)

Gene pool The sum of all alleles carried by the members of a population; the total genetic variability present in any population. (32)

Gene regulation The process that controls each cell's gene activity. (10)

Gene therapy The process of altering genes to combat an inherited defect; when techniques are perfected, genes will be inserted into somatic or germ cells to correct an inherited disease. (9)

Genetic code The specific sequence of three nucleotides in mRNA that encodes an individual amino acid in protein. For example, the code for methionine is AUG. (9,10)

Genetic map A drawing that locates the relative positions of genes on chromosomes. (8)

Genetic variation The phenomenon resulting from the reshuffling of chromosomes during meiosis, which results in offspring that are genetically distinct from their parents or siblings. (7)

Genome A haploid set of chromosomes with the genes they carry. (10)

Genotype The genetic makeup of an individual, including all genes on both sets of chromosomes in a diploid cell. (8)

Genus In taxonomy, a group of very similar organisms of common descent and sharing similar physical features. (14)

Germ cell A sperm cell or ovum (egg cell), the haploid gametes produced by individuals that fuse to form a new individual. (12)

Germination The sprouting, or initiation of development, of a new plant from a seed or spore. (29)

Gibberellin A specific hormone that regulates plant growth. (30)

Gill A specialized structure for gas exchange in water, usually a thin-walled projection from some part of the external body surface, having a rich capillary bed and a large surface area for gas exchange. (21)

Gill slits Pairs of openings that penetrate the back of the mouth from the inside to the outside, used for filtering food or obtaining oxygen. (18)

Gland A group of cells organized into a discrete secretory organ. (25)

Glial cell A nonneural cell of the nervous system that surrounds the neurons and provides them with protection and nutrients. (26)

Glucagon A hormone secreted by the pancreas that acts to lower the concentration of sugar in the blood. (25)

Glycogen A polysaccharide made up of branched chains of glucose; the principal source of stored energy in animals, found mainly in the liver and muscles. (2)

Glycolysis The initial breaking apart of a glucose molecule, resulting in the release of energy in the form of two ATP molecules. The series of reactions does not require the presence of oxygen to occur. (5)

Golgi complex In eukaryotic cells, a collection of flat sacs that processes proteins for export from the cell, or for shunting to different parts of the cell. (3)

Gonad An animal reproductive organ that generates gametes. (12)

Grassland A biome with seasonal extremes of wet and dry weather; contains many grass species and no trees, and borders desert regions. (36)

Greenhouse effect The result of a buildup of carbon dioxide in the atmosphere (for example, through the burning of fossil fuels) in which carbon dioxide traps solar heat beneath the atmospheric layers, which leads to increased global temperatures and changes in climatic patterns. (6, 36)

Ground tissue system Plant tissues that provide support and store starch; consists of parenchyma, colenchyma, and sclerenchyma. (29)

Growth factor A protein produced in certain cells that diffuses throughout the body and speeds up or slows down the cell cycle in other cells. (7)

Gymnosperm (Gr. *Gymnos*, naked + *sperma*, seed) Conifers and their allies; primitive seed plants whose seeds are not enclosed in an ovary; the group has many fossil representatives. (16)

Habitat The area within a species' range where the organism can actually live. (33, 34)

Habituation A progressive decrease in the strength of a behavioral response to a constantly applied weak stimulus, when the stimulus has no negative effects. (26)

Haploid Describes cell that contains one copy of a chromosome set. (7)

Hardy-Weinberg principle In population genetics, a rule developed independently by the English mathematician Hardy and the German biologist Weinberg. The rule states that in the absence of any outside forces, the frequency of each allele in a population will not change over generations. (32)

Heat The random commotion of atoms and molecules. (4)

Heredity The science of inheritance and variation. (8)

Heterotroph (Gr. *heteros*, different + *trophos*, feeder) An organism, such as a human and most other animals, that must take in preformed nutrients. (5, 14)

Heterozygote (Gr. *heteros*, different + *zygote*, pair) An individual having two different alleles for a specific trait. (8)

Hierarchy of life The levels of biological order and organization represented by all living organisms, from the microscopic level to the level of large groups in their environments. (1)

Hinged jaw A jaw that opens wide to allow consumption of large food chunks; derived from gill support bones in the early fishes. (18)

Hippocampus A brain structure located within the cerebrum, involved in the formation of long-term memory. (27)

Histocompatibility proteins (Gr. *histos*, tissue) Tissue-typing proteins in the cells of every individual, capable of distinguishing molecules that are part of the individual's body from invading molecules; each individual has a unique array of the proteins, which work in association with T-cell antigen receptors. (22)

Homeostasis The maintenance of a constant internal milieu despite fluctuations in the external environment. (19)

Homo sapiens The genus and species designating modern humans which first appeared between 30,000 and 100,000 years ago. (18)

Homozygote Having two identical alleles for a specific trait. (8)

Hormone A chemical messenger, usually a peptide or a steroid, produced in one part of the body and transported in the blood to another region, where it exerts an effect. (25)

Hybrid An offspring resulting from the mating between individuals of two different genetic makeups or two different species. (8, 32)

Hydrogen bond A type of weak molecular bond in which a partially negatively charged atom (oxygen or nitrogen) bonds with the partial positive charge on a hydrogen atom when the hydrogen atom is already participating in a covalent bond. (2)

Hypha (pl. hyphae) A long, thin filament of a fungal cell. (16)

Hypothalamus A collection of nerve cells at the base of the brain just below the cerebral hemispheres, responsible for the integration and correlation of many neural and endocrine functions. (25)

Hypothesis A potential answer to a question or solution to a problem. All hypotheses are provisional working ideas that are accepted for the present but may be rejected in the future if the evidence collected is not consistent with the idea. (1)

Immune system The network of cells, tissues, and organs that defends the body against invaders. (22)

Immunological memory The phenomenon by which memory cells form in response to exposure to antigens, thus helping an organism resist disease; the basis for vaccinations and booster shots. (22)

Imprinting The recognition, response, and attachment of young animals to a particular adult or object; in nature the attachment of young to their parents. Animals may imprint on other objects or animals if they are exposed to that animal or object at the sensitive stage; for example, young birds may imprint on humans. (37)

Induction The process by which one set of cells causes another group of cells to develop in a specific way. (12)

Inner cell mass The several layers of cells within the blastocyst that become the embryo. (12)

Insect A small arthropod with specializations for sensing and rapid movement. An insect's head bears one pair of antennae; the thorax has three pairs of legs and usually one or two pairs of wings. Insects are the most successful and largest class of animals on earth. (17)

Insight learning The process of acquiring knowledge through reasoning. (37)

Instinct Any behavioral trait that is innate, that is, not learned. Instinctive behavior is usually not changed by experience during an animal's lifetime. (37)

Insulin A hormone made in the pancreas that helps to regulate the breakdown of sugar. (25)

Interneuron A nerve cell that relays messages between other nerve cells. (26)

Interphase The growth period between cell divisions in a cell. During this period, the cell conducts its normal activities and DNA replication takes place in preparation for the next cell division. (7)

Interspecific competition Competition for resources—e.g., food or space—between individuals of different species. (34)

Intertidal zone The region extending along the coastline between high and low tide level. (36)

Intracellular fluid The fluid contained within a cell's outer membrane; it is rich in ions and dissolved materials. (4)

Introns Sequences of DNA in a eukaryotic gene that are excised from RNA and so are not translated into proteins. (10)

Invertebrate An animal that does not have a backbone. (17)

In vitro fertilization The fertilization of an egg by a sperm under laboratory conditions, literally, "in glass." (13)

Ion An atom that has gained or lost an electron, thereby attaining a positive or negative electrical charge. (2)

Isotope An alternative form of an element having the same atomic number but a different atomic mass due to the different number of neutrons present in the nucleus. Some isotopes are unstable and emit radiation. (2)

Joint A point of contact between two bones. (28)

J-shaped curve The shape of a graph representing a population that has undergone exponential growth because there were no factors limiting its reproductive potential. (33)

Kidney The main excretory organ of the vertebrate body; filters blood to remove nitrogenous wastes and regulates the balance of water and solutes in blood plasma. (24)

Kin selection Form of natural selection in which an individual increases its fitness by helping relatives, who share its genes, to reproduce. (37)

Kingdom The broadest taxonomic designation of living organisms; most biologists recognize five kingdoms: Monera, Protista, Fungi, Plantae, and Animalia. (1, 14)

Krebs cycle The first stage of aerobic respiration, in which a two-carbon fragment is completely broken down into carbon dioxide. (5)

Lateral meristem Area of actively growing cells in the stems and roots of plants that cause these areas to thicken. (29)

Learning An adaptive and enduring change in an individual's behavior based on personal experiences in the environment. (37)

Legume A member of the pea family that often contains nitrogen-fixing bacteria in root nodules; includes beans, alfalfa, clover, and lupine. (31)

Leukocyte White blood cell; important in the body's defense of immune system. (21)

Lichen An association between a fungus and an alga that live symbiotically together. (16)

Life cycle The process by which a fertilized egg develops into an adult organism that produces eggs and/or sperm. These unite with other eggs and sperm to form new fertilized eggs, which grow to new adults, which produce new fertilized eggs, thereby continuing the cycle. (7)

Life history strategy The way an organism allocates energy to growth, survival, or reproduction. (33)

Ligament A band of connective tissue that links bone to bone. (28)

Light-dependent (light-trapping) reactions The first stage in photosynthesis, driven by light energy. Electrons that trap the sun's energy pass the energy to high-energy carriers such as ATP or NADPH, where it is stored in chemical bonds. (6)

Light-independent (sugar-building) reactions The second stage of photosynthesis, also called the Calvin-Benson cycle, which does not require light. During the six steps of the cycle, carbon is fixed and carbohydrates are formed. (6).

Lipid A biological molecule that contains long-chain hydrocarbons and does not dissolve in water; includes fatty acids, fats, and waxes. (2)

Lipid bilayer The two-layered sheet of phospholipids that makes up the plasma membrane of cells and is mainly responsible for the membrane's barrier function. (3)

Liver Large, lobed gland in the body, destroys blood cells, stores glycogen, disperses glucose to the bloodstream, and produces bile. (23)

Loop of Henle U-shaped region of the vertebrate kidney tubules chiefly responsible for reabsorption of water and salts from the filtrate by diffusion. (24)

Lung The principal air-breathing organ of most land animals. (18)

Lymph A yellowish fluid derived from tissue fluid and transported in special vessels to the blood; contains proteins, dead cells, and sometimes bacteria and viruses. (20)

Lymphatic system In vertebrates, a collective term for a system of vessels carrying lymph, fluid that has been forced out of the capillaries, back to the bloodstream. (20)

Lymph node A bean-shaped organ located along lymphatic vessels that filters out dead cells and debris; also harbors white blood cells that attack bacteria and other foreign matter. (20)

Lymphocyte A white blood cell formed in bone marrow and lymph tissue; is active in the immune responses that protect the body against infectious diseases. There are two main classes of lymphocytes: B lymphocytes, involved in antibody formation, and T lymphocytes, involved in cell-mediated immunity. (22)

Lysosome A spherical vesicle within the cell that contains powerful digestive enzymes. (3)

Macroevolution A term for large evolutionary changes in the phenotype or appearance of organisms. These changes are usually great enough to put the descendants into another family order, phylum, or other category. (32)

Macronutrient (Gr. *makros*, large + L. *nutrire*, to nourish) An inorganic chemical element such as nitrogen, potassium, calcium, magnesium, and sulfur, required in large amounts for plant growth. (31)

Macrophage A white cell in the immune system that engulfs invaders and consumes debris when the lymphocytes have completed their task. (22)

Malignant Describes cancerous cells or a mass of cells (tumor) that divide uncontrollably. (12)

Mammal A warm-blooded vertebrate that has a four-chambered heart, body hair or fur, and a placenta and produces milk to suckle its young. (18)

Marsupial A mammal having a pouch in which it carries its young, which are born in a small and underdeveloped state. Found extensively in Australia, with few representatives in America; includes kangaroos, opossums, and koala bears. (18)

Meiosis The type of cell division that occurs during gamete formation; the diploid parent cell divides twice, giving rise to four cells, each of which is haploid, that is, has half the number of chromosome as the rest of the cells in the body. (7)

Menstrual cycle The cyclic reproductive capacities of human and other primate females. Generally a 28-day cycle characterized by a gradual thickening of the lining of the uterus (womb) and the maturation of a follicle from which a mature egg is released. If the egg is not fertilized, the inner lining of the uterus is shed, a process known as menstruation, and the cycle starts again. In humans the menstrual cycle normally starts between the ages of 12 and 14 years. (13)

Meristem Plant tissue containing cells that can continually divide throughout the plant's life. (29)

Messenger RNA (mRNA) An RNA molecule that carries the information to make a specific polypeptide. mRNA is transcribed from structural genes and is translated into protein on the ribosomes. (10)

Metabolism (Gr. *metabole*, to change) The sum of all the chemical reactions that take place within the body; includes photosynthesis, respiration, digestion, and the synthesis of organic molecules. (1, 4)

Metamorphosis (Gr. *meta*, after + *morphe*, form + *osis*, state of) Process in which there is a marked change in morphology during postembryonic development; in insects the complete change in body form that takes place inside a cocoon; the individual enters the cocoon as a larva, such as a caterpillar, and emerges as an adult, a butterfly. Also refers to the change from a tadpole to a frog in amphibians. (25)

Metastasize The process by which malignant cells spread throughout the body. (12)

Microevolution A term used to describe small evolutionary changes within a species. These changes only slightly alter the phenotype or appearance of individuals and do not place the descendants in different taxonomic categories. (32)

Micronutrient (Gr. *mikros*, small + L. *nutrire*, to nourish) A nutrient required in small amounts for plant growth; includes the minerals iron, copper, zinc, and chlorine. (31)

Mimicry The phenomenon in which one organism evolves to resemble another; often a nonpoisonous species evolves to resemble a poisonous one and so avoids being eaten. (34)

Mineral A specific chemical element required by the body in small amounts. (23)

Mitochondrion (pl. mitochondria) Organelle in the cell that provides energy for the cell's activities. Mitochondria are the sites of oxidative (aerobic) respiration, and almost all of the ATP of nonphotosynthetic eukaryotic cells is produced in the mitochondria. (3)

Mitosis Somatic cell division; the process of nuclear division in which replicated chromosomes separate to form two genetically identical daughter nuclei. Mitosis is usually accompanied by cytokinesis and results in two daughter cells, each having the same chromosome number and genetic composition as the parent cell. (7)

Molecule Two or more identical or different atoms linked by covalent, hydrogen, or other bonds. (2)

Mollusk An invertebrate in the phylum Mollusca with bilateral symmetry, a head and gut tube, a true body cavity, gills, and an open circulatory system; includes snails, clams, octopuses, squids, and slugs. (17)

Monera The kingdom comprising single-celled prokaryotes such as bacteria. (1, 14, 15)

Monomer A single part or subunit of a polymer. (2)

Monotreme An egg-laying mammal that also has many primitive or reptilian features. The only living forms are the spiny anteater and the duck-billed platypus. (18)

Morphogenesis The growth, shaping, and arrangement of the organs and tissues of the body. (12)

Motor cortex The region of the cerebral cortex responsible for controlling muscular movements; the left side of the cortex controls muscles on the right side of the body and vice versa. (27)

Motor neuron A neuron that sends messages from the brain or spinal cord to muscles or secretory glands. (26)

Movement Self-propelled motion exhibited by living organisms. (1)

Mutualism A symbiotic relationship in which both species benefit. (34)

Mutation Any heritable change in the base sequence of an organism's DNA; can occur in a single base pair or in large portion of a chromosome. (9)

Myelin sheath Insulating sheath around neurons that speeds impulse travel; composed of the lipid membranes of Schwann cells. (26)

Myofibril A fibril composed of actin and myosin found in the cytoplasm of muscle cells that contracts the muscle. (28)

Nasal cavity One of the two chambers that open into the throat, where entering air is warmed and humidified. (21)

Natural killer cell A type of lymphocyte that helps to guard the body against cancer cells; also probably attacks virus-infected cells. (22)

Natural selection The evolutionary process by which the increased survival and reproduction of the best adapted individuals in a population results in an increase in some allele frequencies and a decrease in others. (1, 32)

Navigation The ability to move from one map point to another. (37)

Nephron The functional unit of the vertebrate kidney; each of the million nephrons in a kidney consists of a glomerulus enclosed by a Bowman's capsule and a long attached tubule. The nephron removes wastes from the blood. (24)

Nerve A group of axons and accompanying cells from many different neurons running in parallel bundles and held together by connective tissue. (26)

Nerve cord A bundle of nerve tissue that runs longitudinally down an animal's body; first seen in flatworms. Appears as paired, ventral cords in invertebrate worms and insects, and as dorsal, hollow tubes of nerve tissue in chordates, becoming the spinal cord in vertebrates. (17, 18)

Nerve impulse A change in ion permeabilities in a neuron's membrane that sweeps down the axon to its terminal, where it excites other cells. (26)

Nerve net The simplest form of nervous organization in animals; consists of nerve cells and fibers interlaced throughout an animals' body. Found in many invertebrates, it permits a diffuse response to stimuli. (27)

Nervous system The integrative network in an animal that serves to coordinate and control all the physiological systems in its body. (26)

Neurons The nerve cells that transmit messages throughout the body; includes the cell body, dendrites, and axons. (26)

Neurotransmitter A chemical that transmits a nerve impulse across a synapse. (26)

Neurulation Developmental stage in an embryo during which the cells differentiate into the brain and spinal cord. (12)

Neutralist An individual who believes that most mutations are "neutral" (neither advantageous nor disadvantageous to an organism's fitness) and that most alleles in natural populations seem to be equivalent; in contrast to a selectionist. (32)

Neutron A subatomic particle with no electrical charge found in the nucleus of an atom. (2)

Niche The manner in which an organism obtains its resources; analogous to its role or job in the community. (34)

Nitrogen cycle The process by which nitrogen circulates from its gaseous phase in the atmosphere to biologically useful compounds in the soil and on plant roots and back to the atmosphere; various species of bacteria fix atmospheric nitrogen into the biologically useful compounds, while other bacterial species release the fixed nitrogen back into the atmosphere. (35)

Nitrogen fixation The process by which certain bacteria (or natural forces, such as lightning) convert gaseous nitrogen from the air to ammonia or nitrate ions, which can then be incorporated into plant compounds. (31, 35)

Nodule A swelling on the roots of legumes that contain nitrogen-fixing bacteria. (31)

Notochord A rod of mesodermal cells in the embryo that gives rise to the backbone; all chordates have a notochord at some point in their life cycle. (12, 18)

Nuclear envelope The bilipid membrane that encloses the nucleus of the cell in eukaryotes. (3)

Nuclear pore A hole that perforates the nuclear membrane. (3)

Nuclear winter A condition resulting from the burning and fires that would be generated during a nuclear war. Clouds of smoke and ash would block out sunlight, resulting in long periods of cold and darkness during which photosynthesis, and with it most life on earth, would be drastically reduced or cease. (36)

Nucleic acid A polymer of nucleotides, for example, DNA and RNA. (2)

Nucleoid The unbounded region of DNA in a prokaryotic cell. (3)

Nucleolus A dark-staining region within the nucleus where ribosomal RNA is formed. (3)

Nucleotide The basic chemical unit of DNA and RNA, consisting of one of four nitrogenous bases linked to a sugar (a deoxyribose or a ribose), which is in turn linked to a phosphate. (2)

Nucleus (atomic) The central core of an atom, containing neutrons and protons. (2)

Nucleus (cell) The central region of eukaryotic cells, which is bounded by a membrane and contains the cell's DNA. (3)

Nutrient Any substance that an organism must obtain from the environment in order to survive and reproduce. (23)

Nutrition The science concerned with determining the amounts and kinds of nutrients needed by the body. (23)

Oil A type of lipid or fat, often in the liquid state; examples are corn, coconut, and olive oils. (2)

Oncogene A cancer gene; often considered a mutant form of a growth-regulating gene that is inappropriately turned on, causing unrestrained cell replication. (11)

Open growth The continuous growth pattern exhibited by a plant throughout its life. (29)

Opposable thumb A first digit that can touch each of the others. (18)

Optic nerve The nerve leading from the eye to the brain. (27)

Order The structural and behavioral organization exhibited by living organisms; for example, the geometric patterns in the facets of a fruit fly's eye or the levels of organization in the hierarchy of life. (1)

Organ A body structure composed of two or more tissues that functions as a unit within the body. (17)

Organism An individual that can independently carry out all life functions. (1, 19)

Organogenesis The formation of organs during embryonic development. (12)

Organ system Two or more interrelated organs that serve a common function. (1, 19)

Orientation The sense of direction-finding; thought to be inborn in some animals. (37)

Osmosis The movement of water through a semipermeable membrane from an area of high concentration to one of low concentration. (4)

Osmoregulation Maintenance of a constant internal salt and water concentration in an organism. (24)

Ovary Egg-producing organ. (13)

Ovulation The release of a mature egg from the ovary by rupture of the follicle. (13)

Pancreas The organ located near the junction of the stomach and small intestine that makes and secretes digestive enzymes, bicarbonate ions, insulin, and glucagon. (23)

Parathyroid gland A set of four small endocrine glands located on the thyroid gland that secrete parathyroid hormone, which controls blood calcium levels. (25)

Parental (P) generation The first generation of individuals used in genetic experiments to test inherited traits; matings between P individuals result in the first filial, or F_1, generation. (8)

Passive transport Movement of substances into or out of a cell by the substances moving from areas of high concentration to areas of low concentration. (4)

Pedigree An ordered diagram of a family's relevant genetic features. (11)

Penis In reptiles and mammals, the male organ of copulation. (13)

Perennial (L. *per*, through + *annus*, year) Plants that live for many years, blooming and setting seed several times before dying. (29)

Phenotype The physical appearance of an organism controlled by its genes interacting with the environment. (8)

Phloem Tubular tissues in plants that transports dissolved sugars and proteins from cells that make or store the nutrients (as in the leaves or roots) to cells that use the nutrients (as in the shoots). (29)

Phospholipid A cellular lipid composed of a phosphate functional group and two fatty acid chains attached to a glycerol molecule; a main component of cell membranes. (2)

Phosphorus cycle The process by which phosphorus circulates from rocks or soil into organisms and back to the soil; can occur locally over short periods or globally over vastly longer periods. (35)

Photon A vibrating particle of light radiation that contains a specific quantity of energy. (4)

Photoperiod Length of light and dark periods each day. (30)

Photoreceptor cell A light-sensitive cell. (27)

Photorespiration A process that occurs in plants when carbon dioxide levels are depleted but oxygen continues to accumulate and the enzyme RuBP carboxylase (Rubisco) fixes oxygen instead of carbon dioxide. (6)

Photosynthesis The metabolic process in which plants trap solar energy and convert it to chemical energy (ATP and NADPH), which in turn is used in the manufacture of sugars from carbon dioxide and water. (6)

Photosystem A special unit of chlorophyll and other pigment molecules associated with proteins and electron acceptors in the thylakoid membrane of a plant cell's chloroplast, where the conversion of light energy to chemical energy takes place. (6)

pH scale A logarithmic scale that measures the hydrogen ion concentration; acids range from pH 1 to 7, water is neutral at a pH of 7, and bases range from pH 7 to 14. (2)

Phyletic gradualism Evolutionary theory that holds that morphological changes occur gradually during evolution and are not always associated with speciation. Distinct from punctuated equilibrium theory. (32)

Physiology The study of the functioning of organs in the body. (19)

Phytochrome Light-sensitive pigment found in plants that absorbs in the red or far-red wavelengths; associated with a number of timing processes such as flowering, dormancy, leaf formation, and seed germination. (30)

Phytoplankton Photosynthetic microorganisms that live near the surface of marine and fresh water. (15)

Pineal gland Endocrine gland located in the roof of the forebrain; involved in measuring day length and controlling reproductive cycles, the onset of puberty, and moods. (25)

Pituitary gland Endocrine gland located beneath the brain; the anterior lobe secretes tropic hormones, growth hormones, and prolactin; the posterior lobe stores and releases oxytocin and ADH. Much of the functioning of the pituitary is under control of the hypothalamus. (25)

Placenta The spongy organ rich in blood vessels by which the developing embryo receives nourishment from the mother. (12, 13)

Plantae The kingdom comprising muticellular, photosynthesizing organisms such as algae, mosses, ferns, and flowering plants. (14)

Plasma The liquid portion of blood or lymph. (20)

Plasma membrane The membrane that surrounds all cells, regulating entry and exit of substances; consists of a single lipid bilayer. (3)

Plasmid Circular piece of DNA that can exist either inside or outside bacterial cells but must reproduce inside a bacterial cell. Plasmids are used extensively in genetic engineering as carriers of foreign genes. (9)

Plastid An organelle found in plants and some protozoa that harvests solar energy and produces and stores carbohydrates or pigments. (3)

Platelets Disk-shaped cell fragments in the blood that are important in blood clotting. (20)

Pleural sac Fluid-filled sac that encases the lungs. (21)

Pollen The male gametophyte of seed plants contained in grains on the anthers. (16)

Pollinator An organism that carries pollen from one plant to another; examples are birds and bees. (16)

Polymer A large molecule made of repeated sequences of similar or identical subunits called monomers. DNA, proteins, and starch are examples of polymers. (2)

Polypeptide (Gr. *polys*, many + *peptin*, to digest) Amino acids joined together by peptide bonds to form long chains. (2)

Polysaccharide A type of carbohydrate made up of many monosaccharides linked together. Glycogen and cellulose are examples of polysaccharides. (2)

Population A group of individuals of the same species living in a particular area. (33)

Population bottleneck A situation arising when only a small number of individuals of a population survive to reproduce; therefore only a small percentage of the original gene pool remains. (32)

Population genetics A set of principles clarifying what happens at the genetic level in populations; derived from the study of how the frequencies of certain alleles in gene pools change over time. (32)

Postsynaptic cell The nerve cell that receives the message crossing a synapse. (26)

Predation The act of a predator species capturing and eating a prey species. (34)

Predator An organism, usually an animal, that obtains its food by eating other living organisms. (34)

Prediction An educated guess about what one will observe in a specific situation if a proposed hypothesis is correct. (1)

Presynaptic cell The neuron that sends a message down its axon and across a synapse to another cell. (26)

Prey A living organism that is eaten by another (predator) organism. (34)

Primary growth Plant growth arising from the apical meristem. (29)

Primate A mammal that has stereoscopic vision, a large brain, an upright gait, and modified teeth and that exhibits extended infant care; includes lemurs, tarsiers, monkeys, apes, and humans. (18)

Principle of independent assortment The second law of heredity, which states that genes on different chromosomes assort independently of each other during the formation of eggs and sperm. (8)

Principle of segregation Mendel's first law of heredity, which states that sexually reproducing diploid organisms have two alleles for each gene and that during gamete formation these two alleles segregate from each other so that the resulting gametes have only one allele of each gene. (8)

Producers Organisms, usually plants, that can produce all their nutritional needs from nonbiological substances. (35)

Product A substance that forms as the result of a chemical reaction. (4)

Progesterone A female steroid hormone produced in the ovaries and involved in the regulation of the menstrual cycle; helps prepare the uterus for implantation of the ovum. (13)

Prokaryotic cell A cell in which the DNA is not bound by a nucleus, such as a bacterium or blue-green algae. Prokaryotic cells are more primitive than eukaryotic cells. (3)

Propagation The process by which an action potential travels from one patch of nerve-cell membrane to an adjacent patch, and then to the next, passing along a nerve impulse. (26)

Prosimian A suborder of the primate order; includes lemurs and tarsiers. (18)

Prostate gland A male reproductive gland that secretes a milky alkaline fluid that enlarges the volume of sperm and seminal fluid and lowers the pH of the mixture. (13)

Protein One of the most fundamental molecules of living organisms, composed of one or more chains of amino acids. A protein typically contains over 100 amino acids and may be composed of more that one peptide chain. (2)

Protein synthesis A two-step process that takes place in cells: first, a strand of DNA is copied (transcribed) to a new strand of messenger RNA; then the RNA message is translated to a specific sequence of amino acids in a polypeptide chain. (10)

Protista The kingdom comprising single-celled eukaryotes such as *Euglena*. (14, 15)

Proton A subatomic particle having a positive charge, found in the nucleus of an atom. (2)

Protozoa Single-celled, animallike organisms in the kingdom Protista that consume other cells or food particles. Examples are *Didinidum* and the sporozoan protist that causes malaria. (15)

Pulmonary artery The Y-shaped artery carrying deoxygenated blood from the heart's right ventricle to the lungs. (20)

Pulmonary circulation Part of the circulatory system in which deoxygenated blood arriving back to the heart from the body is pumped to the lungs, where the blood receives oxygen. Reoxygenated blood is then pumped back to the heart. (20)

Punctuated equilibrium Evolutionary theory that holds that morphological changes occur rapidly in time; during speciation these changes occur in small populations with the resulting new species being distinct from the ancestral form. After speciation, species retain much the same form until extinction; distinct from the phyletic gradualism theory. (32)

Pure breeding Individuals that, when self-fertilized, give rise to progeny identical to themselves. (8)

Quantitative trait A trait controlled by several interacting genes rather than by a single gene's pair of alleles. Examples are traits that vary in a measurable quantity, such as an individual's height or the amount of milk produced by a cow in liters. (8)

Radial body plan A circular body plan having a central axis from which structures radiate outward like the spokes of a wheel. (17)

Radiate The process by which a single species diverges into a number of new species as a result of environmental pressures.

Reabsorption The process by which useful molecules from the urine are absorbed back into the blood from the kidney's nephrons. (24)

Reactant A starting substance for a chemical reaction. (4)

Receptor A cell protein of a specific shape that binds to a particular chemical. (25)

Recessive An allele or corresponding phenotypic trait that is only expressed in the homozygote. (8)

Recombinant DNA technology The techniques involved in excising DNA from one genome and inserting it into a foreign genome. (9)

Recombination The reshuffling of normal and mutated genes on chromosome pairs, which creates new combinations of alleles; occurs mainly during meiosis. (7, 9)

Rectum The short tube located at the end of the colon. (23)

Red blood cell An indented disk-shaped cell that contains hemoglobin, circulates in the blood, and transports oxygen to the tissues. (4, 20)

Reflex arc An automatic reaction, involving only a few neurons and requiring no input from the brain, in which a motor response quickly follows a sensory stimulus. (26)

Regulator cell A cell that detects perturbations in the environment and emits a hormone in response. (25)

Reinforcer A reward or punishment used to reinforce learned behavior in trial-and-error learning. (37)

Releaser A specific environmental signal that triggers a fixed-action behavior; can be any information picked up by the sense organs. (37)

Reproductive isolating mechanism (RIM) Any structural, behavioral, or biochemical feature that prevents individuals of a species from successfully breeding with individuals of another species. (32)

Reptile A cold-blooded, scaly, lung-breathing vertebrate that lays eggs that usually have a shell; includes crocodiles, lizards, and tortoises. (18)

Resource partitioning The process of dividing resources that enables species with similar requirements to use the same resources in different areas, at different times, or in different ways. (34)

Respiration (L. *respirare*, to breathe) The exchange of oxygen and carbon dioxide between cells and the environment. The oxidative breakdown and release of energy from fuel molecules. (21)

Responsiveness A living organism's ability to respond to cues in its environment, such as the presence of food or enemies, and then react in ways that help maintain its body intact. (1)

Restriction fragment length polymorphism (RFLP) Genetic variation between individuals in the lengths of DNA that can occur when a piece of DNA is cut with a restriction enzyme. (11)

Retina (L. a small net) Multilayered region that lines the back of the eyeball and contains light-sensitive cells. (27)

Ribosomal RNA (rRNA) RNA molecules that are structural components of the ribosomes and involved in protein synthesis. (10)

Ribosome A structure in the cells that provides a site for protein synthesis and enzymes that link amino acids to each other in polypeptide formation. Ribosomes may lie freely in the cell or attach to the membranes of the endoplasmic reticulum. (3)

Rod A photoreceptor cell in an animal's eye that is sensitive to low levels of light but unable to distinguish color. (27)

Root Branching structure of a plant that grows downward into the soil; roots anchor the plant and absorb and transport water and mineral nutrients. (29)

Roundworm An invertebrate in the phylum Nematoda with bilateral symmetry, a head and gut tube, a false body cavity, and a hydroskeleton; extremely common in soils and as parasites on other animals and plants. (17)

Saccharide The sugar subunits of carbohydrates. (2)

Saliva A clear substance made of mucus and water and containing enzymes that break down carbohydrates; secreted by three glands in the mouth. (23)

Savanna A biome with year-long warmth and an extended dry season, containing mainly grasses and stunted, widely spaced trees; located in Africa, South America, Australia, and Southeast Asia. (36)

Scientific method The process by which scientific research is carried out; involves (1) asking a question or identifying a problem to be solved, based on observations of the natural world; (2) proposing a hypothesis; (3) making a prediction of what will be observed in a specific situation if the hypothesis is correct; and (4) testing the prediction by performing an experiment. (1)

Secondary growth Enlarged diameter of a stem or root resulting from cell divisions in the lateral meristem. (29)

Second filial (F_2) generation Offspring arising from self-fertilization or a cross between individuals of the first filial (F_1) generation. (8)

Second law of thermodynamics The second energy law, which states that because energy conversions from one form to another are not 100 percent efficient, all systems tend toward greater states of disorder. (4)

Secretion The process by which ions and some drugs are removed from the blood and secreted into the urine in the kidney's nephrons; makes drug testing possible. (24)

Seed Product of a fertilized ovule of a seed plant, generally consisting of an embryo with its food reserves enclosed in a protective coat. (16)

Segmented worm An invertebrate in the phylum Annelida with bilateral symmetry, a head, a gut tube, a body cavity, and segmentation; includes earthworms and leeches. (17)

Selection The process by which certain genes and/or traits in organisms increase or decrease in frequency within a population as a result of environmental pressures; concept can be extended to account for evolution of early cells from precursor molecules. (14, 32)

Selectionist An individual who believes that natural selection is the primary force in evolution. (32)

Self-tolerance The lack of immune-system responsiveness to one's own cells and molecules. (22)

Semen Sperm and its surrounding seminal fluid. (13)

Semicircular canal A fluid-filled tube that functions as an organ of balance and motion detection. (27)

Semipermeable membrane A membrane that allows water molecules to pass freely but prevents the passage of large molecules across its surface. (4)

Senescence A condition characterized by a profound decline in cell and organ function. (13, 30)

Sense organ A group of specialized cells that receive a stimulus and convert it to a kind of energy that can trigger a nerve impulse. (27)

Sensitization A strong behavioral response to a mild stimulus following a strong, harmful stimulus; the opposite of *habituation*. (26)

Sensory cortex The region of the cerebral cortex that registers and integrates sensations from various parts of the body; the left side of the cortex receives sensations from the right side of the body and vice versa. (27)

Sensory neuron A nerve cell that receives information from the external or internal environment and transmits this information to the brain or spinal cord. (26)

Sex chromosome One of the pair of chromosomes involved in sex determination, such as the X and Y chromosomes in humans. (8)

Sexual reproduction The process by which two gametes of opposite mating types (usually male and female) fuse to form a new, unique individual. (7)

Sexual selection A type of selection in which individuals select a mate based on physical or behavioral characteristics without regard for the value of that characteristic in the struggle for survival. (32, 37)

Shoot The above-ground part of a plant; includes the stem, branches, leaves, flowers, and fruit. (29)

Skeletal muscle Muscle consisting of elongate, striated muscle cells; voluntary muscle. (28)

Skeleton The rigid body support to which muscles attach and apply force; can be external, as in insects, or internal, as in mammals. (17, 18, 28)

Skull Bony plates that cover and protect the brain. (18)

Smooth muscle Muscle consisting of spindle-shaped, unstriated muscle cells; involuntary muscle found in the digestive and circulatory systems. (28)

Social insects Insects that live in large colonies with labor divided among subgroups that differ in appearance and behavior; examples are termites, ants, wasps, and bees. (17)

Sociobiology The study of social behavior. (37)

Soil A mixture of organic and inorganic particles consisting of sand, silt, clay, and humus. (31)

Solute A substance that has been dissolved in a solvent. (2)

Solvent A substance capable of dissolving other molecules. (2)

Soma The main cell body of a neuron. (26)

Somatic cell (Gr. *soma*, body) A cell in an animal that is not a germ cell. (12)

Speciation Emergence of a new species; thought to occur mainly as a result of populations becoming geographically isolated from each other and evolving in different directions. (32)

Species A group of organisms whose members have the same structural traits and who can interbreed with each other. (14)

Species richness The diversity or total number of species found in a community. (34)

Specific immune response The process by which white blood cells incapacitate specific invading organisms, such as bacteria, while leaving nontargeted cells of the body untouched. (22)

Spinal cord A tube of nerve tissue that runs the length of a chordate animal's back just above the notochord and serves to integrate the body's movements and sensations. (18)

Sponge Invertebrate animal in the phylum Porifera with an asymmetrical saclike body that filters and consumes food from the surrounding water; these animals have no evolutionary descendants. (17)

Spore A reproductive cell that divides mitotically to produce a new individual. In prokaryotes, a resistant cell capable of surviving harsh conditions and of germinating when conditions are once again favorable. (15, 16)

Sporophyte (Gr. *spora*, seed + *phyton*, plant) The diploid, spore-producing stage of many plants. (16)

S-shaped curve The shape of a graph representing a population whose growth has leveled off at its carrying capacity because of competition for limited resources or other factors. (33)

Starch A type of polysaccharide that serves as an energy source for plants and animals. (2)

Stem The erect part of a plant that supports the plant vertically and acts as a central corridor for the transport of water, minerals, sugar, and other substances. (16)

Stem cell A generative cell that retains the ability to divide throughout life. (12)

Stereoscopic vision Vision based on overlapping visual fields, distinguished by good depth perception, which allows animals to discriminate distances. (18)

Stoma (pl. stomata) A tiny opening on a leaf's surface through which carbon dioxide is taken in and water and oxygen are released; stomata are most numerous on the undersides of leaves. (29)

Stomach The elastic, J-shaped bag located at the end of the esophagus; stores food for later processing and mixes food with digestive enzymes. (23)

Substrate A specific reactant that fits into the active site of an enzyme. (4)

Succession A progression of communities in a particular habitat, each one replacing the other until a persistent, stable climax community is established. (34)

Synapse The region of communication between two neurons. (26)

Systemic circulation Part of the circulatory system that takes oxygenated blood to the body and deoxygenated blood back to the heart. (20)

Tail An appendage extending beyond the anus supported by a notochord or vertebral column; found in all chordates at some point during their development. (18)

Target cell A cell that receives a hormone and responds by carrying out a specific cellular activity that helps the organism adjust to the original perturbation. (25)

Taxonomy (Gr. *taxix*, arrangement + *nomos*, law) The science of classifying organisms into different categories. (14)

T cell (T lymphocyte) A cell involved in the body's cellular immune response that passes through the thymus for processing and maturation. Some T cells kill foreign cells directly, while some regulate the activities of other lymphocytes. (22)

Tendon Strap of connective tissue that connects bone to muscle. (28)

Territoriality A form of behavior in which an individual defends its living or feeding space against any intruders. (37)

Testis One of the male reproductive organs; produces sperm and sex hormones. (13)

Testosterone A male steroid hormone related to cholesterol that stimulates sperm production. (13)

Thalamus A forebrain region just above the hypothalamus that serves as a relay system to the surrounding cerebrum. (27)

Thoracic cavity The cavity that contains the lungs and the heart. (21)

Thymus A gland in the neck or thorax of many vertebrates; makes and stores lymphocytes in addition to secreting hormones. (20, 22)

Thyroid gland Large endocrine gland regulating the body's growth and use of energy by secreting thyroxine. (25)

Thyroxine An iodine-containing hormone that governs both metabolic and growth rates and stimulates nervous system function. (25)

Tidal ventilation Type of respiration exhibited by mammals, including humans, and amphibians; air flows in and out of the lungs through the same set of hollow tubes, rather like the ebb and flow of the tide. (21)

Tissue A group of cells of the same type performing the same function within the body. (17)

Tissue culture In plants, the process of growing new, identical plantlets from somatic cells. (31)

Trachea The "windpipe," the major airway leading into the lungs. (21)

Transcription (L. *trans*, across + *scribere*, to write) The formation of RNA from a single strand of a DNA molecule. (10)

Transfer RNA (tRNA) A small RNA molecule that translates a codon into an amino acid during protein synthesis. (10)

Translation The conversion of the information on a strand of messenger RNA into a sequence of amino acids. (10)

Translocation In plants, the transport of solutes in phloem cells. (31)

Transpiration Loss of water from plants by evaporation, mainly through the stomata on stems and leaves. (31)

Trial-and-error learning A form of learning in which an animal associates a response with a reward or punishment; also called operant conditioning. (37)

Triglyceride A molecule of fat or oil consisting of three fatty acids attached to a glycerol. (2)

Trophoblast A single layer of cells forming the outer layer of the hollow blastocyst in early embryonic development; will burrow into the mother's uterine lining and give rise to the placenta. (12)

Tropism (Gr. *trope*, turning) The movement of a plant in response to external or internal stimuli, for example, phototropism, the response of plants to light. (30)

Unity The condition of being similar to other living organisms, for example, the fact that all living organisms have DNA as their hereditary material. (1)

Urea An organic compound formed in the vertebrate liver; the principal form in which mammals and some fish dispose of nitrogenous wastes. (24)

Ureter One of two long tubes that carry urine from the kidneys to the bladder. (24)

Urethra (Gr. from *ourine*, to urinate) A tube that carries urine and releases it to the outside; in males this tube also carries sperm. (13, 24)

Uric acid An insoluble, white, crystalline compound derived from ammonia; the principal excretory product in birds, reptiles, and insects. (24)

Urination The process by which urine exits the body through the urethra. (24)

Urine The nitrogen-containing waste fluid filtered from the blood by the kidney and stored in the bladder. (24)

Uterus (L. womb) Thick-walled chamber where the embryo develops. (13)

Vagina A hollow, muscular tube that receives the penis during copulation and through which the fetus passes during birth. (13)

Vascular plant A plant with internal transport tubes for food and water; includes ferns, horsetails, club mosses, gymnosperms, and flowering plants. (16)

Vascular tissue system Plant tissue that conducts fluid throughout the plant and helps strengthen roots, stems, and leaves; consists of xylem and phloem cells. (29)

Vas deferens (L. *vas*, a vessel + *defere*, to carry down) A tube that carries sperm from the epididymis to the ejaculatory duct in the penis. (13)

Vein (1) In animals, large, thin-walled blood vessel that brings blood from the body to the heart. (20) (2) In plant leaves, bundles of xylem and phloem cells that form a branching pattern in dicots and run parallel to each other in monocots. (29)

Ventilation The process by which the lungs are filled and emptied, achieved by breathing in and out. (21)

Vertebra (pl. vertebrae) One of a series of interlocking bones that make up the backbone of vertebrates. (18)

Vertebral column The backbone; a series of stacked vertebrae that encase and protect the spinal cord and support the trunk. (28)

Vertebrate An animal that possesses a backbone made of body segments known as vertebrae. (17)

Virus Infectious agent consisting of RNA or DNA encased in a protein coat; is incapable of metabolism or reproduction without a host cell. (15)

Vitamin (L. *vita*, life + *amine*, of chemical origin) Organic compound needed in small amounts for growth and metabolism. (23)

Warm-blooded Refers to animals that are capable of maintaining a constant internal temperature despite environmental changes. (18)

Warning coloration Brightly colored body patterns on poisonous prey species that warn potential predators to avoid them. (34)

Water cycle The process by which water circulates from the atmosphere to the earth's surface as rain or snow and back to the atmosphere by evaporation from puddles, ponds, rivers, and oceans and by transpiration from plants. (35)

Weather The condition of the atmosphere at any particular place and time, including its humidity, wind speed, temperature, and precipitation. (36)

White blood cell A colorless cell that circulates in blood and lymph and helps defend the body against invasion by microorganisms and other foreign materials. (20, 22)

Wood Dead xylem cells that make up stem tissue in saplings, bushes, and trees. (29)

X chromosome One of the two sex chromosomes that determine gender. In humans, females have two identical X chromosomes and males have an X and a Y chromosome. (8)

X-linked Refers to genes carried on the X chromosome. (11)

Xylem Tubular tissue in plants that transports water and minerals from the soil through the roots, stems, and leaves; dead xylem cells form wood. (29)

Y chromosome One of the two sex chromosomes that determine gender. In humans, only males have a Y chromosome as well as an X, while females have two identical X chromosomes. (8)

Yolk Part of an egg cell, or ovum, containing rich stores of lipids, carbohydrates, and special proteins that nourish the embryo during development. (12)

Zygote (Gr. *zygotos*, paired together) A fertilized egg, which is a diploid cell that can either develop into a diploid individual by a series of mitotic divisions or undergo meiosis and develop into a haploid individual. (7)

CREDITS AND ACKNOWLEDGMENTS

CHAPTER 1 Figure 1.1: Michael Fogden/Animals, Animals. Figure 1.2: (a) Marcello Bertinetti/Photo Researchers; (b) Martha Cooper/Photo Researchers; (d) Joseph Nettis/Photo Researchers; (e) Tom Carroll. Figure 1.3: (a) Eric Shabtach, University of Oregon, Eugene; (b) Peter L. Altken/Photo Researchers; (c) George Whitely/Photo Researchers. Figure 1.4: Joyce Poole. Figure 1.6: (a) Science Source/Photo Researchers; (b) M. Abbey/Visuals Unlimited; (c) E. R. Degginger/Animals, Animals; (d) John Postlethwait, University of Oregon, Eugene; (e) Art Wolfe. Figure 1.7: Kim Taylor/Bruce Coleman, Inc. Figure 1.8: John Postlethwait. Figure 1.9: Wolfgang Baylor/Bruce Coleman, Inc. Figure 1.10: Stephen Dalton/Animals, Animals. Figure 1.11a: Marion Patterson/Photo Researchers. Figure 1.13: A. Maslowski/Visuals Unlimited. Figure 1.14: (a) Science Source/Photo Researchers; (b) Animals, Animals; (c) Zig Leszczynski/Animals, Animals. Figure 1.15: B. W. Matthews, University of Oregon, Eugene. Figure 1.16: (a) Bruce Coleman, Inc.; (b) Richard Kolar/Earth Scenes. Figure 1.17: (a) The Bettmann Archive; (b) The Bettmann Archive. Figure 1.19: C. W. Myers, American Museum of Natural History. Figure 1.20: N. Myres/Bruce Coleman, Inc. Figure 1.21: Zig Leszczynski/Animals, Animals. Figure 1.22: (a) Frans Lanting; (b) G. Prance/Visuals Unlimited. Page 19 (4): Bruce Coleman, Inc. Part One: George J. Wilder/Visuals Unlimited.

CHAPTER 2 Figure 2.1: Toby Kaninger/Photo Researchers. Figure 2.3: (a) Alan Pitcairn/Grant Heilman Photography; (b) Toby Kaninger/Photo Researchers. Page 26 (Figure 2): Science Source/Photo Researchers. Figure 2.8: Tom Branch/Photo Researchers. Figure 2.9: Eric V. Gravé/Photo Researchers. Figure 2.10: Grant Heilman Photography. Figure 2.12: (a) Raymond A. Mendez/Animals, Animals; (b) Susan McCartney/Photo Researchers. Figure 2.14: M. A. Chappel/Animals, Animals. Figure 2.19: (a) Frank Oberle/Bruce Coleman, Inc.; (b) Grant Heilman. Figure 2.20: (top left) Science Source/Photo Researchers; (bottom left) Manfred Kage/Peter Arnold, Inc.; (bottom right) Science Source/Photo Researchers. Figure 2.23: Warren Uzzle. Figure 2.25: (b) (top left) Robert P. Apkarian, Yerkes Regional Primate Center, Emory University; (bottom left) Science Source/Photo Researchers; (top right) Helmut Gritscher/Peter Arnold, Inc; (bottom right) Harry Rogers/Photo Researchers. Page 42 (Figure 1): R. F. Thomas. Figure 2.26c: B. W. Matthews.

CHAPTER 3 Figure 3.1: Makio Murayama/Biological Photo Service. Figure 3.2: (a) The Granger Collection; (b) The Bettmann Archive. Figure 3.3: (a) The Granger Collection; (b) Culver Pictures. Figure 3.4: (a) K. O. Stetter, Universität Regensberg; (b) Science Source/Photo Researchers. Figure 3.6: Tony Brain/Photo Researchers. Figure 3.7: Stephen J. Kraseman/Peter Arnold, Inc. Figure 3.8: Richard Kessel, University of Iowa. Figure 3.9: Eric Shabtach. Figure 3.10: The Cystic Fibrosis Foundation. Figure 3.11: M. M. Perry and A. B. Gilbert, *J. Cell Science*, 39 (1979): 357–372. Figure 3.12: (a) Science Source/Photo Researchers; (b) Courtesy of Mark S. Ladinsky and Richard McIntosh, Dept. of Molecular, Cellular, and Developmental Biology, University of Colorado, Boulder; (c) Daniel Branton, Harvard University. Figure 3.14: (a–c) Klaus Weber and Mary Osborn, Max Planck Institüt. Figure 3.15: (a, b) D. W. Fawcett/Photo Researchers; (c) D. W. Fawcett and D. Friend/Photo Researchers. Figure 3.16: Gordon Gahan/Photo Researchers. Figure 3.17: Hans Pfletschinger/Peter Arnold, Inc. Page 61: (a) John Walsh/Photo Researchers; (b) Tony Brain/Photo Researchers; (c) Science Source/Photo Researchers. Figure 3.18: Eric Shabtach. Figure 3.19: Jeremy Burgess/Photo Researchers. Figure 3.20: (a, b) D. Patterson/Photo Researchers. Figure 3.21: Science Source/Photo Researchers. Figure 3.22: Manfred Kage/Peter Arnold, Inc. Figure 3.26: Robert Trelstad.

CHAPTER 4 Figure 4.1: Science Source/Photo Researchers. Figure 4.10: Biological Photo Service. Figure 4.14: B. W. Matthews. Figure 4.15: David Madison. Figure 4.18: (a–c) Grant Heilman. Figure 4.19: M. J. Haigney. Figure 4.20: (a, b) John Postlethwait. Page 85 (Figure 1): Leland C. Clark, Jr., Professor of Pediatrics, Children's Hospital Medical Center, Cincinnati, Ohio. Figure 4.22: Paulo Bonino/Photo Researchers.

CHAPTER 5 Figure 5.1: Hubert Schriebel. Figure 5.6: Steve Dalton/Animals, Animals. Figure 5.9: Marty and Kate Denny/PhotoEdit. Figure 5.10: Joe Munroe/Photo Researchers. Figure 5.11: (a) Linford Christe/Photo Researchers; Peter Marlow/Sygma. Figure 5.13: (a, b) Nancy Kennaway, Oregon Health Science University. Figure 5.18: Deni Bown.

CHAPTER 6 Figure 6.1: Thomas Houland/Grant Heilman Photography. Figure 6.4: Jeremy Burgess/Photo Researchers. Figure 6.7: Grant Heilman. Figure 6.10: James Klingbeil/Third Coast Stock Photos. Part Two: Wardene Wisser/Bruce Coleman, Inc.

CHAPTER 7 Figure 7.1: Bill Ellvey. Figure 7.4: (a–e) Martin Rogers/Prism. Figure 7.5: Runk/Schoenberger/Grant Heilman Photography. Figure 7.9: W. E. Engler. Figure 7.10: (d, e) Mark S. Ladinsky and Richard McIntosh. Figure 7.11: Science Source/Grant Heilman and Richard McIntosh. Figure 7.11: Science Source/Photo Researchers. Figure 7.12: The Bettmann Archive. Figure 7.14b: Custom Medical Stock Photo. Page 127 (Figure 1): Gary Groves/Avon Skin Care Labs. Figure 7.16: (a–c) Mark McKenna. Figure 7.17: Jeff Rotman/Peter Arnold, Inc. Figure 7.19: (a) Grant Heilman; (b) Peter Arnold, Inc. Figure 7.21: Richard Hutchings/Photo Researchers. Figure 7.22: Dova Vargas/The Image Works.

CHAPTER 8 Figure 8.1: John Chellman/Animals, Animals. Figure 8.2: *Journal of Heredity*, V. 63 (No. 5), Sept./Oct., 1972. Figure 8.3a: S. N. Postlethwait. Figure 8.14: Harry Howard, University of Oregon, Eugene. Figure 8.18: Hank Morgan/Rainbow. Figure 8.19: P. Barry Levy/Profiles West. Figure 8.20: John Watney/Photo Researchers. Figure 8.21: Fritz Prenzel/Animals, Animals.

CHAPTER 9 Figure 9.1: R. Langridge/Rainbow. Figure 9.3: Science Source/Photo Researchers. Figure 9.5b: K. Kleinschmidt et al., *Biochem Biophysic Acta*, 61 (1962Z):857. Figure 9.6: Hank Morgan/Photo Researchers. Figure 9.7: Grant Heilman. Figure 9.9: Harvard University. Figure 9.11: Science Source/Photo Researchers. Figure 9.12: B. Kavenof. Figure 9.13: (a) Carolina Biological Supply; (b) W. E. Engler. Figure 9.16: Scott Cunningham/*Sports Illustrated*. Figure 9.17: (a) Wide World; (b) V. Chandler and H. Howard, University of Oregon, Eugene. Figure 9.19: B. W. Matthews. Figure 9.20: Monsanto Chemical Co. Figure 9.22: Nathan T. Wright/Bruce Coleman, Inc.

CHAPTER 10 Figure 10.1: *Oxford Textbook of Medicine*, D. J. Weatherall, ed., Oxford University Press, 1944. Figure 10.5: (a) Gwen Fidler/Comstock; (b) Alexander Lowry/Photo Researchers. Figure 10.8: (a) S. J. Kraseman/Photo Researchers; (b) Richard Kolar/Animals, Animals. Figure 10.11: *Oxford Textbook of Medicine*, p. 25. Figure 10.14a: Peter Arnold, Inc. Page 184 (Figure 1): Fred McDonald. Figure 10.16b: Custom Medical Stock Photo.

CHAPTER 11 Figure 11.1: Milupa Corporation. Figure 11.2: Thomas Ives/Comstock. Figure 11.3: D. D. Whitney. Figure 11.5: Science Source/Photo Researchers. Figure 11.7: J. D. Spillane; reprinted from *The Metabolic Basis of Inherited Disease*, J. B. Stanbury, ed., 2:1262, Fig. 52.2. Figure 11.8: (a, b) Wide World. Figure 11.9b: Michael Mackett, Patterson Laboratory, Manchester, England. Figure 11.10: American Psychological Association. Figure 11.14: Steven A. Rosenberg, National Cancer Institute. Page 201: (1) Art Resource; (2) Richard Hutchings/Photo Researchers; (4) Science Source/Photo Researchers; (5) Federal Bureau of Investigation; (6) Martha Cooper/Peter Arnold, Inc.

CHAPTER 12 Figure 12.1: Adam Felsenfeld, University of Oregon, Eugene. Figure 12.3: (a) Scott Camazine/Photo Researchers; (b) Michael Fogden/Animals, Animals. Figure 12.4a: G. Shatten/Photo Researchers. Figure 12.5: Science Source/Photo Researchers. Figure 12.6a: G. Shatten/Photo Researchers. Figure 12.8: Patricia Clarco-Gillam, University of California, San Francisco. Figure 12.9: John Gerhart. Figure 12.13: (a, b) Science Source/Photo Researchers. Figure 12.15b: Science Source/

Photo Researchers. Figure 12.17: (a, b) K. Vogel, University of Oregon, Eugene. Figure 12.18: G. Steven Martin.

CHAPTER 13 Figure 13.1: Wide World. Figure 13.5d: Manfred Kage/Peter Arnold, Inc. Figure 13.7: Hank Morgan/Time-Life. Figuree 13.8: Science Source/Photo Researchers. Figure 13.10: (a–c) Science Source/Photo Researchers. Figure 13.11: Leonard McComb/Life Picture Service. Figure 13.12: Hugo Lagerkrantz and Theodore A. Slotkin, "The Stress of Being Born," *Scientific American*, 254 (No. 4), April, 1986:101. Figure 13.13. (a–d): Wide World. Figure 13.15: Wide World. Figure 13.17: Richard Hutchings/PhotoEdit. Part Three: Ken W. Davis/Tom Stack & Associates.

CHAPTER 14 Figure 14.1: Alexander Zender, Swiss Federal Institute of Technology. Figure 14.3: (a, b) Chesley Bonestell, painting. Figure 14.4: Science Source/Photo Researchers. Figure 14.5: Roger Ressmeyer/Starlight. Figure 14.7: Woods Hole Oceanographic Institution. Figure 14.8: Pieter Cullis and Michael Hope, University of British Columbia. Figure 14.9: Fred Bavendam/Peter Arnold, Inc.

CHAPTER 15 Figure 15.1: Science Source/Photo Researchers. Figure 15.2: Tony Brain/Photo Researchers. Figure 15.3: (a) Science Source/Photo Researchers; (b) David Phillips/Visuals Unlimited; (c) John D. Cunningham/Visuals Unlimited; (d) Science Source/Photo Researchers. Figure 15.4d: Science Source/Photo Researchers. Figure 15.5: T. J. Beveridge/Biological Photo Service. Figure 15.6: Tony Brain/Photo Researchers. Figure 15.7: Biological Photo Service. Figure 15.8b: Biological Photo Service. Figure 15.9b: Peter Arnold, Inc. Figure 15.10: Ken Graham/Bruce Coleman, Inc. Figure 15.11: Arthur J. Olson, Research Institute of Scripps Clinic. Figure 15.12: Kichara Homata/PhotoNats. Figure 15.14: (a) Biological Photo Service; (b) John D. Cunningham/Visuals Unlimited. Figure 15.15: (a) Eric Gravé/Photo Researchers; (b) Science Source/Photo Researchers. Figure 15.16c: Science Source/Photo Researchers. Figure 15.17: (top) F. J. R. Taylor and G. Gaines, Institute of Oceanography, University of British Columbia, Vancouver; (bottom) Science Source/Photo Researchers. Figure 15.8: Manfred Kage/Peter Arnold, Inc. Figure 15.19: Science Source/Photo Researchers. Figure 15.20: Stephen P. Parker/Photo Researchers.

CHAPTER 16 Figure 16.1: Peter Arnold, Inc. Figure 16.2: Bruce Coleman, Inc. Figure 16.5: D. Marx/Visuals Unlimited. Figure 16.6: S. N. Postlethwait. Figure 16.7: Nichlos Devore III/Bruce Coleman, Inc. Figure 16.8: W. J. Weber/Visuals Unlimited. Figure 16.9: (a) Peter Arnold, Inc.; (b) George H. Morrison/Grant Heilman Photography. Figure 16.10: Rod Planck. Figure 16.13: Breck P. Kent/Animals, Animals. Figure 16.14: Bob Evans/Peter Arnold, Inc. Figure 16.15: Science Source/Photo Researchers. Figure 16.17: S. N. Postlethwait. Figure 16.18: (b, c) Biological Photo Service. Figure 16.20: (a) S. N. Postlethwait; (b) Runk/'Schoenberger/Grant Heilman Photography. Figure 16.21: Weyerhauser, Inc. Figure 16.22: (a) Tom J. Ulrich/Visuals Unlimited; (b) John Colwell/Grant Heilman Photography. Figure 16.24: A. Nelson/Animals, Animals. Page 219 (4): Rijksmuseum, Amsterdam.

CHAPTER 17 Figure 17.1: Glenn D. Prestwich. Figure 17.3: (top) S. Conway Morris; (bottom) Marianne Collins. Figure 17.4a: Nancy Seton/Photo Researchers. Figure 17.5: (a) Tom Branch/Photo Researchers; (b) Glenn Oliver/Visuals Unlimited; (c) Steve Early/Animals, Animals; (d) Biological Photo Service. Figure 17.7: Michael Abbey/Photo Researchers. Figure 17.8: U.S. Army Institute of Pathology. Figure 17.9: Ed Reschke/Peter Arnold, Inc. Figure 17.11: Dianora Niccolini/Medichrome. Figure 17.13: (a) Breck P. Kent/Animals, Animals; (b) Biological Photo Service. Figure 17.14: Breck P. Kent/Animals, Animals. Figure 17.15: G. J. Bernard/Animals, Animals. Figure 17.16: L. C. Lockwood/Animals, Animals. Figure 17.19: (a) K. G. Lucas/Biological Photo Service; (b) John Gerlach/Visuals Unlimited. Figure 17.21: Eric Shabtach. Figure 17.22: (a) D. Ellis/Visuals Unlimited; (b) Robert Redden/Animals, Animals; (c) Karl Maslawski/Animals, Animals; (d) Biological Photo Service. Figure 17.23: Robert F. Myers.

CHAPTER 18 Figure 18.1: Gerard Lacz/Peter Arnold, Inc. Figure 18.5: John D. Cunningham/Visuals Unlimited. Figure 18.6: Tom Stack. Figure 18.7: (a) Zig Leszczynski/Animals, Animals; (b) L. L. T. Rhodes/Animals, Animals. Figure 18.8: Fred MacDonald. Figure 18.9: (a) Zig Leszczynski/Animals, Animals; (b) Tom McHugh/Photo Researchers. Figure 18.10: E. R. Degginger/Earth Scenes. Figure 18.11: (a) John Nees/Animals, Animals; (b) Ron Garrison, San Diego Zoo. Figure 18.13: (a) Bruce Coleman, Inc.; (b) San Diego Zoo. Figure 18.15: (a) Science Source/Photo Researchers; (b) John Chellman/Animals, Animals; (c) Leonard LeRue III/Visuals Unlimited; (d) Ted Levin/Animals, Animals. Figure 18.17: Michael Dick/Animals, Animals. Figure 18.18: Micky Gibson/Animals, Animals. Figure 18.19: M. Amsterdam/Animals, Animals. Figure 18.20: (a) San Diego Zoo; (b) Ron Garrison/San Diego Zoo; (c) Lawrence Migdale/Photo Researchers. Figure 18.22: (a–e) *Discover* 9/86:920. Figure 18.13: David L. Brill, Courtesy of National Geographic Society. Figure 18.24: Tom McHugh/Photo Researchers. Figure 18.25: Smithsonian Institution. Figure 18.26: Maurice Wilson, from Sonia Cole, *The Neolithic Revolution*, Plate 1, British Museum. Page 235 (5): Evidan. Part 4: Michael Fogden, Bruce Coleman, Inc.

CHAPTER 19 Figure 19.1: Richard Sears/Earth Views. Figure 19.2: Tom J. Ulrich/Visuals Unlimited. Figure 19.4: (top left) John D. Cunningham/Visuals Unlimited; (top right) Biological Photo Service; (bottom) Tom Stack; (bottom) Science Source/Photo Researchers. Page 342 (Figure 1): Science Source/Photo Researchers. Figure 19.10: G. C. Kelly/Photo Researchers. Figure 19.11: (a) Milton H. Tierney, Jr./Visuals Unlimited; (b) Stephen J. Krasemann/Photo Researchers.

CHAPTER 20 Figure 20.1: Tim Davis/Photo Researchers. Figure 20.3: Richard G. Kessel and Randy H. Kardon. Figure 20.6c: Richard G. Kessel and Randy H. Kardon. Figure 20.8: Jerry L. Ferrara/Photo Researchers. Page 354: (left and right) East Jefferson General Hospital, New Orleans. Figure 20.9: (a) Karen Pruess; (b) Alan Carey/The Image Works. Figure 20.10: David M. Phillips/Visuals Unlimited.

CHAPTER 21 Figure 21.1: Jonathan T. Wright/Bruce Coleman, Inc. Figure 21.2: Tom McHugh/Photo Researchers. Figure 21.3: Victor H. Hutchison/Visuals Unlimited. Figure 21.5: Harry Howard. Figure 21.8: (a) Michael Gabridge/Visuals Unlimited; (b) Martin M. Rotker. Figure 21.11: Bill O'Conner/Peter Arnold, Inc.

CHAPTER 22 Figure 22.1: Science Source/Photo Researchers. Figure 22.2: Boehringer Ingelheim International. Figure 22.4: David Davies, Howard Hughes Medical Institute. Figure 22.5: Custom Medical Stock Photo. Figure 22.9: Wide World. Figure 22.10: Center for Disease Control. Figure 22.11b: Center for Disease Control. Figure 22.12b: David Weintraub/Photo Researchers. Figure 22.13: James Stevenson/Photo Researchers. Page 384 (5): Dennis Brach/Black Star.

CHAPTER 23 Figure 23.1: Keith Hillett/Animals, Animals. Figure 23.2: Ken Balcomb/Earthviews. Figure 23.5: Jerard Vandystadt/Photo Researchers. Page 396 (Figure 1): D. M. Phillips. Page 402 (1): Art Resource.

CHAPTER 24 Figure 24.1: Tom McHugh/Photo Researchers. Figure 24.4: Animals, Animals. Figure 24.7b: Peter Andrews. Department of Anatomy, Georgetown University. Figure 24.9: David Madison/Duomo. Figure 24.10: (a) Richard LaVal/Animals, Animals; (b) R. Ingo-Riepl. Page 412 (Figure 1): Richard Hutchings/Photo Researchers. Figure 24.13: John Postlethwait.

CHAPTER 25 Figure 25.1: (a) William D. Griffin/Animals, Animals; (b) Breck P. Kent/Animals, Animals. Figure 25.6: Elizabeth Crews/The Image Works. Figure 25.7: Lester Bergman & Associates. Figure 25.9: Lester Bergman & Associates. Page 426 (Figure 1): Bob Daemrich/The Image Works.

CHAPTER 26 Figure 26.1: Brookhaven National Laboratories. Figure 26.4c: Carol Readhead et al. "Expression of a Myelin Basic Protein Gene in Transgenic Shiverer Mice: Correction of the Dismyelinating Phenotype," *Cell* 48 (2/27/87):703. Figure 26.6a: Reprinted from *Human Physiology*, 2e, by Stuart Ira Fox, William C. Brown, Publishers: Dubuque, Iowa, 1987. Figure 26.7: Frank R. George, National Institute of Drug Abuse.

CHAPTER 27 Figure 27.1: Stephen Dalton/Photo Researchers. Figure 27.4: (a, b) *Discover*, 11/82:95. Figure 27.5c: Lenhart Nilsson. Figure

27.6: (a, b) Richard G. Kessel and Randy H. Kardon. Figure 27.8: Jonathan T. Wright/Bruce Coleman, Inc. Figure 27.10: Science Source/Photo Researchers. Figure 27.12: (a, b) Visuals Unlimited. Page 456 (3): J. Lethmate Ibbenbürch.

CHAPTER 28 Figure 28.1: Animals, Animals. Figure 28.2: Rodger Jackman/Animals, Animals. Figure 28.3: Michael Fogden/Animals, Animals. Figure 28.6: (a) Gene Shih and Richard G. Kessel; (b) Fred Hossler/Visuals Unlimited. Figure 28.7: Martin M. Rotker. Figure 28.8: Wide World. Figure 28.13: (a) Focus on Sports; (b) J. D. MacDougall, MacMaster University, Ontario, Canada; (c) Bonie Kamin/Comstock. Page 467: (Figure 1a) Allen Russel/Profiles West; (b) Focus on Sports. Figure 28.14: (a) John Postlethwait; (b) Janet Hopson; (c) Ruth Veres. Page 470: (1) *The Illustrations from the Works of Andreas Vesalius of Brussels*, New York: Dover Publications, 1973: 87, 95; (5) Art Resource. Part Five: Richard H. Gross.

CHAPTER 29 Figure 29.1: S. N. Postlethwait. Figure 29.3: (a) John D. Cunningham, Visuals Unlimited; (b) Runk/Schoenberger/Grant Heilman Photography; (c) George H. Harrison/Grant Heilman Photography. Figure 29.4: (b) Eric Shabtach; (c) George Wilder/Visuals Unlimited; (d) (left) from *Wood Structure and Identification*, H. A. Core, W. A. Cote, and A. C. Day, Syracuse University Press, Syracuse, New York, 1976; (d) (right) SKA. Figure 29.6: (a) Biological Photo Service; (b) Dick Thomas/Visuals Unlimited. Figure 29.7: (a) Harry Howard; (b) Science Source/Photo Researchers. Figure 29.8: S. N. Postlethwait. Figure 29.9: (a) William J. Weber/Visuals Unlimited; (b) Breck P. Kent/Animals, Animals. Figure 29.10: Robert Waaland/Biological Photo Service. Figure 29.11: (a) Runk/Schoenberger/Grant Heilman Photography; (b) Richard Kessel and Gene Shih. Page 480 (Figure 1): Noel Vietmeyer. Figure 29.12: (a–c) Harry Howard. Figure 29.13: (a, b) R. F. Everett, University of Wisconsin, Madison. Figure 29.14c: John D. Cunningham. Figure 29.15: (a) David Newman/Visuals Unlimited; (b) John D. Cunningham/Visuals Unlimited. Figure 29.16: (a) S. N. Postlethwait; (b) Lefever/Grushow/Grant Heilman Photography. (c) Thomas Eisner. Figure 29.17: S. N. Postlethwait. Page 487 (7): Art Resource.

CHAPTER 30 Figure 30.1: Library of Congress. Figure 30.2: (a) International Rice Research Institute; (b) Sylvan H. Wittwer, Michigan State University, East Lansing. Figure 30.3: David Newman/Visuals Unlimited. Figure 30.4: S. N. Postlethwait. Figure 30.6: Breck P. Kent/Animals, Animals. Figure 30.7: Frank B. Salisbury. Figure 30.8a: Mark McKenna. Figure 30.9a: David Newman/Visuals Unlimited. Figure 30.10: Loren McIntyre. Figure 30.12: (a) David M. Doody/Tom Stack & Associates; (b) J. P. Nitsch. Figure 30.13: Larry Lefever/Grant Heilman Photography. Figure 30.14: Ray Ellis/Photo Researchers. Page 501 (3): Tate Gallery, London.

CHAPTER 31 Figure 31.1: G. I. Bernard/Animals, Animals. Figure 31.2: John D. Cunningham/Visuals Unlimited. Figure 31.4: Barry L. Runk/Grant Heilman Photography. Figure 31.5b: Mark McKenna. Figure 31.6: Jeremy Burgess/Photo Researchers. Figure 31.7: (a) C. P. Vance/Visuals Unlimited; (b) Jeremy Burgess/Photo Researchers. Figure 31.8: (a–c) Jack Kelly Clark, University of California Pest Management Project. Figure 31.12: Eric Shabtach. Figure 31.14: (a–c) Monsanto Chemical Co. Figure 31.15: DNAP. Page 516 (6): Donald R. Helinski, University of California, San Diego. Part 6: A. Peter Margosian/PhotoNats.

CHAPTER 32 Figure 32.1: Gregory G. Dimijian. Figure 32.5: from Victor A. McKusick, *Human Genetics*, 2e, Prentice Hall, Englewood Cliffs, NJ, 1969:49. Figure 32.6: (a) John Trager/Visuals Unlimited; (b) Visuals Unlimited. Figure 32.7: (a, b) M. W. Tweedie/Photo Researchers. Figure 32.8: *Journal of Heredity*, V. 5. No. 11, 1914. Figure 32.9: (a) T. Delevoryas, University of Texas; (b) Photo Researchers. Figure 32.10 (top and bottom) Animals, Animals. Figure 32.11: G. Perkins/Visuals Unlimited. Figure 32.13: Victor McKusick, Johns Hopkins University. Figure 32.14a: Patti Murphy/Animals, Animals. Figure 32.15: (a) Ken Cole/Animals, Animals; (b) Bradley Smith/Animals, Animals; (c) E. R. Degginger/Animals, Animals. Figure 32.19: Visuals Unlimited. Page 537 (6): Larry Reynolds.

CHAPTER 33 Figure 33.1: Helga Teives, Arizona State Museum. Figure 33.3: (a) Visuals Unlimited; (b) M. Zohary. Figure 33.11: David C. Fritts/Animals, Animals. Figure 33.12: W. E. Ruth/Bruce Coleman, Inc. Figure 33.14: (a) Renee Purse/Photo Researchers; (b) R. S. Virdee/Grant Heilman Photography. Figure 33.15: (b) B. M. Fagen; (c) Noel D. Vietmeyer; (d) The Bettmann Archive. Figure 33.18: Wendell Metzen/Bruce Coleman, Inc. Page 551 (6): Peter Turnley/Gamma-Liaison.

CHAPTER 34 Figure 34.1: Gilbert Grant/Photo Researchers; (inset) Brooking Tatum/Visuals Unlimited. Figure 34.5: Bob McKeever/Tom Stack & Associates. Figure 34.7: (a) David Pearson/Visuals Unlimited; (b) Arthus-Bertrand/Photo Researchers; (c) B. H. Brattstrom. Figure 34.11: (a, b) Reprinted from A. P. Dodd, *The Biological Campaign Against the Prickly Pear*, Prickly Pear Board, Government Printer, Brisbane, Australia, 1940. Figure 34.12: William Curtsinger/Photo Researchers. Figure 34.13: (a) J. Alcock/Visuals Unlimited; (b) D. W. Gotshall/Visuals Unlimited. Figure 34.14: (a) Susan Leavines/Photo Researchers; (b) Thomas Eisner. Figure 34.15: (a) W. J. Weber/Visuals Unlimited; (b) John D. Cunningham/Visuals Unlimited. Figure 34.17: (a, b) Richard Gross. Figure 34.18: (a) Steve McCutcheon/Visuals Unlimited; (b) Steve Coombs/Photo Researchers; (c) Keven and Betty Collins/Visuals Unlimited. Page 564 (Figure 1): E. R. Degginger/Animals, Animals. Figure 34.19: (a) R. T. Domingo/Visuals Unlimited; (b) George Stevens, NOAA. Page 566 (Figure 1): David Simberloff, Florida State University. Page 568 (3) Lew Eatherton/Photo Researchers; (4) David Wells/The Image Works; (5) Norman Myers/Bruce Coleman, Inc.

CHAPTER 35 Figure 35.1: Woods Hole Oceanographic Institute. Figure 35.2: Woods Hole Oceanographic Institute. Figure 35.6: Robert Pierce, U.S. Dept. of Agriculture. Fiure 35.14: John D. Cunningham/Visuals Unlimited. Page 582 (4): The Image Works.

CHAPTER 36 Figure 36.1: Fritz Trillmach. Figure 36.7a: Gregory G. Dimijian. Figure 36.8: Terry G. Murphey/Animals, Animals. Figure 36.9: Michael Fogden/Animals, Animals. Figure 36.10: William J. Weber/Visuals Unlimited. Figure 36.11: James R. Fisher/Photo Researchers. Figure 36.12: Peter B. Kaplan/Photo Researchers. Figure 36.13: John Shawn. Figure 36.14: Bill Curtsinger/Photo Researchers. Figure 36.17: Joel Arrington/Visuals Unlimited. Figure 36.19: Carl Rossler/Animals, Animals. Page 599 (5): D. H. Davies/Bruce Coleman, Inc.

CHAPTER 37 Figure 37.1: Gordon Langsbury/Bruce Coleman, Inc. Figure 37.2: W. A. Banaszewski/Visuals Unlimited. Figure 37.3: Zig Leszczynski/Animals, Animals. Figure 37.4: Louis Darling. Figure 37.5: Nina Leen/Time-Life. Figure 37.6: The Bettmann Archive. Figure 37.7: Susan Kuklin/Photo Researchers. Figure 37.8: (a) Ron Austing/Photo Researchers; (b) François Gohier/Photo Researchers. Figure 37.10: R. H. Armstrong/Animals, Animals. Figure 37.11: William E. Townsend, Jr./Photo Researchers. Figure 37.14: Charles Palek/Animals, Animals. Figure 37.15: Scott Camazine/Photo Researchers. Figure 37.16: Tom Cajocob. Figure 37.17: Bob Sacha. Page 614: (5, left) Rijksmuseum, Amsterdam; (5, right) Art Resource.

INDEX

Pages on which definitions or main discussions of topics appear are indicated by **boldface**.

Pages containing illustrations or tables are indicated by *italics*.

Metric–English and English–Metric Conversions

Prefixes Used with Units (metric)

Prefix	Symbol	Value	Equivalents	Example
Kilo-	k	1,000, or 10^3	1 kilometer (km) = 1×10^3 m	(six tenths of a mile)
Centi-	c	1/100, or 10^{-2}	1 centimeter (cm) = 0.01 m	1/100 m (about the width of your little finger)
Milli-	m	1/1,000, or 10^{-3}	1 millimeter (mm) = 0.001m	1/1,000 m (less than the width of a letter on this page)
Micro-	μ	1/1,000,000, or 10^{-6}	1 micrometer (μm) = 1×10^{-6} m	1/1,000,000 m (1/100 the thickness of a page in this book)
Nano-	n	1/1,000,000,000, or 10^{-9}	1 nanometer (nm) = 1×10^{-9} m	1/1,000,000,000 m (1/3 the size of a small protein; myoglobin is about 3 nm across)

Length

Metric	=	English (USA)
millimeter (0.001 m)	=	0.039 in
centimeter (0.01 m)	=	0.39 in
meter	=	3.28 ft, 39.37 in
kilometer (1×10^3 m)	=	0.62 mi, 1,091 yd, 3,273 ft

English (USA)	=	Metric
inch	=	2.54 cm
foot	=	0.30 m, 30.48 cm
yard	=	0.91 m, 91.4 cm
mile (statute) (5,280 ft)	=	1.61 km, 1,609 m

Practice Examples:

A foot–long hotdog is 30.5 cm long.
A 32-inch softball bat is 0.8 m long.
A 100-yard long football field is 91 m long.
When driving at 55 mph, you will travel
   88.6 km in an hour.

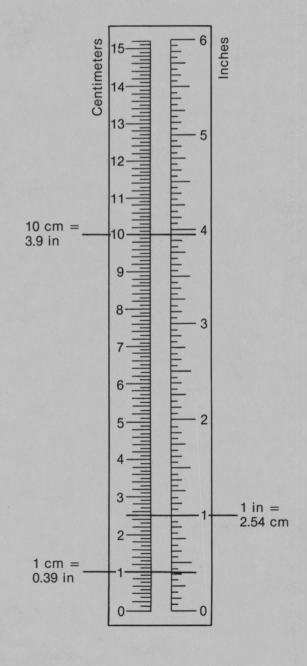

10 cm = 3.9 in

1 in = 2.54 cm

1 cm = 0.39 in